SOLIDS (SPACE FIGURES):

L = Lateral Area; T (or S) = Total (Surface) Area; V = Volume

Parallelepiped (box):

$$T = 2\ell w + 2\ell h + 2wh$$
$$V = \ell wh$$

Right Prism:

$$L = hP$$
$$T = L + 2B$$
$$V = Bh$$

Regular Pyramid:

$$L = \frac{1}{2}\ell P$$
$$\ell^2 = a^2 + h^2$$
$$T = L + B$$
$$V = \frac{1}{3}Bh$$

Right Circular Cylinder:

$$L = 2\pi rh$$
$$T = 2\pi rh + 2\pi r^2$$
$$V = \pi r^2 h$$

Right Circular Cone:

$$L = \pi r\ell$$
$$\ell^2 = r^2 + h^2$$
$$T = \pi r\ell + \pi r^2$$
$$V = \frac{1}{3}\pi r^2 h$$

Sphere:

$$S = 4\pi r^2$$
$$V = \frac{4}{3}\pi r^3$$

Miscellaneous:

Euler's Equation: $V + F = E + 2$

ANALYTIC GEOMETRY:

Cartesian Plane

Distance:
$$d = \sqrt{(x_2 - x_1)^2 + (y_2 - y_1)^2}$$

Midpoint:
$$M = \left(\frac{x_1 + x_2}{2}, \frac{y_1 + y_2}{2}\right)$$

Slope: $m = \dfrac{y_2 - y_1}{x_2 - x_1}, x_1 \neq x_2$

Parallel Lines:
$$\ell_1 \| \ell_2 \leftrightarrow m_1 = m_2$$
Perpendicular Lines:
$$\ell_1 \perp \ell_2 \leftrightarrow m_1 \cdot m_2 = -1$$

Equations of a Line:
  Slope-Intercept: $y = mx + b$
  Point-Slope: $y - y_1 = m(x - x_1)$
  General: $Ax + By = C$

Cartesian Space

Distance: $d = \sqrt{(x_2 - x_1)^2 + (y_2 - y_1)^2 + (z_2 - z_1)^2}$

Midpoint: $M = \left(\dfrac{x_1 + x_2}{2}, \dfrac{y_1 + y_2}{2}, \dfrac{z_1 + z_2}{2}\right)$

Equations of a Line:
  Vector Form: $(x, y, z) = (x_1, y_1, z_1) + n(a, b, c)$
  Point Form: $(x, y, z) = (x_1 + na, y_1 + nb, z_1 + nc)$

Equation of a Plane:
  $Ax + By + Cz = D$

TRIGONOMETRY:

Right Triangle:

$$\sin \theta = \frac{opposite}{hypotenuse} = \frac{a}{c}$$

$$\cos \theta = \frac{adjacent}{hypotenuse} = \frac{b}{c}$$

$$\tan \theta = \frac{opposite}{adjacent} = \frac{a}{b}$$

$$\sin^2\theta + \cos^2\theta = 1$$

Triangle:

$$A = \frac{1}{2}bc \sin \alpha$$

$$\frac{\sin \alpha}{a} = \frac{\sin \beta}{b} = \frac{\sin \gamma}{c}$$

$$c^2 = a^2 + b^2 - 2ab \cos \gamma$$
or
$$\cos \gamma = \frac{a^2 + b^2 - c^2}{2ab}$$

6E

Elementary Geometry

for College Students

6E

Elementary Geometry

for College Students

Daniel C. Alexander
Parkland College, Professor Emeritus

Geralyn M. Koeberlein
Mahomet-Seymour High School,
Mathematics Department Chair, Retired

CENGAGE
Learning·

Australia • Brazil • Mexico • Singapore • United Kingdom • United States

Elementary Geometry for College Students,
Sixth Edition
Daniel C. Alexander
Geralyn M. Koeberlein

Product Director: Liz Covello

Product Team Manager: Richard Stratton

Product Assistant: Jennifer Cordoba

Media Developer: Lynh Pham

Marketing Manager: Gordon Lee

Content Project Manager: Cheryll Linthicum

Art Director: Vernon Boes

Manufacturing Planner: Becky Cross

Rights Acquisitions Specialist: Roberta Broyer

Production Service and Compositor:
    Graphic World Inc.

Photo Researcher: PreMedia Global

Text Researcher: PreMedia Global

Illustrator: Jade Myers, Matrix Art Services;
    diacriTech

Text and Cover Designer: Terri Wright

Cover Image: © Sara Meijer/Alamy

For product information and technology assistance, contact us at
Cengage Learning Customer & Sales Support, 1-800-354-9706.

For permission to use material from this text or product,
submit all requests online at **www.cengage.com/permissions.**
Further permissions questions can be e-mailed to
permissionrequest@cengage.com.

Library of Congress Control Number: 2013942407

ISBN-13: 978-1-285-19569-8

ISBN-10: 1-285-19569-8

Cengage Learning
200 First Stamford Place, 4th Floor
Stamford, CT 06902
USA

Cengage Learning is a leading provider of customized learning solutions with office locations around the globe, including Singapore, the United Kingdom, Australia, Mexico, Brazil, and Japan. Locate your local office at **www.cengage.com/global.**

Cengage Learning products are represented in Canada by Nelson Education, Ltd.

To learn more about Cengage Learning Solutions, visit **www.cengage.com.**

Purchase any of our products at your local college store or at our preferred online store **www.cengagebrain.com.**

Printed in China
2 3 4 5 6 7 18 17 16 15

This edition is dedicated to our spouses, children, and grandchildren.
Dan Alexander and Geralyn Koeberlein

LETTER FROM THE AUTHOR

As of the late 1980s, the geometry textbooks written for the elementary college student contained flaws in reasoning as well as errors and even contradictions. In those textbooks, there seemed to be an overall lack of geometric figures that are essential to visual reasoning. Using my collected notes and problems, I developed an outline for the geometry textbook that would present the necessary topics in a logical order. Providing descriptions and explanations that students could read and comprehend, students were able to learn the vocabulary of geometry, recognize visual relationships, solve problems, and even create some proofs of theorems as well. The textbook would have to provide several exercises, many of them serving as building blocks that would transition the student to mid-range and challenging problems and applications. Without doubt, earlier editions of this textbook have evolved so that any improvements would advance my early goals. In time, the Interactive Companion (for added practice and reinforcement) was incorporated into the Student Study Guide. Both the technology for geometry and the applications of geometry are evident throughout this textbook.

As authors, Geralyn Koeberlein and I feel quite strongly that we have always accurately and completely addressed the fundamental concepts of geometry, as suggested by a number of professional mathematical associations. However, many of the changes in the sixth edition are attributable to current users and reviewers of the fifth edition. For instance, we chose to include an increased discussion of the parabola as well as a new section dealing with three-dimensional coordinate geometry. As always, we present new topics concisely and with easily understood explanations.

We continue to include the visual explanations of theorems that are enabled by accurate and well-labeled figures. Comparable to the guidance provided by roadway signs and GPS systems, the geometric figures found in this textbook provide guidance for student readers too. Thus, students will have the tools, figures, and background to "see" results intuitively, explore relationships inductively, and establish principles deductively.

We believe that explanations are a necessary component of the geometry textbook. As well as forcing us to look back (to review), these "proofs" are learning experiences within themselves. Our textbook presents these proofs in the most compact and understandable form that we can find. As the reader will discover, we provide suggestions and insights into the construction of a proof whenever possible.

Writing this textbook for college students, I have incorporated my philosophy for the teaching of geometry. The student who is willing to study geometry and to accept the responsibility and challenges herein will be well-prepared for advanced mathematical endeavors and will also have developed skills of logic that are useful in other disciplines.

Daniel C. Alexander

Contents

10 Analytic Geometry 435

11 Introduction to Trigonometry 489

APPENDICES 529

ANSWERS 565

As authors, we aim to help the students of the sixth edition of *Elementary Geometry for College Students* become familiar with the terminology of geometry, explore (and perhaps discover) geometric principles, strengthen their skills in deductive reasoning, and gain new skills in problem solving, particularly in the area of geometry-based applications. Our style of writing enables novices of geometric study to open doors, refreshes the memories of students who have had an earlier introduction to the subject, provides a different perspective for other students, and even encourages and directs those who might someday be teaching the subject matter.

As any classroom teacher of geometry would expect, we have developed this textbook in a logical order and with features that are intuitive, informative, and motivational; as a result, students are enabled to achieve the goals cited in the preceding paragraph. For this textbook to be an effective learning tool, it is imperative that it contain a multitude of figures and illustrations; of course, this approach follows from the assumption that "A picture is worth a thousand words."

We are well aware that the completion of a proof is quite the challenge for the student of geometry. With this in mind, we seek initially to have the student recognize and appreciate the role of proof throughout the development of our geometry textbook. To achieve this end, the student will initially be asked to follow the flow of a given proof. In turn, the student should be able to supply missing pieces (statements and reasons) for the proof, thus recognizing a need to both order and justify one's claims. Of course, the ultimate goal is that the student actually generate a geometric proof, beginning with writing a lower-level proof and then progressing toward creating a higher-end proof.

For completeness, convenience, and compactness, we provide proof in a variety of forms: the two-column proof, the paragraph proof, and the "picture" proof. A student's actual creation of the two-column proof demonstrates the student's understanding; that is, it becomes evident that the student can both order and justify conclusions toward a desired end. Our belief is that such accomplishments in the ability to reason extend themselves to other disciplines; for instance, successful students will likely improve paragraph writing in a composition class by improving the order, flow, and even the justification of their claims. Also, the elements of logic found in the study of geometry may very well enable students who are also or will be enrolled in a computer science class to create more powerful and more compact subroutines in their computer codes.

In each edition, we have continued to be inspired and guided by both the National Council of Teachers of Mathematics (NCTM) and the American Mathematical Association of Two-Year Colleges (AMATYC). Of course, we encourage suggestions for content and improvement in this textbook from those who are current users.

OUTCOMES FOR THE STUDENT

- Mastery of the essential concepts of geometry, for intellectual and vocational needs
- Preparation of the transfer student for further study of mathematics and geometry at the senior-level institution
- Understanding of the step-by-step reasoning necessary to fully develop a mathematical system such as geometry
- Enhancement of one's interest in geometry through discovery activities, features, and solutions to exercises

FEATURES OF THE SIXTH EDITION

- Inclusion of approximately 150 new exercises, many of a challenging nature
- Increased uniformity in the steps outlining construction techniques
- Creation of a new Section 10.6 that discusses analytic geometry in three dimensions
- Extension of the feature *Strategy for Proof*, which provides insight into the development of proofs of geometric theorems
- Expanded coverage of parabolas
- Extension of the Discover activities
- Extension of Appendix A.4 of the Fifth Edition—now Appendixes A.4, Factoring and Quadratic Equations, and A.5, The Quadratic Formula and Square Root Properties of the Sixth Edition

TRUSTED FEATURES

Full-color format aids in the development of concepts, solutions, and investigations through application of color to all figures and graphs. The authors have continued the introduction of color to all figures to ensure that it is both accurate and instructionally meaningful.

Reminders found in the text margins provide a convenient recall mechanism.

Discover activities emphasize the importance of induction in the development of geometry.

Geometry in Nature and **Geometry in the Real World** illustrate geometry found in everyday life.

Overviews found in chapter-ending material organize important properties and other information from the chapter.

An **Index of Applications** calls attention to the practical applications of geometry.

A **Glossary of Terms** at the end of the textbook provides a quick reference of geometry terms.

Chapter-opening photographs highlight subject matter for each chapter.

Warnings are provided so that students might avoid common pitfalls.

Chapter Summaries review the chapter, preview the chapter to follow, and provide a list of important concepts found in the current chapter.

Perspective on History boxes provide students with biographical sketches and background leading to geometric discoveries.

Perspective on Applications boxes explore classical applications and proofs.

Chapter Reviews provide numerous practice problems to help solidify student understanding of chapter concepts.

Chapter Tests provide students the opportunity to prepare for exams.

Formula pages at the front of the book list important formulas with relevant art to illustrate.

Reference pages at the back of the book summarize the important abbreviations and symbols used in the textbook.

STUDENT RESOURCES

Student Study Guide with Solutions Manual (978-1-285-19681-7) provides worked-out solutions to select odd-numbered problems from the text as well as new Interactive Exercise sets for additional review. Select solutions for the additional Interactive Exercise sets are provided within the study guide. Complete solutions are available on the instructors website.

Text-Specific DVDs (978-1-285-19687-9), hosted by Dana Mosely, provide professionally produced content that covers key topics of the text, offering a valuable resource to augment classroom instruction or independent study and review.

The Geometers Sketchpad CD-ROM (978-0-618-76840-0) helps you construct and measure geometric figures, explore properties and form conjectures, and create polished homework assignments and presentations. This CD-ROM is a must-have resource for your classes.

STUDENT WEBSITE

Visit **www.cengagebrain.com** to access additional course materials and companion resources. At the CengageBrain.com home page, search for the ISBN of your title (from the back cover of your book) using the search box at the top of the page. This will take you to the product page, where free companion resources can be found.

INSTRUCTOR RESOURCES

Instructor's Solutions Manual provides solutions to all the exercises in the book, alternatives for order of presentation of the topics included, and suggestions for teaching each topic.

Cognero (Included on the Instructor Companion Website) Cengage Learning Testing Powered by Cognero is a flexible, online system that allows you to author, edit, and manage test bank content from multiple Cengage Learning solutions; create multiple test versions in an instant; and deliver tests from your LMS, your classroom, or wherever you want.

Text-Specific DVDs (978-1-285-19687-9), hosted by Dana Mosely, provide professionally produced content that covers key topics of the text, offering a valuable resource to augment classroom instruction or independent study and review.

The **Solution Builder** allows instructors to create customizable solutions that they can print out to distribute or post as needed. This is a convenient and expedient way to deliver solutions to specific homework sets.

The **Geometer's Sketchpad CD-ROM** (978-0-618-76840-0) helps users to construct and measure geometric figures, explore properties and form conjectures, and create polished homework assignments and presentations. This CD-ROM is a must-have resource for your classes.

INSTRUCTOR WEBSITE

Everything you need for your course in one place! This collection of book-specific lecture and class tools formerly found on the PowerLecture CD, is available online at **www.cengage.com/login**. Here you can access and download PowerPoint presentations, images, and more.

ACKNOWLEDGMENTS

The authors wish to thank Marc Bove, Rita Lombard, and Maureen Ross at Cengage for their efforts in making the sixth edition of the textbook a reality. Also instrumental in the process were Amy Simpson of Graphic World Inc. and Megan Guiney at diacriTech, Inc. In addition, we would like to recognize and thank those who made earlier editions of this textbook possible: Beth Dahlke, Theresa Grutz, Florence Powers, Dawn Nuttall, Lynn Cox, Melissa Parkin, and Noel Kamm.

We express our gratitude to reviewers of the current edition and previous editions:

Houssam M. Al-Basha, *Lincoln Land Community College*
Paul Allen, *University of Alabama*
Jane C. Beatie, *University of South Carolina at Aiken*
Dr. Michaela Bertch, *Cathedral Catholic High School*
Steven Blasberg, *West Valley College*
Barbara Brown, *Anoka Ramsey Community College*
Patricia Clark, *Indiana State University*
Joyce Cutler, *Framingham State College*
Walter Czarnec, *Framingham State College*
Kahroline di Passero, *Moorpark Community College*
Darwin G. Dorn, *University of Wisconsin–Washington County*
William W. Durand, *Henderson State University*
Zoltan Fischer, *Minneapolis Community and Technical College*
Kathryn E. Godshalk, *Cypress College*
Chris Graham, *Mt. San Antonio Community College*
Sharon Gronberg, *Southwest Texas State University*
Geoff Hagopian, *College of the Desert*
Edith Hays, *Texas Woman's University*
Ben L. Hill, *Lane Community College*
George L. Holloway, *Los Angeles Valley College*
Leigh Hollyer, *University of Houston–Central Campus*
Tracy Hoy, *College of Lake County*
Greg Jamison, *Fresno City College*
Josephine G. Lane, *Eastern Kentucky University*
John C. Longnecker, *University of Northern Iowa*
Erin C. Martin, *Parkland College*
Nicholas Martin, *Shepherd College*
Jill McKenney, *Lane Community College*
James R. McKinney, *Cal Poly at Pomona*
Iris C. McMurtry, *Motlow State Community College*
Jack Morrell, *Atlanta Metropolitan State College*
Michael Naylor, *Western Washington University*
Maurice Ngo, *Chabot College*
Ellen L. Rebold, *Brookdale Community College*
Lauri Semarne, *Los Angeles, California*
Patty Shovanec, *Texas Technical University*
Marvin Stick, *University of Massachusetts–Lowell*
Joseph F. Stokes, *Western Kentucky University*
Kay Stroope, *Phillips Community College–University of Arkansas*
Dr. John Stroyls, *Georgia Southwestern State University*
Karen R. Swick, *Palm Beach Atlantic College*
Steven L. Thomassin, *Ventura College*
Bettie A. Truitt, *Black Hawk College*
Jean A. Vrechek, *Sacramento City College*
Tom Zerger, *Saginaw Valley State University*

In the sixth edition of *Elementary Geometry for College Students*, the topics that comprise a minimal course include most of Chapters 1–6 and Chapter 8. For a complete basic course, coverage of Chapters 1–8 is recommended. Some sections that can be treated as optional in formulating a course description include the following:

- Section 2.6 Symmetry and Transformations
- Section 3.4 Basic Constructions Justified
- Section 3.5 Inequalities in a Triangle
- Section 5.6 Segments Divided Proportionally
- Section 6.4 Some Constructions and Inequalities for the Circle
- Section 7.1 Locus of Points
- Section 7.2 Concurrence of Lines
- Section 7.3 More About Regular Polygons
- Section 8.5 More Area Relationships in the Circle
- Section 10.6 The Three-Dimensional Coordinate System

Given that this textbook is utilized for three-, four-, and five-hour courses, the following flowchart depicts possible orders in which the textbook can be used. As suggested by the preceding paragraph, it is possible to treat certain sections as optional.

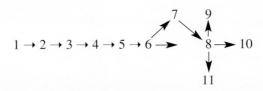

For students who need further review of related algebraic topics, consider these topics found in Appendix A:

A.1: Algebraic Expressions
A.2: Formulas and Equations
A.3: Inequalities
A.4: Factoring and Quadratic Equations
A.5: The Quadratic Formula and Square Root Properties

Sections A.4 and A.5 include these methods of solving quadratic equations: the factoring method, the square roots method, and the Quadratic Formula.

Logic appendices can be found at the textbook website. These include:

Logic Appendix 1: Truth Tables
Logic Appendix 2: Valid Arguments

Daniel C. Alexander and Geralyn M. Koeberlein

Index of Applications

6E

Elementary Geometry

for College Students

Doug Byrnes/Corbis

Chapter 1

Line and Angle Relationships

Magical! In geometry, figures can be conceived to create an illusion. Known as the Bridge of Aspiration, this passageway was conceptualized by the Wilkinson Eyre Architects. It connects the Royal Opera House and the Royal Ballet School in Covent Garden in London, England. In this geometric design, 23 square portals are each rotated slightly in order to create the illusion of a twisted passage. Although a visual inspection of the bridge might have one think that people would have to walk on walls to cross Floral Street below, it is easy to walk upright. The architectural design successfully creates the fluidity, grace, and spirit of the dance. This chapter opens with a discussion of the types of reasoning: intuition, induction, and deduction. Additional topics found in Chapter 1 and useful for design include tools of geometry, such as the ruler, protractor, and compass. By considering relationships between lines and angles, the remainder of the chapter begins the logical development of geometry. For the geometry student needing an algebra review, several topics are found in Appendix A. Other topics are developed as needed.

Additional video explanations of concepts, sample problems, and applications are available on DVD.

1.1 Sets, Statements, and Reasoning

KEY CONCEPTS

Statement	Conclusion	Law of Detachment
Variable	Reasoning	Set
Conjunction	Intuition	Subset
Disjunction	Induction	Venn Diagram
Negation	Deduction	Intersection
Implication	Argument (Valid	Union
(Conditional)	and Invalid)	
Hypothesis		

SETS

A **set** is any collection of objects, all of which are known as the *elements* of the set. The statement $A = \{1, 2, 3\}$ is read, "A is the set of elements 1, 2, and 3." In geometry, geometric figures such as lines and angles are actually sets of points.

Where $A = \{1, 2, 3\}$ and $B = \{\text{counting numbers}\}$, A is a *subset* of B because each element in A is also in B; in symbols, $A \subseteq B$. In Chapter 2, we will discover that $T = \{$all triangles$\}$ is a subset of $P = \{$all polygons$\}$; that is, $T \subseteq P$.

STATEMENTS

DEFINITION

A **statement** is a set of words and/or symbols that collectively make a claim that can be classified as true or false.

Figure 1.1

EXAMPLE 1

Classify each of the following as a true statement, a false statement, or neither.

1. $4 + 3 = 7$
2. An angle has two sides. (See Figure 1.1.)
3. Robert E. Lee played shortstop for the Yankees.
4. $7 < 3$ (This is read "7 is less than 3.")
5. Look out!

SOLUTION 1 and 2 are true statements; 3 and 4 are false statements; 5 is not a statement.

Some statements contain one or more *variables*; a **variable** is a letter that represents a number. The claim "$x + 5 = 6$" is called an *open sentence* or *open statement* because it can be classified as true or false, depending on the replacement value of x. For instance, $x + 5 = 6$ is true if $x = 1$; for x not equal to 1, $x + 5 = 6$ is false. Some statements containing variables are classified as true because they are true for all replacements. Consider the Commutative Property of Addition, usually stated in the form $a + b = b + a$. In words, this property states that the same result is obtained when two numbers are added in either order; for instance, when $a = 4$ and $b = 7$, it follows that $4 + 7 = 7 + 4$.

The **negation** of a given statement P makes a claim opposite that of the original statement. If the given statement is true, its negation is false, and vice versa. If P is a statement, we use $\sim P$ (which is read "not P") to indicate its negation.

EXAMPLE 2

Give the negation of each statement.

a) $4 + 3 = 7$ **b)** All fish can swim

SOLUTION
a) $4 + 3 \neq 7$ ($\neq$ means "is not equal to.")
b) Some fish cannot swim. (To negate "All fish can swim," we say that at least one fish cannot swim.)

TABLE 1.1 *The Conjunction*		
P	**Q**	**P and Q**
T	T	T
T	F	F
F	T	F
F	F	F

A *compound* statement is formed by combining other statements used as "building blocks." In such cases, we may use letters such as P and Q to represent simple statements. For example, the letter P may refer to the statement "$4 + 3 = 7$," and the letter Q to the statement "Babe Ruth was a U.S. president." The statement "$4 + 3 = 7$ *and* Babe Ruth was a U.S. president" has the form *P and Q*, and is known as the **conjunction** of P and Q. The statement "$4 + 3 = 7$ *or* Babe Ruth was a U.S. president" has the form *P or Q*, and is known as the **disjunction** of statement P and statement Q. A conjunction is true only when P and Q are *both* true. A disjunction is false only when P and Q are *both* false. See Tables 1.1 and 1.2.

TABLE 1.2 *The Disjunction*		
P	**Q**	**P or Q**
T	T	T
T	F	T
F	T	T
F	F	F

EXAMPLE 3

Assume that statement P and statement Q are both true.

$$P: 4 + 3 = 7$$
$$Q: \text{An angle has two sides.}$$

Classify the following statements as true or false.

1. $4 + 3 \neq 7$ and an angle has two sides.
2. $4 + 3 \neq 7$ or an angle has two sides.

SOLUTION Statement 1 is false because the conjunction has the form "F and T."
Statement 2 is true because the disjunction has the form "F or T."

The statement "If P, then Q," known as a **conditional statement** (or **implication**), is classified as true or false as a whole. A statement of this form can be written in equivalent forms; for instance, the conditional statement, "If an angle is a right angle, then it measures 90 degrees" is equivalent to the statement, "All right angles measure 90 degrees."

EXAMPLE 4

Classify each conditional statement as true or false.

1. If an animal is a fish, then it can swim. (States, "All fish can swim.")
2. If two sides of a triangle are equal in length, then two angles of the triangle are equal in measure. (See Figure 1.2 below.)

Figure 1.2

3. If Wendell studies, then he will receive an A on the test.

SOLUTION Statements 1 and 2 are true. Statement 3 is false; Wendell may study yet not receive an A.

In the conditional statement "If *P*, then *Q*," *P* is the **hypothesis** and *Q* is the **conclusion.** In statement 2 of Example 4, we have

Hypothesis: Two sides of a triangle are equal in length.

Conclusion: Two angles of the triangle are equal in measure.

SSG **EXS. 1–7** For the true statement "If *P*, then *Q*," the hypothetical situation described in *P* implies the conclusion described in *Q*. This type of statement is often used in reasoning, so we turn our attention to this matter.

REASONING

Success in the study of geometry requires vocabulary development, attention to detail and order, supporting claims, and thinking. **Reasoning** is a process based on experience and principles that allows one to arrive at a conclusion. The following types of reasoning are used to develop mathematical principles.

1.	Intuition	An inspiration leading to the statement of a theory
2.	Induction	An organized effort to test and validate the theory
3.	Deduction	A formal argument that proves the tested theory

➤ Intuition

We are often inspired to think and say, "It occurs to me that. . . ." With **intuition,** a sudden insight allows one to make a statement without applying any formal reasoning. When intuition is used, we sometimes err by "jumping" to conclusions. In a cartoon, the character having the "bright idea" (using intuition) is shown with a light bulb next to her or his head.

> **EXAMPLE 5**
>
> Figure 1.3 is called a *regular pentagon* because its five sides have equal lengths and its five interior angles have equal measures. What do you suspect is true of the lengths of the dashed parts of lines from *B* to *E* and from *B* to *D*?
>
> **SOLUTION** Intuition suggests that the lengths of the dashed parts of lines (known as *diagonals* of the pentagon) are the same.
> **NOTE 1:** Using induction (and a *ruler*), we can verify that this claim is true. We will discuss measurement with the ruler in more detail in Section 1.2.
> **NOTE 2:** Using methods found in Chapter 3, we could use deduction to prove that the two diagonals do indeed have the same length.

Figure 1.3

The role intuition plays in formulating mathematical thoughts is truly significant. But to have an idea is not enough! Testing a theory may lead to a revision of the theory or even to its total rejection. If a theory stands up to testing, it moves one step closer to becoming mathematical law.

➤ Induction

We often use specific observations and experiments to draw a general conclusion. This type of reasoning is called **induction.** As you would expect, the observation/experimentation process is common in laboratory and clinical settings. Chemists, physicists, doctors, psychologists, weather forecasters, and many others use collected data as a basis for drawing conclusions. In our study of geometry, the inductive process generally has us use the ruler or the *protractor* (to measure angles).

EXAMPLE 6

While in a grocery store, you examine several 6-oz cartons of yogurt. Although the flavors and brands differ, each carton is priced at 75 cents. What do you conclude?

CONCLUSION Every 6-oz carton of yogurt in the store costs 75 cents.

As you may already know (see Figure 1.2), a figure with three straight sides is called a *triangle*.

EXAMPLE 7

In a geometry class, you have been asked to measure the three interior angles of each triangle in Figure 1.4. You discover that triangles I, II, and IV have two angles (as marked) that have equal measures. What may you conclude?

CONCLUSION The triangles that have two sides of equal length also have two angles of equal measure.

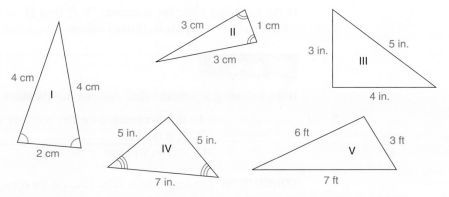

Figure 1.4

NOTE: The protractor, used to support the conclusion above, will be discussed in Section 1.2.

▶ Deduction

> **DEFINITION**
>
> **Deduction** is the type of reasoning in which the knowledge and acceptance of selected assumptions guarantee the truth of a particular conclusion.

In Example 8, we illustrate a **valid argument,** a form of deductive reasoning used frequently in the development of geometry. In this form, at least two statements are treated as facts; these assumptions are called the *premises* of the argument. On the basis of the premises, a particular *conclusion* must follow. This form of deduction is called the **Law of Detachment.**

EXAMPLE 8

If you accept the following statements 1 and 2 as true, what must you conclude?

1. If a student plays on the Rockville High School boys' varsity basketball team, then he is a talented athlete.
2. Todd plays on the Rockville High School boys' varsity basketball team.

CONCLUSION Todd is a talented athlete.

To more easily recognize this pattern for deductive reasoning, we use letters to represent statements in the following generalization.

LAW OF DETACHMENT

Let P and Q represent simple statements, and assume that statements 1 and 2 are true. Then a valid argument having conclusion C has the form

$$\left. \begin{array}{l} \textbf{1.} \text{ If } P, \text{ then } Q \\ \textbf{2.} \ P \end{array} \right\} \quad \text{premises}$$

$$\left. \text{C.} \ \therefore Q \quad \right\} \quad \text{conclusion}$$

NOTE: The symbol $\therefore$ means "therefore."

In the preceding form, the statement "If P, then Q" is often read "P implies Q." That is, when P is known to be true, Q must follow.

EXAMPLE 9

Is the following argument valid? Assume that premises 1 and 2 are true.

1. If it is raining, then Tim will stay in the house.
2. It is raining.
C. $\therefore$ Tim will stay in the house.

CONCLUSION The argument is valid because the form of the argument is

1. If P, then Q
2. P
C. $\therefore Q$

with $P = $ "It is raining," and $Q = $ "Tim will stay in the house."

EXAMPLE 10

Is the following argument valid? Assume that premises 1 and 2 are true.

1. If a man lives in London, then he lives in England.
2. William lives in England.
C. $\therefore$ William lives in London.

CONCLUSION The argument is not valid. Here, $P = $ "A man lives in London," and $Q = $ "A man lives in England." Thus, the form of this argument is

1. If P, then Q
2. Q
C. $\therefore P$

To represent a valid argument, the Law of Detachment would require that the first statement has the form "If Q, then P." Even though statement Q is true, it does not enable us to draw a valid conclusion about P. Of course, if William lives in England, he *might* live in London; but he might instead live in Liverpool, Manchester, Coventry, or any of countless other places in England. Each of these possibilities is a **counterexample** disproving the validity of the argument. Remember that deductive reasoning is concerned with reaching conclusions that *must be true,* given the truth of the premises.

VALID ARGUMENT	INVALID ARGUMENT
1. If P, then Q	**1.** If P, then Q
2. P	**2.** Q
C. $\therefore Q$	C. $\therefore P$

We will use deductive reasoning throughout our work in geometry. For example, suppose that you know these two facts:

SSG **EXS. 8–12**

    1. If an angle is a right angle, then it measures $90°$.
    2. Angle A is a right angle.

Because the form found in statements 1 and 2 matches the form of the valid argument, you may draw the following conclusion.

    C. Angle A measures $90°$.

VENN DIAGRAMS

Sets of objects are often represented by geometric figures known as *Venn Diagrams*. Their creator, John Venn, was an Englishman who lived from 1834 to 1923. In a Venn Diagram, each set is represented by a closed (bounded) figure such as a circle or rectangle. If statements P and Q of the conditional statement "If P, then Q" are represented by sets of objects P and Q, respectively, then the Law of Detachment can be justified by a geometric argument. When a Venn Diagram is used to represent the statement "If P, then Q," it is absolutely necessary that circle P lies in circle Q; that is, P is a *subset* of Q. (See Figure 1.5.)

If P, then Q.

Figure 1.5

EXAMPLE 11

Use Venn Diagrams to verify Example 8.

SOLUTION Let B = students on the Rockville High varsity boys' basketball team. Let A = people who are talented athletes.

To represent the statement "If a basketball player (B), then a talented athlete (A)," we show B within A. In Figure 1.6 we use point T to represent Todd, a person on the basketball team (T in B). With point T also in circle A, we conclude that "Todd is a talented athlete."

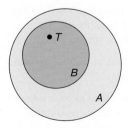

Figure 1.6

The statement "If P, then Q" is sometimes expressed in the form "All P are Q." For instance, the conditional statement of Examples 8 and 11 can be written "All Rockville High School basketball players are talented athletes." Venn Diagrams can also be used to demonstrate that the argument of Example 10 is not valid. To show the invalidity of the argument in Example 10, one must show that an object in Q may *not* lie in circle P. (See Figure 1.5.)

The compound statements known as the conjunction and the disjunction can also be related to the intersection and union of sets, relationships that can be illustrated by the use of Venn Diagrams. For the Venn Diagram, we assume that the sets *P* and *Q* may have elements in common. (See Figure 1.7.)

The elements common to *P* and *Q* form the **intersection** of *P* and *Q*, which is written $P \cap Q$. This set, $P \cap Q$, is the set of all elements in *both P* **and** *Q*. The elements that are in *P*, in *Q*, or in both form the **union** of *P* and *Q*, which is written $P \cup Q$. This set, $P \cup Q$, is the set of elements in *P* **or** *Q*.

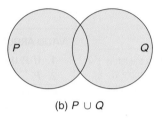

(a) $P \cap Q$ (b) $P \cup Q$

SSG **EXS. 13–15**   Figure 1.7

Exercises 1.1

In Exercises 1 and 2, which sentences are statements? If a sentence is a statement, classify it as true or false.

1. a) Where do you live?
    b) $4 + 7 \neq 5$.
    c) Washington was the first U.S. president.
    d) $x + 3 = 7$ when $x = 5$.

2. a) Chicago is located in the state of Illinois.
    b) Get out of here!
    c) $x < 6$ (read as "*x* is less than 6") when $x = 10$.
    d) Babe Ruth is remembered as a great football player.

In Exercises 3 and 4, give the negation of each statement.

3. a) Christopher Columbus crossed the Atlantic Ocean.
    b) All jokes are funny.

4. a) No one likes me.
    b) Angle 1 is a right angle.

In Exercises 5 to 10, classify each statement as simple, conditional, a conjunction, or a disjunction.

5. If Alice plays, the volleyball team will win.

6. Alice played and the team won.

7. The first-place trophy is beautiful.

8. An integer is odd or it is even.

9. Matthew is playing shortstop.

10. You will be in trouble if you don't change your ways.

In Exercises 11 to 18, state the hypothesis and the conclusion of each statement.

11. If you go to the game, then you will have a great time.

12. If two chords of a circle have equal lengths, then the arcs of the chords are congruent.

13. If the diagonals of a parallelogram are perpendicular, then the parallelogram is a rhombus.

14. If $\frac{a}{b} = \frac{c}{d}$, where $b \neq 0$ and $d \neq 0$, then $a \cdot d = b \cdot c$.

15. Corresponding angles are congruent if two parallel lines are cut by a transversal.

16. Vertical angles are congruent when two lines intersect.

17. All squares are rectangles.

18. Base angles of an isosceles triangle are congruent.

In Exercises 19 to 24, classify each statement as true or false.

19. If a number is divisible by 6, then it is divisible by 3.

20. Rain is wet and snow is cold.

21. Rain is wet or snow is cold.

22. If Jim lives in Idaho, then he lives in Boise.

23. Triangles are round or circles are square.

24. Triangles are square or circles are round.

In Exercises 25 to 32, name the type of reasoning (if any) used.

25. While participating in an Easter egg hunt, Sarah notices that each of the seven eggs she has found is numbered. Sarah concludes that all eggs used for the hunt are numbered.

26. You walk into your geometry class, look at the teacher, and conclude that you will have a quiz today.

27. Albert knows the rule "If a number is added to each side of an equation, then the new equation has the same solution set as the given equation." Given the equation $x - 5 = 7$, Albert concludes that $x = 12$.

28. You believe that "Anyone who plays major league baseball is a talented athlete." Knowing that Duane Gibson has just been called up to the major leagues, you conclude that Duane Gibson is a talented athlete.

29. As a handcuffed man is brought into the police station, you glance at him and say to your friend, "That fellow looks guilty to me."

30. While judging a science fair project, Mr. Cange finds that each of the first 5 projects is outstanding and concludes that all 10 will be outstanding.

31. You know the rule "If a person lives in the Santa Rosa Junior College district, then he or she will receive a tuition break at Santa Rosa." Emma tells you that she has received a tuition break. You conclude that she resides in the Santa Rosa Junior College district.

32. As Mrs. Gibson enters the doctor's waiting room, she concludes that it will be a long wait.

In Exercises 33 to 36, use intuition to state a conclusion.

33. You are told that the opposite angles formed when two lines cross are **vertical angles**. In the figure, angles 1 and 2 are vertical angles. Conclusion?

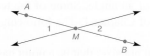

Exercises 33, 34

34. In the figure, point *M* is called the **midpoint** of line segment *AB*. Conclusion?

35. The two triangles shown are **similar** to each other. Conclusion?

36. Observe (but do not measure) the following angles. Conclusion?

In Exercises 37 to 40, use induction to state a conclusion.

37. Several movies directed by Lawrence Garrison have won Academy Awards, and many others have received nominations. His latest work, *A Prisoner of Society*, is to be released next week. Conclusion?

38. On Monday, Matt says to you, "Andy hit his little sister at school today." On Tuesday, Matt informs you, "Andy threw his math book into the wastebasket during class." On Wednesday, Matt tells you, "Because Andy was throwing peas in the school cafeteria, he was sent to the principal's office." Conclusion?

39. While searching for a classroom, Tom stopped at an instructor's office to ask directions. On the office bookshelves are books titled *Intermediate Algebra, Calculus, Modern Geometry, Linear Algebra,* and *Differential Equations.* Conclusion?

40. At a friend's house, you see several food items, including apples, pears, grapes, oranges, and bananas. Conclusion?

In Exercises 41 to 50, use deduction to state a conclusion, if possible.

41. If the sum of the measures of two angles is 90°, then these angles are called "complementary." Angle 1 measures 27° and angle 2 measures 63°. Conclusion?

42. If a person attends college, then he or she will be a success in life. Kathy Jones attends Dade County Community College. Conclusion?

43. All mathematics teachers have a strange sense of humor. Alex is a mathematics teacher. Conclusion?

44. All mathematics teachers have a strange sense of humor. Alex has a strange sense of humor. Conclusion?

45. If Stewart Powers is elected president, then every family will have an automobile. Every family has an automobile. Conclusion?

46. If Tabby is meowing, then she is hungry. Tabby is hungry. Conclusion?

47. If a person is involved in politics, then that person will be in the public eye. June Jesse has been elected to the Missouri state senate. Conclusion?

48. If a student is enrolled in a literature course, then he or she will work very hard. Bram Spiegel digs ditches by hand six days a week. Conclusion?

49. If a person is rich and famous, then he or she is happy. Marilyn is wealthy and well known. Conclusion?

50. If you study hard and hire a tutor, then you will make an A in this course. You make an A in this course. Conclusion?

In Exercises 51 to 54, use Venn Diagrams to determine whether the argument is valid or not valid.

51. 1) If an animal is a cat, then it makes a "meow" sound.
    2) Tipper is a cat.
    C) Then Tipper makes a "meow" sound.

52. 1) If an animal is a cat, then it makes a "meow" sound.
    2) Tipper makes a "meow" sound.
    C) Then Tipper is a cat.

53. 1) All Boy Scouts serve the United States of America.
    2) Sean serves the United States of America.
    C) Sean is a Boy Scout.

54. 1) All Boy Scouts serve the United States of America.
    2) Sean is a Boy Scout.
    C) Sean serves the United States of America.

55. Where $A = \{1,2,3\}$ and $B = \{2,4,6,8\}$, classify each of the following as true or false.
    a) $A \cap B = \{2\}$
    b) $A \cup B = \{1,2,3,4,6,8\}$
    c) $A \subseteq B$

In Exercises 56 and 57, P is a true statement, while Q and R are false statements. Classify each of the following statements as true or false.

56. a) (P and Q) or R
b) (P or Q) and R

57. a) (P and Q) or ~R
b) (P or Q) and ~R

1.2 Informal Geometry and Measurement

KEY CONCEPTS

Point	Midpoint	Perpendicular
Line	Congruence	Compass
Plane	Protractor	Constructions
Collinear Points	Parallel	Circle
Line Segment	Bisect	Arc
Betweenness of Points	Intersect	Radius

In geometry, the terms *point, line,* and *plane* are described but not defined. Other concepts that are accepted intuitively, but never defined, include the *straightness* of a line, the *flatness* of a plane, the notion that a point on a line lies *between* two other points on the line, and the notion that a point lies in the *interior* or *exterior* of an angle. Some of the terms found in this section are formally defined in later sections of Chapter 1. The following are descriptions of some of the undefined terms.

A **point,** which is represented by a dot, has location but not size; that is, a point has no dimensions. An uppercase italic letter is used to name a point. Figure 1.8 shows points *A, B,* and *C.*

The second undefined geometric term is **line.** A line is an infinite set of points. Given any two points on a line, there is always a point that lies between them on that line. Lines have a quality of "straightness" that is not defined but assumed. Given several points on a line, these points form a straight path. Whereas a point has no dimensions, a line is one-dimensional; that is, the distance between any two points on a given line can be measured. Line *AB*, represented symbolically by $\overleftrightarrow{AB}$, extends infinitely far in opposite directions, as suggested by the arrows on the line. A line may also be represented by a single lowercase letter. Figures 1.9(a) and (b) show the lines *AB* and *m*. When a lowercase letter is used to name a line, the line symbol is omitted; that is, $\overleftrightarrow{AB}$ and *m* can name the same line.

Note the position of point *X* on $\overleftrightarrow{AB}$ in Figure 1.9(c). When three points such as *A, X,* and *B* are on the same line, they are said to be **collinear.** In the order shown, which is symbolized *A-X-B* or *B-X-A*, point *X* is said to be *between A* and *B*. When a drawing is not provided, the notation *A-B-C* means that these points are collinear, with *B* between *A* and *C*. When a drawing is provided, we assume that all points in the drawing that appear to be collinear *are* collinear, *unless otherwise stated.* Figure 1.9(d) shows that *A, B,* and *C* are collinear; in Figure 1.8, points *A, B,* and *C* are *noncollinear.*

At this time, we informally introduce some terms that will be formally defined later. You have probably encountered the terms *angle, triangle,* and *rectangle* many times. An example of each is shown in Figure 1.10.

Figure 1.8

Figure 1.9

Angle *ABC*
(a)

Triangle *DEF*
(b)

Rectangle *WXYZ*
(c)

Figure 1.10

Using symbols, we refer to Figures 1.10(a), (b), and (c) as $\angle ABC$, $\triangle DEF$, and $\square WXYZ$, respectively. Some caution must be used in naming figures; although the angle in Figure 1.10(a) can be called $\angle CBA$, it is incorrect to describe the angle as $\angle ACB$ because that order implies a path from point A to point C to point B . . . a different angle! In $\angle ABC$, the point B at which the sides meet is called the **vertex** of the angle. Because there is no confusion regarding the angle described, $\angle ABC$ is also known as $\angle B$ (using only the vertex) or as $\angle 1$. The points D, E, and F at which the sides of $\triangle DEF$ (also called $\triangle DFE$, $\triangle EFD$, etc.) meet are called the *vertices* (plural of *vertex*) of the triangle. Similarly, W, X, Y, and Z are the vertices of the rectangle; the vertices are named in an order that traces the rectangle.

A **line segment** is part of a line. It consists of two distinct points on the line and all points between them. (See Figure 1.11.) Using symbols, we indicate the line segment by $\overline{BC}$; note that $\overline{BC}$ is a set of points but is not a number. We use BC (omitting the segment symbol) to indicate the *length* of this line segment; thus, BC is a number. The sides of a triangle or rectangle are line segments.

Figure 1.11

EXAMPLE 1

Can the rectangle in Figure 1.10(c) be named **a)** $\square XYZW$? **b)** $\square WYXZ$?

SOLUTION
    a) Yes, because the points taken in this order trace the figure.
    b) No; for example, $\overline{WY}$ is not a side of the rectangle.

MEASURING LINE SEGMENTS

The instrument used to measure a line segment is a scaled straightedge such as a *ruler,* a *yardstick,* or a *meter stick.* Line segment RS ($\overline{RS}$ in symbols) in Figure 1.12 measures 5 centimeters. Because we express the length of $\overline{RS}$ by RS (with no bar), we write $RS = 5$ cm.

> To find the length of a line segment using a ruler:
>
>   **1.** Place the ruler so that "0" corresponds to one endpoint of the line segment.
>   **2.** Read the length of the line segment by reading the number at the remaining endpoint of the line segment.

Because manufactured measuring devices such as the ruler, yardstick, and meter stick may lack perfection or be misread, there is a margin of error each time one is used. In Figure 1.12, for instance, RS may actually measure 5.02 cm (and that could be rounded from 5.023 cm, etc.). Measurements are approximate, not perfect.

Figure 1.12

In Example 2, a ruler (not drawn to scale) is shown in Figure 1.13. In the drawing, the distance between consecutive marks on the ruler corresponds to 1 inch. The measure of a line segment is known as *linear measure.*

EXAMPLE 2

In rectangle *ABCD* of Figure 1.13, the line segments $\overline{AC}$ and $\overline{BD}$ shown are the diagonals of the rectangle. How do the lengths of the diagonals compare?

Figure 1.13

SOLUTION As shown on the ruler, $AC = 10''$. As intuition suggests, the lengths of the diagonals are the same, so it follows that $BD = 10''$.

NOTE: In linear measure, $10''$ means 10 inches, and $10'$ means 10 feet.

Figure 1.14

In Figure 1.14, point *B* lies between *A* and *C* on $\overline{AC}$. If $AB = BC$, then *B* is the **midpoint** of $\overline{AC}$. When $AB = BC$, the geometric figures $\overline{AB}$ and $\overline{BC}$ are said to be **congruent;** in effect, geometric figures are congruent when one can be placed over the other (a perfect match). Numerical lengths may be equal, but the actual line segments (geometric figures) are congruent. The symbol for congruence is $\cong$; thus, $\overline{AB} \cong \overline{BC}$ if *B* is the midpoint of $\overline{AC}$. Example 3 emphasizes the relationship between $\overline{AB}$, $\overline{BC}$, and $\overline{AC}$ when *B* lies between *A* and *C*.

EXAMPLE 3

SSG **EXS. 1–8**

In Figure 1.15, the lengths of $\overline{AB}$ and $\overline{BC}$ are $AB = 4$ and $BC = 8$. What is *AC*, the length of $\overline{AC}$?

Figure 1.15

SOLUTION As intuition suggests, the length of $\overline{AC}$ equals $AB + BC$. Thus, $AC = 4 + 8 = 12$.

MEASURING ANGLES

Discover

The word geometry means the measure (from *metry*) of the earth (from *geo*). Words that contain *meter* also suggest the measure of some quantity. What is measured by each of the following objects?

odometer, pedometer, thermometer, altimeter, clinometer, anemometer

ANSWER
Vehicle mileage, distance walked, temperature, altitude, angle of inclination, wind speed

Although we formally define an angle in Section 1.4, we consider it intuitively at this time.

An angle's measure depends not on the lengths of its sides but on the amount of opening between its sides. In Figure 1.16, the arrows on the angles' sides suggest that the sides extend indefinitely.

Figure 1.16

Figure 1.17

The instrument shown in Figure 1.17 (and used in the measurement of angles) is a **protractor.** For example, we would express the measure of ∠*RST* by writing m∠*RST* = 50°; this statement is read, "The measure of ∠*RST* is 50 degrees." Measuring the angles in Figure 1.16 with a protractor, we find that m∠*B* = 55° and m∠1 = 90°. If the degree symbol is missing, the measure is understood to be in degrees; thus, m∠1 = 90.

In practice, the protractor shown will measure an angle that is greater than 0° but less than or equal to 180°.

To find the degree measure of an angle using a protractor:

1. Place the notch of the protractor at the point where the sides of the angle meet (the vertex of the angle). See point S in Figure 1.18.
2. Place the edge of the protractor along a side of the angle so that the scale reads "0." See point *T* in Figure 1.18 where we use "0" on the outer scale.
3. Using the same (outer) scale, read the angle size by reading the degree measure that corresponds to the second side of the angle.

EXAMPLE 4

For Figure 1.18, find the measure of ∠*RST*.

Figure 1.18

SOLUTION Using the protractor, we find that the measure of angle *RST* is 31°.
(In symbols, m∠*RST* = 31° or m∠*RST* = 31.)

Some protractors show a full 360°; such a protractor is used to measure an angle whose measure is between 0° and 360°. An angle whose measure is between 180° and 360° is known as a *reflex angle*.

Just as measurement with a ruler is not perfect, neither is measurement with a protractor.

The lines on a sheet of paper in a notebook are *parallel*. Informally, **parallel** lines lie on the same page and will not cross over each other even if they are extended indefinitely. We say that lines ℓ and *m* in Figure 1.19(a) are parallel; note here the use of a lowercase letter to name a line. We say that line segments are parallel if they are parts of parallel lines; if $\overleftrightarrow{RS}$ is parallel to $\overleftrightarrow{MN}$, then $\overline{RS}$ is parallel to $\overline{MN}$ in Figure 1.19(b).

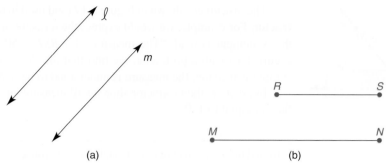

(a)   (b)

Figure 1.19

For $A = \{1, 2, 3\}$ and $B = \{6, 8, 10\}$, there are no common elements; for this reason, we say that the intersection of A and B is the **empty set** (symbol is $\varnothing$). Just as $A \cap B = \varnothing$, the parallel lines in Figure 1.19(a) are characterized by $\ell \cap m = \varnothing$.

EXAMPLE 5

In Figure 1.20, the sides of angles ABC and DEF are parallel ($\overline{AB}$ to $\overline{DE}$ and $\overline{BC}$ to $\overline{EF}$). Use a protractor to decide whether these angles have equal measures.

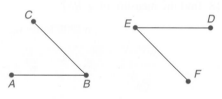

Figure 1.20

SOLUTION The angles have equal measures. Both measure 44°.

Two angles with equal measures are said to be *congruent*. In Figure 1.20, we see that $\angle ABC \cong \angle DEF$. In Figure 1.21, $\angle ABC \cong \angle CBD$.

In Figure 1.21, angle ABD has been separated into smaller angles ABC and CBD; if the two smaller angles are congruent (have equal measures), then angle ABD has been *bisected*. In general, the word **bisect** means to separate a line segment (or an angle) into two parts of equal measure; similarly, the word **trisect** means that the line segment (or angle) is separated into three parts of equal measure.

Any angle having a 180° measure is called a **straight angle,** an angle whose sides are in opposite directions. See straight angle RST in Figure 1.22(a). When a straight angle is bisected, as shown in Figure 1.22(b), the two angles formed are **right angles** (each measures 90°).

When two lines have a point in common, as in Figure 1.23, they are said to **intersect.** When two lines intersect and form congruent adjacent angles, they are said to be **perpendicular.** In Figure 1.23, lines r and t are perpendicular if $\angle 1 \cong \angle 2$ or $\angle 2 \cong \angle 3$, and so on.

Figure 1.21

Figure 1.22

Figure 1.23

EXAMPLE 6

In Figure 1.23, suppose that **a)** $\angle 3 \cong \angle 4$ **b)** $\angle 1 \cong \angle 4$ **c)** $\angle 1 \cong \angle 3$. Are lines r and t perpendicular?

SSG EXS. 9–13

SOLUTION a) Yes **b)** Yes **c)** No

Figure 1.24

CONSTRUCTIONS

Another tool used in geometry is the **compass.** This instrument, shown in Figure 1.24, is used to draw circles and parts of circles known as *arcs.*

The ancient Greeks insisted that only two tools (a compass and a straightedge) be used for geometric **constructions,** which were idealized drawings assuming perfection in the use of these tools. The compass was used to create "perfect" circles and for marking off segments of "equal" length. The straightedge could be used to draw a straight line through two designated points.

A **circle** is the set of all points in a plane that are at a given distance from a particular point (known as the "center" of the circle). The part of a circle between any two of its points is known as an **arc.** Any line segment joining the center to a point on the circle is a **radius** (plural: *radii*) of the circle. See Figure 1.25.

Construction 1, which follows, is quite basic and depends only on using arcs of the same radius length to construct line segments of the same length. The arcs are created by using a compass. Construction 2 is more difficult to perform and explain, so we will delay its explanation to a later chapter (see Section 3.4).

Figure 1.25

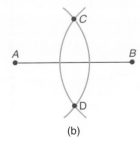

(a)

(b)

CONSTRUCTION 1 To construct a segment congruent to a given segment.

GIVEN: $\overline{AB}$ in Figure 1.26(a).

CONSTRUCT: $\overline{CD}$ on line m so that $\overline{CD} \cong \overline{AB}$ (or $CD = AB$)

CONSTRUCTION: With your compass open to the length of $\overline{AB}$, place the stationary point of the compass at C and mark off a length equal to AB at point D, as shown in Figure 1.26(b). Then $CD = AB$.

(a)

(b)

Figure 1.26

The following construction is shown step by step in Figure 1.27. Intuition suggests that point M in Figure 1.27(c) is the midpoint of $\overline{AB}$.

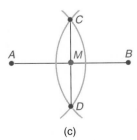

(c)

Figure 1.27

SSG EXS. 14–17

CONSTRUCTION 2 To construct the midpoint M of a given line segment AB.

GIVEN: $\overline{AB}$ in Figure 1.27(a)

CONSTRUCT: M on $\overline{AB}$ so that $AM = MB$

CONSTRUCTION: Figure 1.27(a): Open your compass to a length greater than one-half of $\overline{AB}$.

Figure 1.27(b): Using A as the center of the arc, mark off an arc that extends both above and below segment AB. With B as the center and keeping the same length of radius, mark off an arc that extends above and below $\overline{AB}$ so that two points (C and D) are determined where the arcs cross.

Figure 1.27(c): Now draw $\overline{CD}$. The point where $\overline{CD}$ crosses $\overline{AB}$ is the midpoint M.

EXAMPLE 7

In Figure 1.28, M is the midpoint of $\overline{AB}$.

Figure 1.28

a) Find AM if $AB = 15$.
b) Find AB if $AM = 4.3$.
c) Find AB if $AM = 2x + 1$.

SOLUTION

a) AM is one-half of AB, so $AM = 7\frac{1}{2}$.
b) AB is twice AM, so $AB = 2(4.3)$ or $AB = 8.6$.
c) AB is twice AM, so $AB = 2(2x + 1)$ or $AB = 4x + 2$.

The technique from algebra used in Example 8 and also needed for Exercises 47 and 48 of this section depends on the following properties of addition and subtraction.

If $a = b$ and $c = d$, then $a + c = b + d$.

Words: Equals added to equals provide equal sums.

Illustration: Since $0.5 = \frac{5}{10}$ and $0.2 = \frac{2}{10}$, it follows that
$0.5 + 0.2 = \frac{5}{10} + \frac{2}{10}$; that is, $0.7 = \frac{7}{10}$.

If $a = b$ and $c = d$, then $a - c = b - d$.

Words: Equals subtracted from equals provide equal differences.

Illustration: Since $0.5 = \frac{5}{10}$ and $0.2 = \frac{2}{10}$, it follows that
$0.5 - 0.2 = \frac{5}{10} - \frac{2}{10}$; that is, $0.3 = \frac{3}{10}$.

EXAMPLE 8

Figure 1.29

In Figure 1.29, point B lies on $\overline{AC}$ between A and C. If $AC = 10$ and AB is 2 units longer than BC, find the length x of $\overline{AB}$ and the length y of $\overline{BC}$.

SOLUTION
Because $AB + BC = AC$, we have $x + y = 10$.
Because $AB - BC = 2$, we have $x - y = 2$.
Adding the left and right sides of these equations, we have

$$\begin{array}{r} x + y = 10 \\ x - y = 2 \\ \hline 2x = 12 \end{array} \quad \text{so } x = 6.$$

If $x = 6$, then $x + y = 10$ becomes $6 + y = 10$ and $y = 4$.

SSG **EXS. 18, 19**   Thus, $AB = 6$ and $BC = 4$.

Exercises 1.2

1. If line segment *AB* and line segment *CD* are drawn to scale, what does intuition tell you about the lengths of these segments?

2. If angles *ABC* and *DEF* were measured with a protractor, what does intuition tell you about the degree measures of these angles?

3. How many endpoints does a line segment have? How many midpoints does a line segment have?

4. Do the points *A, B,* and *C* appear to be collinear?

   Exercises 4–6

5. How many lines can be drawn that contain both points *A* and *B*? How many lines can be drawn that contain points *A, B,* and *C*?

6. Consider noncollinear points *A, B,* and *C*. If each line must contain two of the points, what is the total number of lines that are determined by these points?

7. Name all the angles in the figure.

8. Which of the following measures can an angle have? 23°, 90°, 200°, 110.5°, −15°

9. Must two different points be collinear? Must three or more points be collinear? Can three or more points be collinear?

10. Which symbol(s) correctly expresses the order in which the points *A, B,* and *X* lie on the given line, *A-X-B* or *A-B-X*?

11. Which symbols correctly name the angle shown? ∠*ABC,* ∠*ACB,* ∠*CBA*

12. A triangle is named △*ABC*. Can it also be named △*ACB*? Can it be named △*BAC*?

13. Consider rectangle *MNPQ*. Can it also be named rectangle *PQMN*? Can it be named rectangle *MNQP*?

14. Suppose ∠*ABC* and ∠*DEF* have the same measure. Which statements are expressed correctly?
    a) m∠*ABC* = m∠*DEF*   b) ∠*ABC* = ∠*DEF*
    c) m∠*ABC* ≅ m∠*DEF*   d) ∠*ABC* ≅ ∠*DEF*

15. Suppose $\overline{AB}$ and $\overline{CD}$ have the same length. Which statements are expressed correctly?
    a) *AB* = *CD*   b) $\overline{AB}$ = $\overline{CD}$
    c) *AB* ≅ *CD*   d) $\overline{AB}$ ≅ $\overline{CD}$

16. When two lines cross (intersect), they have exactly one point in common. In the drawing, what is the point of intersection? How do the measures of ∠1 and ∠2 compare?

17. Judging from the ruler shown (not to scale), estimate the measure of each line segment.
    a) *AB*   b) *CD*

    Exercises 17, 18

18. Judging from the ruler, estimate the measure of each line segment.
    a) *EF*   b) *GH*

19. Judging from the protractor provided, estimate the measure of each angle to the nearest multiple of 5° (e.g., 20°, 25°, 30°, etc.).
    a) m∠1   b) m∠2

    Exercises 19, 20

20. Using the drawing for Exercise 19, estimate the measure of each angle to the nearest multiple of 5° (e.g., 20°, 25°, 30°, etc.).
a) m∠3
b) m∠4

21. Consider the square at the right, *RSTV*. It has four right angles and four sides of the same length. How are sides $\overline{RS}$ and $\overline{ST}$ related? How are sides $\overline{RS}$ and $\overline{VT}$ related?

22. Square *RSTV* has diagonals $\overline{RT}$ and $\overline{SV}$ (not shown). If the diagonals are drawn, how will their lengths compare? Do the diagonals of a square appear to be perpendicular?

Exercises 21, 22

23. Use a compass to draw a circle. Draw a radius, a line segment that connects the center to a point on the circle. Measure the length of the radius. Draw other radii and find their lengths. How do the lengths of the radii compare?

24. Use a compass to draw a circle of radius 1 inch. Draw a chord, a line segment that joins two points on the circle. Draw other chords and measure their lengths. What is the largest possible length of a chord in this circle?

25. The sides of the pair of angles are parallel. Are ∠1 and ∠2 congruent?

26. The sides of the pair of angles are parallel. Are ∠3 and ∠4 congruent?

27. The sides of the pair of angles are perpendicular. Are ∠5 and ∠6 congruent?

28. The sides of the pair of angles are perpendicular. Are ∠7 and ∠8 congruent?

29. On a piece of paper, use your compass to construct a triangle that has two sides of the same length. Cut the triangle out of the paper and fold the triangle in half so that the congruent sides coincide (one lies over the other). What seems to be true of two angles of that triangle?

30. On a piece of paper, use your protractor to draw a triangle that has two angles of the same measure. Cut the triangle out of the paper and fold the triangle in half so that the angles of equal measure coincide (one lies over the other). What seems to be true of two of the sides of that triangle?

31. A trapezoid is a four-sided figure that contains one pair of parallel sides. Which sides of the trapezoid *MNPQ* appear to be parallel?

32. In the rectangle shown, what is true of the lengths of each pair of opposite sides?

33. A line segment is bisected if its two parts have the same length. Which line segment, $\overline{AB}$ or $\overline{CD}$, is bisected at point *X*?

34. An angle is bisected if its two parts have the same measure. Use three letters to name the angle that is bisected.

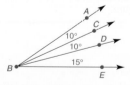

In Exercises 35 to 38, with A-B-C on $\overline{AC}$, it follows that AB + BC = AC.

Exercises 35–38

35. Find *AC* if *AB* = 9 and *BC* = 13.

36. Find *AB* if *AC* = 25 and *BC* = 11.

37. Find *x* if *AB* = *x*, *BC* = *x* + 3, and *AC* = 21.

38. Find an expression for *AC* (the length of $\overline{AC}$) if *AB* = *x* and *BC* = *y*.

39. ∠ABC is a straight angle. Using your protractor, you can show that m∠1 + m∠2 = 180°. Find m∠1 if m∠2 = 56°.

Exercises 39, 40

40. Find m∠1 if m∠1 = 2x and m∠2 = x.

(HINT: See Exercise 39.)

In Exercises 41 to 44, m∠1 + m∠2 = m∠ABC.

41. Find m∠ABC if m∠1 = 32° and m∠2 = 39°.

42. Find m∠1 if m∠ABC = 68° and m∠1 = m∠2.

43. Find x if m∠1 = x, m∠2 = 2x + 3, and m∠ABC = 72°.

44. Find an expression for m∠ABC if m∠1 = x and m∠2 = y.

Exercises 41–44

45. A compass was used to mark off three congruent segments, $\overline{AB}$, $\overline{BC}$, and $\overline{CD}$. Thus, $\overline{AD}$ has been trisected at points B and C. If AD = 32.7, how long is $\overline{AB}$?

46. Use your compass and straightedge to bisect $\overline{EF}$.

*47. In the figure, m∠1 = x and m∠2 = y. If x − y = 24°, find x and y.

(HINT: m∠1 + m∠2 = 180°.)

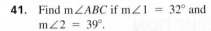

*48. In the drawing, m∠1 = x and m∠2 = y. If m∠RSV = 67° and x − y = 17°, find x and y.

(HINT: m∠1 + m∠2 = m∠RSV.)

For Exercises 49 and 50, use the following information. Relative to its point of departure or some other point of reference, the angle that is used to locate the position of a ship or airplane is called its bearing. The bearing may also be used to describe the direction in which the airplane or ship is moving. By using an angle between 0° and 90°, a bearing is measured from the North-South line toward the East or West. In the diagram, airplane A (which is 250 miles from Chicago's O'Hare airport's control tower) has a bearing of S 53° W.

49. Find the bearing of airplane B relative to the control tower.

50. Find the bearing of airplane C relative to the control tower.

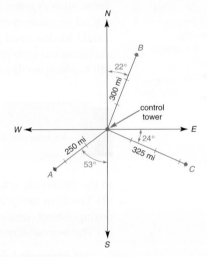

Exercises 49, 50

1.3	**Early Definitions and Postulates**

KEY CONCEPTS	Mathematical System	Congruent Segments	Parallel Lines
	Axiom or Postulate	Midpoint of a Line	Plane
	Theorem	Segment	Coplanar Points
	Ruler Postulate	Ray	Space
	Distance	Opposite Rays	
	Segment-Addition Postulate	Intersection of Two Geometric Figures	

A MATHEMATICAL SYSTEM

Like algebra, the branch of mathematics called geometry is a **mathematical system.** The formal study of a mathematical system begins with undefined terms. Building on this foundation, we can then define additional terms. Once the terminology is sufficiently developed, certain properties (characteristics) of the system become apparent. These properties are known as **axioms** or **postulates** of the system; more generally, such statements are called **assumptions** in that they are assumed to be true. Once we have developed a vocabulary and accepted certain postulates, many principles follow logically as we apply deductive methods. These statements can be proved and are called **theorems.** The following box summarizes the components of a mathematical system (sometimes called a logical system or deductive system).

FOUR PARTS OF A MATHEMATICAL SYSTEM

1. Undefined terms ⎫
2. Defined terms ⎭ vocabulary

3. Axioms or postulates ⎫
4. Theorems ⎭ principles

Discover

Although we cannot actually define *line* and *plane*, we can compare them in the following analogy. Please complete: A _?_ is to *straight* as a _?_ is to *flat*.

ANSWERS
line; plane

CHARACTERISTICS OF A GOOD DEFINITION

Terms such as *point, line,* and *plane* are classified as undefined because they do not fit into any set or category that has been previously determined. Terms that *are* defined, however, should be described precisely. *But what is a good definition?* A good definition is like a mathematical equation written using words. A good definition must possess four characteristics, which we illustrate with a term that we will redefine at a later time.

DEFINITION

An **isosceles triangle** is a triangle that has two congruent sides.

In the definition, notice that: (1) The term being defined—*isosceles triangle*—is named. (2) The term being defined is placed into a larger category (a type of *triangle*). (3) The distinguishing quality (that two sides of the triangle are congruent) is included. (4) The *reversibility* of the definition is illustrated by these statements:

"If a triangle is isosceles, then it has two congruent sides."
"If a triangle has two congruent sides, then it is an isosceles triangle."

CHARACTERISTICS OF A GOOD DEFINITION

1. It names the term being defined.
2. It places the term into a set or category.
3. It distinguishes the defined term from other terms without providing unnecessary facts.
4. It is reversible.

The reversibility of a definition is achieved by using the phrase "if and only if." For instance, we could define *congruent angles* by saying "Two angles are congruent if and only if these angles have equal measures." The "if and only if" statement has the following dual meaning:

"If two angles are congruent, then they have equal measures."
"If two angles have equal measures, then they are congruent."

Figure 1.30

When represented by a Venn Diagram, the definition above would relate set $C = \{$congruent angles$\}$ to set $E = \{$angles with equal measures$\}$ as shown in Figure 1.30. The sets C and E are identical and are known as **equivalent sets.**

Once undefined terms have been described, they become the building blocks for other terminology. In this textbook, primary terms are defined within boxes, whereas related terms are often boldfaced and defined within statements. Consider the following definition (see Figure 1.31).

Figure 1.31

> **DEFINITION**
>
> A **line segment** is the part of a line that consists of two points, known as endpoints, and all points between them.

Geometry in the Real World

On the road map, driving distances between towns are shown. In traveling from town *A* to town *D*, which path traverses the least distance?
Solution *A* to *E*, *E* to *C*, *C* to *D*:
10 + 4 + 5 = 19

EXAMPLE 1

State the four characteristics of a good definition of the term "line segment."

1. The term being defined, *line segment*, is clearly present in the definition.
2. A line segment is defined as part of a line (a category).
3. The definition distinguishes the line segment as a specific part of a line.
4. The definition is reversible.
   i) A line segment is the part of a line between and including two points.
   ii) The part of a line between and including two points is a line segment.

INITIAL POSTULATES

Recall that a postulate is a statement that is assumed to be true.

> **POSTULATE 1**
>
> Through two distinct points, there is exactly one line.

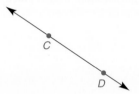

Figure 1.32

Postulate 1 is sometimes stated in the form "Two points determine a line." See Figure 1.32, in which points *C* and *D* determine exactly one line, namely $\overleftrightarrow{CD}$. Of course, Postulate 1 also implies that there is a unique line segment determined by two distinct points used as endpoints. Recall Figure 1.31, in which points *A* and *B* determine $\overline{AB}$.

NOTE: In geometry, the reference numbers used with postulates (as in Postulate 1) need not be memorized.

EXAMPLE 2

In Figure 1.33, how many distinct lines can be drawn through
  a) point *A*?
  b) both points *A* and *B* at the same time?
  c) all points *A, B,* and *C* at the same time?

SOLUTION
  a) An infinite (countless) number
  b) Exactly one
  c) No line contains all three points.

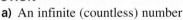

Figure 1.33

Recall from Section 1.2 that the symbol for line segment *AB*, named by its endpoints, is $\overline{AB}$. Omission of the bar from $\overline{AB}$, as in *AB*, means that we are considering the *length* of the segment. These symbols are summarized in Table 1.3.

TABLE 1.3

Symbol	Words for Symbol	Geometric Figure
$\overleftrightarrow{AB}$	Line AB	A ●————————● B (with arrows)
$\overline{AB}$	Line segment AB	A ●————————● B
AB	Length of segment AB	A number

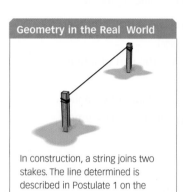
A ruler is used to measure the length of a line segment such as $\overline{AB}$. This length may be represented by AB or BA (the order of A and B is not important). However, AB must be a positive number.

POSTULATE 2 ■ Ruler Postulate

The measure of any line segment is a unique positive number.

We wish to call attention to the term *unique* and to the general notion of uniqueness. The Ruler Postulate implies the following:

1. There exists a number measure for each line segment.
2. Only *one* measure is permissible.

Characteristics 1 and 2 are both necessary for uniqueness! Other phrases that may replace the term *unique* include

One and only one

Exactly one

One and no more than one

A more accurate claim than the commonly heard statement "The shortest distance between two points is a straight line" is found in the following definition.

DEFINITION

The **distance** between two points A and B is the length of the line segment $\overline{AB}$ that joins the two points.

A ●—————●————————● B
⠀⠀⠀⠀X

Figure 1.34

As we saw in Section 1.2, there is a relationship between the lengths of the line segments determined in Figure 1.34. This relationship is stated in the third postulate. The title and meaning of the postulate are equally important! The title "Segment-Addition Postulate" will be cited frequently in later sections.

POSTULATE 3 ■ Segment-Addition Postulate

If X is a point of $\overline{AB}$ and A-X-B, then $AX + XB = AB$.

Technology Exploration

Use software if available.

1. Draw line segment $\overline{XY}$.
2. Choose point P on $\overline{XY}$.
3. Measure $\overline{XP}$, $\overline{PY}$, and $\overline{XY}$.
4. Show that $XP + PY = XY$.

EXAMPLE 3

In Figure 1.34, find AB if
⠀⠀⠀**a)** $AX = 7.32$ and $XB = 6.19$.⠀⠀⠀⠀⠀⠀**b)** $AX = 2x + 3$ and $XB = 3x - 7$.

SOLUTION

a) $AB = 7.32 + 6.19$, so $AB = 13.51$.

b) $AB = (2x + 3) + (3x - 7)$, so $AB = 5x - 4$.

> **DEFINITION**
>
> Congruent ($\cong$) line **segments** are two line segments that have the same length.

Figure 1.35

In general, geometric figures that can be made to coincide (fit perfectly one on top of the other) are said to be **congruent.** The symbol $\cong$ is a combination of the symbol ~, which means that the figures have the same shape, and $=$, which means that the corresponding parts of the figures have the same measure. In Figure 1.35, $\overline{AB} \cong \overline{CD}$, but $\overline{AB} \not\cong \overline{EF}$ (meaning that $\overline{AB}$ and $\overline{EF}$ are not congruent). Does it appear that $\overline{CD} \cong \overline{EF}$?

> **EXAMPLE 4**

In the U.S. system of measures, 1 foot = 12 inches. If $AB = 2.5$ feet and $CD = 2$ feet 6 inches, are $\overline{AB}$ and $\overline{CD}$ congruent?

SOLUTION Yes, $\overline{AB} \cong \overline{CD}$ because 2.5 feet = 2 feet + 0.5 feet or 2 feet + 0.5(12 inches), or 2 feet 6 inches.

> **DEFINITION**
>
> The **midpoint** of a line segment is the point that separates the line segment into two congruent parts.

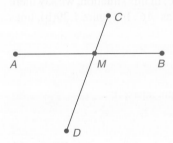

Figure 1.36

In Figure 1.36, if A, M, and B are collinear and $\overline{AM} \cong \overline{MB}$, then M is the **midpoint** of $\overline{AB}$. Equivalently, M is the midpoint of $\overline{AB}$ if $AM = MB$. Also, if $\overline{AM} \cong \overline{MB}$, then $\overline{CD}$ is described as a **bisector** of $\overline{AB}$.

If M is the midpoint of $\overline{AB}$ in Figure 1.36, we can draw any of these conclusions:

$$AM = MB \qquad MB = \tfrac{1}{2}(AB) \qquad AB = 2\,(MB)$$
$$AM = \tfrac{1}{2}(AB) \qquad AB = 2\,(AM)$$

> **EXAMPLE 5**

GIVEN: M is the midpoint of $\overline{EF}$ (not shown). $EM = 3x + 9$ and $MF = x + 17$

FIND: x, EM, and MF

SOLUTION Because M is the midpoint of $\overline{EF}$, $EM = MF$. Then

$$3x + 9 = x + 17$$
$$2x + 9 = 17$$
$$2x = 8$$
$$x = 4$$

By substitution, $EM = 3(4) + 9 = 12 + 9 = 21$ and $MF = 4 + 17 = 21$. Thus, $x = 4$ while $EM = MF = 21$.

In geometry, the word **union** is used to describe the joining or combining of two figures or sets of points.

Discover

Assume that M is the midpoint of $\overline{AB}$ in Figure 1.36. Can you also conclude that M is the midpoint of $\overline{CD}$?

ANSWER
ON

> **DEFINITION**
>
> **Ray** AB, denoted by $\overrightarrow{AB}$, is the union of $\overline{AB}$ and all points X on $\overleftrightarrow{AB}$ such that B is between A and X.

In Figure 1.37, $\overleftrightarrow{AB}$, $\overrightarrow{AB}$, and $\overrightarrow{BA}$ are shown in that order; note that $\overrightarrow{AB}$ and $\overrightarrow{BA}$ are not the same ray.

Figure 1.37

Opposite rays are two rays with a common endpoint; also, the union of opposite rays is a straight line. In Figure 1.39(a), $\overrightarrow{BA}$ and $\overrightarrow{BC}$ are opposite rays.

The **intersection** of two geometric figures is the set of points that the two figures have in common. In everyday life, the intersection of Bradley Avenue and Neil Street is the part of the roadway that the two roads have in common (Figure 1.38).

> **POSTULATE 4**
>
> If two lines intersect, they intersect at a point.

When two lines share two (or more) points, the lines coincide; in this situation, we say there is only one line. In Figure 1.39(a), $\overleftrightarrow{AB}$ and $\overleftrightarrow{BC}$ are the same as $\overleftrightarrow{AC}$. In Figure 1.39(b), lines ℓ and m intersect at point P.

> **DEFINITION**
>
> **Parallel lines** are lines that lie in the same plane but do not intersect.

In Figure 1.40, suppose that ℓ and n are parallel; in symbols, $\ell \parallel n$ and $\ell \cap n = \varnothing$. However, ℓ and m are not parallel because they intersect at point A; so $\ell \nparallel m$ and $\ell \cap m = A$.

Figure 1.38

(a)

(b)

Figure 1.39

 EXS. 5–12

> **EXAMPLE 6**

In Figure 1.40, $\ell \parallel n$. What is the intersection of

a) lines n and m?
b) lines ℓ and n?

SOLUTION

a) Point B
b) Parallel lines do not intersect.

Figure 1.40

Another undefined term in geometry is **plane.** A plane is two-dimensional; that is, it has infinite length and infinite width but no thickness. Except for its limited size, a flat surface such as the top of a table could be used as an example of a plane. An uppercase letter can be used to name a plane. Because a plane (like a line) is infinite, we can show only a portion of the plane or planes, as in Figure 1.41.

Planes *R* and *S*   Planes *T* and *V*

Figure 1.41

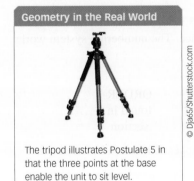

Figure 1.42

A plane is two-dimensional, consists of an infinite number of points, and contains an infinite number of lines. Two distinct points may determine (or "fix") a line; likewise, exactly three noncollinear points determine a plane. Just as collinear points lie on the same line, **coplanar points** lie in the same plane. In Figure 1.42, points *B*, *C*, *D*, and *E* are coplanar, whereas *A*, *B*, *C*, and *D* are noncoplanar.

In this book, points shown in figures are generally assumed to be coplanar unless otherwise stated. For instance, points *A*, *B*, *C*, *D*, and *E* are coplanar in Figure 1.43(a), as are points *F*, *G*, *H*, *J*, and *K* in Figure 1.43(b).

(a)   (b)

Figure 1.43

POSTULATE 5

Through three noncollinear points, there is exactly one plane.

On the basis of Postulate 5, we can see why a three-legged table sits evenly but a four-legged table would "wobble" if the legs were of unequal length.

Space is the set of all possible points. It is three-dimensional, having qualities of length, width, and depth. When two planes intersect in space, their intersection is a line. An opened greeting card suggests this relationship, as does Figure 1.44(a). This notion gives rise to our next postulate.

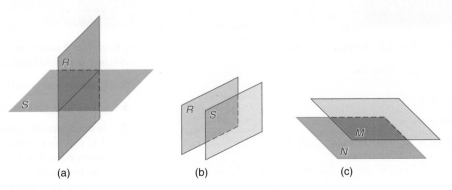

(a)   (b)   (c)

Figure 1.44

Discover

During a baseball game, the catcher and the third baseman follow the path of a foul pop fly toward the grandstand. Does it appear that there is a play on the baseball?

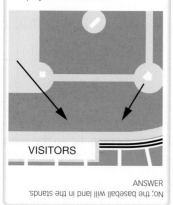

VISITORS

ANSWER
No; the baseball will land in the stands.

 EXS. 13–16

A   M   B

Figure 1.45

 EXS. 17–20

POSTULATE 6

If two distinct planes intersect, then their intersection is a line.

The intersection of two planes is infinite because it is a line. [See Figure 1.44(a) on page 25.] If two planes do not intersect, then they are **parallel.** The parallel **vertical** planes *R* and *S* in Figure 1.44(b) may remind you of the opposite walls of your classroom. The parallel **horizontal** planes *M* and *N* in Figure 1.44(c) suggest the relationship between the ceiling and the floor.

Imagine a plane and two points of that plane, say points *A* and *B*. Now think of the line containing the two points and the relationship of $\overleftrightarrow{AB}$ to the plane. Perhaps your conclusion can be summed up as follows.

POSTULATE 7

Given two distinct points in a plane, the line containing these points also lies in the plane.

Because the uniqueness of the midpoint of a line segment can be justified, we call the following statement a theorem. The "proof" of the theorem is found in Section 2.2.

THEOREM 1.3.1

The midpoint of a line segment is unique.

If *M* is the midpoint of $\overline{AB}$ in Figure 1.45, then no other point can separate $\overline{AB}$ into two congruent parts. The proof of this theorem is based on the Ruler Postulate. *M* is *the* point that is located $\frac{1}{2}(AB)$ units from *A* (and from *B*).

The numbering system used to identify Theorem 1.3.1 need not be memorized. However, this theorem number may be used in a later reference. The numbering system works as follows:

1	3	1
CHAPTER	SECTION	ORDER
where	where	found in
found	found	section

A summary of the theorems presented in this textbook appears at the end of the book.

Exercises 1.3

In Exercises 1 and 2, complete the statement.

A   B   C

Exercises 1, 2

1. $AB + BC =$ __?__

2. If $AB = BC$, then *B* is the __?__ of $\overline{AC}$.

In Exercises 3 and 4, use the fact that 1 foot = 12 inches.

3. Convert 6.25 feet to a measure in inches.

4. Convert 52 inches to a measure in feet and inches.

In Exercises 5 and 6, use the fact that 1 meter ≈ 3.28 feet (measure is approximate).

5. Convert $\frac{1}{2}$ meter to feet.

6. Convert 16.4 feet to meters.

7. In the figure, the 15-mile road from *A* to *C* is under construction. A detour from *A* to *B* of 5 miles and then from *B* to *C* of 13 miles must be taken. How much farther is the "detour" from *A* to *C* than the road from *A* to *C*?

Exercises 7, 8

8. A cross-country runner jogs at a rate of 15 feet per second. If she runs 300 feet from A to B, 450 feet from B to C, and then 600 feet from C back to A, how long will it take her to return to point A? See figure for Exercise 7.

In Exercises 9 to 28, use the drawings as needed to answer the following questions.

9. Name three points that appear to be
   a) collinear.   b) noncollinear.

10. How many lines can be drawn through
   a) point A?
   b) points A and B?
   c) points A, B, and C?
   d) points A, B, and D?

Exercises 9, 10

11. Give the meanings of $\overleftrightarrow{CD}$, $\overline{CD}$, CD, and $\overrightarrow{CD}$.

12. Explain the difference, if any, between
   a) $\overleftrightarrow{CD}$ and $\overleftrightarrow{DC}$.   c) $\overline{CD}$ and DC.
   b) $\overline{CD}$ and $\overline{DC}$.   d) $\overrightarrow{CD}$ and $\overrightarrow{DC}$.

13. Name two lines that appear to be
   a) parallel.   b) nonparallel.

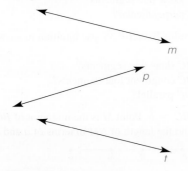

14. Classify as true or false:
   a) $AB + BC = AD$ d) $AB + BC + CD = AD$
   b) $AD - CD = AB$ e) $AB = BC$
   c) $AD - CD = AC$

Exercises 14–17

15. *Given:* M is the midpoint of $\overline{AB}$
              $AM = 2x + 1$ and $MB = 3x - 2$
    Find: x and AM

16. *Given:* M is the midpoint of $\overline{AB}$
              $AM = 2(x + 1)$ and $MB = 3(x - 2)$
    Find: x and AB

17. *Given:* $AM = 2x + 1$, $MB = 3x + 2$, and
              $AB = 6x - 4$
    Find: x and AB

18. Can a segment bisect a line? a segment? Can a line bisect a segment? a line?

19. In the figure, name
   a) two opposite rays.
   b) two rays that are not opposite.

20. Suppose that (a) point C lies in plane X and (b) point D lies in plane X. What can you conclude regarding $\overleftrightarrow{CD}$?

21. Make a sketch of
   a) two intersecting lines that are perpendicular.
   b) two intersecting lines that are *not* perpendicular.
   c) two parallel lines.

22. Make a sketch of
   a) two intersecting planes.
   b) two parallel planes.
   c) two parallel planes intersected by a third plane that is not parallel to the first or the second plane.

23. Suppose that (a) planes M and N intersect, (b) point A lies in both planes M and N, and (c) point B lies in both planes M and N. What can you conclude regarding $\overleftrightarrow{AB}$?

24. Suppose that (a) points A, B, and C are collinear and (b) $AB > AC$. Which point can you conclude *cannot* lie between the other two?

25. Suppose that points A, R, and V are collinear. If $AR = 7$ and $RV = 5$, then which point cannot possibly lie between the other two?

26. Points A, B, C, and D are coplanar; B, C, and D are collinear; point E is not in plane M. How many planes contain
   a) points A, B, and C?
   b) points B, C, and D?
   c) points A, B, C, and D?
   d) points A, B, C, and E?

27. Using the number line provided, name the point that
   a) is the midpoint of $\overline{AE}$.
   b) is the endpoint of a segment of length 4, if the other endpoint is point G.
   c) has a distance from B equal to $3(AC)$.

Exercises 27, 28

28. Consider the figure for Exercise 27. Given that B is the midpoint of $\overline{AC}$ and C is the midpoint of $\overline{BD}$, what can you conclude about the lengths of
   a) $\overline{AB}$ and $\overline{CD}$?   c) $\overline{AC}$ and $\overline{CD}$?
   b) $\overline{AC}$ and $\overline{BD}$?

In Exercises 29 to 32, use only a compass and a straightedge to complete each construction.

29. *Given:* $\overline{AB}$ and $\overline{CD}$ ($AB > CD$)
    Construct: $\overline{MN}$ on line ℓ so that $MN = AB + CD$

Exercises 29, 30

30. *Given:* $\overline{AB}$ and $\overline{CD}$ ($AB > CD$)
    Construct: $\overline{EF}$ on line ℓ so that $EF = AB - CD$

31. *Given:* $\overline{AB}$ as shown in the figure
   Construct: $\overline{PQ}$ on line *n* so that $PQ = 3(AB)$

Exercises 31, 32

32. *Given:* $\overline{AB}$ as shown in the figure
   Construct: $\overline{TV}$ on line *n* so that $TV = \frac{1}{2}(AB)$

33. Can you use the construction for the midpoint of a segment to divide a line segment into
   a)   three congruent parts?   c)   six congruent parts?
   b)   four congruent parts?   d)   eight congruent parts?

34. Generalize your findings in Exercise 33.

35. Consider points *A, B, C,* and *D*, no three of which are collinear. Using two points at a time (such as *A* and *B*), how many lines are determined by these points?

36. Consider noncoplanar points *A, B, C,* and *D*. Using three points at a time (such as *A, B,* and *C*), how many planes are determined by these points?

37. Line ℓ is parallel to plane *P* (that is, it will not intersect *P* even if extended). Line *m* intersects line ℓ. What can you conclude about *m* and *P*?

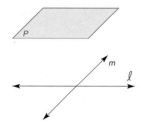

38. $\overleftrightarrow{AB}$ and $\overleftrightarrow{EF}$ are said to be **skew** lines because they neither intersect nor are parallel. How many planes are determined by
   a)   parallel lines *AB* and *DC*?
   b)   intersecting lines *AB* and *BC*?
   c)   skew lines *AB* and *EF*?
   d)   lines *AB, BC,* and *DC*?
   e)   points *A, B,* and *F*?
   f)   points *A, C,* and *H*?
   g)   points *A, C, F,* and *H*?

Exercises 38–40

39. In the "box" shown for Exercise 38, use intuition to answer each question.
   a) Are $\overline{AB}$ and $\overline{DC}$ parallel?
   b) Are $\overline{AB}$ and $\overline{FE}$ skew line segments?
   c) Are $\overline{AB}$ and $\overline{FE}$ perpendicular?

40. In the "box" shown for Exercise 38, use intuition to answer each question.
   a)   Are $\overline{AG}$ and $\overline{BC}$ skew line segments?
   b)   Are $\overline{AG}$ and $\overline{BC}$ congruent line segments?
   c)   Are $\overline{GF}$ and $\overline{DC}$ parallel?

*41. Let $AB = a$ and $BC = b$. Point *M* is the midpoint of $\overline{BC}$. If $AN = \frac{2}{3}(AB)$, find the length of $\overline{NM}$ in terms of *a* and *b*.

1.4   Angles and Their Relationships

KEY CONCEPTS	Angle: Sides of Angle, Vertex of Angle	Angle-Addition Postulate	Complementary Angles Supplementary Angles
	Protractor Postulate	Adjacent Angles	Vertical Angles
	Acute, Right, Obtuse, Straight, and Reflex Angles	Congruent Angles Bisector of an Angle	

This section introduces you to the language of angles. Recall from Sections 1.1 and 1.3 that the word *union* means that two sets or figures are joined.

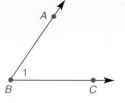

Figure 1.46

> **DEFINITION**
>
> An **angle** is the union of two rays that share a common endpoint.

The preceding definition is illustrated in Figure 1.46, in which $\overrightarrow{BA}$ and $\overrightarrow{BC}$ have the common endpoint *B*.

As shown, this angle is represented by $\angle ABC$ or $\angle CBA$. The rays BA and BC are known as the **sides** of the angle. B, the common endpoint of these rays, is known as the **vertex** of the angle. When three letters are used to name an angle, the vertex is always named in the middle. Recall that a single letter or numeral may be used to name the angle. The angle in Figure 1.46 may be described as $\angle B$ (the vertex of the angle) or as $\angle 1$. In set notation, we see that $\angle B = \overrightarrow{BA} \cup \overrightarrow{BC}$.

POSTULATE 8 ■ Protractor Postulate

The measure of an angle is a unique positive number.

NOTE: In Chapters 1 to 10, the measures of most angles will be between 0° and 180°, including 180°. Angles with measures between 180° and 360° are introduced in this section; these angles are not often encountered in our study of geometry.

TYPES OF ANGLES

An angle whose measure is less than 90° is an **acute angle.** If the angle's measure is exactly 90°, the angle is a **right angle.** If the angle's measure is between 90° and 180°, the angle is **obtuse.** An angle whose measure is exactly 180° is a **straight angle;** alternatively, a straight angle is one whose sides form opposite rays (a straight line). A **reflex angle** is one whose measure is between 180° and 360°. See Table 1.4 on page 30.

Figure 1.47

In Figure 1.47, $\angle ABC$ contains the noncollinear points A, B, and C. Unless otherwise stated or indicated (by an arc), $\angle ABC$ is an acute angle. The three points A, B, and C also determine a plane. The plane containing $\angle ABC$ is separated into three subsets by the angle:

Points such as D are said to be in the *interior* of $\angle ABC$.

Points such as E are said to be *on* $\angle ABC$.

Points such as F are said to be in the *exterior* of $\angle ABC$.

With this description, it is possible to state the Angle-Addition Postulate, which is the counterpart of the Segment-Addition Postulate! Consider Figure 1.48 as you read Postulate 9.

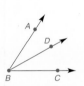

Figure 1.48

POSTULATE 9 ■ Angle-Addition Postulate

If a point D lies in the interior of an angle ABC, then $m\angle ABD + m\angle DBC = m\angle ABC$.

Technology Exploration

Use software if available.
1. Draw ∠RST.
2. Through point V in the interior of ∠RST, draw $\overrightarrow{SV}$.
3. Measure ∠RST, ∠RSV, and ∠VST.
4. Show that m∠RSV + m∠VST = m∠RST.

TABLE 1.4
Angles

Angle	Example
Acute (1)	m∠1 = 23°
Right (2)	m∠2 = 90°
Obtuse (3)	m∠3 = 112°
Straight (4)	m∠4 = 180°
Reflex (5)	m∠5 = 337°

NOTE: An arc is necessary in indicating a reflex angle, and it can be used to indicate a straight angle as well.

Discover

When greater accuracy is needed in angle measurement, a degree can be divided into 60 minutes. In symbols, 1° = 60′. Convert 22.5° to degrees and minutes.

ANSWER
22° 30′,

EXAMPLE 1

Use Figure 1.48 on page 29 to find m∠ABC if:

a) m∠ABD = 27° and m∠DBC = 42°
b) m∠ABD = x° and m∠DBC = $(2x - 3)$°

SOLUTION
a) Using the Angle-Addition Postulate,
m∠ABC = m∠ABD + m∠DBC. That is, m∠ABC = 27° + 42° = 69°.
b) m∠ABC = m∠ABD + m∠DBC = x° + $(2x - 3)$° = $(3x - 3)$°.

Discover

An index card can be used to categorize the types of angles displayed. In each sketch, an index card is placed over an angle. A dashed ray indicates that a side is hidden. What type of angle is shown in each figure? (Note the placement of the card in each figure.)

One edge of the index card coincides with both of the angle's sides	Sides of the angle coincide with two edges of the card	Card hides the second side of the angle	Card exposes the second side of the angle

ANSWER
Obtuse angle   Acute angle   Right Angle   Straight Angle

SSG EXS. 1–6

CLASSIFYING PAIRS OF ANGLES

Many angle relationships involve exactly two angles (a pair)—never more than two angles and never less than two angles!

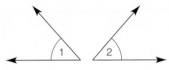

Figure 1.48

In Figure 1.48, $\angle ABD$ and $\angle DBC$ are said to be *adjacent* angles. In this description, the term *adjacent* means that angles lie "next to" each other; in everyday life, one might say that the Subway sandwich shop is adjacent to the Baskin-Robbins ice cream shop. When two angles are adjacent, they have a common vertex and a common side between them. In Figure 1.48, $\angle ABC$ and $\angle ABD$ are not adjacent because they have interior points in common; notice that the common side $(\overrightarrow{BA})$ does not lie between $\angle ABC$ and $\angle ABD$.

> **DEFINITION**
>
> Two angles are **adjacent** (adj. $\angle$s) if they have a common vertex and a common side between them.

We now recall the meaning of *congruent* angles.

> **DEFINITION**
>
> **Congruent angles** ($\cong \angle$**s**) are two angles with the same measure.

Congruent angles must coincide when one is placed over the other. (Do not consider that the sides appear to have different lengths; remember that rays are infinite in length!) In symbols, $\angle 1 \cong \angle 2$ if $m\angle 1 = m\angle 2$. In Figure 1.49, as well as in later figures, similar markings (arcs) indicate that two angles are congruent; thus, $\angle 1 \cong \angle 2$.

Figure 1.49

EXAMPLE 2

GIVEN: $\angle 1 \cong \angle 2$
$m\angle 1 = 2x + 15$
$m\angle 2 = 3x - 2$

FIND: x

SOLUTION $\angle 1 \cong \angle 2$ means $m\angle 1 = m\angle 2$. Therefore,

$$2x + 15 = 3x - 2$$
$$17 = x \quad \text{or} \quad x = 17$$

NOTE: $m\angle 1 = 2(17) + 15 = 49°$ and $m\angle 2 = 3(17) - 2 = 49°$.

> **DEFINITION**
>
> The **bisector** of an angle is the ray that separates the given angle into two congruent angles.

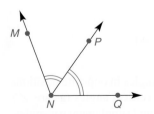

Figure 1.50

With P in the interior of $\angle MNQ$ so that $\angle MNP \cong \angle PNQ$, $\overrightarrow{NP}$ is said to **bisect** $\angle MNQ$. Equivalently, $\overrightarrow{NP}$ is the bisector or angle-bisector of $\angle MNQ$. On the basis of Figure 1.50, possible consequences of the definition of bisector of an angle are

$m\angle MNP = m\angle PNQ \quad m\angle MNQ = 2(m\angle PNQ) \quad m\angle MNQ = 2(m\angle MNP)$
$m\angle PNQ = \frac{1}{2}(m\angle MNQ) \quad m\angle MNP = \frac{1}{2}(m\angle MNQ)$

> **DEFINITION**
>
> Two angles are **complementary** if the sum of their measures is 90°. Each angle in the pair is known as the **complement** of the other angle.

Angles with measures of 37° and 53° are complementary. The 37° angle is the complement of the 53° angle, and vice versa. If the measures of two angles are x and y and it is known that $x + y = 90°$, then these two angles are complementary.

> **DEFINITION**
>
> Two angles are **supplementary** if the sum of their measures is 180°. Each angle in the pair is known as the **supplement** of the other angle.

Discover

In a game of billiards, a ball ricochets off an end bumper to create congruent angles *BCA* and *DCE*. If m∠*BCD* = 34°, find m∠*DCE*.

ANSWER
73°

EXAMPLE 3

Given that m∠1 = 29°, find:
a) the complement x of ∠1   **b)** the supplement y of ∠1

SOLUTION
   a) $x + 29 = 90$, so $x = 61°$; complement $= 61°$
   b) $y + 29 = 180$, so $y = 151°$; supplement $= 151°$

EXAMPLE 4

GIVEN: ∠P and ∠Q are complementary, where

$$m\angle P = \frac{x}{2} \quad \text{and} \quad m\angle Q = \frac{x}{3}$$

FIND: x, m∠P, and m∠Q

SOLUTION

$$m\angle P + m\angle Q = 90$$
$$\frac{x}{2} + \frac{x}{3} = 90$$

Multiplying by 6 (the least common denominator, or LCD, of 2 and 3), we have

$$6 \cdot \frac{x}{2} + 6 \cdot \frac{x}{3} = 6 \cdot 90$$
$$3x + 2x = 540$$
$$5x = 540$$
$$x = 108$$

$$m\angle P = \frac{x}{2} = \frac{108}{2} = 54°$$

$$m\angle Q = \frac{x}{3} = \frac{108}{3} = 36°$$

NOTE: m∠P = 54° and m∠Q = 36°, so their sum is exactly 90°.

Figure 1.51

When two straight lines intersect, the pairs of nonadjacent angles in opposite positions are known as **vertical angles.** In Figure 1.51, ∠5 and ∠6 are vertical angles (as are ∠7 and ∠8). In addition, ∠5 and ∠7 can be described as adjacent and supplementary angles, as can ∠5 and ∠8. If m∠7 = 30°, what is m∠5 and what is m∠8? It is true in general that vertical angles are congruent, and we will prove this in Example 3 of Section 1.6. We apply this property in Example 5 of this section.

Recall the Addition and Subtraction Properties of Equality: If $a = b$ and $c = d$, then $a \pm c = b \pm d$. These principles can be used in solving a system of equations, such as the following:

SSG **EXS. 7–12**

$$x + y = 5$$
$$\underline{2x - y = 7}$$
$$3x \quad\quad = 12 \quad \text{(left and right sides are added)}$$
$$x = 4$$

We can substitute 4 for x in either equation to solve for y:

$$x + y = 5$$
$$4 + y = 5 \quad \text{(by substitution)}$$
$$y = 1$$

If $x = 4$ and $y = 1$, then $x + y = 5$ and $2x - y = 7$.

When each term in an equation is multiplied by the same nonzero number, the solutions of the equation are not changed. For instance, the equations $2x - 3 = 7$ and $6x - 9 = 21$ (each term multiplied by 3) both have the solution $x = 5$. Likewise, the values of x and y that make the equation $4x + y = 180$ true also make the equation $16x + 4y = 720$ (each term multiplied by 4) true. We use this method in Example 5.

EXAMPLE 5

GIVEN: In Figure 1.51 on page 32, ℓ and m intersect so that

$$m\angle 5 = 2x + 2y$$
$$m\angle 8 = 2x - y$$
$$m\angle 6 = 4x - 2y$$

FIND: x and y

SOLUTION $\angle 5$ and $\angle 8$ are supplementary (adjacent and exterior sides form a straight angle). Therefore, $m\angle 5 + m\angle 8 = 180$. $\angle 5$ and $\angle 6$ are congruent (vertical). Therefore, $m\angle 5 = m\angle 6$. Consequently, we have

$$(2x + 2y) + (2x - y) = 180 \quad \text{(supplementary } \angle\text{s 5 and 8)}$$
$$2x + 2y = 4x - 2y \quad \text{($\cong \angle$s 5 and 6)}$$

Simplifying, $$4x + y = 180$$
$$2x - 4y = 0$$

Using the Multiplication Property of Equality, we multiply the equation $4x + y = 180$ by 4. Then the equivalent system allows us to eliminate variable y by addition.

$$16x + 4y = 720$$
$$\underline{2x - 4y = 0}$$
$$18x = 720 \quad \text{(adding left, right sides)}$$
$$x = 40$$

Using the equation $4x + y = 180$, it follows that

$$4(40) + y = 180$$
$$160 + y = 180$$
$$y = 20$$

Summarizing, $x = 40$ and $y = 20$.

NOTE: It follows that $m\angle 5 = 120°$, $m\angle 8 = 60°$, and $m\angle 6 = 120°$.

Discover

On the grid shown, points are uniformly spaced. Name two congruent line segments in figure (a). Name two congruent angles in figure (b).

(a)   (b)

ANSWER

(a) $\overline{AB} \cong \overline{CD}$ (b) $\angle A \cong \angle D$

CONSTRUCTIONS WITH ANGLES

In Section 1.2, we considered Constructions 1 and 2 with line segments. Now consider two constructions that involve angle concepts. In Section 3.4, it will become clear why these methods are valid. However, intuition suggests that the techniques are appropriate.

(a)

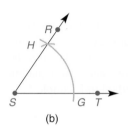

(b)

CONSTRUCTION 3 To construct an angle congruent to a given angle.

GIVEN: ∠RST in Figure 1.52(a)

CONSTRUCT: With $\overrightarrow{PQ}$ as one side, ∠NPQ ≅ ∠RST

CONSTRUCTION: Figure 1.52(b): With a compass, mark an arc to intersect both sides of ∠RST at points G and H.

Figure 1.52(c): Without changing the radius, mark an arc to intersect $\overrightarrow{PQ}$ at K and the "would-be" second side of ∠NPQ.

Figure 1.52(b): Now mark an arc to measure the distance from G to H.

Figure 1.52(d): Using the same radius as in the preceding step, mark an arc with K as center to intersect the would-be second side of the desired angle. Now draw the ray from P through the point of intersection of the two arcs.

The resulting angle (∠NPQ) is the one desired, as we will prove in Section 3.4, Example 1.

Just as a line segment can be bisected, so can an angle. This takes us to a fourth construction method.

(c)

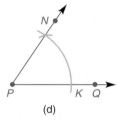

(d)

Figure 1.52

SSG **EXS. 13–20**

CONSTRUCTION 4 To construct the bisector of a given angle.

GIVEN: ∠PRT in Figure 1.53(a)

CONSTRUCT: $\overrightarrow{RS}$ so that ∠PRS ≅ ∠SRT

(a)

(b)

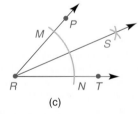

(c)

Figure 1.53

CONSTRUCTION: Figure 1.53(b): Using a compass, mark an arc to intersect the sides of ∠PRT at points M and N.

Figure 1.53(c): Now, with M and N as centers, mark off two arcs with equal radii to intersect at point S in the interior of ∠PRT, as shown. Now draw ray RS, the desired angle bisector.

Reasoning from the definition of an angle bisector, the Angle-Addition Postulate, and the Protractor Postulate, we can justify the following theorem.

THEOREM 1.4.1

There is one and only one bisector for a given angle.

This theorem is often stated, "The bisector of an angle is unique." This statement is proved in Example 5 of Section 2.2.

Exercises **1.4**

1. What type of angle has the given measure?
   a) 47°   b) 90°   c) 137.3°

2. What type of angle has the given measure?
   a) 115°   b) 180°   c) 36°

3. What *relationship*, if any, exists between two angles
   a) with measures of 37° and 53°?
   b) with measures of 37° and 143°?

4. What *relationship*, if any, exists between two angles
   a) with equal measures?
   b) that have the same vertex and a common side between
   them?

*In Exercises 5 to 8, describe in one word the relationship
between the angles.*

5. ∠*ABD* and ∠*DBC*   6. ∠7 and ∠8

  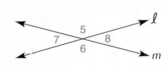

7. ∠1 and ∠2   8. ∠3 and ∠4

*Use drawings as needed to answer each of the following
questions.*

9. Must two rays with a common endpoint be coplanar? Must
   three rays with a common endpoint be coplanar?

10. Suppose that $\overrightarrow{AB}$, $\overrightarrow{AC}$, $\overrightarrow{AD}$, $\overrightarrow{AE}$, and $\overrightarrow{AF}$ are coplanar.

   Exercises 10–13

   Classify the following as true or false:
   a) m∠*BAC* + m∠*CAD* = m∠*BAD*
   b) ∠*BAC* ≅ ∠*CAD*
   c) m∠*BAE* − m∠*DAE* = m∠*BAC*
   d) ∠*BAC* and ∠*DAE* are adjacent
   e) m∠*BAC* + m∠*CAD* + m∠*DAE* = m∠*BAE*

11. Without using a protractor, name the type of angle repre-
    sented by:
    a) ∠*BAE*   b) ∠*FAD*   c) ∠*BAC*   d) ∠*FAE*

12. What, if anything, is wrong with the claim
    m∠*FAB* + m∠*BAE* = m∠*FAE*?

13. ∠*FAC* and ∠*CAD* are adjacent and $\overrightarrow{AF}$ and $\overrightarrow{AD}$ are oppo-
    site rays. What can you conclude about ∠*FAC* and ∠*CAD*?

For Exercises 14 and 15, let m∠1 = x and m∠2 = y.

14. Using variables *x* and *y*, write an equation that expresses the
    fact that ∠1 and ∠2 are:
    a) supplementary   b) congruent

15. Using variables *x* and *y*, write an equation that expresses the
    fact that ∠1 and ∠2 are:
    a) complementary   b) vertical

16. *Given:*   m∠*RST* = 39°
              m∠*TSV* = 23°
    Find: m∠*RSV*

    Exercises 16–24

17. *Given:*   m∠*RSV* = 59°
              m∠*TSV* = 17°
    Find: m∠*RST*

18. *Given:*   m∠*RST* = 2*x* + 9
              m∠*TSV* = 3*x* − 2
              m∠*RSV* = 67°
    Find: *x*

19. *Given:*   m∠*RST* = 2*x* − 10
              m∠*TSV* = *x* + 6
              m∠*RSV* = 4(*x* − 6)
    Find: *x* and m∠*RSV*

20. *Given:*   m∠*RST* = 5(*x* + 1) − 3
              m∠*TSV* = 4(*x* − 2) + 3
              m∠*RSV* = 4(2*x* + 3) − 7
    Find: *x* and m∠*RSV*

21. *Given:*   m∠*RST* = $\frac{x}{2}$
              m∠*TSV* = $\frac{x}{4}$
              m∠*RSV* = 45°
    Find: *x* and m∠*RST*

22. *Given:*   m∠*RST* = $\frac{2x}{3}$
              m∠*TSV* = $\frac{x}{2}$
              m∠*RSV* = 49°
    Find: *x* and m∠*TSV*

For Exercises 23 and 24, see figure on page 35.

23. *Given:* $\overrightarrow{ST}$ bisects $\angle RSV$
$m\angle RST = x + y$
$m\angle TSV = 2x - 2y$
$m\angle RSV = 64°$
Find: x and y

24. *Given:* $\overrightarrow{ST}$ bisects $\angle RSV$
$m\angle RST = 2x + 3y$
$m\angle TSV = 3x - y + 2$
$m\angle RSV = 80°$
Find: x and y

25. *Given:* $\overleftrightarrow{AB}$ and $\overleftrightarrow{AC}$ in plane P as shown
$\overleftrightarrow{AD}$ intersects P at point A
$\angle CAB \cong \angle DAC$
$\angle DAC \cong \angle DAB$
What can you conclude?

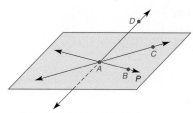

26. Two angles are complementary. One angle is 12° larger than the other. Using two variables x and y, find the size of each angle by solving a system of equations.

27. Two angles are supplementary. One angle is 24° more than twice the other. Using two variables x and y, find the measure of each angle.

28. For two complementary angles, find an expression for the measure of the second angle if the measure of the first is:
a) $x°$
b) $(3x - 12)°$
c) $(2x + 5y)°$

29. Suppose that two angles are supplementary. Find expressions for the supplements, using the expressions provided in Exercise 28, parts (a) to (c).

30. On the protractor shown, $\overrightarrow{NP}$ bisects $\angle MNQ$. Find x.

Exercises 30, 31

31. On the protractor shown for Exercise 30, $\angle MNP$ and $\angle PNQ$ are complementary. Find x.

32. Classify as true or false:
a) If points P and Q lie in the interior of $\angle ABC$, then $\overline{PQ}$ lies in the interior of $\angle ABC$.
b) If points P and Q lie in the interior of $\angle ABC$, then $\overleftrightarrow{PQ}$ lies in the interior of $\angle ABC$.
c) If points P and Q lie in the interior of $\angle ABC$, then $\overrightarrow{PQ}$ lies in the interior of $\angle ABC$.

In Exercises 33 to 40, use only a compass and a straightedge to perform the indicated constructions.

Exercises 33–35

33. *Given:* Obtuse $\angle MRP$
Construct: With $\overrightarrow{OA}$ as one side, an angle $\cong \angle MRP$

34. *Given:* Obtuse $\angle MRP$
Construct: $\overrightarrow{RS}$, the angle bisector of $\angle MRP$

35. *Given:* Obtuse $\angle MRP$
Construct: Rays $\overrightarrow{RS}$, $\overrightarrow{RT}$, and $\overrightarrow{RU}$ so that $\angle MRP$ is divided into four $\cong$ angles

36. *Given:* Straight $\angle DEF$
Construct: A right angle with vertex at E

(HINT: Use Construction 4.)

37. Draw a triangle with three acute angles. Construct angle bisectors for each of the three angles. On the basis of the appearance of your construction, what seems to be true?

38. *Given:* Acute $\angle 1$ and $\overline{AB}$
Construct: Triangle ABC with $\angle A \cong \angle 1$, $\angle B \cong \angle 1$, and side $\overline{AB}$

39. What seems to be true of two of the sides in the triangle you constructed in Exercise 38?

40. *Given:* Straight $\angle ABC$ and $\overrightarrow{BD}$
Construct: Bisectors of $\angle ABD$ and $\angle DBC$
What type of angle is formed by the bisectors of the two angles?

41. Refer to the circle with center *O*.
    a) Use a protractor to find m∠*B*.
    b) Use a protractor to find m∠*D*.
    c) Compare results in parts (a) and (b).

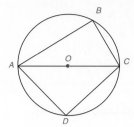

42. If m∠*TSV* = 38°, m∠*USW* = 40°, and m∠*TSW* = 61°, find m∠*USV*.

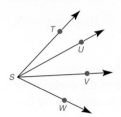

Exercises 42, 43

43. If m∠*TSU* = *x* + 2*z*, m∠*USV* = *x* − *z*, and m∠*VSW* = 2*x* − *z*, find *x* if m∠*TSW* = 60. Also, find *z* if m∠*USW* = 3*x* − 6.

44. Refer to the circle with center *P*.
    a) Use a protractor to find m∠1.
    b) Use a protractor to find m∠2.
    c) Compare results in parts (a) and (b).

45. On the hanging sign, the three angles (∠*ABD*, ∠*ABC*, and ∠*DBC*) at vertex *B* have the sum of measures 360°. If m∠*DBC* = 90° and $\overrightarrow{BA}$ bisects the indicated reflex angle, find m∠*ABC*.

46. With 0 < *x* < 90, an acute angle has measure *x*. Find the difference between the measure of its supplement and the measure of its complement.

1.5 Introduction to Geometric Proof

KEY CONCEPTS   Algebraic Properties   Proof   Given Problem and Prove Statement

To believe certain geometric principles, it is necessary to have proof. This section introduces some guidelines for establishing the proof of these geometric properties. Several examples are offered to help you develop your own proofs. In the beginning, the form of proof will be a two-column proof, with statements in the left column and reasons in the right column. But where do the statements and reasons come from?

To deal with this question, you must ask "What" is known (Given) and "Why" the conclusion (Prove) should follow from this information. In correctly piecing together a proof, you will usually scratch out several conclusions, discarding some and reordering the rest. Each conclusion must be justified by citing the Given (hypothesis), a previously stated definition or postulate, or a theorem previously proved.

Selected properties from algebra are often used as reasons to justify statements. For instance, we use the Addition Property of Equality to justify adding the same number to each side of an equation. Reasons found in a proof often include the properties found in Tables 1.5 and 1.6 on page 38.

Reminder

Additional properties and techniques of algebra are found in Appendix A.

TABLE 1.5
Properties of Equality (a, b, and c are real numbers)

Addition Property of Equality:	If $a = b$, then $a + c = b + c$.
Subtraction Property of Equality:	If $a = b$, then $a - c = b - c$.
Multiplication Property of Equality:	If $a = b$, then $a \cdot c = b \cdot c$.
Division Property of Equality:	If $a = b$ and $c \neq 0$, then $\dfrac{a}{c} = \dfrac{b}{c}$.

As we discover in Example 1, some properties can be used interchangeably.

EXAMPLE 1

Which property of equality justifies each conclusion?
    a) If $2x - 3 = 7$, then $2x = 10$.   **b)** If $2x = 10$, then $x = 5$.

SOLUTION
    a) Addition Property of Equality; added 3 to each side of the equation.
    b) Multiplication Property of Equality; multiplied each side of the equation by $\frac{1}{2}$.
        OR Division Property of Equality; divided each side of the equation by 2.

TABLE 1.6
Further Algebraic Properties of Equality (a, b, and c are real numbers)

Reflexive Property:	$a = a$.
Symmetric Property:	If $a = b$, then $b = a$.
Distributive Property:	$a(b + c) = a \cdot b + a \cdot c$.
Substitution Property:	If $a = b$, then a replaces b in any equation.
Transitive Property:	If $a = b$ and $b = c$, then $a = c$.

Before considering geometric proof, we study algebraic proof in Examples 2 and 3. Each statement in the proof is supported by the reason *why* we can make that statement (claim). The first claim in the proof is the *Given* statement, and the sequence of steps must conclude with a final statement representing the claim to be proved (called the *Prove statement*).

In Example 2, we construct the algebraic proof of the claim, "If $2x - 3 = 7$, then $x = 5$." Where P represents the statement "$2x - 3 = 7$," and R represents "$x = 5$," the theorem has the form "If P, then R." We also use letter Q to name the intermediate conclusion "$2x = 10$." Using the letters P, Q, and R, we show the logical development for the proof at the left. This logical format *will not* be provided in future proofs.

EXAMPLE 2

    GIVEN: $2x - 3 = 7$
    PROVE: $x = 5$

PROOF

Logical Format	Statements	Reasons
P	1. $2x - 3 = 7$	1. Given
If P, then Q	2. $2x - 3 + 3 = 7 + 3$	2. Addition Property of Equality
Q	3. $2x = 10$	3. Substitution
If Q, then R	4. $\dfrac{2x}{2} = \dfrac{10}{2}$	4. Division Property of Equality
R	5. $x = 5$	5. Substitution

SSG **EXS. 1–4**

Study Example 3. Then cover the reasons and provide the reason for each statement. In turn, with statements covered, find the statement corresponding to each reason.

EXAMPLE 3

GIVEN: $2(x - 3) + 4 = 10$

PROVE: $x = 6$

PROOF

Statements	Reasons
1. $2(x - 3) + 4 = 10$	1. Given
2. $2x - 6 + 4 = 10$	2. Distributive Property
3. $2x - 2 = 10$	3. Substitution
4. $2x = 12$	4. Addition Property of Equality
5. $x = 6$	5. Division Property of Equality

NOTE 1: Alternatively, Step 5 could use the reason Multiplication Property of Equality (multiply by $\frac{1}{2}$).

NOTE 2: The fifth step is the final step because the Prove statement ($x = 6$) has been made and justified.

SSG **EXS. 5–7**

Discover

In the diagram, the wooden trim pieces are mitered (cut at an angle) to be equal and to form a right angle when placed together. Use the properties of algebra to explain why the measures of $\angle 1$ and $\angle 2$ are both 45°. What you have done is an informal "proof."

ANSWER

oslɐ °St = 2∠ɯ uǝɥ┴ ˙°St = ⌊∠ɯ ʇɐɥʇ ǝǝs
ǝʍ 'Z ʎq ƃuᴉpᴉʌᴉp 'puɐ ˙°06 = ⌊∠ɯ · Z 'snɥ┴
˙°06 = Z∠ɯ + ⌊∠ɯ ʇɐɥʇ ǝǝs ǝʍ 'Z∠ɯ
= ⌊∠ɯ ǝsnɐɔǝꓭ ˙°06 = Z∠ɯ + ⌊∠ɯ

The Discover activity at the left suggests that formal geometric proofs also exist. The typical format for a problem requiring geometric proof is

GIVEN: _____   DRAWING

PROVE: _____

Consider this problem:

GIVEN: A-P-B on $\overline{AB}$ (Figure 1.54)

PROVE: $AP = AB - PB$

Figure 1.54

First consider the Drawing (Figure 1.54), and relate it to any additional information described by the Given. Then consider the Prove statement. Do you understand the claim, and does it seem reasonable? If it seems reasonable, intermediate claims can be ordered and supported to form the contents of the proof. Because a proof must begin with the Given and conclude with the Prove, the proof of the preceding problem has this form:

PROOF	
Statements	**Reasons**
1. $A\text{-}P\text{-}B$ on $\overline{AB}$	1. Given
2. ?	2. ?
.	.
.	.
.	.
?. $AP = AB - PB$	?. ?

To construct the preceding proof, you must deduce from the Drawing and the Given that

$$AP + PB = AB$$

In turn, you may conclude (through subtraction) that $AP = AB - PB$. The complete proof problem will have the appearance of Example 4, which follows the first of several "Strategy for Proof" features used in this textbook.

> **STRATEGY FOR PROOF ■ The First Line of Proof**
>
> *General Rule:* The first *statement* of the proof includes the "Given" information; also, the first *reason* is Given.
> *Illustration:* See the first line in the proof of Example 4.

EXAMPLE 4

A   P   B

Figure 1.55

GIVEN: $A\text{-}P\text{-}B$ on $\overline{AB}$ (Figure 1.55)
PROVE: $AP = AB - PB$

PROOF	
Statements	**Reasons**
1. $A\text{-}P\text{-}B$ on $\overline{AB}$	1. Given
2. $AP + PB = AB$	2. Segment-Addition Postulate
3. $AP = AB - PB$	3. Subtraction Property of Equality

SSG   **EXS. 8–10**

Some properties of inequality (see Table 1.7) are useful in geometric proof.

TABLE 1.7	
Properties of Inequality (a, b, and c are real numbers)	
Addition Property of Inequality:	If $a > b$, then $a + c > b + c$.
	If $a < b$, then $a + c < b + c$.
Subtraction Property of Inequality:	If $a > b$, then $a - c > b - c$.
	If $a < b$, then $a - c < b - c$.

SAMPLE PROOFS

Consider Figure 1.56 and this problem:

GIVEN: $MN > PQ$

PROVE: $MP > NQ$

M   N   P   Q

Figure 1.56

To understand the situation, first study the Drawing (Figure 1.56) and the related Given. Then read the Prove with reference to the Drawing. What may be confusing here is that the Given involves *MN* and *PQ*, whereas the Prove involves *MP* and *NQ*. However, this is easily remedied through the addition of *NP* to each side of the inequality $MN > PQ$; see Step 2 in the proof of Example 5.

M   N   P   Q

Figure 1.57

EXAMPLE 5

GIVEN: $MN > PQ$ (Figure 1.57)
PROVE: $MP > NQ$

PROOF

Statements	Reasons
1. $MN > PQ$	1. Given
2. $MN + NP > NP + PQ$	2. Addition Property of Inequality
3. $MN + NP = MP$ and $NP + PQ = NQ$	3. Segment-Addition Postulate
4. $MP > NQ$	4. Substitution

NOTE: The final reason may come as a surprise. However, the Substitution Axiom of Equality allows you to replace a quantity with its equal in *any* statement—including an inequality! See Appendix A.3 for more information.

STRATEGY FOR PROOF ■ The Last Statement of the Proof

General Rule: The final *statement* of the proof is the "Prove" statement.
Illustration: See the last statement in the proof of Example 6.

EXAMPLE 6

Study this proof, noting the order of the statements and reasons.

GIVEN: $\overrightarrow{ST}$ bisects $\angle RSU$
$\overrightarrow{SV}$ bisects $\angle USW$ (Figure 1.58)

PROVE: $m\angle RST + m\angle VSW = m\angle TSV$

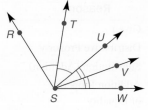

Figure 1.58

PROOF

Statements	Reasons
1. $\overrightarrow{ST}$ bisects $\angle RSU$	1. Given
2. $m\angle RST = m\angle TSU$	2. If an angle is bisected, then the measures of the resulting angles are equal.
3. $\overrightarrow{SV}$ bisects $\angle USW$	3. Same as reason 1
4. $m\angle VSW = m\angle USV$	4. Same as reason 2
5. $m\angle RST + m\angle VSW = m\angle TSU + m\angle USV$	5. Addition Property of Equality (use the equations from statements 2 and 4)
6. $m\angle TSU + m\angle USV = m\angle TSV$	6. Angle-Addition Postulate
7. $m\angle RST + m\angle VSW = m\angle TSV$	7. Substitution

SSG **EXS. 11, 12**

Exercises 1.5

In Exercises 1 to 6, which property justifies the conclusion of the statement?

1. If $2x = 12$, then $x = 6$.

2. If $x + x = 12$, then $2x = 12$.

3. If $x + 5 = 12$, then $x = 7$.

4. If $x - 5 = 12$, then $x = 17$.

5. If $\frac{x}{5} = 3$, then $x = 15$.

6. If $3x - 2 = 13$, then $3x = 15$.

In Exercises 7 to 10, state the property or definition that justifies the conclusion (the "then" clause).

7. Given that $\angle$s 1 and 2 are supplementary, then $m\angle 1 + m\angle 2 = 180°$.

8. Given that $m\angle 3 + m\angle 4 = 180°$, then $\angle$s 3 and 4 are supplementary.

9. Given $\angle RSV$ and $\overrightarrow{ST}$ as shown, then $m\angle RST + m\angle TSV = m\angle RSV$.

10. Given that $m\angle RST = m\angle TSV$, then $\overrightarrow{ST}$ bisects $\angle RSV$.

Exercises 9, 10

In Exercises 11 to 22, use the Given information to draw a conclusion based on the stated property or definition.

Exercises 11, 12

11. *Given:* A-M-B; Segment-Addition Postulate

12. *Given:* M is the midpoint of $\overline{AB}$; definition of midpoint

13. *Given:* $m\angle 1 = m\angle 2$; definition of angle bisector

14. *Given:* $\overrightarrow{EG}$ bisects $\angle DEF$; definition of angle bisector

15. *Given:* $\angle$s 1 and 2 are complementary; definition of complementary angles

16. *Given:* $m\angle 1 + m\angle 2 = 90°$; definition of complementary angles

Exercises 13–16

17. *Given:* $2x - 3 = 7$; Addition Property of Equality

18. *Given:* $3x = 21$; Division Property of Equality

19. *Given:* $7x + 5 - 3 = 30$; Substitution Property of Equality

20. *Given:* $\frac{1}{2} = 0.5$ and $0.5 = 50\%$; Transitive Property of Equality

21. *Given:* $3(2x - 1) = 27$; Distributive Property

22. *Given:* $\frac{x}{5} = -4$; Multiplication Property of Equality

In Exercises 23 and 24, fill in the missing reasons for the algebraic proof.

23. *Given:* $3(x - 5) = 21$
    Prove: $x = 12$

PROOF	
Statements	**Reasons**
1. $3(x - 5) = 21$	1. ?
2. $3x - 15 = 21$	2. ?
3. $\qquad 3x = 36$	3. ?
4. $\qquad x = 12$	4. ?

24. *Given:* $2x + 9 = 3$
    Prove: $x = -3$

PROOF	
Statements	**Reasons**
1. $2x + 9 = 3$	1. ?
2. $\qquad 2x = -6$	2. ?
3. $\qquad x = -3$	3. ?

In Exercises 25 and 26, fill in the missing statements for the algebraic proof.

25. *Given:* $2(x + 3) - 7 = 11$
    Prove: $x = 6$

PROOF	
Statements	**Reasons**
1. ?	1. Given
2. ?	2. Distributive Property
3. ?	3. Substitution (Addition)
4. ?	4. Addition Property of Equality
5. ?	5. Division Property of Equality

26. *Given:* $\frac{x}{5} + 3 = 9$
    Prove: $x = 30$

PROOF	
Statements	**Reasons**
1. ?	1. Given
2. ?	2. Subtraction Property of Equality
3. ?	3. Multiplication Property of Equality

In Exercises 27 to 30, fill in the missing reasons for each geometric proof.

27. *Given:*   D-E-F on $\overleftrightarrow{DF}$

      Prove: $DE = DF - EF$ *Exercises 27, 28*

PROOF

Statements	Reasons
1. D-E-F on $\overleftrightarrow{DF}$	1. ?
2. $DE + EF = DF$	2. ?
3. $DE = DF - EF$	3. ?

28. *Given:*   E is the midpoint of $\overline{DF}$

      Prove: $DE = \frac{1}{2}(DF)$

PROOF

Statements	Reasons
1. E is the midpoint of $\overline{DF}$	1. ?
2. $DE = EF$	2. ?
3. $DE + EF = DF$	3. ?
4. $DE + DE = DF$	4. ?
5. $2(DE) = DF$	5. ?
6. $DE = \frac{1}{2}(DF)$	6. ?

29. *Given:* $\overrightarrow{BD}$ bisects $\angle ABC$

      Prove: $m\angle ABD = \frac{1}{2}(m\angle ABC)$

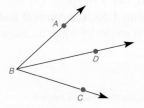

Exercises 29, 30

PROOF

Statements	Reasons
1. $\overrightarrow{BD}$ bisects $\angle ABC$	1. ?
2. $m\angle ABD = m\angle DBC$	2. ?
3. $m\angle ABD + m\angle DBC = m\angle ABC$	3. ?
4. $m\angle ABD + m\angle ABD = m\angle ABC$	4. ?
5. $2(m\angle ABD) = m\angle ABC$	5. ?
6. $m\angle ABD = \frac{1}{2}(m\angle ABC)$	6. ?

30. *Given:* $\angle ABC$ and $\overrightarrow{BD}$ (See figure for Exercise 29.)

      Prove: $m\angle ABD = m\angle ABC - m\angle DBC$

PROOF

Statements	Reasons
1. $\angle ABC$ and $\overrightarrow{BD}$	1. ?
2. $m\angle ABD + m\angle DBC = m\angle ABC$	2. ?
3. $m\angle ABD = m\angle ABC - m\angle DBC$	3. ?

In Exercises 31 and 32, fill in the missing statements and reasons.

31. *Given:*   M-N-P-Q on $\overline{MQ}$

      Prove: $MN + NP + PQ = MQ$

PROOF

Statements	Reasons
1. ?	1. ?
2. $MN + NQ = MQ$	2. ?
3. $NP + PQ = NQ$	3. ?
4. ?	4. Substitution Property of Equality

32. *Given:* $\angle TSW$ with $\overrightarrow{SU}$ and $\overrightarrow{SV}$

      Prove:

      $m\angle TSW = m\angle TSU + m\angle USV + m\angle VSW$

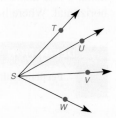

PROOF

Statements	Reasons
1. ?	1. ?
2. $m\angle TSW = m\angle TSU + m\angle USW$	2. ?
3. $m\angle USW = m\angle USV + m\angle VSW$	3. ?
4. ?	4. Substitution Property of Equality

33. When the Distributive Property is written in its *symmetric* form, it reads $a \cdot b + a \cdot c = a(b + c)$. Use this form to rewrite $5x + 5y$.

34. Another form of the Distributive Property (see Exercise 33) reads $b \cdot a + c \cdot a = (b + c)a$. Use this form to rewrite $5x + 7x$. Then simplify.

35. The Multiplication Property of Inequality requires that we *reverse* the inequality symbol when multiplying by a *negative* number. Given that $-7 < 5$, form the inequality that results when we multiply each side by -2.

36. The Division Property of Inequality requires that we *reverse* the inequality symbol when dividing by a *negative* number. Given that $12 > -4$, form the inequality that results when we divide each side by -4.

37. Provide reasons for this proof. "If $a = b$ and $c = d$, then $a + c = b + d$."

PROOF	
Statements	**Reasons**
1. $a = b$	1. ?
2. $a + c = b + c$	2. ?
3. $c = d$	3. ?
4. $a + c = b + d$	4. ?

38. Write a proof for: "If $a = b$ and $c = d$, then $a - c = b - d$."

(HINT: Use Exercise 37 as a guide.)

1.6   Relationships: Perpendicular Lines

KEY CONCEPTS

Vertical Line(s)	Relations: Reflexive,	Equivalence Relation
Horizontal Line(s)	Symmetric, and	Perpendicular Bisector
Perpendicular Lines	Transitive Properties	of a Line Segment

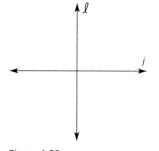

Figure 1.59

Informally, a **vertical** line is one that extends up and down, like a flagpole. On the other hand, a line that extends left to right is **horizontal.** In Figure 1.59, ℓ is vertical and j is horizontal. Where lines ℓ and j intersect, they appear to form angles of equal measure.

> **DEFINITION**
>
> **Perpendicular lines** are two lines that meet to form congruent adjacent angles.

Perpendicular lines do not have to be vertical and horizontal. In Figure 1.60, the slanted lines m and p are perpendicular ($m \perp p$). As in Figure 1.60, a small square is often placed in the opening of an angle formed by perpendicular lines.

Example 1 provides a formal proof of the relationship between perpendicular lines and right angles. Study this proof, noting the order of the statements and reasons. The numbers in parentheses to the left of the statements refer to the earlier statement(s) of the proof upon which the new statement is based.

Figure 1.60

> **STRATEGY FOR PROOF ■ The Drawing for the Proof**
>
> *General Rule:* Make a drawing that accurately characterizes the "Given" information.
> *Illustration:* For the proof of Example 1, see Figure 1.61.

> **THEOREM 1.6.1**
>
> If two lines are perpendicular, then they meet to form right angles.

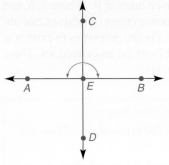

Figure 1.61

EXAMPLE 1

GIVEN: $\overleftrightarrow{AB} \perp \overleftrightarrow{CD}$, intersecting at E (See Figure 1.61)

PROVE: $\angle AEC$ is a right angle

PROOF	
Statements	**Reasons**
1. $\overleftrightarrow{AB} \perp \overleftrightarrow{CD}$, intersecting at E	1. Given
(1) 2. $\angle AEC \cong \angle CEB$	2. Perpendicular lines meet to form congruent adjacent angles (Definition)
(2) 3. $m\angle AEC = m\angle CEB$	3. If two angles are congruent, their measures are equal
4. $\angle AEB$ is a straight angle and $m\angle AEB = 180°$	4. Measure of a straight angle equals $180°$
5. $m\angle AEC + m\angle CEB = m\angle AEB$	5. Angle-Addition Postulate
(4), (5) 6. $m\angle AEC + m\angle CEB = 180°$	6. Substitution
(3), (6) 7. $m\angle AEC + m\angle AEC = 180°$ or $2 \cdot m\angle AEC = 180°$	7. Substitution
(7) 8. $m\angle AEC = 90°$	8. Division Property of Equality
(8) 9. $\angle AEC$ is a right angle	9. If the measure of an angle is $90°$, then the angle is a right angle

EXAMPLE 2

In Figure 1.61, find the sum of $m\angle AEC + m\angle CEB + m\angle BED + m\angle DEA$.

SOLUTION
Because each angle of the sum measures $90°$, the total is $4 \cdot 90°$ or $360°$.

In general, "The sum of the measures of the nonoverlapping adjacent angles about a point is $360°$."

RELATIONS

The relationship between perpendicular lines suggests the more general, but undefined, mathematical concept of **relation.** In general, a relation "connects" two elements of an associated set of objects. Table 1.8 provides several examples of the concept of a relation R.

SSG **EXS. 1, 2**

TABLE 1.8		
Relation R	**Objects Related**	**Example of Relationship**
is equal to	numbers	$2 + 3 = 5$
is greater than	numbers	$7 > 5$
is perpendicular to	lines	$\ell \perp m$
is complementary to	angles	$\angle 1$ is comp. to $\angle 2$
is congruent to	line segments	$\overline{AB} \cong \overline{CD}$
is a brother of	people	Matt is a brother of Phil

There are three special properties that may exist for a given relation R. Where a, b, and c are objects associated with relation R, the properties consider one object (reflexive), two objects in either order (symmetric), or three objects (transitive). For the properties to exist, it is necessary that the statements be true for all objects selected from the associated set. These properties are generalized, and specific examples are given below:

Reflexive property: aRa ($5 = 5$; equality of numbers has a reflexive property)

Symmetric property: If aRb, then bRa. (If $\ell \perp m$, then $m \perp \ell$; perpendicularity of lines has a symmetric property)

Transitive property: If aRb and bRc, then aRc. (If $\angle 1 \cong \angle 2$ and $\angle 2 \cong \angle 3$, then $\angle 1 \cong \angle 3$; congruence of angles has a transitive property)

Geometry in Nature

An icicle formed from freezing
water assumes a vertical path.

© Karel Brož/Shutterstock.com

EXAMPLE 2

Does the relation "is less than" for numbers have a reflexive property? a symmetric property? a transitive property?

SOLUTION Because "$2 < 2$" is false, there is *no* reflexive property.
"If $2 < 5$, then $5 < 2$" is also false; there is *no* symmetric property.
"If $2 < 5$ and $5 < 9$, then $2 < 9$" is true; there is a transitive property.
NOTE: The same results are obtained for choices other than 2, 5, and 9.

Congruence of angles (or of line segments) is closely tied to equality of angle measures (or line segment measures) by the definition of congruence.

PROPERTIES FOR THE CONGRUENCE OF ANGLES	
Reflexive:	$\angle 1 \cong \angle 1$; an angle is congruent to itself.
Symmetric:	If $\angle 1 \cong \angle 2$, then $\angle 2 \cong \angle 1$.
Transitive:	If $\angle 1 \cong \angle 2$ and $\angle 2 \cong \angle 3$, then $\angle 1 \cong \angle 3$.

 EXS. 3–9

Any relation (such as congruence of angles) that has reflexive, symmetric, and transitive properties is known as an *equivalence relation.* In later chapters, we will see that *congruence of triangles* and *similarity of triangles* also have reflexive, symmetric, and transitive properties; therefore, these relations are also equivalence relations.

Returning to the formulation of a proof, the final example in this section is based on the fact that vertical angles are congruent when two lines intersect. See Figure 1.62(a). Because there are two pairs of congruent angles, the Prove could be stated

Figure 1.62(a)

Prove: $\angle 1 \cong \angle 3$ and $\angle 2 \cong \angle 4$

Such a conclusion is a conjunction and would be proved if both congruences were established. For simplicity, the Prove of Example 3 is stated

Prove: $\angle 2 \cong \angle 4$

Study this proof of Theorem 1.6.2, noting the order of the statements and reasons.

THEOREM 1.6.2

If two lines intersect, then the vertical angles formed are congruent.

Figure 1.62(b)

EXAMPLE 3

GIVEN: $\overleftrightarrow{AC}$ intersects $\overleftrightarrow{BD}$ at O [See Figure 1.62(b).]

PROVE: $\angle 2 \cong \angle 4$

PROOF

Statements	Reasons
1. $\overleftrightarrow{AC}$ intersects $\overleftrightarrow{BD}$ at O	1. Given
2. $\angle$s AOC and DOB are straight $\angle$s, with m$\angle AOC = 180$ and m$\angle DOB = 180$	2. The measure of a straight angle is 180°
3. m$\angle DOB$ = m$\angle AOC$	3. Substitution
4. m$\angle 1$ + m$\angle 4$ = m$\angle DOB$ and m$\angle 1$ + m$\angle 2$ = m$\angle AOC$	4. Angle-Addition Postulate
5. m$\angle 1$ + m$\angle 4$ = m$\angle 1$ + m$\angle 2$	5. Substitution
6. m$\angle 4$ = m$\angle 2$	6. Subtraction Property of Equality
7. $\angle 4 \cong \angle 2$	7. If two angles are equal in measure, the angles are congruent
8. $\angle 2 \cong \angle 4$	8. Symmetric Property of Congruence of Angles

In the preceding proof, there is no need to reorder the congruent angles from statement 7 to statement 8 because congruence of angles is symmetric; in the later work, statement 7 will be written to match the Prove statement even if the previous line does not have the same order. The same type of thinking applies to proving lines perpendicular or parallel: The order is simply not important!

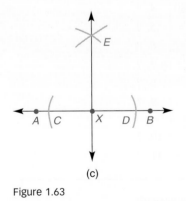

Figure 1.63

CONSTRUCTIONS LEADING TO PERPENDICULAR LINES

Construction 2 in Section 1.2 determined not only the midpoint of $\overline{AB}$ but also that of the **perpendicular bisector** of $\overline{AB}$. In many instances, we need the line perpendicular to another line at a point other than the midpoint of a segment.

CONSTRUCTION 5 To construct the line perpendicular to a given line at a specified point on the given line.

GIVEN: $\overleftrightarrow{AB}$ with point X in Figure 1.63(a)

CONSTRUCT: A line $\overleftrightarrow{EX}$, so that $\overleftrightarrow{EX} \perp \overleftrightarrow{AB}$

CONSTRUCTION: Figure 1.63(b): Using X as the center, mark off arcs of equal radii on each side of X to intersect $\overleftrightarrow{AB}$ at C and D.

Figure 1.63(c): Now, using C and D as centers, mark off arcs of equal radii with a length greater than XD so that these arcs intersect either above (as shown) or below $\overleftrightarrow{AB}$.

Calling the point of intersection E, draw $\overleftrightarrow{EX}$, which is the desired line; that is, $\overleftrightarrow{EX} \perp \overleftrightarrow{AB}$.

The theorem that Construction 5 is based on is a consequence of the Protractor Postulate, and we state it without proof.

THEOREM 1.6.3

In a plane, there is exactly one line perpendicular to a given line at any point on the line.

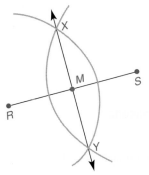

Figure 1.64

SSG **EXS. 10–14**

Construction 2, which was used to locate the midpoint of a line segment in Section 1.2, is also the method for constructing the perpendicular bisector of a line segment. In Figure 1.64, $\overleftrightarrow{XY}$ is the perpendicular bisector of $\overline{RS}$. The following theorem can be proved by methods developed later in this book.

> **THEOREM 1.6.4**
>
> The perpendicular bisector of a line segment is unique.

Exercises 1.6

In Exercises 1 and 2, supply reasons.

1. *Given:* $\angle 1 \cong \angle 3$
   Prove: $\angle MOP \cong \angle NOQ$

PROOF

Statements	Reasons
1. $\angle 1 \cong \angle 3$	1. ?
2. $m\angle 1 = m\angle 3$	2. ?
3. $m\angle 1 + m\angle 2 = m\angle MOP$ and $m\angle 2 + m\angle 3 = m\angle NOQ$	3. ?
4. $m\angle 1 + m\angle 2 = m\angle 2 + m\angle 3$	4. ?
5. $m\angle MOP = m\angle NOQ$	5. ?
6. $\angle MOP \cong \angle NOQ$	6. ?

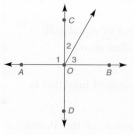

Exercise 2

2. *Given:* $\overleftrightarrow{AB}$ intersects $\overleftrightarrow{CD}$ at O so that $\angle 1$ is a right $\angle$ (Use the figure following Exercise 1.)
   Prove: $\angle 2$ and $\angle 3$ are complementary

PROOF

Statements	Reasons
1. $\overleftrightarrow{AB}$ intersects $\overleftrightarrow{CD}$ at O	1. ?
2. $\angle AOB$ is a straight $\angle$, so $m\angle AOB = 180$	2. ?
3. $m\angle 1 + m\angle COB = m\angle AOB$	3. ?
4. $m\angle 1 + m\angle COB = 180$	4. ?
5. $\angle 1$ is a right angle	5. ?
6. $m\angle 1 = 90$	6. ?
7. $90 + m\angle COB = 180$	7. ?
8. $m\angle COB = 90$	8. ?
9. $m\angle 2 + m\angle 3 = m\angle COB$	9. ?
10. $m\angle 2 + m\angle 3 = 90$	10. ?
11. $\angle 2$ and $\angle 3$ are complementary	11. ?

In Exercises 3 and 4, supply statements.

3. *Given:* $\angle 1 \cong \angle 2$ and $\angle 2 \cong \angle 3$
   Prove: $\angle 1 \cong \angle 3$
   (Use the figure following Exercise 1.)

Exercise 3

PROOF

Statements	Reasons
1. ?	1. Given
2. ?	2. Transitive Property of Congruence

4. *Given:* $m\angle AOB = m\angle 1$
   $m\angle BOC = m\angle 1$
   Prove: $\overrightarrow{OB}$ bisects $\angle AOC$

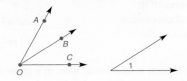

PROOF

Statements	Reasons
1. ?	1. Given
2. ?	2. Substitution
3. ?	3. Angles with equal measures are congruent
4. ?	4. If a ray divides an angle into two congruent angles, then the ray bisects the angle

In Exercises 5 to 9, use a compass and a straightedge to complete the constructions.

5. *Given:* Point N on line s
   Construct: Line m through N so that $m \perp s$

6. *Given:* $\overrightarrow{OA}$
   Construct: Right angle BOA

   (HINT: *Use a straightedge to extend* $\overrightarrow{OA}$ *to the left.*)

7. *Given:* Line ℓ containing point A
   Construct: A 45° angle with vertex at A

8. *Given:* $\overline{AB}$
   Construct: The perpendicular bisector of $\overline{AB}$

9. *Given:* Triangle ABC
   Construct: The perpendicular bisectors of sides $\overline{AB}$, $\overline{AC}$, and $\overline{BC}$

10. Draw a conclusion based on the results of Exercise 9.

In Exercises 11 and 12, provide the missing statements and reasons.

11. *Given:* ∠s 1 and 3 are complementary
    ∠s 2 and 3 are complementary
    Prove: $\angle 1 \cong \angle 2$

PROOF

Statements	Reasons
1. ∠s 1 and 3 are complementary; ∠s 2 and 3 are complementary	1. ?
2. $m\angle 1 + m\angle 3 = 90$; $m\angle 2 + m\angle 3 = 90$	2. The sum of the measures of complementary ∠s is 90
(2) 3. $m\angle 1 + m\angle 3 = m\angle 2 + m\angle 3$	3. ?
4. ?	4. Subtraction Property of Equality
(4) 5. ?	5. If two ∠s are = in measure, they are ≅

12. *Given:* $\angle 1 \cong \angle 2$; $\angle 3 \cong \angle 4$
    ∠s 2 and 3 are complementary
    Prove: ∠s 1 and 4 are complementary

PROOF

Statements	Reasons
1. $\angle 1 \cong \angle 2$ and $\angle 3 \cong \angle 4$	1. ?
2. ? and ?	2. If two ∠s are ≅, then their measures are equal
3. ∠s 2 and 3 are complementary	3. ?
(3) 4. ?	4. The sum of the measures of complementary ∠s is 90
(2), (4) 5. $m\angle 1 + m\angle 4 = 90$	5. ?
6. ?	6. If the sum of the measures of two angles is 90, then the angles are complementary

13. Does the relation "is perpendicular to" have a reflexive property (consider line ℓ)? a symmetric property (consider lines ℓ and m)? a transitive property (consider lines ℓ, m, and n)?

14. Does the relation "is greater than" have a reflexive property (consider real number a)? a symmetric property (consider real numbers a and b)? a transitive property (consider real numbers a, b, and c)?

15. Does the relation "is complementary to" for angles have a reflexive property (consider one angle)? a symmetric property (consider two angles)? a transitive property (consider three angles)?

16. Does the relation "is less than" for numbers have a reflexive property (consider one number)? a symmetric property (consider two numbers)? a transitive property (consider three numbers)?

17. Does the relation "is a brother of" have a reflexive property (consider one male)? a symmetric property (consider two males)? a transitive property (consider three males)?

18. Does the relation "is in love with" have a reflexive property (consider one person)? a symmetric property (consider two people)? a transitive property (consider three people)?

19. This textbook has used numerous symbols and abbreviations. In this exercise, indicate what word is represented or abbreviated by each of the following:
a) $\perp$ b) $\angle$s   c) supp.   d) rt.   e) m$\angle$1

20. This textbook has used numerous symbols and abbreviations. In this exercise, indicate what word is represented or abbreviated by each of the following:
a) post.   b) $\cup$ c) $\varnothing$ d) <   e) pt.

21. This textbook has used numerous symbols and abbreviations. In this exercise, indicate what word is represented or abbreviated by each of the following:
a) adj.   b) comp.   c) $\overrightarrow{AB}$ d) $\cong$ e) vert.

22. If there were no understood restriction to lines in a plane in Theorem 1.6.3, the theorem would be false. Explain why the following statement is false: "In space, there is exactly one line perpendicular to a given line at any point on the line."

23. Prove the Extended Segment Addition Property by using the Drawing, the Given, and the Prove that follow.
Given: M-N-P-Q on $\overline{MQ}$
Prove: $MN + NP + PQ = MQ$

24. The Segment-Addition Postulate can be generalized as follows: "The length of a line segment equals the sum of the lengths of its parts." State a general conclusion about AE based on the following figure.

$A \quad B \quad C \quad D \quad E$

25. Prove the Extended Angle Addition Property by using the Drawing, the Given, and the Prove that follow.
Given: $\angle TSW$ with $\overrightarrow{SU}$ and $\overrightarrow{SV}$
Prove: m$\angle TSW$ = m$\angle TSU$ + m$\angle USV$ + m$\angle VSW$

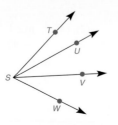

26. The Angle-Addition Postulate can be generalized as follows: "The measure of an angle equals the sum of the measures of its parts." State a general conclusion about m$\angle GHK$ based on the figure shown.

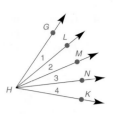

27. If there were no understood restriction to lines in a plane in Theorem 1.6.4, the theorem would be false. Explain why the following statement is false: "In space, the perpendicular bisector of a line segment is unique."

*28. In the proof below, provide the missing reasons.
Given: $\angle 1$ and $\angle 2$ are complementary
$\angle 1$ is acute
Prove: $\angle 2$ is also acute

PROOF

	Statements	Reasons
	1. $\angle 1$ and $\angle 2$ are complementary	1. ?
(1)	2. m$\angle 1$ + m$\angle 2$ = 90	2. ?
	3. $\angle 1$ is acute	3. ?
(3)	4. Where m$\angle 1$ = x, $0 < x < 90$	4. ?
(2)	5. x + m$\angle 2$ = 90	5. ?
(5)	6. m$\angle 2$ = 90 − x	6. ?
(4)	7. $-x < 0 < 90 - x$	7. ?
(7)	8. $90 - x < 90 < 180 - x$	8. ?
(7), (8)	9. $0 < 90 - x < 90$	9. ?
(6), (9)	10. $0 <$ m$\angle 2 < 90$	10. ?
(10)	11. $\angle 2$ is acute	11. ?

29. Without writing a proof, explain the conclusion for the following problem.
Given: $\overleftrightarrow{AB}$
Prove: m$\angle 1$ + m$\angle 2$ + m$\angle 3$ + m$\angle 4$ = 180

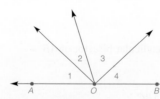

Exercises 29, 30

30. Without writing a proof, explain the conclusion for the following problem.
Given: $\overleftrightarrow{AB}$, $\angle$s 2 and 3 are complementary.
Prove: $\angle$s 1 and 4 are complementary.

(HINT: Use the result from Exercise 29.)

1.7   The Formal Proof of a Theorem

Recall from Section 1.3 that statements that can be proved are called *theorems*. To understand the formal proof of a theorem, we begin by considering the terms *hypothesis* and *conclusion*. The hypothesis of a statement describes the given situation (Given), whereas the conclusion describes what you need to establish (Prove). When a statement has the form "If H, then C," the hypothesis is H and the conclusion is C. Some theorems must be reworded to fit into "If . . . , then . . ." form so that the hypothesis and conclusion are easy to recognize.

EXAMPLE 1

Give the hypothesis H and conclusion C for each of these statements.

a) If two lines intersect, then the vertical angles formed are congruent.

b) All right angles are congruent.

c) Parallel lines do not intersect.

d) Lines are perpendicular when they meet to form congruent adjacent angles.

SOLUTION

a) As is   H: Two lines intersect.
                    C: The vertical angles formed are congruent.

b) Reworded   If two angles are right angles, then these angles are congruent.
                    H: Two angles are right angles.
                    C: The angles are congruent.

c) Reworded   If two lines are parallel, then these lines do not intersect.
                    H: Two lines are parallel.
                    C: The lines do not intersect.

d) Reordered   When (if) two lines meet to form congruent adjacent angles, these
                    lines are perpendicular.
                    H: Two lines meet to form congruent adjacent angles.
                    C: The lines are perpendicular.

Why do we need to distinguish between the hypothesis and the conclusion? For a theorem, the hypothesis determines the Given and the Drawing. The Given provides a description of the Drawing's known characteristics. The conclusion (Prove) determines the relationship that you wish to establish in the Drawing.

SSG   **EXS. 1–3**

THE WRITTEN PARTS OF A FORMAL PROOF

The five necessary parts of a formal proof are listed in the following box in the order in which they should be developed.

ESSENTIAL PARTS OF THE FORMAL PROOF OF A THEOREM

1. *Statement:* States the theorem to be proved.
2. *Drawing:* Represents the hypothesis of the theorem.
3. *Given:* Describes the Drawing according to the information found in the hypothesis of the theorem.
4. *Prove:* Describes the Drawing according to the claim made in the conclusion of the theorem.
5. *Proof:* Orders a list of claims (Statements) and justifications (Reasons), beginning with the Given and ending with the Prove; there must be a logical flow in this Proof.

The most difficult aspect of a formal proof is the thinking process that must take place between parts 4 and 5. This game plan or analysis involves deducing and ordering conclusions based on the given situation. One must be somewhat like a lawyer, selecting the claims that help prove the case while discarding those that are superfluous. In the process of ordering the statements, it may be beneficial to think in reverse order, like so:

The Prove statement would be true if what else were true?

The final proof must be arranged in an order that allows one to reason from an earlier statement to a later claim by using deduction (perhaps several times). Where principle P has the form "If H, then C," the logical order follows.

$$
\begin{array}{lll}
\text{H: hypothesis} & \longleftarrow & \textbf{statement of proof} \\
\underline{\text{P: principle}} & \longleftarrow & \textbf{reason of proof} \\
\therefore \text{C: conclusion} & \longleftarrow & \textbf{next statement in proof}
\end{array}
$$

Consider the following theorem, which was proved in Example 1 of Section 1.6.

THEOREM 1.6.1

If two lines are perpendicular, then they meet to form right angles.

EXAMPLE 2

Write the parts of the formal proof of Theorem 1.6.1.

SOLUTION

1. State the theorem.
   If two lines are perpendicular, then they meet to form right angles.
2. The hypothesis is H: Two lines are perpendicular.
   Make a Drawing to fit this description.
   (See Figure 1.65.)
3. Write the Given statement, using the Drawing and based on the hypothesis H: Two lines are ⊥.
   Given: $\overleftrightarrow{AB} \perp \overleftrightarrow{CD}$ intersecting at E
4. Write the Prove statement, using the Drawing and based on the conclusion C: They meet to form right angles.
   Prove: $\angle AEC$ is a right angle.
5. Construct the Proof. This formal proof is found in Example 1, Section 1.6.

Figure 1.65

SSG **EXS. 4, 5**

CONVERSE OF A STATEMENT

The converse of the statement "If *P*, then *Q*" is "If *Q*, then *P*." That is, the converse of a given statement interchanges its hypothesis and conclusion. Consider the following:

Statement: If a person lives in London, then that person lives in England.

Converse: If a person lives in England, then that person lives in London.

As shown above, the given statement is true, whereas its converse is false. Sometimes the converse of a true statement is also true. In fact, Example 3 presents the formal proof of Theorem 1.7.1, which is the converse of Theorem 1.6.1.

Once a theorem has been proved, it may be cited thereafter as a reason in future proofs. Thus, any theorem found in this section can be used for justification in proof problems found in later sections.

The proof that follows is nearly complete! It is difficult to provide a complete formal proof that explains the "how to" and simultaneously presents the final polished form. Example 3 illustrates the polished proof. You do not see the thought process and the scratch paper needed to piece this puzzle together.

The proof of a theorem is not unique! For instance, students' Drawings need not match, even though the same relationships should be indicated. Certainly, different letters are likely to be chosen for the Drawing that illustrates the hypothesis.

> **Warning**
>
> You should not make a drawing that embeds qualities beyond those described in the hypothesis; nor should your drawing indicate fewer qualities than the hypothesis prescribes!

THEOREM 1.7.1

If two lines meet to form a right angle, then these lines are perpendicular.

EXAMPLE 3

Give a formal proof for Theorem 1.7.1.

If two lines meet to form a right angle, then these lines are perpendicular.

GIVEN: $\overleftrightarrow{AB}$ and $\overleftrightarrow{CD}$ intersect at *E* so that $\angle AEC$ is a right angle (Figure 1.66)

PROVE: $\overleftrightarrow{AB} \perp \overleftrightarrow{CD}$

Figure 1.66

PROOF

	Statements	Reasons
	1. $\overleftrightarrow{AB}$ and $\overleftrightarrow{CD}$ intersect so that $\angle AEC$ is a right angle	1. Given
	2. $m\angle AEC = 90$	2. If an $\angle$ is a right $\angle$, its measure is 90
	3. $\angle AEB$ is a straight $\angle$, so $m\angle AEB = 180$	3. If an $\angle$ is a straight $\angle$, its measure is 180
	4. $m\angle AEC + m\angle CEB = m\angle AEB$	4. Angle-Addition Postulate
(2), (3), (4)	5. $90 + m\angle CEB = 180$	5. Substitution
(5)	6. $m\angle CEB = 90$	6. Subtraction Property of Equality
(2), (6)	7. $m\angle AEC = m\angle CEB$	7. Substitution
	8. $\angle AEC \cong \angle CEB$	8. If two $\angle$s have $=$ measures, the $\angle$s are $\cong$
	9. $\overleftrightarrow{AB} \perp \overleftrightarrow{CD}$	9. If two lines form $\cong$ adjacent $\angle$s, these lines are $\perp$

Because perpendicular lines lead to right angles, and conversely, a square (see Figure 1.66) may be used to indicate perpendicular lines or a right angle.

SSG EXS. 6–8

Several additional theorems are now stated, the proofs of which are left as exercises. This list contains theorems that are quite useful when cited as reasons in later proofs. A formal proof is provided only for Theorem 1.7.6.

THEOREM 1.7.2

If two angles are complementary to the same angle (or to congruent angles), then these angles are congruent.

See Exercise 27 for a drawing describing Theorem 1.7.2.

THEOREM 1.7.3

If two angles are supplementary to the same angle (or to congruent angles), then these angles are congruent.

See Exercise 28 for a drawing describing Theorem 1.7.3.

THEOREM 1.7.4

Any two right angles are congruent.

THEOREM 1.7.5

If the exterior sides of two adjacent acute angles form perpendicular rays, then these angles are complementary.

For Theorem 1.7.5, we create an informal proof called a picture proof. Although such a proof is less detailed, the impact of the explanation is the same! This is the first of several "picture proofs" found in this textbook. In Figure 1.67, the square is used to indicate that $\overrightarrow{BA} \perp \overrightarrow{BC}$

PICTURE PROOF OF THEOREM 1.7.5

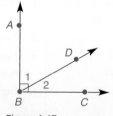

Figure 1.67

Given: $\overrightarrow{BA} \perp \overrightarrow{BC}$

Prove: $\angle 1$ and $\angle 2$ are complementary

Proof: With $\overrightarrow{BA} \perp \overrightarrow{BC}$, we see that $\angle 1$ and $\angle 2$ are parts of a right angle.

Then $m\angle 1 + m\angle 2 = 90°$, so $\angle 1$ and $\angle 2$ are complementary.

STRATEGY FOR PROOF ■ The Final Reason in the Proof

General Rule: The last reason explains why the last *statement* must be true. Never write the word "Prove" for any reason in a proof.

Illustration: The final reason in the proof of Theorem 1.7.6 is the definition of supplementary angles: If the sum of measures of two angles is 180°, the angles are supplementary.

<table>
<tr><td>

Technology Exploration

Use computer software if available.

1. Draw $\overleftrightarrow{EG}$ containing point F. Also draw $\overrightarrow{FH}$ as in Figure 1.68.

2. Measure ∠3 and ∠4.

3. Show that m∠3 + m∠4 = 180°. (Answer may not be perfect.)

</td></tr>
</table>

EXAMPLE 4

Study the formal proof of Theorem 1.7.6.

> **THEOREM 1.7.6**
>
> If the exterior sides of two adjacent angles form a straight line, then these angles are supplementary.

GIVEN: ∠3 and ∠4 and $\overleftrightarrow{EG}$ (Figure 1.68)

PROVE: ∠3 and ∠4 are supplementary

Figure 1.68

SSG **EXS. 9–12**

PROOF

Statements	Reasons
1. ∠3 and ∠4 and $\overleftrightarrow{EG}$	1. Given
2. m∠3 + m∠4 = m∠EFG	2. Angle-Addition Postulate
3. ∠EFG is a straight angle	3. If the sides of an ∠ are opposite rays, it is a straight ∠
4. m∠EFG = 180	4. The measure of a straight ∠ is 180
5. m∠3 + m∠4 = 180	5. Substitution
6. ∠3 and ∠4 are supplementary	6. If the sum of the measures of two ∠s is 180, the ∠s are supplementary

The final two theorems in this section are stated for convenience. We suggest that the student make drawings to illustrate Theorem 1.7.7 and Theorem 1.7.8.

> **THEOREM 1.7.7**
>
> If two line segments are congruent, then their midpoints separate these segments into four congruent segments.

> **THEOREM 1.7.8**
>
> If two angles are congruent, then their bisectors separate these angles into four congruent angles.

SSG **EXS. 13, 14**

Exercises 1.7

In Exercises 1 to 6, state the hypothesis H and the conclusion C for each statement.

1. If a line segment is bisected, then each of the equal segments has half the length of the original segment.

2. If two sides of a triangle are congruent, then the triangle is isosceles.

3. All squares are quadrilaterals.

4. Every regular polygon has congruent interior angles.

5. Two angles are congruent if each is a right angle.

6. The lengths of corresponding sides of similar polygons are proportional.

7. Name, in order, the five parts of the formal proof of a theorem.

8. Which part (hypothesis or conclusion) of a theorem determines the
   a) Drawing?
   b) Given?
   c) Prove?

9. Which part (Given or Prove) of the proof depends upon the
   a) hypothesis of theorem?
   b) conclusion of theorem?

10. Which of the following can be cited as a reason in a proof?
    a) Given   c) Definition
    b) Prove   d) Postulate

11. When can a theorem be cited as a "reason" for a proof?

12. Based upon the hypothesis of a theorem, do the drawings of different students have to be identical (same names for vertices, etc.)?

For each theorem stated in Exercises 13 to 18, make a Drawing. On the basis of your Drawing, write a Given and a Prove for the theorem.

13. If two lines are perpendicular, then these lines meet to form a right angle.

14. If two lines meet to form a right angle, then these lines are perpendicular.

15. If two angles are complementary to the same angle, then these angles are congruent.

16. If two angles are supplementary to the same angle, then these angles are congruent.

17. If two lines intersect, then the vertical angles formed are congruent.

18. Any two right angles are congruent.

In Exercises 19 to 26, use the drawing in which $\overleftrightarrow{AC}$ intersects $\overleftrightarrow{DB}$ at point O.

19. If m∠1 = 125°, find m∠2, m∠3, and m∠4.

20. If m∠2 = 47°, find m∠1, m∠3, and m∠4.

21. If m∠1 = 3x + 10 and m∠3 = 4x − 30, find x and m∠1.

22. If m∠2 = 6x + 8 and m∠4 = 7x, find x and m∠2.

23. If m∠1 = 2x and m∠2 = x, find x and m∠1.

24. If m∠2 = x + 15 and m∠3 = 2x, find x and m∠2.

25. If m∠2 = $\frac{x}{2}$ − 10 and m∠3 = $\frac{x}{3}$ + 40, find x and m∠2.

26. If m∠1 = x + 20 and m∠4 = $\frac{x}{3}$, find x and m∠4.

In Exercises 27 to 35, complete the formal proof of each theorem.

27. If two angles are complementary to the same angle, then these angles are congruent.
    Given: ∠1 is comp. to ∠3
                 ∠2 is comp. to ∠3
    Prove: ∠1 ≅ ∠2

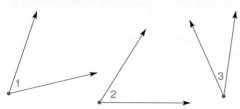

PROOF	
Statements	**Reasons**
1. ∠1 is comp. to ∠3 ∠2 is comp. to ∠3	1. ?
2. m∠1 + m∠3 = 90 m∠2 + m∠3 = 90	2. ?
3. m∠1 + m∠3 = m∠2 + m∠3	3. ?
4. m∠1 = m∠2	4. ?
5. ∠1 ≅ ∠2	5. ?

28. If two angles are supplementary to the same angle, then these angles are congruent.
    Given: ∠1 is supp. to ∠2
                 ∠3 is supp. to ∠2
    Prove: ∠1 ≅ ∠3

    (HINT: See Exercise 27 for help.)

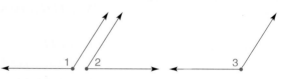

Exercise 28

29. If two lines intersect, the vertical angles formed are congruent.

30. Any two right angles are congruent.

31. If the exterior sides of two adjacent acute angles form perpendicular rays, then these angles are complementary.
    Given: $\overrightarrow{BA} \perp \overrightarrow{BC}$
    Prove: ∠1 is comp. to ∠2

PROOF	
Statements	**Reasons**
1. $\overrightarrow{BA} \perp \overrightarrow{BC}$	1. ?
2. ?	2. If two rays are ⊥, then they meet to form a rt. ∠
3. m∠ABC = 90	3. ?
4. m∠ABC = m∠1 + m∠2	4. ?
5. m∠1 + m∠2 = 90	5. Substitution
6. ?	6. If the sum of the measures of two angles is 90, then the angles are complementary

32. If two line segments are congruent, then their midpoints separate these segments into four congruent segments.

Given: $\overline{AB} \cong \overline{DC}$
M is the midpoint of $\overline{AB}$
N is the midpoint of $\overline{DC}$
Prove: $\overline{AM} \cong \overline{MB} \cong \overline{DN} \cong \overline{NC}$

33. If two angles are congruent, then their bisectors separate these angles into four congruent angles.

Given: $\angle ABC \cong \angle EFG$
$\overrightarrow{BD}$ bisects $\angle ABC$
$\overrightarrow{FH}$ bisects $\angle EFG$
Prove: $\angle 1 \cong \angle 2 \cong \angle 3 \cong \angle 4$

34. The bisectors of two adjacent supplementary angles form a right angle.

Given: $\angle ABC$ is supp. to $\angle CBD$
$\overrightarrow{BE}$ bisects $\angle ABC$
$\overrightarrow{BF}$ bisects $\angle CBD$
Prove: $\angle EBF$ is a right angle

35. The supplement of an acute angle is an obtuse angle.

(HINT: Use Exercise 28 of Section 1.6 as a guide.)

PERSPECTIVE ON HISTORY

THE DEVELOPMENT OF GEOMETRY

One of the first written accounts of geometric knowledge appears in the Rhind papyrus, a collection of documents that date back to more than 1000 years before Christ. In this document, Ahmes (an Egyptian scribe) describes how north-south and east-west lines were redrawn following the overflow of the Nile River. Astronomy was used to lay out the north-south line. The rest was done by people known as "rope-fasteners." By tying knots in a rope, it was possible to separate the rope into segments with lengths that were in the ratio 3 to 4 to 5. The knots were fastened at stakes in such a way that a right triangle would be formed. In Figure 1.69, the right angle is formed so that one side (of length 4, as shown) lies in the north-south line, and the second side (of length 3, as shown) lies in the east-west line.

Figure 1.69

The principle that was used by the rope-fasteners is known as the Pythagorean Theorem. However, we also know that the ancient Chinese were aware of this relationship. That is, the Pythagorean Theorem was known and applied many centuries before the time of Pythagoras (the Greek mathematician for whom the theorem is named).

Ahmes describes other facts of geometry that were known to the Egyptians. Perhaps the most impressive of these facts was that their approximation of π was 3.1604. To four decimal places of accuracy, we know today that the correct value of π is 3.1416.

Like the Egyptians, the Chinese treated geometry in a very practical way. In their constructions and designs, the Chinese used the rule (ruler), the square, the compass, and the level. Unlike the Egyptians and the Chinese, the Greeks formalized and expanded the knowledge base of geometry by pursuing it as an intellectual endeavor.

According to the Greek scribe Proclus (about 50 B.C.), Thales (625–547 B.C.) first established deductive proofs for several of the known theorems of geometry. Proclus also notes that it was Euclid (330–275 B.C.) who collected, summarized, ordered, and verified the vast quantity of knowledge of geometry in his time. Euclid's work *Elements* was the first textbook of geometry. Much of what was found in *Elements* is the core knowledge of geometry and thus can be found in this textbook as well.

PERSPECTIVE ON APPLICATIONS

PATTERNS

In much of the study of mathematics, we seek patterns related to the set of counting numbers $N = \{1,2,3,4,5, \ldots\}$. Some of these patterns are geometric and are given special names that reflect the configuration of sets of points. For instance, the set of *square numbers* is shown geometrically in Figure 1.70 and, of course, corresponds to the numbers 1, 4, 9, 16, ...

Figure 1.70

EXAMPLE 1

Find the fourth number in the pattern of triangular numbers shown in Figure 1.71(a).

1	3	6	?
(1 point)	(3 points)	(6 points)	(? points)

Figure 1.71(a)

SOLUTION Adding a row of 4 points at the bottom, we have the diagram shown in Figure 1.71(b), which contains 10 points. The fourth triangular number is 10.

Figure 1.71(b)

Some patterns of geometry lead to principles known as postulates and theorems. One of the principles that we will explore in the next example is based on the total number of *diagonals* found in a polygon with a given number of sides. A diagonal of a polygon (many-sided figure) joins two non-consecutive vertices of the polygon together. Of course, joining any two vertices of a triangle will determine a side; thus, a triangle has no diagonals. In Example 2, both the number of sides of the polygon and the number of diagonals are shown.

EXAMPLE 2

Find the total number of diagonals for a polygon of 6 sides.

3 sides	4 sides	5 sides	6 sides
0 diagonals	2 diagonals	5 diagonals	? diagonals

Figure 1.72(a)

SOLUTION By drawing all possible diagonals as shown in Figure 1.72(b) and counting them, we find that there are a total of 9 diagonals!

Figure 1.72(b)

Certain geometric patterns are used to test students, as in testing for intelligence (IQ) or on college admissions tests. A simple example might have you predict the next (fourth) figure in the pattern of squares shown in Figure 1.73(a).

Figure 1.73(a)

We rotate the square once more to obtain the fourth figure as shown in Figure 1.73(b).

Figure 1.73(b)

EXAMPLE 3

Midpoints of the sides of a *square* are used to generate new figures in the sequence shown in Figure 1.74(a). Draw the fourth figure.

Figure 1.74(a)

SOLUTION By continuing to add and join midpoints in the third figure, we form a figure like the one shown in Figure 1.74(b).

Figure 1.74(b)

Note that each new figure within the previous figure is also a square!

Summary

A Look Back at Chapter 1

Our goal in this chapter has been to introduce geometry. We discussed the types of reasoning that are used to develop geometric relationships. The use of the tools of measurement (ruler and protractor) was described. We encountered the four elements of a mathematical system: undefined terms, definitions, postulates, and theorems. The undefined terms were needed to lay the foundation for defining new terms. The postulates were needed to lay the foundation for the theorems we proved here and for the theorems that lie ahead. Constructions presented in this chapter included the bisector of an angle and the perpendicular to a line at a point on the line.

A Look Ahead to Chapter 2

The theorems we will prove in the next chapter are based on a postulate known as the Parallel Postulate. A new method of proof, called indirect proof, will be introduced; it will be used in later chapters. Although many of the theorems in Chapter 2 deal with parallel lines, several theorems in the chapter deal with the angles of a polygon. Symmetry and transformations will be discussed.

Key Concepts

1.1

Statement • Variable • Conjunction • Disjunction • Negation • Implication (Conditional) • Hypothesis • Conclusion • Reasoning • Intuition • Induction • Deduction • Argument (Valid and Invalid) • Law of Detachment • Set • Subsets • Venn Diagram • Intersection • Union

1.2

Point • Line • Plane • Collinear Points • Vertex • Line Segment • Betweenness of Points • Midpoint • Congruence

• Protractor • Parallel • Bisect • Intersect • Perpendicular • Compass • Constructions • Circle • Arc • Radius

1.3

Mathematical System • Axiom or Postulate • Assumption • Theorem • Ruler Postulate • Distance • Segment-Addition Postulate • Congruent Segments • Midpoint of a Line Segment • Ray • Opposite Rays • Intersection of Two Geometric Figures • Parallel Lines • Plane • Coplanar Points • Space

1.4

Angle • Sides of Angle • Vertex of Angle • Protractor Postulate • Acute, Right, Obtuse, Straight, and Reflex Angles • Angle-Addition Postulate • Adjacent Angles • Congruent Angles • Bisector of an Angle • Complementary Angles • Supplementary Angles • Vertical Angles

1.5

Algebraic Properties • Proof • Given Problem and Prove Statement

1.6

Vertical and Horizontal Line(s) • Perpendicular Lines • Relations • Reflexive, Symmetric, and Transitive Properties • Equivalence Relation • Perpendicular Bisector of a Line Segment

1.7

Formal Proof of a Theorem • Converse of a Theorem • Picture Proof of a Theorem

Overview ■ Chapter 1

Line and Line Segment Relationships

Figure	Relationship	Symbols
	Parallel lines (and segments)	$\ell \parallel m$ or $\overleftrightarrow{AB} \parallel \overleftrightarrow{CD}$; $\overline{AB} \parallel \overline{CD}$
	Intersecting lines	$\overleftrightarrow{EF} \cap \overleftrightarrow{GH} = K$
	Perpendicular lines (*t* shown vertical, *v* shown horizontal)	$t \perp v$
	Congruent line segments	$\overline{MN} \cong \overline{PQ}$; $MN = PQ$
	Point *B* between *A* and *C* on $\overline{AC}$	A-B-C; $AB + BC = AC$
	Point *M* the midpoint of $\overline{PQ}$	$\overline{PM} \cong \overline{MQ}$; $PM = MQ$; $PM = \frac{1}{2}(PQ)$; $PQ = 2(PM)$

Angle Classification (One Angle)

Figure	Type	Angle Measure
	Acute angle	$0° < m\angle 1 < 90°$
	Right angle	$m\angle 2 = 90°$

Angle Classification (One Angle)

Figure	Type	Angle Measure
	Obtuse angle	$90° < m\angle 3 < 180°$
	Straight angle	$m\angle 4 = 180°$
	Reflex angle	$180° < m\angle 5 < 360°$

Angle Relationships (Two Angles)

Figure	Relationship	Symbols
	Congruent angles	$\angle 1 \cong \angle 2$; $m\angle 1 = m\angle 2$
	Adjacent angles	$m\angle 3 + m\angle 4 = m\angle ABC$
	Bisector of angle ($\overrightarrow{HK}$ bisects $\angle GHJ$)	$\angle 5 \cong \angle 6$; $m\angle 5 = m\angle 6$; $m\angle 5 = \frac{1}{2}(m\angle GHJ)$ $m\angle GHJ = 2(m\angle 5)$
	Complementary angles	$m\angle 7 + m\angle 8 = 90°$
	Supplementary angles	$m\angle 9 + m\angle 10 = 180°$
	Vertical angles ($\angle 11$ and $\angle 12$; $\angle 13$ and $\angle 14$)	$\angle 11 \cong \angle 12$; $\angle 13 \cong \angle 14$

Chapter 1 Review Exercises

1. Name the four components of a mathematical system.

2. Name three types of reasoning.

3. Name the four characteristics of a good definition.

In Review Exercises 4 to 6, name the type of reasoning illustrated.

4. While watching the pitcher warm up, Phillip thinks, "I'll be able to hit against him."

5. Laura is away at camp. On the first day, her mother brings her additional clothing. On the second day, her mother brings her another pair of shoes. On the third day, her mother brings her cookies. Laura concludes that her mother will bring her something on the fourth day.

6. Sarah knows the rule "A number (not 0) divided by itself equals 1." The teacher asks Sarah, "What is 5 divided by 5?" Sarah says, "The answer is 1."

In Review Exercises 7 and 8, state the hypothesis and conclusion for each statement.

7. If the diagonals of a trapezoid are equal in length, then the trapezoid is isosceles.

8. The diagonals of a parallelogram are congruent if the parallelogram is a rectangle.

In Review Exercises 9 to 11, draw a valid conclusion where possible.

9. 1. If a person has a good job, then that person has a college degree.
   2. Henry has a college degree.
   C. ∴ ?

10. 1. If a person has a good job, then that person has a college degree.
    2. Jody Smithers has a good job.
    C. ∴ ?

11. 1. If the measure of an angle is $90°$, then that angle is a right angle.
    2. Angle A has a measure of $90°$.
    C. ∴ ?

12. A, B, and C are three points on a line. $AC = 8$, $BC = 4$, and $AB = 12$. Which point must be between the other two points?

13. Use three letters to name the angle shown. Also use one letter to name the same angle. Decide whether the angle measure is less than $90°$, equal to $90°$, or greater than $90°$.

14. Figure MNPQ is a rhombus. Draw diagonals $\overline{MP}$ and $\overline{QN}$ of the rhombus. How do $\overline{MP}$ and $\overline{QN}$ appear to be related?

In Review Exercises 15 to 17, sketch and label the figures described.

15. Points A, B, C, and D are coplanar. A, B, and C are the only three of these points that are collinear.

16. Line ℓ intersects plane X at point P.

17. Plane M contains intersecting lines j and k.

18. On the basis of appearance, what *type of angle* is shown?

(a)   (b)

19. On the basis of appearance, what *type of angle* is shown?

(a)   (b)

20. *Given:* $\overrightarrow{BD}$ bisects $\angle ABC$
    $m\angle ABD = 2x + 15$
    $m\angle DBC = 3x + 5$
    Find: $m\angle ABC$

21. *Given:* $m\angle ABD = 2x + 5$
    $m\angle DBC = 3x - 4$
    $m\angle ABC = 86°$
    Find: $m\angle DBC$

Exercises 20, 21

22. *Given:* $AM = 3x - 1$
    $MB = 4x - 5$
    M is the midpoint of $\overline{AB}$
    Find: AB

Exercises 22, 23

23. *Given:* $AM = 4x - 4$
    $MB = 5x + 2$
    $AB = 25$
    Find: MB

24. *Given:* D is the midpoint of $\overline{AC}$
    $\overline{AC} \cong \overline{BC}$
    $CD = 2x + 5$
    $BC = x + 28$
    Find: AC

25. *Given:* $m\angle 3 = 7x - 21$
           $m\angle 4 = 3x + 7$
     Find: $m\angle FMH$

26. *Given:* $m\angle FMH = 4x + 1$
           $m\angle 4 = x + 4$ *Exercises 25–27*
     Find: $m\angle 4$

27. In the figure, find:
a) $\overleftrightarrow{KH} \cap \overleftrightarrow{FJ}$
b) $\overrightarrow{MJ} \cup \overrightarrow{MH}$
c) $\angle KMJ \cap \angle JMH$
d) $\overrightarrow{MK} \cup \overrightarrow{MH}$

28. *Given:* $\angle EFG$ is a right angle
           $m\angle HFG = 2x - 6$
           $m\angle EFH = 3 \cdot m\angle HFG$
     Find: $m\angle EFH$

29. Two angles are supplementary. One angle is 40° more than four times the other. Find the measures of the two angles.

30. a) Write an expression for the perimeter of the triangle shown.

(HINT: Add the lengths of the sides.)

b) If the perimeter is 32 centimeters, find the value of x.
c) Find the length of each side of the triangle.

31. The sum of the measures of all three angles of the triangle in Review Exercise 30 is 180°. If the sum of the measures of angles 1 and 2 is more than 130°, what can you conclude about the measure of angle 3?

32. Susan wants to have a 4-ft board with some pegs on it. She wants to leave 6 in. on each end and 4 in. between pegs. How many pegs will fit on the board?

(HINT: If n represents the number of pegs, then $(n - 1)$ represents the number of equal spaces.)

State whether the sentences in Review Exercises 33 to 37 are always true (A), sometimes true (S), or never true (N).

33. If $AM = MB$, then A, M, and B are collinear.

34. If two angles are congruent, then they are right angles.

35. The bisectors of vertical angles are opposite rays.

36. Complementary angles are congruent.

37. The supplement of an obtuse angle is another obtuse angle.

38. Fill in the missing statements or reasons.
     Given: $\angle 1 \cong \angle P$
           $\angle 4 \cong \angle P$
           $\overrightarrow{VP}$ bisects $\angle RVO$
     Prove: $\angle TVP \cong \angle MVP$

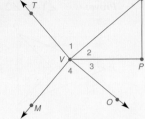

PROOF	
Statements	**Reasons**
1. $\angle 1 \cong \angle P$	1. Given
2. ?	2. Given
(1), (2) 3. ?	3. Transitive Prop. of $\cong$
(3) 4. $m\angle 1 = m\angle 4$	4. ?
5. $\overrightarrow{VP}$ bisects $\angle RVO$	5. ?
6. ?	6. If a ray bisects an $\angle$, it forms two $\angle$s of equal measure
(4), (6) 7. ?	7. Addition Prop. of Equality
8. $m\angle 1 + m\angle 2 = m\angle TVP$; $m\angle 4 + m\angle 3 = m\angle MVP$	8. ?
(7), (8) 9. $m\angle TVP = m\angle MVP$	9. ?
10. ?	10. If two $\angle$s are $=$ in measure, then they are $\cong$

Write two-column proofs for Review Exercises 39 to 46.

Exercises 39–41

39. *Given:* $\overline{KF} \perp \overline{FH}$
           $\angle JHF$ is a right $\angle$
     Prove: $\angle KFH \cong \angle JHF$

40. *Given:* $\overline{KH} \cong \overline{FJ}$
           G is the midpoint of both $\overline{KH}$ and $\overline{FJ}$
     Prove: $\overline{KG} \cong \overline{GJ}$

41. *Given:* $\overline{KF} \perp \overline{FH}$
     Prove: $\angle KFJ$ is comp. to $\angle JFH$

42. *Given:*   ∠1 is comp. to ∠M
                     ∠2 is comp. to ∠M
   Prove: ∠1 ≅ ∠2

Exercises 42, 43

43. *Given:*   ∠MOP ≅ ∠MPO
                     $\overrightarrow{OR}$ bisects ∠MOP
                     $\overrightarrow{PR}$ bisects ∠MPO
   Prove: ∠1 ≅ ∠2

For Review Exercise 44, see the figure that follows Review Exercise 45.

44. *Given:*   ∠4 ≅ ∠6
   Prove: ∠5 ≅ ∠6

45. *Given:*   Figure as shown
   Prove: ∠4 is supp. to ∠2

Exercises 44–46

46. *Given:*   ∠3 is supp. to ∠5
                     ∠4 is supp. to ∠6
   Prove: ∠3 ≅ ∠6

47. *Given:* $\overline{VP}$
   Construct: $\overline{VW}$ such that
                  VW = 4 · VP

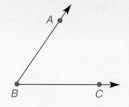

48. Construct a 135° angle.

49. *Given:*   Triangle *PQR*
   Construct: The three angle bisectors
   What did you discover about the three angle bisectors of this triangle?

50. *Given:* $\overline{AB}$, $\overline{BC}$, and ∠B as shown in Review Exercise 51.
   Construct: Triangle *ABC*

51. *Given:*   m∠B = 50°
   Construct: An angle whose measure is 20°

52. If m∠1 = 90°, find the measure of reflex angle 2.

Chapter **1**   Test

1. Which type of reasoning is illustrated below? _____
   Because it has rained the previous four days, Annie concludes that it will rain again today.

2. Given ∠ABC (as shown), provide a second correct method for naming this angle. _____

3. Using the Segment-Addition Postulate, state a conclusion regarding the accompanying figure. _____

A ————————— P ——— B

4. Complete each postulate:
   a) If two lines intersect, they intersect in a _____
   b) If two planes intersect, they intersect in a _____

5. Given that *x* is the measure of an angle, name the type of angle when:
   a) x = 90° _____   b) 90° < x < 180° _____

6. What word would describe two angles
   a) whose sum of measures is equal to 180°? _____
   b) that have equal measures? _____

7. Given that $\overrightarrow{NP}$ bisects $\angle MNQ$, state a conclusion involving
   m$\angle MNP$ and m$\angle PNQ$.
   _____

8. Complete each theorem:
   a) If two lines are perpendicular, they meet to form
      _____ angles.
   b) If the exterior sides of two adjacent angles form a
      straight line, these angles are _____

9. State the conclusion for the following deductive argument.
   (1) If you study geometry, then you will develop reasoning
       skills.
   (2) Kianna is studying geometry this semester.
   (C) _____

Questions 10, 11

10. In the figure, A-B-C-D and M is the midpoint of $\overline{AB}$. If
   $AB = 6.4$ inches and $BD = 7.2$ inches, find MD. _____

11. In the figure, $AB = x$, $BD = x + 5$, and $AD = 27$.
   Find: a) x _____   b) BD _____

Questions 12, 13

12. In the figure, m$\angle EFG = 68°$ and m$\angle 3 = 33°$.
   Find m$\angle 4$. _____

13. In the figure, m$\angle 3 = x$ and m$\angle 4 = 2x - 3$.
   If m$\angle EFG = 69°$, find:
   a) x _____   b) m$\angle 4$ _____

Questions 14–16

14. *Lines ℓ and m intersect at point P.*
   If m$\angle 1 = 43°$, find:
   a) m$\angle 2$ _____   b) m$\angle 3$ _____

15. If m$\angle 1 = 2x - 3$ and m$\angle 3 = 3x - 28$, find:
   a) x _____   b) m$\angle 1$ _____

16. If m$\angle 1 = 2x - 3$ and m$\angle 2 = 6x - 1$, find:
   a) x _____   b) m$\angle 2$ _____

17. $\angle$s 3 and 4 (not shown) are complementary. Where
   m$\angle 3 = x$ and m$\angle 4 = y$, write an equation using
   variables x and y.
   _____

18. Construct the angle bisector of obtuse angle *RST*.

19. Construct the perpendicular bisector of $\overline{AB}$.

*In Exercises 20 to 22, complete the missing statements/reasons
for each proof.*

20. *Given:* M-N-P-Q on $\overline{MQ}$
   Prove: $MN + NP + PQ = MQ$

PROOF	
Statements	**Reasons**
1. M-N-P-Q on $\overline{MQ}$	1. _____
2. $MN + NQ = MQ$	2. _____
3. $NP + PQ = NQ$	3. _____
4. $MN + NP + PQ = MQ$	4. _____

21. *Given:* $2x - 3 = 17$
   Prove: $x = 10$

PROOF	
Statements	**Reasons**
1. _____	1. Given
2. _____	2. Addition Property of Equality
3. _____	3. Division Property of Equality

22. *Given:* ∠*ABC* is a right angle;
$\overrightarrow{BD}$ bisects ∠*ABC*

Prove: m∠1 = 45°

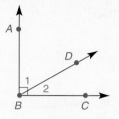

23. Obtuse angle *ABC* is bisected by $\overrightarrow{BD}$ and is trisected by $\overrightarrow{BE}$ and $\overrightarrow{BF}$. If m∠*EBD* = 18°, find m∠*ABC*. _____

PROOF

Statements	Reasons
1. ∠*ABC* is a right angle	1. _____
2. m∠*ABC* = _____	2. Definition of a right angle
3. m∠1 + m∠2 = m∠*ABC*	3. _____
4. m∠1 + m∠2 = _____	4. Substitution Property of Equality
5. $\overrightarrow{BD}$ bisects ∠*ABC*	5. _____
6. m∠1 = m∠2	6. _____
7. m∠1 + m∠1 = 90° or 2 · m∠1 = 90°	7. _____
8. _____	8. Division Property of Equality

© John Coletti/The Image Bank/Getty Images

Chapter 2

Parallel Lines

Breathtaking! The widest cable-stayed bridge in the world, the Leonard P. Zakim Bridge (also known as the Bunker Hill Bridge) lies at the north end of Boston, Massachusetts. Lying above the Charles River, this modern design bridge was dedicated in 2002. Cables for the bridge are parallel or nearly parallel to each other. The vertical towers above the bridge are perpendicular to the bridge floor. In this chapter, we consider relationships among parallel and perpendicular lines. Thanks to the line relationships, we can establish a most important fact regarding angle measures for the triangle in Section 2.4. Another look at the Bunker Hill Bridge suggests the use of symmetry, a topic that is given considerable attention in Section 2.6.

Additional video explanations of concepts, sample problems, and applications are available on DVD.

2.1 The Parallel Postulate and Special Angles

KEY CONCEPTS

Perpendicular Lines	Parallel Postulate	Corresponding Angles
Perpendicular Planes	Transversal	Alternate Interior
Parallel Lines	Interior Angles	Angles
Parallel Planes	Exterior Angles	Alternate Exterior
		Angles

PERPENDICULAR LINES

By definition, two lines (or segments or rays) are perpendicular if they meet to form congruent adjacent angles. Using this definition, we proved the theorem stating that "perpendicular lines meet to form right angles." We can also say that two rays or line segments are perpendicular if they are parts of perpendicular lines. We now consider a method for constructing a line perpendicular to a given line.

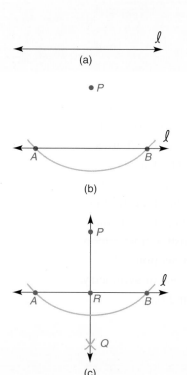

(a)

(b)

(c)

Figure 2.1

> **CONSTRUCTION 6** To construct the line that is perpendicular to a given line from a point not on the given line.
>
> GIVEN: In Figure 2.1(a), line ℓ and point P not on ℓ
> CONSTRUCT: $\overleftrightarrow{PQ} \perp \ell$
> CONSTRUCTION: Figure 2.1(b): With P as the center, open the compass to a length great enough to intersect ℓ in two points A and B.
>   Figure 2.1(c): With A and B as centers, mark off arcs of equal radii (using the same compass opening) to intersect at a point Q, as shown.
>   Draw $\overleftrightarrow{PQ}$ to complete the desired line.

In this construction, $\angle PRA$ and $\angle PRB$ are right angles. Greater accuracy is achieved if the arcs drawn from A and B intersect on the opposite side of line ℓ from point P.

Construction 6 suggests a uniqueness relationship that can be proved.

THEOREM 2.1.1

From a point not on a given line, there is exactly one line perpendicular to the given line.

The term *perpendicular* includes line-ray, line-plane, and plane-plane relationships. In Figure 2.1(c), $\overrightarrow{RP} \perp \ell$. The drawings in Figure 2.2 on page 69 indicate two perpendicular lines, a line perpendicular to a plane, and two perpendicular planes.

PARALLEL LINES

Just as the word *perpendicular* can relate lines and planes, the word *parallel* can also be used to describe relationships among lines and planes. However, parallel lines must lie in the same plane, as the following definition emphasizes.

(a) $\ell \perp m$ (b) $\ell \perp P$ (c) $R \perp P$

Figure 2.2

DEFINITION

Parallel lines are lines in the same plane that do not intersect.

More generally, two lines in a plane, a line and a plane, or two planes are parallel if they do not intersect. Figure 2.3 illustrates possible applications of the word *parallel*.

(a) $r \parallel s$ (b) $r \parallel T$ (c) $T \parallel V$
$r \cap s = \varnothing$ $r \cap T = \varnothing$ $T \cap V = \varnothing$

Figure 2.3

In Figure 2.4, two parallel planes M and N are both intersected by a third plane G. How must the lines of intersection, a and b, be related?

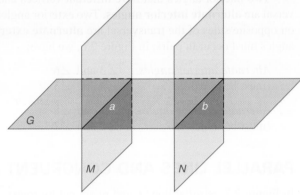

SSG **EXS. 1–3**   Figure 2.4

EUCLIDEAN GEOMETRY

Warning

Two lines that do not meet may be skew (rather than parallel). In the figure, the lines suggested by straw *a* and straw *c* are skew.

The type of geometry found in this textbook is known as *Euclidean geometry*. In this geometry, we interpret a plane as a flat, two-dimensional surface in which the line segment joining any two points of the plane lies entirely within the plane. Whereas the postulate that follows characterizes Euclidean geometry, the Perspective on Applications section near the end of this chapter discusses alternative geometries. Postulate 10, the Euclidean Parallel Postulate, is easy to accept because of the way we perceive a plane.

> ### POSTULATE 10 ■ Parallel Postulate
>
> Through a point not on a line, exactly one line is parallel to the given line.

Consider Figure 2.5, in which line *m* and point *P* (with *P* not on *m*) both lie in plane *R*. It seems reasonable that exactly one line can be drawn through *P* parallel to line *m*. The method of construction for the unique line through *P* parallel to *m* is provided in Section 2.3.

A **transversal** is a line that intersects two (or more) other lines at distinct points; all of the lines lie in the same plane. In Figure 2.6, *t* is a transversal for lines *r* and *s*. Angles that are formed between *r* and *s* are **interior angles;** those outside *r* and *s* are **exterior angles.** Relative to Figure 2.6, we have

Figure 2.5

Interior angles: $\angle 3, \angle 4, \angle 5, \angle 6$

Exterior angles: $\angle 1, \angle 2, \angle 7, \angle 8$

Consider the angles in Figure 2.6 that are formed when lines are cut by a transversal. Two angles that lie in the same relative positions (such as *above* and *left*) are called **corresponding angles** for these lines. In Figure 2.6, $\angle 1$ and $\angle 5$ are corresponding angles; each angle is *above* the line and to the *left* of the transversal that together form the angle. As shown in Figure 2.6, we have

Corresponding angles:	$\angle 1$ and $\angle 5$	above left
(must be in pairs)	$\angle 3$ and $\angle 7$	below left
	$\angle 2$ and $\angle 6$	above right
	$\angle 4$ and $\angle 8$	below right

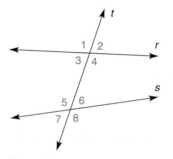

Figure 2.6

Two interior angles that have different vertices and lie on opposite sides of the transversal are **alternate interior angles.** Two exterior angles that have different vertices and lie on opposite sides of the transversal are **alternate exterior angles.** Both types of alternate angles must occur in pairs; in Figure 2.6, we have:

Alternate interior angles:	$\angle 3$ and $\angle 6$
(must be in pairs)	$\angle 4$ and $\angle 5$

Alternate exterior angles:	$\angle 1$ and $\angle 8$
(must be in pairs)	$\angle 2$ and $\angle 7$

SSG **EXS. 4–6**

PARALLEL LINES AND CONGRUENT ANGLES

In Figure 2.7, *parallel* lines ℓ and *m* are cut by transversal *v*. If a protractor were used to measure $\angle 1$ and $\angle 5$, these corresponding angles would have equal measures; that is, they would be congruent. Similarly, any other pair of corresponding angles would be congruent as long as $\ell \parallel m$.

> ### POSTULATE 11
>
> If two parallel lines are cut by a transversal, then the pairs of corresponding angles are congruent.

Figure 2.7

EXAMPLE 1

In Figure 2.7, $\ell \parallel m$ and $m\angle 1 = 117°$. Find:

a) $m\angle 2$ **c)** $m\angle 4$
b) $m\angle 5$ **d)** $m\angle 8$

SOLUTION

a) $m\angle 2 = 63°$ supplementary to $\angle 1$
b) $m\angle 5 = 117°$ corresponding to $\angle 1$
c) $m\angle 4 = 117°$ vertical to $\angle 1$
d) $m\angle 8 = 117°$ corresponding to $\angle 4$ [found in part (c)]

Several theorems follow from Postulate 11; for some of these theorems, formal proofs are provided. Study the proofs and be able to state all the theorems. You can cite the theorems that have been proven as reasons in subsequent proofs.

THEOREM 2.1.2

If two parallel lines are cut by a transversal, then the pairs of alternate interior angles are congruent.

Technology Exploration

Use computer software if available.
1. Draw $\overleftrightarrow{AB} \parallel \overleftrightarrow{CD}$.
2. Draw transversal $\overleftrightarrow{EF}$.
3. By numbering the angles as in Figure 2.8, find the measures of all eight angles.
4. Show that pairs of corresponding angles are congruent.

GIVEN: $a \parallel b$ in Figure 2.8
         Transversal k

PROVE: $\angle 3 \cong \angle 6$

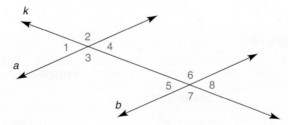

Figure 2.8

PROOF	
Statements	**Reasons**
1. $a \parallel b$; transversal k	1. Given
2. $\angle 2 \cong \angle 6$	2. If two $\parallel$ lines are cut by a transversal, corresponding $\angle$s are $\cong$
3. $\angle 3 \cong \angle 2$	3. If two lines intersect, vertical $\angle$s formed are $\cong$
4. $\angle 3 \cong \angle 6$	4. Transitive (of $\cong$)

Although we did not establish that alternate interior angles 4 and 5 are congruent, it is easy to prove that these are congruent because they are supplements to $\angle 3$ and $\angle 6$. A theorem that is similar to Theorem 2.1.2 follows, but the proof is left as Exercise 28.

THEOREM 2.1.3

If two parallel lines are cut by a transversal, then the pairs of alternate exterior angles are congruent.

PARALLEL LINES AND SUPPLEMENTARY ANGLES

When two parallel lines are cut by a transversal, it can be shown that the two interior angles on the same side of the transversal are supplementary. A similar claim can be made for the pair of exterior angles on the same side of the transversal.

> **STRATEGY FOR PROOF** ■ **Using Substitution in a Proof Statement**
>
> *General Rule:* In an equation, an expression can replace its equal.
> *Illustration:* See statements 3, 6, and 7 in the proof of Theorem 2.1.4. Note that m∠1 (found in statement 3) is substituted for m∠2 in statement 6 to obtain statement 7.

Discover

In the figure below, ℓ ∥ m with transversal *v*. Use your protractor to determine the relationship between ∠4 and ∠6.

ANSWER
ʎɹɐʇuǝɯǝlddnS

> **THEOREM 2.1.4**
>
> If two parallel lines are cut by a transversal, then the pairs of interior angles on the same side of the transversal are supplementary.

GIVEN: In Figure 2.9, $\overleftrightarrow{TV} \parallel \overleftrightarrow{WY}$ with transversal $\overleftrightarrow{RS}$

PROVE: ∠1 and ∠3 are supplementary

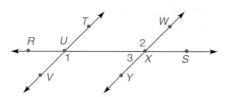

Figure 2.9

PROOF

Statements	Reasons
1. $\overleftrightarrow{TV} \parallel \overleftrightarrow{WY}$; transversal $\overleftrightarrow{RS}$	1. Given
2. ∠1 ≅ ∠2	2. If two ∥ lines are cut by a transversal, alternate interior ∠s are ≅
3. m∠1 = m∠2	3. If two ∠s are ≅, their measures are =
4. ∠WXY is a straight ∠, so m∠WXY 5 1808	4. If an ∠ is a straight ∠, its measure is 180°
5. m∠2 1 m∠3 5 m∠WXY	5. Angle-Addition Postulate
6. m∠2 1 m∠3 5 1808	6. Substitution
7. m∠1 1 m∠3 5 1808	7. Substitution
8. ∠1 and ∠3 are supplementary	8. If the sum of measures of two ∠s is 180°, the ∠s are supplementary

The proof of the following theorem is left as an exercise.

> **THEOREM 2.1.5**
>
> If two parallel lines are cut by a transversal, then the pairs of exterior angles on the same side of the transversal are supplementary.

SSG   **EXS. 7–11**

The remaining examples in this section illustrate methods from algebra and deal with the angles formed when two parallel lines are cut by a transversal.

EXAMPLE 2

GIVEN: $\overleftrightarrow{TV} \parallel \overleftrightarrow{WY}$ with transversal $\overleftrightarrow{RS}$

$\quad\quad\quad$ $m\angle RUV = (x + 4)(x - 3)$

$\quad\quad\quad$ $m\angle WXS = x^2 - 3$

FIND: x

SOLUTION $\angle RUV$ and $\angle WXS$ are alternate exterior angles, so they are congruent. Then $m\angle RUV = m\angle WXS$. Therefore,

$$(x + 4)(x - 3) = x^2 - 3$$
$$x^2 + x - 12 = x^2 - 3$$
$$x - 12 = -3$$
$$x = 9$$

NOTE: $\angle RUV$ and $\angle WXS$ both measure $78°$ when $x = 9$.

In Figure 2.10, lines r and s are known to be parallel; thus, $\angle 1 \cong \angle 5$, since these are corresponding angles for r and s with transversal ℓ.

For ℓ and m of Figure 2.10 to be parallel as well, name two angles that would have to be congruent. If we think of line s as a transversal, $\angle 5$ would have to be congruent to $\angle 9$, since these are corresponding angles for ℓ and m cut by transversal s.

For Example 3, recall that two equations are necessary to solve a problem in two variables.

SSG   **EXS. 12, 13**

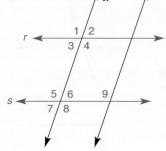

Figure 2.10

EXAMPLE 3

GIVEN: In Figure 2.10, $r \parallel s$ and transversal ℓ

$\quad\quad\quad$ $m\angle 3 = 4x + y$

$\quad\quad\quad$ $m\angle 5 = 6x + 5y$

$\quad\quad\quad$ $m\angle 6 = 5x - 2y$

FIND: x and y

SOLUTION $\angle 3$ and $\angle 6$ are congruent alternate interior angles; also, $\angle 3$ and $\angle 5$ are supplementary angles according to Theorem 2.1.4. These facts lead to the following system of equations:

$$4x + y = 5x - 2y$$
$$(4x + y) + (6x + 5y) = 180$$

The equations shown above can be simplified as follows

$$x - 3y = 0$$
$$10x + 6y = 180$$

After we divide each term of the second equation by 2, the system becomes

$$x - 3y = 0$$
$$5x + 3y = 90$$

Addition leads to the equation $6x = 90$, so $x = 15$. Substituting 15 for x into the equation $x - 3y = 0$, we have

$$15 - 3y = 0$$
$$-3y = -15$$
$$y = 5$$

Our solution, $x = 15$ and $y = 5$, yields the following angle measures:

$$m\angle 3 = 65°$$
$$m\angle 5 = 115°$$
$$m\angle 6 = 65°$$

NOTE: For an alternative solution, the equation $x - 3y = 0$ could be multiplied by 2 to obtain $2x - 6y = 0$. Then the equations $2x - 6y = 0$ and $10x + 6y = 180$ could be added to eliminate y.

Note that the angle measures determined in Example 3 are consistent with the required relationships for the angles named in Figure 2.10. For instance, $m\angle 3 + m\angle 5 = 180°$, and we see that interior angles on the same side of the transversal are indeed supplementary.

Exercises 2.1

For Exercises 1 to 4, $\ell \parallel m$ with transversal v.

1. If $m\angle 1 = 108°$, find:
   a) $m\angle 5$ b) $m\angle 7$

2. If $m\angle 3 = 71°$, find:
   a) $m\angle 5$ b) $m\angle 6$

3. If $m\angle 2 = 68.3°$, find:
   a) $m\angle 3$ b) $m\angle 6$

4. If $m\angle 4 = 110.8°$, find:
   a) $m\angle 5$ b) $m\angle 8$

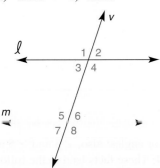

Use drawings, as needed, to answer each question.

5. Does the relation "is parallel to" have a
   a) reflexive property? (consider a line m)
   b) symmetric property? (consider lines m and n in a plane)
   c) transitive property? (consider coplanar lines m, n, and q)

6. In a plane, $\ell \perp m$ and $t \perp m$. By appearance, how are ℓ and t related?

7. Suppose that $r \parallel s$. Both interior angles ($\angle 4$ and $\angle 6$) have been bisected. Using intuition, what appears to be true of $\angle 9$ formed by the bisectors?

8. Make a sketch to represent two planes that are
   a) parallel.
   b) perpendicular.

9. Suppose that r is parallel to s and $m\angle 2 = 87°$. Find:
   a) $m\angle 3$ c) $m\angle 1$
   b) $m\angle 6$ d) $m\angle 7$

10. In Euclidean geometry, how many lines can be drawn through a point P not on a line ℓ that are
    a) parallel to line ℓ?
    b) perpendicular to line ℓ?

11. Lines r and s are cut by transversal t. Which angle
    a) corresponds to $\angle 1$?
    b) is the alternate interior $\angle$ for $\angle 4$?
    c) is the alternate exterior $\angle$ for $\angle 1$?
    d) is the other interior angle on the same side of transversal t as $\angle 3$?

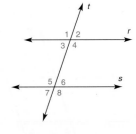

12. $\overrightarrow{AD} \parallel \overrightarrow{BC}$, $\overrightarrow{AB} \parallel \overrightarrow{DC}$, and $m\angle A = 92°$. Find:
    a) $m\angle B$
    b) $m\angle C$
    c) $m\angle D$

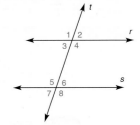

13. $\ell \parallel m$, with transversal t, and $\overrightarrow{OQ}$ bisects $\angle MON$. If $m\angle 1 = 112°$, find the following:
    a) $m\angle 2$
    b) $m\angle 4$
    c) $m\angle 5$
    d) $m\angle MOQ$

14. *Given:* $\ell \parallel m$
    Transversal t
    $m\angle 1 = 4x + 2$
    $m\angle 6 = 4x - 2$
    Find: x and $m\angle 5$

Exercises 13, 14

15. *Given:* $m \parallel n$
          Transversal k
          $m\angle 3 = x^2 - 3x$
          $m\angle 6 = (x + 4)(x - 5)$
    Find: x and $m\angle 4$

16. *Given:* $m \parallel n$
          Transversal k
          $m\angle 1 = 5x + y$
          $m\angle 2 = 3x + y$
          $m\angle 8 = 3x + 5y$
    Find: x, y, and $m\angle 8$

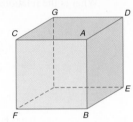

Exercises 15–17

17. *Given:* $m \parallel n$
          Transversal k
          $m\angle 3 = 6x + y$
          $m\angle 5 = 8x + 2y$
          $m\angle 6 = 4x + 7y$
    Find: x, y, and $m\angle 7$

18. In the three-dimensional figure, $\overline{CA} \perp \overline{AB}$ and $\overline{BE} \perp \overline{AB}$. Are $\overleftrightarrow{CA}$ and $\overleftrightarrow{BE}$ parallel to each other? (Compare with Exercise 6.)

19. *Given:* $\ell \parallel m$ and $\angle 3 \cong \angle 4$
    Prove: $\angle 1 \cong \angle 4$
    (See figure below.)

PROOF

Statements	Reasons
1. $\ell \parallel m$	1. ?
2. $\angle 1 \cong \angle 2$	2. ?
3. $\angle 2 \cong \angle 3$	3. ?
4. ?	4. Given
5. ?	5. Transitive of $\cong$

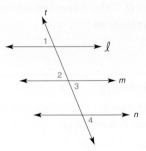

Exercises 19, 20

20. *Given:* $\ell \parallel m$ and $m \parallel n$ (See figure for Exercise 19.)
    Prove: $\angle 1 \cong \angle 4$

PROOF

Statements	Reasons
1. $\ell \parallel m$	1. ?
2. $\angle 1 \cong \angle 2$	2. ?
3. $\angle 2 \cong \angle 3$	3. ?
4. ?	4. Given
5. $\angle 3 \cong \angle 4$	5. ?
6. ?	6. ?

21. *Given:* $\overleftrightarrow{CE} \parallel \overleftrightarrow{DF}$
          Transversal $\overleftrightarrow{AB}$
          $\overrightarrow{CX}$ bisects $\angle ACE$
          $\overrightarrow{DE}$ bisects $\angle CDF$
    Prove: $\angle 1 \cong \angle 3$

Exercises 21, 22

22. *Given:* $\overleftrightarrow{CE} \parallel \overleftrightarrow{DF}$
          Transversal $\overleftrightarrow{AB}$
          $\overrightarrow{DE}$ bisects $\angle CDF$
    Prove: $\angle 3 \cong \angle 6$

23. *Given:* $r \parallel s$
          Transversal t
          $\angle 1$ is a right $\angle$
    Prove: $\angle 2$ is a right $\angle$

Exercises 23, 26

24. *Given:* $\overleftrightarrow{AB} \parallel \overleftrightarrow{DE}$
          $m\angle BAC = 42°$
          $m\angle EDC = 54°$
    Find: $m\angle ACD$

    (*HINT: There is a line through C parallel to both $\overleftrightarrow{AB}$ and $\overleftrightarrow{DE}$.*)

Exercises 24, 25

25. *Given:* $\overleftrightarrow{AB} \parallel \overleftrightarrow{DE}$
          $m\angle BAC + m\angle CDE = 93°$
    Find: $m\angle ACD$

    (*See "Hint" in Exercise 24.*)

26. *Given:* $r \parallel s$, $r \perp t$ (See figure for Exercise 23.)
    Prove: $s \perp t$

27. In triangle ABC, line t is drawn through vertex A in such a way that $t \parallel \overline{BC}$.
    a) Which pairs of $\angle$s are $\cong$?
    b) What is the sum of $m\angle 1$, $m\angle 4$, and $m\angle 5$?
    c) What is the sum of measures of the $\angle$s of $\triangle ABC$?

In Exercises 28 to 30, write a formal proof of each theorem.

28. If two parallel lines are cut by a transversal, then the pairs of alternate exterior angles are congruent.

29. If two parallel lines are cut by a transversal, then the pairs of exterior angles on the same side of the transversal are supplementary.

30. If a transversal is perpendicular to one of two parallel lines, then it is also perpendicular to the other line.

31. Suppose that two lines are cut by a transversal in such a way that the pairs of corresponding angles are not congruent. Can those two lines be parallel?

32. *Given:*   Line ℓ and point P not on ℓ
       Construct: $\overleftrightarrow{PQ} \perp \ell$

33. *Given:*   Triangle ABC with three acute angles
       Construct: $\overline{BD} \perp \overline{AC}$, with D on $\overline{AC}$.

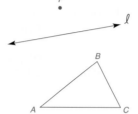

34. *Given:*   Triangle MNQ with obtuse $\angle MNQ$
       Construct: $\overline{NE} \perp \overline{MQ}$, with E on $\overline{MQ}$.

35. *Given:*   Triangle MNQ with obtuse $\angle MNQ$
       Construct: $\overline{MR} \perp \overline{NQ}$
       (*HINT: Extend* $\overline{NQ}$.)

Exercises 34, 35

36. *Given:*   A line m and a point T not on m

Suppose that you do the following:
i)   Construct a perpendicular line r from T to line m.
ii)   Construct a line s perpendicular to line r at point T.
What is the relationship between lines s and m?

2.2   Indirect Proof

KEY CONCEPTS	Conditional	Contrapositive	Indirect Proof
	Converse	Law of Negative	
	Inverse	Inference	

Let $P \rightarrow Q$ represent the *conditional* statement "If P, then Q." The following statements are related to this conditional statement (also called an *implication*).

NOTE: Recall that $\sim P$ represents the negation of P.

Conditional (or Implication)	$P \rightarrow Q$	If P, then Q.
Converse of Conditional	$Q \rightarrow P$	If Q, then P.
Inverse of Conditional	$\sim P \rightarrow \sim Q$	If not P, then not Q.
Contrapositive of Conditional	$\sim Q \rightarrow \sim P$	If not Q, then not P.

Consider the following conditional statement.

> *If Tom lives in San Diego, then he lives in California.*

This true conditional statement has the following related statements:

> *Converse:* If Tom lives in California, then he lives in San Diego. (false)

> *Inverse:* If Tom does not live in San Diego, then he does not live in California. (false)

> *Contrapositive:* If Tom does not live in California, then he does not live in San Diego. (true)

In general, the conditional statement and its contrapositive are either both true or both false! Similarly, the converse and the inverse are also either both true or both false.

EXAMPLE 1

For the conditional statement that follows, give the converse, the inverse, and the contrapositive. Then classify each as true or false.

If two angles are vertical angles, then they are congruent angles.

SOLUTION

CONVERSE: If two angles are congruent angles, then they are vertical angles. (false)

INVERSE: If two angles are not vertical angles, then they are not congruent angles. (false)

CONTRAPOSITIVE: If two angles are not congruent angles, then they are not vertical angles. (true)

Figure 2.11

> "If P, then Q" and "If not Q, then not P" are equivalent.

Venn Diagrams can be used to explain why the conditional statement $P \rightarrow Q$ and its contrapositive $\sim Q \rightarrow \sim P$ are equivalent. The relationship "If P, then Q" is represented in Figure 2.11. Note that if any point is selected outside of Q (that is $\sim Q$), then it cannot possibly lie in set P (thus, $\sim P$).

THE LAW OF NEGATIVE INFERENCE (CONTRAPOSITION)

SSG **EXS. 1, 2**

Consider the following circumstances, and accept each premise as true:

1. If Matt cleans his room, then he will go to the movie. ($P \rightarrow Q$)
2. Matt does not get to go to the movie. ($\sim Q$)

What can you conclude? You should have deduced that Matt did not clean his room; if he had, he would have gone to the movie. This "backdoor" reasoning is based on the fact that the truth of $P \rightarrow Q$ implies the truth of $\sim Q \rightarrow \sim P$.

LAW OF NEGATIVE INFERENCE (CONTRAPOSITION)
1. $P \rightarrow Q$
2. $\sim Q$
C. $\therefore \sim P$

EXAMPLE 2

Use the Law of Negative Inference to draw a valid conclusion for this argument.

1. If the weather is nice Friday, we will go on a picnic.
2. We did not go on a picnic Friday.
C. $\therefore$?

SOLUTION

The weather was not nice Friday.

Like the Law of Detachment from Section 1.1, the Law of Negative Inference (Law of Contraposition) is a form of deduction. Whereas the Law of Detachment characterizes the method of "direct proof" found in preceding sections, the Law of Negative Inference characterizes the method of proof known as **indirect proof.**

INDIRECT PROOF

SSG ▶ **EXS. 3, 4**

You will need to know when to use the indirect method of proof. Often the theorem to be proved has the form $P \rightarrow Q$, in which Q is a negation and denies some claim. For instance, an indirect proof might be best if Q reads in one of these ways:

c is *not* equal to d

ℓ is *not* perpendicular to m

However, we will see in Example 5 of this section that the indirect method can be used to prove that line ℓ is parallel to line m. Indirect proof is also used for proving existence and uniqueness theorems; see Example 6.

The method of indirect proof is illustrated in Example 3. All indirect proofs in this book are given in paragraph form (as are some of the direct proofs).

In any paragraph proof, each statement must still be justified. Because of the need to order your statements properly, writing any type of proof may have a positive impact on the essays you write for your other classes!

Geometry in the Real World

When the bubble displayed on the level is not centered, the board used in construction is neither vertical nor horizontal.

EXAMPLE 3

GIVEN: In Figure 2.12, $\overrightarrow{BA}$ is *not* perpendicular to $\overrightarrow{BD}$

PROVE: $\angle 1$ and $\angle 2$ are *not* complementary

PROOF: Suppose that $\angle 1$ and $\angle 2$ *are* complementary. Then $m\angle 1 + m\angle 2 = 90°$ because the sum of the measures of two complementary $\angle$s is 90. We also know that $m\angle 1 + m\angle 2 = m\angle ABD$ by the Angle-Addition Postulate. In turn, $m\angle ABD = 90°$ by substitution. Then $\angle ABD$ is a right angle. In turn, $\overrightarrow{BA} \perp \overrightarrow{BD}$. But this contradicts the given hypothesis; therefore, the supposition must be false, and it follows that $\angle 1$ and $\angle 2$ are not complementary.

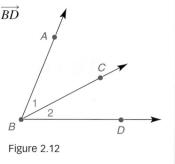

Figure 2.12

In Example 3 and in all indirect proofs, the first statement takes the form

Suppose/Assume the exact opposite of the Prove Statement.

By its very nature, such a statement cannot be supported even though every other statement in the proof can be justified; thus, when a contradiction is reached, the finger of blame points to the supposition. Having reached a contradiction, we may say that the claim involving $\sim Q$ has failed and is false; in effect, the double negative $\sim(\sim Q)$ is equivalent to Q. Thus, our only recourse is to conclude that Q is true. Following is an outline of this technique.

SSG **EXS. 5–7**

> **STRATEGY FOR PROOF ■ Method of Indirect Proof**
>
> To prove the statement $P \rightarrow Q$ or to complete the proof problem of the form
>
> > Given: P
> >
> > Prove: Q
>
> by the indirect method, use the following steps:
>
>   **1.** Suppose that $\sim Q$ is true.
>   **2.** Reason from the supposition until you reach a contradiction.
>   **3.** Note that the supposition claiming that $\sim Q$ is true must be false and that Q must therefore be true.
>
> Step 3 completes the proof.

The contradiction found in an indirect proof often takes the form "Q and $\sim Q$," which cannot be true. Thus, the assumed statement $\sim Q$ has forced the conclusion $\sim P$, asserting that $\sim Q \rightarrow \sim P$ is true. Then the desired theorem $P \rightarrow Q$ (which is equivalent to the contrapositive of $\sim Q \rightarrow \sim P$) is also true.

> **STRATEGY FOR PROOF ■ The First Line of an Indirect Proof**
>
> *General Rule:* The first *statement* of an indirect proof is generally "Suppose/Assume the *opposite* of the Prove statement."
>
> *Illustration:* See Example 4, which begins "Assume that $\ell \parallel m$."

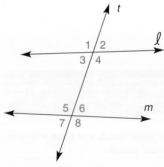

Figure 2.13

SSG **EXS. 8, 9**

EXAMPLE 4

Complete a formal proof of the following theorem:

> *If two lines are cut by a transversal so that corresponding angles are not congruent, then the two lines are not parallel.*

GIVEN: In Figure 2.13, ℓ and m are cut by transversal t
     $\angle 1 \not\cong \angle 5$

PROVE: $\ell \not\parallel m$

PROOF: Assume that $\ell \parallel m$. When these lines are cut by transversal t, any two corresponding angles (including $\angle 1$ and $\angle 5$) are congruent. But $\angle 1 \not\cong \angle 5$ by hypothesis. Thus, the assumed statement, which claims that $\ell \parallel m$, must be false. It follows that $\ell \not\parallel m$.

The versatility of the indirect proof is shown in the final examples of this section. The indirect proofs preceding Example 5 contain a negation in the conclusion (Prove); the proofs in the final illustrations use the indirect method to arrive at a positive conclusion.

EXAMPLE 5

GIVEN: In Figure 2.14, plane T intersects parallel planes P and Q in lines ℓ and m, respectively

PROVE: $\ell \parallel m$

PROOF: Assume that ℓ is not parallel to m. Then ℓ and m intersect at some point A. But if so, point A must be on both planes P and Q, which means that planes P and Q intersect; but P and Q are parallel by hypothesis. Therefore, the assumption that ℓ and m are not parallel must be false, and it follows that $\ell \parallel m$.

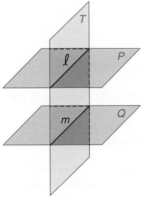

Figure 2.14

Indirect proofs are also used to establish uniqueness theorems, as Example 6 illustrates.

EXAMPLE 6

Prove the statement "The bisector of an angle is unique."

GIVEN: In Figure 2.15(a), $\overrightarrow{BD}$ bisects $\angle ABC$

PROVE: $\overrightarrow{BD}$ is the only bisector for $\angle ABC$

PROOF: Suppose that $\overrightarrow{BE}$ [as shown in Figure 2.15(b)] is also a bisector of $\angle ABC$ and that $m\angle ABE = \frac{1}{2}m\angle ABC$. Given that $\overrightarrow{BD}$ bisects $\angle ABC$, it follows that $m\angle ABD = \frac{1}{2}m\angle ABC$.

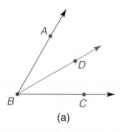

(a)   (b)

Figure 2.15

By the Angle-Addition Postulate, $m\angle ABD = m\angle ABE + m\angle EBD$. By substitution, $\frac{1}{2}m\angle ABC = \frac{1}{2}m\angle ABC + m\angle EBD$; but then $m\angle EBD = 0$ by subtraction. An angle with a measure of 0 contradicts the Protractor Postulate, which states that the measure of an angle is a unique positive number. Therefore, the assumed statement must be false, and it follows that the bisector of an angle is unique.

SSG **EX. 10**

Exercises 2.2

In Exercises 1 to 4, write the converse, the inverse, and the contrapositive of each statement. When possible, classify the statement as true or false.

1. If Juan wins the state lottery, then he will be rich.

2. If $x > 2$, then $x \neq 0$.

3. Two angles are complementary if the sum of their measures is 90°.

4. In a plane, if two lines are not perpendicular to the same line, then these lines are not parallel.

In Exercises 5 to 10, draw a conclusion where possible.

5. 1. If two triangles are congruent, then the triangles are similar.
2. Triangles *ABC* and *DEF* are not congruent.
C. ∴ ?

6. 1. If two triangles are congruent, then the triangles are similar.
2. Triangles *ABC* and *DEF* are not similar.
C. ∴ ?

7. 1. If Alice plays in the volleyball match, our team will win.
2. Our team lost the volleyball match.
C. ∴ ?

8. 1. If you send the package on Tuesday, it will arrive on Thursday.
2. The package arrived on Friday.
C. ∴ ?

9. 1. If $x > 3$, then $x = 5$.
2. $x > 3$
C. ∴ ?

10. 1. If $x > 3$, then $x = 5$.
2. $x \neq 5$
C. ∴ ?

11. Which of the following statements would you prove by the indirect method?
a) In triangle *ABC*, if $m\angle A > m\angle B$, then $AC \neq BC$.
b) If alternate exterior $\angle 1 \not\cong$ alternate exterior $\angle 8$, then ℓ is not parallel to *m*.
c) If $(x + 2) \cdot (x - 3) = 0$, then $x = -2$ or $x = 3$.
d) If two sides of a triangle are congruent, then the two angles opposite these sides are also congruent.
e) The perpendicular bisector of a line segment is unique.

In Exercises 12 to 14, write the first statement of the indirect proof of the given statement.

12. If $AC > BC$ in $\triangle ABC$, then $m\angle B \neq m\angle A$.

13. If ℓ is not parallel to m, then $\angle 1 \not\cong \angle 2$.

14. If $\overline{AB}$ is not perpendicular to $\overline{BC}$, then $\angle ABC$ is not a right angle.

For Exercises 15 to 18, the given statement is true. Write an equivalent (but more compact) statement that must be true.

15. If $\angle A$ and $\angle B$ are not congruent, then $\angle A$ and $\angle B$ are not vertical angles.

16. If lines ℓ and m are not perpendicular, then the angles formed by ℓ and m are not right angles.

17. If all sides of a triangle are not congruent, then the triangle is not an equilateral triangle.

18. If no two sides of a quadrilateral (figure with four sides) are parallel, then the quadrilateral is not a trapezoid.

In Exercises 19 and 20, state a conclusion for the argument. Statements 1 and 2 are true.

19. 1. If the areas of two triangles are not equal, then the two triangles are not congruent.
2. Triangle ABC is congruent to triangle DEF.
C. ∴ ?

20. 1. If two triangles do not have the same shape, then the triangles are not similar.
2. Triangle RST is similar to triangle XYZ.
C. ∴ ?

21. A periscope uses an indirect method of observation. This instrument allows one to see what would otherwise be obstructed. Mirrors are located (see $\overline{AB}$ and $\overline{CD}$ in the drawing) so that an image is reflected twice. How are $\overline{AB}$ and $\overline{CD}$ related to each other?

22. Some stores use an indirect method of observation. The purpose may be for safety (to avoid collisions) or to foil the attempts of would-be shoplifters. In this situation, a mirror (see $\overline{EF}$ in the drawing) is placed at the intersection of two aisles as shown. An observer at point P can then see any movement along the indicated aisle. In the sketch, what is the measure of $\angle GEF$?

In Exercises 23 to 34, give the indirect proof for each problem or statement.

23. Given: $\angle 1 \not\cong \angle 5$
Prove: $r \not\parallel s$

24. Given: $\angle ABD \not\cong \angle DBC$
Prove: $\overrightarrow{BD}$ does not bisect $\angle ABC$

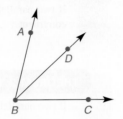

25. Given: $m\angle 3 > m\angle 4$
Prove: $\overleftrightarrow{FH}$ is not $\perp$ to $\overleftrightarrow{EG}$

26. Given: $MB > BC$
$AM = CD$
Prove: B is not the midpoint of $\overline{AD}$

27. If two angles are not congruent, then these angles are not vertical angles.

28. If $x^2 \neq 25$, then $x \neq 5$.

29. If alternate interior angles are not congruent when two lines are cut by a transversal, then the lines are not parallel.

30. If a and b are positive numbers, then $\sqrt{a^2 + b^2} \neq a + b$.

31. The midpoint of a line segment is unique.

32. There is exactly one line perpendicular to a given line at a point on the line.

*33. In a plane, if two lines are parallel to a third line, then the two lines are parallel to each other.

*34. In a plane, if two lines are intersected by a transversal so that the corresponding angles are congruent, then the lines are parallel.

2.3 Proving Lines Parallel

KEY CONCEPTS   Proving Lines Parallel

For this section, here is a quick review of the relevant postulate and theorems from Section 2.1. Each theorem has the hypothesis "If two parallel lines are cut by a transversal"; each theorem has a conclusion involving an angle relationship.

> **POSTULATE 11**
>
> If two parallel lines are cut by a transversal, then the pairs of corresponding angles are congruent.

> **THEOREM 2.1.2**
>
> If two parallel lines are cut by a transversal, then the pairs of alternate interior angles are congruent.

> **THEOREM 2.1.3**
>
> If two parallel lines are cut by a transversal, then the pairs of alternate exterior angles are congruent.

> **THEOREM 2.1.4**
>
> If two parallel lines are cut by a transversal, then the pairs of interior angles on the same side of the transversal are supplementary.

> **THEOREM 2.1.5**
>
> If two parallel lines are cut by a transversal, then the pairs of exterior angles on the same side of the transversal are supplementary.

SSG   **EXS. 1, 2**

Suppose that we wish to prove that two lines are parallel rather than to establish an angle relationship (as the previous statements do). Such a theorem would take the form "If . . . , then these lines are parallel." At present, the only method we have of proving lines parallel is based on the definition of parallel lines. Establishing the conditions of the definition (that coplanar lines do *not* intersect) is virtually impossible! Thus, we begin to develop methods for proving that lines in a plane are parallel by proving Theorem 2.3.1 by the indirect method. Counterparts of Theorems 2.1.2–2.1.5, namely, Theorems 2.3.2–2.3.5, are proved directly but depend on Theorem 2.3.1. Except for Theorem 2.3.6, the theorems of this section require coplanar lines.

> **THEOREM 2.3.1**
>
> If two lines are cut by a transversal so that two corresponding angles are congruent, then these lines are parallel.

Figure 2.16

GIVEN: ℓ and m cut by transversal t
$\angle 1 \cong \angle 2$ (See Figure 2.16)

PROVE: $\ell \parallel m$

PROOF: Suppose that $\ell \not\parallel m$. Then a line r can be drawn through point P that is parallel to m; this follows from the Parallel Postulate. If $r \parallel m$, then $\angle 3 \cong \angle 2$ because these angles correspond. But $\angle 1 \cong \angle 2$ by hypothesis. Now $\angle 3 \cong \angle 1$ by the Transitive Property of Congruence; therefore, $m\angle 3 = m\angle 1$. But $m\angle 3 + m\angle 4 = m\angle 1$. (See Figure 2.16.) Substituting $m\angle 1$ for $m\angle 3$ leads to $m\angle 1 + m\angle 4 = m\angle 1$; and by subtraction, $m\angle 4 = 0$. This contradicts the Protractor Postulate, which states that the measure of any angle must be a positive number. Then r and ℓ must coincide, and it follows that $\ell \parallel m$.

Each claim in Theorems 2.3.2–2.3.5 is the converse of its counterpart in Section 2.1, and each claim provides a method for proving that lines are parallel.

THEOREM 2.3.2

If two lines are cut by a transversal so that two alternate interior angles are congruent, then these lines are parallel.

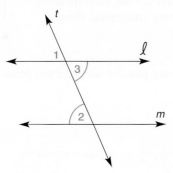

Figure 2.17

GIVEN: Lines ℓ and m and transversal t
$\angle 2 \cong \angle 3$ (See Figure 2.17.)

PROVE: $\ell \parallel m$

PLAN FOR THE PROOF: Show that $\angle 1 \cong \angle 2$ (corresponding angles). Then apply Theorem 2.3.1, in which $\cong$ corresponding $\angle$s imply parallel lines.

PROOF	
Statements	**Reasons**
1. ℓ and m; trans. t; $\angle 2 \cong \angle 3$	1. Given
2. $\angle 1 \cong \angle 3$	2. If two lines intersect, vertical $\angle$s are $\cong$
3. $\angle 1 \cong \angle 2$	3. Transitive Property of Congruence
4. $\ell \parallel m$	4. If two lines are cut by a transversal so that corr. $\angle$s are $\cong$, then these lines are parallel

The following theorem is proved in a manner much like the proof of Theorem 2.3.2. The proof is left as an exercise.

THEOREM 2.3.3

If two lines are cut by a transversal so that two alternate exterior angles are congruent, then these lines are parallel.

In a more complex drawing, it may be difficult to decide which lines are parallel because of congruent angles. Consider Figure 2.18 on page 84. Suppose that $\angle 1 \cong \angle 3$. Which lines must be parallel? The resulting confusion (it appears that a may be parallel to b and c may be parallel to d) can be overcome by asking, "Which lines help form $\angle 1$ and $\angle 3$?" In this case, $\angle 1$ and $\angle 3$ are formed by lines a and b with c as the transversal. Thus, $a \parallel b$.

Discover

When a staircase is designed, "stringers" are cut for each side of the stairs as shown. How are angles 1 and 3 related? How are angles 1 and 2 related?

ANSWER
ʎɹɐʇuǝɯǝldɯoↃ 'ʇuǝnɹᵷuoↃ

Figure 2.18

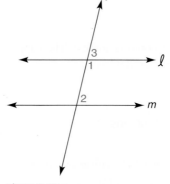

Figure 2.19

EXAMPLE 1

In Figure 2.18, which lines must be parallel if $\angle 3 \cong \angle 8$?

SOLUTION $\angle 3$ and $\angle 8$ are the alternate exterior angles formed when lines c and d are cut by transversal b. Thus, $c \parallel d$.

EXAMPLE 2

In Figure 2.18, $m\angle 3 = 94°$. Find $m\angle 5$ so that $c \parallel d$.

SOLUTION With b as a transversal for lines c and d, $\angle 3$ and $\angle 5$ are corresponding angles. Then c would be parallel to d if $\angle 3$ and $\angle 5$ were congruent. Thus, $m\angle 5 = 94°$.

Theorems 2.3.4 and 2.3.5 enable us to prove that lines are parallel when certain pairs of angles are supplementary.

THEOREM 2.3.4

If two lines are cut by a transversal so that two interior angles on the same side of the transversal are supplementary, then these lines are parallel.

EXAMPLE 3

Prove Theorem 2.3.4. (See Figure 2.19.)

GIVEN: Lines ℓ and m; transversal t
$\angle 1$ is supplementary to $\angle 2$

PROVE: $\ell \parallel m$

PROOF

Statements	Reasons
1. ℓ and m; trans. t; $\angle 1$ is supp. to $\angle 2$	1. Given
2. $\angle 1$ is supp. to $\angle 3$	2. If the exterior sides of two adjacent $\angle$s form a straight line, these $\angle$s are supplementary
3. $\angle 2 \cong \angle 3$	3. If two $\angle$s are supp. to the same $\angle$, they are $\cong$
4. $\ell \parallel m$	4. If two lines are cut by a transversal so that corr. $\angle$s are $\cong$, then these lines are parallel

The proof of Theorem 2.3.5 is similar to that of Theorem 2.3.4. The proof is left as an exercise.

THEOREM 2.3.5

If two lines are cut by a transversal so that two exterior angles on the same side of the transversal are supplementary, then the lines are parallel.

Figure 2.20

EXAMPLE 4

In Figure 2.20, which line segments must be parallel if $\angle B$ and $\angle C$ are supplementary?

SOLUTION Again, the solution lies in the question "Which line segments form $\angle B$ and $\angle C$?" With $\overline{BC}$ as a transversal, $\angle B$ and $\angle C$ are formed by $\overline{AB}$ and $\overline{DC}$. Because the supplementary interior angles B and C lie on the same side of transversal $\overline{BC}$, it follows that $\overline{AB} \parallel \overline{DC}$.

We include two final theorems that provide additional means of proving that lines are parallel. The proof of Theorem 2.3.6 (see Exercise 33) requires an auxiliary line (a transversal).

Discover

On the grid shown, points are uniformly spaced. Name two parallel line segments in figure (a). In figure (b), what relationship exists between $\overline{RS}$ and $\overline{TV}$?

(a)   (b)

ANSWER
(a) $\overline{AB} \parallel \overline{EF}$ (b) perpendicular

THEOREM 2.3.6

If two lines are each parallel to a third line, then these lines are parallel to each other.

Theorem 2.3.6 is true even if the three lines described are not coplanar. In Theorem 2.3.7, the lines must be coplanar; in Example 5, we prove Theorem 2.3.7.

THEOREM 2.3.7

If two coplanar lines are each perpendicular to a third line, then these lines are parallel to each other.

STRATEGY FOR PROOF ■ Proving That Lines Are Parallel

General Rule: The proof of Theorem 2.3.7 depends upon establishing the condition found in one of the Theorems 2.3.1–2.3.6.

Illustration: In Example 5, we establish congruent corresponding angles in statement 3 so that lines are parallel by Theorem 2.3.1.

SSG **EXS. 3–8**

Figure 2.21

EXAMPLE 5

GIVEN: $\overleftrightarrow{AC} \perp \overleftrightarrow{BE}$ and $\overleftrightarrow{DF} \perp \overleftrightarrow{BE}$ (See Figure 2.21.)
PROVE: $\overleftrightarrow{AC} \parallel \overleftrightarrow{DF}$

PROOF	
Statements	**Reasons**
1. $\overleftrightarrow{AC} \perp \overleftrightarrow{BE}$ and $\overleftrightarrow{DF} \perp \overleftrightarrow{BE}$	1. Given
2. $\angle$s 1 and 2 are rt. $\angle$s	2. If two lines are perpendicular, they meet to form right $\angle$s
3. $\angle 1 \cong \angle 2$	3. All right angles are $\cong$
4. $\overleftrightarrow{AC} \parallel \overleftrightarrow{DF}$	4. If two lines are cut by a transversal so that corr. $\angle$s are $\cong$, then these lines are parallel

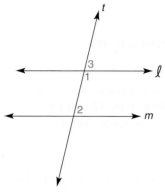

Figure 2.22

EXAMPLE 6

GIVEN: m∠1 = 7x and m∠2 = 5x (See Figure 2.22.)

FIND: x, so that ℓ will be parallel to m

SOLUTION For ℓ to be parallel to m, ∠s 1 and 2 would have to be supplementary. This follows from Theorem 2.3.4 because ∠s 1 and 2 are interior angles on the same side of transversal t. Then

$$7x + 5x = 180$$
$$12x = 180$$
$$x = 15$$

NOTE: With m∠1 = 105° and m∠2 = 75°, we see that ∠1 and ∠2 are supplementary. Then ℓ ∥ m.

Construction 7 depends on Theorem 2.3.1, which is restated below.

THEOREM 2.3.1

If two lines are cut by a transversal so that two corresponding angles are congruent, then these lines are parallel.

SSG EXS. 9–16

CONSTRUCTION 7 *To construct the line parallel to a given line from a point not on that line.*

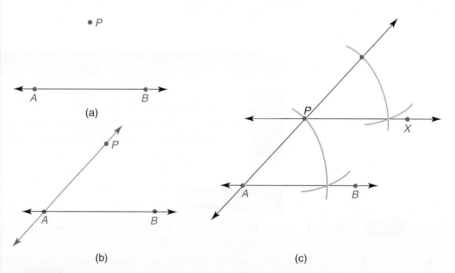

Figure 2.23

GIVEN: $\overleftrightarrow{AB}$ and point P not on $\overleftrightarrow{AB}$, as in Figure 2.23(a)

CONSTRUCT: The line through point P parallel to $\overleftrightarrow{AB}$

CONSTRUCTION: Figure 2.23(b): Draw a line (to become a transversal) through point P and some point on $\overleftrightarrow{AB}$. For convenience, we choose point A and draw $\overleftrightarrow{AP}$.

Figure 2.23(c): Using P as the vertex, construct the angle that corresponds to ∠PAB so that this angle is congruent to ∠PAB. It may be necessary to extend $\overleftrightarrow{AP}$ upward to accomplish this. $\overleftrightarrow{PX}$ is the desired line parallel to $\overleftrightarrow{AB}$.

SSG EX. 17

Exercises 2.3

In Exercises 1 to 6, ℓ and m are cut by transversal v. On the basis of the information given, determine whether ℓ must be parallel to m.

1. $m\angle 1 = 107°$ and
$m\angle 5 = 107°$

2. $m\angle 2 = 65°$ and
$m\angle 7 = 65°$

3. $m\angle 1 = 106°$ and $m\angle 7 = 76°$

4. $m\angle 1 = 106°$ and
$m\angle 4 = 106°$

5. $m\angle 3 = 113.5°$ and
$m\angle 5 = 67.5°$

6. $m\angle 6 = 71.4°$ and $m\angle 7 = 71.4°$

Exercises 1–6

In Exercises 7 to 16, name the lines (if any) that must be parallel under the given conditions.

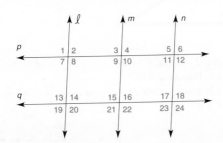

Exercises 7–16

7. $\angle 1 \cong \angle 20$

8. $\angle 3 \cong \angle 10$

9. $\angle 9 \cong \angle 14$

10. $\angle 7 \cong \angle 11$

11. $\ell \perp p$ and $n \perp p$

12. $\ell \parallel m$ and $m \parallel n$

13. $\ell \perp p$ and $m \perp q$

14. $\angle 8$ and $\angle 9$ are supplementary.

15. $m\angle 8 = 110°$, $p \parallel q$, and $m\angle 18 = 70°$

16. The bisectors of $\angle 9$ and $\angle 21$ are parallel.

In Exercises 17 and 18, complete each proof by filling in the missing statements and reasons.

17. *Given:* $\angle 1$ and $\angle 2$ are complementary
$\angle 3$ and $\angle 1$ are complementary
Prove: $\overline{BC} \parallel \overline{DE}$

PROOF

Statements	Reasons
1. $\angle$s 1 and 2 are comp.; $\angle$s 3 and 1 are comp.	1. ?
2. $\angle 2 \cong \angle 3$	2. ?
3. ?	3. If two lines are cut by a transversal so that corr. $\angle$s are $\cong$, the lines are $\parallel$

18. *Given:* $\ell \parallel m$
$\angle 3 \cong \angle 4$
Prove: $\ell \parallel n$

PROOF

Statements	Reasons
1. $\ell \parallel m$	1. ?
2. $\angle 1 \cong \angle 2$	2. ?
3. $\angle 2 \cong \angle 3$	3. If two lines intersect, the vertical $\angle$s formed are $\cong$
4. ?	4. Given
5. $\angle 1 \cong \angle 4$	5. Transitive Prop. of $\cong$
6. ?	6. ?

In Exercises 19 to 22, complete the proof.

19. *Given:* $\overline{AD} \perp \overline{DC}$
$\overline{BC} \perp \overline{DC}$
Prove: $\overline{AD} \parallel \overline{BC}$

20. *Given:* $\angle 1 \cong \angle 3$
$\angle 2 \cong \angle 4$
Prove: $\overleftrightarrow{CD} \parallel \overleftrightarrow{EF}$

21. *Given:* $\overrightarrow{DE}$ bisects $\angle CDA$
$\angle 3 \cong \angle 1$
Prove: $\overline{ED} \parallel \overline{AB}$

22. *Given:* $\overline{XY} \parallel \overline{WZ}$
$\angle 1 \cong \angle 2$
Prove: $\overline{MN} \parallel \overline{XY}$

In Exercises 23 to 30, determine the value of x so that line ℓ will be parallel to line m.

23. $m\angle 4 = 5x$
$m\angle 5 = 4(x + 5)$

24. $m\angle 2 = 4x + 3$
$m\angle 7 = 5(x - 3)$

25. $m\angle 3 = \frac{x}{2}$
$m\angle 5 = x$

26. $m\angle 1 = \frac{x}{2} + 35$
$m\angle 5 = \frac{3x}{4}$

27. $m\angle 6 = x^2 - 9$
$m\angle 2 = x(x - 1)$

28. $m\angle 4 = 2x^2 - 3x + 6$
$m\angle 5 = 2x(x - 1) - 2$

29. $m\angle 3 = (x + 1)(x + 4)$
$m\angle 5 = 16(x + 3) - (x^2 - 2)$

30. $m\angle 2 = (x^2 - 1)(x + 1)$
$m\angle 8 = 185 - x^2(x + 1)$

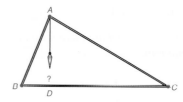

Exercises 23–30

In Exercises 31 to 33, give a formal proof for each theorem.

31. If two lines are cut by a transversal so that a pair of alternate exterior angles are congruent, then these lines are parallel.

32. If two lines are cut by a transversal so that a pair of exterior angles on the same side of the transversal are supplementary, then these lines are parallel.

33. If two lines are each parallel to the same line, then these lines are parallel to each other. (Assume three coplanar lines.)

34. Explain why the statement in Exercise 33 remains true even if the three lines are not coplanar.

35. Given that point *P* does *not* lie on line ℓ, construct the line through point *P* that is parallel to line ℓ.

36. Given that point *Q* does not lie on $\overline{AB}$, construct the line through point *Q* that is parallel to $\overline{AB}$.

37. A carpenter drops a plumb line from point *A* to $\overline{BC}$. Assuming that $\overline{BC}$ is horizontal, the point *D* at which the plumb line intersects $\overline{BC}$ will determine the vertical line segment $\overline{AD}$. Use a construction to locate point *D*.

38. *Given:* $m\angle 2 + m\angle 3 = 90°$
$\overrightarrow{BE}$ bisects $\angle ABC$
$\overrightarrow{CE}$ bisects $\angle BCD$
Prove: $\ell \parallel n$

2.4   The Angles of a Triangle

KEY CONCEPTS

Triangles	Equilateral Triangle	Determined
Vertices	Acute Triangle	Underdetermined
Sides of a Triangle	Obtuse Triangle	Overdetermined
Interior and Exterior	Right Triangle	Corollary
of a Triangle	Equiangular Triangle	Exterior Angle of a
Scalene Triangle	Auxiliary Line	Triangle
Isosceles Triangle		

In geometry, the word *union* means that figures are joined or combined.

DEFINITION

A **triangle** (symbol $\triangle$) is the union of three line segments that are determined by three noncollinear points.

Figure 2.24

The triangle shown in Figure 2.24 is known as $\triangle ABC$, or $\triangle BCA$, etc. (any order of letters A, B, and C can be used). Each point A, B, and C is a **vertex** of the triangle; collectively, these three points are the **vertices** of the triangle. $\overline{AB}$, $\overline{BC}$, and $\overline{AC}$ are the **sides** of the triangle. Point D is in the **interior** of the triangle; point E is on the triangle; and point F is in the **exterior** of the triangle.

Triangles may be categorized by the lengths of their sides. Table 2.1 presents each type of triangle, the relationship among its sides, and a drawing in which congruent sides are marked. You should become familiar with the types of triangles found in both Table 2.1 and Table 2.2.

TABLE 2.1
Triangles Classified by Congruent Sides

Type		Number of Congruent Sides
Scalene		None
Isosceles		Two
Equilateral		Three

Triangles may also be classified according to the measures of their angles as shown in Table 2.2.

TABLE 2.2
Triangles Classified by Angles

Type		Angle(s)	Type		Angle(s)
Acute		All angles acute	Right		One right angle
Obtuse		One obtuse angle	Equiangular		All angles congruent

EXAMPLE 1

SSG   EXS. 1–7

In $\triangle HJK$ (not shown), $HJ = 4$, $JK = 4$, and $m\angle J = 90°$. Describe completely the type of triangle represented.

SOLUTION $\triangle HJK$ is a right isosceles triangle, or $\triangle HJK$ is an isosceles right triangle.

In an earlier exercise, it was suggested that the sum of the measures of the three interior angles of a triangle is 180°. This is proved through the use of an **auxiliary** (or helping) **line.** When an auxiliary line is added to the drawing for a proof, a *justification* must be given for the existence of that line. Justifications include statements such as

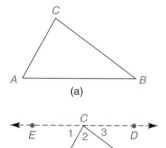

(a)

(b)

Figure 2.25

Technology Exploration

Use computer software, if available.

1. Draw △ABC.
2. Measure ∠A, ∠B, and ∠C.
3. Show that m∠A + m∠B + m∠C = 180°.

(Answer may not be "perfect.")

There is exactly one line through two distinct points.

An angle has exactly one bisector.

There is only one line perpendicular to another line at a point on that line.

When an auxiliary line is introduced into a proof, the original drawing is redrawn for the sake of clarity. Each auxiliary figure must be **determined**, but not **underdetermined** or **overdetermined.** A figure is underdetermined when more than one figure is possible. On the other extreme, a figure is overdetermined when it is impossible for the drawing to include *all* conditions described.

> ### THEOREM 2.4.1
>
> In a triangle, the sum of the measures of the interior angles is 180°.

The first statement in the following "picture proof" establishes the auxiliary line that is used. The auxiliary line is justified by the Parallel Postulate.

PICTURE PROOF OF THEOREM 2.4.1

GIVEN: △ABC in Figure 2.25(a)

PROVE: m∠A + m∠B + m∠C = 180°

PROOF: Through C in Figure 2.25(b), draw $\overleftrightarrow{ED} \parallel \overline{AB}$.

We see that m∠1 + m∠2 + m∠3 = 180°.

But m∠1 = m∠A and m∠3 = m∠B (alternate interior angles).

Then m∠A + m∠B + m∠C = 180° in Figure 2.25(a).

The notions of equality of angle measures and congruence of angles are at times used interchangeably within a proof, as in the preceding "picture proof."

> ### EXAMPLE 2
>
> In △RST (not shown), m∠R = 45° and m∠S = 64°. Find *m∠T*.
>
> **SOLUTION** In △RST, m∠R + m∠S + m∠T = 180°, so
> 45° + 64° + m∠T = 180°. Thus, 109° + m∠T = 180° and m∠T = 71°.

A theorem that follows directly from a previous theorem is known as a **corollary** of that theorem. Corollaries, like theorems, must be proved before they can be used. These proofs are often brief, but they depend on the related theorem. Some corollaries of Theorem 2.4.1 follow. We suggest that the student make a drawing to illustrate each corollary.

> ### COROLLARY 2.4.2
>
> Each angle of an equiangular triangle measures 60°.

> ### COROLLARY 2.4.3
>
> The acute angles of a right triangle are complementary.

SSG ▶ EXS. 8–12

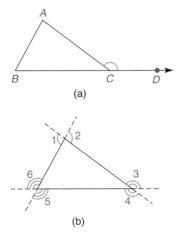
STRATEGY FOR PROOF ■ Proving a Corollary

General Rule: The proof of a corollary is completed by using the theorem upon which the corollary depends.

Illustration: Using $\triangle NMQ$ of Example 3, the proof of Corollary 2.4.3 depends on the fact that $m\angle M + m\angle N + m\angle Q = 180°$. With $m\angle M = 90°$, it follows that $m\angle N + m\angle Q = 90°$.

EXAMPLE 3

Given: $\angle M$ is a right angle in $\triangle NMQ$ (not shown); $m\angle N = 57°$

Find: $m\angle Q$

SOLUTION
According to Corollary 2.4.3, the acute $\angle$s of a right triangle are complementary. Then

$$m\angle N + m\angle Q = 90°$$
$$57° + m\angle Q = 90°$$
$$m\angle Q = 33°$$

COROLLARY 2.4.4

If two angles of one triangle are congruent to two angles of another triangle, then the third angles are also congruent.

The following example illustrates Corollary 2.4.4.

EXAMPLE 4

In $\triangle RST$ and $\triangle XYZ$ (triangles not shown), $m\angle R = m\angle X = 52°$. Also, $m\angle S = m\angle Y = 59°$.

a) Find $m\angle T$.   b) Find $m\angle Z$.   c) Is $\angle T \cong \angle Z$?

SOLUTION
a) $m\angle R + m\angle S + m\angle T = 180°$
$52° + 59° + m\angle T = 180°$
$111° + m\angle T = 180°$
$m\angle T = 69°$
b) Using $m\angle X + m\angle Y + m\angle Z = 180°$, we also find that $m\angle Z = 69°$.
c) Yes, $\angle T \cong \angle Z$ (both measure 69°).

When the sides of a triangle are extended, each angle that is formed by a side and an extension of the adjacent side is an **exterior angle** of the triangle. With *B-C-D* in Figure 2.26(a), $\angle ACD$ is an exterior angle of $\triangle ABC$; for a triangle, there are a total of six exterior angles—two at each vertex. [See Figure 2.26(b).]

In Figure 2.26(a), $\angle A$ and $\angle B$ are the two *nonadjacent* interior angles for exterior $\angle ACD$. These angles (*A* and *B*) are sometimes called *remote* interior angles for exterior $\angle ACD$. Of course, $\angle ACB$ is the adjacent interior angle for $\angle ACD$.

COROLLARY 2.4.5

The measure of an exterior angle of a triangle equals the sum of the measures of the two nonadjacent interior angles.

Figure 2.26

According to Corollary 2.4.5, $m\angle 1 = m\angle S + m\angle T$ in Figure 2.27.

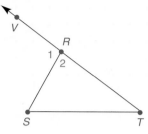

EXAMPLE 5

GIVEN: In Figure 2.27,
$$m\angle 1 = x^2 + 2x$$
$$m\angle S = x^2 - 2x$$
$$m\angle T = 3x + 10$$

FIND: x

SOLUTION Applying Corollary 2.4.5,

$$m\angle 1 = m\angle S + m\angle T$$
$$x^2 + 2x = (x^2 - 2x) + (3x + 10)$$
$$x^2 + 2x = x^2 + x + 10$$
$$2x = x + 10$$
$$x = 10$$

Figure 2.27

Check: With $x = 10$, $m\angle 1 = 120°$, $m\angle S = 80°$, and $m\angle T = 40°$; so $120 = 80 + 40$, which satisfies the conditions of Corollary 2.4.5.

SSG **EXS. 13–19**

Exercises 2.4

In Exercises 1 to 4, refer to △ABC. On the basis of the infor-mation given, determine the measure of the remaining angle(s) of the triangle.

1. $m\angle A = 63°$ and $m\angle B = 42°$

2. $m\angle B = 39°$ and $m\angle C = 82°$

3. $m\angle A = m\angle C = 67°$

Exercises 1–6

4. $m\angle B = 42°$ and $m\angle A = m\angle C$

5. Describe the auxiliary line (segment) as determined, overdetermined, or underdetermined.
  a) Draw the line through vertex C of △ABC.
  b) Through vertex C, draw the line parallel to $\overline{AB}$.
  c) With M the midpoint of $\overline{AB}$, draw $\overline{CM}$ perpendicular to $\overline{AB}$.

6. Describe the auxiliary line (segment) as determined, overdetermined, or underdetermined.
  a) Through vertex B of △ABC, draw $\overleftrightarrow{AB} \perp \overline{AC}$.
  b) Draw the line that contains A, B, and C.
  c) Draw the line that contains M, the midpoint of $\overline{AB}$.

In Exercises 7 and 8, classify the triangle (not shown) by considering the lengths of its sides.

7. a) All sides of △ABC are of the same length.
  b) In △DEF, $DE = 6$, $EF = 6$, and $DF = 8$.

8. a) In △XYZ, $\overline{XY} \cong \overline{YZ}$.
  b) In △RST, $RS = 6$, $ST = 7$, and $RT = 8$.

In Exercises 9 and 10, classify the triangle (not shown) by considering the measures of its angles.

9. a) All angles of △ABC measure 60°.
  b) In △DEF, $m\angle D = 40°$ and $m\angle E = 50°$.

10. a) In △XYZ, $m\angle X = 123°$.
  b) In △RST, $m\angle R = 45°$, $m\angle S = 65°$, and $m\angle T = 70°$.

In Exercises 11 and 12, make drawings as needed.

11. Suppose that for △ABC and △MNQ, you know that $\angle A \cong \angle M$ and $\angle B \cong \angle N$. Explain why $\angle C \cong \angle Q$.

12. Suppose that T is a point on side $\overline{PQ}$ of △PQR. Also, $\overrightarrow{RT}$ bisects $\angle PRQ$, and $\angle P \cong \angle Q$. If $\angle 1$ and $\angle 2$ are the angles formed when $\overrightarrow{RT}$ intersects $\overline{PQ}$, explain why $\angle 1 \cong \angle 2$.

In Exercises 13 to 15, $j \parallel k$ and △ABC.

13. *Given:* $m\angle 3 = 50°$
 $m\angle 4 = 72°$
 Find: $m\angle 1$, $m\angle 2$, and $m\angle 5$

14. *Given:* $m\angle 3 = 55°$
 $m\angle 2 = 74°$
 Find: $m\angle 1$, $m\angle 4$, and $m\angle 5$

Exercises 13–15

15. *Given:* $m\angle 1 = 122.3°$, $m\angle 5 = 41.5°$
 Find: $m\angle 2$, $m\angle 3$, and $m\angle 4$

16. *Given:* $\overline{MN} \perp \overline{NQ}$ and ∠s as shown
Find: x, y, and z

17. *Given:* $\overline{AB} \parallel \overline{DC}$
$\overrightarrow{DB}$ bisects ∠ADC
m∠A = 110°
Find: m∠3

Exercises 17, 18

18. *Given:* $\overline{AB} \parallel \overline{DC}$
$\overrightarrow{DB}$ bisects ∠ADC
m∠1 = 36°
Find: m∠A

19. *Given:* △ABC with B-D-E-C
m∠3 = m∠4 = 30°
m∠1 = m∠2 = 70°
Find: m∠B

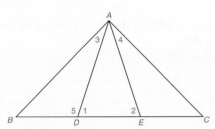

Exercises 19–22

20. *Given:* △ABC with B-D-E-C
m∠1 = 2x
m∠3 = x
Find: m∠B in terms of x

21. *Given:* △ADE with m∠1 = m∠2 = x
m∠DAE = $\frac{x}{2}$
Find: x, m∠1, and m∠DAE

22. *Given:* △ABC with m∠B = m∠C = $\frac{x}{2}$
m∠BAC = x
Find: x, m∠BAC, and m∠B

23. Consider any triangle and one exterior angle at each vertex. What is the sum of the measures of the three exterior angles of the triangle?

24. *Given:* Right △ABC with
right ∠C
m∠1 = 7x + 4
m∠2 = 5x + 2
Find: x

Exercises 24–27

For Exercises 25 to 27, see the figure for Exercise 24.

25. *Given:* m∠1 = x, m∠2 = y, m∠3 = 3x
Find: x and y

26. *Given:* m∠1 = x, m∠2 = $\frac{x}{2}$
Find: x

27. *Given:* m∠1 = $\frac{x}{2}$, m∠2 = $\frac{x}{3}$
Find: x

28. *Given:* m∠1 = 8(x + 2)
m∠3 = 5x − 3
m∠5 = 5(x + 1) − 2
Find: x

29. *Given:* m∠1 = x
m∠2 = 4y
m∠3 = 2y
m∠4 = 2x − y − 40
Find: x, y, and m∠5

Exercises 28, 29

30. *Given:* Equiangular △RST
$\overrightarrow{RV}$ bisects ∠SRT
Prove: △RVS is a right △

31. *Given:* $\overline{MN}$ and $\overline{PQ}$ intersect
at K; ∠M ≅ ∠Q
Prove: ∠P ≅ ∠N

32. The sum of the measures of two angles of a triangle equals the measure of the third (largest) angle. What type of triangle is described?

33. Draw, if possible, an
a) isosceles obtuse triangle.
b) equilateral right triangle.

34. Draw, if possible, a
a) right scalene triangle.
b) triangle having both a right angle and an obtuse angle.

35. Along a straight shoreline, two houses are located at points H and M. The houses are 5000 feet apart. A small island lies in view of both houses, with angles as indicated. Find m∠I.

36. An airplane has leveled off (is flying horizontally) at an altitude of 12,000 feet. Its pilot can see each of two farmhouses at points R and T in front of the plane. With angle measures as indicated, find m∠R.

37. On a map, three Los Angeles suburbs are located at points N (Newport Beach), P (Pomona), and B (Burbank). With angle measures as indicated, determine m$\angle N$ and m$\angle P$.

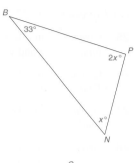

38. The roofline of a house shows the shape of right triangle ABC with m$\angle C = 90°$. If the measure of $\angle CAB$ is 24° larger than the measure of $\angle CBA$, then how large is each angle?

39. A lamppost has a design such that m$\angle C = 110°$ and $\angle A \cong \angle B$. Find m$\angle A$ and m$\angle B$.

40. For the lamppost of Exercise 39, suppose that m$\angle A = $ m$\angle B$ and that m$\angle C = 3($m$\angle A)$. Find m$\angle A$, m$\angle B$, and m$\angle C$.

41. The triangular symbol on the "PLAY" button of a DVD has congruent angles at M and N. If m$\angle P = 30°$, what are the measures of angle M and angle N?

42. A polygon with four sides is called a *quadrilateral*. Consider the figure and the dashed auxiliary line. What is the sum of the measures of the four interior angles of this (or any other) quadrilateral?

Exercises 39, 40

43. Explain why the following statement is true.
Each interior angle of an equiangular triangle measures 60°.

44. Explain why the following statement is true.
The acute angles of a right triangle are complementary.

In Exercises 45 to 47, write a formal proof for each corollary.

45. The measure of an exterior angle of a triangle equals the sum of the measures of the two nonadjacent interior angles.

46. If two angles of one triangle are congruent to two angles of another triangle, then the third angles are also congruent.

47. Use an indirect proof to establish the following theorem: A triangle cannot have more than one right angle.

48. *Given:* $\overleftrightarrow{AB}$, $\overrightarrow{DE}$, and $\overleftrightarrow{CF}$
$\overleftrightarrow{AB} \parallel \overrightarrow{DE}$
$\overrightarrow{CG}$ bisects $\angle BCF$
$\overrightarrow{FG}$ bisects $\angle CFE$
Prove: $\angle G$ is a right angle

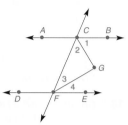

*49. *Given:* $\overrightarrow{NQ}$ bisects $\angle MNP$
$\overrightarrow{PQ}$ bisects $\angle MPR$
m$\angle Q = 42°$
Find: m$\angle M$

*50. *Given:* In rt. $\triangle ABC$, $\overline{AD}$ bisects $\angle CAB$ and $\overline{BF}$ bisects $\angle ABC$.
Find: m$\angle FED$

2.5 Convex Polygons

KEY CONCEPTS

Convex Polygons (Triangle, Quadrilateral, Pentagon, Hexagon, Heptagon, Octagon, Nonagon, Decagon)

Concave Polygon
Diagonals of a Polygon
Regular Polygon

Equilateral Polygon
Equiangular Polygon
Polygram

DEFINITION

A **polygon** is a closed plane figure whose sides are line segments that intersect only at the endpoints.

Most polygons considered in this textbook are **convex;** the angle measures of convex polygons are between 0° and 180°. Some convex polygons are shown in Figure 2.28; those in Figure 2.29 are **concave.** A line segment joining two points of a concave polygon can contain points in the exterior of the polygon. Thus, a concave polygon always has at least one reflex angle. Figure 2.30 shows some figures that aren't polygons at all!

Convex Polygons

Figure 2.28

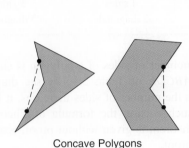

Concave Polygons

Figure 2.29

Not Polygons

Figure 2.30

Table 2.3 categorizes polygons by their number of sides.

TABLE 2.3

Polygon	Number of Sides	Polygon	Number of Sides
Triangle	3	Heptagon	7
Quadrilateral	4	Octagon	8
Pentagon	5	Nonagon	9
Hexagon	6	Decagon	10

With Venn Diagrams, the set of all objects under consideration is called the **universe.** If P = {all polygons} is the universe, then we can describe sets T = {triangles} and Q = {quadrilaterals} as subsets that lie within the universe P. Sets T and Q are described as **disjoint** because they have no elements in common. See Figure 2.31.

Figure 2.31

DIAGONALS OF A POLYGON

A **diagonal** of a polygon is a line segment that joins two nonconsecutive vertices.

Figure 2.32 shows heptagon $ABCDEFG$ for which $\angle GAB$, $\angle B$, and $\angle BCD$ are some of the interior angles and $\angle 1$, $\angle 2$, and $\angle 3$ are some of the exterior angles. Because they join consecutive vertices, $\overline{AB}$, $\overline{BC}$, and $\overline{CD}$ are some of the sides of the heptagon. Because a diagonal joins nonconsecutive vertices of $ABCDEFG$, $\overline{AC}$, $\overline{AD}$, and $\overline{AE}$ are among the many diagonals of the polygon.

Table 2.4 illustrates selected polygons by number of sides and the corresponding total number of diagonals for each type.

Figure 2.32

TABLE 2.4

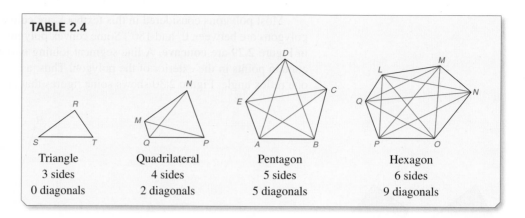

Triangle	Quadrilateral	Pentagon	Hexagon
3 sides	4 sides	5 sides	6 sides
0 diagonals	2 diagonals	5 diagonals	9 diagonals

When the number of sides of a polygon is small, we can list all diagonals by name. For pentagon $ABCDE$ of Table 2.4, we see diagonals $\overline{AC}$, $\overline{AD}$, $\overline{BD}$, $\overline{BE}$, and $\overline{CE}$—a total of five. As the number of sides increases, it becomes more difficult to count all the diagonals. In such a case, the formula of Theorem 2.5.1 is most convenient to use. Although this theorem is given without proof, Exercise 39 of this section provides some insight for the proof.

THEOREM 2.5.1

The total number of diagonals D in a polygon of n sides is given by the formula $D = \frac{n(n-3)}{2}$.

Theorem 2.5.1 reaffirms the fact that a triangle has no diagonals; when $n = 3$, $D = \frac{3(3-3)}{2} = 0$. We also apply this theorem in Example 1.

EXAMPLE 1

Find (a) the number of diagonals for any pentagon (b) the type of polygon that has 9 diagonals.

SSG **EXS. 1–5**

SOLUTION (a) For a pentagon, $n = 5$. Then $D = \frac{5(5-3)}{2} = \frac{5(2)}{2} = 5$. Thus, the pentagon has 5 diagonals.

(b)
$$\frac{n(n-3)}{2} = 9$$
$$\frac{n^2 - 3n}{2} = 9$$
$$n^2 - 3n = 18$$
$$n^2 - 3n - 18 = 0$$
$$(n-6)(n+3) = 0$$
$$n - 6 = 0 \quad \text{or} \quad n + 3 = 0$$
$$n = 6 \quad \text{or} \quad n = -3 \text{ (discard)}$$

When $n = 6$, the polygon is a hexagon.

SUM OF THE INTERIOR ANGLES OF A POLYGON

The following theorem provides the formula for the sum of the interior angles of any polygon.

Reminder

The sum of the interior angles of a triangle is 180°.

THEOREM 2.5.2

The sum S of the measures of the interior angles of a polygon with n sides is given by $S = (n - 2) \cdot 180°$. Note that $n > 2$ for any polygon.

Let us consider an informal proof of Theorem 2.5.2 for the special case of a pentagon. The proof would change for a polygon of a different number of sides but only by the number of triangles into which the polygon can be separated. Although Theorem 2.5.2 is also true for concave polygons, we consider the proof only for the case of the convex polygon.

PROOF

Consider the pentagon *ABCDE* in Figure 2.33, with auxiliary segments (diagonals from one vertex) as shown.

The equations that follow are based upon the sum of the interior angles in triangles *ABC*, *ACD*, and *ADE*. Adding columns of angle measures, we have

$$
\begin{array}{l}
m\angle 1 + \qquad\qquad m\angle 2 + m\angle 3 = 180° \\
\quad m\angle 6 + m\angle 5 \qquad\quad + m\angle 4 = 180° \\
\underline{m\angle 8 + m\angle 9 + m\angle 7 \qquad\qquad\quad = 180°} \\
m\angle E + m\angle A + m\angle D + m\angle B + m\angle C = 540°
\end{array}
$$

For pentagon *ABCDE*, in which $n = 5$, the sum of the measures of the interior angles is $(5 - 2) \cdot 180°$, which equals 540°.

When drawing diagonals from one vertex of a polygon of n sides, we always form $(n - 2)$ triangles. The sum of the measures of the interior angles always equals $(n - 2) \cdot 180°$.

Figure 2.33

EXAMPLE 2

Find the sum of the measures of the interior angles of a hexagon. Then find the measure of each interior angle of an equiangular hexagon.

SOLUTION For the hexagon, $n = 6$, so the sum of the measures of the interior angles is $S = (6 - 2) \cdot 180°$ or $4(180°)$ or $720°$.

In an equiangular hexagon, each of the six interior angles measures $\frac{720°}{6}$, or 120°.

EXAMPLE 3

Find the number of sides in a polygon whose sum of measures for its interior angles is 2160°.

SOLUTION Here $S = 2160$ in the formula of Theorem 2.5.2. Because $(n - 2) \cdot 180 = 2160$, we have $180n - 360 = 2160$.

Then $180n = 2520$
$n = 14$

SSG **EXS. 6–9**

The polygon has 14 sides.

REGULAR POLYGONS

On page 98, Figure 2.34 shows polygons that are, respectively, (a) **equilateral**, (b) **equiangular**, and (c) **regular** (both sides and angles are congruent). Note the dashes that indicate congruent sides and the arcs that indicate congruent angles.

(a)   (b)   (c)

Figure 2.34

A **regular polygon** is a polygon that is both equilateral and equiangular.

The polygon in Figure 2.34(c) is a *regular pentagon*. Other examples of regular polygons include the equilateral triangle and the square. In Chapter 3, we will prove that any equilateral triangle is also equiangular.

Based upon the formula $S = (n - 2) \cdot 180°$ from Theorem 2.5.2, there is also a formula for the measure of each interior angle of a regular polygon having n sides. It applies to equiangular polygons as well.

COROLLARY 2.5.3

The measure I of each interior angle of a regular polygon or equiangular polygon of n sides is $I = \frac{(n - 2) \cdot 180°}{n}$.

Figure 2.35

EXAMPLE 4

Find the measure of each interior angle of a ceramic floor tile in the shape of an equiangular octagon (Figure 2.35).

SOLUTION For an octagon, $n = 8$. Applying Corollary 2.5.3,

$$I = \frac{(8 - 2) \cdot 180}{8}$$

$$= \frac{6 \cdot 180}{8}$$

$$= \frac{1080}{8}, \text{ so } I = 135°$$

Each interior angle of the tile measures 135°.

NOTE: For the octagonal tiles of Example 4, small squares are used as "fillers" to cover the floor. The pattern, known as a tessellation, is found in Section 8.3.

EXAMPLE 5

Each interior angle of a certain regular polygon has a measure of 144°. Find its number of sides, and identify the type of polygon it is.

SOLUTION Let n be the number of sides the polygon has. All n of the interior angles are equal in measure.

The measure of each interior angle is given by

$$I = \frac{(n - 2) \cdot 180}{n} \text{ where } I = 144$$

SSG **EXS. 10–12**   Then

$$\frac{(n-2)\cdot 180}{n} = 144$$

$$(n-2)\cdot 180 = 144n \qquad \text{(multiplying by } n\text{)}$$

$$180n - 360 = 144n$$

$$36n = 360$$

$$n = 10$$

With 10 sides, the polygon is a regular decagon.

A second corollary to Theorem 2.5.2 concerns the sum of the interior angles of any quadrilateral. For the proof, we simply let $n = 4$ in the formula $S = (n-2)\cdot 180°$. Then $S = (4-2)\cdot 180° = 2\cdot 180° = 360°$. Also, see the Discover at the left.

Discover

From a paper quadrilateral, cut the angles from the "corners." Now place the angles so that they have the same vertex and do **not** overlap. What is the sum of measures of the four angles?

ANSWER
360°

> ### COROLLARY 2.5.4
>
> The sum of the measures of the four interior angles of a quadrilateral is 360°.

On the basis of Corollary 2.5.4, it is clearly the case that each interior angle of a square or rectangle measures 90°.

The following interesting corollary to Theorem 2.5.2 can be established through algebra.

> ### COROLLARY 2.5.5
>
> The sum of the measures of the exterior angles of a polygon, one at each vertex, is 360°.

We now consider an algebraic proof for Corollary 2.5.5.

PROOF

A polygon of n sides has n interior angles and n exterior angles, if one is considered at each vertex. As shown in Figure 2.36, these interior and exterior angles may be grouped into pairs of supplementary angles such as $\angle 1$ and $\angle 1'$ (read $\angle 1$ "prime"). Because there are n pairs of angles, the sum of the measures of all pairs is $180 \cdot n$ degrees.

Of course, the sum of the measures of the interior angles is $(n-2)\cdot 180°$.

In words, we have

$$\begin{array}{ccc}\text{Sum of Measures} & \text{Sum of Measures} & \text{Sum of Measures of All}\\ \text{of Interior Angles} + \text{of Exterior Angles} = \text{Supplementary Pairs}\end{array}$$

Figure 2.36

Where S represents the sum of the measures of the exterior angles,

$$(n-2)\cdot 180 + S = 180n$$

$$180n - 360 + S = 180n$$

$$-360 + S = 0$$

$$\therefore S = 360$$

The next corollary follows from Corollary 2.5.5. The claim made in Corollary 2.5.6 is applied in Example 6.

> ### COROLLARY 2.5.6
>
> The measure E of each exterior angle of a regular polygon or equiangular polygon of n sides is $E = \frac{360°}{n}$.

EXAMPLE 6

Use Corollary 2.5.6 to find the number of sides of a regular polygon if each interior angle measures 144°. (Note that we are repeating Example 5.)

SOLUTION If each interior angle measures 144°, then each exterior angle measures 36° (they are supplementary, because exterior sides of these adjacent angles form a straight line).
Now each of the n exterior angles has the measure

$$\frac{360°}{n}$$

In this case, $\frac{360}{n} = 36$, and it follows that $36n = 360$, so $n = 10$. The polygon (a decagon) has 10 sides.

POLYGRAMS

 EXS. 13, 14

A **polygram** is the star-shaped figure that results when the sides of convex polygons with five or more sides are extended. When the polygon is regular, the resulting polygram is also regular—that is, the interior acute angles are congruent, the interior reflex angles are congruent, and all sides are congruent. The names of polygrams come from the names of the polygons whose sides were extended. Figure 2.37 shows a pentagram, a hexagram, and an octagram. With congruent angles and sides indicated, these figures are **regular polygrams.**

Geometry in Nature

The starfish has the shape of a pentagram.

Pentagram

Hexagram

Octagram

 EXS. 15–17

Figure 2.37

Exercises 2.5

For Exercises 1 and 2, consider a group of regular polygons.

1. As the number of sides of a regular polygon increases, does each interior angle increase or decrease in measure?

2. As the number of sides of a regular polygon increases, does each exterior angle increase or decrease in measure?

3. *Given:* $\overline{AB} \parallel \overline{DC}, \overline{AD} \parallel \overline{BC},$ $\overline{AE} \parallel \overline{FC},$ with angle measures as indicated
   Find: x, y, and z

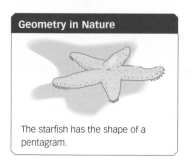

4. In pentagon *ABCDE* with $\angle B \cong \angle D \cong \angle E$, find the measure of interior angle *D*.

5. Find the total number of diagonals for a polygon of n sides if:
   a) $n = 5$ b) $n = 10$

6. Find the total number of diagonals for a polygon of n sides if:
   a) $n = 6$ b) $n = 8$

7. Find the sum of the measures of the interior angles of a polygon of n sides if:
   a) $n = 5$ b) $n = 10$

8. Find the sum of the measures of the interior angles of a polygon of n sides if:
   a) $n = 6$
   b) $n = 8$

9. Find the measure of each interior angle of a regular polygon of n sides if:
   a) $n = 4$
   b) $n = 12$

10. Find the measure of each interior angle of a regular polygon of n sides if:
    a) $n = 6$
    b) $n = 10$

11. Find the measure of each exterior angle of a regular polygon of n sides if:
    a) $n = 4$
    b) $n = 12$

12. Find the measure of each exterior angle of a regular polygon of n sides if:
    a) $n = 6$
    b) $n = 10$

13. Find the number of sides for a polygon whose sum of the measures of its interior angles is:
    a) $900°$
    b) $1260°$

14. Find the number of sides for a polygon whose sum of the measures of its interior angles is:
    a) $1980°$
    b) $2340°$

15. Find the number of sides for a regular polygon whose measure of each interior angle is:
    a) $108°$
    b) $144°$

16. Find the number of sides for a regular polygon whose measure of each interior angle is:
    a) $150°$
    b) $168°$

17. Find the number of sides for a regular polygon whose exterior angles each measure:
    a) $24°$
    b) $18°$

18. Find the number of sides for a regular polygon whose exterior angles each measure:
    a) $45°$
    b) $9°$

19. What is the measure of each interior angle of a stop sign?

20. Lug bolts are equally spaced about the wheel to form the equal angles shown in the figure. What is the measure of each of the equal obtuse angles?

In Exercises 21 to 26, with P = {all polygons} as the universe, draw a Venn Diagram to represent the relationship between these sets. Describe a subset relationship, if one exists. Are the sets described disjoint or equivalent? Do the sets intersect?

21. $T = \{\text{triangles}\}$; $I = \{\text{isosceles triangles}\}$

22. $R = \{\text{right triangles}\}$; $S = \{\text{scalene triangles}\}$

23. $A = \{\text{acute triangles}\}$; $S = \{\text{scalene triangles}\}$

24. $Q = \{\text{quadrilaterals}\}$; $L = \{\text{equilateral polygons}\}$

25. $H = \{\text{hexagons}\}$; $O = \{\text{octagons}\}$

26. $T = \{\text{triangles}\}$; $Q = \{\text{quadrilaterals}\}$

27. *Given:* Quadrilateral $RSTQ$ with exterior $\angle$s at R and T
    Prove: $m\angle 1 + m\angle 2 = m\angle 3 + m\angle 4$

28. *Given:* Regular hexagon $ABCDEF$ with diagonal $\overline{AC}$ and exterior $\angle 1$
    Prove: $m\angle 2 + m\angle 3 = m\angle 1$

29. *Given:* Quadrilateral $RSTV$ with diagonals $\overline{RT}$ and $\overline{SV}$ intersecting at W
    Prove: $m\angle 1 + m\angle 2 = m\angle 3 + m\angle 4$

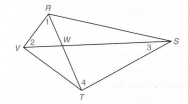

30. *Given:* Quadrilateral $ABCD$ with $\overline{BA} \perp \overline{AD}$ and $\overline{BC} \perp \overline{DC}$
    Prove: $\angle$s B and D are supplementary

31. A father wishes to make a baseball home plate for his son to use in practicing pitching. Find the size of each of the equal angles if the home plate is modeled on the one in (a) and if it is modeled on the one in (b).

(a)

(b)

32. The adjacent interior and exterior angles of a polygon are supplementary, as indicated in the drawing. Assume that you know that the measure of each interior angle of a regular polygon is $\frac{(n-2)180}{n}$.
    a) Express the measure of each exterior angle as the supplement of the interior angle.
    b) Simplify the expression in part (a) to show that each exterior angle has a measure of $\frac{360}{n}$.

33. Find the measure of each (a) acute interior angle of a regular pentagram (b) reflex interior angle of the pentagram.

34. Find the measure of each (a) acute interior angle of a regular octagram (b) reflex interior angle of the octagram.

35. Consider any regular polygon; find and join (in order) the midpoints of the sides. What does intuition tell you about the resulting polygon?

36. Consider a regular hexagon *RSTUVW*. What does intuition tell you about △*RTV*, the result of drawing diagonals $\overline{RT}$, $\overline{TV}$, and $\overline{VR}$?

37. The face of a clock has the shape of a regular polygon with 12 sides. What is the measure of the angle formed by two consecutive sides?

38. The top surface of a picnic table is in the shape of a regular hexagon. What is the measure of the angle formed by two consecutive sides?

*39. Consider a polygon of *n* sides determined by the *n* noncollinear vertices *A*, *B*, *C*, *D*, and so on.

a) Choose any vertex of the polygon. To how many of the remaining vertices of the polygon can the selected vertex be joined to form a diagonal?

b) Considering that each of the *n* vertices in (a) can be joined to any one of the remaining (*n* − 3) vertices to form diagonals, the product *n*(*n* − 3) appears to represent the total number of diagonals possible. However, this number includes duplications, such as $\overline{AC}$ and $\overline{CA}$. What expression actually represents *D*, the total number of diagonals in a polygon of *n* sides?

40. For the concave quadrilateral *ABCD*, explain why the sum of the interior angles is 360°.

(HINT: Draw $\overline{BD}$.)

Exercises 40, 41

41. If m∠*A* = 20°, m∠*B* = 88°, and m∠*C* = 31°, find the measure of the reflex angle at vertex *D*.

(HINT: See Exercise 40.)

42. Is it possible for a polygon to have the following sum of measures for its interior angles?
a) 600°
b) 720°

43. Is it possible for a regular polygon to have the following measures for each interior angle?
a) 96°
b) 140°

44. Draw a concave hexagon that has:
a) one interior reflex angle.
b) two interior reflex angles.

45. Draw a concave pentagon that has:
a) one interior reflex angle.
b) two interior reflex angles.

46. For concave pentagon *ABCDE*, find the measure of the reflex angle at vertex *E* if m∠*A* = m∠*D* = *x*, m∠*B* = m∠*C* = 2*x*, and m∠*E* = 4*x*.

(HINT: ∠E is the indicated reflex angle.)

47. For concave hexagon *HJKLMN*, m∠*H* = *y* and the measure of the reflex angle at *N* is 2(*y* + 10). Find the measure of the interior angle at vertex *N*.

(HINT: Note congruences in the figure.)

2.6	**Symmetry and Transformations**

KEY CONCEPTS	Symmetry Line of Symmetry Axis of Symmetry Point Symmetry	Transformations Slides Translations Reflections	Rotations Angle of Rotation Center of Rotation

LINE SYMMETRY

In the figure below, rectangle *ABCD* is said to have *symmetry with respect to line ℓ* because each point to the left of the *line of symmetry* or *axis of symmetry* has a corresponding point to the right; for instance, *X* and *Y* are *corresponding points*.

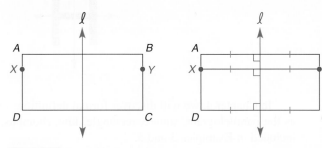

Figure 2.38

> **DEFINITION**
>
> A figure has *symmetry with respect to a line ℓ* if for every point *X* on the figure, there is a second point *Y* on the figure for which ℓ is the perpendicular bisector of $\overline{XY}$.

In particular, *ABCD* of Figure 2.38 has *horizontal symmetry* with respect to line ℓ; that is, a vertical axis of symmetry leads to a pairing of corresponding points on a horizontal line. In Example 1, we see that a horizontal axis leads to *vertical symmetry* for points on □*ABCD*.

Geometry in Nature

© aislutsky/Shutterstock.com

Like many of nature's creations, the butterfly displays line symmetry.

EXAMPLE 1

Rectangle *ABCD* in Figure 2.38 has a second line of symmetry. Draw this horizontal line (or axis) for which there is *vertical symmetry*.

SOLUTION Line *m* (determined by the midpoints of $\overline{AD}$ and $\overline{BC}$) is the desired line of symmetry. As shown in Figure 2.39(b), *R* and *S* are located symmetrically with respect to line *m*.

(a)   (b)

Figure 2.39

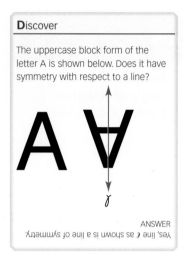
EXAMPLE 2

a) Which letter(s) shown below has (have) a line of symmetry?
b) Which letter(s) has (have) more than one line of symmetry?

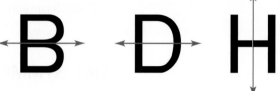

B   D   F   G   H

SOLUTION

a) B, D, and H as shown

b) H as shown

In Chapter 4, we will discover formal definitions of the types of quadrilaterals known as the parallelogram, square, rectangle, kite, rhombus, and rectangle. Some of these are included in Examples 3 and 5.

EXAMPLE 3

a) Which geometric figures have at least one line of symmetry?
b) Which geometric figures have more than one line of symmetry?

      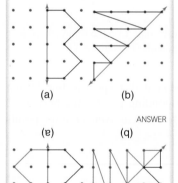

Isosceles Triangle   Square   Quadrilateral   Regular Pentagon

Figure 2.40(a)

SOLUTION

a) The isosceles triangle, square, and the regular pentagon all have at least one line of symmetry.
b) The square and regular pentagon have more than one line of symmetry; each figure is shown with two lines of symmetry. (There are actually four lines of symmetry for the square and five lines of symmetry for the regular pentagon.)

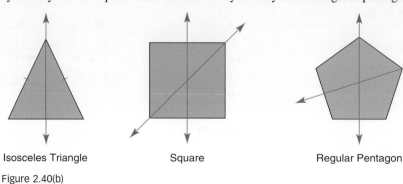

Isosceles Triangle   Square   Regular Pentagon

Figure 2.40(b)

SSG **EXS. 1–4**

POINT SYMMETRY

In Figure 2.41, rectangle *ABCD* is also said to have *symmetry with respect to a point*. As shown, point *P* is determined by the intersection of the diagonals of ▱*ABCD*.

Figure 2.41

DEFINITION

A figure has **symmetry with respect to point *P*** if for every point *M* on the figure, there is a second point *N* on the figure for which point *P* is the midpoint of $\overline{MN}$.

On the basis of this definition, each point on ▱*ABCD* in Figure 2.41(a) has a corresponding point that is the same distance from *P* but which lies in the opposite direction from *P*. In Figure 2.41(b), *M* and *N* are a pair of corresponding points. Even though a figure may have multiple lines of symmetry, a figure can have only one point of symmetry. Thus, the point of symmetry (when one exists) is unique.

EXAMPLE 4

Which letter(s) shown below have point symmetry?

M   N   P   S   X

N S X

SOLUTION *N*, *S*, and *X* as shown all have point symmetry.

EXAMPLE 5

Which geometric figures in Figure 2.42(a) have point symmetry?

Isosceles Triangle	Square	Rhombus	Regular Pentagon	Regular Hexagon

Figure 2.42(a)

SOLUTION Only the square, the rhombus, and the regular hexagon have point symmetry. In the regular pentagon, consider the "centrally" located point *P* and note that *AP* ≠ *PM*.

Square YES	Rhombus YES	Regular Hexagon YES	Regular Pentagon NO

SSG   **EXS. 5–8**   Figure 2.42(b)

TRANSFORMATIONS

In the following material, we will generate new figures from old figures by association of points. In particular, the *transformations* included in this textbook will preserve the shape and size of the given figure; in other words, these transformations preserve lengths and angle measures and thus lead to a second figure that is *congruent* to the given figure. The types of transformations included are (1) the *slide* or *translation*, (2) the *reflection*, and (3) the *rotation*. Each of these types of transformations is also called an *isometry*, which translates to "same measure."

❯❯ Slides (Translations)

With this type of transformation, every point of the original figure is associated with a second point by locating it through a movement of a fixed length and direction. In Figure 2.43, △*ABC* is translated to the second triangle (its image △*DEF*) by sliding each point through the distance and in the direction that takes point *A* to point *D*. The background grid is not necessary to demonstrate the slide, but it lends credibility to our claim that the same length and direction have been used to locate each point.

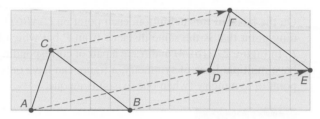

Figure 2.43

Slide △*XYZ* horizontally as shown in Figure 2.44 to form △*RST*. In this example, the distance (length of the slide) is *XR*.

SOLUTION

Figure 2.44

In Example 6, △*XYZ* ≅ △*RTS*. In every slide, the given figure and the produced figure (its *image*) are necessarily congruent. In Example 6, the correspondence of vertices is given by *X*↔*R*, *Y*↔*T*, and *Z*↔*S*.

Where *A*↔*E*, complete the slide of quadrilateral *ABCD* to form quadrilateral *EFGH*. Indicate the correspondence among the remaining vertices. See the figure at the top of page 107.

SOLUTION $B \leftrightarrow F$, $C \leftrightarrow G$, and $D \leftrightarrow H$.

Figure 2.45

❯ Reflections

With the reflection, every point of the original figure is reflected across a line in such a way as to make the given line a line of symmetry. Each pair of corresponding points will lie on opposite sides of the line of reflection and at like distances. In Figure 2.46, obtuse triangle MNP is reflected across the vertical line $\overleftrightarrow{AB}$ to produce the image $\triangle GHK$. The vertex N of the given obtuse angle corresponds to the vertex H of the obtuse angle in the image triangle. With the vertical line as the axis of reflection, a drawing such as Figure 2.46 is sometimes called a *horizontal reflection*, since the image lies to the right of the given figure. It is possible for the line of reflection to be horizontal or oblique (slanted).

Figure 2.46

EXAMPLE 8

Draw the reflection of right $\triangle ABC$

a) across line ℓ to form $\triangle XYZ$.

b) across line m to form $\triangle PQR$.

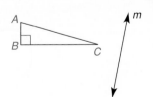

SOLUTION Shown in Figure 2.47

Figure 2.47

With the horizontal axis (line) of reflection, the reflection in Example 8(a) is often called a *vertical reflection*. In the vertical reflection of Figure 2.47(a), the image lies below the given figure. In Example 9, we use a side of the given figure as the line (line segment) of reflection. This reflection is neither horizontal nor vertical.

EXAMPLE 9

Draw the reflection of △*ABC* across side $\overline{BC}$ to form △*DBC*. How are △*ABC* and △*DBC* related in Figure 2.48?

SOLUTION

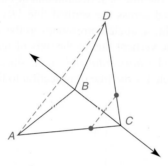

Figure 2.48

The triangles are congruent; also, notice that *D*↔*A*, *B*↔*B*, and *C*↔*C*.

EXAMPLE 10

Complete the figure produced by a reflection across the given line. See the solution in Figure 2.49.

SOLUTION

Figure 2.49

⟩ Rotations

In this transformation, every point of the given figure leads to a point (its image) by rotation about a given point through a prescribed angle measure. In Figure 2.50, ray *AB* rotates about point *A* clockwise through an angle of 30° to produce the image ray *AC*. This has the same appearance as the second hand of a clock over a five-second period of time. In this figure, A↔A and B↔C. As shown in Figure 2.50, the *angle of rotation* measures 30°; also the *center of rotation* is point A.

Figure 2.50

EXAMPLE 11

In Figure 2.51, square *WXYZ* has been rotated counterclockwise about its center *C*, (intersection of diagonals) through an angle of rotation of 45° to form congruent square *QMNP*. What is the name of the eight-pointed geometric figure that is formed by the two intersecting squares?

SOLUTION

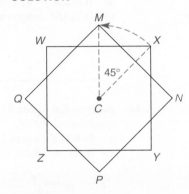

Figure 2.51

The eight-pointed figure formed is a regular octagram.

EXAMPLE 12

Shown in Figure 2.52 are the uppercase A, line ℓ, and point *O*. Which of the pairs of transformations produce the original figure?

   a) The letter A is reflected across ℓ, and that image is reflected across ℓ again.
   b) The letter A is reflected across ℓ, and that image is rotated clockwise 60° about point *O*.
   c) The letter A is rotated 180° about *O*, followed by another 180° rotation about *O*.

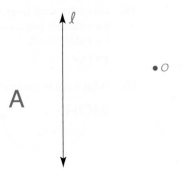

SSG **EXS. 9–14**

Figure 2.52

SOLUTION (a) and (c)

Exercises 2.6

1. Which letters have symmetry with respect to a line?

   M   N   P   T   X

2. Which letters have symmetry with respect to a line?

   I   K   S   V   Z

3. Which letters have symmetry with respect to a point?

   M   N   P   T   X

4. Which letters have symmetry with respect to a point?

   I   K   S   V   Z

5. Which geometric figures have symmetry with respect to at least one line?

   a)   b)   c)

6. Which geometric figures have symmetry with respect to at least one line?

   a)   b)   c)

7. Which geometric figures have symmetry with respect to a point?

   a)   b)   c)

8. Which geometric figures have symmetry with respect to a point?

   a)   b)   c)

9. Which words have a vertical line of symmetry?

   DAD   MOM   NUN   EYE

10. Which words have a vertical line of symmetry?

    WOW   BUB   MAM   EVE

11. Complete each figure so that it has symmetry with respect to line ℓ.

    a)   b)

12. Complete each figure so that it has symmetry with respect to line m.

    a)   b)

13. Complete each figure so that it reflects across line ℓ.

    a)   b)

14. Complete each figure so that it reflects across line m.

    a)   b)

15. Suppose that $\triangle ABC$ slides to the right to the position of $\triangle DEF$.

    a) If $m\angle A = 63°$, find $m\angle D$.   b) Is $\overline{AC} \cong \overline{DF}$?
    c) Is $\triangle ABC$ congruent to $\triangle DEF$?
    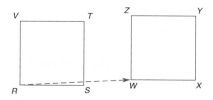

16. Suppose that square $RSTV$ slides point for point to form quadrilateral $WXYZ$.

    a) Is $WXYZ$ a square?   b) Is $RSTV \cong WXYZ$?
    c) If $RS = 1.8$ cm, find WX.

17. Given that the vertical line is a line of symmetry, complete each letter to discover the hidden word.

18. Given that the horizontal line is a line of symmetry, complete each letter to discover the hidden word.

    DECODE

19. Given that each letter has symmetry with respect to the indicated point, complete each letter to discover the hidden word.

    SIX

20. What word is produced by a 180° rotation about the point?

    MOH   ·

21. What word is produced by a 180° rotation about the point?

22. What word is produced by a 360° rotation about the point?

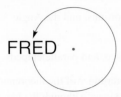

23. In which direction (clockwise or counterclockwise) will pulley 1 rotate if pulley 2 rotates in the clockwise direction?

a)   b)

24. In which direction (clockwise or counterclockwise) will gear 1 rotate if gear 2 rotates in the clockwise direction?

a)   b)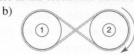

25. Considering that the consecutive dials on the electric meter rotate in opposite directions, what is the current reading in kilowatt hours of usage?

KWH

26. Considering that the consecutive dials on the natural gas meter rotate in opposite directions, what is the current reading in cubic feet of usage?

Cu FT

27. Describe the type(s) of symmetry displayed by each of these automobile logos.

a) Toyota

The Toyota brand and logos as well as Toyota model names are trademarks of Toyota Motor

b) Chevrolet

This logo appears as a courtesy of General Motors Corporation

c) Volkswagen

Volkswagen Group of America, Inc.

28. Describe the type(s) of symmetry displayed by each of these department store logos.

a) Kmart

The Kmart logo is a registered trademark of Sears Brands, LLC.

b) Target

Target and the Bullseye Design are registered trademarks of Target Brands, Inc. All rights reserved.

c) Bergner's

Bergner's

29. Given a figure, which of the following pairs of transformations leads to an image that repeats the original figure?
    a) Figure slides 10 cm to the right *twice*.
    b) Figure is reflected about a vertical line *twice*.
    c) Figure is rotated clockwise about a point 180° *twice*.
    d) Figure is rotated clockwise about a point 90° *twice*.

30. Given a figure, which of the following pairs of transformations leads to an image that repeats the original figure?
    a) Figure slides 10 cm to the right, followed by slide of 10 cm to the left.
    b) Figure is reflected about the same horizontal line *twice*.
    c) Figure is rotated clockwise about a point 120° *twice*.
    d) Figure is rotated clockwise about a point 360° *twice*.

31. A regular hexagon is rotated about a centrally located point (as shown). How many rotations are needed to repeat the given hexagon vertex for vertex, if the angle of rotation is
    a) 30°?   b) 60°?   c) 90°?   d) 240°?

32. A regular octagon is rotated about a centrally located point (as shown). How many rotations are needed to repeat the given octagon vertex for vertex, if the angle of rotation is
    a) 10°?   b) 45°?   c) 90°?   d) 120°?

33. $\angle A'B'C'$ is the image of $\angle ABC$ following the reflection of $\angle ABC$ across line ℓ. If m$\angle A'B'C' = \frac{x}{5} + 20$ and m$\angle ABC = \frac{x}{2} + 5$, find x.

34. $\angle X'YZ'$ is the image of $\angle XYZ$ following a 100° counterclockwise rotation of $\angle XYZ$ about point Y. If m$\angle XYZ = \frac{5x}{6}$ and m$\angle X'YZ' = 130°$, find x.

35. Hexagon $ABCB'A'D$ is determined when the open figure with vertices A, B, C, and D is reflected across $\overline{DC}$.
    a) How many diagonals does $ABCB'A'D$ have?
    b) How many of the diagonals from part (a) lie in the exterior of the hexagon?

36. Rectangle $BCAC'$ is formed when the right triangle ABC is rotated as shown.
    a) What is the measure of the angle of rotation?
    b) What point is the center of this rotation?

PERSPECTIVE ON HISTORY

SKETCH OF EUCLID

Names often associated with the early development of Greek mathematics, beginning in approximately 600 B.C., include Thales, Pythagoras, Archimedes, Appolonius, Diophantus, Eratosthenes, and Heron. However, the name most often associated with traditional geometry is that of Euclid, who lived around 300 B.C.

Euclid, himself a Greek, was asked to head the mathematics department at the University of Alexandria (in Egypt), which was the center of Greek learning. It is believed that Euclid told Ptolemy (the local ruler) that "There is no royal road to geometry," in response to Ptolemy's request for a quick and easy knowledge of the subject.

Euclid's best-known work is the *Elements*, a systematic treatment of geometry with some algebra and number theory. That work, which consists of 13 volumes, has dominated the study of geometry for more than 2000 years. Most secondary-level geometry courses, even today, are based on Euclid's *Elements* and in particular on these volumes:

Book I: Triangles and congruence, parallels, quadrilaterals, the Pythagorean theorem, and area relationships

Book III: Circles, chords, secants, tangents, and angle measurement

Book IV: Constructions and regular polygons

Book VI: Similar triangles, proportions, and the Angle Bisector theorem

Book XI: Lines and planes in space, and parallelepipeds

One of Euclid's theorems was a forerunner of the theorem of trigonometry known as the Law of Cosines. Although the law is difficult to understand now, it will make sense to you later. As stated by Euclid, "In an obtuse-angled triangle, the square of the side opposite the obtuse angle equals the sum of the squares of the other two sides and the product of one side and the projection of the other upon it."

While it is believed that Euclid was a great teacher, he is also recognized as a great mathematician and as the first author of an elaborate textbook. In Chapter 2 of *this* textbook, Euclid's Parallel Postulate has been central to our study of plane geometry.

PERSPECTIVE ON APPLICATIONS

NON-EUCLIDEAN GEOMETRIES

The geometry we present in this book is often described as Euclidean geometry. A non-Euclidean geometry is a geometry characterized by the existence of at least one contradiction of a Euclidean geometry postulate. To appreciate this subject, you need to realize the importance of the word *plane* in the Parallel Postulate. Thus, the Parallel Postulate is now restated.

PARALLEL POSTULATE

In a plane, through a point not on a line, exactly one line is parallel to the given line.

The Parallel Postulate characterizes a course in plane geometry; it corresponds to the theory that "the earth is flat." On a small scale (most applications aren't global), the theory works well and serves the needs of carpenters, designers, and most engineers.

To begin the move to a different geometry, consider the surface of a **sphere** (like the earth). See Figure 2.53. By definition, a sphere is the set of all points in space that are at a fixed distance from a given point. If a line segment on the surface of the

sphere is extended to form a line, it becomes a great circle (like the equator of the earth). Each line in this geometry, known as *spherical geometry*, is the intersection of a plane containing the center of the sphere with the sphere.

(a) ℓ and *m* are lines in spherical geometry

(b) These circles are *not* lines in spherical geometry

Figure 2.53

Spherical (or elliptical) geometry is actually a model of Riemannian geometry, named in honor of Georg F. B. Riemann (1826–1866), the German mathematician responsible for the next postulate. The Reimannian Postulate is not numbered in this book, because it does not characterize Euclidean geometry.

Through a point not on a line, there are no lines parallel to the given line.

To understand the Reimannian Postulate, consider a sphere (Figure 2.54) containing line ℓ and point P not on ℓ. Any line drawn through point P must intersect ℓ in two points. To see this develop, follow the frames in Figure 2.55, which depict an attempt to draw a line parallel to ℓ through point P.

(a)

(b)

Figure 2.54

Consider the natural extension to Riemannian geometry of the claim that the shortest distance between two points is a straight line. For the sake of efficiency and common sense, a person traveling from New York City to London will follow the path of a line as it is known in spherical geometry. As you might guess, this concept is used to chart international flights between cities. In Euclidean geometry, the claim suggests that a person tunnel under the earth's surface from one city to the other.

A second type of non-Euclidean geometry is attributed to the works of a German, Karl F. Gauss (1777–1855), a Russian, Nikolai Lobachevski (1793–1856), and a Hungarian, Johann Bolyai (1802–1862). The postulate for this system of non-Euclidean geometry is as follows:

LOBACHEVSKIAN POSTULATE

Through a point not on a line, there are infinitely many lines parallel to the given line.

This form of non-Euclidean geometry is termed *hyperbolic geometry*. Rather than using the plane or sphere as the surface

for study, mathematicians use a saddle-like surface known as a **hyperbolic paraboloid**. (See Figure 2.56.) A line ℓ is the intersection of a plane with this surface. Clearly, more than one plane can intersect this surface to form a line containing P that does not intersect ℓ. In fact, an infinite number of planes intersect the surface in an infinite number of lines parallel to ℓ and containing P. Table 2.5 compares the three types of geometry.

(a) Small part of surface of the sphere

(b) Line through P "parallel" to ℓ on larger part of surface

(c) Line through P shown to intersect ℓ on larger portion of surface

(d) All of line ℓ and the line through P shown on entire sphere

Figure 2.55

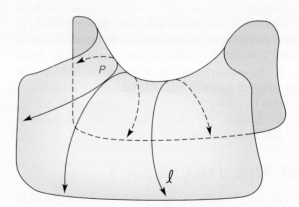

Figure 2.56

TABLE 2.5
Comparison of Types of Geometry

Postulate	Model	Line	Number of Lines Through P Parallel to ℓ
Parallel (Euclidean)	Plane geometry	Intersection of plane with plane	One
Riemannian	Spherical geometry	Intersection of plane with sphere (plane contains center of sphere)	None
Lobachevskian	Hyperbolic geometry	Intersection of plane with hyperbolic paraboloid	Infinitely many

Summary

A Look Back at Chapter 2

One goal of this chapter has been to prove several theorems based on the postulate, "If two parallel lines are cut by a transversal, then the pairs of corresponding angles are congruent." The method of indirect proof was introduced as a basis for proving lines parallel if a pair of corresponding angles are congruent. Several methods of proving lines parallel were then demonstrated by the direct method. The Parallel Postulate was used to prove that the sum of the measures of the interior angles of a triangle is 180°. Several corollaries followed naturally from this theorem. A formula was then developed for finding the sum of the measures of the interior angles of any polygon. The chapter closed with a discussion of symmetry and transformations.

A Look Ahead to Chapter 3

In the next chapter, the concept of congruence will be extended to triangles, and several methods of proving triangles congruent will be developed. Several theorems dealing with inequality relationships in a triangle will also be proved. The Pythagorean Theorem will be introduced.

Key Concepts

2.1

Perpendicular Lines • Perpendicular Planes • Parallel Lines • Parallel Planes • Parallel Postulate • Transversal • Interior Angles • Exterior Angles • Corresponding Angles • Alternate Interior Angles • Alternate Exterior Angles

2.2

Conditional • Converse • Inverse • Contrapositive • Law of Negative Inference • Indirect Proof

2.3

Proving Lines Parallel

2.4

Triangles • Vertices • Sides of a Triangle • Interior and Exterior of a Triangle • Scalene Triangle • Isosceles Triangle • Equilateral Triangle • Acute Triangle • Obtuse Triangle • Right Triangle • Equiangular Triangle • Auxiliary Line • Determined • Underdetermined • Overdetermined • Corollary • Exterior Angle of a Triangle

2.5

Convex Polygons (Triangle, Quadrilateral, Pentagon, Hexagon, Heptagon, Octagon, Nonagon, Decagon) • Concave Polygon • Diagonals of a Polygon • Regular Polygon • Equilateral Polygon • Equiangular Polygon • Polygram

2.6

Symmetry • Line of Symmetry • Axis of Symmetry • Point Symmetry • Transformations • Slides • Translations • Reflections • Rotations • Angle of Rotation • Center of Rotation

Overview ■ Chapter 2

Parallel Lines and Transversal

Figure	Relationship	Symbols
	$\ell \parallel m$	Corresponding $\angle$s $\cong$; $\angle 1 \cong \angle 5$, $\angle 2 \cong \angle 6$, etc.
		Alternate interior $\angle$s $\cong$; $\angle 3 \cong \angle 6$ and $\angle 4 \cong \angle 5$
		Alternate exterior $\angle$s $\cong$; $\angle 1 \cong \angle 8$ and $\angle 2 \cong \angle 7$
		Supplementary $\angle$s; $m\angle 3 + m\angle 5 = 180°$; $m\angle 1 + m\angle 7 = 180°$, etc.

Triangles Classified by Sides

Figure	Type	Number of Congruent Sides
	Scalene	None
	Isosceles	Two
	Equilateral	Three

Triangles Classified by Angles

Figure	Type	Angle(s)
	Acute	Three acute angles
	Right	One right angle
	Obtuse	One obtuse angle
	Equiangular	Three congruent angles

Polygons: Sum *S* of All Interior Angles

Figure	Type of Polygon	Sum of Interior Angle(s)
	Triangle	$S = 180°$
	Quadrilaterial	$S = 360°$
	Polygon with n sides	$S = (n - 2) \cdot 180°$

Polygons: Sum *S* of All Exterior Angles; *D* Is the Total Number of Diagonals

Figure	Type of Polygon	Relationships
	Polygon with n sides	$S = 360°$ $D = \dfrac{n(n - 3)}{2}$

Symmetry

Figure	Type of Symmetry	Figure Redrawn to Display Symmetry
Z	Point	Z
D	Line	←[D]→

Chapter 2 Review Exercises

1. If m∠1 = m∠2, which lines are parallel?

(a)

(b)

2. *Given:* m∠13 = 70°
   Find: m∠3

3. *Given:* m∠9 = 2x + 17
         m∠11 = 5x − 94
   Find: x

a∥b and c∥d

Exercises 2, 3

4. *Given:* m∠B = 75°,
         m∠DCE = 50°
   Find: m∠D and m∠DEF

5. *Given:* m∠DCA = 130°
         m∠BAC = 2x + y
         m∠BCE = 150°
         m∠DEC = 2x − y
   Find: x and y

AB∥CD and BC∥DE

Exercises 4, 5

6. *Given:* AC ∥ DF
         AE ∥ BF
         m∠AEF = 3y
         m∠BFE = x + 45
         m∠FBC = 2x + 15
   Find: x and y

Exercises 6–11

For Review Exercises 7 to 11, use the given information to name the segments that must be parallel. If there are no such segments, write "none." Assume A-B-C and D-E-F. (Use the drawing from Exercise 6.)

7. ∠3 ≅ ∠11

8. ∠4 ≅ ∠5

9. ∠7 ≅ ∠10

10. ∠6 ≅ ∠9

11. ∠8 ≅ ∠5 ≅ ∠3

For Review Exercises 12 to 15, find the values of x and y.

12.

a∥b

13.

14.

a∥b

15.

16. *Given:* m∠1 = x^2 − 12
         m∠4 = x(x − 2)
   Find: x so that $\overrightarrow{AB} \parallel \overrightarrow{CD}$

17. *Given:* $\overline{AB} \parallel \overline{CD}$
         m∠2 = x^2 − 3x + 4
         m∠1 = 17x − x^2 − 5
         m∠ACE = 111°
   Find: m∠3, m∠4, and m∠5

Exercises 16, 17

18. *Given:* $\overline{DC} \parallel \overline{AB}$
         ∠A ≅ ∠C
         m∠A = 3x + y
         m∠D = 5x + 10
         m∠C = 5y + 20
   Find: m∠B

For Review Exercises 19 to 24, decide whether the statements are always true (A), sometimes true (S), or never true (N).

19. An isosceles triangle is a right triangle.

20. An equilateral triangle is a right triangle.

21. A scalene triangle is an isosceles triangle.

22. An obtuse triangle is an isosceles triangle.

23. A right triangle has two congruent angles.

24. A right triangle has two complementary angles.

25. Complete the following table for regular polygons.

Number of sides	8	12	20		
Measure of each exterior ∠				24	36
Measure of each interior ∠				157.5	178
Number of diagonals					

For Review Exercises 26 to 29, sketch, if possible, the polygon described.

26. A quadrilateral that is equiangular but not equilateral

27. A quadrilateral that is equilateral but not equiangular

28. A triangle that is equilateral but not equiangular

29. A hexagon that is equilateral but not equiangular

For Review Exercises 30 and 31, write the converse, inverse, and contrapositive of each statement.

30. If two angles are right angles, then the angles are congruent.

31. If it is not raining, then I am happy.

32. Which statement—the converse, the inverse, or the contrapositive—always has the same truth or falsity as a given implication?

33. *Given:* $\overline{AB} \parallel \overline{CF}$
  $\angle 2 \cong \angle 3$
  Prove: $\angle 1 \cong \angle 3$

34. *Given:* $\angle 1$ is complementary to $\angle 2$; $\angle 2$ is complementary to $\angle 3$
  Prove: $\overline{BD} \parallel \overline{AE}$

35. *Given::* $\overline{BE} \perp \overline{DA}$
  $\overline{CD} \perp \overline{DA}$
  Prove: $\angle 1 \cong \angle 2$

36. *Given:* $\angle A \cong \angle C$
  $\overline{DC} \parallel \overline{AB}$
  Prove: $\overline{DA} \parallel \overline{CB}$

For Review Exercises 37 and 38, give the first statement for an indirect proof.

37. If $x^2 + 7x + 12 \neq 0$, then $x \neq -3$

38. If two angles of a triangle are not congruent, then the sides opposite those angles are not congruent.

39. *Given:* $m \not\parallel n$
  Prove: $\angle 1 \not\cong \angle 2$

40. *Given:* $\angle 1 \not\cong \angle 3$
  Prove: $m \not\parallel n$

Exercises 39, 40

41. Construct the line through C parallel to $\overline{AB}$.

42. Construct an equilateral triangle ABC with side $\overline{AB}$.

43. Which block letters have
  a) line symmetry (at least one axis)?
  b) point symmetry?

B   H   J   S   W

44. Which figures have
  a) line symmetry (at least one axis)?
  b) point symmetry?

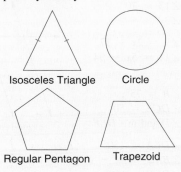

Isosceles Triangle   Circle

Regular Pentagon   Trapezoid

45. When $\triangle ABC$ slides to its image $\triangle DEF$, how are $\triangle ABC$ and $\triangle DEF$ related?

46. Complete the drawing so that the figure is reflected across
  a) line ℓ.   b) line m.

47. Through what approximate angle of rotation must a baseball pitcher turn when throwing to first base rather than home plate?

Chapter **2**   **Test**

1. Consider the figure shown at the right.

   a) Name the angle that corresponds to ∠1.

   _____
   b) Name the alternate interior angle for ∠6.

   _____

2. In the accompanying figure, m∠2 = 68°, m∠8 = 112°, and m∠9 = 110°.

   a) Which lines (*r* and *s* OR ℓ and *m*) must be parallel? _____
   b) Which pair of lines (*r* and *s* OR ℓ and *m*) cannot be parallel? _____

3. To prove a theorem of the form "If *P*, then *Q*" by the indirect method, the first line of the proof should read: Suppose that _____ is true.

4. Assuming that statements 1 and 2 are true, draw a valid conclusion if possible.
   1. If two angles are both right angles, then the angles are congruent.
   2. ∠*R* and ∠*S* are not congruent.
   C. ∴? _____

5. Let all of the lines named be coplanar. Make a drawing to reach a conclusion.
   a) If *r* ∥ *s* and *s* ∥ *t*, then _____.
   b) If *a* ⊥ *b* and *b* ⊥ *c*, then _____.

6. Through point *A*, construct the line that is perpendicular to line ℓ.

7. For △*ABC*, find m∠*B* if
   a) m∠*A* = 65° and m∠*C* = 79°.

   _____
   b) m∠*A* = 2*x*, m∠*B* = *x*, and m∠*C* = 2*x* + 15. _____

8. a) What word describes a polygon with five sides? _____
   b) How many diagonals does a polygon with five sides have? _____

9. a) Given that the polygon shown has six congruent angles, this polygon is known as a(n)

   _____ _____.
   b) What is the measure of each of the congruent interior angles?

   _____

10. Consider the block letters A, D, N, O, and X. Which type of symmetry (line symmetry, point symmetry, both types, or neither type) is illustrated by each letter?
    A _____   D _____
    N _____   O _____
    X _____

11. Which type of transformation (slide, reflection, or rotation) is illustrated?
    a) _____   b) _____   c) _____

    (a)

    (b)
    (c)

12. In the figure shown, suppose that $\overline{AB} \parallel \overline{DC}$ and $\overline{AD} \parallel \overline{BC}$. If m∠1 = 82° and m∠4 = 37°, find m∠*C*. _____

    Exercises 12, 13

13. If m∠1 = *x* + 28 and m∠2 = 2*x* − 26, find the value *x* for which it follows that $\overline{AB} \parallel \overline{DC}$. _____

14. In the figure shown, suppose that ray *CD* bisects exterior angle ∠*ACE* of △*ABC*. If m∠1 = 70° and m∠2 = 30°, find m∠4.

    _____
    Exercises 14, 15

15. In the figure shown, ∠*ACE* is an exterior angle of △*ABC*. If $\overrightarrow{CD} \parallel \overrightarrow{BA}$, m∠1 = 2(m∠2), and m∠*ACE* = 117°, find the measure of ∠1. _____

In Exercises 16 and 18, complete the missing statements or reasons for each proof.

16. *Given:* ∠1 ≅ ∠2
∠3 ≅ ∠4
Prove: ℓ ∥ n

PROOF

Statements	Reasons
1. ∠1 ≅ ∠2 and ∠3 ≅ ∠4	1. _____
2. _____	2. If two lines intersect, the vertical ∠s are ≅
3. ∠1 ≅ 4	3. _____
4. _____	4. If two lines are cut by a transversal so that a pair of alternate exterior ∠s are ≅, the lines are ∥

17. Use an indirect proof to complete the following proof.
Given: △MNQ with m∠N = 120°
Prove: ∠M and ∠Q are not complementary

18. *Given:* In △ABC, m∠C = 90°
Prove: ∠1 and ∠2 are complementary

PROOF

Statements	Reasons
1. △ABC, m∠C = 90°	1. _____
2. m∠1 + m∠2 + m∠C = _____	2. The sum of ∠s of a △ is 180°
3. _____	3. Substitution Prop. of Equality
4. m∠1 + m∠2 = ___	4. Subtraction Prop. of Equality
5. _____	5. _____

19. In △XYZ, ∠XYZ is trisected by $\overline{YW}$ and $\overline{YV}$. With angle measures as shown, find m∠Z. _____

Photodisc/Allan Baxter/Getty Images

Chapter **3**

Triangles

Majestic! In Statue Square of Hong Kong, the Bank of China (the structure shown at the left in the photograph above) rises 1209 feet above the square. Designed by I. M. Pei (who studied at the Massachusetts Institute of Technology and also graduated from the Harvard Graduate School of Design), the Bank of China displays many triangles of the same shape and size. Such triangles, known as congruent triangles, are also displayed in the Ferris wheel found in Exercise 43 of Section 3.3. While Chapter 3 is devoted to the study of triangle types and their characteristics, the properties of triangles developed herein also provide a much-needed framework for the study of quadrilaterals found in Chapter 4.

Additional video explanations of concepts, sample problems, and applications are available on DVD.

3.1   Congruent Triangles

KEY CONCEPTS

Congruent Triangles	Included Side	Symmetric and
SSS	Included Angle	Transitive Properties
SAS	Reflexive Property of	of Congruence
ASA	Congruence (Identity)	
AAS		

Two triangles are **congruent** if one coincides with (fits perfectly over) the other. In Figure 3.1, we say that $\triangle ABC \cong \triangle DEF$ if these congruences hold:

$$\angle A \cong \angle D, \angle B \cong \angle E, \angle C \cong \angle F, \overline{AB} \cong \overline{DE}, \overline{BC} \cong \overline{EF}, \text{ and } \overline{AC} \cong \overline{DF}$$

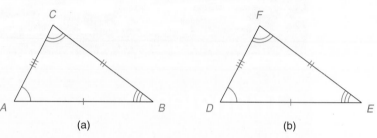

(a)   (b)

Figure 3.1

From the indicated congruences, we also say that vertex A corresponds to vertex D, as does B to E and C to F. In symbols, the correspondences are represented by

$$A \leftrightarrow D, \qquad B \leftrightarrow E, \qquad \text{and} \qquad C \leftrightarrow F.$$

In Section 2.6, we used a slide transformation on $\triangle ABC$ to form its image $\triangle DEF$.

The claim $\triangle MNQ \cong \triangle RST$ orders corresponding vertices of the triangles (not shown), so we can conclude from this statement that

$$M \leftrightarrow R, \qquad N \leftrightarrow S, \qquad \text{and} \qquad Q \leftrightarrow T.$$

This correspondence of vertices implies the congruence of corresponding parts such as $\angle M \cong \angle R$ and $\overline{NQ} \cong \overline{ST}$. Conversely, if the correspondence of vertices of two congruent triangles is $M \leftrightarrow R$, $N \leftrightarrow S$, and $Q \leftrightarrow T$, we order these vertices to make the claims $\triangle MNQ \cong \triangle RST$, $\triangle NQM \cong \triangle STR$, and so on.

EXAMPLE 1

For two congruent triangles, the correspondence of vertices is given by $A \leftrightarrow D$, $B \leftrightarrow E$, and $C \leftrightarrow F$. Complete each statement:

a) $\triangle BCA \cong$?   **b)** $\triangle DEF \cong$?

SOLUTION With due attention to the order of corresponding vertices, we have

a) $\triangle BCA \cong \triangle EFD$ **b)** $\triangle DEF \cong \triangle ABC$

DEFINITION

Two triangles are **congruent** if the six parts of the first triangle are congruent to the six corresponding parts of the second triangle.

As always, any definition is reversible! If two triangles are known to be congruent, we may conclude that the corresponding parts are congruent. Moreover, if the six pairs of parts are known to be congruent, then so are the triangles! From the congruent parts indicated in Figure 3.2, we can conclude that $\triangle MNQ \cong \triangle RST$. Using the terminology introduced in Section 2.6 and Figure 3.2, $\triangle TSR$ is the reflection of $\triangle QNM$ across a vertical line (not shown) that lies midway between the two triangles.

Following Figure 3.2 are some of the properties of congruent triangles that are useful in later proofs and explanations. In the figure, notice that corresponding sides lie *opposite* corresponding angles and conversely.

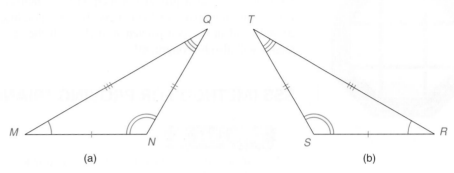

(a)   (b)

Figure 3.2

---

1. $\triangle ABC \cong \triangle ABC$ (Reflexive Property of Congruence)
2. If $\triangle ABC \cong \triangle DEF$, then $\triangle DEF \cong \triangle ABC$. (Symmetric Property of Congruence)
3. If $\triangle ABC \cong \triangle DEF$ and $\triangle DEF \cong \triangle GHI$, then $\triangle ABC \cong \triangle GHI$. (Transitive Property of Congruence)

SSG **EXS. 1, 2**

On the basis of the properties above, we see that the "congruence of triangles" is an equivalence relation.

It would be difficult to establish that triangles were congruent if all six pairs of congruent parts had to first be verified. Fortunately, it is possible to prove triangles congruent by establishing fewer than six pairs of congruences. To suggest a first method, consider the construction in Example 2.

EXAMPLE 2

Construct a triangle whose sides have the lengths of the segments provided in Figure 3.3(a).

SOLUTION Figure 3.3(b): Choose $\overline{AB}$ as the first side of the triangle (the choice of $\overline{AB}$ is arbitrary) and mark its length as shown.

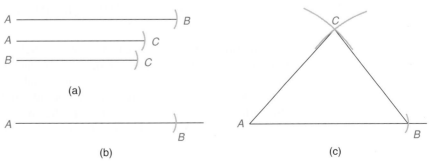

(a)

(b)   (c)

Figure 3.3

Figure 3.3(c): Using the left endpoint A, mark off an arc of length equal to that of $\overline{AC}$. Now mark off an arc the length of $\overline{BC}$ from the right endpoint B so that these arcs intersect at C, the third vertex of the triangle. Joining point C to A and then to B completes the desired triangle.

Look again at Example 2. If a "different" triangle were constructed by choosing $\overline{AC}$ to be the first side, it would be congruent to the triangle shown; however, it might be necessary to flip or rotate it to have corresponding vertices match. The objective of Example 2 is that it provides a method for establishing the congruence of triangles by using only three pairs of their parts. If corresponding angles are measured in the given triangle and in the constructed triangle with the same lengths for sides, these pairs of angles will also be congruent!

SSS (METHOD FOR PROVING TRIANGLES CONGRUENT)

POSTULATE 12

If the three sides of one triangle are congruent to the three sides of a second triangle, then the triangles are congruent (SSS).

The designation SSS will be cited as a reason in the proof that follows. Each letter of SSS refers to a *pair* of congruent sides.

Geometry in the Real World

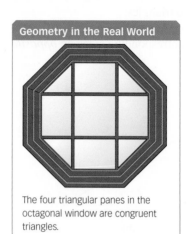

The four triangular panes in the octagonal window are congruent triangles.

EXAMPLE 3

GIVEN: $\overline{AB}$ and $\overline{CD}$ bisect each other at M
$\overline{AC} \cong \overline{DB}$
(See Figure 3.4.)

PROVE: $\triangle AMC \cong \triangle BMD$

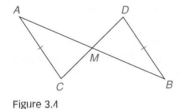

Figure 3.4

PROOF

Statements	Reasons
1. $\overline{AB}$ and $\overline{CD}$ bisect each other at M	1. Given
2. $\overline{AM} \cong \overline{MB}$ $\overline{CM} \cong \overline{MD}$	2. If a segment is bisected, the segments formed are $\cong$
3. $\overline{AC} \cong \overline{DB}$	3. Given
4. $\triangle AMC \cong \triangle BMD$	4. SSS

NOTE 1: In Steps 2 and 3, the three pairs of sides were shown to be congruent; thus, SSS is cited as the reason that justifies why $\triangle AMC \cong \triangle BMD$.

NOTE 2: $\triangle BMD$ is the image determined by the clockwise or counterclockwise rotation of $\triangle AMC$ about point M through a 180° angle of rotation.

The two sides that form an angle of a triangle are said to **include that angle** of the triangle. In $\triangle TUV$ in Figure 3.5(a), sides $\overline{TU}$ and $\overline{TV}$ form $\angle T$; therefore, $\overline{TU}$ and $\overline{TV}$ include $\angle T$. In turn, $\angle T$ is said to be the included angle for $\overline{TU}$ and $\overline{TV}$. Similarly, any two angles of a triangle must have a common side, and these two angles are said to **include that side**. In $\triangle TUV$, $\angle U$ and $\angle T$ share the common side $\overline{UT}$; therefore, $\angle U$ and $\angle T$ include the side $\overline{UT}$; equivalently, $\overline{UT}$ is the side included by $\angle U$ and $\angle T$.

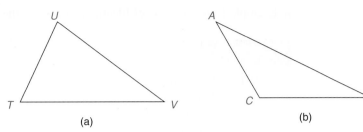

Figure 3.5

Informally, the term *include* names the part of a triangle that is "between" two other named parts. In a triangle, two sides include an angle, while two angles include a side. The two sides that include the angle actually *form* the angle. The two angles that include a side *share* that side.

EXAMPLE 4

In $\triangle ABC$ of Figure 3.5(b):

a) Which angle is included by $\overline{AC}$ and $\overline{CB}$?
b) Which sides include $\angle B$?
c) What is the included side for $\angle A$ and $\angle B$?
d) Which angles include $\overline{CB}$?

SOLUTION

a) $\angle C$ (because it is formed by $\overline{AC}$ and $\overline{CB}$)
b) $\overline{AB}$ and $\overline{BC}$ (because these form $\angle B$)
c) $\overline{AB}$ (because it is the common side for $\angle A$ and $\angle B$)
d) $\angle C$ and $\angle B$ (because $\overline{CB}$ is a side of each angle)

SAS (METHOD FOR PROVING TRIANGLES CONGRUENT)

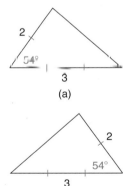

Figure 3.6

A second way of establishing that two triangles are congruent involves showing that two sides and the included angle of one triangle are congruent to two sides and the included angle of a second triangle. If two people each draw a triangle so that two of the sides measure 2 cm and 3 cm and their included angle measures 54°, then those triangles are congruent. See Figure 3.6.

POSTULATE 13

If two sides and the included angle of one triangle are congruent to two sides and the included angle of a second triangle, then the triangles are congruent (SAS).

The order of the letters SAS in Postulate 13 helps us to remember that the two sides that are named have the angle "between" them; that is, the two sides referred to by S and S form the angle, represented by A.

In Example 5 on page 126, the two triangles to be proved congruent share a common side; the statement $\overline{PN} \cong \overline{PN}$ is justified by the Reflexive Property of Congruence, which is conveniently expressed as **Identity.**

DEFINITION

In a proof, **Identity** (also known as the Reflexive Property of Congruence) is the reason cited when verifying that a line segment or an angle is congruent to itself.

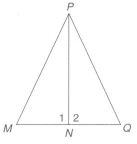

Figure 3.7

In Example 5, note the use of Identity and SAS as the final reasons.

EXAMPLE 5

GIVEN: $\overline{PN} \perp \overline{MQ}$
$\overline{MN} \cong \overline{NQ}$
(See Figure 3.7.)

PROVE: $\triangle PNM \cong \triangle PNQ$

PROOF

Statements	Reasons
1. $\overline{PN} \perp \overline{MQ}$	1. Given
2. $\angle 1 \cong \angle 2$	2. If two lines are $\perp$, they meet to form $\cong$ adjacent $\angle$s
3. $\overline{MN} \cong \overline{NQ}$	3. Given
4. $\overline{PN} \cong \overline{PN}$	4. Identity (or Reflexive)
5. $\triangle PNM \cong \triangle PNQ$	5. SAS

NOTE: In $\triangle PNM$, $\overline{MN}$ (Step 3) and $\overline{PN}$ (Step 4) include $\angle 1$; similarly, $\overline{NQ}$ and $\overline{PN}$ include $\angle 2$ in $\triangle PNQ$. Thus, SAS is used to verify that $\triangle PNM \cong \triangle PNQ$ in reason 5.

SSG EXS. 3–6

ASA (METHOD FOR PROVING TRIANGLES CONGRUENT)

The next method for proving triangles congruent requires a combination of two angles and the included side. If two triangles are drawn so that two angles measure 33° and 47° while their included side measures 5 centimeters, then these triangles must be congruent. See the figure below.

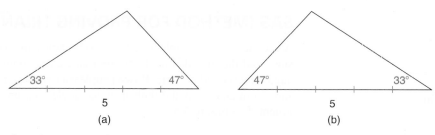

Figure 3.8

POSTULATE 14

If two angles and the included side of one triangle are congruent to two angles and the included side of a second triangle, then the triangles are congruent (ASA).

Although the method in Postulate 14 is written compactly as ASA, you must use caution as you write these abbreviations that verify that triangles are congruent! For example, ASA refers to two angles and the included side, whereas SAS refers to two sides and the included angle. For us to apply any postulate, the specific conditions described in it must be satisfied.

SSS, SAS, and ASA are all valid methods of proving triangles congruent, but SSA is *not* a method and *cannot* be used. In Figure 3.9 on page 127, the two triangles are marked to demonstrate the SSA relationship, yet the two triangles are *not* congruent.

(a)   (b)

Figure 3.9

Another combination that cannot be used to prove triangles congruent is AAA. See Figure 3.10. Three pairs of congruent angles in two triangles do not guarantee three pairs of congruent sides!

In Example 6, the triangles to be proved congruent overlap (see Figure 3.11). To clarify relationships, the triangles have been redrawn separately in Figure 3.12. In Figure 3.12, the parts indicated as congruent are established as congruent in the proof. For statement 3, Identity (or Reflexive) is used to justify that an angle is congruent to itself.

Figure 3.10

SSG **EXS. 7–11**

Figure 3.11

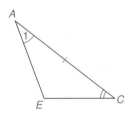

Figure 3.12

EXAMPLE 6

GIVEN: $\overline{AC} \cong \overline{DC}$
          $\angle 1 \cong \angle 2$
          (See Figure 3.11.)

PROVE: $\triangle ACE \cong \triangle DCB$

PROOF

Statements	Reasons
1. $\overline{AC} \cong \overline{DC}$ (See Figure 3.12.)	1. Given
2. $\angle 1 \cong \angle 2$	2. Given
3. $\angle C \cong \angle C$	3. Identity
4. $\triangle ACE \cong \triangle DCB$	4. ASA

Next we consider the AAS theorem; this theorem can be proved by applying the ASA postulate.

AAS (METHOD FOR PROVING TRIANGLES CONGRUENT)

THEOREM 3.1.1

If two angles and a nonincluded side of one triangle are congruent to two angles and a nonincluded side of a second triangle, then the triangles are congruent (AAS).

GIVEN: $\angle T \cong \angle K$, $\angle S \cong \angle J$, and $\overline{SR} \cong \overline{HJ}$ (See Figure 3.13.)

PROVE: $\triangle TSR \cong \triangle KJH$

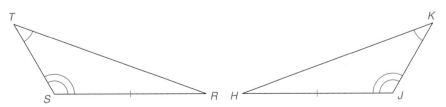

Figure 3.13

PROOF

Statements	Reasons
1. $\angle T \cong \angle K$ $\angle S \cong \angle J$	1. Given
2. $\angle R \cong \angle H$	2. If two $\angle$s of one $\triangle$ are $\cong$ to two $\angle$s of another $\triangle$, then the third $\angle$s are also congruent
3. $\overline{SR} \cong \overline{HJ}$	3. Given
4. $\triangle TSR \cong \triangle KJH$	4. ASA

STRATEGY FOR PROOF ■ Proving That Two Triangles Are Congruent

General Rule: Methods of proof (possible final reasons) available in Section 3.1 are SSS, SAS, ASA, and AAS.

Illustration: See Exercises 9–12 of this section.

SSG **EXS. 12–14**

Exercises 3.1

In Exercises 1 to 4, consider the congruent triangles shown.

1. For the triangles shown, we can express their congruence with the statement $\triangle ABC \cong \triangle FED$. By reordering the vertices, express this congruence with a different statement.

Exercises 1–4

2. With corresponding angles indicated, the triangles shown are congruent. Find values for *a, b,* and *c*.

3. With corresponding angles indicated, find m$\angle A$ if m$\angle F = 72°$.

4. With corresponding angles indicated, find m$\angle E$ if m$\angle A = 57°$ and m$\angle C = 85°$.

5. Consider $\triangle ABC$ and $\triangle ABD$ in the figure shown. By the reason Identity, $\angle A \cong \angle A$ and $\overline{AB} \cong \overline{AB}$.
(a) If $\overline{BC} \cong \overline{BD}$ can you prove that $\triangle ABC \cong \triangle ABD$?
(b) If yes in part (a), by what reason are the triangles congruent?

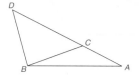

6. In a right triangle, the sides that form the right angle are the legs; the longest side (opposite the right angle) is the hypotenuse. Some textbooks say that when two right triangles have congruent pairs of legs, the right triangles are congruent by the reason LL. In our work, LL is just a special case of one of the postulates in this section. Which postulate is that?

7. In $\triangle ABC$, the midpoints of the sides are joined. (a) What does intuition suggest regarding the relationship between $\triangle AED$ and $\triangle FDE$? (We will prove this relationship later.) (b) What does intuition suggest regarding $\triangle AED$ and $\triangle EBF$?

8. (a) Suppose that you wish to prove that $\triangle RST \cong \triangle SRV$. Using the reason Identity, name one pair of corresponding parts that are congruent.

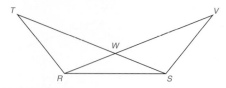

(b) Suppose you wish to prove that $\triangle RWT \cong \triangle SWV$. Considering the figure, name one pair of corresponding angles of these triangles that must be congruent.

In Exercises 9 to 12, congruent parts are indicated by like dashes (sides) or arcs (angles). State which method (SSS, SAS, ASA, or AAS) would be used to prove the two triangles congruent.

9.

10.

11.

12.
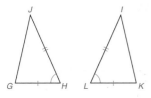

In Exercises 13 to 18, use only the given information to state the reason why △ABC ≅ △DBC. Redraw the figure and use marks like those used in Exercises 9 to 12.

Exercises 13–18

13. ∠A ≅ ∠D, $\overline{AB}$ ≅ $\overline{BD}$, and ∠1 ≅ ∠2

14. ∠A ≅ ∠D, $\overline{AC}$ ≅ $\overline{CD}$, and B is the midpoint of $\overline{AD}$

15. ∠A ≅ ∠D, $\overline{AC}$ ≅ $\overline{CD}$, and $\overrightarrow{CB}$ bisects ∠ACD

16. ∠A ≅ ∠D, $\overline{AC}$ ≅ $\overline{CD}$, and $\overline{AB}$ ≅ $\overline{BD}$

17. $\overline{AC}$ ≅ $\overline{CD}$, $\overline{AB}$ ≅ $\overline{BD}$, and $\overline{CB}$ ≅ $\overline{CB}$ (by Identity)

18. ∠1 and ∠2 are right ∠s, $\overline{AB}$ ≅ $\overline{BD}$, and ∠A ≅ ∠D

In Exercises 19 and 20, the triangles to be proved congruent have been redrawn separately. Congruent parts are marked.

a) *Name an additional pair of parts that are congruent by using the reason Identity.*

b) *Considering the congruent parts, state the reason why the triangles must be congruent.*

19. △ABC ≅ △AED

20. △MNP ≅ △MQP
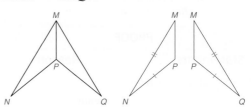

In Exercises 21 to 24, the triangles named can be proven congruent. Considering the congruent pairs marked, name the additional pair of parts that must be congruent in order to use the method named.

21. SAS

△ABD ≅ △CBE

22. ASA

△WVY ≅ △ZVX

23. SSS

△MNO ≅ △OPM

24. AAS

$\triangle EFG \cong \triangle JHG$

In Exercises 25 and 26, complete each proof. Use the figure shown below.

25. *Given:* $\overline{AB} \cong \overline{CD}$ and $\overline{AD} \cong \overline{CB}$
   Prove: $\triangle ABC \cong \triangle CDA$

PROOF	
Statements	**Reasons**
1. $\overline{AB} \cong \overline{CD}$ and $\overline{AD} \cong \overline{CB}$	1. ?
2. ?	2. Identity
3. $\triangle ABC \cong \triangle CDA$	3. ?

Exercises 25, 26

26. *Given:* $\overline{DC} \parallel \overline{AB}$ and $\overline{AD} \parallel \overline{BC}$
   Prove: $\triangle ABC \cong \triangle CDA$

PROOF	
Statements	**Reasons**
1. $\overline{DC} \parallel \overline{AB}$	1. ?
2. $\angle DCA \cong \angle BAC$	2. ?
3. ?	3. Given
4. ?	4. If two ∥ lines are cut by a transversal, alt. int. ∠s are ≅
5. $\overline{AC} \cong \overline{AC}$	5. ?
6. ?	6. ASA

In Exercises 27 to 32, use SSS, SAS, ASA, or AAS to prove that the triangles are congruent.

27. *Given:* $\overrightarrow{PQ}$ bisects $\angle MPN$
   $\overline{MP} \cong \overline{NP}$
   Prove: $\triangle MQP \cong \triangle NQP$

28. *Given:* $\overline{PQ} \perp \overline{MN}$ and
   $\angle 1 \cong \angle 2$
   Prove: $\triangle MQP \cong \triangle NQP$

Exercises 27, 28

29. *Given:* $\overline{AB} \perp \overline{BC}$ and $\overline{AB} \perp \overline{BD}$
   $\overline{BC} \cong \overline{BD}$
   Prove: $\triangle ABC \cong \triangle ABD$

30. *Given:* $\overline{PN}$ bisects $\overline{MQ}$
   $\angle M$ and $\angle Q$ are right angles
   Prove: $\triangle PQR \cong \triangle NMR$

31. *Given:* $\angle VRS \cong \angle TSR$ and $\overline{RV} \cong \overline{TS}$
   Prove: $\triangle RST \cong \triangle SRV$

Exercises 31, 32

32. *Given:* $\overline{VS} \cong \overline{TR}$ and $\angle TRS \cong \angle VSR$
   Prove: $\triangle RST \cong \triangle SRV$

In Exercises 33 to 36, the methods to be used are SSS, SAS, ASA, and AAS.

33. Given that $\triangle RST \cong \triangle RVU$, does it follow that $\triangle RSU$ is also congruent to $\triangle RVT$? Name the method, if any, used in arriving at this conclusion.

Exercises 33, 34

34. Given that $\angle S \cong \angle V$ and $\overline{ST} \cong \overline{UV}$, does it follow that $\triangle RST \cong \triangle RVU$? Which method, if any, did you use?

35. Given that $\angle A \cong \angle E$ and $\angle B \cong \angle D$, does it follow that $\triangle ABC \cong \triangle EDC$? If so, cite the method used in arriving at this conclusion.

Exercises 35, 36

36. Given that $\angle A \cong \angle E$ and $\overline{BC} \cong \overline{DC}$, does it follow that $\triangle ABC \cong \triangle EDC$? Cite the method, if any, used in reaching this conclusion. (See the figure for Exercise 35.)

37. In quadrilateral $ABCD$, $\overline{AC}$ and $\overline{BD}$ are perpendicular bisectors of each other. Name all triangles that are congruent to:
   a) $\triangle ABE$　　　b) $\triangle ABC$　　　c) $\triangle ABD$

38. In $\triangle ABC$ and $\triangle DEF$, you know that $\angle A \cong \angle D$, $\angle C \cong \angle F$, and $\overline{AB} \cong \overline{DE}$. Before concluding that the triangles are congruent by ASA, you need to show that $\angle B \cong \angle E$. State the postulate or theorem that allows you to confirm this statement ($\angle B \cong \angle E$).

In Exercises 39 and 40, complete each proof.

39. *Given:*　　Plane M
　　　　　　　　C is the midpoint of $\overline{EB}$
　　　　　　　　$\overline{AD} \perp \overline{BE}$ and $\overline{AB} \parallel \overline{ED}$
　　Prove:　　$\triangle ABC \cong \triangle DEC$

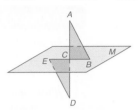

40. *Given:*　　$\overline{SP} \cong \overline{SQ}$ and $\overline{ST} \cong \overline{SV}$
　　Prove:　　$\triangle SPV \cong \triangle SQT$ and $\triangle TPQ \cong \triangle VQP$

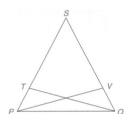

41. *Given:*　　$\angle ABC$; $\overline{RS}$ is the perpendicular bisector of $\overline{AB}$; $\overline{RT}$ is the perpendicular bisector of $\overline{BC}$.
　　Prove:　　$\overline{AR} \cong \overline{RC}$

3.2 **Corresponding Parts of Congruent Triangles**

KEY CONCEPTS　　CPCTC　　　　　　　　　　HL
　　　　　　　　　　Hypotenuse and Legs　　　Pythagorean Theorem
　　　　　　　　　　　of a Right Triangle　　　Square Roots Property

Recall that the definition of congruent triangles states that *all* six parts (three sides and three angles) of one triangle are congruent respectively to the six corresponding parts of the second triangle. If we have proved that $\triangle ABC \cong \triangle DEF$ by SAS (the congruent parts are marked in Figure 3.14), then we can draw further conclusions such as $\angle C \cong \angle F$ and $\overline{AC} \cong \overline{DF}$. The following reason (CPCTC) is often cited for drawing such conclusions and is based on the definition of congruent triangles.

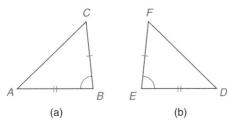

　　　　(a)　　　　　　　　　　　(b)

Figure 3.14

CPCTC

Corresponding parts of congruent triangles are congruent.

STRATEGY FOR PROOF ■ Using CPCTC

General Rule: In a proof, two triangles must be proven congruent *before* CPCTC can be used to verify that another pair of sides or angles of these triangles are also congruent.

Illustration: In the proof of Example 1, statement 5 (triangles congruent) must be stated before we conclude that $\overline{TZ} \cong \overline{VZ}$ by CPCTC.

Figure 3.15

EXAMPLE 1

GIVEN: $\overrightarrow{WZ}$ bisects $\angle TWV$
$\overline{WT} \cong \overline{WV}$
(See Figure 3.15.)

PROVE: $\overline{TZ} \cong \overline{VZ}$

PROOF

Statements	Reasons
1. $\overrightarrow{WZ}$ bisects $\angle TWV$	1. Given
2. $\angle TWZ \cong \angle VWZ$	2. The bisector of an angle separates it into two $\cong \angle s$
3. $\overline{WT} \cong \overline{WV}$	3. Given
4. $\overline{WZ} \cong \overline{WZ}$	4. Identity
5. $\triangle TWZ \cong \triangle VWZ$	5. SAS
6. $\overline{TZ} \cong \overline{VZ}$	6. CPCTC

In Example 1, we could just as easily have used CPCTC to prove that two angles are congruent. If we had been asked to prove that $\angle T \cong \angle V$, then the final statement of the proof would have read

6. $\angle T \cong \angle V$	6. CPCTC

We can take the proof in Example 1 a step further by proving triangles congruent, using CPCTC, and finally reaching another conclusion such as parallel or perpendicular lines. In Example 1, suppose we had been asked to prove that $\overline{WZ}$ bisects $\overline{TV}$. Then Steps 1–6 of Example 1 would have remained as they are, and a seventh step of the proof would have read

7. $\overline{WZ}$ bisects $\overline{TV}$	7. If a line segment is divided into two $\cong$ parts, then it has been bisected

Reminder

CPCTC means "**C**orresponding **P**arts of **C**ongruent **T**riangles are **C**ongruent."

STRATEGY FOR PROOF ■ **Proofs That Involve Congruent Triangles**

In our study of triangles, we will establish three types of conclusions:

1. *Proving triangles congruent,* such as $\triangle TWZ \cong \triangle VWZ$
2. *Proving corresponding parts of congruent triangles are congruent,* such as $\overline{TZ} \cong \overline{VZ}$. (Note that two $\triangle$s have to be proved $\cong$ before CPCTC can be used.)
3. *Establishing a further relationship,* such as $\overline{WZ}$ bisects $\overline{TV}$. (Note that we must establish that two $\triangle$s are $\cong$ and also apply CPCTC before this goal can be reached.)

Little has been said about a "plan for proof," but every geometry student and teacher must have a plan before a proof can be completed. Though we generally do not write the "plan," we demonstrate the technique in Example 2.

EXAMPLE 2

Figure 3.16

GIVEN: $\overline{ZW} \cong \overline{YX}$
$\overline{ZY} \cong \overline{WX}$
(See Figure 3.16.)

PROVE: $\overline{ZY} \parallel \overline{WX}$

PLAN FOR PROOF: By showing that $\triangle ZWX \cong \triangle XYZ$, we can show that $\angle 1 \cong \angle 2$ by CPCTC. Then $\angle$s 1 and 2 are congruent alternate interior angles for $\overline{ZY}$ and $\overline{WX}$ with transversal $\overline{ZX}$; thus, $\overline{ZY}$ and $\overline{WX}$ must be parallel.

PROOF

Statements	Reasons
1. $\overline{ZW} \cong \overline{YX}; \overline{ZY} \cong \overline{WX}$	1. Given
2. $\overline{ZX} \cong \overline{ZX}$	2. Identity
3. $\triangle ZWX \cong \triangle XYZ$	3. SSS
4. $\angle 1 \cong \angle 2$	4. CPCTC
5. $\overline{ZY} \parallel \overline{WX}$	5. If two lines are cut by a transversal so that the alt. int. $\angle$s are $\cong$, these lines are $\parallel$

SSG EXS. 4–6

SUGGESTIONS FOR PROVING TRIANGLES CONGRUENT

Because many proofs depend upon establishing congruent triangles, we offer the following suggestions.

Figure 3.17

STRATEGY FOR PROOF ■ **Drawings Used to Prove Triangles Congruent**

Suggestions for a proof that involves congruent triangles:

1. Mark the figures systematically, using:
   a) A *square* in the opening of each right angle
   b) The same number of *dashes* on congruent sides
   c) The same number of *arcs* inside congruent angles
2. Trace the triangles to be proved congruent in different colors.
3. If the triangles overlap, draw them separately.

NOTE: In Figure 3.17, consider the like markings.

SSG EXS. 7–9

Figure 3.18

RIGHT TRIANGLES

In a right triangle, the side opposite the right angle is the **hypotenuse** of that triangle, and the sides of the right angle are the **legs** of the right triangle. These parts of a right triangle are illustrated in Figure 3.18.

Another method for proving triangles congruent is the HL method, which applies exclusively to right triangles. In HL, H refers to hypotenuse and L refers to leg. The proof of this method will be delayed until Section 5.4.

HL (METHOD FOR PROVING TRIANGLES CONGRUENT)

Figure 3.19

> **THEOREM 3.2.1**
>
> If the hypotenuse and a leg of one right triangle are congruent to the hypotenuse and a leg of a second right triangle, then the triangles are congruent (HL).

The relationships described in Theorem 3.2.1 (HL) are illustrated as marked in Figure 3.19. In Example 3, the construction based upon HL leads to a unique right triangle.

EXAMPLE 3

GIVEN: $\overline{AB}$ and $\overline{CA}$ in Figure 3.20(a); note that $AB > CA$.

CONSTRUCT: The right triangle with hypotenuse of length equal to AB and one leg of length equal to CA

SOLUTION Figure 3.20(b): Construct $\overleftrightarrow{CQ}$ perpendicular to $\overleftrightarrow{EF}$ at point C.

Figure 3.20(c): Now mark off the length of $\overline{CA}$ on $\overleftrightarrow{CQ}$. (See page 135.)

Figure 3.20

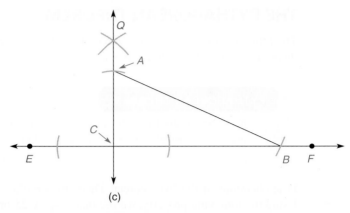

(c)

Figure 3.20

Figure 3.20(c): Finally, with point A as center, mark off an arc with its length equal to that of $\overline{AB}$ as shown. $\triangle ABC$ is the desired right $\triangle$.

We apply the HL Theorem in Example 4.

EXAMPLE 4

GIVEN: $\overline{AB} \perp \overline{CD}$; $\overline{AC} \cong \overline{AD}$

PROVE: $\triangle ABC \cong \triangle ABD$ and $\overline{CB} \cong \overline{DB}$

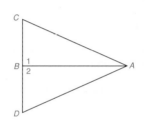

PROOF	
Statements	**Reasons**
1. $\overline{AB} \perp \overline{CD}$	1. Given
2. $\angle 1$ and $\angle 2$ are right angles	2. If two lines are perpendicular, they form right angles
3. $\overline{AC} \cong \overline{AD}$	3. Given
4. $\overline{AB} \cong \overline{AB}$	4. Identity
5. $\triangle ABC \cong \triangle ABD$	5. HL
6. $\overline{CB} \cong \overline{DB}$	6. CPCTC

In Example 5, we emphasize the expanded list of methods for proving triangles congruent. The list includes SSS, SAS, ASA, AAS, and HL.

EXAMPLE 5

Cite the reason why the right triangles $\triangle ABC$ and $\triangle ECD$ in Figure 3.21 are congruent if:

a) $\overline{AB} \cong \overline{EC}$ and $\overline{AC} \cong \overline{ED}$
b) $\angle A \cong \angle E$ and C is the midpoint of $\overline{BD}$
c) $\overline{BC} \cong \overline{CD}$ and $\angle 1 \cong \angle 2$
d) $\overline{AB} \cong \overline{EC}$ and $\overline{EC}$ bisects $\overline{BD}$

SOLUTION

a) HL   b) AAS   c) ASA   d) SAS

Geometry in the Real World

In the manufacturing process, the parts of many machines must be congruent. The two sides of the hinge shown are congruent.

A   E

B   C   D

Figure 3.21

SSG **EXS. 10–11**

THE PYTHAGOREAN THEOREM

The following theorem can be applied only when a triangle is a right triangle. Proof of the theorem is delayed until Section 5.4.

PYTHAGOREAN THEOREM

The square of the length (c) of the hypotenuse of a right triangle equals the sum of squares of the lengths (a and b) of the legs of the right triangle; that is, $c^2 = a^2 + b^2$.

In applications of the Pythagorean Theorem, we often find statements such as $c^2 = 25$. Using the following property, we see that $c = \sqrt{25}$ or $c = 5$.

SQUARE ROOTS PROPERTY

Let x represent the length of a line segment, and let p represent a positive number. If $x^2 = p$, then $x = \sqrt{p}$.

The *square root of p*, symbolized $\sqrt{p}$, represents the number that when multiplied times itself equals p. As we indicated earlier, $\sqrt{25} = 5$ because $5\sqrt{5} = 25$. When a square root is not exact, a calculator can be used to find its approximate value, where the symbol $\approx$ means "is equal to approximately," $\sqrt{22} \approx 4.69$ because $4.69\sqrt{4.69} = 21.9961 \approx 22$.

EXAMPLE 6

Find the length of the third side of the right triangle. See the figure below.

a) Find c if $a = 6$ and $b = 8$.
b) Find b if $a = 7$ and $c = 10$.

SOLUTION

a) $c^2 = a^2 + b^2$, so $c^2 = 6^2 + 8^2$
or $c^2 = 36 + 64 = 100$.
Then $c = \sqrt{100} = 10$.

b) $c^2 = a^2 + b^2$, so $10^2 = 7^2 + b^2$
or $100 = 49 + b^2$. Subtracting yields
$b^2 = 51$, so $b = \sqrt{51} \approx 7.14$.

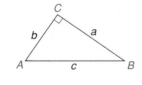

SSG **EXS. 12–14**

Discover

On the pegboard shown, each vertical (and horizontal) space between consecutive pegs measures one unit. Apply the Pythagorean Theorem to find the perimeter (sum of the lengths of the sides) of each polygon.

(a)   (b)

ANSWER
(b) $8 + 4\sqrt{2}$ units
(a) $6 + 4\sqrt{2} + \sqrt{5} + \sqrt{17}$ units

Exercises 3.2

In Exercises 1 to 4, state the reason (SSS, SAS, ASA, AAS, or HL) why the triangles are congruent.

1. *Given:* $\angle 1 \cong \angle 2$
$\angle CAB \cong \angle DAB$
Prove: $\triangle CAB \cong \triangle DAB$

2. *Given:* $\angle CAB \cong \angle DAB$
$\overline{AC} \cong \overline{AD}$
Prove: $\triangle CAB \cong \triangle DAB$

Exercise 1, 2

3. *Given:* $\angle M$ and $\angle R$
are right angles
$\overline{MN} \cong \overline{QR}$
$\overline{MP} \cong \overline{RP}$
Prove: $\triangle MNP \cong \triangle QRP$

4. *Given:* P is the midpoint of
$\overline{MR}$ and $\angle N \cong \angle Q$
Prove: $\triangle MNP \cong \triangle QRP$

Exercise 3, 4

In Exercises 5 to 12, plan and write the two-column proof for each problem.

5. *Given:* ∠1 and ∠2 are right ∠s
           $\overline{CA} \cong \overline{DA}$
       Prove: △ABC ≅ △ABD

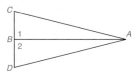

Exercises 5, 6

6. *Given:* ∠1 and ∠2 are right ∠s
           $\overrightarrow{AB}$ bisects ∠CAD
       Prove: △ABC ≅ △ABD

7. *Given:* P is the midpoint
           of both $\overline{MR}$ and $\overline{NQ}$
       Prove: △MNP ≅ △RQP

8. *Given:* $\overline{MN} \parallel \overline{QR}$
           $\overline{MN} \cong \overline{QR}$
       Prove: △MNP ≅ △RQP

Exercises 7, 8

9. *Given:* ∠R and ∠V are
           right ∠s
           ∠1 ≅ ∠2
       Prove: △RST ≅ △VST

10. *Given:* ∠1 ≅ ∠2
            ∠3 ≅ ∠4
        Prove: △RST ≅ △VST

11. *Given:* $\overline{SR} \cong \overline{SV}$
            $\overline{RT} \cong \overline{VT}$
        Prove: △RST ≅ △VST

Exercises 9–12

12. *Given:* ∠R and ∠V are right ∠s
            $\overline{RT} \cong \overline{VT}$
        Prove: △RST ≅ △VST

In Exercises 13 and 14, complete each proof.

13. *Given:* $\overline{HJ} \perp \overline{KL}$ and $\overline{HK} \cong \overline{HL}$ (See figure below.)
        Prove: $\overline{KJ} \cong \overline{JL}$

PROOF

Statements	Reasons
1. $\overline{HJ} \perp \overline{KL}$ and $\overline{HK} \cong \overline{HL}$	1. ?
2. ∠s HJK and HJL are rt. ∠s	2. ?
3. $\overline{HJ} \cong \overline{HJ}$	3. ?
4. ?	4. HL
5. ?	5. CPCTC

14. In Exercise 13, you can add a Step 6 to prove that "J is the midpoint of $\overline{KL}$." What reason would you use to establish this conclusion?

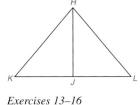

Exercises 13–16

15. *Given:* $\overrightarrow{HJ}$ bisects ∠KHL (See figure for exercise 13.)
            $\overline{HJ} \perp \overline{KL}$
        Prove: ∠K ≅ ∠L

PROOF

Statements	Reasons
1. ?	1. Given
2. ∠JHK ≅ ∠JHL	2. ?
3. $\overline{HJ} \perp \overline{KL}$	3. ?
4. ∠HJK ≅ ∠HJL	4. ?
5. ?	5. Identity
6. ?	6. ASA
7. ∠K ≅ ∠L	7. ?

16. In Exercise 15, you can delete Steps 5 and 6 and still prove that "∠K ≅ ∠L." What reason would you use to establish that ∠K ≅ ∠L in the shorter proof?

In Exercises 17 to 20, first prove that triangles are congruent, and then use CPCTC.

17. *Given:* ∠P and ∠R are right ∠s
            M is the midpoint of $\overline{PR}$
        Prove: ∠N ≅ ∠Q

Exercises 17, 18

18. *Given:* M is the midpoint of $\overline{NQ}$
            $\overline{NP} \parallel \overline{RQ}$ with transversals $\overline{PR}$ and $\overline{NQ}$
        Prove: $\overline{NP} \cong \overline{QR}$

19. *Given:* ∠1 and ∠2 are right ∠s
            H is the midpoint of $\overline{FK}$
            $\overline{FG} \parallel \overline{HJ}$
        Prove: $\overline{FG} \cong \overline{HJ}$

20. *Given:* $\overline{DE} \perp \overline{EF}$ and $\overline{CB} \perp \overline{AB}$
            $\overline{AB} \parallel \overline{FE}$
            $\overline{AC} \cong \overline{FD}$
        Prove: $\overline{EF} \cong \overline{BA}$

In Exercises 21 to 26, △ABC is a right triangle. Use the given information to find the length of the third side of the triangle.

21. $a = 4$ and $b = 3$

22. $a = 12$ and $b = 5$

23. $a = 15$ and $c = 17$

24. $b = 6$ and $c = 10$

25. $a = 5$ and $b = 4$

26. $a = 7$ and $c = 8$

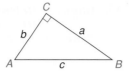

In Exercises 27 to 29, prove the indicated relationship.

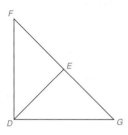

Exercises 27–29

27. *Given:* $\overline{DF} \cong \overline{DG}$ and $\overline{FE} \cong \overline{EG}$
Prove: $\overrightarrow{DE}$ bisects $\angle FDG$

28. *Given:* $\overrightarrow{DE}$ bisects $\angle FDG$
$\angle F \cong \angle G$
Prove: E is the midpoint of $\overline{FG}$

29. *Given:* E is the midpoint of $\overline{FG}$
$\overline{DF} \cong \overline{DG}$
Prove: $\overline{DE} \perp \overline{FG}$

In Exercises 30 to 32, draw the triangles that are to be shown congruent separately. Then complete the proof.

Exercises 30–32

30. *Given:* $\angle MQP$ and $\angle NPQ$ are rt. $\angle$s
$\overline{MQ} \cong \overline{NP}$
Prove: $\overline{MP} \cong \overline{NQ}$

(*HINT: Show* $\triangle MQP \cong \triangle NPQ$.)

31. *Given:* $\angle 1 \cong \angle 2$ and $\overline{MN} \cong \overline{QP}$
Prove: $\overline{MQ} \parallel \overline{NP}$

(*HINT: Show* $\triangle NMP \cong \triangle QPM$.)

32. *Given:* $\overline{MN} \parallel \overline{QP}$ and $\overline{MQ} \parallel \overline{NP}$
Prove: $\overline{MQ} \cong \overline{NP}$

(*HINT: Show* $\triangle MQP \cong \triangle PNM$.)

33. *Given:* $\overrightarrow{RW}$ bisects $\angle SRU$
$\overline{RS} \cong \overline{RU}$
Prove: $\triangle TRU \cong \triangle VRS$

(*HINT: First show that* $\triangle RSW \cong \triangle RUW$.)

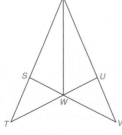

Exercise 33

34. *Given:* $\overline{DB} \perp \overline{BC}$ and $\overline{CE} \perp \overline{DE}$
$\overline{AB} \cong \overline{AE}$
Prove: $\triangle BDC \cong \triangle ECD$

(*HINT: First show that* $\triangle ACE \cong \triangle ADB$.)

35. In the roof truss shown, $AB = 8$ and m$\angle HAF = 37°$. Find:
a) AH b) m$\angle BAD$ c) m$\angle ADB$

(*HINT: The design of the roof truss displays line symmetry.*)

36. In the support system of the bridge shown, $AC = 6$ ft and m$\angle ABC = 28°$. Find:
a) m$\angle RST$ b) m$\angle ABD$ c) BS

(*HINT: The smaller triangles shown in the figure are all congruent to each other.*)

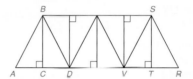

37. As a car moves along the roadway in a mountain pass, it passes through a horizontal run of 750 feet and through a vertical rise of 45 feet. To the nearest foot, how far does the car move along the roadway?

38. Because of construction along the road from A to B, Alinna drives 5 miles from A to C and then 12 miles from C to B. How much farther did Alinna travel by using the alternative route from A to B?

39. *Given:* Regular pentagon $ABCDE$ with diagonals $\overline{BE}$ and $\overline{BD}$
Prove: $\overline{BE} \cong \overline{BD}$

(*HINT: First prove* $\triangle ABE \cong \triangle CBD$.)

40. In the figure with regular pentagon $ABCDE$, do $\overrightarrow{BE}$ and $\overrightarrow{BD}$ trisect $\angle ABC$?

(*HINT: m$\angle ABE$ = m$\angle AEB$.*)

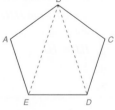

Exercises 39, 40

<table>
<tr><td rowspan="5" style="vertical-align:top">**3.3**</td><td colspan="4">**Isosceles Triangles**</td></tr>
</table>

KEY CONCEPTS	Isosceles Triangle
	Vertex, Legs, and Base
	  of an Isosceles
	  Triangle
	Base Angles
	Vertex Angle
		Angle Bisector
		Median
		Altitude
		Perpendicular Bisector
		Auxiliary Line
			Determined,
			  Undetermined,
			  Overdetermined
			Equilateral and
			  Equiangular Triangles
			Perimeter

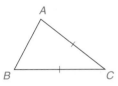

Figure 3.22

In an isosceles triangle, the two sides of equal length are **legs,** and the third side is the **base.** See Figure 3.22. The point at which the two legs meet is the **vertex** of the triangle, and the angle formed by the legs (and opposite the base) is the **vertex angle.** The two remaining angles are **base angles.**

If $\overline{AC} \cong \overline{BC}$ in Figure 3.23, then $\triangle ABC$ is isosceles with legs $\overline{AC}$ and $\overline{BC}$, base $\overline{AB}$, vertex C, vertex angle C, and base angles at A and B. With $\overline{AC} \cong \overline{BC}$, we see that the base $\overline{AB}$ of this isosceles triangle is not necessarily the "bottom" side.

Helping lines known as **auxiliary lines** are needed to prove many theorems. To this end, we consider some of the lines (line segments) that may prove helpful. Each angle of a triangle has a unique **angle bisector;** this may be indicated by a ray or segment from the vertex of the bisected angle. See Figure 3.24(a). Just as an angle bisector begins at the vertex of an angle, the **median** also joins a vertex to the midpoint of the opposite side. See Figure 3.24(b). Generally, the median from a vertex of a triangle is not the same as the angle bisector from that vertex. An **altitude** is a line segment drawn from a vertex to the opposite side so that it is perpendicular to the opposite side. See Figure 3.24(c). Finally, the **perpendicular bisector** of a side of a triangle is shown as a line in Figure 3.24(d). A segment or ray could also perpendicularly bisect a side of the triangle.

Figure 3.23

$\angle 1 \cong \angle 2$, so $\overrightarrow{AD}$ is the angle bisector of $\angle BAC$ in $\triangle ABC$

(a)

M is the midpoint of $\overline{BC}$, so $\overline{AM}$ is the median from A to $\overline{BC}$

(b)

$\overline{AE} \perp \overline{BC}$, so $\overline{AE}$ is the altitude of $\triangle ABC$ from vertex A to $\overline{BC}$

(c)

M is the midpoint of $\overline{BC}$ and $\overleftrightarrow{FM} \perp \overline{BC}$, so $\overleftrightarrow{FM}$ is the perpendicular bisector of side $\overline{BC}$ in $\triangle ABC$

(d)

Figure 3.24

In Figure 3.25, $\overrightarrow{AD}$ is the bisector of $\angle BAC$; $\overline{AE}$ is the altitude from A to $\overline{BC}$; M is the midpoint of $\overline{BC}$; $\overline{AM}$ is the median from A to $\overline{BC}$; and $\overleftrightarrow{FM}$ is the perpendicular bisector of $\overline{BC}$.

An altitude can actually lie in the exterior of a triangle. In obtuse $\triangle RST$ of Figure 3.26 on page 140, the altitude from R must be drawn to an extension of side $\overline{ST}$. Later we will use the length h of the altitude $\overline{RH}$ and the length b of side $\overline{ST}$ in the following formula for the area of a triangle:

$$A = \frac{1}{2}bh$$

Any angle bisector and any median necessarily lie in the interior of the triangle.

Figure 3.25

Figure 3.26

Figure 3.27

Each triangle has three altitudes—one from each vertex. As shown for △ABC in Figure 3.27, the three altitudes seem to meet at a common point.

We now consider the proof of a statement that involves the corresponding altitudes of congruent triangles; corresponding altitudes are those drawn to corresponding sides of the congruent triangles.

> ## THEOREM 3.3.1
>
> Corresponding altitudes of congruent triangles are congruent.

GIVEN: △ABC ≅ △RST
        Altitudes $\overline{CD}$ to $\overline{AB}$ and $\overline{TV}$ to $\overline{RS}$

PROVE: $\overline{CD} \cong \overline{TV}$

PROOF	
Statements	**Reasons**
1. △ABC ≅ △RST Altitudes $\overline{CD}$ to $\overline{AB}$ and $\overline{TV}$ to $\overline{RS}$	1. Given
2. $\overline{CD} \perp \overline{AB}$ and $\overline{TV} \perp \overline{RS}$	2. An altitude of a △ is the line segment from one vertex drawn ⊥ to the opposite side
3. ∠CDA and ∠TVR are right ∠s	3. If two lines are ⊥, they form right ∠s
4. ∠CDA ≅ ∠TVR	4. All right angles are ≅
5. $\overline{AC} \cong \overline{RT}$ and ∠A ≅ ∠R	5. CPCTC (from △ABC ≅ △RST)
6. △CDA ≅ △TVR	6. AAS
7. $\overline{CD} \cong \overline{TV}$	7. CPCTC

Each triangle has three medians—one from each vertex to the midpoint of the opposite side. As the medians are drawn for △DEF in Figure 3.28(a), it appears that the three medians intersect at a point.

Each triangle has three angle bisectors—one for each of the three angles. As these are shown for △MNP in Figure 3.28(b), it appears that the three angle bisectors have a point in common.

Each triangle has three perpendicular bisectors for its sides; these are shown for △RST in Figure 3.28(c). Like the altitudes, medians, and angle bisectors, the perpendicular bisectors of the sides also appear to meet at a single point.

(a)

(b)

(c)

Figure 3.28

The angle bisectors (like the medians) of a triangle *always* meet in the interior of the triangle. However, the altitudes (like the perpendicular bisectors of the sides) can meet in the exterior of the triangle; see Figure 3.28(c). The points of intersection described in this

SSG **EXS. 1–6**

paragraph and the preceding three paragraphs will be given greater attention in Chapter 7.

The Discover activity at the left opens the doors to further discoveries.

In Figure 3.29, the bisector of the vertex angle of isosceles △ABC is a line (segment) of symmetry for △ABC.

Discover

Using a sheet of construction paper, cut out an isosceles triangle. Now use your compass to bisect the vertex angle. Fold along the angle bisector to form two smaller triangles. How are the smaller triangles related?

ANSWER
They are congruent.

EXAMPLE 1

Give a formal proof of Theorem 3.3.2.

THEOREM 3.3.2

The bisector of the vertex angle of an isosceles triangle separates the triangle into two congruent triangles.

Figure 3.29

GIVEN: Isosceles △ABC, with $\overline{AB} \cong \overline{BC}$
$\overrightarrow{BD}$ bisects ∠ABC
(See Figure 3.29.)

PROVE: $\triangle ABD \cong \triangle CBD$

PROOF	
Statements	**Reasons**
1. Isosceles △ABC with $\overline{AB} \cong \overline{BC}$	1. Given
2. $\overrightarrow{BD}$ bisects ∠ABC	2. Given
3. ∠1 ≅ ∠2	3. The bisector of an ∠ separates it into two ≅ ∠s
4. $\overline{BD} \cong \overline{BD}$	4. Identity
5. $\triangle ABD \cong \triangle CBD$	5. SAS

(a)

(b)

(c)

Figure 3.30

Recall from Section 2.4 that an auxiliary figure must be determined. Consider Figure 3.30 and the following three descriptions, which are coded **D** for *determined*, **U** for *underdetermined*, and **O** for *overdetermined*:

D: Draw a line segment from A perpendicular to $\overline{BC}$ so that the terminal point is on $\overline{BC}$. [*Determined* because the line from A perpendicular to $\overline{BC}$ is unique; see Figure 3.30(a).]

U: Draw a line segment from A to $\overline{BC}$ so that the terminal point is on $\overline{BC}$. [*Underdetermined* because many line segments are possible; see Figure 3.30(b).]

O: Draw a line segment from A perpendicular to $\overline{BC}$ so that it bisects $\overline{BC}$. [*Overdetermined* because the line segment from A drawn perpendicular to $\overline{BC}$ will not contain the midpoint M of $\overline{BC}$; see Figure 3.30(c).]

For Example 2, an auxiliary segment will be needed. As you study the following proof, note the uniqueness of the segment and its justification (reason 2) in the proof.

STRATEGY FOR PROOF ■ Using an Auxillary Line

General Rule: An early *statement* of the proof establishes the "helping line," such as *the* altitude or *the* angle bisector.

Illustration: See the second line in the proof of Example 2. The chosen angle bisector leads to congruent triangles, which enable us to complete the proof.

EXAMPLE 2

Give a formal proof of Theorem 3.3.3.

THEOREM 3.3.3

If two sides of a triangle are congruent, then the angles opposite these sides are also congruent.

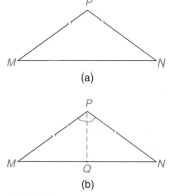

(a)

(b)

Figure 3.31

GIVEN: Isosceles $\triangle MNP$
with $\overline{MP} \cong \overline{NP}$
[See Figure 3.31(a).]

PROVE: $\angle M \cong \angle N$

NOTE: Figure 3.31(b) shows the auxiliary segment, the bisector of $\angle MPN$.

PROOF

Statements	Reasons
1. Isosceles $\triangle MNP$ with $\overline{MP} \cong \overline{NP}$	1. Given
2. Draw $\angle$ bisector $\overrightarrow{PQ}$ from P to $\overline{MN}$	2. Every angle has one and only one bisector
3. $\triangle MPQ \cong \triangle NPQ$	3. The bisector of the vertex angle of an isosceles $\triangle$ separates it into two $\cong$ $\triangle$s
4. $\angle M \cong \angle N$	4. CPCTC

Discover

Using construction paper and scissors, cut out an isosceles triangle MNP with $\overline{MP} \cong \overline{PN}$. Fold it so that $\angle M$ coincides with $\angle N$. What can you conclude?

ANSWER
$N\angle \cong W\angle$

For the proof of Theorem 3.3.3, a different angle bisector (such as the bisector of $\angle M$) would not lead to congruent triangles; that is, the choice of auxiliary line must lead to the desired outcome!

Theorem 3.3.3 is sometimes stated, "The base angles of an isosceles triangle are congruent." We apply this theorem in Example 3.

EXAMPLE 3

Find the size of each angle of the isosceles triangle shown in Figure 3.32 on page 143 if:

a) $m\angle 1 = 36°$
b) The measure of each base angle is $5°$ less than twice the measure of the vertex angle

SOLUTION

a) $m\angle 1 + m\angle 2 + m\angle 3 = 180°$. Since $m\angle 1 = 36°$ and $\angle 2$ and $\angle 3$ are $\cong$, we have

$$36 + 2(m\angle 2) = 180$$
$$2(m\angle 2) = 144$$
$$m\angle 2 = 72$$

Thus, $m\angle 1 = 36°$ and $m\angle 2 = m\angle 3 = 72°$.

Figure 3.33

b) Let the vertex angle measure be given by x. Then the size of each base angle is $2x - 5$. Because the sum of the measures is 180°,

$$x + (2x - 5) + (2x - 5) = 180$$
$$5x - 10 = 180$$
$$5x = 190$$
$$x = 38$$
$$2x - 5 = 2(38) - 5 = 76 - 5 = 71$$

Therefore, $\text{m}\angle 1 = 38°$ and $\text{m}\angle 2 = \text{m}\angle 3 = 71°$.

Figure 3.32

In some instances, a carpenter may want to get a quick, accurate measurement without having to go get his or her tools. Suppose that the carpenter's square shown in Figure 3.33 is handy but that a miter box is not nearby. If two marks are made at lengths of 4 inches from the corner of the square and these are then joined, what size angle is determined? You should see that each angle indicated by an arc measures 45°.

Example 4 shows us that the converse of the theorem "The base angles of an isosceles triangle are congruent" is also true. However, see the accompanying Warning.

> **W**arning
>
> The converse of an "If, then" statement is not always true.

EXAMPLE 4

Study the picture proof of Theorem 3.3.4.

> **THEOREM 3.3.4**
>
> If two angles of a triangle are congruent, then the sides opposite these angles are also congruent.

Figure 3.34

SSG **EXS. 7–17**

PICTURE PROOF OF THEOREM 3.3.4

GIVEN: $\triangle TUV$ with $\angle T \cong \angle U$ [See Figure 3.34(a).]

PROVE: $\overline{VU} \cong \overline{VT}$

PROOF: Drawing $\overline{VP} \perp \overline{TU}$ [see Figure 3.34(b)], we see that $\triangle VPT \cong \triangle VPU$ by AAS. Now $\overline{VU} \cong \overline{VT}$ by CPCTC.

A consequence of Theorem 3.3.4 is that "A triangle with two congruent angles must be an isosceles triangle."

When all three sides of a triangle are congruent, the triangle is **equilateral.** If all three angles are congruent, then the triangle is **equiangular.** Theorems 3.3.3 and 3.3.4 can be used to prove that the sets {equilateral triangles} and {equiangular triangles} are equivalent.

> **COROLLARY 3.3.5**
>
> An equilateral triangle is also equiangular.

> **COROLLARY 3.3.6**
>
> An equiangular triangle is also equilateral.

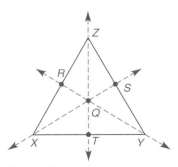

Figure 3.35

An equilateral (or equiangular) triangle such as $\triangle XYZ$ has line symmetry with respect to each of the three axes shown in Figure 3.35. In the figure, R, S, and T are the midpoints of the sides. Because it can be shown that $QZ > QT$, $\triangle XYZ$ does not have point symmetry with respect to point Q (or any other point).

Figure 3.36

> **DEFINITION**
>
> The **perimeter** of a triangle is the sum of the lengths of its sides. Thus, if a, b, and c are the lengths of the three sides, then the perimeter P is given by $P = a + b + c$. (See Figure 3.36.)

EXAMPLE 5

GIVEN: $\angle B \cong \angle C$
  $AB = 5.3$ and $BC = 3.6$

FIND: The perimeter of $\triangle ABC$

SOLUTION If $\angle B \cong \angle C$, then $AC = AB = 5.3$. Therefore,

$$P = a + b + c$$
$$P = 3.6 + 5.3 + 5.3$$
$$P = 14.2$$

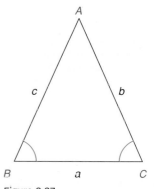

Figure 3.37

SSG **EXS. 18–22**

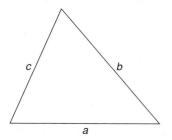

Geometry in the Real World

Braces that create triangles are used to provide stability for a bookcase. The triangle is called a rigid figure.

EXAMPLE 6

The perimeter of $\triangle ABC$ (in Figure 3.37) is 47. If $AB = x$, $AC = x + 1$, and $BC = x - 5$, find x, AB, AC, and BC.

SOLUTION

$$P = AB + AC + BC$$
$$47 = x + (x + 1) + (x - 5)$$
$$47 = 3x - 4$$
$$3x = 51$$
$$x = 17$$

Thus, $AB = 17$, $AC = 18$, and $BC = 12$.

Many of the properties of triangles that were investigated in earlier sections of this chapter are summarized in Table 3.1 on page 145.

TABLE 3.1
Selected Properties of Triangles

	Scalene	Isosceles	Equilateral (equiangular)	Acute	Right	Obtuse
Sides	No two are ≅	Exactly two are ≅	All three are ≅	Possibly two or three ≅ sides	Possibly two ≅ sides; $c^2 = a^2 + b^2$	Possibly two ≅ sides
Angles	Sum of ∠s is 180°	Sum of ∠s is 180°; two ∠s ≅	Sum of ∠s is 180°; three ≅ 60° ∠s	All ∠s acute; sum of ∠s is 180°; possibly two or three ≅ ∠s	One right ∠; sum of ∠s is 180°; possibly two ≅ 45° ∠s; acute ∠s are complementary	One obtuse ∠; sum of ∠s is 180°; possibly two ≅ acute ∠s

Exercises 3.3

For Exercises 1 to 8, use the accompanying drawing.

1. If $\overline{VU} \cong \overline{VT}$, what type of triangle is △*VTU*?

2. If $\overline{VU} \cong \overline{VT}$, which angles of △*VTU* are congruent?

3. If ∠*T* ≅ ∠*U*, which sides of △*VTU* are congruent?

4. If $\overline{VU} \cong \overline{VT}$, *VU* = 10, and *TU* = 8, what is the perimeter of △*VTU*?

5. If $\overline{VU} \cong \overline{VT}$ and m∠*T* = 69°, find m∠*U*.

6. If $\overline{VU} \cong \overline{VT}$ and m∠*T* = 69°, find m∠*V*.

7. If $\overline{VU} \cong \overline{VT}$ and m∠*T* = 72°, find m∠*V*.

8. If $\overline{VU} \cong \overline{VT}$ and m∠*V* = 40°, find m∠*T*.

Exercises 1–8

In Exercises 9 to 12, determine whether the sets have a subset relationship. Are the two sets disjoint or equivalent? Do the sets intersect?

9. *L* = {equilateral triangles}; *E* = {equiangular triangles}

10. *S* = {triangles with two ≅ sides}; *A* = {triangles with two ≅ ∠s}

11. *R* = {right triangles}; *O* = {obtuse triangles}

12. *I* = {isosceles triangles}; *R* = {right triangles}

In Exercises 13 to 18, describe the line segment as determined, underdetermined, or overdetermined. Use the accompanying drawing for reference.

Exercises 13–18

13. Draw a line segment through point *A*.

14. Draw a line segment with endpoints *A* and *B*.

15. Draw a line segment $\overline{AB}$ parallel to line *m*.

16. Draw a line segment $\overline{AB}$ perpendicular to *m*.

17. Draw a line from *A* perpendicular to *m*.

18. Draw $\overline{AB}$ so that line *m* bisects $\overline{AB}$.

19. Given that △*ABC* is isosceles with $\overline{AB} \cong \overline{AC}$, give the reason why △*ABD* ≅ △*ACD* if *D* lies on $\overline{BC}$ and:
    (a) $\overline{AD}$ is an altitude of △*ABC*.
    (b) $\overline{AD}$ is a median of △*ABC*.
    (c) $\overline{AD}$ is the bisector of ∠*BAC*.

20. Is it possible for a triangle to be:
    (a) an acute isosceles triangle?
    (b) an obtuse isosceles triangle?
    (c) an equiangular isosceles triangle?

21. A surveyor knows that a lot has the shape of an isosceles triangle. If the vertex angle measures 70° and each equal side is 160 ft long, what measure does each of the base angles have?

22. In concave quadrilateral $ABCD$, the angle at A measures 40°. $\triangle ABD$ is isosceles, $\overrightarrow{BC}$ bisects $\angle ABD$, and $\overrightarrow{DC}$ bisects $\angle ADB$. What are the measures of $\angle ABC$, $\angle ADC$, and $\angle 1$?

In Exercises 23 to 28, use arithmetic or algebra as needed to find the measures indicated. Note the use of dashes on equal sides of the given isosceles triangles.

23. Find $m\angle 1$ and $m\angle 2$ if $m\angle 3 = 68°$.

24. If $m\angle 3 = 68°$, find $m\angle 4$, the angle formed by the bisectors of $\angle 3$ and $\angle 2$.

25. Find the measure of $\angle 5$, which is formed by the bisectors of $\angle 1$ and $\angle 3$. Again let $m\angle 3 = 68°$.

26. Find an expression for the measure of $\angle 5$ if $m\angle 3 = 2x$ and the line segments shown bisect the angles of the isosceles triangle.

Exercises 24–26

27. In isosceles $\triangle ABC$ with vertex A (not shown), each base angle is 12° larger than the vertex angle. Find the measure of each angle.

28. In isosceles $\triangle ABC$ (not shown), vertex angle A is 5° more than one-half of base angle B. Find the size of each angle of the triangle.

In Exercises 29 to 32, suppose that $\overline{BC}$ is the base of isosceles $\triangle ABC$ (not shown).

29. Find the perimeter of $\triangle ABC$ if $AB = 8$ and $BC = 10$.

30. Find AB if the perimeter of $\triangle ABC$ is 36.4 and $BC = 14.6$.

31. Find x if the perimeter of $\triangle ABC$ is 40, $AB = x$, and $BC = x + 4$.

32. Find x if the perimeter of $\triangle ABC$ is 68, $AB = x$, and $BC = 1.4x$.

33. Suppose that $\triangle ABC \cong \triangle DEF$. Also, $\overline{AX}$ bisects $\angle CAB$ and $\overline{DY}$ bisects $\angle FDE$. Are these corresponding angle bisectors of congruent triangles congruent?

Exercises 33, 34

34. Suppose that $\triangle ABC \cong \triangle DEF$, $\overline{AX}$ is the median from A to $\overline{BC}$, and $\overline{DY}$ is the median from D to $\overline{EF}$. Are these corresponding medians of congruent triangles congruent?

In Exercises 35 and 36, complete each proof using the drawing below.

35. *Given:* $\angle 3 \cong \angle 1$
     Prove: $\overline{AB} \cong \overline{AC}$

Exercises 35, 36

PROOF

Statements	Reasons
1. $\angle 3 \cong \angle 1$	1. ?
2. ?	2. If two lines intersect, the vertical $\angle$s formed are $\cong$
3. ?	3. Transitive Property of Congruence
4. $\overline{AB} \cong \overline{AC}$	4. ?

36. *Given:* $\overline{AB} \cong \overline{AC}$
     Prove: $\angle 6 \cong \angle 7$

PROOF

Statements	Reasons
1. ?	1. Given
2. $\angle 2 \cong \angle 1$	2. ?
3. $\angle 2$ and $\angle 6$ are supplementary; $\angle 1$ and $\angle 7$ are supplementary	3. ?
4. ?	4. If two $\angle$s are supplementary to $\cong$ $\angle$s, they are $\cong$ to each other

In Exercises 37 to 39, complete each proof.

37. *Given:* $\angle 1 \cong \angle 3$
        $\overline{RU} \cong \overline{VU}$
     Prove: $\triangle STU$ is isosceles
     (HINT: First show that $\triangle RUS \cong \triangle VUT$.)

38. *Given:* $\overline{WY} \cong \overline{WZ}$
     M is the midpoint
     of $\overline{YZ}$
     $\overline{MX} \perp \overline{WY}$
     $\overline{MT} \perp \overline{WZ}$
 Prove: $\overline{MX} \cong \overline{MT}$

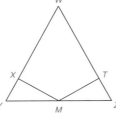

39. *Given:* Isosceles △*MNP*
     with vertex *P*
     Isosceles △*MNQ*
     with vertex *Q*
 Prove: △*MQP* ≅ △*NQP*

40. In isosceles triangle *BAT*,
 $\overline{AB} \cong \overline{AT}$. Also, $\overline{BR} \cong \overline{BT} \cong \overline{AR}$.
 If *AB* = 12.3 and *AR* = 7.6, find
 the perimeter of:
 a) △*BAT*
 b) △*ARB*
 c) △*RBT*

41. In △*BAT*, $\overline{BR} \cong \overline{BT} \cong \overline{AR}$,
 and m∠*RBT* = 20°. Find:
 a) m∠*T*
 b) m∠*ARB*
 c) m∠*A*

Exercises 40, 41

42. In △*PMN*, $\overline{PM} \cong \overline{PN}$, $\overrightarrow{MB}$ bisects ∠*PMN*, and $\overrightarrow{NA}$ bisects
 ∠*PNM*. If m∠*P* = 36°, name all isosceles triangles shown
 in the drawing.

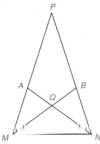

43. △*ABC* lies in the structural support system of the Ferris
 wheel. If m∠*A* = 30° and *AB* = *AC* = 20 ft, find the
 measures of ∠*B* and ∠*C*.

In Exercises 44 to 46, explain why each statement is true.

44. The altitude from the vertex of an isosceles triangle is also
 the median to the base of the triangle.

45. The bisector of the vertex angle of an isosceles triangle
 bisects the base.

46. The bisectors of the base angles of an isosceles triangle,
 together with the base, form an isosceles triangle.

*47. *Given:* In the figure, $\overline{XZ} \cong \overline{YZ}$ and *Z* is the midpoint
     of $\overline{XW}$.

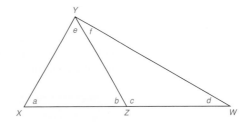

 Prove: △*XYW* is a right triangle with m∠*XYW* = 90°.
 (HINT: Let m∠*X* = *a*.)

*48. *Given:* In the figure, *a* = *e* = 66°. Also, $\overline{YZ} \cong \overline{ZW}$.
     If *YW* = 14.3 in. and *YZ* = 7.8 in., find the
     perimeter of △*XYW* to the nearest tenth of
     an inch.

3.4	**Basic Constructions Justified**

KEY CONCEPTS	Justifying Constructions

In earlier sections, the construction methods that were introduced were presented
intuitively. In this section, we justify these construction methods and apply them in further
constructions. The justification of the method is a "proof" that demonstrates that the
construction accomplished its purpose. See Example 1.

EXAMPLE 1

Justify the method for constructing an angle congruent to a given angle.

GIVEN: $\angle ABC$
$\overline{BD} \cong \overline{BE} \cong \overline{ST} \cong \overline{SR}$ (by construction)
$\overline{DE} \cong \overline{TR}$ (by construction)

PROVE: $\angle B \cong \angle S$

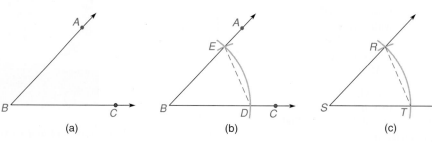

Figure 3.38

PROOF	
Statements	**Reasons**
1. $\angle ABC$; $\overline{BD} \cong \overline{BE} \cong \overline{ST} \cong \overline{SR}$	1. Given
2. $\overline{DE} \cong \overline{TR}$	2. Given
3. $\triangle EBD \cong \triangle RST$	3. SSS
4. $\angle B \cong \angle S$	4. CPCTC

In Example 2, we will apply the construction method that was justified in Example 1. Our goal is to construct an isosceles triangle that contains an obtuse angle. It is necessary that the congruent sides include the obtuse angle.

(a)

(b)

Figure 3.39

EXAMPLE 2

Construct an isosceles triangle in which obtuse $\angle A$ is included by two sides of length a [see Figure 3.39(a)].

SOLUTION Construct an angle congruent to $\angle A$. From A, mark off arcs of length a at points B and C as shown in Figure 3.39(b). Join B to C to complete $\triangle ABC$.

SSG **EXS. 1–2**

In Example 3, we recall the method of construction used to bisect an angle. Although the technique is illustrated, the objective here is to justify the method.

EXAMPLE 3

Justify the method for constructing the bisector of an angle. Provide the missing reasons in the proof.

GIVEN: $\angle XYZ$
$\overline{YM} \cong \overline{YN}$ (by construction)
$\overline{MW} \cong \overline{NW}$ (by construction)
(See Figure 3.40 on page 149.)

PROVE: $\overrightarrow{YW}$ bisects $\angle XYZ$

(a)

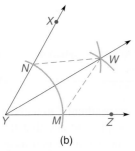

(b)

Figure 3.40

PROOF	
Statements	**Reasons**
1. ∠XYZ; $\overline{YM} \cong \overline{YN}$ and $\overline{MW} \cong \overline{NW}$	1. ?
2. $\overline{YW} \cong \overline{YW}$	2. ?
3. △YMW ≅ △YNW	3. ?
4. ∠MYW ≅ ∠NYW	4. ?
5. $\overrightarrow{YW}$ bisects ∠XYZ	5. ?

The angle bisector method can be used to construct angles of certain measures. For instance, if a right angle has been constructed, then an angle of measure 45° can be constructed by bisecting the 90° angle. In Example 4, we construct an angle of measure 30°.

EXAMPLE 4

Construct an angle that measures 30°.

SOLUTION Figures 3.41(a) and (b): We begin by constructing an equilateral (and therefore equiangular) triangle. To accomplish this, mark off a line segment of length a. From the endpoints of this line segment, mark off arcs with the same radius length a. The point of intersection determines the third vertex of this equilateral triangle, whose angles measure 60° each.
Figure 3.41(c): By constructing the bisector of one angle, we determine an angle that measures 30°.

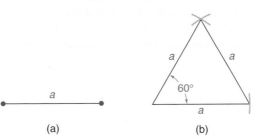

(a)   (b)   (c)

Figure 3.41

In Example 5, we justify the method for constructing a line perpendicular to a given line from a point not on that line. In the example, point P lies above line ℓ.

EXAMPLE 5

GIVEN: P not on ℓ
$\overline{PA} \cong \overline{PB}$ (by construction)
$\overline{AQ} \cong \overline{BQ}$ (by construction)
(See Figure 3.42.)

PROVE: $\overline{PQ} \perp \overline{AB}$

Provide the missing statements and reasons in the proof on page 150.

(a)

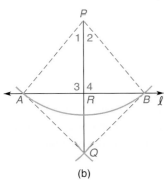

(b)

Figure 3.42

PROOF

Statements	Reasons
1. P not on ℓ $\overline{PA} \cong \overline{PB}$ and $\overline{AQ} \cong \overline{BQ}$	1. ?
2. $\overline{PQ} \cong \overline{PQ}$	2. ?
3. $\triangle PAQ \cong \triangle PBQ$	3. ?
4. $\angle 1 \cong \angle 2$	4. ?
5. $\overline{PR} \cong \overline{PR}$	5. ?
6. $\triangle PRA \cong \triangle PRB$	6. ?
7. $\angle 3 \cong \angle 4$	7. ?
8. ?	8. If two lines meet to form $\cong$ adjacent $\angle$s, these lines are $\perp$

In Example 6, we recall the method for constructing the line perpendicular to a given line at a point on the line. We illustrate the technique in the example and ask that the student justify the method in Exercise 29. In Example 6, we construct an angle that measures 45°.

EXAMPLE 6

Construct an angle that measures 45°.

SOLUTION Figure 3.43(a): We begin by constructing a line segment perpendicular to a given line ℓ at a point P on that line.

Figure 3.43(b): Next we bisect one of the right angles that was determined. The bisector forms an angle whose measure is 45°.

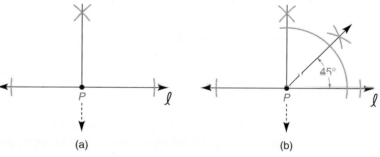

(a)   (b)

SSG **EXS. 3–5**   Figure 3.43

As we saw in Example 4, constructing an equilateral triangle is fairly simple. It is also possible to construct other regular polygons, such as a square or a regular hexagon. In the following box, we recall some facts that will help us to perform such constructions.

> To construct a regular polygon with n sides:
>
> **1.** Each interior angle must measure $I = \frac{(n-2)180}{n}$ degrees; alternatively, each exterior angle must measure $E = \frac{360}{n}$ degrees.
> **2.** All sides must be congruent.

EXAMPLE 7

Construct a regular hexagon having sides of length *a*.

SOLUTION Figure 3.44(a): We begin by marking off a line segment of length *a*.
Figure 3.44(b): Each exterior angle of the hexagon ($n = 6$) must measure
$E = \frac{360}{6} = 60°$; then each interior angle measures 120°. We construct an
equilateral triangle (all sides measure *a*) so that a 60° exterior angle is formed.
Figure 3.44(c): Again marking off an arc of length *a* for the second side, we
construct another exterior angle of measure 60°.
Figure 3.44(d): This procedure is continued until the regular hexagon *ABCDEF*
is determined.

(SSG) **EXS. 6–7**   Figure 3.44

Exercises **3.4**

In Exercises 1 to 6, use line segments of given lengths a, b, and c to perform the constructions.

Exercises 1–6

1. Construct a line segment of length 2*b*.

2. Construct a line segment of length $b + c$.

3. Construct a line segment of length $\frac{1}{2}c$.

4. Construct a line segment of length $a - b$.

5. Construct a triangle with sides of lengths *a*, *b*, and *c*.

6. Construct an isosceles triangle with a base of length *b* and legs of length *a*.

In Exercises 7 to 12, use the angles provided to perform the constructions.

Exercises 7–12

7. Construct an angle that is congruent to acute ∠*A*.

8. Construct an angle that is congruent to obtuse ∠*B*.

9. Construct an angle that has one-half the measure of ∠*A*.

10. Construct an angle that has a measure equal to
m∠*B* − m∠*A*.

11. Construct an angle that has twice the measure of ∠*A*.

12. Construct an angle whose measure averages the measures of ∠*A* and ∠*B*.

In Exercises 13 and 14, use the angles and lengths of sides provided to construct the triangle described.

13. Construct the triangle that has sides of lengths *r* and *t* with included angle *S*.

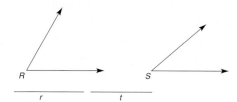

Exercises 13, 14

14. Construct the triangle that has a side of length *t* included by angles *R* and *S*.

In Exercises 15 to 18, construct angles having the given measures.

15. 90° and then 45°

16. 60° and then 30°

17. 30° and then 15°

18. 45° and then 105°

(HINT: 105° = 45° + 60°)

19. Describe how you would construct an angle measuring 22.5°.

20. Describe how you would construct an angle measuring 75°.

21. Construct the complement of the acute angle *Q* shown.

22. Construct the supplement of the obtuse angle *T* shown.

In Exercises 23 to 26, use line segments of lengths a and c as shown.

23. Construct the right triangle with hypotenuse of length *c* and a leg of length *a*.

c ———————

a ————

Exercises 23–26

24. Construct an isosceles triangle with base of length *c* and altitude of length *a*.

(HINT: The altitude lies on the perpendicular bisector of the base.)

25. Construct an isosceles triangle with a vertex angle of 30° and each leg of length *c*.

26. Construct a right triangle with base angles of 45° and hypotenuse of length *c*.

In Exercises 27 and 28, use the given angle R and the line segment of length b.

27. Construct the right triangle in which acute angle *R* has a side (one leg of the triangle) of length *b*.

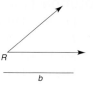

Exercises 27, 28

28. Construct an isosceles triangle with base of length *b* and congruent base angles having the measure of angle *R*.

29. Complete the justification of the construction of the line perpendicular to a given line at a point on that line.
Given: Line *m*, with point *P* on *m*
$\overline{PQ} \cong \overline{PR}$ (by construction)
$\overline{QS} \cong \overline{RS}$ (by construction)
Prove: $\overleftrightarrow{SP} \perp m$

30. Complete the justification of the construction of the perpendicular bisector of a line segment.
Given: $\overline{AB}$ with $\overline{AC} \cong \overline{BC} \cong \overline{AD} \cong \overline{BD}$ (by construction)
Prove: $\overline{AM} \cong \overline{MB}$ and $\overleftrightarrow{CD} \perp \overline{AB}$

31. To construct a regular hexagon, what measure would be necessary for each interior angle? Construct an angle of that measure.

32. To construct a regular octagon, what measure would be necessary for each interior angle? Construct an angle of that measure.

33. To construct a regular dodecagon (12 sides), what measure would be necessary for each interior angle? Construct an angle of that measure.

34. Draw an acute triangle and construct the three medians of the triangle. Do the medians appear to meet at a common point?

35. Draw an obtuse triangle and construct the three altitudes of the triangle. Do the altitudes appear to meet at a common point?

(HINT: In the construction of two of the altitudes, sides need to be extended.)

36. Draw a right triangle and construct the angle bisectors of the triangle. Do the angle bisectors appear to meet at a common point?

37. Draw an obtuse triangle and construct the three perpendicular bisectors of its sides. Do the perpendicular bisectors of the three sides appear to meet at a common point?

38. Construct an equilateral triangle and its three altitudes. What does intuition tell you about the three medians, the three angle bisectors, and the three perpendicular bisectors of the sides of that triangle?

39. A carpenter has placed a square over an angle in such a manner that $\overline{AB} \cong \overline{AC}$ and $\overline{BD} \cong \overline{CD}$. In the drawing, what can you conclude about the location of point D?

*40. In right triangle ABC, $m\angle C = 90°$. Also, $BC = a$, $CA = b$, and $AB = c$. Construct the bisector of $\angle B$ so that it intersects $\overline{CA}$ at point D. Now construct $\overline{DE}$ perpendicular to $\overline{AB}$ with E on $\overline{AB}$. In terms of a, b, and c, find the length of $\overline{EA}$.

3.5	Inequalities in a Triangle

| **KEY CONCEPTS** | Lemma | Inequality of Sides and Angles in a Triangle | The Triangle Inequality |

Important inequality relationships exist among the measured parts of a triangle. To establish some of these, we recall and apply some facts from both algebra and geometry. A more in-depth review of inequalities can be found in Appendix A, Section A.3.

> **DEFINITION**
>
> Let a and b be real numbers. **a > b** (read "a is greater than b") if and only if there is a positive number p for which $a = b + p$.

SSG EXS. 1–3

For instance, $9 > 4$, because there is the positive number 5 for which $9 = 4 + 5$. Because $5 + 2 = 7$, we also know that $7 > 2$ and $7 > 5$. In geometry, let A-B-C on $\overline{AC}$ so that $AB + BC = AC$; then $AC > AB$, because BC is a positive number. The statement $a < b$ (read "a is less than b") is true when $a + p = b$ for some positive number p. When $a < b$, it is also true that $b > a$.

LEMMAS (HELPING THEOREMS)

The following theorems can be used to prove relationships found later in this textbook. In their role as "helping" theorems, each of the five statements that follow is called a **lemma.**

> **LEMMA 3.5.1**
>
> If B is between A and C on $\overline{AC}$, then $AC > AB$ and $AC > BC$. (The measure of a line segment is greater than the measure of any of its parts. See Figure 3.45.)

Figure 3.45

PROOF

By the Segment-Addition Postulate, $AC = AB + BC$. According to the Ruler Postulate, $BC > 0$ (meaning BC is positive); it follows that $AC > AB$. Similarly, $AC > BC$. These relationships follow directly from the definition of $a > b$. ∎

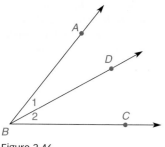

Figure 3.46

LEMMA 3.5.2

If $\overrightarrow{BD}$ separates $\angle ABC$ into two parts ($\angle 1$ and $\angle 2$), then $m\angle ABC > m\angle 1$ and $m\angle ABC > m\angle 2$. (The measure of an angle is greater than the measure of any of its parts. See Figure 3.46.)

PROOF

By the Angle-Addition Postulate, $m\angle ABC = m\angle 1 + m\angle 2$. Using the Protractor Postulate, $m\angle 2 > 0$; it follows that $m\angle ABC > m\angle 1$. Similarly, $m\angle ABC > m\angle 2$. ■

Figure 3.47

LEMMA 3.5.3

If $\angle 3$ is an exterior angle of a triangle and $\angle 1$ and $\angle 2$ are the nonadjacent interior angles, then $m\angle 3 > m\angle 1$ and $m\angle 3 > m\angle 2$. (The measure of an exterior angle of a triangle is greater than the measure of either nonadjacent interior angle. See Figure 3.47.)

PROOF

Because the measure of an exterior angle of a triangle equals the sum of measures of the two nonadjacent interior angles, $m\angle 3 = m\angle 1 + m\angle 2$. It follows that $m\angle 3 > m\angle 1$ and $m\angle 3 > m\angle 2$. ■

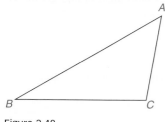

Figure 3.48

LEMMA 3.5.4

In $\triangle ABC$, if $\angle C$ is a right angle or an obtuse angle, then $m\angle C > m\angle A$ and $m\angle C > m\angle B$. (If a triangle contains a right or an obtuse angle, then the measure of this angle is greater than the measure of either of the remaining angles. See Figure 3.48.)

PROOF

In $\triangle ABC$, $m\angle A + m\angle B + m\angle C = 180°$. With $\angle C$ being a right angle or an obtuse angle, $m\angle C \geq 90°$; it follows that $m\angle A + m\angle B \leq 90°$. Then $m\angle A < 90°$ and $m\angle B < 90°$. Thus, $m\angle C > m\angle A$ and $m\angle C > m\angle B$. ■

The following theorem (also a lemma) is used in Example 1. Its proof (not given) depends on the definition of "is greater than."

LEMMA 3.5.5 ■ Addition Property of Inequality

If $a > b$ and $c > d$, then $a + c > b + d$.

Geometry in the Real World

A carpenter's "plumb" determines the shortest distance to a horizontal line. A vertical brace provides structural support for the roof.

EXAMPLE 1

Give a paragraph proof for the following problem.

GIVEN: $AB > CD$ and $BC > DE$

PROVE: $AC > CE$

Figure 3.49

PROOF: If $AB > CD$ and $BC > DE$, then $AB + BC > CD + DE$ by Lemma 3.5.5. But $AB + BC = AC$ and $CD + DE = CE$ by the Segment-Addition Postulate. Using substitution, it follows that $AC > CE$. ■

The paragraph proof in Example 1 could have been written in this standard format.

PROOF

Statements	Reasons
1. $AB > CD$ and $BC > DE$	1. Given
2. $AB + BC > CD + DE$	2. Lemma 3.5.5
3. $AB + BC = AC$ and $CD + DE = CE$	3. Segment-Addition Postulate
4. $AC > CE$	4. Substitution

 EXS. 4–8

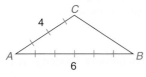

Figure 3.50

The paragraph proof and the two-column proof of Example 1 are equivalent. In either format, statements must be ordered and justified.

The remaining theorems are the "heart" of this section. Before studying the theorem and its proof, it is a good idea to visualize each theorem. Many statements of inequality are intuitive; that is, they are easy to believe even though they may not be easily proved.

Study Theorem 3.5.6 and consider Figure 3.50, in which it appears that $m\angle C > m\angle B$.

THEOREM 3.5.6

If one side of a triangle is longer than a second side, then the measure of the angle opposite the longer side is greater than the measure of the angle opposite the shorter side.

(a)

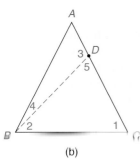

(b)

Figure 3.51

EXAMPLE 2

Provide a paragraph proof of Theorem 3.5.6.

GIVEN: $\triangle ABC$, with $AC > BC$ [See Figure 3.51(a).]

PROVE: $m\angle B > m\angle A$

PROOF: Given $\triangle ABC$ with $AC > BC$. We use the Ruler Postulate to locate point D on $\overline{AC}$ so that $\overline{CD} \cong \overline{BC}$ in Figure 3.51(b). Now $m\angle 2 = m\angle 5$ in the isosceles triangle BDC. By Lemma 3.5.2, $m\angle ABC > m\angle 2$; therefore, $m\angle ABC > m\angle 5$ (*) by substitution. By Lemma 3.5.3, $m\angle 5 > m\angle A$ (*) because ∠5 is an exterior angle of △ADB. Using the two starred statements, we can conclude by the Transitive Property of Inequality that $m\angle ABC > m\angle A$; that is, $m\angle B > m\angle A$ in Figure 3.51(a). ∎

The relationship described in Theorem 3.5.6 extends, of course, to all sides and all angles of a triangle. That is, the largest of the three angles of a triangle is opposite the longest side, and the smallest angle is opposite the shortest side.

Technology Exploration

Use computer software if available.

1. Draw a $\triangle ABC$ with $\overline{AB}$ as the longest side.

2. Measure $\angle A$, $\angle B$, and $\angle C$.

3. Show that $\angle C$ has the greatest measure.

EXAMPLE 3

Given that the three sides of $\triangle ABC$ (not shown) are $AB = 4$, $BC = 5$, and $AC = 6$, arrange the three angles of the triangle by size.

SOLUTION Because $AC > BC > AB$, the largest angle of $\triangle ABC$ is $\angle B$ because it lies opposite the longest side $\overline{AC}$. The angle intermediate in size is $\angle A$, which lies opposite $\overline{BC}$. The smallest angle is $\angle C$, which lies opposite the shortest side, $\overline{AB}$. Thus, the order of the angles by size is

$$m\angle B > m\angle A > m\angle C$$

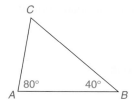

Figure 3.52

Discover

Using construction paper and a protractor, draw △RST so that m∠R = 75°, m∠S = 60°, and m∠T = 45°. Measure the length of each side.

a) Which side is longest?

b) Which side is shortest?

ANSWER S

(ə) *RS̲* (ɥ) *S̲T*

The converse of Theorem 3.5.6 is also true. It is necessary, however, to use an indirect proof to establish the converse. Recall that this method of proof begins by supposing the opposite of what we want to show. Because this assumption leads to a contradiction, the assumption must be false and the desired claim is therefore true.

Study Theorem 3.5.7 and consider Figure 3.52, in which m∠A = 80° and m∠B = 40°. Compare the lengths of the sides opposite ∠A and ∠B. It appears that the longer side lies opposite the larger angle; that is, it appears that $BC > AC$.

THEOREM 3.5.7

If the measure of one angle of a triangle is greater than the measure of a second angle, then the side opposite the larger angle is longer than the side opposite the smaller angle.

The proof of Theorem 3.5.7 depends on a fact known as the Trichotomy Property. It is stated below.

THE TRICHOTOMY PROPERTY

Given real numbers a and b, one and only one of the following statements can be true.

$$a > b, \quad a = b, \quad or \quad a < b$$

EXAMPLE 4

Prove Theorem 3.5.7 by using an indirect approach.

GIVEN: △ABC with m∠B > m∠A (See Figure 3.53.)

PROVE: $AC > BC$

PROOF: Given △ABC with m∠B > m∠A, assume that $AC \leq BC$. But if $AC = BC$, then m∠B = m∠A, which contradicts the hypothesis. Also, if $AC < BC$, then it follows by Theorem 3.5.6 that m∠B < m∠A, which also contradicts the hypothesis. Thus, the assumed statement must be false; applying the Trichotomy Property, it follows that $AC > BC$.

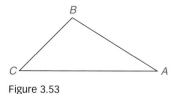

Figure 3.53

EXAMPLE 5

Given △RST (not shown) in which m∠R = 80° and m∠S = 55°, write an extended inequality that compares the lengths of the three sides.

SOLUTION Because the sum of angles of △RST is 180°, we can show that m∠T = 45°. With m∠R > m∠S > m∠T, it follows that the sides opposite these ∠s are unequal in the same order. That is,

$$ST > RT > SR.$$

The following corollary is a consequence of Theorem 3.5.7.

SSG **EXS. 9–12**

COROLLARY 3.5.8

The perpendicular line segment from a point to a line is the shortest line segment that can be drawn from the point to the line.

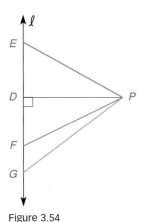

Figure 3.54

SSG **EXS. 13–14**

Figure 3.55

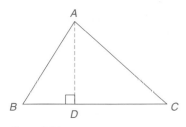

Figure 3.56

In Figure 3.54, $PD < PE$, $PD < PF$, and $PD < PG$. In every case, $\overline{PD}$ lies opposite an acute angle of a triangle, whereas the second segment always lies opposite a right angle of that triangle (necessarily the largest angle of the triangle involved). With $\overline{PD} \perp \ell$, we say that PD is the *distance* from P to ℓ.

Corollary 3.5.8 can easily be extended to three dimensions.

> **COROLLARY 3.5.9**
>
> The perpendicular line segment from a point to a plane is the shortest line segment that can be drawn from the point to the plane.

In Figure 3.55, $\overline{PD}$ is a leg of each right triangle shown. With $\overline{PE}$ the hypotenuse of $\triangle PDE$, $\overline{PF}$ the hypotenuse of $\triangle PDF$, and $\overline{PG}$ the hypotenuse of $\triangle PDG$, the length of $\overline{PD}$ is less than that of $\overline{PE}$, $\overline{PF}$, $\overline{PG}$, or any other line segment joining point P to a point in plane R. With $\overline{PD} \perp$ plane R, the length of $\overline{PD}$ is the *distance* from point P to plane R.

Our final theorem shows that no side of a triangle can have a length greater than or equal to the sum of the lengths of the other two sides. In the proof, the relationship is validated for only one of three possible inequalities. Theorem 3.5.10 is often called the Triangle Inequality. See Figure 3.56.

> **THEOREM 3.5.10 ■ Triangle Inequality**
>
> The sum of the lengths of any two sides of a triangle is greater than the length of the third side.

GIVEN: $\triangle ABC$

PROVE: $BA + CA > BC$

PROOF: Draw $\overline{AD} \perp \overline{BC}$. Because the shortest segment from a point to $\overline{AD}$ is the perpendicular segment, $BA > BD$ and $CA > CD$. Using Lemma 3.5.5, we add the inequalities; $BA + CA > BD + CD$. By the Segment-Addition Postulate, the sum $BD + CD$ can be replaced by BC to yield $BA + CA > BC$. ◾

The following statement is an alternative and expanded form of Theorem 3.5.10. If a, b, and c are the lengths of the sides of a triangle and c is the length of any side, then $a < b + c$ and $c < a + b$; this implies that $a - b < c$ and $c < a + b$, which is equivalent to $a - b < c < a + b$. This leads to an alternative form of Theorem 3.5.10.

> **THEOREM 3.5.10 ■ Triangle Inequality**
>
> The length of any side of a triangle must lie between the sum and difference of the lengths of the other two sides.

> **EXAMPLE 6**

Can a triangle have sides of the following lengths?

    a) 3, 4, and 5
    b) 3, 4, and 7
    c) 3, 4, and 8
    d) 3, 4, and x

Discover: The Magic Triangle

To create a particular *Magic Triangle*, we place the numbers 1, 2, 3, 4, 5, and 6 along the sides of a triangle so that the sum on each side totals 9.

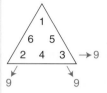

By using the numbers 1, 2, 3, 4, . . ., and 9, complete the *Magic Triangle* in such a way that the sum of all the 4 numbers on each side totals 17.

ANSWER

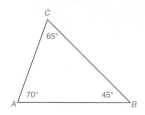

SOLUTION

a) Yes, because the sum of the lengths of two sides is greater than the length of the third side.

b) No, because $3 + 4 = 7$

c) No, because $3 + 4 < 8$

d) Yes, if $4 - 3 < x < 4 + 3$ or $1 < x < 7$

The alternative form of Theorem 3.5.10 was used in part (d) of Example 6 to show that the length of the third side must be between 1 and 7.

Our final example illustrates a practical application of inequality relationships in triangles.

EXAMPLE 7

On a map, firefighters are located at points A and B. A fire has broken out at point C. Which group of firefighters is nearer the location of the fire? (See Figure 3.57.)

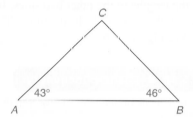

Figure 3.57

SOLUTION With $m\angle A = 43°$ and $m\angle B = 46°$, the side opposite $\angle B$ has a greater length than the side opposite $\angle A$; that is, $AC > BC$. Because the distance from B to C is less than the distance from A to C, the firefighters at site B should arrive at the fire located at C first.

NOTE: In Example 7, we assume that highways from A and B (to C) are equally accessible.

SSG **EXS. 15–18**

Exercises 3.5

In Exercises 1 to 10, classify each statement as true or false.

1. $\overline{AB}$ is the longest side of $\triangle ABC$.

Exercises 1, 2

2. $AB < BC$

3. $DB > AB$

Exercises 3, 4

4. Because $m\angle A = m\angle B$, it follows that $DA = DC$.

5. $m\angle A + m\angle B = m\angle C$

Exercises 5, 6

6. $m\angle A > m\angle B$

7. $DF > DE + EF$

Exercises 7, 8

8. If $\overrightarrow{DG}$ is the bisector of $\angle EDF$, then $DG > DE$.

9. $DA > AC$

Exercises 9, 10

10. $CE = ED$

11. Is it possible to draw a triangle whose angles measure
   a) $100°$, $100°$, and $60°$?
   b) $45°$, $45°$, and $90°$?

12. Is it possible to draw a triangle whose angles measure
   a) $80°$, $80°$, and $50°$?
   b) $50°$, $50°$, and $80°$?

13. Is it possible to draw a triangle whose sides measure
   a) 8, 9, and 10?
   b) 8, 9, and 17?
   c) 8, 9, and 18?

14. Is it possible to draw a triangle whose sides measure
   a) 7, 7, and 14?
   b) 6, 7, and 14?
   c) 6, 7, and 8?

In Exercises 15 to 18, describe the triangle ($\triangle XYZ$, not shown) as scalene, isosceles, or equilateral. Also, is the triangle acute, right, or obtuse?

15. $m\angle X = 43°$ and $m\angle Y = 47°$

16. $m\angle X = 60°$ and $\angle Y \cong \angle Z$

17. $m\angle X = m\angle Y = 40°$

18. $m\angle X = 70°$ and $m\angle Y = 40°$

19. Two of the sides of an isosceles triangle have lengths of 10 cm and 4 cm. Which length must be the length of the base?

20. The sides of a right triangle have lengths of 6 cm, 8 cm, and 10 cm. Which length is that of the hypotenuse?

21. A triangle is both isosceles *and* acute. If one angle of the triangle measures $36°$, what is the measure of the largest angle(s) of the triangle? What is the measure of the smallest angle(s) of the triangle?

22. One of the angles of an isosceles triangle measures $96°$. What is the measure of the largest angle(s) of the triangle? What is the measure of the smallest angle(s) of the triangle?

23. NASA in Huntsville, Alabama (at point H), has called a manufacturer for parts needed as soon as possible. NASA will, in fact, send a courier for the necessary equipment. The manufacturer has two distribution centers located in nearby Tennessee—one in Nashville (at point N) and the other in Jackson (at point J). Using the angle measurements indicated on the accompanying map, determine to which town the courier should be dispatched to obtain the needed parts.

24. A tornado has just struck a small Kansas community at point T. There are Red Cross units stationed in both Salina (at point S) and Wichita (at point W). Using the angle measurements indicated on the accompanying map, determine which Red Cross unit would reach the victims first. (Assume that both units have the same mode of travel and accessible roadways available.)

In Exercises 25 and 26, complete each proof shown on page 160.

25. Given: $m\angle ABC > m\angle DBE$
              $m\angle CBD > m\angle EBF$
   Prove: $m\angle ABD > m\angle DBF$

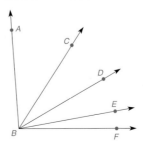

PROOF

Statements	Reasons
1. ?	1. Given
2. m∠ABC + m∠CBD > m∠DBE + m∠EBF	2. Addition Property of Inequality
3. m∠ABD = m∠ABC + m∠CBD and m∠DBF = m∠DBE + m∠EBF	3. ?
4. ?	4. Substitution

26. *Given:*   Equilateral △ABC and D-B-C
    Prove: DA > AC

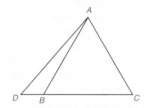

PROOF

Statements	Reasons
1. ?	1. Given
2. △ABC is equiangular, so m∠ABC = m∠C	2. ?
3. m∠ABC > m∠D (∠D of △ABD)	3. The measure of an ext. ∠ of a △ is greater than the measure of either nonadjacent int. ∠
4. ?	4. Substitution
5. ?	5. ?

In Exercises 27 and 28, construct proofs.

27. *Given:*   Quadrilateral RSTU with diagonal $\overline{US}$
    ∠R and ∠TUS are right ∠s
    Prove: TS > UR

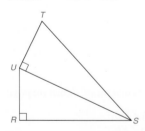

28. *Given:*   Quadrilateral ABCD with $\overline{AB} \cong \overline{DE}$
    Prove: DC > AB

29. For △ABC and △DEF (not shown), suppose that $\overline{AC} \cong \overline{DF}$, $\overline{AB} \cong \overline{DE}$, and m∠A < m∠D. Draw a conclusion regarding the lengths of $\overline{BC}$ and $\overline{EF}$.

30. In △MNP (not shown), point Q lies on $\overline{NP}$ so that $\overrightarrow{MQ}$ bisects ∠NMP. If MN < MP, draw a conclusion about the relative lengths of $\overline{NQ}$ and $\overline{QP}$.

In Exercises 31 to 34, apply a form of Theorem 3.5.10.

31. The sides of a triangle have lengths of 4, 6, and x. Write an inequality that states the possible values of x.

32. The sides of a triangle have lengths of 7, 13, and x. As in Exercise 31, write an inequality that describes the possible values of x.

33. If the lengths of two sides of a triangle are represented by 2x + 5 and 3x + 7 (in which x is positive), describe in terms of x the possible lengths of the third side whose length is represented by y.

34. Prove by the indirect method: "The length of a diagonal of a square is not equal in length to the length of any of the sides of the square."

35. Prove by the indirect method:
    Given: △MPN is not isosceles
    Prove: PM ≠ PN

36. Prove by the indirect method:
    Given: Scalene △XYZ in which $\overline{ZW}$ bisects ∠XYZ (point W lies on $\overline{XY}$).
    Prove: $\overline{ZW}$ is not perpendicular to $\overline{XY}$

In Exercises 37 and 38, prove each theorem.

37. The length of the median from the vertex of an isosceles triangle is less than the length of either of the legs.

38. The length of an altitude of an acute triangle is less than the length of either side containing the same vertex as the altitude.

SKETCH OF ARCHIMEDES

Whereas Euclid (see Perspective on History, Chapter 2) was a great teacher and wrote so that the majority might understand the principles of geometry, Archimedes wrote only for the very well-educated mathematicians and scientists of his day. Archimedes (287–212 B.C.) wrote on such topics as the measure of the circle, the quadrature of the parabola, and spirals. In his works, Archimedes found a very good approximation of π. His other geometric works included investigations of conic sections and spirals, and he also wrote about physics. He was a great inventor and is probably remembered more for his inventions than for his writings.

Several historical events concerning the life of Archimedes have been substantiated, and one account involves his detection of a dishonest goldsmith. In that story, Archimedes was called upon to determine whether the crown that had been ordered by the king was constructed entirely of gold. By applying the principle of hydrostatics (which he had discovered), Archimedes established that the goldsmith had not constructed the crown entirely of gold. (The principle of hydrostatics states that an object placed in a fluid displaces an amount of fluid equal in weight to the amount of weight the object loses while submerged.)

One of his inventions is known as Archimedes' screw. This device allows water to flow from one level to a higher level so that, for example, holds of ships can be emptied of water. Archimedes' screw was used in Egypt to drain fields when the Nile River overflowed its banks.

When Syracuse (where Archimedes lived) came under siege by the Romans, Archimedes designed a long-range catapult that was so effective that Syracuse was able to fight off the powerful Roman army for three years before being overcome.

One report concerning the inventiveness of Archimedes has been treated as false, because his result has not been duplicated. It was said that he designed a wall of mirrors that could focus and reflect the sun's heat with such intensity as to set fire to Roman ships at sea. Because recent experiments with concave mirrors have failed to produce such intense heat, this account is difficult to believe.

Archimedes eventually died at the hands of a Roman soldier, even though the Roman army had been given orders not to harm him. After his death, the Romans honored his brilliance with a tremendous monument displaying the figure of a sphere inscribed in a right circular cylinder.

PASCAL'S TRIANGLE

Blaise Pascal (1623–1662) was a French mathematician who contributed to several areas of mathematics, including conic sections, calculus, and the invention of a calculating machine. But Pascal's name is most often associated with the array of numbers known as Pascal's Triangle, which follows:

$$
\begin{array}{ccccccccc}
 & & & & 1 & & & & \\
 & & & 1 & & 1 & & & \\
 & & 1 & & 2 & & 1 & & \\
 & 1 & & 3 & & 3 & & 1 & \\
1 & & 4 & & 6 & & 4 & & 1
\end{array}
$$

Each row of entries in Pascal's Triangle begins and ends with the number 1. Intermediate entries in each row are found by the addition of the upper-left and upper-right entries of the preceding row. The row following 1 4 6 4 1 has the form

$$1 \quad 5 \quad 10 \quad 10 \quad 5 \quad 1$$

Applications of Pascal's Triangle include the counting of subsets of a given set, which we will consider in the following paragraph. While we do not pursue this notion, Pascal's Triangle is also useful in the algebraic expansion of a binomial to a power such as $(a + b)^2$, which equals $a^2 + 2ab + b^2$. Notice that the multipliers in the product found with exponent 2 are 1 2 1,

from a row of Pascal's Triangle. In fact, the expansion $(a + b)^3$ leads to $a^3 + 3a^2b + 3ab^2 + b^3$, in which the multipliers (also known as coefficients) take the form 1 3 3 1, a row of Pascal's Triangle.

SUBSETS OF A GIVEN SET

A subset of a given set is a set formed from choices of elements from the given set. Because a subset of a set with n elements can have from 0 to n elements, we find that Pascal's Triangle provides a count of the number of subsets containing a given counting number of elements.

Pascal's Triangle	Set	Number of Elements	Subsets of the Set	Number of Subsets
1	$\varnothing$	0	$\varnothing$	1
1 1	$\{a\}$	1	$\varnothing, \{a\}$ 1 + 1 subsets	2
1 2 1	$\{a, b\}$	2	$\varnothing, \{a\}, \{b\}, \{a, b\}$ 1 + 2 + 1 subsets	4
1 3 3 1	$\{a, b, c\}$	3	$\varnothing, \{a\}, \{b\}, \{c\}, \{a, b\},$ $\{a, c\}, \{b, c\}, \{a, b, c\}$ 1 + 3 + 3 + 1 subsets	8

1 subset of 0 elements, **3** subsets of 1 element each, **3** subsets of 2 elements each, **1** subset of 3 elements

In algebra, it is shown that $2^0 = 1$; not by coincidence, the set $\varnothing$, which has 0 elements, has 1 subset. Just as $2^1 = 2$, the set $\{a\}$ which has 1 element, has 2 subsets. The pattern continues so that a set with 2 elements has $2^2 = 4$ subsets and a set with 3 elements has $2^3 = 8$ subsets. A quick examination suggests this fact:

The total number of subsets for a set with n elements is 2^n.

The entries of the fifth row of Pascal's Triangle correspond to the numbers of subsets of the four-element set $\{a, b, c, d\}$; of course, the subsets of $\{a, b, c, d\}$ must have 0 elements, 1 element each, 2 elements each, 3 elements each, or 4 elements each. Based upon the preceding principle, there will be a total of $2^4 = 16$ subsets for $\{a, b, c, d\}$.

EXAMPLE 1

List all 16 subsets of the set $\{a, b, c, d\}$ by considering the fifth row of Pascal's Triangle, namely 1   4   6   4   1. Notice also that $1 + 4 + 6 + 4 + 1$ must equal 16.

SOLUTION $\varnothing$, $\{a\}$, $\{b\}$, $\{c\}$, $\{d\}$, $\{a, b\}$, $\{a, c\}$, $\{a, d\}$, $\{b, c\}$, $\{b, d\}$, $\{c, d\}$, $\{a, b, c\}$, $\{a, b, d\}$, $\{a, c, d\}$, $\{b, c, d\}$, $\{a, b, c, d\}$.

EXAMPLE 2

Find the number of subsets for a set with six elements.

SOLUTION   The number of subsets is 2^6, or 64.

Looking back at Example 1, we notice that the number of subsets of the four-element set $\{a, b, c, d\}$ is $1 + 4 + 6 + 4 + 1$, which equals 16, or 2^4. The preceding principle can be restated in the following equivalent form:

The sum of the entries in row n of Pascal's Triangle is 2^{n-1}.

EXAMPLE 3

The sixth row of Pascal's Triangle is 1   5   10   10   5   1. Use the principle above to find the sum of the entries of this row.

SOLUTION   With $n = 6$, it follows that $n - 1 = 5$. Then $1 + 5 + 10 + 10 + 5 + 1 = 2^5$, or 32.

NOTE: There are 32 subsets for a set containing five elements; consider $\{a, b, c, d, e\}$.

In closing, we note that only a few of the principles based upon Pascal's Triangle have been explored in this Perspective on Application!

Summary

A Look Back at Chapter 3

In this chapter, we considered several methods for proving triangles congruent. We explored properties of isosceles triangles and justified construction methods of earlier chapters. Inequality relationships for the sides and angles of a triangle were also investigated.

A Look Ahead to Chapter 4

In the next chapter, we use properties of triangles to develop the properties of quadrilaterals. We consider several special types of quadrilaterals, including the parallelogram, kite, rhombus, and trapezoid.

Key Concepts

3.1

Congruent Triangles • SSS, SAS, ASA, AAS • Included Side, Included Angle • Reflexive Property of Congruence (Identity) • Symmetric and Transitive Properties of Congruence

3.2

CPCTC • Hypotenuse and Legs of a Right Triangle • HL • Pythagorean Theorem • Square Roots Property

3.3

Isosceles Triangle • Vertex, Legs, and Base of an Isosceles Triangle • Base Angles • Vertex Angle • Angle Bisector • Median • Altitude • Perpendicular Bisector • Auxiliary Line • Determined, Underdetermined, Overdetermined • Equilateral and Equiangular Triangles • Perimeter

3.4

Justifying Constructions

3.5

Lemma • Inequality of Sides and Angles in a Triangle • The Triangle Inequality

Overview ■ Chapter 3

Methods of Proving Triangles Congruent: △ABC ≅ △DEF

Figure (Note Marks)	Method	Steps Needed in Proof
	SSS	$\overline{AB} \cong \overline{DE}$, $\overline{AC} \cong \overline{DF}$, and $\overline{BC} \cong \overline{EF}$
	SAS	$\overline{AB} \cong \overline{DE}$, $\angle A \cong \angle D$, and $\overline{AC} \cong \overline{DF}$
	ASA	$\angle A \cong \angle D$, $\overline{AC} \cong \overline{DF}$, and $\angle C \cong \angle F$
	AAS	$\angle A \cong \angle D$, $\angle C \cong \angle F$, and $\overline{BC} \cong \overline{EF}$
	HL	$\angle A$ and $\angle D$ are rt. $\angle$s, $\overline{AC} \cong \overline{DF}$, and $\overline{BC} \cong \overline{EF}$

Special Relationships

Figure	Relationship	Conclusion
	Pythagorean Theorem	$c^2 = a^2 + b^2$
	$\overline{DF} \cong \overline{EF}$ (two $\cong$ sides)	$\angle E \cong \angle D$ (opposite $\angle$s $\cong$)
	$\angle D \cong \angle E$ (two $\cong$ angles)	$\overline{EF} \cong \overline{DF}$ (opposite sides $\cong$)

Inequality Relationships in a Triangle

Figure	Relationship	Conclusion
	$ST > RS$	m$\angle R >$ m$\angle T$ (opposite angles)
	m$\angle Y >$ m$\angle X$	$XZ > YZ$ (opposite sides)
	Triangle Inequality	$b + c > a$ $a + c > b$ $a + b > c$
	Triangle Inequality Alternate	$c - b < a < c + b$ $a - c < b < a + c$ $a - b < c < a + b$

Chapter **3**   Review Exercises

1. *Given:* $\angle AEB \cong \angle DEC$
 $\overline{AE} \cong \overline{ED}$
 Prove: $\triangle AEB \cong \triangle DEC$

2. *Given:* $\overline{AB} \cong \overline{EF}$
 $\overline{AC} \cong \overline{DF}$
 $\angle 1 \cong \angle 2$
 Prove: $\angle B \cong \angle E$

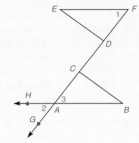

3. *Given:* $\overline{AD}$ bisects $\overline{BC}$
 $\overline{AB} \perp \overline{BC}$
 $\overline{DC} \perp \overline{BC}$
 Prove: $\overline{AE} \cong \overline{ED}$

4. *Given:* $\overline{OA} \cong \overline{OB}$
 $\overline{OC}$ is the median to $\overline{AB}$
 Prove: $\overline{OC} \perp \overline{AB}$

5. *Given:* $\overline{AB} \cong \overline{DE}$
 $\overline{AB} \parallel \overline{DE}$
 $\overline{AC} \cong \overline{DF}$
 Prove: $\overline{BC} \parallel \overline{FE}$

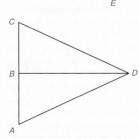

6. *Given:* B is the midpoint of $\overline{AC}$
 $\overline{BD} \perp \overline{AC}$
 Prove: $\triangle ADC$ is isosceles

7. *Given:* $\overline{JM} \perp \overline{GM}$ and $\overline{GK} \perp \overline{KJ}$
 $\overline{GH} \cong \overline{HJ}$
 Prove: $\overline{GM} \cong \overline{JK}$

8. *Given:* $\overline{TN} \cong \overline{TR}$
 $\overline{TO} \perp \overline{NP}$
 $\overline{TS} \perp \overline{PR}$
 $\overline{TO} \cong \overline{TS}$
 Prove: $\angle N \cong \angle R$

9. *Given:* $\overline{YZ}$ is the base of an isosceles triangle; $\overrightarrow{XA} \parallel \overline{YZ}$
 Prove: $\angle 1 \cong \angle 2$

10. *Given:* $\overline{AB} \parallel \overline{DC}$
 $\overline{AB} \cong \overline{DC}$
 C is the midpoint of $\overline{BE}$
 Prove: $\overline{AC} \parallel \overline{DE}$

11. *Given:* $\angle BAD \cong \angle CDA$
 $\overline{AB} \cong \overline{CD}$
 Prove: $\overline{AE} \cong \overline{ED}$

(*HINT: Prove* $\triangle BAD \cong \triangle CDA$ *first.*)

12. *Given:* $\overline{BE}$ is the altitude to $\overline{AC}$
 $\overline{AD}$ is the altitude to $\overline{CE}$
 $\overline{BC} \cong \overline{CD}$
 Prove: $\overline{BE} \cong \overline{AD}$

(*HINT: Prove* $\triangle CBE \cong \triangle CDA$.)

13. *Given:* $\overline{AB} \cong \overline{CD}$
$\angle BAD \cong \angle CDA$
Prove: $\triangle AED$ is isosceles

(HINT: Prove

14. *Given:* $\overrightarrow{AC}$ bisects $\angle BAD$
Prove: $AD > CD$

15. In $\triangle PQR$ (not shown), m$\angle P = 67°$ and m$\angle Q = 23°$.
a) Name the shortest side.
b) Name the longest side.

16. In $\triangle ABC$ (not shown), m$\angle A = 40°$ and m$\angle B = 65°$.
List the sides in order of their lengths, starting with the
smallest side.

17. In $\triangle PQR$ (not shown), $PQ = 1.5$, $PR = 2$, and
$QR = 2.5$. List the angles in order of size, starting with
the smallest angle.

18. Name the longest line segment shown in quadrilateral
ABCD.

19. Which of the following can be the lengths of the sides of a
triangle?
a) 3, 6, 9
b) 4, 5, 8
c) 2, 3, 8

20. Two sides of a triangle have lengths 15 and 20. The length
of the third side can be any number between __?__ and __?__.

21. *Given:* $\overline{DB} \perp \overline{AC}$
$\overline{AD} \cong \overline{DC}$
m$\angle C = 70°$
Find: m$\angle ADB$

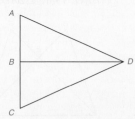

22. *Given:* $\overline{AB} \cong \overline{BC}$
$\angle DAC \cong \angle BCD$
m$\angle B = 50°$
Find: m$\angle ADC$

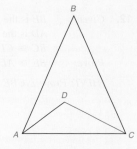

23. *Given:* $\triangle ABC$ is isosceles with base $\overline{AB}$
m$\angle 2 = 3x + 10$
m$\angle 4 = \frac{5}{2}x + 18$
Find: m$\angle C$

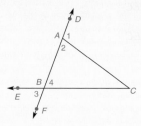

Exercises 23, 24

24. *Given:* $\triangle ABC$ with perimeter 40
$AB = 10$
$BC = x + 6$
$AC = 2x - 3$
Find: Whether $\triangle ABC$ is scalene, isosceles, or
equilateral

25. *Given:* $\triangle ABC$ is isosceles with base $\overline{AB}$
$AB = y + 7$
$BC = 3y + 5$
$AC = 9 - y$
Find: Whether $\triangle ABC$ is also equilateral

Exercises 25, 26

26. *Given:* $\overline{AC}$ and $\overline{BC}$ are the legs of isosceles $\triangle ABC$
m$\angle 1 = 5x$
m$\angle 3 = 2x + 12$
Find: m$\angle 2$

27. Construct an angle that measures 75°.

28. Construct a right triangle that has acute angle A and
hypotenuse of length c.

29. Construct a second isosceles triangle in which the base
angles are half as large as the base angles of the given
isosceles triangle.

$\angle CAD \cong \angle BDA$

1. It is given that $\triangle ABC \cong \triangle DEF$ (triangles not shown).
   a) If m$\angle A = 37°$ and m$\angle E = 68°$, find m$\angle F$. ____
   b) If $AB = 7.3$ cm, $BC = 4.7$ cm, and $AC = 6.3$ cm, find EF. ____

2. Consider $\triangle XYZ$ (not shown).
   a) Which side is included by $\angle X$ and $\angle Y$? ____
   b) Which angle is included by sides $\overline{XY}$ and $\overline{YZ}$? ____

3. State the reason (SSS, SAS, ASA, AAS, or HL) why the triangles are congruent. Note the marks that indicate congruent parts.

   $\triangle RVS \cong \triangle RTS$ ____ $\triangle XMW \cong \triangle MYZ$ ____

4. Write the statement that is represented by the acronym CPCTC. _____

5. With congruent parts marked, are the two triangles congruent? Answer YES or NO.
   $\triangle ABC$ and $\triangle DAC$ ____
   $\triangle RSM$ and $\triangle WVM$ ____

6. With $\triangle ABD \cong \triangle CBE$ and A-D-E-C, does it necessarily follow that $\triangle AEB$ and $\triangle CDB$ are congruent? Answer YES or NO. ____

$\triangle ABD \cong \triangle CBE$

7. In $\triangle ABC$, m$\angle C = 90°$. Find:
   a) c if $a = 8$ and $b = 6$ ____
   b) b if $a = 6$ and $c = 8$ ____

8. $\overline{CM}$ is the median for $\triangle ABC$ from vertex C to side $\overline{AB}$.
   a) Name two line segments that must be congruent.
      ____
   b) Is $\angle 1$ necessarily congruent to $\angle 2$? ____

9. In $\triangle TUV$, $\overline{TV} \cong \overline{UV}$.
   a) If m$\angle T = 71°$, find m$\angle V$.
      ____
   b) If m$\angle T = 7x + 2$ and m$\angle U = 9(x - 2)$, find m$\angle V$.
      ____

Exercises 9, 10

10. In $\triangle TUV$, $\angle T \cong \angle U$.
    a) If $VT = 7.6$ inches and $TU = 4.3$ inches, find VU.
       ____
    b) If $VT = 4x + 1$, $TU = 2x$ and $VU = 6x - 10$, find the perimeter of $\triangle TUV$. ____

    (HINT: Find the value of x.)

11. Show all arcs in the following construction.
    a) Construct an angle that measures 60°.
    b) Using the result from part (a), construct an angle that measures 30°.

12. Show all arcs in the following construction. Construct an isoceles right triangle in which each leg has the length of line segment $\overline{AB}$.

    _____
    A B

13. In $\triangle ABC$, m$\angle C = 46°$, and m$\angle B = 93°$.
    a) Name the shortest side of $\triangle ABC$.
       ____
    b) Name the longest side of $\triangle ABC$. ____

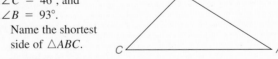

14. In $\triangle TUV$ (not shown), $TU > TV > VU$. Write a three-part inequality that compares the measures of the three angles of $\triangle TUV$. ____

15. In the figure, $\angle A$ is a right angle, $AD = 4$, $DE = 3$, $AB = 5$, and $BC = 2$. Of the two line segments $\overline{DC}$ and $\overline{EB}$, which one is longer? ____

16. Given △ABC, draw the triangle that results when △ABC is rotated clockwise 180° about M, the midpoint of $\overline{AC}$. Let D name the image of point B. In these congruent triangles, which side of △CDA corresponds to side $\overline{BC}$ of △ABC? _____

17. Complete all statements and reasons for the following proof problem.

Given: ∠R and ∠V are right angles; ∠1 ≅ ∠2

Prove: △RST ≅ △VST

PROOF

Statements	Reasons

18. Complete the missing statements and reasons in the following proof.

Given: △RUV; ∠R ≅ ∠V and ∠1 ≅ ∠3

Prove: △STU is an isosceles triangle

PROOF

Statements	Reasons
1. △RUV; ∠R ≅ ∠V	1. _____
2. $\overline{UV}$ ≅ $\overline{UR}$	2. _____
3. _____	3. Given
4. △RSU ≅ △VTU	4. _____
5. _____	5. CPCTC
6. _____	6. If 2 sides of a △ are ≅, this triangle is an isosceles triangle.

19. The perimeter of an isosceles triangle is 32 cm. If the length of the altitude drawn to the base is 8 cm, how long is each leg of the isosceles triangle? _____

© Richard A. Cooke/CORBIS.

Chapter 4

Quadrilaterals

Comforting! Designed by architect Frank Lloyd Wright (1867–1959), this private home is nestled among the trees in the Bear Run Nature Preserve of southwestern Pennsylvania. Known as Fallingwater, this house was constructed in the 1930s. The geometric figure that dominates the homes designed by Wright is the quadrilateral. In this chapter, we consider numerous types of quadrilaterals— among them the parallelogram, the rhombus, and the trapezoid. Also, the language and properties for each type of quadrilateral are developed. Each type of quadrilateral has its own properties and applications. Many of these real-world applications can be found in the examples and exercises of Chapter 4.

Additional video explanations of concepts, sample problems, and applications are available on DVD.

4.1 | Properties of a Parallelogram

KEY CONCEPTS

Quadrilateral
Skew
Quadrilateral

Parallelogram
Diagonals of a
Parallelogram

Altitudes of a
Parallelogram

A **quadrilateral** is a polygon that has exactly four sides. Unless otherwise stated, the term *quadrilateral* refers to a plane figure such as *ABCD* in Figure 4.1(a), in which the line segment sides lie within a single plane. When two sides of the quadrilateral are skew (not coplanar), as with *MNPQ* in Figure 4.1(b), that quadrilateral is said to be **skew.** Thus, *MNPQ* is a skew quadrilateral. In this textbook, we generally consider quadrilaterals whose sides are coplanar.

(a)   (b)

Figure 4.1

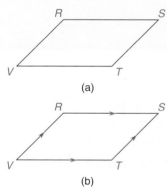

R   S

V   T

(a)

R   S

V   T

(b)

Figure 4.2

DEFINITION

A **parallelogram** is a quadrilateral in which both pairs of opposite sides are parallel.

In Figure 4.2(a), if $\overline{RS} \parallel \overline{VT}$ and $\overline{RV} \parallel \overline{ST}$, then *RSTV* is a parallelogram. We sometimes indicate parallel lines (and line segments) by showing arrows in the same direction. Thus, Figure 4.2(b) emphasizes the fact that *RSTV* is a parallelogram. Because the symbol for parallelogram is $\square$, the quadrilateral in Figure 4.2(b) is $\square RSTV$. The set $P = \{parallelograms\}$ is a subset of $Q = \{quadrilaterals\}$; that is, $P \subseteq Q$.

The Discover activity at the left hints at many of the theorems of this section.

Discover

From a standard sheet of construction paper, cut out a parallelogram as shown. Then cut along one diagonal. How are the two triangles that are formed related?

ANSWER

They are congruent.

EXAMPLE 1

Give a formal proof of Theorem 4.1.1.

THEOREM 4.1.1

A diagonal of a parallelogram separates it into two congruent triangles.

GIVEN: $\square ABCD$ with diagonal $\overline{AC}$ (See Figure 4.3 on page 171.)

PROVE: $\triangle ACD \cong \triangle CAB$

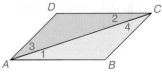

Figure 4.3

PROOF	
Statements	**Reasons**
1. $\square ABCD$	1. Given
2. $\overline{AB} \parallel \overline{CD}$	2. The opposite sides of a $\square$ are $\parallel$ (definition)
3. $\angle 1 \cong \angle 2$	3. If two $\parallel$ lines are cut by a transversal, the alternate interior $\angle$s are congruent
4. $\overline{AD} \parallel \overline{BC}$	4. Same as reason 2
5. $\angle 3 \cong \angle 4$	5. Same as reason 3
6. $\overline{AC} \cong \overline{AC}$	6. Identity
7. $\triangle ACD \cong \triangle CAB$	7. ASA

> **STRATEGY FOR PROOF ■ Using Congruent Triangles**
>
> *General Rule:* To prove that parts of a quadrilateral are congruent, we often use a diagonal as an auxiliary line to prove that triangles are congruent. Then we apply CPCTC.
>
> *Illustration:* This strategy is used in the proof of Corollaries 4.1.2 and 4.1.3. In the proof of Corollary 4.1.4, we do not need the auxiliary line.

Reminder

The sum of the measures of the interior angles of a quadrilateral is 360°.

The proofs of corollaries 4.1.2 to 4.1.5 are left as exercises for the student.

COROLLARY 4.1.2

The opposite angles of a parallelogram are congruent.

According to Corollary 4.1.2, $\angle B \cong \angle D$ and $\angle DAB \cong \angle BCD$ in Figure 4.3.

COROLLARY 4.1.3

The opposite sides of a parallelogram are congruent.

According to Corollary 4.1.3, $\overline{AD} \cong \overline{BC}$ and $\overline{AB} \cong \overline{DC}$ in Figure 4.3.

COROLLARY 4.1.4

The diagonals of a parallelogram bisect each other.

If diagonal $\overline{BD}$ were added to Figure 4.3, diagonals $\overline{AC}$ and $\overline{DB}$ would bisect each other.

Recall Theorem 2.1.4: "If two parallel lines are cut by a transversal, then the interior angles on the same side of the transversal are supplementary." A corollary of that theorem is stated next.

SSG **EXS. 1–6**

COROLLARY 4.1.5

Two consecutive angles of a parallelogram are supplementary.

In the figure for Example 2, $\angle R$ and $\angle S$ are supplementary. Other pairs of supplementary angles are $\angle S$ and $\angle T$, $\angle T$ and $\angle V$, and $\angle R$ and $\angle V$.

EXAMPLE 2

In $\square RSTV$, m$\angle S = 42°$, $ST = 5.3$ cm, and $VT = 8.1$ cm. Find:

a) m$\angle V$ **b)** m$\angle T$ **c)** RV **d)** RS

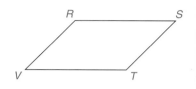

SOLUTION

a) m$\angle V = 42°$; $\angle V \cong \angle S$
because these are opposite $\angle$s of $\square RSTV$.

b) m$\angle T = 138°$; $\angle T$ and $\angle S$ are supplementary because these angles are consecutive angles of $\square RSTV$.

c) $RV = 5.3$ cm; $\overline{RV} \cong \overline{ST}$ because these are opposite sides of $\square RSTV$.

d) $RS = 8.1$ cm; $\overline{RS} \cong \overline{VT}$, also a pair of opposite sides of $\square RSTV$.

Example 3 illustrates Theorem 4.1.6, the fact that two parallel lines are everywhere equidistant. In general, the phrase *distance between two parallel lines* refers to the length of the perpendicular segment between the two parallel lines. These concepts will provide insight into the definition of altitude of a parallelogram.

STRATEGY FOR PROOF ■ Separating the Given Information

General Rule: When only part of the "Given" information leads to an important conclusion, it may be separated (for emphasis) from other Given facts in the statements of the proof.

Illustration: See lines 1 and 2 in the proof of Example 3. Notice that the Given facts found in statement 2 lead to statement 3.

THEOREM 4.1.6

Two parallel lines are everywhere equidistant.

Figure 4.4

EXAMPLE 3

GIVEN: $\overleftrightarrow{AB} \parallel \overleftrightarrow{CD}$
$\overline{AC} \perp \overleftrightarrow{CD}$ and $\overline{BD} \perp \overleftrightarrow{CD}$
(See Figure 4.4.)
PROVE: $\overline{AC} \cong \overline{BD}$

PROOF	
Statements	**Reasons**
1. $\overleftrightarrow{AB} \parallel \overleftrightarrow{CD}$	1. Given
2. $\overline{AC} \perp \overleftrightarrow{CD}$ and $\overline{BD} \perp \overleftrightarrow{CD}$	2. Given
3. $\overline{AC} \parallel \overline{BD}$	3. If two lines are $\perp$ to the same line, they are parallel to each other
4. $ABDC$ is a $\square$	4. If both pairs of opposite sides of a quadrilateral are $\parallel$, the quadrilateral is a $\square$
5. $\overline{AC} \cong \overline{BD}$	5. Opposite sides of a $\square$ are congruent

In Example 3, we used the definition of a parallelogram to prove that a particular quadrilateral was a parallelogram; however, there are other ways of establishing that a given quadrilateral is a parallelogram. We will investigate those methods in Section 4.2.

> **DEFINITION**
>
> An **altitude** of a parallelogram is a line segment drawn from one side so that it is perpendicular to the nonadjacent side (or to an extension of that side).

Figure 4.5

For ☐*RSTV*, $\overline{RW}$ and $\overline{SX}$ are altitudes to side $\overline{VT}$ (or to side $\overline{RS}$), as shown in Figure 4.5(a). With respect to side $\overline{RS}$, sometimes called base $\overline{RS}$, the length *RW* (or *SX*) is the *height* of *RSTV*. Similarly, in Figure 4.5(b), $\overline{TY}$ and $\overline{SZ}$ are altitudes to side $\overline{RV}$ (or to side $\overline{ST}$). Also, the length *TY* (or *ZS*) is called the *height* of parallelogram *RSTV* with respect to side $\overline{ST}$ (or $\overline{RV}$).

Next we consider an inequality relationship for the parallelogram. To develop this relationship, we need to investigate an inequality involving two triangles. We will use, but not prove, the following relationship found in Lemma 4.1.7.

> **LEMMA 4.1.7**
>
> If two sides of one triangle are congruent to two sides of a second triangle and the measure of the included angle of the first triangle is greater than the measure of the included angle of the second, then the length of the side opposite the included angle of the first triangle is greater than the length of the side opposite the included angle of the second.

> **Discover**
>
> On one piece of paper, draw a triangle (△*ABC*) so that *AB* = 3, *BC* = 5, and m∠*B* = 110°. Then draw △*DEF*, in which *DE* = 3, *EF* = 5, and m∠*E* = 50°. Which is longer, $\overline{AC}$ or $\overline{DF}$?
>
> ANSWER
>

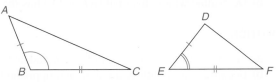

Figure 4.6

GIVEN: $\overline{AB} \cong \overline{DE}$ and $\overline{BC} \cong \overline{EF}$; m∠*B* > m∠*E* (See Figure 4.6.)

PROVE: *AC* > *DF*

Now we can compare the lengths of the diagonals of a parallelogram. For a parallelogram having no right angles, two consecutive angles are unequal but supplementary; thus, one angle of the parallelogram will be acute and the consecutive angle will be obtuse. In Figure 4.7(a) on page 174, ☐*ABCD* has acute angle *A* and obtuse angle *D*. In Figure 4.7(b), diagonal $\overline{AC}$ lies opposite the obtuse angle *ADC* in △*ACD*, and diagonal $\overline{BD}$ lies opposite the acute angle *DAB* in △*ABD*. In Figures 4.7(c) and (d), we have taken △*ACD* and △*ABD* from ☐*ABCD* of Figure 4.7(b). Note that the two sides of the triangles that include ∠*A* and ∠*D* are congruent. Also, note that $\overline{AC}$ (opposite obtuse ∠*D*) is longer than $\overline{DB}$ (opposite acute ∠*A*).

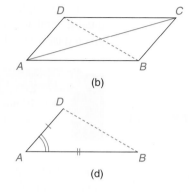

Figure 4.7

On the basis of Lemma 4.1.7 and the preceding discussion, we have the following theorem.

THEOREM 4.1.8

In a parallelogram with unequal pairs of consecutive angles, the longer diagonal lies opposite the obtuse angle.

EXAMPLE 4

In parallelogram *RSTV* (not shown), m∠*R* = 67°.

a) Find the measure of ∠*S*.
b) Determine which diagonal ($\overline{RT}$ or $\overline{SV}$) has the greater length.

SOLUTION

a) m∠*S* = 180° − 67° = 113° (∠*R* and ∠*S* are supplementary).
b) Because ∠*S* is obtuse, the diagonal opposite this angle is longer; that is, $\overline{RT}$ is the longer diagonal.

We use an indirect approach to solve Example 5.

EXAMPLE 5

In parallelogram *ABCD* (not shown), $\overline{AC}$ and $\overline{BD}$ are diagonals, and $BD > AC$. Determine which angles of the parallelogram are obtuse and which angles are acute.

SOLUTION Because the longer diagonal $\overline{BD}$ lies opposite angles *A* and *C*, these angles are obtuse. The remaining angles *B* and *D* are necessarily acute.

SSG EXS. 12–15

Our next example uses algebra to relate angle sizes and diagonal lengths.

EXAMPLE 6

In □*MNPQ* in Figure 4.8, m∠*M* = 2(*x* + 10) and m∠*Q* = 3*x* − 10. Determine which diagonal would be longer, $\overline{QN}$ or $\overline{MP}$.

Figure 4.8

SOLUTION Consecutive angles *M* and *Q* are supplementary, so m∠*M* + m∠*Q* = 180°.

$$2(x + 10) + (3x - 10) = 180$$
$$2x + 20 + 3x - 10 = 180$$
$$5x + 10 = 180 \rightarrow 5x = 170 \rightarrow x = 34$$

Then m∠M = 2(34 + 10) = 88°, whereas m∠Q = 3(34) − 10 = 92°. Because m∠Q > m∠M, diagonal $\overline{MP}$ (opposite ∠Q) would be longer than $\overline{QN}$ (opposite ∠M).

SPEED AND DIRECTION OF AIRCRAFT

For the application in Example 7, we will indicate the velocity of an airplane or of the wind by drawing a directed arrow. In each case, a scale is used on a grid in which a north-south line meets an east-west line at right angles. Consider the sketches in Figure 4.9 and read their descriptions.

Plane travels due north at 400 mph

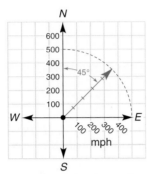

Plane travels at 500 mph in the direction N 45° E

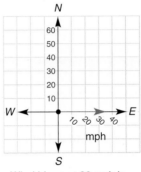

Wind blows at 30 mph in the direction west to east

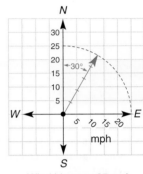

Wind blows at 25 mph in the direction N 30° E

Figure 4.9

In some scientific applications, such as Example 7, a parallelogram can be used to determine the solution to the problem. For instance, the Parallelogram Law enables us to determine the resulting speed and direction of an airplane when the velocity of the airplane and that of the wind are considered together. In Figure 4.10, the arrows representing the two velocities are placed head-to-tail from the point of origin. Because the order of the two velocities is reversible, the drawing leads to a parallelogram. In the parallelogram, it is the length and direction of the indicated diagonal that solve the problem. In Example 7, accuracy is critical in scaling the drawing that represents the problem. Otherwise, the ruler and protractor will give poor results in your answer.

Figure 4.10

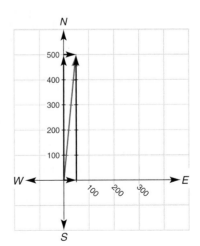

Figure 4.11

NOTE: In Example 7, kph means *kilometers per hour.*

EXAMPLE 7

An airplane travels due north at 500 kph. If the wind blows at 50 kph from west to east, what are the resulting speed and direction of the plane? See Figure 4.11 on page 175.

SOLUTION Using a ruler to measure the diagonal of the parallelogram, we find that the length corresponds to a speed of approximately 505 kph. Using a protractor, we find that the direction is approximately N 6° E.

SSG **EXS. 16–17**   **NOTE:** The actual speed is approximately 502.5 kph, while the direction is N 5.7° E.

Exercises 4.1

1. *ABCD* is a parallelogram.
   a) Using a ruler, compare the lengths of sides $\overline{AB}$ and $\overline{DC}$.
   b) Using a protractor, compare the measures of ∠*A* and ∠*C*.

Exercises 1, 2

2. *ABCD* is a parallelogram.
   a) Using a ruler, compare the lengths of $\overline{AD}$ and $\overline{BC}$.
   b) Using a protractor, compare the measures of ∠*B* and ∠*D*.

3. *MNPQ* is a parallelogram. Suppose that $MQ = 5$, $MN = 8$, and m∠*M* = 110°. Find:
   a) *QP*   c) m∠*Q*
   b) *NP*   d) m∠*P*

Exercises 3, 4

4. *MNPQ* is a parallelogram. Suppose that $MQ = 12.7$, $MN = 17.9$, and m∠*M* = 122°. Find:
   a) *QP*   c) m∠*Q*
   b) *NP*   d) m∠*P*

For Exercises 5 to 8, MNPQ is a parallelogram with diagonals $\overline{QN}$ and $\overline{MP}$.

Exercises 5–8

5. a) If $QN = 12.8$, find *QR*.
     b) If $MR = 5.3$, find *MP*.

6. a) If $QR = 7.3$, find *RN*.
     b) If $MP = 10.6$, find *RP*.

7. If $QR = 2x + 3$ and $RN = x + 7$, find *QR*, *RN*, and *QN*.

8. If $MR = 5(a + 7)$ and $MP = 12a + 34$, find *MR*, *RP*, and *MP*.

9. Given that $AB = 3x + 2$, $BC = 4x + 1$, and $CD = 5x - 2$, find the length of each side of ▱*ABCD*.

Exercises 9–16

10. Given that m∠*A* = $2x + 3$ and m∠*C* = $3x - 27$, find the measure of each angle of ▱*ABCD*.

11. Given that m∠*A* = $2x + 3$ and m∠*B* = $3x - 23$, find the measure of each angle of ▱*ABCD*.

12. Given that m∠*A* = $\frac{2x}{5}$ and m∠*B* = $\frac{x}{2}$, find the measure of each angle of ▱*ABCD*.

13. Given that m∠*A* = $\frac{2x}{3}$ and m∠*C* = $\frac{x}{2} + 20$, find the measure of each angle of ▱*ABCD*.

For Exercises 14 to 16, see the figure for Exercise 9.

14. Given that m$\angle A = 2x + y$, m$\angle B = 2x + 3y - 20$, and m$\angle C = 3x - y + 16$, find the measure of each angle of $\square ABCD$.

15. Assuming that m$\angle B >$ m$\angle A$ in $\square ABCD$, which diagonal ($\overline{AC}$ or $\overline{BD}$) would be longer?

16. Suppose that diagonals $\overline{AC}$ and $\overline{BD}$ of $\square ABCD$ are drawn and that $AC > BD$. Which angle ($\angle A$ or $\angle B$) would have the greater measure?

In Exercises 17 and 18, consider $\square RSTV$ with $\overline{VX} \perp \overline{RS}$ and $\overline{VY} \perp \overline{ST}$.

17. a) Which line segment is the altitude of $\square RSTV$ with respect to base $\overline{ST}$?

b) Which number is the height of $\square RSTV$ with respect to base $\overline{ST}$?

Exercises 17, 18

18. a) Which line segment is the altitude of $\square RSTV$ with respect to base $\overline{RS}$?
b) Which number is the height of $\square RSTV$ with respect to base $\overline{RS}$?

In Exercises 19 to 22, classify each statement as true or false. In Exercises 19 and 20, recall that the symbol $\subseteq$ means "is a subset of."

19. Where $Q = \{$quadrilaterals$\}$ and $P = \{$polygons$\}$, $Q \subseteq P$.

20. Where $Q = \{$quadrilaterals$\}$ and $P = \{$parallelograms$\}$, $Q \subseteq P$.

21. A parallelogram has point symmetry about the point where its two diagonals intersect.

22. A parallelogram has line symmetry and either diagonal is an axis of symmetry.

23. In quadrilateral $RSTV$, the midpoints of consecutive sides are joined in order. Try drawing other quadrilaterals and joining their midpoints. What can you conclude about the resulting quadrilateral in each case?

24. In quadrilateral $ABCD$, the midpoints of opposite sides are joined to form two intersecting segments. Try drawing other quadrilaterals and joining their opposite midpoints. What can you conclude about these segments in each case?

25. Quadrilateral $ABCD$ has $\overline{AB} \cong \overline{DC}$ and $\overline{AD} \cong \overline{BC}$. Using intuition, what type of quadrilateral is $ABCD$?

26. Quadrilateral $RSTV$ has $\overline{RS} \cong \overline{TV}$ and $\overline{RS} \parallel \overline{TV}$. Using intuition, what type of quadrilateral is $RSTV$?

In Exercises 27 to 30, use the definition of a parallelogram to complete each proof.

27. *Given:* $\overline{RS} \parallel \overline{VT}, \overline{RV} \perp \overline{VT}$, and $\overline{ST} \perp \overline{VT}$
Prove: $RSTV$ is a parallelogram

PROOF	
Statements	**Reasons**
1. $\overline{RS} \parallel \overline{VT}$	1. ?
2. ?	2. Given
3. ?	3. If two lines are $\perp$ to the same line, they are $\parallel$ to each other
4. ?	4. If both pairs of opposite sides of a quadrilateral are $\parallel$, the quad. is a $\square$

28. *Given:* $\overline{WX} \parallel \overline{ZY}$ and $\angle$s Z and Y are supplementary
Prove: $WXYZ$ is a parallelogram

PROOF	
Statements	**Reasons**
1. $\overline{WX} \parallel \overline{ZY}$	1. ?
2. ?	2. Given
3. ?	3. If two lines are cut by a transversal so that int. $\angle$s on the same side of the trans. are supplementary, these lines are $\parallel$
4. ?	4. If both pairs of opposite sides of a quadrilateral are $\parallel$, the quad. is a $\square$

29. *Given:* Parallelogram *RSTV*; also $\overline{XY} \parallel \overline{VT}$
Prove: $\angle 1 \cong \angle S$
Plan: First show that *RSYX* is a parallelogram.

30. *Given:* Parallelogram *ABCD* with $\overline{DE} \perp \overline{AB}$ and $\overline{FB} \perp \overline{AB}$
Prove: $\overline{DE} \cong \overline{FB}$
Plan: First show that *DEBF* is a parallelogram.

In Exercises 31 to 34, write a formal proof of each theorem or corollary.

31. The opposite angles of a parallelogram are congruent.

32. The opposite sides of a parallelogram are congruent.

33. The diagonals of a parallelogram bisect each other.

34. The consecutive angles of a parallelogram are supplementary.

35. The bisectors of two consecutive angles of □*HJKL* are shown. What can you conclude regarding $\angle P$?

36. When the bisectors of two consecutive angles of a parallelogram meet at a point on the remaining side, what type of triangle is:
a) △*DEC*? b) △*ADE*? c) △*BCE*?

37. Draw parallelogram *RSTV* with $m\angle R = 70°$ and $m\angle S = 110°$. Which diagonal of □*RSTV* has the greater length?

38. Draw parallelogram *RSTV* so that the diagonals have the lengths $RT = 5$ and $SV = 4$. Which two angles of □*RSTV* have the greater measure?

39. The following problem is based on the Parallelogram Law. In the scaled drawing, each unit corresponds to 50 mph. A small airplane travels due east at 250 mph. The wind is blowing at 50 mph in the direction due north. Using the scale provided, determine the approximate length of the indicated diagonal and use it to determine the speed of the airplane in miles per hour.

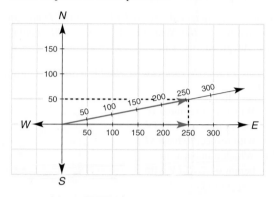

Exercises 39, 40

40. In the drawing for Exercise 39, the bearing (direction) in which the airplane travels is described as north $x°$ east, where x is the measure of the angle from the north axis toward the east axis. Using a protractor, find the approximate bearing of the airplane.

41. Two streets meet to form an obtuse angle at point *B*. On that corner, the newly poured foundation for a building takes the shape of a parallelogram. Which diagonal, $\overline{AC}$ or $\overline{BD}$, is longer?

Exercises 41, 42

42. To test the accuracy of the foundation's measurements, lines (strings) are joined from opposite corners of the building's foundation. How should the strings that are represented by $\overline{AC}$ and $\overline{BD}$ be related?

43. For a quadrilateral *ABCD*, the measures of its angles are $m\angle A = x + 16$, $m\angle B = 2(x + 1)$, $m\angle C = \frac{3}{2}x - 11$, and $m\angle D = \frac{7}{3}x - 16$. Determine the measure of each angle of *ABCD* and whether *ABCD* is a parallelogram.

*44. *Prove:* In a parallelogram, the sum of squares of the lengths of its diagonals is equal to the sum of squares of the lengths of its sides.

4.2 | The Parallelogram and Kite

KEY CONCEPTS Quadrilaterals That Are   Rectangle
Parallelograms   Kite

The quadrilaterals discussed in this section have two pairs of congruent sides.

THE PARALLELOGRAM

In Section 4.1, we sought to develop the properties of parallelograms. In this section, we prove that quadrilaterals with certain characteristics must be parallelograms.

> **STRATEGY FOR PROOF ■ The "Bottom Up" Approach to Proof**
>
> *General Rule:* This method answers the question, "Why would the last statement be true?" The answer often provides insight into the statement(s) preceding the last statement.
>
> *Illustration:* In line 8 of Example 1, we state that *RSTV* is a parallelogram by definition. With $\overline{RS} \parallel \overline{VT}$ in line 1, we need to show that $\overline{RV} \parallel \overline{ST}$ (shown in line 7).

EXAMPLE 1

Give a formal proof of Theorem 4.2.1.

> **THEOREM 4.2.1**
>
> If two sides of a quadrilateral are both congruent and parallel, then the quadrilateral is a parallelogram.

(a)

(b)

Figure 4.12

GIVEN: In Figure 4.12(a), $\overline{RS} \parallel \overline{VT}$ and $\overline{RS} \cong \overline{VT}$
PROVE: *RSTV* is a $\square$

PROOF

Statements	Reasons
1. $\overline{RS} \parallel \overline{VT}$ and $\overline{RS} \cong \overline{VT}$	1. Given
2. Draw diagonal $\overline{VS}$, as in Figure 4.12(b)	2. Exactly one line passes through two points
3. $\overline{VS} \cong \overline{VS}$	3. Identity
4. $\angle RSV \cong \angle SVT$	4. If two $\parallel$ lines are cut by a transversal, alternate interior $\angle$s are $\cong$
5. $\triangle RSV \cong \triangle TVS$	5. SAS
6. $\angle RVS \cong \angle VST$	6. CPCTC
7. $\overline{RV} \parallel \overline{ST}$	7. If two lines are cut by a transversal so that alternate interior $\angle$s are $\cong$, these lines are $\parallel$
8. *RSTV* is a $\square$	8. If both pairs of opposite sides of a quadrilateral are $\parallel$, the quadrilateral is a parallelogram

Figure 4.13

SSG EXS. 1–4

Consider the Discover activity at the left. Through it, we discover conditions that force a quadrilateral to be a parallelogram. This activity also leads to the following theorem; proof of the theorem is left to the student.

> ### THEOREM 4.2.2
>
> If both pairs of opposite sides of a quadrilateral are congruent, then the quadrilateral is a parallelogram.

Another quality of quadrilaterals that determines a parallelogram is found in Theorem 4.2.3. Its proof is also left to the student.

> ### THEOREM 4.2.3
>
> If the diagonals of a quadrilateral bisect each other, then the quadrilateral is a parallelogram.

To understand Theorem 4.2.3, connect vertices A, B, C, and D of Figure 4.13 to form quadrilateral $ABCD$. Now Theorem 4.2.2 can be used to show that $ABCD$ is a parallelogram.

When a figure is drawn to represent the hypothesis of a theorem, we should not include more conditions than the hypothesis states. Relative to Theorem 4.2.3, if we drew two diagonals that not only bisected each other but also were equal in length, then the quadrilateral would be the special type of parallelogram known as a **rectangle.** We will deal with rectangles in the next section.

THE KITE

The next quadrilateral type that we consider is known as a *kite*, a quadrilateral that gets its name from the child's toy pictured at the left. In the construction of the kite, there are two pairs of congruent *adjacent* sides. See Figure 4.14(a) on page 181.

> ### DEFINITION
>
> A **kite** is a quadrilateral with two distinct pairs of congruent adjacent sides.

See the Discover activity at the lower left.

The word *distinct* is used in the definition of kite to clarify that the kite does not have four congruent sides. The word *adjacent* is necessary for the quadrilateral to be a kite; if the word adjacent were replaced by the word *opposite*, the quadrilateral described would be a parallelogram.

> ### THEOREM 4.2.4
>
> In a kite, one pair of opposite angles are congruent.

In Example 2, we verify Theorem 4.2.4 by proving that $\angle B \cong \angle D$ in Figure 4.14(a) on page 181. With congruent sides as marked, $\angle A \not\cong \angle C$ in kite $ABCD$.

EXAMPLE 2

Complete the proof of Theorem 4.2.4.

GIVEN: Kite $ABCD$ with congruent sides as marked. [See Figure 4.14(a).]

PROVE: $\angle B \cong \angle D$

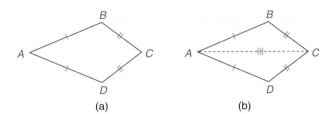

(a)   (b)

Figure 4.14

PROOF	
Statements	**Reasons**
1. Kite *ABCD*	1. ?
2. $\overline{BC} \cong \overline{CD}$ and $\overline{AB} \cong \overline{AD}$	2. A kite has two pairs of ≅ adjacent sides
3. Draw $\overline{AC}$ [Figure 4.14(b)]	3. Through two points, there is exactly one line
4. $\overline{AC} \cong \overline{AC}$	4. ?
5. $\triangle ACD \cong \triangle ACB$	5. ?
6. ?	6. CPCTC

SSG **EXS. 5–10**

The proof of Theorem 4.2.4 required auxiliary diagonal $\overline{AC}$; if drawn, diagonal $\overline{BD}$ would not determine two congruent triangles—thus, no help!

When observing an old barn or shed, we often see that it has begun to lean. Unlike a triangle, which is rigid in shape [Figure 4.15(a)] and bends only when broken, a quadrilateral [Figure 4.15(b)] does *not* provide the same level of strength and stability. In the construction of a house, bridge, building, or swing set [Figure 4.15(c)], note the use of wooden or metal triangles as braces.

(a)

(b)

(c)

Figure 4.15

The brace in the swing set in Figure 4.15(c) suggests the following theorem.

> **THEOREM 4.2.5**
>
> The line segment that joins the midpoints of two sides of a triangle is parallel to the third side and has a length equal to one-half the length of the third side.

Refer to Figure 4.16(a) on page 182, in which *M* is the midpoint of $\overline{AB}$, while *N* is the midpoint of $\overline{AC}$. Theorem 4.2.5 claims that $\overline{MN} \parallel \overline{BC}$ and $MN = \frac{1}{2}(BC)$. We will prove the first part of this theorem but leave the second part as an exercise.

The line segment that joins the midpoints of two sides of a triangle is parallel to the third side of the triangle.

GIVEN: In Figure 4.16(a), △ABC with midpoints M and N of $\overline{AB}$ and $\overline{AC}$, respectively

PROVE: $\overline{MN} \parallel \overline{BC}$

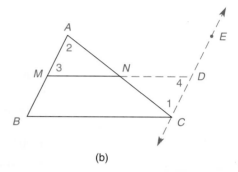

(a)   (b)

Figure 4.16

PROOF

Statements	Reasons
1. △*ABC*, with midpoints *M* and *N* of $\overline{AB}$ and $\overline{AC}$, respectively	1. Given
2. Through *C*, construct $\overleftrightarrow{CE} \parallel \overline{AB}$, as in Figure 4.16(b)	2. Parallel Postulate
3. Extend $\overline{MN}$ to meet $\overleftrightarrow{CE}$ at *D*, as in Figure 4.16(b)	3. Exactly one line passes through two points
4. $\overline{AM} \cong \overline{MB}$ and $\overline{AN} \cong \overline{NC}$	4. The midpoint of a segment divides it into ≅ segments
5. ∠1 ≅ ∠2 and ∠4 ≅ ∠3	5. If two ∥ lines are cut by a transversal, alternate interior ∠s are ≅
6. △*ANM* ≅ △*CND*	6. AAS
7. $\overline{AM} \cong \overline{DC}$	7. CPCTC
8. $\overline{MB} \cong \overline{DC}$	8. Transitive (both are ≅ to $\overline{AM}$)
9. Quadrilateral *BMDC* is a ▱	9. If two sides of a quadrilateral are both ≅ and ∥, the quadrilateral is a parallelogram
10. $\overline{MN} \parallel \overline{BC}$	10. Opposite sides of a ▱ are ∥

In the preceding proof, we needed to show that a quadrilateral having certain characteristics is a parallelogram.

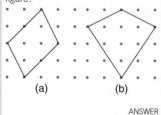
STRATEGY FOR PROOF ■ Proving That a Quadrilateral Is a Parallelogram

General Rule: Methods for proof include the definition of a parallelogram as well as Theorems 4.2.1, 4.2.2, and 4.2.3.

Illustration: In the proof of Theorem 4.2.5, statements 2 and 8 allowed the conclusion in statement 9 (used Theorem 4.2.1).

Theorem 4.2.5 also asserts the following statement. The proof is left to the student.

> The line segment that joins the midpoints of two sides of a triangle has a length equal to one-half the length of the third side.

With midpoints M and N in Figure 4.17, we see that $MN = \frac{1}{2}(ST)$.

EXAMPLE 3

In $\triangle RST$ in Figure 4.17, M and N are the midpoints of $\overline{RS}$ and $\overline{RT}$, respectively.

  a) If $ST = 12.7$, find MN.
  b) If $MN = 15.8$, find ST.

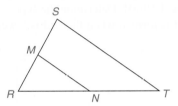

Figure 4.17

Discover

Draw a triangle $\triangle ABC$ with midpoints D of $\overline{CA}$ and E of $\overline{CB}$. Cut out $\triangle CDE$ and place it at the base $\overline{AB}$. By sliding $\overline{DE}$ along $\overline{AB}$, what do you find?

ANSWER

$DE = \frac{1}{2}(AB)$ or $AB = 2(DE)$

SOLUTION

  a) $MN = \frac{1}{2}(ST)$, so $MN = \frac{1}{2}(12.7) = 6.35$.

  b) $MN = \frac{1}{2}(ST)$, so $15.8 = \frac{1}{2}(ST)$. Multiplying by 2,
    we find that $ST = 31.6$.

EXAMPLE 4

GIVEN: $\triangle ABC$ in Figure 4.18, with D the midpoint of $\overline{AC}$ and E the midpoint of $\overline{BC}$; $DE = 2x + 1$; $AB = 5x - 1$
FIND: x, DE, and AB

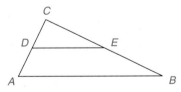

Figure 4.18

SOLUTION Applying Theorem 4.2.5,

$$DE = \frac{1}{2}(AB)$$

so

$$2x + 1 = \frac{1}{2}(5x - 1)$$

Multiplying by 2, we have

$$4x + 2 = 5x - 1$$
$$3 = x$$

Therefore, $DE = 2 \cdot 3 + 1 = 7$. Similarly, $AB = 5 \cdot 3 - 1 = 14$.

SSG **EXS. 11–15**   **NOTE:** In Example 4, a check shows that $DE = \frac{1}{2}(AB)$.

In the final example of this section, we consider product design. Also see related Exercises 17 and 18 of this section.

EXAMPLE 5

In a studio apartment, there is a Murphy bed that folds down from the wall. In the vertical position, the design shows drop-down legs of equal length; that is, $AB = CD$ [see Figure 4.19(a)]. Determine the type of quadrilateral $ABDC$, shown in Figure 4.19(b), that is formed when the bed is lowered to a horizontal position.

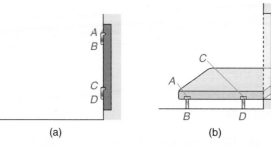

(a)   (b)

Figure 4.19

SOLUTION See Figure 4.19(a). Because $AB = CD$, it follows that $AB + BC = BC + CD$; here, BC was added to each side of the equation. But $AB + BC = AC$ and $BC + CD = BD$. Thus, $AC = BD$ by substitution.

In Figure 4.19(b), we know that $AB = CD$ and $AC = BD$. Because both pairs of opposite sides of the quadrilateral are congruent, $ABDC$ is a parallelogram.

NOTE: In Section 4.3, we will be able to show that $ABDC$ of Figure 4.19(b) is a *rectangle* (a special type of parallelogram).

Exercises 4.2

1. a) As shown, must quadrilateral $ABCD$ be a parallelogram?
   b) Given the lengths of the sides as shown, is the measure of $\angle A$ unique?

2. a) As shown, must $RSTV$ be a parallelogram?
   b) With measures as shown, is it necessary that $RS = 8$?

3. In the drawing, suppose that $\overline{WY}$ and $\overline{XZ}$ bisect each other. What type of quadrilateral is $WXYZ$?

Exercises 3, 4

4. In the drawing, suppose that $\overline{ZX}$ is the perpendicular bisector of $\overline{WY}$. What type of quadrilateral is $WXYZ$?

5. A carpenter lays out boards of lengths 8 ft, 8 ft, 4 ft, and 4 ft by placing them end to end.
   a) If these are joined at the ends to form a quadrilateral that has the 8-ft pieces connected in order, what type of quadrilateral is formed?
   b) If these are joined at the ends to form a quadrilateral that has the 4-ft and 8-ft pieces alternating, what type of quadrilateral is formed?

6. A carpenter joins four boards of lengths 6 ft, 6 ft, 4 ft, and 4 ft, in that order, to form quadrilateral *ABCD* as shown.
   a) What type of quadrilateral is formed?
   b) How are angles *B* and *D* related?

7. In parallelogram *ABCD* (not shown), $AB = 8$, $m\angle B = 110°$, and $BC = 5$. Which diagonal has the greater length?

8. In quadrilateral *WXYZ*, the measures of selected angles are shown.
   a) What type of quadrilateral is *WXYZ*?
   b) Which diagonal of the quadrilateral has the greater length?

9. In △*ABC*, *M* and *N* are midpoints of $\overline{AC}$ and $\overline{BC}$, respectively. If $AB = 12.36$, how long is $\overline{MN}$?

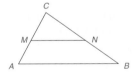

Exercises 9, 10

10. In △*ABC*, *M* and *N* are midpoints of $\overline{AC}$ and $\overline{BC}$, respectively. If $MN = 7.65$, how long is $\overline{AB}$?

In Exercises 11 to 14, assume that X, Y, and Z are midpoints of the sides of △RST.

11. If $RS = 12$, $ST = 14$, and $RT = 16$, find:
    a) *XY*   b) *XZ*   c) *YZ*

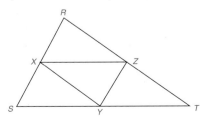

Exercises 11–14

For Exercises 12 to 14, see the figure for Exercise 11.

12. If $XY = 6$, $YZ = 8$, and $XZ = 10$, find:
    a) *RS*   b) *ST*   c) *RT*

13. If the perimeter (sum of the lengths of all three sides) of △*RST* is 20, what is the perimeter of △*XYZ*?

14. If the perimeter (sum of the lengths of all three sides) of △*XYZ* is 12.7, what is the perimeter of △*RST*?

15. Consider any kite.
    a) Does it have line symmetry? If so, describe an axis of symmetry.
    b) Does it have point symmetry? If so, describe the point of symmetry.

16. Consider any parallelogram.
    a) Does it have line symmetry? If so, describe an axis of symmetry.
    b) Does it have point symmetry? If so, describe the point of symmetry.

17. For compactness, the drop-down wheels of a stretcher (or gurney) are folded under it as shown. In order for the board's upper surface to be parallel to the ground when the wheels are dropped, what relationship must exist between $\overline{AB}$ and $\overline{CD}$?

18. For compactness, the drop-down legs of an ironing board fold up under the board. A sliding mechanism at point *A* and the legs being connected at common midpoint *M* cause the board's upper surface to be parallel to the floor. How are $\overline{AB}$ and $\overline{CD}$ related?

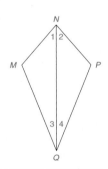

In Exercises 19 to 24, complete each proof.

19. Given: $\angle 1 \cong \angle 2$ and $\angle 3 \cong \angle 4$
    Prove: *MNPQ* is a kite

PROOF	
Statements	**Reasons**
1. $\angle 1 \cong \angle 2$ and $\angle 3 \cong \angle 4$	1. ?
2. $\overline{NQ} \cong \overline{NQ}$	2. ?
3. ?	3. ASA
4. $\overline{MN} \cong \overline{PN}$ and $\overline{MQ} \cong \overline{PQ}$	4. ?
5. ?	5. If a quadrilateral has two pairs of $\cong$ adjacent sides, it is a kite

20. *Given:* Quadrilateral *ABCD*, with midpoints *E, F, G,* and *H* of the sides

Prove: $\overline{EF} \parallel \overline{HG}$

PROOF

Statements	Reasons
1. ?	1. Given
2. Draw $\overline{AC}$	2. Through two points, there is one line
3. In $\triangle ABC$, $\overline{EF} \parallel \overline{AC}$ and in $\triangle ADC$, $\overline{HG} \parallel \overline{AC}$	3. ?
4. ?	4. If two lines are $\parallel$ to the same line, these lines are $\parallel$ to each other

21. *Given:* *M-Q-T* and *P-Q-R* such that *MNPQ* and *QRST* are $\square$s

Prove: $\angle N \cong \angle S$

22. *Given:* $\square WXYZ$ with diagonals $\overline{WY}$ and $\overline{XZ}$

Prove: $\triangle WMX \cong \triangle YMZ$

23. *Given:* Kite *HJKL* with diagonal $\overline{HK}$

Prove: $\overrightarrow{HK}$ bisects $\angle LHJ$

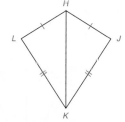

24. *Given:* $\square MNPQ$, with *T* the midpoint of $\overline{MN}$ and *S* the midpoint of $\overline{QP}$

Prove: $\triangle QMS \cong \triangle NPT$ and *MSPT* is a $\square$

In Exercises 25 to 28, write a formal proof of each theorem or corollary.

25. If both pairs of opposite sides of a quadrilateral are congruent, then the quadrilateral is a parallelogram.

26. If the diagonals of a quadrilateral bisect each other, then the quadrilateral is a parallelogram.

27. In a kite, one diagonal is the perpendicular bisector of the other diagonal.

28. One diagonal of a kite bisects two of the angles of the kite.

In Exercises 29 to 31, M and N are the midpoints of sides $\overline{RS}$ and $\overline{RT}$ of $\triangle RST$, respectively.

29. *Given:* $MN = 2y - 3$
$ST = 3y$
Find: *y, MN,* and *ST*

30. *Given:* $MN = x^2 + 5$
$ST = x(2x + 5)$
Find: *x, MN,* and *ST*

31. *Given:* $RM = RN = 2x + 1$
$ST = 5x - 3$
$m\angle R = 60°$
Find: *x, RM,* and *ST*

Exercises 29–31

For Exercises 32 to 35, consider kite ABCD with $\overline{AB} \cong \overline{AD}$ and $\overline{BC} \cong \overline{DC}$.

32. For kite *ABCD*, $m\angle B = \frac{3x}{2} + 2$ and $m\angle D = \frac{9x}{4} - 3$. Find *x*.

33. For kite *ABCD*, $m\angle C = m\angle B - 30$ while $m\angle A = m\angle B - 50$. Find $m\angle B$

34. For kite *ABCD*, $AB = BC + 5$. If the perimeter of *ABCD* is 59.2, find *BC*.

35. For kite *ABCD*, $AB = \frac{x}{6} + 5$, $AD = \frac{x}{3} + 3$, and $BC = x - 2$. Find the perimeter of *ABCD*.

36. *RSTV* is a kite, with $\overline{RS} \perp \overline{ST}$ and $\overline{RV} \perp \overline{VT}$. If $m\angle STV = 40°$, how large is the angle formed
a) by the bisectors of $\angle RST$ and $\angle STV$?
b) by the bisectors of $\angle SRV$ and $\angle RST$?

37. In concave kite *ABCD*, there is an interior angle at vertex *B* that is a reflex angle. Given that $m\angle A = m\angle C = m\angle D = 30°$, find the measure of the indicated reflex angle.

38. If the length of side $\overline{AB}$ (for kite *ABCD*) is 6 in., find the length of $\overline{AC}$ (not shown). Recall that $m\angle A = m\angle C = m\angle D = 30°$.

Exercises 37, 38

*39. Prove that the segment that joins the midpoints of two sides of a triangle has a length equal to one-half the length of the third side.

(*HINT: In the drawing, $\overline{MN}$ is extended to D, a point on $\overline{CD}$. Also, $\overline{CD}$ is parallel to $\overline{AB}$.*)

*40. Prove that when the midpoints of consecutive sides of a quadrilateral are joined in order, the resulting quadrilateral is a parallelogram.

4.3	The Rectangle, Square, and Rhombus

KEY CONCEPTS	Rectangle Square	Rhombus Pythagorean Theorem

Figure 4.20

THE RECTANGLE

In this section, we investigate special parallelograms. The first of these is the rectangle (symbol □ and abbreviated "rect."), which is defined as follows:

> **DEFINITION**
>
> A **rectangle** is a parallelogram that has a right angle. (See Figure 4.20.)

Any reader who is familiar with the rectangle may be confused by the fact that the preceding definition calls for only one right angle. Because a rectangle is a parallelogram by definition, the fact that a rectangle has four right angles is easily proved by applying Corollaries 4.1.3 and 4.1.5; in order, these corollaries remind us that a parallelogram has "opposite angles that are congruent" and "consecutive angles that are supplementary." The proof of Corollary 4.3.1 is left to the student.

> **COROLLARY 4.3.1**
>
> All angles of a rectangle are right angles.

The following theorem is true for rectangles, but not for parallelograms in general.

> **COROLLARY 4.3.2**
>
> The diagonals of a rectangle are congruent.

Reminder

A rectangle is a parallelogram. Thus, it has all the properties of a parallelogram plus some properties of its own.

NOTE: To follow the flow of the proof in Example 1, it may be best to draw triangles *NMQ* and *PQM* of Figure 4.21 on page 188 separately.

EXAMPLE 1

Complete a proof of Theorem 4.3.2. Use Figure 4.21.

GIVEN: ▱*MNPQ* with diagonals $\overline{MP}$ and $\overline{NQ}$

PROVE: $\overline{MP} \cong \overline{NQ}$

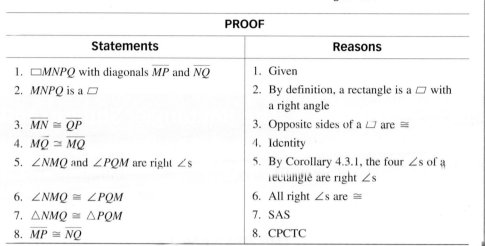

Figure 4.21

PROOF

Statements	Reasons
1. ▱*MNPQ* with diagonals $\overline{MP}$ and $\overline{NQ}$	1. Given
2. *MNPQ* is a ▱	2. By definition, a rectangle is a ▱ with a right angle
3. $\overline{MN} \cong \overline{QP}$	3. Opposite sides of a ▱ are ≅
4. $\overline{MQ} \cong \overline{MQ}$	4. Identity
5. ∠*NMQ* and ∠*PQM* are right ∠s	5. By Corollary 4.3.1, the four ∠s of a rectangle are right ∠s
6. ∠*NMQ* ≅ ∠*PQM*	6. All right ∠s are ≅
7. △*NMQ* ≅ △*PQM*	7. SAS
8. $\overline{MP} \cong \overline{NQ}$	8. CPCTC

SSG **EXS. 1–4**

Square *ABCD*

Figure 4.22

THE SQUARE

All rectangles are parallelograms; some parallelograms are rectangles; and some rectangles are *squares*.

DEFINITION

A **square** is a rectangle that has two congruent adjacent sides. (See Figure 4.22.)

COROLLARY 4.3.3

All sides of a square are congruent.

SSG **EXS. 5–7**

Because a square is a type of rectangle, it has four right angles and its diagonals are congruent. Because a square is also a parallelogram, its opposite sides are parallel. For any square, we can show that the diagonals are perpendicular; see Exercise 32.

In Chapter 8, we measure area in "square units." For the calculation of area, we count the number of congruent squares that fit inside a geometric region.

THE RHOMBUS

The next type of quadrilateral we consider is the rhombus. The plural of the word *rhombus* is *rhombi* (pronounced rhŏm-bī).

DEFINITION

A **rhombus** is a parallelogram with two congruent adjacent sides. (See Figure 4.23.)

Geometry in the Real World

80"

72"

A carpenter has installed the frame for a patio door. To check the frame for "square-ness," he measures the lengths of the two diagonals of the rectangular opening...to be sure that the lengths are equal and that the opening is actually a rectangle.

Figure 4.23

In Figure 4.23, the adjacent sides $\overline{AB}$ and $\overline{AD}$ of rhombus *ABCD* are marked congruent. Because a rhombus is a type of parallelogram, it is also necessary that $\overline{AB} \cong \overline{DC}$ and $\overline{AD} \cong \overline{BC}$. Thus, we have Corollary 4.3.4.

COROLLARY 4.3.4

All sides of a rhombus are congruent.

We will use Corollary 4.3.4 in the proof of the following theorem.

THEOREM 4.3.5

The diagonals of a rhombus are perpendicular.

To visualize Theorem 4.3.5, see Figure 4.24(a).

Geometry in the Real World

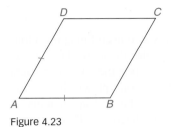

The jack used in changing an automobile tire illustrates the shape of a rhombus.

EXAMPLE 2

Study the picture proof of Theorem 4.3.5. In the proof, pairs of triangles are congruent by the reason SSS.

PICTURE PROOF OF THEOREM 4.3.5

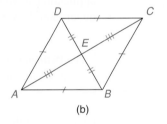

Figure 4.24

GIVEN: Rhombus *ABCD*, with diagonals $\overline{AC}$ and $\overline{DB}$ [see Figure 4.24(a)].

PROVE: $AC \perp DB$

PROOF: Rhombus *ABCD* has congruent sides as indicated. The diagonals of rhombus *ABCD* (a type of parallelogram) bisect each other. By SSS, the four small triangles are congruent; thus, each angle at vertex *E* is a right angle. Then $\overline{AC} \perp \overline{DB}$.

Discover

Sketch regular hexagon *RSTVWX*. Draw diagonals $\overline{RT}$ and $\overline{XV}$. What type of quadrilateral is *RTVX*?

ANSWER
Rectangle

An alternative definition of *square* is "A square is a rhombus whose adjacent sides form a right angle." Therefore, a further property of a square is that its diagonals are perpendicular. Because the square and the rhombus are both types of parallelograms, we have the following consequence.

 EXS. 8–11

COROLLARY 4.3.6

The diagonals of a rhombus (or square) are perpendicular bisectors of each other.

THE PYTHAGOREAN THEOREM

The Pythagorean Theorem, which deals with right triangles, is also useful in applications involving quadrilaterals that have right angles. In antiquity, the theorem claimed that "the square upon the hypotenuse equals the sum of the squares upon the legs of the right triangle." See Figure 4.25(a). This interpretation involves the area concept, which we study in a later chapter. By counting squares in Figure 4.25(a), one sees that 25 "square units" is the sum of 9 and 16 square units. Our interpretation of the Pythagorean Theorem uses number (length) relationships.

Figure 4.25

The Pythagorean Theorem will be proved in Section 5.4. Although it was introduced in Section 3.2, we restate the Pythagorean Theorem for convenience and then review its application to the *right* triangle in Example 3. When right angle relationships exist in quadrilaterals, we often apply the "rule of Pythagoras" as well; see Examples 4, 5, and 6.

> **The Pythagorean Theorem** In a right triangle with hypotenuse of length c and legs of lengths a and b, it follows that $c^2 = a^2 + b^2$.

Provided that the lengths of two of the sides of a right triangle are known, the Pythagorean Theorem can be applied to determine the length of the third side. In Example 3, we seek the length of the hypotenuse in a right triangle whose lengths of legs are known. When we are using the Pythagorean Theorem, c must represent the length of the hypotenuse; however, either leg can be chosen for length a (or b).

EXAMPLE 3

What is the length of the hypotenuse of a right triangle whose legs measure 6 in. and 8 in.? (See Figure 4.26.)

SOLUTION

$$c^2 = a^2 + b^2$$
$$c^2 = 6^2 + 8^2$$
$$c^2 = 36 + 64 \rightarrow c^2 = 100 \rightarrow c = 10 \text{ in.}$$

Figure 4.26

In the following example, the diagonal of a rectangle separates it into two congruent right triangles. As shown in Figure 4.27, the diagonal of the rectangle is the hypotenuse of each right triangle formed by the diagonal.

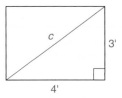

Figure 4.27

EXAMPLE 4

What is the length of the diagonal in a rectangle whose sides measure 3 ft and 4 ft?

SOLUTION For each triangle in Figure 4.27, $c^2 = a^2 + b^2$ becomes $c^2 = 3^2 + 4^2$ or $c^2 = 9 + 16$. Then $c^2 = 25$, so $c = 5$. The length of the diagonal is 5 ft. ∎

In Example 5, we apply Corollary 4.3.6, "The diagonals of the rhombus are perpendicular bisectors of each other."

EXAMPLE 5

What is the length of each side of a rhombus whose diagonals measure 10 cm and 24 cm? (See Figure 4.28.)

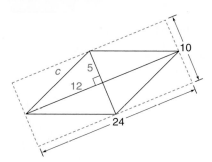

Figure 4.28

SOLUTION The diagonals of a rhombus are perpendicular bisectors of each other. Thus, the diagonals separate the rhombus shown into four congruent right triangles, each of which has legs of lengths 5 cm and 12 cm. For each triangle, $c^2 = a^2 + b^2$ becomes $c^2 = 5^2 + 12^2$, or $c^2 = 25 + 144$. Then $c^2 = 169$, so $c = 13$. The length of each side of the rhombus is 13 cm. ∎

Discovery: The Magic Square

To create a certain *Magic Square*, we place the numbers 1, 2, 3, 4,..., and 9 in the square configuration in such a way that the sum of the 3 numbers along each side *and* each diagonal has the sum of 15.

2	7	6
9	5	1
4	3	8

By using the numbers 3, 4, 5,..., and 11, complete the Magic Square so that the sum of any 3 numbers on a side or a diagonal is 21.

?	?	?
?	7	?
?	?	4

ANSWER

4	11	6
9	7	5
8	3	10

SSG **EXS. 12–14**

EXAMPLE 6

On a softball diamond (actually a square), the distance along the base paths is 60 ft. Using the triangle in Figure 4.29, find the distance from home plate to second base.

Figure 4.29

SOLUTION Using $c^2 = a^2 + b^2$, we have
$$c^2 = 60^2 + 60^2$$
$$c^2 = 7200$$
Then
$$c = \sqrt{7200} \qquad \text{or} \qquad c \approx 84.85 \text{ ft.}$$

Discover

A logo is a geometric symbol or emblem that represents a company. The very sight of the symbol serves as advertising for the company or corporation. Many logos are derived from common geometric shapes. Which company is represented by these symbols?

The sides of an equilateral triangle are trisected and then connected as shown, and finally the middle sections are erased.

The vertices of a regular pentagon are joined to the "center" of the polygon as shown.

A square is superimposed on and centered over a long and narrow parallelogram as shown. Interior line segments are then eliminated.

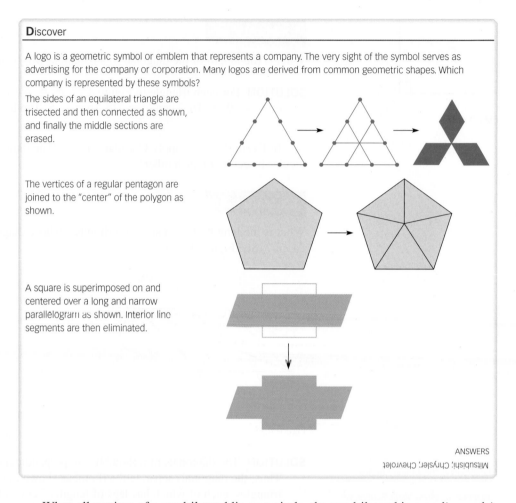

ANSWERS

Mitsubishi; Chrysler; Chevrolet

Reminder

A circle is a set of points in a plane that are at a fixed distance from a point known as the center of the circle.

When all vertices of a quadrilateral lie on a circle, the quadrilateral is a *cyclic quadrilateral*. As it happens, all rectangles are cyclic quadrilaterals, but no rhombus is a cyclic quadrilateral. The key factor in determining whether a quadrilateral is cyclic lies in the fact that the diagonals must intersect at a point that is equidistant from all four vertices. In Figure 4.30(a), rectangle *ABCD* is cyclic because *A, B, C,* and *D* all lie on the circle. However, rhombus *WXYZ* in Figure 4.30(b) is *not* cyclic because *X* and *Z* cannot lie on the circle when *W* and *Y* do lie on the circle.

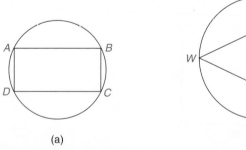

(a)   (b)

Figure 4.30

EXAMPLE 7

For cyclic rectangle *ABCD*, *AB* = 8. Diagonal $\overline{DB}$ of the rectangle contains the center of the circle and *DB* = 10. Find the perimeter of *ABCD* shown in Figure 4.31.

SOLUTION $AB = DC = 8$. Let $AD = b$; applying the Pythagorean Theorem with right triangle ABD, we find that $10^2 = 8^2 + b^2$.

Then $100 = 64 + b^2$ and $b^2 = 36$, so $b = \sqrt{36}$ or 6. In turn, $AD = BC = 6$. The perimeter of $ABCD$ is $2(8) + 2(6) = 16 + 12 = 28$.

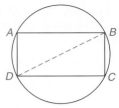

Figure 4.31

Exercises 4.3

1. Being as specific as possible, name the *type* of quadrilateral that
   a) has four congruent sides.
   b) is a parallelogram with a right angle.

2. Being as specific as possible, name the *type* of quadrilateral that
   a) has two pairs of congruent adjacent sides.
   b) has two pairs of congruent opposite sides.

3. Being as specific as possible, name the *type* of parallelogram that
   a) has congruent diagonals.
   b) has perpendicular diagonals.

4. Being as specific as possible, name the *type* of rhombus that
   a) has all angles congruent.
   b) has congruent diagonals.

5. If the diagonals of a parallelogram are perpendicular and congruent, what can you conclude regarding the parallelogram?

6. If the diagonals of a quadrilateral are perpendicular bisectors of each other (but not congruent), what can you conclude regarding the quadrilateral?

7. A line segment joins the midpoints of two opposite sides of a rectangle as shown. What can you conclude regarding $\overline{MN}$ and MN?

In Exercises 8 to 10, use the properties of rectangles to solve each problem. Rectangle ABCD is shown in the figure.

Exercises 8–10

8. *Given:* $AB = 5$ and $BC = 12$
   Find: CD, AD, and AC (not shown)

In Exercises 9 and 10, see the figure for Exercise 8.

9. *Given:* $AB = 2x + 7$, $BC = 3x + 4$, and $CD = 3x + 2$
   Find: x and DA

10. *Given:* $AB = x + y$, $BC = x + 2y$, $CD = 2x - y - 1$, and $DA = 3x - 3y + 1$
    Find: x and y

In Exercises 11 to 14, consider □MNPQ with diagonals $\overline{MP}$ and $\overline{NQ}$. When the answer is not a whole number, leave a square root answer.

11. If $MQ = 6$ and $MN = 8$, find NQ and MP.

12. If $QP = 9$ and $NP = 6$, find NQ and MP.

13. If $NP = 7$ and $MP = 11$, find QP and MN.

14. If $QP = 15$ and $MP = 17$, find MQ and NP.

Exercises 11–14

In Exercises 15 to 18, consider rhombus ABCD with diagonals $\overline{AC}$ and $\overline{DB}$. When the answer is not a whole number, leave a square root answer.

15. If $AE = 5$ and $DE = 4$, find AD.

16. If $AE = 6$ and $EB = 5$, find AB.

17. If $AC = 10$ and $DB = 6$, find AD.

18. If $AC = 14$ and $DB = 10$, find BC.

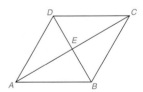

Exercises 15–18

19. *Given:* □$ABCD$ (not shown) with $AB = 8$ and $BC = 6$; M and N are the midpoints of sides $\overline{AB}$ and $\overline{BC}$, respectively.
    Find: MN

20. *Given:* Rhombus $RSTV$ (not shown) with diagonals $\overline{RT}$ and $\overline{SV}$ so that $RT = 8$ and $SV = 6$
    Find: RS, the length of a side

For Exercises 21 and 22, let P = {parallelograms},
R = {rectangles}, and H = {rhombi}. Classify as true
or false:

21. $H \subseteq P$ and $R \subseteq P$

22. $R \cup H = P$ and $R \cap H = \varnothing$

In Exercises 23 and 24, supply the missing statements and
reasons.

23. *Given:* Quadrilateral *PQST*
with midpoints
A, *B*, *C*, and *D* of
the sides
Prove: *ABCD* is a ▱

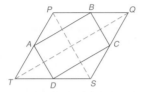

PROOF	
Statements	**Reasons**
1. Quadrilateral *PQST* with midpoints *A*, *B*, *C*, and *D* of the sides	1. ?
2. Draw $\overline{TQ}$	2. Through two points, there is one line
3. $\overline{AB} \parallel \overline{TQ}$ in △*TPQ*	3. The line joining the midpoints of two sides of a triangle is ∥ to the third side
4. $\overline{DC} \parallel \overline{TQ}$ in △*TSQ*	4. ?
5. $\overline{AB} \parallel \overline{DC}$	5. ?
6. Draw $\overline{PS}$	6. ?
7. $\overline{AD} \parallel \overline{PS}$ in △*TSP*	7. ?
8. $\overline{BC} \parallel \overline{PS}$ in △*PSQ*	8. ?
9. $\overline{AD} \parallel \overline{BC}$	9. ?
10. ?	10. If both pairs of opposite sides of a quadrilateral are ∥, the quad. is a ▱

24. *Given:* ▱*WXYZ* with
diagonals $\overline{WY}$
and $\overline{XZ}$
Prove: $\angle 1 \cong \angle 2$

PROOF	
Statements	**Reasons**
1. ?	1. Given
2. ?	2. The diagonals of a rectangle are ≅
3. $\overline{WZ} \cong \overline{XY}$	3. The opposite sides of a rectangle are ≅
4. $\overline{ZY} \cong \overline{ZY}$	4. ?
5. △*XZY* ≅ △*WYZ*	5. ?
6. ?	6. ?

25. Which type(s) of quadrilateral(s) is(are) necessarily cyclic?
a) A square   b) A parallelogram

26. Which type(s) of quadrilateral(s) is(are) necessarily cyclic?
a) A kite   b) A rectangle

27. Find the perimeter of the cyclic quadrilateral shown.

28. Find the perimeter of the square shown.

In Exercises 29 to 31, explain why each statement is true.

29. All angles of a rectangle are right angles.

30. All sides of a rhombus are congruent.

31. All sides of a square are congruent.

In Exercises 32 to 37, write a formal proof of each theorem.

32. The diagonals of a square are perpendicular.

33. A diagonal of a rhombus bisects two angles of the rhombus.

34. If the diagonals of a parallelogram are congruent, the parallelogram is a rectangle.

35. If the diagonals of a parallelogram are perpendicular, the parallelogram is a rhombus.

36. If the diagonals of a parallelogram are congruent and perpendicular, the parallelogram is a square.

37. If the midpoints of the sides of a rectangle are joined in order, the quadrilateral formed is a rhombus.

In Exercises 38 and 39, you will need to use the square root
($\sqrt{\ }$) function of your calculator.

38. A wall that is 12 ft long by 8 ft high has a triangular brace along the diagonal. Use a calculator to approximate the length of the brace to the nearest tenth of a foot.

39. A walk-up ramp moves horizontally 20 ft while rising 4 ft. Use a calculator to approximate its length to the nearest tenth of a foot.

40. a) Argue that the midpoint of the hypotenuse of a right triangle is equidistant from the three vertices of the triangle. Use the fact that the congruent diagonals of a rectangle bisect each other. Be sure to provide a drawing.
b) Use the relationship from part (a) to find *CM*, the length of the median to the hypotenuse of right △*ABC*, in which m∠*C* = 90°, *AC* = 6, and *BC* = 8.

41. Two sets of rails (railroad tracks are equally spaced) intersect but not at right angles. Being as specific as possible, indicate what type of quadrilateral *WXYZ* is formed.

42. In square *ABCD* (not shown), point *E* lies on side $\overline{DC}$. If *AB* = 8 and *AE* = 10, find *BE*.

43. In square *ABCD* (not shown), point *E* lies in the interior of *ABCD* in such a way that △*ABE* is an equilateral triangle. Find m∠*DEC*.

44. The sides of square *ABCD* are trisected at the indicated points. If *AB* = 3, find the perimeter of
(a) quadrilateral *EGIK*.
(b) quadrilateral *EHIL*.

4.4	**The Trapezoid**

KEY CONCEPTS	Trapezoid Bases Legs	Base Angles Median	Isosceles Trapezoid Right Trapezoid

> **DEFINITION**
>
> A **trapezoid** is a quadrilateral with exactly two parallel sides.

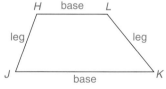

Figure 4.32

Consider Figure 4.32. If $\overline{HL} \parallel \overline{JK}$, then *HJKL* is a trapezoid. The abbreviation for the word trapezoid is "trap." The parallel sides $\overline{HL}$ and $\overline{JK}$ of trapezoid *HJKL* are its **bases,** and the nonparallel sides $\overline{HJ}$ and $\overline{LK}$ are its **legs.** Because ∠*J* and ∠*K* both have base $\overline{JK}$ for a side, they are a pair of **base angles** of the trapezoid; ∠*H* and ∠*L* are also a pair of base angles because $\overline{HL}$ is a base.

When the midpoints of the two legs of a trapezoid are joined, the resulting line segment is known as the **median** of the trapezoid. Given that *M* and *N* are the midpoints of the legs $\overline{HJ}$ and $\overline{LK}$ in trapezoid *HJKL*, $\overline{MN}$ is the median of the trapezoid. [See Figure 4.33(a)].

If the two legs of a trapezoid are congruent, the trapezoid is known as an **isosceles trapezoid.** In Figure 4.33(b), *RSTV* is an **isosceles trapezoid** because $\overline{RS} \parallel \overline{VT}$ and $\overline{RV} \cong \overline{ST}$.

(a)

(b)

(c)

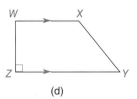

(d)

Figure 4.33

Every trapezoid contains two pairs of consecutive interior angles that are supplementary. Each of these pairs of angles is formed when parallel lines are cut by a transversal. In Figure 4.33(c) on the previous page, $\overline{HL} \parallel \overline{JK}$; thus, angles H and J are supplementary, as are angles L and K. See the "Reminder" at the left.

Given that $\overline{WX} \parallel \overline{ZY}$ and that $\angle Z$ is a right angle in Figure 4.33(d), trapezoid $WXYZ$ is a **right trapezoid**. Based upon the "Reminder," we conclude that $\angle W$ is a right angle as well.

EXAMPLE 1

In Figure 4.32 on page 195, $\overline{HL} \parallel \overline{JK}$. Suppose that $m\angle H = 107°$ and $m\angle K = 58°$. Find $m\angle J$ and $m\angle L$.

SOLUTION Because $\overline{HL} \parallel \overline{JK}$, $\angle$s H and J are supplementary angles, as are $\angle$s L and K. Then $m\angle H + m\angle J = 180$ and $m\angle L + m\angle K = 180$. Substitution leads to $107 + m\angle J = 180$ and $m\angle L + 58 = 180$, so $m\angle J = 73°$ and $m\angle L = 122°$.

DEFINITION

An **altitude** of a trapezoid is a line segment from one base of the trapezoid perpendicular to the opposite base (or to an extension of that base).

In Figure 4.34, $\overline{HX}$, $\overline{LY}$, $\overline{JP}$, and $\overline{KQ}$ are altitudes of trapezoid $HJKL$. The length of any altitude of $HJKL$ is called the *height* of the trapezoid.

Figure 4.34

SSG EXS. 1–6

Theorems 4.4.1 and 4.4.2 involve isosceles trapezoids.

THEOREM 4.4.1

The base angles of an isosceles trapezoid are congruent.

EXAMPLE 2

Study the picture proof of Theorem 4.4.1.

PICTURE PROOF OF THEOREM 4.4.1

(a)

(b)

Figure 4.35

GIVEN: Trapezoid $RSTV$ with $\overline{RV} \cong \overline{ST}$ and $\overline{RS} \parallel \overline{VT}$ [See Figure 4.35(a)].

PROVE: $\angle V \cong \angle T$ and $\angle R \cong \angle S$

PROOF: By drawing $\overline{RY} \perp \overline{VT}$ and $\overline{SZ} \perp \overline{VT}$, we see that $\overline{RY} \cong \overline{SZ}$ (Theorem 4.1.6). By HL, $\triangle RYV \cong \triangle SZT$ so $\angle V \cong \angle T$ (CPCTC). $\angle R \cong \angle S$ in Figure 4.35(a) because these angles are supplementary to congruent angles ($\angle V$ and $\angle T$).

Geometry in the Real World

Some of the glass panels and trim pieces of the light fixture are isosceles trapezoids. Other glass panels are pentagons.

The following statement is a corollary of Theorem 4.4.1. Its proof is left to the student.

COROLLARY 4.4.2

The diagonals of an isosceles trapezoid are congruent.

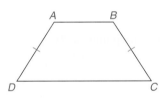

Figure 4.36

Given the isosceles trapezoid $ABCD$ in Figure 4.36, diagonals $\overline{AC}$ and $\overline{BD}$ (if drawn) would necessarily be congruent; that is, $\overline{AC} \cong \overline{BD}$.

EXAMPLE 3

In Figure 4.36, $\overline{AB} \parallel \overline{DC}$ and $\overline{AD} \cong \overline{BC}$.

a) Find the measures of the angles of $ABCD$ if m$\angle A = 12x + 30$ and m$\angle B = 10x + 46$.
b) Find the length of each diagonal (not shown) if it is known that $AC = 2y - 5$ and $BD = 19 - y$.

SOLUTION

a) Applying Theorem 4.4.1, m$\angle A$ = m$\angle B$. Then $12x + 30 = 10x + 46$, so $2x = 16$ and $x = 8$. Then m$\angle A = 12(8) + 30$ or $126°$, and m$\angle B = 10(8) + 46$ or $126°$. Subtracting $(180 - 126 = 54)$, we determine the supplements of $\angle$s A and B. That is, m$\angle C$ = m$\angle D = 54°$.
b) By Corollary 4.4.2, $\overline{AC} \cong \overline{BD}$, so $2y - 5 = 19 - y$. Then $3y = 24$ and $y = 8$. Thus, $AC = 2(8) - 5 = 11$. Also $BD = 19 - 8 = 11$.

For completeness, we state two further properties of the isosceles trapezoid.

1. An isosceles trapezoid has line symmetry; this line (or axis) of symmetry is the perpendicular bisector of either base.
2. An isosceles trapezoid is cyclic; the center of the circle containing all four vertices of the trapezoid is the point of intersection of the perpendicular bisectors of any two consecutive sides (or of the two legs).

Discover

An ancient Chinese puzzle known as a *tangram* begins with the specified separation of a square into seven precise smaller shapes known as *tans*. As shown, the tans are a smaller square, five right triangles, and a parallelogram. Each puzzle challenges the player to rearrange all the tans to form recognizable two-dimensional shapes:

  geometric shapes: a trapezoid, a rectangle, a parallelogram, etc.

  animal shapes: a cat, a rabbit, a horse, etc.

  object shapes: a shoe, a teapot, a person walking, etc.

In researching (Googling) the tangram, you may discover this topic and challenge to be of great personal interest.

The proof of the following theorem is left as Exercise 33. We apply Theorem 4.4.3 in Examples 4 and 5.

The length of the median of a trapezoid equals one-half the sum of the lengths of the two bases.

NOTE: The length of the median of a trapezoid is the "average" of the lengths of the bases. Where m is the length of the median and b_1 and b_2 are the lengths of the bases, $m = \frac{1}{2}(b_1 + b_2)$; equivalently, $m = \frac{b_1 + b_2}{2}$.

Figure 4.37

EXAMPLE 4

In trapezoid $RSTV$ in Figure 4.37, $\overline{RS} \parallel \overline{VT}$ and M and N are the midpoints of $\overline{RV}$ and $\overline{TS}$, respectively. Find the length of median $\overline{MN}$ if $RS = 12$ and $VT = 18$.

SOLUTION Using Theorem 4.4.3, $MN = \frac{1}{2}(RS + VT)$, so $MN = \frac{1}{2}(12 + 18)$, or $MN = \frac{1}{2}(30)$. Thus, $MN = 15$

EXAMPLE 5

In trapezoid $RSTV$ in Figure 4.37, $\overline{RS} \parallel \overline{VT}$ and M and N are the midpoints of $\overline{RV}$ and $\overline{TS}$, respectively. Find MN, RS, and VT if $RS = 2x$, $MN = 3x - 5$, and $VT = 2x + 10$.

SOLUTION Using Theorem 4.4.3, we have $MN = \frac{1}{2}(RS + VT)$, so

$$3x - 5 = \frac{1}{2}[2x + (2x + 10)] \qquad \text{or} \qquad 3x - 5 = \frac{1}{2}(4x + 10)$$

Then $3x - 5 = 2x + 5$ and $x = 10$. Now $RS = 2x = 2(10)$, so $RS = 20$. Also, $MN = 3x - 5 = 3(10) - 5$, therefore, $MN = 25$. Finally, $VT = 2x + 10$; therefore, $VT = 2(10) + 10 = 30$.

NOTE: As a check, $MN = \frac{1}{2}(RS + VT)$ leads to the true statement $25 = \frac{1}{2}(20 + 30)$.

The median of a trapezoid is parallel to each base.

The proof of Theorem 4.4.4 is left as Exercise 28. If $\overline{MN}$ is the median of trapezoid $RSTV$ in Figure 4.37, then $\overline{MN} \parallel \overline{RS}$ and $\overline{MN} \parallel \overline{VT}$.

Theorems 4.4.5 and 4.4.6 enable us to show that a trapezoid with certain characteristics is an isosceles trapezoid. We state these theorems as follows:

SSG **EXS. 7–12**

If two base angles of a trapezoid are congruent, the trapezoid is an isosceles trapezoid.

Figure 4.38

Consider the following plan for proving Theorem 4.4.5. See Figure 4.38.

GIVEN: Trapezoid $RSTV$ with $\overline{RS} \parallel \overline{VT}$ and $\angle V \cong \angle T$

PROVE: $RSTV$ is an isosceles trapezoid

PLAN: Draw auxiliary altitudes $\overline{RX}$ and $\overline{SY}$ of trapezoid $RSTV$. Because $\overline{RX} \cong \overline{SY}$ by Theorem 4.1.6, we can show that $\triangle RXV \cong \triangle SYT$ by AAS. Then $\overline{RV} \cong \overline{ST}$ by CPCTC and $RSTV$ is an isosceles trapezoid.

THEOREM 4.4.6

If the diagonals of a trapezoid are congruent, the trapezoid is an isosceles trapezoid.

(a)

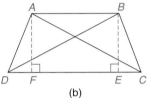

(b)

Figure 4.39

Theorem 4.4.6 has a lengthy proof, for which we have provided a sketch.

GIVEN: Trapezoid $ABCD$ with $\overline{AB} \parallel \overline{DC}$ and $\overline{AC} \cong \overline{DB}$ [See Figure 4.39(a).]

PROVE: $ABCD$ is an isosceles trapezoid.

PLAN: Draw $\overline{AF} \perp \overline{DC}$ and $\overline{BE} \perp \overline{DC}$ in Figure 4.39(b). By Theorem 4.1.6, $\overline{AF} \cong \overline{BE}$. Then $\triangle AFC \cong \triangle BED$ by HL. In turn, $\angle ACD \cong \angle BDC$ by CPCTC. With $\overline{DC} \cong \overline{DC}$ by Identity, $\triangle ACD \cong \triangle BDC$ by SAS. By CPCTC, $\overline{AD} \cong \overline{BC}$. Then trapezoid $ABCD$ is isosceles.

In Figure 4.40, lines a, b, and c are parallel. If $\overline{AB} \cong \overline{BC}$, we claim that $\overline{DE} \cong \overline{EF}$. See Theorem 4.4.7 and its "PLAN" for proof.

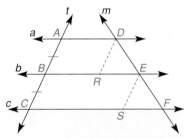

Figure 4.40

THEOREM 4.4.7

If three (or more) parallel lines intercept congruent line segments on one transversal, then they intercept congruent line segments on any transversal.

SSG EXS. 13–16

GIVEN: Parallel lines a, b, and c cut by transversal t so that $\overline{AB} \cong \overline{BC}$; also transversal m in Figure 4.40

PROVE: $\overline{DE} \cong \overline{EF}$

PLAN: Through D and E, draw $\overline{DR} \parallel \overleftrightarrow{AB}$ and $\overline{ES} \parallel \overleftrightarrow{AB}$. In each $\square$ formed, $\overline{DR} \cong \overline{AB}$ and $\overline{ES} \cong \overline{BC}$. Given $\overline{AB} \cong \overline{BC}$, it follows that $\overline{DR} \cong \overline{ES}$. By AAS, we can show $\triangle DER \cong \triangle EFS$; then $\overline{DE} \cong \overline{EF}$ by CPCTC.

EXAMPLE 6

In Figure 4.40, $a \parallel b \parallel c$. If $AB = BC = 7.2$ and $DE = 8.4$, find EF.

SSG EXS. 16, 17

SOLUTION Applying Theorem 4.4.7, $DE = EF$. It follows that $EF = 8.4$.

Based upon Theorem 4.4.7, we can construct three (or more) congruent line segments on a given line segment. In Example 7, we trisect a line segment.

EXAMPLE 7

Construct points X and Y so that $\overline{AB}$ in Figure 4.41(a) is trisected.

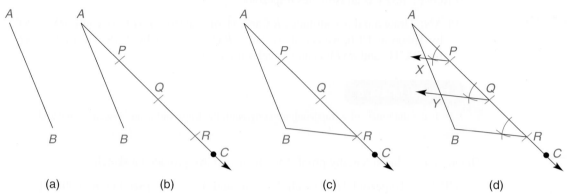

(a)   (b)   (c)   (d)

Figure 4.41

SOLUTION

Figures 4.41(a) and (b): Draw an auxiliary ray $\overrightarrow{AC}$. Now use the compass to mark off congruent arcs at points P, Q, and R. Then $\overline{AP} \cong \overline{PQ} \cong \overline{QR}$.

Figure 4.41(c): Join point R of Figure 4.41(b) to point B.

Figure 4.41(d): Construct angles at Q and P that are congruent to $\angle BRQ$. See arcs of Figure 4.41(d) and note that $\overline{BR} \parallel \overline{YQ} \parallel \overline{XP}$; according to Theorem 4.4.7, $\overline{AX} \cong \overline{XY} \cong \overline{YB}$.

Exercises 4.4

1. Find the measures of the remaining angles of trapezoid $ABCD$ (not shown) if $\overline{AB} \parallel \overline{DC}$ and m$\angle A = 58°$ and m$\angle C = 125°$.

2. Find the measures of the remaining angles of trapezoid $ABCD$ (not shown) if $\overline{AB} \parallel \overline{DC}$ and m$\angle B = 63°$ and m$\angle D = 118°$.

3. What *type* of trapezoid
   a) has congruent diagonals?
   b) has congruent base angles?

4. What *type* of quadrilateral is formed when the midpoints of the sides of an isosceles trapezoid are joined in order?

5. Given isosceles trapezoid $ABCD$, find
   (a) AC, if $BD = 12.3$ cm.
   (b) x, if $AC = 3(x + 7)$ and $BD = 9(x - 6)$.

6. In trapezoid $ABCD$, $\overline{MN}$ is the median. Without writing a formal proof, explain why $MN = \frac{1}{2}(AB + DC)$.

7. If $\angle H$ and $\angle J$ are supplementary, what type of quadrilateral is $HJKL$?

Exercises 7–8

8. If $\angle H$ and $\angle J$ are supplementary in $HJKL$, are $\angle K$ and $\angle L$ necessarily supplementary also?

For Exercises 9 and 10, consider isosceles trapezoid RSTV with $\overline{RS} \parallel \overline{VT}$ and midpoints M, N, P, and Q of the sides.

9. Would *RSTV* have symmetry with respect to
a) $\overleftrightarrow{MP}$? b) $\overleftrightarrow{QN}$?

Exercises 9, 10

10. a) Does $QN = \frac{1}{2}(RS + VT)$?
b) Does $MP = \frac{1}{2}(RV + ST)$?

In Exercises 11 to 16, the drawing shows trapezoid ABCD with $\overline{AB} \parallel \overline{DC}$; also, M and N are midpoints of $\overline{AD}$ and $\overline{BC}$, respectively.

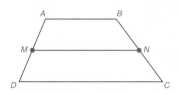

Exercises 11–16

11. *Given:* $AB = 7.3$ and $DC = 12.1$
 Find: MN

12. *Given:* $MN = 6.3$ and $DC = 7.5$
 Find: AB

13. *Given:* $AB = 8.2$ and $MN = 9.5$
 Find: DC

14. *Given:* $AB = 7x + 5$, $DC = 4x - 2$, and
    $MN = 5x + 3$
 Find: x

15. *Given:* $AB = 6x + 5$ and $DC = 8x - 1$
 Find: MN, in terms of x

16. *Given:* $AB = x + 3y + 4$ and $DC = 3x + 5y - 2$
 Find: MN, in terms of x and y

17. *Given:* $ABCD$ is an isosceles trapezoid
 Prove: $\triangle ABE$ is isosceles

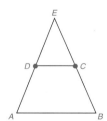

Exercises 17, 18

18. *Given:* Isosceles $\triangle ABE$ with $\overline{AE} \cong \overline{BE}$; also, D and C are midpoints of $\overline{AE}$ and $\overline{BE}$, respectively
 Prove: $ABCD$ is an isosceles trapezoid

19. In isosceles trapezoid *WXYZ* with bases $\overline{ZY}$ and $\overline{WX}$, $ZY = 8$, $YX = 10$, and $WX = 20$. Find height h (the length of $\overline{ZD}$ or $\overline{YE}$).

20. In trapezoid *WXYZ* with bases $\overline{ZY}$ and $\overline{WX}$, $ZY = 12$, $YX = 10$, $WZ = 17$, and $ZD = 8$. Find the length of base $\overline{WX}$.

Exercises 19, 20

21. In isosceles trapezoid *MNPQ* with $\overline{MN} \parallel \overline{QP}$, diagonal $\overline{MP} \perp \overline{MQ}$. If $PQ = 13$ and $NP = 5$, how long is diagonal $\overline{MP}$?

22. In trapezoid *RSTV*, $\overline{RV} \parallel \overline{ST}$, $m\angle SRV = 90°$, and M and N are midpoints of the nonparallel sides.
a) What type of triangle is $\triangle RMN$?
b) If $ST = 13$, $RV = 17$, and $RS = 16$, how long is $\overline{RN}$?
c) What type of triangle is $\triangle RVN$?

[HINT: Use lengths of sides found in part (b).]

23. Each vertical section of a suspension bridge is in the shape of a trapezoid. For additional support, a vertical cable is placed midway as shown. If the two vertical columns shown have heights of 20 ft and 24 ft and the section is 10 ft wide, what will the height of the cable be?

24. The state of Nevada approximates the shape of a trapezoid with these dimensions for boundaries: 340 miles on the north, 515 miles on the east, 435 miles on the south, and 225 miles on the west. If A and B are points located midway across the north and south boundaries, what is the approximate distance from A to B?

25. In the figure, $a \parallel b \parallel c$ and B is the midpoint of $\overline{AC}$. If $AB = 2x + 3$, $BC = x + 7$, and $DE = 3x + 2$, find the length of $\overline{EF}$.

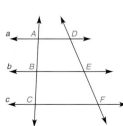

26. In the figure, $a \parallel b \parallel c$ and B is the midpoint of $\overline{AC}$.
If $AB = 2x + 3y$,
$BC = x + y + 7$,
$DE = 2x + 3y + 3$, and
$EF = 5x - y + 2$, find x and y.

Exercises 25, 26

In Exercises 27 to 33, complete a formal proof.

27. The diagonals of an isosceles trapezoid are congruent.

28. The median of a trapezoid is parallel to each base.

29. If two consecutive angles of a quadrilateral are supplementary, the quadrilateral is a trapezoid.

30. If two base angles of a trapezoid are congruent, the trapezoid is an isosceles trapezoid.

31. If three parallel lines intercept congruent segments on one transversal, then they intercept congruent segments on any transversal.

32. If the midpoints of the sides of an isosceles trapezoid are joined in order, then the quadrilateral formed is a rhombus.

33. *Given:* $\overline{EF}$ is the median of trapezoid $ABCD$
Prove: $EF = \frac{1}{2}(AB + DC)$

(*HINT: Using Theorem 4.4.7, show that M is the midpoint of $\overline{AC}$. For $\triangle ADC$ and $\triangle CBA$, apply Theorem 4.2.5.*)

Exercises 33–35

For Exercises 34 and 35, $\overline{EF}$ is the median of trapezoid $ABCD$ in the figure above.

34. Suppose that $AB = 12.8$ and $DC = 18.4$. Find:
a) MF c) EF
b) EM d) Whether $EF = \frac{1}{2}(AB + DC)$

35. Suppose that $EM = 7.1$ and $MF = 3.5$. Find:
a) AB c) EF
b) DC d) Whether $EF = \frac{1}{2}(AB + DC)$

36. *Given:* $\overline{AB} \parallel \overline{DC}$
$m\angle A = m\angle B = 56°$
$\overline{CE} \parallel \overline{DA}$ and $\overrightarrow{CF}$
bisects $\angle DCB$
Find: $m\angle FCE$

37. In a gambrel style roof, the gable end of a barn has the shape of an isosceles trapezoid surmounted by an isosceles triangle. If $AE = 30$ ft and $BD = 24$ ft, find:
a) AS b) VD c) CD d) DE

38. Successive steps on a ladder form isosceles trapezoids with the sides. $AH = 2$ ft and $BI = 2.125$ ft.
a) Find GN, the width of the bottom step
b) Which step is the median of the trapezoid with bases $\overline{AH}$ and $\overline{GN}$?

39. The vertical sidewall of an in-ground pool that is 24 ft in length has the shape of a right trapezoid. What is the depth of the pool in the middle?

40. For the in-ground pool shown in Exercise 39, find the length of the sloped bottom from point D to point C.

41. With $\overline{MN} \parallel \overline{QP}$ and $\overline{MQ} \perp \overline{QP}$, $MNPQ$ is a right trapezoid. Find
a) $m\angle P$, if $m\angle MNP - m\angle M = 31°$.
b) the length of $\overline{NR}$, if $MN = 6$ in., $NP = 5$ in., and $QP = 9$ in.

42. With $\overline{MN} \parallel \overline{QP}$ and $\angle M \cong \angle Q$, $MNPQ$ is a right trapezoid. Find
a) $m\angle P$, if $m\angle MNP - m\angle P = 54°$.
b) the length of side $\overline{NP}$, if $MN = 15$ cm, $MQ = 12$ cm, and $PQ = 20$ cm

*43. In trap. $ABCD$ (not shown), $m\angle A = \frac{x}{2} + 10$, $m\angle B = \frac{x}{3} + 50$, and $m\angle C = \frac{x}{5} + 50$. Find all possible values of x.

*44. In trap. $ABCD$, $\overline{BC} \perp \overline{AB}$ and $\overline{BC} \perp \overline{CD}$. If $DA = 17$, $AB = 6$, and $BC = 8$, find the perimeter of $\triangle DAC$.

*45. Draw and then trisect $\overline{AB}$. Use the construction method found in Example 7.

PERSPECTIVE ON HISTORY

SKETCH OF THALES

One of the most significant contributors to the development of geometry was the Greek mathematician Thales of Miletus (625–547 B.C.). Thales is credited with being the "Father of Geometry" because he was the first person to organize geometric thought and utilize the deductive method as a means of verifying propositions (theorems). It is not surprising that Thales made original discoveries in geometry. Just as significant as his discoveries was Thales' persistence in verifying the claims of his predecessors. In this textbook, you will find that propositions such as these are only a portion of those that can be attributed to Thales:

> Chapter 1: If two straight lines intersect, the opposite
> (vertical) angles formed are equal.
> Chapter 3: The base angles of an isosceles triangle
> are equal.
> Chapter 5: The lengths of the sides of similar triangles
> are proportional.
> Chapter 6: An angle inscribed in a semicircle is a
> right angle.

Thales' knowledge of geometry was matched by the wisdom that he displayed in everyday affairs. For example, he is known to have measured the height of the Great Pyramid of Egypt by comparing the lengths of the shadows cast by the pyramid and by his own staff. Thales also used his insights into geometry to measure the distances from the land to ships at sea.

Perhaps the most interesting story concerning Thales was one related by Aesop (famous for fables). It seems that Thales was on his way to market with his beasts of burden carrying saddlebags filled with salt. Quite by accident, one of the mules discovered that rolling in the stream where he was led to drink greatly reduced this load; of course, this was due to the dissolving of salt in the saddlebags. On subsequent trips, the same mule continued to lighten his load by rolling in the water. Thales soon realized the need to do something (anything!) to modify the mule's behavior. When preparing for the next trip, Thales filled the offensive mule's saddlebags with sponges. When the mule took his usual dive, he found that his load was heavier than ever. Soon the mule realized the need to keep the saddlebags out of the water. In this way, it is said that Thales discouraged the mule from allowing the precious salt to dissolve during later trips to market.

PERSPECTIVE ON APPLICATIONS

SQUARE NUMBERS AS SUMS

In algebra, there is a principle that is generally "proved" by a quite sophisticated method known as mathematical induction. However, verification of the principle is much simpler when provided a geometric justification.

In the following paragraphs, we:

1. State the principle
2. Illustrate the principle
3. Provide the geometric justification for the principle

> Where n is a counting number, the sum of the first n positive odd counting numbers is n^2.

The principle stated above is illustrated for various choices of n.

> Where $n = 1, 1 = 1^2$.
> Where $n = 2, 1 + 3 = 2^2$, or 4.
> Where $n = 3, 1 + 3 + 5 = 3^2$, or 9.
> Where $n = 4, 1 + 3 + 5 + 7 = 4^2$, or 16.

The geometric explanation for this principle utilizes a *wrap-around* effect. Study the diagrams in Figure 4.42.

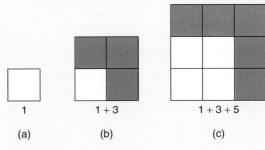

    1   1 + 3   1 + 3 + 5

  (a)   (b)   (c)

Figure 4.42

Given a unit square (one with sides of length 1), we build a second square by wrapping 3 unit squares around the first unit square; in Figure 4.42(b), the "wrap-around" is indicated by 3 shaded squares. Now for the second square (sides of length 2), we form the next square by wrapping 5 unit squares around this square; see Figure 4.42(c).

The next figure in the sequence of squares illustrates that

$$1 + 3 + 5 + 7 = 4^2, \text{ or } 16$$

In the "wrap-around," we emphasize that the next number in the sum is an odd number. The "wrap-around" approach adds $2 \times 3 + 1$, or 7 unit squares in Figure 4.43. When building each sequential square, we always add an odd number of unit squares as in Figure 4.43.

Figure 4.43

PROBLEM

Use the following principle to answer each question:

Where n is a counting number, the sum of the first n positive odd counting numbers is n^2.

a) Find the sum of the first five positive odd integers; that is, find $1 + 3 + 5 + 7 + 9$.

b) Find the sum of the first six positive odd integers.

c) How many positive odd integers were added to obtain the sum 81?

SOLUTIONS

a) 5^2, or 25   **b)** 6^2, or 36   **c)** 9, because $9^2 = 81$

Summary

A Look Back at Chapter 4

The goal of this chapter has been to develop the properties of quadrilaterals, including special types of quadrilaterals such as the parallelogram, rectangle, and trapezoid. The Overview of Chapter 4 on page 205 summarizes the properties of quadrilaterals.

A Look Ahead to Chapter 5

In the next chapter, similarity will be defined for all polygons, with an emphasis on triangles. The Pythagorean Theorem, which we applied in Chapter 4, will be proved in Chapter 5. Special right triangles will be discussed.

Key Concepts

4.1

Quadrilateral • Skew Quadrilateral • Parallelogram • Diagonals of a Parallelogram • Altitudes of a Parallelogram

4.2

Quadrilaterals That Are Parallelograms • Rectangle • Kite

4.3

Rectangle • Square • Rhombus • Pythagorean Theorem

4.4

Trapezoid • Bases • Legs • Base Angles • Median • Isosceles Trapezoid • Right Trapezoid

Overview ■ Chapter 4

Properties of Quadrilaterals

	Parallelo-gram	Rectangle	Rhombus	Square	Kite	Trapezoid	Isosceles Trapezoid
Congruent sides	Both pairs of opposite sides	Both pairs of opposite sides	All four sides	All four sides	Both pairs of adjacent sides	Possible; also see isosceles trapezoid	Pair of legs: possibly 3 sides
Parallel sides	Both pairs of opposite sides	Both pairs of opposite sides	Both pairs of opposite sides	Both pairs of opposite sides	Generally none	Pair of bases	Pair of bases
Perpendicular sides	If the parallelogram is a rectangle or square	Consecutive pairs	If rhombus is a square	Consecutive pairs	Possible	Possibly a right trapezoid	Generally none
Congruent angles	Both pairs of opposite angles	All four angles	Both pairs of opposite angles	All four angles	One pair of opposite angles	Possible; also see isosceles trapezoid	Each pair of base angles
Supplementary angles	All pairs of consecutive angles	Any two angles	All pairs of consecutive angles	Any two angles	Possibly two pairs	Each pair of leg angles	Each pair of leg angles
Diagonal relationships	Bisect each other	Congruent; bisect each other	Perpendicular; bisect each other and interior angles	Congruent; perpendicular; bisect each other and interior angles	Perpendicular; one bisects other and two interior angles	Intersect	Congruent

Chapter 4   Review Exercises

State whether the statements in Review Exercises 1 to 12 are always true (A), sometimes true (S), or never true (N).

1. A square is a rectangle.

2. If two of the angles of a trapezoid are congruent, then the trapezoid is isosceles.

3. The diagonals of a trapezoid bisect each other.

4. The diagonals of a parallelogram are perpendicular.

5. A rectangle is a square.

6. The diagonals of a square are perpendicular.

7. Two consecutive angles of a parallelogram are supplementary.

8. Opposite angles of a rhombus are congruent.

9. The diagonals of a rectangle are congruent.

10. The four sides of a kite are congruent.

11. The diagonals of a parallelogram are congruent.

12. The diagonals of a kite are perpendicular bisectors of each other.

13. *Given:* $\square ABCD$
    $CD = 2x + 3$
    $BC = 5x - 4$
    Perimeter of $\square ABCD = 96$ cm
    Find: The lengths of the sides of $\square ABCD$

Exercises 13, 14

14. *Given:* $\square ABCD$
    $m\angle A = 2x + 6$
    $m\angle B = x + 24$
    Find: $m\angle C$

15. The diagonals of $\square ABCD$ (not shown) are perpendicular. If one diagonal has a length of 10 and the other diagonal has a length of 24, find the perimeter of the parallelogram.

16. *Given:* $\square MNOP$
    $m\angle M = 4x$
    $m\angle O = 2x + 50$
    Find: $m\angle M$ and $m\angle P$

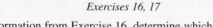

Exercises 16, 17

17. Using the information from Exercise 16, determine which diagonal ($\overline{MO}$ or $\overline{PN}$) would be longer.

18. In quadrilateral $ABCD$, M is the midpoint only of $\overline{BD}$ and $\overline{AC} \perp \overline{DB}$ at M. What special type of quadrilateral is $ABCD$?

19. In isosceles trapezoid $DEFG$, $\overline{DE} \parallel \overline{GF}$ and $m\angle D = 108°$. Find the measures of the other angles in the trapezoid.

20. One base of a trapezoid has a length of 12.3 cm and the length of the other base is 17.5 cm. Find the length of the median of the trapezoid.

21. In trapezoid $MNOP$, $\overline{MN} \parallel \overline{PO}$ and R and S are the midpoints of $\overline{MP}$ and $\overline{NO}$, respectively. Find the lengths of the bases if $RS = 15$, $MN = 3x + 2$, and $PO = 2x - 7$.

In Review Exercises 22 to 24, M and N are the midpoints of $\overline{FJ}$ and $\overline{FH}$, respectively.

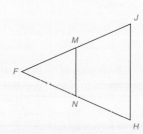

Exercises 22–24

22. *Given:*   Isosceles $\triangle FJH$ with
    $\overline{FJ} \cong \overline{FH}$
    $FM = 2y + 3$
    $NH = 5y - 9$
    $JH = 2y$
    Find: The perimeter of $\triangle FMN$

23. *Given:* $JH = 12$
    $m\angle J = 80°$
    $m\angle F = 60°$
    Prove: MN, $m\angle FMN$, $m\angle FNM$

24. *Given:* $MN = x^2 + 6$
    $JH = 2x(x + 2)$
    Prove: x, MN, JH

25. *Given:* $ABCD$ is a $\square$
    $\overline{AF} \cong \overline{CE}$
    Prove: $\overline{DF} \parallel \overline{EB}$

26. *Given:* $ABEF$ is a rectangle
    $BCDE$ is a rectangle
    $\overline{FE} \cong \overline{ED}$
    Prove: $\overline{AE} \cong \overline{BD}$ and $\overline{AE} \parallel \overline{BD}$

27. *Given:* $\overline{DE}$ is a median of $\triangle ADC$
$\overline{BE} \cong \overline{FD}$
$\overline{EF} \cong \overline{FD}$

Prove: $ABCF$ is a $\square$

28. *Given:* $\triangle FAB \cong \triangle HCD$
$\triangle EAD \cong \triangle GCB$

Prove: $ABCD$ is a $\square$

29. *Given:* $ABCD$ is a parallelogram
$\overline{DC} \cong \overline{BN}$
$\angle 3 \cong \angle 4$

Prove: $ABCD$ is a rhombus

30. *Given:* $\triangle TWX$ is isosceles, with base $\overline{WX}$
$\overline{RY} \parallel \overline{WX}$

Prove: $RWXY$ is an isosceles trapezoid
(See the figure at the top of the next column.)

31. Construct rhombus $ABCD$, given these lengths for the diagonals.

A ├──────────┤ C

B ├──────────────┤ D

32. Draw rectangle $ABCD$ with $AB = 5$ and $BC = 12$. Include diagonals $\overline{AC}$ and $\overline{BD}$.
a) How are $\overline{AB}$ and $\overline{BC}$ related?
b) Find the length of diagonal $\overline{AC}$.

33. Draw rhombus $WXYZ$ with diagonals $\overline{WY}$ and $\overline{XZ}$. Let $\overline{WY}$ name the longer diagonal.
a) How are diagonals $\overline{WY}$ and $\overline{XZ}$ related?
b) If $WX = 17$ and $XZ = 16$, find the length of diagonal $\overline{WY}$.

34. Considering parallelograms, kites, rectangles, squares, rhombi, trapezoids, and isosceles trapezoids, which figures have
a) line symmetry?
b) point symmetry?

35. What type of quadrilateral is formed when the triangle is reflected across the indicated side?
a) Isosceles $\triangle ABC$ across $\overline{BC}$
b) Obtuse $\triangle XYZ$ across $\overline{XY}$

Chapter 4 | Test

1. Consider $\square ABCD$ as shown.
a) How are $\angle A$ and $\angle C$ related? _____
b) How are $\angle A$ and $\angle B$ related? _____

2. In $\square RSTV$ (not shown), $RS = 5.3$ cm and $ST = 4.1$ cm. Find the perimeter of $RSTV$. _____

3. In $\square ABCD$, $AD = 5$ and $DC = 9$. If the altitude from vertex D to $\overline{AB}$ has length 4 (that is, $DE = 4$), find the length of EB. _____

4. In $\square RSTV$, $m\angle S = 57°$. Which diagonal ($\overline{VS}$ or $\overline{RT}$) would have the greater length? _____

Exercises 4, 5

5. In $\square RSTV$, $VT = 3x - 1$, $TS = 2x + 1$, and $RS = 4(x - 2)$. Find the value of x. _____

6. Complete each statement:
a) If a quadrilateral has two pairs of congruent *adjacent* sides, then the quadrilateral is a(n) _____.
b) If a quadrilateral has two pairs of congruent *opposite* sides, then the quadrilateral is a(n) _____.

7. Complete each statement:
a) In □*RSTV*, $\overline{RW}$ is the _____ from vertex *R* to base $\overline{VT}$.

b) If altitude $\overline{RW}$ of figure (a) is congruent to altitude $\overline{TY}$ of figure (b), then □*RSTV* must also be a(n) _____.

8. In △*ABC*, *M* is the midpoint of $\overline{AB}$ and *N* is the midpoint of $\overline{AC}$.
a) How are line segments $\overline{MN}$ and $\overline{BC}$ related? _____
b) Use an equation to state how the lengths *MN* and *BC* are related. _____

Exercises 8–10

9. In △*ABC*, *M* is the midpoint of $\overline{AB}$ and *N* is the midpoint of $\overline{AC}$. If *MN* = 7.6 cm, find *BC*. _____

10. In △*ABC*, *M* is the midpoint of $\overline{AB}$ and *N* is the midpoint of $\overline{AC}$. If *MN* = 3*x* − 11 and *BC* = 4*x* + 24, find the value of *x*. _____

11. In rectangle *ABCD*, *AD* = 12 and *DC* = 5. Find the length of diagonal $\overline{AC}$ (not shown). _____

12. In trapezoid *RSTV*, $\overline{RS} \parallel \overline{VT}$.
a) Which sides are the legs of *RSTV*? _____
b) Name two angles that are supplementary. _____

13. In trapezoid *RSTV*, $\overline{RS} \parallel \overline{VT}$ and $\overline{MN}$ is the median. Find the length *MN* if *RS* = 12.4 in. and *VT* = 16.2 in. _____

Exercises 13, 14

14. In trapezoid *RSTV* of Exercise 13, $\overline{RS} \parallel \overline{VT}$ and $\overline{MN}$ is the median. Find *x* if *VT* = 2*x* + 9, *MN* = 6*x* − 13, and *RS* = 15. _____

15. Complete the proof of the following theorem:
"In a kite, one pair of opposite angles are congruent."
Given: Kite *ABCD*; $\overline{AB} \cong \overline{AD}$ and $\overline{BC} \cong \overline{DC}$
Prove: ∠*B* ≅ ∠*D*

(a)   (b)

PROOF	
Statements	**Reasons**
1. _____	1. _____
2. Draw $\overline{AC}$.	2. Through two points, there is exactly one line
3. _____	3. Identity
4. △*ACD* ≅ △*ACB*	4. _____
5. _____	5. _____

16. Complete the proof of the following theorem:
"The diagonals of an isosceles trapezoid are congruent."
Given: Trapezoid *ABCD* with $\overline{AB} \parallel \overline{DC}$ and $\overline{AD} \cong \overline{BC}$
Prove: $\overline{AC} \cong \overline{DB}$

PROOF	
Statements	**Reasons**
1. _____	1. _____
2. ∠*ADC* ≅ ∠*BCD*	2. Base ∠s of an isosceles trapezoid are _____
3. *DC* ≅ *DC*	3. _____
4. △*ADC* ≅ △*BCD*	4. _____
5. _____	5. CPCTC

17. In kite *RSTV*, *RS* = 2*x* − 4, *ST* = *x* − 1, *TV* = *y* − 3, and *RV* = *y*. Find the perimeter of *RSTV*.

Dmitry Fischer/Shutterstock.com

Chapter 5

Similar Triangles

Talented! The handiwork of a skillful craftsman, these Russian nesting dolls have the same shape but different sizes. Because of their design, each doll can be placed within another so that they all nest together. These painted dolls are said to be similar in shape. In nature, water lily pads have the same shape but different sizes. In the every-day world, cylindrical containers found on grocery store shelves may have the same shape but different sizes. In all these situations, one figure is merely an enlargement of the other; in geometry, we say that two such figures are *similar*. Further illustrations of both two-dimensional and three-dimensional similar figures can be found in Sections 5.2 and 5.3. The solutions for some applications in this and later chapters lead to quadratic equations. A review of the methods that are used to solve quadratic equations can be found in Appendix A.4 and Appendix A.5 of this textbook.

Additional video explanations of concepts, sample problems, and applications are available on DVD.

5.1 Ratios, Rates, and Proportions

KEY CONCEPTS	Ratio	Means	Extended Ratio
	Rate	Means-Extremes	Extended Proportion
	Proportion	Property	
	Extremes	Geometric Mean	

The concepts and techniques discussed in Section 5.1 are often necessary for managing the algebraic and geometric applications found in this chapter and beyond.

A **ratio** is the quotient $\frac{a}{b}$ (where $b \neq 0$) that provides a comparison between the numbers a and b. Because every fraction indicates a division, every fraction represents a ratio. Read "a to b," the ratio is sometimes written in the form $a{:}b$. The numbers a and b are often called the *terms* of the ratio.

It is generally preferable to express the ratio in simplified form (lowest terms), so the ratio 6 to 8 would be reduced (in fraction form) from $\frac{6}{8}$ to $\frac{3}{4}$. If units of measure are found in a ratio, these units must be **commensurable** (convertible to the same unit of measure). When simplifying the ratio of two quantities that are expressed in the same unit, we eliminate the common unit in the process. If two quantities cannot be compared because no common unit of measure is possible, the quantities are said to be **incommensurable.**

> **Reminder**
>
> Units are neither needed nor desirable in a simplified ratio.

EXAMPLE 1

Find the simplified form of each ratio:

a) 12 to 20

b) 12 in. to 24 in.

c) 12 in. to 3 ft (**NOTE:** 1 ft = 12 in.)

d) 5 lb to 20 oz (**NOTE:** 1 lb = 16 oz)

e) 5 lb to 2 ft

f) 4 m to 30 cm (**NOTE:** 1 m = 100 cm)

Geometry in the Real World

At a grocery store, the cost per unit is a rate that allows the consumer to know which brand is more expensive.

SOLUTION

a) $\dfrac{12}{20} = \dfrac{3}{5}$

b) $\dfrac{12 \text{ in.}}{24 \text{ in.}} = \dfrac{12}{24} = \dfrac{1}{2}$

c) $\dfrac{12 \text{ in.}}{3 \text{ ft}} = \dfrac{12 \text{ in.}}{3(12 \text{ in.})} = \dfrac{12 \text{ in.}}{36 \text{ in.}} = \dfrac{1}{3}$

d) $\dfrac{5 \text{ lb}}{20 \text{ oz}} = \dfrac{5(16 \text{ oz})}{20 \text{ oz}} = \dfrac{80 \text{ oz}}{20 \text{ oz}} = \dfrac{4}{1}$

e) $\dfrac{5 \text{ lb}}{2 \text{ ft}}$ is incommensurable!

f) $\dfrac{4 \text{ m}}{30 \text{ cm}} = \dfrac{4(100 \text{ cm})}{30 \text{ cm}} = \dfrac{400 \text{ cm}}{30 \text{ cm}} = \dfrac{40}{3}$

A **rate** is a quotient that relates two quantities that are incommensurable. If an automobile can travel 300 miles along an interstate while consuming 10 gallons of gasoline, then its consumption *rate* is $\frac{300 \text{ miles}}{10 \text{ gallons}}$. In simplified form, the consumption rate is $\frac{30 \text{ mi}}{\text{gal}}$, which is read as "30 miles per gallon" and abbreviated 30 mpg.

EXAMPLE 2

Simplify each rate. Units are necessary in each answer.

a) $\dfrac{120 \text{ miles}}{5 \text{ gallons}}$

b) $\dfrac{100 \text{ meters}}{10 \text{ seconds}}$

c) $\dfrac{12 \text{ teaspoons}}{2 \text{ quarts}}$

d) $\dfrac{\$18.45}{5 \text{ gal}}$

SOLUTION

a) $\dfrac{120 \text{ mi}}{5 \text{ gal}} = \dfrac{24 \text{ mi}}{\text{gal}}$ or 24 mpg

b) $\dfrac{100 \text{ m}}{10 \text{ s}} = \dfrac{10 \text{ m}}{\text{s}}$

c) $\dfrac{12 \text{ teaspoons}}{2 \text{ quarts}} = \dfrac{6 \text{ teaspoons}}{\text{quart}}$

d) $\dfrac{\$18.45}{5 \text{ gal}} = \dfrac{\$3.69}{\text{gal}}$

SSG **EXS. 1–2**

A **proportion** is a statement that equates two ratios or two rates. Thus, $\frac{a}{b} = \frac{c}{d}$ is a proportion and may be read as "*a* is to *b* as *c* is to *d*." In the order read, *a* is the *first term* of the proportion, *b* is the *second term*, *c* is the *third term*, and *d* is the *fourth term*. The first and last terms (*a* and *d*) of the proportion are the **extremes**, whereas the second and third terms (*b* and *c*) are the **means**.

The following property is extremely convenient for solving proportions.

PROPERTY 1 ■ Means-Extremes Property

In a proportion, the product of the means equals the product of the extremes; that is, if $\frac{a}{b} = \frac{c}{d}$ (where $b \neq 0$ and $d \neq 0$), then $a \cdot d = b \cdot c$.

Because a proportion is a statement, it could be true or false. The truth or falsity of a proportion can be "tested" by applying the Means-Extremes Property. In the false proportion $\frac{9}{12} = \frac{2}{3}$, it is obvious that $9 \cdot 3 \neq 12 \cdot 2$; on the other hand, the truth of the statement $\frac{9}{12} = \frac{3}{4}$ is evident from the fact that $9 \cdot 4 = 12 \cdot 3$. Henceforth, any proportion given in this text is intended to be a true proportion.

EXAMPLE 3

Use the Means-Extremes Property to solve each proportion for *x*.

a) $\dfrac{4}{8} = \dfrac{5}{12}$ c) $\dfrac{3}{x} = \dfrac{x}{2}$ e) $\dfrac{x+3}{5} = \dfrac{4}{x-1}$

b) $\dfrac{x+1}{9} = \dfrac{x-3}{3}$ d) $\dfrac{x+3}{3} = \dfrac{9}{x-3}$

SOLUTION

a) $x \cdot 12 = 8 \cdot 5$ (Means-Extremes Property)
$12x = 40$
$x = \dfrac{40}{12} = \dfrac{10}{3}$

b) $3(x + 1) = 9(x - 3)$ (Means-Extremes Property)
$3x + 3 = 9x - 27$
$30 = 6x$
$x = 5$

c) $3 \cdot 2 = x \cdot x$ (Means-Extremes Property)
$x^2 = 6$
$x = \pm\sqrt{6} \approx \pm 2.45$

d) $(x + 3)(x - 3) = 3 \cdot 9$ (Means-Extremes Property)

$$x^2 - 9 = 27$$
$$x^2 - 36 = 0$$
$$(x + 6)(x - 6) = 0 \qquad \text{(using factoring)}$$

$x + 6 = 0$	or	$x - 6 = 0$
$x = -6$	or	$x = 6$

e) $(x + 2)(x - 1) = 5 \cdot 4$ (Means-Extremes Property)

$$x^2 + x - 2 = 20$$
$$x^2 + x - 22 = 0$$
$$x = \frac{-b \pm \sqrt{b^2 - 4ac}}{2a} \qquad \text{(using Quadratic Formula; see Appendix A.5)}$$
$$= \frac{-1 \pm \sqrt{(1)^2 - 4(1)(-22)}}{2(1)}$$
$$= \frac{-1 \pm \sqrt{1 + 88}}{2} = \frac{-1 \pm \sqrt{89}}{2}, \text{ where}$$
$$\frac{-1 + \sqrt{89}}{2} \approx 4.22 \text{ while } \frac{-1 - \sqrt{89}}{2} \approx -5.22$$

In application problems involving proportions, it is essential to order the related quantities in each ratio or rate. The first step in the solution of Example 4 illustrates the care that must be taken in forming the proportion for an application. Because of consistency, units may be eliminated in the actual proportion.

EXAMPLE 4

If an automobile can travel 90 mi on 4 gal of gasoline, how far can it travel on 6 gal of gasoline?

SOLUTION By form,

$$\frac{\text{number miles first trip}}{\text{number gallons first trip}} = \frac{\text{number miles second trip}}{\text{number gallons second trip}}$$

Where x represents the number of miles traveled on the second trip, we have

$$\frac{90}{4} = \frac{x}{6}$$
$$4x = 540$$
$$x = 135$$

Thus, the car can travel 135 mi on 6 gal of gasoline.

DEFINITION

The nonzero number b is the **geometric mean** of a and c if $\frac{a}{b} = \frac{b}{c}$ or $\frac{c}{b} = \frac{b}{a}$.

In either proportion of the definition above, the second and third terms (the geometric mean) must be identical. For example, 6 and -6 are the geometric means of 4 and 9 because $\frac{4}{6} = \frac{6}{9}$ and $\frac{4}{-6} = \frac{-6}{9}$. Because applications in geometry generally require positive solutions, we usually seek only the positive geometric mean of a and c.

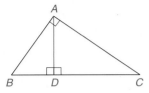

Figure 5.1

SSG **EXS. 3–6**

EXAMPLE 5

In Figure 5.1, AD is the geometric mean of BD and DC. If $BC = 10$ and $BD = 4$, determine AD.

SOLUTION By the definition of geometric mean, $\frac{BD}{AD} = \frac{AD}{DC}$. Because $DC = BC - BD$, we know that $DC = 10 - 4 = 6$. Where x is the length of $\overline{AD}$, the proportion above becomes

$$\frac{4}{x} = \frac{x}{6}$$

Applying the Means-Extremes Property,

$$x^2 = 24$$
$$x = \pm\sqrt{24} = \pm\sqrt{4 \cdot 6} = \pm\sqrt{4} \cdot \sqrt{6} = \pm 2\sqrt{6}$$

To have a permissible length for $\overline{AD}$, the solution must be positive. Thus, $AD = 2\sqrt{6}$ so that $AD \approx 4.90$.

An **extended ratio** compares more than two quantities and must be expressed in a form such as $a{:}b{:}c$ or $d{:}e{:}f{:}g$. If you know that the angles of a triangle are 90°, 60°, and 30°, then the ratio that compares these measures is 90:60:30, or 3:2:1 (because 90, 60, and 30 have the greatest common factor of 30).

PROPERTY OF EXTENDED RATIOS

Unknown quantities in the ratio $a{:}b{:}c{:}d$ can be represented by ax, bx, cx, and dx.

We apply the Property of Extended Ratios in Example 6.

EXAMPLE 6

Suppose that the perimeter of a quadrilateral is 70 and the lengths of the sides are in the ratio 2:3:4:5. Find the measure of each side.

SOLUTION We represent the lengths of the sides by $2x$, $3x$, $4x$, and $5x$. Then

$$2x + 3x + 4x + 5x = 70$$
$$14x = 70$$
$$x = 5$$

Because $2x = 10$, $3x = 15$, $4x = 20$, and $5x = 25$, the lengths of the sides of the quadrilateral are 10, 15, 20, and 25.

It is possible to solve certain problems in more ways than one, as illustrated in the next example. However, the solution is unique and is not altered by the method chosen.

EXAMPLE 7

The measures of two complementary angles are in the ratio 2 to 3. Find the measure of each angle.

SOLUTION Let the first of the complementary angles have measure x; then the second angle has the measure $90 - x$. Thus, we have

$$\frac{x}{90 - x} = \frac{2}{3}$$

Using the Means-Extremes Property, we have

$$3x = 2(90 - x)$$
$$3x = 180 - 2x$$
$$5x = 180$$
$$x = 36$$

Therefore, $90 - x = 54$ so the angles have measures of $36°$ and $54°$.

ALTERNATIVE SOLUTION Because the measures of the angles are in the ratio 2:3, we represent their measures by $2x$ and $3x$. Because the angles are complementary,

$$2x + 3x = 90$$
$$5x = 90$$
$$x = 18$$

SSG **EXS. 7–9**   Now $2x = 36$ and $3x = 54$, so the measures of the two angles are $36°$ and $54°$.

The remaining properties of proportions are theorems that can be proven by applying the Means-Extremes Property. See Exercises 40 and 41.

STRATEGY FOR PROOF ■ Proving Properties of Proportions

General Rule: To prove these theorems, apply the Means-Extremes Property as well as the Addition, Subtraction, Multiplication, and Division Properties of Equality.

Illustration: Proving the first part of Property 3 begins with the addition of 1 to each side of the proportion $\frac{a}{b} = \frac{c}{d}$.

PROPERTY 2 ■ Alternative Forms of Proportions

In a proportion, the means or the extremes (or both) may be interchanged; that is, if $\frac{a}{b} = \frac{c}{d}$ (where a, b, c, and d are nonzero), then $\frac{a}{c} = \frac{b}{d}, \frac{d}{b} = \frac{c}{a}$, and $\frac{d}{c} = \frac{b}{a}$.

Given the proportion $\frac{2}{3} = \frac{8}{12}$, Property 2 leads to conclusions such as

1. $\dfrac{2}{8} = \dfrac{3}{12}$ (means interchanged)

2. $\dfrac{12}{3} = \dfrac{8}{2}$ (extremes interchanged)

3. $\dfrac{3}{2} = \dfrac{12}{8}$ (both sides inverted; the result obtained when both means and extremes are interchanged)

PROPERTY 3 ■ Sum and Difference Properties of a Proportion

If $\frac{a}{b} = \frac{c}{d}$ (where $b \neq 0$ and $d \neq 0$), then $\frac{a + b}{b} = \frac{c + d}{d}$ and $\frac{a - b}{b} = \frac{c - d}{d}$.

With Property 3, we can add (or subtract) each denominator to (from) the corresponding numerator in order to obtain a valid proportion.

Given the proportion $\frac{2}{3} = \frac{8}{12}$, the Sum and Difference Property leads to conclusions such as

1. $\dfrac{2 + 3}{3} = \dfrac{8 + 12}{12}$ (each side simplifies to $\frac{5}{3}$)

2. $\dfrac{2 - 3}{3} = \dfrac{8 - 12}{12}$ (each side simplifies to $-\frac{1}{3}$)

SSG EXS. 10, 11

Just as there are extended ratios, there are also **extended proportions** such as

$$\frac{a}{b} = \frac{c}{d} = \frac{e}{f} = \cdots$$

Suggested by different numbers of servings of a particular recipe, the statement below is an extended proportion comparing numbers of eggs to numbers of cups of milk:

$$\frac{2 \text{ eggs}}{3 \text{ cups}} = \frac{4 \text{ eggs}}{6 \text{ cups}} = \frac{6 \text{ eggs}}{9 \text{ cups}}$$

EXAMPLE 8

In the triangles shown in Figure 5.2, $\dfrac{AB}{DE} = \dfrac{AC}{DF} = \dfrac{BC}{EF}$. Find the lengths of $\overline{DF}$ and $\overline{EF}$.

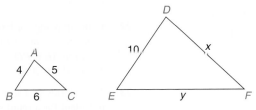

Figure 5.2

SOLUTION Substituting into the proportion $\dfrac{AB}{DE} = \dfrac{AC}{DF} = \dfrac{BC}{EF}$, we have

$$\frac{4}{10} = \frac{5}{x} = \frac{6}{y}$$

From the equation

$$\frac{4}{10} = \frac{5}{x}$$

it follows that $4x = 50$ and that $x = DF = 12.5$. Using the equation

$$\frac{4}{10} = \frac{6}{y}$$

SSG EXS. 12, 13   we find that $4y = 60$, so $y = EF = 15$.

Discover

THE GOLDEN RATIO

It is believed that the "ideal" rectangle is determined when a square can be removed in such a way as to leave a smaller rectangle with the same shape as the original rectangle. As we shall find, the rectangles are known as *similar* in shape. Upon removal of the square, the similarity in the shapes of the rectangles requires that $\frac{W}{L} = \frac{L-W}{W}$. To discover the relationship between L and W, we choose $W = 1$ and solve the equation $\frac{1}{L} = \frac{L-1}{1}$ for L. The solution is $L = \frac{1+\sqrt{5}}{2}$. The ratio comparing length to width is known as the golden ratio. Because $L = \frac{1+\sqrt{5}}{2} \approx 1.62$ when $W = 1$, the *ideal* rectangle has a length that is approximately 1.62 times its width; that is, $L \approx 1.62W$.

Exercises 5.1

In Exercises 1 to 4, give the ratios in simplified form.

1. a) 12 to 15   c) 1 ft to 18 in.
    b) 12 in. to 15 in.   d) 1 ft to 18 oz

2. a) 20 to 36   c) 20 oz to 2 lb (1 lb = 16 oz)
    b) 24 oz to 52 oz   d) 2 lb to 20 oz

3. a) 15:24   c) 2 m:150 cm (1 m = 100 cm)
    b) 2 ft:2 yd (1 yd = 3 ft)   d) 2 m:1 lb

4. a) 24:32   c) 150 cm:2 m
    b) 12 in.:2 yd   d) 1 gal:24 mi

In Exercises 5 to 14, find the value of x in each proportion.

5. a) $\dfrac{x}{4} = \dfrac{9}{12}$ b) $\dfrac{7}{x} = \dfrac{21}{24}$

6. a) $\dfrac{x - 1}{10} = \dfrac{3}{5}$ b) $\dfrac{x + 1}{6} = \dfrac{10}{12}$

7. a) $\dfrac{x - 3}{8} = \dfrac{x + 3}{24}$ b) $\dfrac{x + 1}{6} = \dfrac{4x - 1}{18}$

8. a) $\dfrac{9}{x} = \dfrac{x}{16}$ b) $\dfrac{32}{x} = \dfrac{x}{2}$

9. a) $\dfrac{x}{4} = \dfrac{7}{x}$ b) $\dfrac{x}{6} = \dfrac{3}{x}$

10. a) $\dfrac{x + 1}{3} = \dfrac{10}{x + 2}$ b) $\dfrac{x - 2}{5} = \dfrac{12}{x + 2}$

11. a) $\dfrac{x + 1}{x} = \dfrac{10}{2x}$ b) $\dfrac{2x + 1}{x + 1} = \dfrac{14}{3x - 1}$

12. a) $\dfrac{x + 1}{2} = \dfrac{7}{x - 1}$ b) $\dfrac{x + 1}{3} = \dfrac{5}{x - 2}$

13. a) $\dfrac{x + 1}{x} = \dfrac{2x}{3}$ b) $\dfrac{x + 1}{x - 1} = \dfrac{2x}{5}$

14. a) $\dfrac{x + 1}{x} = \dfrac{x}{x - 1}$ b) $\dfrac{x + 2}{x} = \dfrac{2x}{x - 2}$

15. Sarah ran the 300-m hurdles in 47.7 sec. In meters per second, find the rate at which Sarah ran. Give the answer to the nearest tenth of a meter per second.

16. Fran has been hired to sew the dance troupe's dresses for the school musical. If $13\frac{1}{3}$ yd of material is needed for the four dresses, find the rate that describes the amount of material needed for each dress.

In Exercises 17 to 22, use proportions to solve each problem.

17. A recipe calls for 4 eggs and 3 cups of milk. To prepare for a larger number of guests, a cook uses 14 eggs. How many cups of milk are needed?

18. If a school secretary copies 168 worksheets for a class of 28 students, how many worksheets must be prepared for a class of 32 students?

19. An electrician installs 25 electrical outlets in a new six-room house. Assuming proportionality, how many outlets should be installed in a new construction having seven rooms? (Round to nearest integer.)

20. The secretarial pool (15 secretaries in all) on one floor of a corporate complex has access to four copy machines. If there are 23 secretaries on a different floor, approximately what number of copy machines should be available? (Assume a proportionality.)

21. Assume that *AD* is the geometric mean of *BD* and *DC* in △*ABC* shown in the accompanying drawing.
a) Find *AD* if *BD* = 6 and *DC* = 8.
b) Find *BD* if *AD* = 6 and *DC* = 8.

Exercises 21, 22

22. In the drawing for Exercise 21, assume that *AB* is the geometric mean of *BD* and *BC*.
a) Find *AB* if *BD* = 6 and *DC* = 10.
b) Find *DC* if *AB* = 10 and *BC* = 15.

23. The salaries of a secretary, a salesperson, and a vice president for a retail sales company are in the ratio 2:3:5. If their combined annual salaries amount to $124,500, what is the annual salary of each?

24. The salaries of a school cook, custodian, and bus driver are in the ratio 2:4:3. If their combined monthly salaries for November total $8,280, what is the monthly salary for each person?

25. If the measures of the angles of a quadrilateral are in the ratio 3:4:5:6, find the measure of each angle.

26. If the measures of the angles of a quadrilateral are in the ratio of 2:3:4:6, find the measure of each angle.

27. The measures of two complementary angles are in the ratio 4:5. Find the measure of each angle, using the two methods shown in Example 7.

28. The measures of two supplementary angles are in the ratio of 2:7. Find the measure of each angle, using the two methods of Example 7.

29. If 1 in. equals 2.54 cm, use a proportion to convert 12 in. to centimeters.
$\left(HINT: \dfrac{2.54 \text{ cm}}{1 \text{ in.}} = \dfrac{x \text{ cm}}{12 \text{ in.}}\right)$

30. If 1 kg equals 2.2 lb, use a proportion to convert 12 pounds to kilograms.

31. For the quadrilaterals shown, $\frac{MN}{WX} = \frac{NP}{XY} = \frac{PQ}{YZ} = \frac{MQ}{WZ}$. If $MN = 7$, $WX = 3$, and $PQ = 6$, find YZ.

Exercises 31, 32

32. For this exercise, use the drawing and extended ratio of Exercise 31. If $NP = 2 \cdot XY$ and $WZ = 3\frac{1}{2}$, find MQ.

33. Two numbers a and b are in the ratio 3:4. If the first number is decreased by 2 and the second is decreased by 1, they are in the ratio 2:3. Find a and b.

34. Two numbers a and b are in the ratio 2:3. If both numbers are decreased by 2, the ratio of the resulting numbers becomes 3:5. Find a and b.

35. If the ratio of the measure of the complement of an angle to the measure of its supplement is 1:3, find the measure of the angle.

36. If the ratio of the measure of the complement of an angle to the measure of its supplement is 1:4, find the measure of the angle.

37. On a blueprint, a 1-in. scale corresponds to 3 ft. To show a room with actual dimensions 12 ft wide by 14 ft long, what dimensions should be shown on the blueprint?

38. To find the golden ratio (see the Discover activity on page 215), solve the equation $\frac{1}{L} = \frac{L-1}{1}$ for L.

(HINT: You will need the Quadratic Formula.)

39. Find
  a) the exact length of an ideal rectangle with width $W = 5$ by solving $\frac{5}{L} = \frac{L-5}{5}$.
  b) the approximate length of an ideal rectangle with width $W = 5$ by using $L \approx 1.62W$.

40. *Prove:* If $\frac{a}{b} = \frac{c}{d}$ (where a, b, c, and d are nonzero), then $\frac{a}{c} = \frac{b}{d}$.

41. *Prove:* If $\frac{a}{b} = \frac{c}{d}$ (where $b \neq 0$ and $d \neq 0$), then $\frac{a+b}{b} = \frac{c+d}{d}$.

5.2 Similar Polygons

KEY CONCEPTS   Similar Polygons   Corresponding Vertices,
                    Congruent Polygons   Angles, and Sides

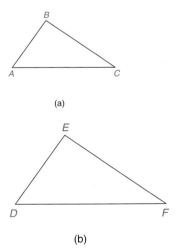

(a)

(b)

Figure 5.3

When two geometric figures have exactly the same shape, they are **similar;** the symbol for "is similar to" is $\sim$. When two figures have the same shape ($\sim$) and all corresponding parts have equal ($=$) measures, the two figures are **congruent** ($\cong$). Note that the symbol for congruence combines the symbols for similarity and equality; that is, congruent polygons always have the same shape and the measures of corresponding parts are equal.

> Two congruent polygons are also similar polygons.

While two-dimensional figures such as $\triangle ABC$ and $\triangle DEF$ in Figure 5.3 can be similar, it is also possible for three-dimensional figures to be similar. Similar orange juice containers are shown in Figures 5.4(a) and 5.4(b) on page 218. Informally, two figures are "similar" if one is an enlargement of the other. Thus, a tuna fish can and an orange juice can are *not* similar, even if both are right-circular cylinders [see Figures 5.4(b) and 5.4(c) on page 218.] We will consider cylinders in greater detail in Chapter 9.

(a)   (b)   (c)

Figure 5.4

In this chapter, the discussion of similarity will generally be limited to plane figures.

For two polygons to be similar, it is necessary that each angle of one polygon be congruent to the corresponding angle of the other. However, the congruence of angles is not sufficient to establish the similarity of polygons. The vertices of the congruent angles are **corresponding vertices** of the similar polygons. Consider Figure 5.5. If ∠*A* in one polygon is congruent to ∠*H* in the second polygon, then vertex *A* corresponds to vertex *H*, and this is symbolized *A*↔*H*; we can indicate that ∠*A* corresponds to ∠*H* by writing ∠*A*↔∠*H*. A pair of angles like ∠*A* and ∠*H* are **corresponding angles,** and the sides determined by consecutive and corresponding vertices are **corresponding sides** of the similar polygons. For instance, if *A*↔*H* and *B*↔*J*, then $\overline{AB}$ corresponds to $\overline{HJ}$.

EXAMPLE 1

Given similar quadrilaterals *ABCD* and *HJKL* with congruent angles as indicated in Figure 5.5, name the vertices, angles, and sides that correspond to each other.

(a)   (b)

Figure 5.5

SOLUTION Because ∠*A* ≅ ∠*H*, it follows that

$$
\begin{array}{lll}
A \leftrightarrow H & \text{and} & \angle A \leftrightarrow \angle H. \text{ Similarly,} \\
B \leftrightarrow J & \text{and} & \angle B \leftrightarrow \angle J \\
C \leftrightarrow K & \text{and} & \angle C \leftrightarrow \angle K \\
D \leftrightarrow L & \text{and} & \angle D \leftrightarrow \angle L
\end{array}
$$

Associating pairs of consecutive and corresponding vertices of similar polygons, we determine the endpoints of the corresponding sides.

$$\overline{AB} \leftrightarrow \overline{HJ}, \quad \overline{BC} \leftrightarrow \overline{JK}, \quad \overline{CD} \leftrightarrow \overline{KL}, \quad \text{and} \quad \overline{AD} \leftrightarrow \overline{HL}$$

With an understanding of the terms corresponding angles and corresponding sides, we can define similar polygons.

DEFINITION

Two polygons are **similar** if and only if two conditions are satisfied:
1. All pairs of corresponding angles are congruent.
2. All pairs of corresponding sides are proportional.

The second condition for similarity requires that the following extended proportion exists for the sides of the similar quadrilaterals of Example 1 on page 218.

$$\frac{AB}{HJ} = \frac{BC}{JK} = \frac{CD}{KL} = \frac{AD}{HL}$$

Note that *both* conditions 1 and 2 for similarity are necessary! Although condition 1 of the definition is satisfied for square *EFGH* and rectangle *RSTU* [see Figures 5.6(a) and (b)], the figures are not similar. That is, one is not an enlargement of the other because the extended proportion comparing the lengths of corresponding sides is not true. On the other hand, condition 2 of the definition is satisfied for square *EFGH* and rhombus *WXYZ* [see Figures 5.6(a) and 5.6(c)], but the figures are not similar because the pairs of corresponding angles are not congruent.

(a)   (b)   (c)

Figure 5.6

EXAMPLE 2

Which figures must be similar?

a) Any two isosceles triangles   **c)** Any two rectangles
b) Any two regular pentagons   **d)** Any two squares

SOLUTION

a) No; ∠ pairs need not be ≅, nor do the pairs of sides need to be proportional.
b) Yes; all angles are congruent (measure 108° each), and all pairs of sides are proportional.
c) Not all angles measure 90°, but the pairs of sides are not necessarily proportional.
d) Yes; all angles measure 90°, and all pairs of sides are proportional.

SSG **EXS. 1–4**

EXAMPLE 3

If △*ABC* ~ △*DEF* in Figure 5.7, use the indicated measures to find the measures of the remaining parts of each of the triangles.

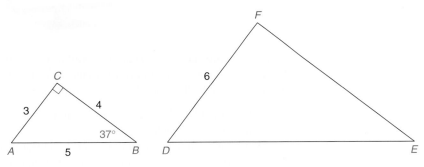

Figure 5.7

SOLUTION Because the sum of the measures of the angles of a triangle is 180°,

$$m\angle A = 180 - (90 + 37) = 53°.$$

Because of the similarity and the correspondence of vertices in Figure 5.7,

$$m\angle D = 53°, \qquad m\angle E = 37°, \qquad \text{and} \qquad m\angle F = 90°.$$

The proportion that relates the lengths of the sides is

$$\frac{AC}{DF} = \frac{CB}{FE} = \frac{AB}{DE} \qquad \text{so} \qquad \frac{3}{6} = \frac{4}{FE} = \frac{5}{DE}.$$

From $\frac{3}{6} = \frac{4}{FE}$, we see that $3 \cdot FE = 6 \cdot 4$ so that $3 \cdot FE = 24$ and $FE = 8$.

From $\frac{3}{6} = \frac{5}{DE}$, we see that $3 \cdot DE = 6 \cdot 5$ so that $3 \cdot DE = 30$ and $DE = 10$.

In any proportion, the ratios can all be inverted; thus, Example 3 could have been solved by using the proportion

$$\frac{DF}{AC} = \frac{FE}{CB} = \frac{DE}{AB}$$

In any extended proportion, the ratios must all be equal to the same constant value. By designating this number (which is often called the "constant of proportionality") by k, we see that

$$\frac{DF}{AC} = k, \qquad \frac{FE}{CB} = k, \text{ and} \qquad \frac{DE}{AB} = k$$

It follows that $DF = k \cdot AC$, $FE = k \cdot CB$, and $DE = k \cdot AB$. In Example 3, this constant of proportionality has the value $k = 2$, which means that the length of each side of the larger triangle was twice the length of the corresponding side of the smaller triangle.

SSG **EXS. 5–10**

In the following application, the constant of proportionality is the number 4.3.

EXAMPLE 4

For television viewers, there is an ideal relationship between the size of the flat panel television and the distance at which the viewer watches it. Where D is the diagonal measure of the television set and V is the distance to the viewer, we express this relationship by the proportion $\frac{D_1}{V_1} = \frac{D_2}{V_2}$, where the ratio between D and V equals the constant of proportionality k. Of course, this also leads to the form $D = kV$. For the following example, it is further noted that flat panel TV sets are commonly manufactured with diagonals that measure 19″, 22″, 26″, 32″, 37″, 42″, and 50″.

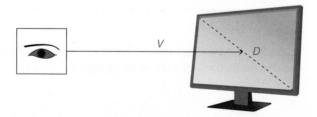

In the following statement, D is measured in inches while V is measured in feet. While D measures the length of the diagonal of the TV and V the distance from the observer to the TV, the ideal relationship is given by $D = 4.3V$.

  a) Find the ideal distance for a person to watch a TV that has a diagonal measuring 37 inches.
  b) A shopper intends to purchase an LCD flat panel set that will be viewed at a distance of approximately 6 feet. What size set should she purchase?

SOLUTION

    a) Using the relationship $D = 4.3V$ and knowing that $D = 37$, we have
$37 = 4.3V$. Then $V = \frac{37}{4.3} \approx 8.6$. That is, the viewer would ideally watch
the TV from a distance of 8.6 feet.

    b) With $D = 4.3V$ and knowing that $V = 6$, we have $D = 4.3(6)$. Then
$D = 25.8$, so the shopper should buy an LCD with a diagonal measuring
26 inches.

Let k represent the constant of proportionality for two similar polygons. When $k > 1$,
the similarity produces an enlarged figure known as a *stretch* or *dilation*. If $0 < k < 1$,
the similarity produces a smaller figure known as *shrink* or *contraction*.

    The constant of proportionality is also used to *scale* a map, a diagram, or a blueprint.
As a consequence, scaling problems can be solved by using proportions.

EXAMPLE 5

On a map, a length of 1 in. represents a distance of 30 mi. On the map, how far apart
should two cities appear if they are actually 140 mi apart along a straight line?

SOLUTION Where $x = $ the map distance desired (in inches),

$$\frac{1}{30} = \frac{x}{140}$$

Then $30x = 140$ and $x = 4\frac{2}{3}$ in.

EXAMPLE 6

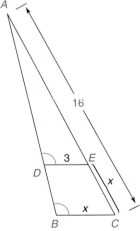

Figure 5.8

In Figure 5.8, $\triangle ABC \sim \triangle ADE$ with $\angle ADE \cong \angle B$. If $DE = 3$, $AC = 16$, and
$EC = BC$, find the length BC.

SOLUTION With $\triangle ABC \sim \triangle ADE$, we have $\frac{DE}{BC} = \frac{AE}{AC}$. With $AC = AE + EC$ and
representing the lengths of the congruent segments ($\overline{EC}$ and $\overline{BC}$) by x, we have

$$16 = AE + x \quad \text{so} \quad AE = 16 - x$$

Substituting into the proportion above, we have

$$\frac{3}{x} = \frac{16 - x}{16}$$

It follows that

$$x(16 - x) = 3 \cdot 16$$
$$16x - x^2 = 48$$
$$x^2 - 16x + 48 = 0$$
$$(x - 4)(x - 12) = 0$$
$$x - 4 = 0 \quad \text{or} \quad x - 12 = 0$$
$$x = 4 \quad \text{or} \quad x = 12$$

Thus, BC equals 4 or 12. Each length is acceptable, but the scaled drawings
differ. See the illustrations in Figure 5.9 on the following page.

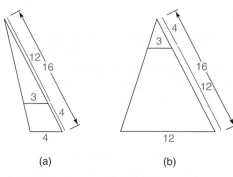

(a) (b)

Figure 5.9

The following example uses a method called *shadow reckoning*. This method of calculating a length dates back more than 2500 years when it was used by the Greek mathematician, Thales, to estimate the height of the pyramids in Egypt. In Figure 5.10, $\triangle ABC \sim \triangle DEF$. Note that $\angle A \cong \angle D$ and $\angle C \cong \angle F$.

EXAMPLE 7

Darnell is curious about the height of a flagpole that stands in front of his school. Darnell, who is 6 ft tall, casts a shadow that he paces off at 9 ft. He walks the length of the shadow of the flagpole, a distance of 30 ft. How tall is the flagpole?

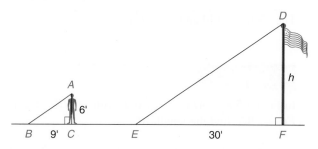

Figure 5.10

SOLUTION In Figure 5.10, $\triangle ABC \sim \triangle DEF$. From similar triangles, we know that $\frac{AC}{DF} = \frac{BC}{EF}$ or $\frac{AC}{BC} = \frac{DF}{EF}$ by interchanging the means.

Where h is the height of the flagpole, substitution into the second proportion leads to

$$\frac{6}{9} = \frac{h}{30} \rightarrow 9h = 180 \rightarrow h = 20$$

The height of the flagpole is 20 ft.

SSG **EXS. 11–13**

Exercises 5.2

1. a) What is true of any pair of corresponding angles of two similar polygons?
   b) What is true of any pairs of corresponding sides of two similar polygons?

2. a) Are any two quadrilaterals similar?
   b) Are any two squares similar?

3. a) Are any two regular pentagons similar?
   b) Are any two equiangular pentagons similar?

4. a) Are any two equilateral hexagons similar?
   b) Are any two regular hexagons similar?

In Exercises 5 and 6, refer to the drawing.

5. a) Given that $A \leftrightarrow X$, $B \leftrightarrow T$, and $C \leftrightarrow N$, write a statement claiming that the triangles shown are similar.
   b) Given that $A \leftrightarrow N$, $C \leftrightarrow X$, and $B \leftrightarrow T$, write a statement claiming that the triangles shown are similar.

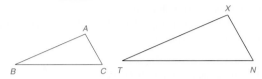

Exercises 5, 6

6. a) If $\triangle ABC \sim \triangle XTN$, which angle of $\triangle ABC$ corresponds to $\angle N$ of $\triangle XTN$?
   b) If $\triangle ABC \sim \triangle XTN$, which side of $\triangle XTN$ corresponds to side $\overline{AC}$ of $\triangle ABC$?

7. A **sphere** is the three-dimensional surface that contains all points in space lying at a fixed distance from a point known as the center of the sphere. Consider the two spheres shown. Are these two spheres similar? Are any two spheres similar? Explain.

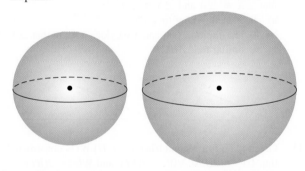

8. Given that rectangle $ABCE$ is similar to rectangle $MNPR$ and that $\triangle CDE \sim \triangle PQR$, what can you conclude regarding pentagon $ABCDE$ and pentagon $MNPQR$?

9. *Given:* $\triangle MNP \sim \triangle QRS$, $m\angle M = 56°$, $m\angle R = 82°$, $MN = 9$, $QR = 6$, $RS = 7$, $MP = 12$
   Find: a) $m\angle N$ c) NP
          b) $m\angle P$ d) QS

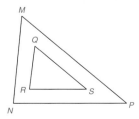

10. *Given:* $\triangle ABC \sim \triangle PRC$, $m\angle A = 67°$, $PC = 5$, $CR = 12$, $PR = 13$, $AB = 26$
   Find: a) $m\angle B$
          b) $m\angle RPC$
          c) AC
          d) CB

11. a) Does the similarity relationship have a **reflexive** property for triangles (and polygons in general)?
   b) Is there a **symmetric** property for the similarity of triangles (and polygons)?
   c) Is there a **transitive** property for the similarity of triangles (and polygons)?

12. Using the names of properties from Exercise 11, identify the property illustrated by each statement:
   a) If $\triangle 1 \sim \triangle 2$, then $\triangle 2 \sim \triangle 1$.
   b) If $\triangle 1 \sim \triangle 2$, $\triangle 2 \sim \triangle 3$, and $\triangle 3 \sim \triangle 4$, then $\triangle 1 \sim \triangle 4$.
   c) $\triangle 1 \sim \triangle 1$

13. In the drawing, $\triangle HJK \sim \triangle FGK$. If $HK = 6$, $KF = 8$, and $HJ = 4$, find FG.

14. In the drawing, $\triangle HJK \sim \triangle FGK$. If $HK = 6$, $KF = 8$, and $FG = 5$, find HJ.

Exercises 13, 14

15. Quadrilateral $ABCD \sim$ quadrilateral $HJKL$. If $m\angle A = 55°$, $m\angle J = 128°$, and $m\angle D = 98°$, find $m\angle K$.

(a)   (b)

Exercises 15–20

16. Quadrilateral $ABCD \sim$ quadrilateral $HJKL$. If $m\angle A = x$, $m\angle J = x + 50$, $m\angle D = x + 35$, and $m\angle K = 2x - 45$, find x.

17. Quadrilateral $ABCD \sim$ quadrilateral $HJKL$. If $AB = 5$, $BC = n$, $HJ = 10$, and $JK = n + 3$, find n.

18. Quadrilateral $ABCD \sim$ quadrilateral $HJKL$. If $m\angle D = 90°$, $AD = 8$, $DC = 6$, and $HL = 12$, find the length of diagonal $\overline{HK}$ (not shown).

19. Quadrilateral $ABCD \sim$ quadrilateral $HJKL$. If $m\angle A = 2x + 4$, $m\angle H = 68°$, and $m\angle D = 3x - 6$, find $m\angle L$.

20. Quadrilateral $ABCD \sim$ quadrilateral $HJKL$. If $m\angle A = m\angle K = 70°$, and $m\angle B = 110°$, what types of quadrilaterals are $ABCD$ and $HJKL$?

In Exercises 21 to 24, △ADE ~ △ABC.

21. Given: $DE = 4, AE = 6, EC = BC$
    Find: BC

22. Given: $DE = 5, AD = 8, DB = BC$
    Find: AB

    (HINT: Find DB first.)

23. Given: $DE = 4, AC = 20,$
          $EC = BC$
    Find: BC

24. Given: $AD = 4, AC = 18,$
          $DB = AE$
    Find: AE

Exercises 21–24

25. Pentagon $ABCDE \sim$ pentagon $GHJKL$ (not shown), $AB = 6$, and $GH = 9$. If the perimeter of $ABCDE$ is 50, find the perimeter of $GHJKL$.

26. Quadrilateral $MNPQ \sim$ quadrilateral $WXYZ$ (not shown), $PQ = 5$, and $YZ = 7$. If the longest side of $MNPQ$ is of length 8, find the length of the longest side of $WXYZ$.

27. A blueprint represents the 72-ft length of a building by a line segment of length 6 in. What length on the blueprint would be used to represent the height of this 30-ft-tall building?

28. A technical drawing shows the $3\frac{1}{2}$-ft lengths of the legs of a baby's swing by line segments 3 in. long. If the diagram should indicate the legs are $2\frac{1}{2}$ ft apart at the base, what length represents this distance on the diagram?

In Exercises 29 to 32, use the fact that triangles are similar.

29. A person who is walking away from a 10-ft lamppost casts a shadow 6 ft long. If the person is at a distance of 10 ft from the lamppost at that moment, what is the person's height?

30. With 100 ft of string out, a kite is 64 ft above ground level. When the girl flying the kite pulls in 40 ft of string, the angle formed by the string and the ground does not change. What is the height of the kite above the ground after the 40 ft of string have been taken in?

31. While admiring a rather tall tree, Fred notes that the shadow of his 6-ft frame has a length of 3 paces. On the level ground, he walks off the complete shadow of the tree in 37 paces. How tall is the tree?

32. As a garage door closes, light is cast 6 ft beyond the base of the door (as shown in the accompanying drawing) by a light fixture that is set in the garage ceiling 10 ft back from the door. If the ceiling of the garage is 10 ft above the floor, how far is the garage door above the floor at the time that light is cast 6 ft beyond the door?

33. In the drawing, $\overleftrightarrow{AB} \parallel \overleftrightarrow{DC} \parallel \overleftrightarrow{EF}$ with transversals ℓ and m. If D and C are the midpoints of $\overline{AE}$ and $\overline{BF}$, respectively, then is trapezoid $ABCD$ similar to trapezoid $DCFE$?

34. In the drawing, $\overleftrightarrow{AB} \parallel \overleftrightarrow{DC} \parallel \overleftrightarrow{EF}$. Suppose that transversals ℓ and m are also parallel. D and C are the midpoints of $\overline{AE}$ and $\overline{BF}$, respectively. Is parallelogram $ABCD$ similar to parallelogram $DCFE$?

Exercises 33, 34

35. Given △ABC, a second triangle (△XTN) is constructed so that $\angle X \cong \angle A$ and $\angle N \cong \angle C$.
    a) Is $\angle T$ congruent to $\angle B$?
    b) Using intuition (appearance), does it seem that △XTN is similar to △ABC?

36. Given △RST, a second triangle (△UVW) is constructed so that $UV = 2(RS)$, $VW = 2(ST)$, and $WU = 2(RT)$.
    a) What is the constant value of the ratios $\frac{UV}{RS}$, $\frac{VW}{ST}$, and $\frac{WU}{RT}$?
    b) Using intuition (appearance), does it seem that △UVW is similar to △RST?

37. Henry watches his 32″ diagonal LCD television while sitting on the couch 7 feet away. When he purchases a TV for a room where he will be watching from a distance of 11 feet, what size TV should he purchase?
    $\left(\text{HINT: Use } \frac{D_1}{V_1} = \frac{D_2}{V_2}. \right)$

38. Use $D = 4.3V$ to find the ideal viewing distance V for a TV set that has a diagonal measuring 42 inches. Answer to the nearest foot.

    (HINT: See Example 4.)

For Exercises 39 and 40, use intuition to form a proportion based on the drawing shown.

*39. △*ABC* has an inscribed rhombus *ARST*. If *AB* = 10 and *AC* = 6, find the length *x* of each side of the rhombus.

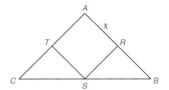

*40. A square with sides of length 2 in. rests (as shown) on a square with sides of length 6 in. Find the perimeter of trapezoid *ABCD*.

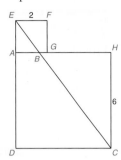

5.3	**Proving Triangles Similar**

KEY CONCEPTS	AAA	CSSTP	SAS ∼
	AA	CASTC	SSS ∼

It is quite difficult to establish a proportionality between the lengths of the corresponding sides of polygons. For this reason, our definition of similar polygons (and therefore of similar triangles) is almost impossible to use as a method of proof. Fortunately, some easier methods for proving triangles similar are available. If two triangles are carefully sketched or constructed so that pairs of angles are congruent, the triangles will have the same shape as shown in Figure 5.11.

△*HJK* ∼ △*SRT*

Figure 5.11

POSTULATE 15

If the three angles of one triangle are congruent to the three angles of a second triangle, then the triangles are similar (AAA).

Corollary 5.3.1 of Postulate 15 follows from Corollary 2.4.4, which states "If two angles of one triangle are congruent to two angles of another triangle, then the third pair of angles *must* also be congruent."

COROLLARY 5.3.1

If two angles of one triangle are congruent to two angles of another triangle, then the triangles are similar (AA).

Rather than use AAA to prove triangles similar, we will use AA in Example 1 and later proofs because it requires fewer steps.

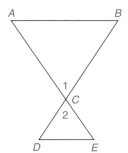

Figure 5.12

EXAMPLE 1

Provide a two-column proof for the following problem.

GIVEN: $\overline{AB} \parallel \overline{DE}$ in Figure 5.12

PROVE: $\triangle ABC \sim \triangle EDC$

PROOF

Statements	Reasons
1. $\overline{AB} \parallel \overline{DE}$	1. Given
2. $\angle A \cong \angle E$	2. If two $\parallel$ lines are cut by a transversal, the alternate interior angles are $\cong$
3. $\angle 1 \cong \angle 2$	3. Vertical angles are $\cong$
4. $\triangle ABC \sim \triangle EDC$	4. AA

> **STRATEGY FOR PROOF ■ Proving That Two Triangles Are Similar**
>
> *General Rule:* Although there *will be* three methods of proof (AA, SAS $\sim$, and SSS $\sim$) for similar triangles, we use AA whenever possible. This leads to a more efficient proof.
>
> *Illustration:* See statements 2 and 3 in the proof of Example 1. Notice that statement 4 follows by the reason AA.

In some instances, we wish to prove a relationship that takes us beyond the similarity of triangles. The following consequences of the definition of similarity are often cited as reasons in a proof.

 EXS. 1–4

> **CSSTP**
>
> Corresponding sides of similar triangles are proportional.

> **CASTC**
>
> Corresponding angles of similar triangles are congruent.

The first fact, abbreviated CSSTP, is used in Example 2. Although the CSSTP statement involves triangles, the corresponding sides of *any* two similar polygons are proportional. That is, the ratio of the lengths of any pair of corresponding sides of one polygon equals the ratio of the lengths of another pair of corresponding sides of the second polygon. The second fact, abbreviated CASTC, is used in Example 4.

> **STRATEGY FOR PROOF ■ Proving a Proportion**
>
> *General Rule:* First prove that triangles are similar. Then apply CSSTP.
>
> *Illustration:* In Example 2, statement 3 verifies that triangles are similar. In turn, CSSTP justifies statement 4 of the proof.

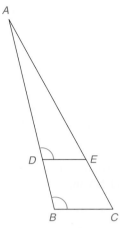

Figure 5.13

Reminder

CSSTP means "**C**orresponding **S**ides of **S**imilar **T**riangles are **P**roportional."

EXAMPLE 2

Complete the following two-column proof.

GIVEN: $\angle ADE \cong \angle B$ in Figure 5.13

PROVE: $\dfrac{DE}{BC} = \dfrac{AE}{AC}$

PROOF	
Statements	**Reasons**
1. $\angle ADE \cong \angle B$	1. Given
2. $\angle A \cong \angle A$	2. Identity
3. $\triangle ADE \sim \triangle ABC$	3. AA
4. $\dfrac{DE}{BC} = \dfrac{AE}{AC}$	4. CSSTP

NOTE: In Step 4, $\overline{DE}$ and $\overline{BC}$ are corresponding sides, as are $\overline{AE}$ and $\overline{AC}$.

THEOREM 5.3.2

The lengths of the corresponding altitudes of similar triangles have the same ratio as the lengths of any pair of corresponding sides.

The proof of Theorem 5.3.2 is left to the student; see Exercise 35. Note that this proof also requires the use of CSSTP.

STRATEGY FOR PROOF ■ Proving Products of Lengths Equal

General Rule: First prove that two triangles are similar. Then form a proportion involving the lengths of corresponding sides. Finally, apply the Means-Extremes Property.

Illustration: See the following proof and Example 3 (an alternative form of the proof).

The paragraph style of proof is generally used in upper-level mathematics classes. These paragraph proofs are no more than modified two-column proofs. Compare the following two-column proof to the paragraph proof found in Example 3.

GIVEN: $\angle M \cong \angle Q$ in Figure 5.14

PROVE: $NP \cdot QR = RP \cdot MN$

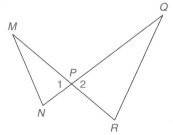

Figure 5.14

PROOF	
Statements	**Reasons**
1. $\angle M \cong \angle Q$	1. Given (hypothesis)
2. $\angle 1 \cong \angle 2$	2. Vertical angles are $\cong$
3. $\triangle MPN \sim \triangle QPR$	3. AA
4. $\dfrac{NP}{RP} = \dfrac{MN}{QR}$	4. CSSTP
5. $NP \cdot QR = RP \cdot MN$	5. Means-Extremes Property

EXAMPLE 3

Use a paragraph proof to complete this problem.

GIVEN: $\angle M \cong \angle Q$ in Figure 5.14 on page 227

PROVE: $NP \cdot QR = RP \cdot MN$

PROOF: By hypothesis, $\angle M \cong \angle Q$. Also, $\angle 1 \cong \angle 2$ by the fact that vertical angles are congruent. Now $\triangle MPN \sim \triangle QPR$ by AA. Using CSSTP, $\frac{NP}{RP} = \frac{MN}{QR}$. Then $NP \cdot QR = RP \cdot MN$ by the Means-Extremes Property.

NOTE: In the proof, the sides selected for the proportion were carefully chosen. The statement to be proved required that we include NP, QR, RP, and MN in the proportion.

SSG EXS. 5–7

In addition to AA, there are other methods that can be used to establish similar triangles. Recall that SAS and SSS name methods for proving that triangles are congruent. To distinguish the following techniques for showing that triangles are similar from the methods for proving that triangles are congruent, we use SAS ~ and SSS ~ to identify the similarity theorems. We will prove SAS ~ in Example 6, but the proof of SSS ~ is at the website for this textbook.

> **Reminder**
>
> CASTC means "**C**orresponding **A**ngles of **S**imilar **T**riangles are **C**ongruent."

THEOREM 5.3.3 (SAS~)

If an angle of one triangle is congruent to an angle of a second triangle and the pairs of sides including the angles are proportional, then the triangles are similar.

Consider this application of Theorem 5.3.3.

Figure 5.15

EXAMPLE 4

In Figure 5.15, $\frac{DG}{DE} = \frac{DH}{DF}$. Also, m$\angle E = x$, m$\angle D = x + 22$, and m$\angle DHG = x - 10$. Find the value of x and the measure of each angle.

SOLUTION With $\angle D \cong \angle D$ (Identity) and $\frac{DG}{DE} = \frac{DH}{DF}$ (Given), $\triangle DGH \sim \triangle DEF$ by SAS~. By CASTC, $\angle F \cong \angle DHG$, so m$\angle F = x - 10$. The sum of angles in $\triangle DEF$ is $x + x + 22 + x - 10 = 180$, so $3x + 12 = 180$. Then $3x = 168$ and $x = 56$. In turn, m$\angle E = \angle DGH = 56°$, m$\angle F = $ m$\angle DHG = 46°$, and m$\angle D = 78°$.

> **Warning**
>
> SSS and SAS prove that triangles are congruent. SSS~ and SAS~ prove that triangles are similar.

THEOREM 5.3.4 (SSS~)

If the three sides of one triangle are proportional to the three corresponding sides of a second triangle, then the triangles are similar.

Along with AA and SAS~, Theorem 5.3.4 (SSS~) provides the third (and final) method of establishing that triangles are similar.

EXAMPLE 5

Which method (AA, SAS ~, or (SSS ~) establishes that $\triangle ABC \sim \triangle XTN$? See Figure 5.16 on the following page.

a) $\angle A \cong \angle X$, $AC = 6$, $XN = 9$, $AB = 8$, and $XT = 12$
b) $AB = 6$, $AC = 4$, $BC = 8$, $XT = 9$, $XN = 6$, and $TN = 12$

SOLUTION

a) SAS~ because $\angle A \cong \angle X$ and $\frac{AC}{XN} = \frac{AB}{XT}$

b) SSS~ because $\frac{AB}{XT} = \frac{AC}{XN} = \frac{BC}{TN}$

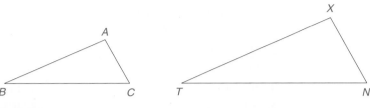

SSG **EXS. 8–10**   Figure 5.16

We close this section by proving a special case of Theorem 5.3.3 (SAS ~). To achieve this goal, we prove a helping theorem by the indirect method. In Figure 5.17(a), we say that sides $\overline{CA}$ and $\overline{CB}$ are divided proportionally by $\overline{DE}$ if $\frac{DA}{CD} = \frac{EB}{CE}$. See the Discover activity at the left.

LEMMA 5.3.5

If a line segment divides two sides of a triangle proportionally, then this line segment is parallel to the third side of the triangle.

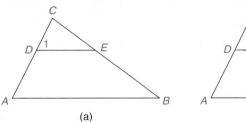

Figure 5.17

GIVEN: $\triangle ABC$ with $\frac{DA}{CD} = \frac{EB}{CE}$ in Figure 5.17(a)

PROVE: $\overline{DE} \parallel \overline{AB}$

PROOF: $\frac{DA}{CD} = \frac{EB}{CE}$ in $\triangle ABC$. Applying Property 3 of Section 5.1, we have $\frac{CD + DA}{CD} = \frac{CE + EB}{CE}$, so $\frac{CA}{CD} = \frac{CB}{CE}$ (*).

Now suppose that $\overline{DE}$ is not parallel to $\overline{AB}$. Through D, we draw $\overline{DF} \parallel \overline{AB}$ in Figure 5.17(b). It follows that $\angle CDF \cong \angle A$. With $\angle C \cong \angle C$, it follows that $\triangle CDF \sim \triangle CAB$ by the reason AA. By CSSTP, $\frac{CA}{CD} = \frac{CB}{CF}$ (**). Using the starred statements and substitution, $\frac{CB}{CE} = \frac{CB}{CF}$ (both ratios are equal to $\frac{CA}{CD}$). Applying the Means-Extremes Property, $CB \cdot CF = CB \cdot CE$; dividing by CB, we find that $CF = CE$. But $CF = CE + EF$, so $CE + EF = CE$ by substitution. By subtraction, $EF = 0$; this statement contradicts the Ruler Postulate. Thus, F must coincide with E; it follows that $\overline{DE} \parallel \overline{AB}$.

In Example 6, we use Lemma 5.3.5 to prove this case of the SAS ~ theorem.

EXAMPLE 6

GIVEN: $\triangle ABC$ and $\triangle DEC$; $\dfrac{CA}{CD} = \dfrac{CB}{CE}$

PROVE: $\triangle ABC \sim \triangle DEC$

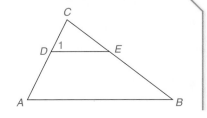

PROOF	
Statements	**Reasons**
1. $\triangle ABC$ and $\triangle DEC$; $\dfrac{CA}{CD} = \dfrac{CB}{CE}$	1. Given
2. $\dfrac{CA - CD}{CD} = \dfrac{CB - CE}{CE}$	2. Difference Property of a Proportion
3. $\dfrac{DA}{CD} = \dfrac{EB}{CE}$	3. Substitution
4. $\overline{DE} \parallel \overline{AB}$	4. Lemma 5.3.5
5. $\angle 1 \cong \angle A$	5. If 2 $\parallel$ lines are cut by a trans., corr. $\angle$s are $\cong$
6. $\angle C \cong \angle C$	6. Identity
7. $\triangle ABC \sim \triangle DEC$	7. AA

SSG **EXS. 11, 12**

In closing, note that any two polygons having n sides are similar if all pairs of corresponding angles are congruent while the lengths of their corresponding pairs of sides are also proportional.

Exercises 5.3

1. What is the acronym that is used to represent the statement "Corresponding angles of similar triangles are congruent?"

2. What is the acronym that is used to represent the statement "Corresponding sides of similar triangles are proportional?"

3. Classify as true or false:
   a) If the vertex angles of two isosceles triangles are congruent, the triangles are similar.
   b) Any two equilateral triangles are similar.

4. Classify as true or false:
   a) If the midpoints of two sides of a triangle are joined, the triangle formed is similar to the original triangle.
   b) Any two isosceles triangles are similar.

In Exercises 5 to 8, name the method (AA, SSS~, or SAS~) that is used to show that the triangles are similar.

5. $WU = \frac{3}{2} \cdot TR$, $WV = \frac{3}{2} \cdot TS$, and $UV = \frac{3}{2} \cdot RS$

Exercises 5–8

6. $\angle T \cong \angle W$ and $\angle R \cong \angle U$

7. $\angle T \cong \angle W$ and $\dfrac{TR}{WU} = \dfrac{TS}{WV}$

8. $\dfrac{TR}{WU} = \dfrac{TS}{WV} = \dfrac{RS}{UV}$

In Exercises 9 to 12, name the method that explains why $\triangle DGH \sim \triangle DEF$. See the figure in the right column.

9. $\dfrac{DG}{DE} = \dfrac{DH}{DF}$

10. $DE = 3 \cdot DG$ and $DF = 3 \cdot DH$

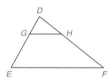

Exercises 9–12

11. $\dfrac{DG}{DE} = \dfrac{DH}{DF} = \dfrac{GH}{EF} = \frac{1}{3}$

12. $\angle DGH \cong \angle DEF$

In Exercises 13 to 16, provide the missing reasons.

13. Given: $\square RSTV$; $\overline{VW} \perp \overline{RS}$; $\overline{VX} \perp \overline{TS}$
   Prove: $\triangle VWR \sim \triangle VXT$

PROOF	
Statements	**Reasons**
1. $\square RSTV$; $\overline{VW} \perp \overline{RS}$; $\overline{VX} \perp \overline{TS}$	1. ?
2. $\angle VWR$ and $\angle VXT$ are rt. $\angle$s	2. ?
3. $\angle VWR \cong \angle VXT$	3. ?
4. $\angle R \cong \angle T$	4. ?
5. $\triangle VWR \sim \triangle VXT$	5. ?

14. *Given:*   △DET and □ABCD
      Prove: △ABE ~ △CTB

PROOF

Statements	Reasons
1. △DET and □ABCD	1. ?
2. $\overline{AB} \parallel \overline{DT}$	2. Opposite sides of a □ are ‖
3. ∠EBA ≅ ∠T	3. ?
4. $\overline{ED} \parallel \overline{CB}$	4. ?
5. ∠E ≅ ∠CBT	5. ?
6. △ABE ~ △CTB	6. ?

15. *Given:*   △ABC; M and N are midpoints of $\overline{AB}$ and $\overline{AC}$, respectively
      Prove: △AMN ~ △ABC

PROOF

Statements	Reasons
1. △ABC; M and N are the midpoints of $\overline{AB}$ and $\overline{AC}$, respectively	1. ?
2. $AM = \frac{1}{2}(AB)$ and $AN = \frac{1}{2}(AC)$	2. ?
3. $MN = \frac{1}{2}(BC)$	3. ?
4. $\frac{AM}{AB} = \frac{1}{2}$, $\frac{AN}{AC} = \frac{1}{2}$, and $\frac{MN}{BC} = \frac{1}{2}$	4. ?
5. $\frac{AM}{AB} = \frac{AN}{AC} = \frac{MN}{BC}$	5. ?
6. △AMN ~ △ABC	6. ?

16. *Given:*   △XYZ with $\overline{XY}$ trisected at P and Q and $\overline{YZ}$ trisected at R and S
      Prove: △XYZ ~ △PYR

PROOF

Statements	Reasons
1. △XYZ; $\overline{XY}$ trisected at P and Q; $\overline{YZ}$ trisected at R and S	1. ?
2. $\frac{YR}{YZ} = \frac{1}{3}$ and $\frac{YP}{YX} = \frac{1}{3}$	2. Definition of trisect
3. $\frac{YR}{YZ} = \frac{YP}{YX}$	3. ?
4. ∠Y ≅ ∠Y	4. ?
5. △XYZ ~ △PYR	5. ?

In Exercises 17 to 24, complete each proof.

17. *Given:* $\overline{MN} \perp \overline{NP}$, $\overline{QR} \perp \overline{RP}$
      Prove: △MNP ~ △QRP

Exercises 17, 18

PROOF

Statements	Reasons
1. ?	1. Given
2. ∠s N and QRP are right ∠s	2. ?
3. ?	3. All right ∠s are ≅
4. ∠P ≅ ∠P	4. ?
5. ?	5. ?

18. *Given:* $\overline{MN} \parallel \overline{QR}$ (See figure for Exercise 17.)
      Prove: △MNP ~ △QRP

PROOF

Statements	Reasons
1. ?	1. Given
2. ∠M ≅ ∠RQP	2. ?
3. ?	3. If two ‖ lines are cut by a transversal, the corresponding ∠s are ≅
4. ?	4. ?

19. *Given:*   ∠H ≅ ∠F
      Prove: △HJK ~ △FGK

Exercises 19, 20

PROOF

Statements	Reasons
1. ?	1. Given
2. ∠HKJ ≅ ∠FKG	2. ?
3. ?	3. ?

20. *Given:* $\overline{HJ} \perp \overline{JF}$, $\overline{HG} \perp \overline{FG}$ (See figure for Exercise 19.)
      Prove: △HJK ~ △FGK

PROOF

Statements	Reasons
1. ?	1. Given
2. ∠s G and J are right ∠s	2. ?
3. ∠G ≅ ∠J	3. ?
4. ∠HKJ ≅ ∠GKF	4. ?
5. ?	5. ?

21. *Given:* $\frac{RQ}{NM} = \frac{RS}{NP} = \frac{QS}{MP}$

Prove: $\angle N \cong \angle R$

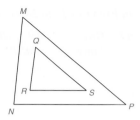

PROOF	
Statements	**Reasons**
1. ?	1. Given
2. ?	2. SSS ~
3. ?	3. CASTC

22. *Given:* $\frac{DG}{DE} = \frac{DH}{DF}$

Prove: $\angle DGH \cong \angle E$

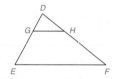

PROOF	
Statements	**Reasons**
1. ?	1. ?
2. $\angle D \cong \angle D$	2. ?
3. $\triangle DGH \sim \triangle DEF$	3. ?
4. ?	4. ?

23. *Given:* $\overline{RS} \parallel \overline{UV}$

Prove: $\frac{RT}{VT} = \frac{RS}{VU}$

PROOF	
Statements	**Reasons**
1. ?	1. ?
2. $\angle R \cong \angle V$ and $\angle S \cong \angle U$	2. ?
3. ?	3. AA
4. ?	4. ?

24. *Given:* $\overline{AB} \parallel \overline{DC}, \overline{AC} \parallel \overline{DE}$

Prove: $\frac{AB}{DC} = \frac{BC}{CE}$

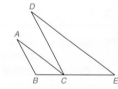

PROOF	
Statements	**Reasons**
1. $\overline{AB} \parallel \overline{DC}$	1. ?
2. ?	2. If 2 ∥ lines are cut by a trans. corr. ∠s are ≅
3. ?	3. Given
4. $\angle ACB \cong \angle E$	4. ?
5. $\triangle ACB \sim \triangle DEC$	5. ?
6. ?	6. ?

In Exercises 25 to 28, $\triangle ABC \sim \triangle DBE$.

Exercises 25–28

25. *Given:* $AC = 8, DE = 6, CB = 6$

Find: EB

(*HINT: Let EB* $= x$, *and solve an equation.*)

26. *Given:* $AC = 10, CB = 12$

E is the midpoint of $\overline{CB}$

Find: DE

27. *Given:* $AC = 10, DE = 8, AD = 4$

Find: DB

28. *Given:* $CB = 12, CE = 4, AD = 5$

Find: DB

29. $\triangle CDE \sim \triangle CBA$ with $\angle CDE \cong \angle B$. If $CD = 10$, $DA = 8$, and $CE = 6$, find EB.

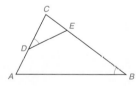

Exercises 29, 30

30. $\triangle CDE \sim \triangle CBA$ with $\angle CDE \cong \angle B$. If $CD = 10$, $CA = 16$, and $EB = 12$, find CE. (See the figure for Exercise 29.)

31. $\triangle ABF \sim \triangle CBD$ with obtuse angles at vertices D and F as indicated. If $m\angle B = 45°$, $m\angle C = x$ and $m\angle AFB = 4x$, find x.

32. $\triangle ABF \sim \triangle CBD$ with obtuse angles at vertices D and F. If $m\angle B = 44°$ and $m\angle A : m\angle CDB = 1:3$, find $m\angle A$.

Exercises 31, 32

In Exercise 33, provide a two-column proof.

33. Given: $\overline{AB} \parallel \overline{DF}, \overline{BD} \parallel \overline{FG}$
Prove: $\triangle ABC \sim \triangle EFG$

In Exercise 34, provide a paragraph proof.

34. Given: $\overline{RS} \perp \overline{AB}, \overline{CB} \perp \overline{AC}$
Prove: $\triangle BSR \sim \triangle BCA$

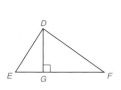

35. Use a two-column proof to prove the following theorem: "The lengths of the corresponding altitudes of similar triangles have the same ratio as the lengths of any pair of corresponding sides."
Given: $\triangle DEF \sim \triangle MNP; \overline{DG}$ and $\overline{MQ}$ are altitudes
Prove: $\dfrac{DG}{MQ} = \dfrac{DE}{MN}$

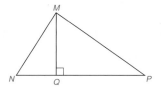

36. Provide a paragraph proof for the following problem.
Given: $\overline{RS} \parallel \overline{YZ}, \overline{RU} \parallel \overline{XZ}$
Prove: $RS \cdot ZX = ZY \cdot RT$

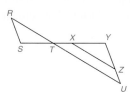

37. Use the result of Exercise 13 to do the following problem. In $\square MNPQ$, $QP = 12$ and $QM = 9$. The length of altitude $\overline{QR}$ (to side $\overline{MN}$) is 6. Find the length of altitude $\overline{QS}$ from Q to $\overline{PN}$.

38. Use the result of Exercise 13 to do the following problem. In $\square ABCD$, $AB = 7$ and $BC = 12$. The length of altitude $\overline{AF}$ (to side $\overline{BC}$) is 5. Find the length of altitude $\overline{AE}$ from A to $\overline{DC}$.

39. The distance across a pond is to be measured indirectly by using similar triangles. If $XY = 160$ ft, $YW = 40$ ft, $TY = 120$ ft, and $WZ = 50$ ft, find XT.

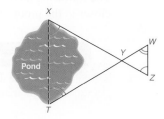

40. In the figure, $\angle ABC \cong \angle ADB$. Find AB if $AD = 2$ and $DC = 6$.

41. Prove that the altitude drawn to the hypotenuse of a right triangle separates the right triangle into two right triangles that are similar to each other and to the original right triangle.

42. Prove that the line segment joining the midpoints of two sides of a triangle determines a triangle that is similar to the original triangle.

43. The vertices of rhombus *ARST* lie on $\triangle ABC$ as shown. Where $AB = c$ and $AC = b$, show that $RS = \dfrac{bc}{b + c}$. (Let $AR = x$.)

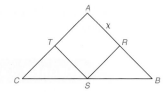

5.4 The Pythagorean Theorem

KEY CONCEPTS   Pythagorean Theorem   Converse of   Pythagorean Triple
                                              Pythagorean Theorem

The following theorem, which was proved in Exercise 41 of Section 5.3, will take us a step closer to the proof of the well-known Pythagorean Theorem.

> **THEOREM 5.4.1**
>
> The altitude drawn to the hypotenuse of a right triangle separates the right triangle into two right triangles that are similar to each other and to the original right triangle.

Theorem 5.4.1 is illustrated by Figure 5.18, in which the right triangle $\triangle ABC$ has its right angle at vertex C so that $\overline{CD}$ is the altitude to hypotenuse $\overline{AB}$. The smaller triangles are shown in Figures 5.18(b) and (c), and the original triangle is shown in Figure 5.18(d). Note the matching arcs indicating congruent angles.

(a)

(b)

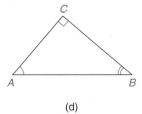
(c)   (d)

Figure 5.18

In Figure 5.18(a), $\overline{AD}$ and $\overline{DB}$ are known as *segments* (parts) of the hypotenuse $\overline{AB}$. Furthermore, $\overline{AD}$ is the segment of the hypotenuse *adjacent* to (next to) leg $\overline{AC}$, and BD is the segment of the hypotenuse *adjacent* to leg $\overline{BC}$. Proof of the following theorem is left as Exercise 43. Compare the statement of Theorem 5.4.2 to the "Prove" statement that follows it.

> **THEOREM 5.4.2**
>
> The length of the altitude to the hypotenuse of a right triangle is the geometric mean of the lengths of the segments of the hypotenuse.

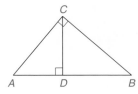

Figure 5.19

GIVEN: $\triangle ABC$ in Figure 5.19, with right $\angle ACB$; $\overline{CD} \perp \overline{AB}$

PROVE: $\dfrac{AD}{CD} = \dfrac{CD}{DB}$

PLAN FOR PROOF: Show that $\triangle ADC \sim \triangle CDB$. Then use CSSTP.

NOTE: In the proportion $\frac{AD}{CD} = \frac{CD}{DB}$, recall that CD is a geometric mean because the second and the third terms are identical.

The proof of the following lemma is left as Exercise 44. Compare the statement of Lemma 5.4.3 to the "Prove" statement that follows it.

Figure 5.20

Figure 5.21

LEMMA 5.4.3

The length of each leg of a right triangle is the geometric mean of the length of the hypotenuse and the length of the segment of the hypotenuse adjacent to that leg.

GIVEN: $\triangle ABC$ with right $\angle ACB$; $\overline{CD} \perp \overline{AB}$ in Figure 5.20.

PROVE: $\dfrac{AB}{AC} = \dfrac{AC}{AD}$

PLAN: Show that $\triangle ADC \sim \triangle ACB$ in Figure 5.21. Then use CSSTP.

NOTE: Although $\overline{AD}$ and $\overline{DB}$ are both segments of the hypotenuse, $\overline{AD}$ is the segment adjacent to $\overline{AC}$. We can also prove that $\frac{AB}{CB} = \frac{CB}{DB}$ by showing that $\triangle CDB \sim \triangle ACB$.

Lemma 5.4.3 opens the doors to a proof of the famous Pythagorean Theorem, one of the most frequently applied relationships in geometry. Although the theorem's title gives credit to the Greek geometer Pythagoras, many other proofs are known, and the ancient Chinese were aware of the relationship before the time of Pythagoras.

SSG **EXS. 1, 2**

THEOREM 5.4.4 ■ Pythagorean Theorem

The square of the length of the hypotenuse of a right triangle is equal to the sum of the squares of the lengths of the legs.

Thus, where c is the length of the hypotenuse and a and b are the lengths of the legs, $c^2 = a^2 + b^2$. See Figure 5.22(a).

GIVEN: In Figure 5.22(a), $\triangle ABC$ with right $\angle C$

PROVE: $c^2 = a^2 + b^2$

PROOF: Draw $\overline{CD} \perp \overline{AB}$, as shown in Figure 5.22(b). Denote $AD = x$ and $DB = y$. By Lemma 5.4.3,

Therefore, $\qquad b^2 = cx \qquad$ and $\qquad a^2 = cy$

Using the Addition Property of Equality, we have

$$a^2 + b^2 = cy + cx = c(y + x)$$

But $y + x = x + y = AD + DB = AB = c$. Thus, $a^2 + b^2 = c(c) = c^2$; equivalently, $c^2 = a^2 + b^2$.

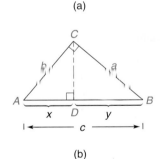

(a)

(b)

Figure 5.22

Discover

A video entitled "The Rule of Pythagoras" is available through Project Mathematics at Cal Tech University in Pasadena, CA. It is well worth watching!

EXAMPLE 1

Given $\triangle RST$ with right $\angle S$ in Figure 5.23, find:

a) RT if $RS = 3$ and $ST = 4$
b) ST if $RS = 6$ and $RT = 9$

SOLUTION With right $\angle S$, the hypotenuse is $\overline{RT}$. Then $RT = c$, $RS = a$, and $ST = b$, as shown.

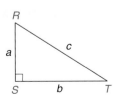

Figure 5.23

a) $3^2 + 4^2 = c^2 \rightarrow 9 + 16 = c^2$
$$c^2 = 25$$
$$c = 5; RT = 5$$
b) $6^2 + b^2 = 9^2 \rightarrow 36 + b^2 = 81$
$$b^2 = 45$$
$$b = \sqrt{45} = \sqrt{9 \cdot 5} = \sqrt{9} \cdot \sqrt{5} = 3\sqrt{5}$$
$$ST = 3\sqrt{5} \approx 6.71$$

SSG **EXS. 3, 4**

The converse of the Pythagorean Theorem is also true.

> ### THEOREM 5.4.5 ■ Converse of the Pythagorean Theorem
>
> If a, b, and c are the lengths of the three sides of a triangle, with c the length of the longest side, and if $c^2 = a^2 + b^2$, then the triangle is a right triangle with the right angle opposite the side of length c.

(a)

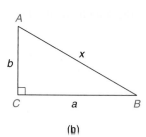

(b)

Figure 5.24

GIVEN: $\triangle RST$ [Figure 5.24(a)] with sides a, b, and c so that $c^2 = a^2 + b^2$

PROVE: $\triangle RST$ is a right triangle.

PROOF: We are given $\triangle RST$ for which $c^2 = a^2 + b^2$. Construct the right $\triangle ABC$, which has legs of lengths a and b and a hypotenuse of length x. [See Figure 5.24(b).] By the Pythagorean Theorem, $x^2 = a^2 + b^2$. By substitution, $x^2 = c^2$ and $x = c$. Thus, $\triangle RTS \cong \triangle ABC$ by SSS. Then $\angle S$ (opposite the side of length c) must be $\cong$ to $\angle C$, the right angle of $\triangle ABC$. Then $\angle S$ is a right angle, and $\triangle RST$ is a right triangle.

We apply Theorem 5.4.5 in Example 2.

EXAMPLE 2

Do the following represent the lengths of the sides of a right triangle?

a) $a = 5, b = 12, c = 13$
b) $a = 15, b = 8, c = 17$
c) $a = 7, b = 9, c = 10$
d) $a = \sqrt{2}, b = \sqrt{3}, c = \sqrt{5}$

SOLUTION

a) Yes. Because $5^2 + 12^2 = 13^2$ (that is, $25 + 144 = 169$), this triangle is a right triangle.
b) Yes. Because $15^2 + 8^2 = 17^2$ (that is, $225 + 64 = 289$), this triangle is a right triangle.
c) No. $7^2 + 9^2 = 49 + 81 = 130$, which is not 10^2 (that is, 100), so this triangle is not a right triangle.
d) Yes. Because $(\sqrt{2})^2 + (\sqrt{3})^2 = (\sqrt{5})^2$ leads to $2 + 3 = 5$, this triangle is a right triangle.

SSG **EXS. 5, 6**

EXAMPLE 3

A ladder 12 ft long is leaning against a wall so that its base is 4 ft from the wall at ground level (see Figure 5.25). How far up the wall does the ladder reach?

Figure 5.25

SOLUTION Applying the Pythagorean Theorem, the desired height h is the length of a leg of the indicated right triangle.

$$4^2 + h^2 = 12^2$$
$$16 + h^2 = 144$$
$$h^2 = 128$$
$$h = \sqrt{128} = \sqrt{64 \cdot 2} = \sqrt{64} \cdot \sqrt{2} = 8\sqrt{2}$$

The height is exactly $h = 8\sqrt{2}$, which is approximately 11.31 ft.

Reminder

The diagonals of a rhombus are perpendicular bisectors of each other.

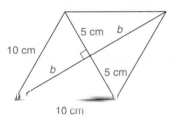

Figure 5.26

EXAMPLE 4

One diagonal of a rhombus has the same length, 10 cm, as each side of the rhombus in Figure 5.26. How long is the other diagonal?

SOLUTION Because the diagonals are perpendicular bisectors of each other, four right △s are formed. For each right △, a side of the rhombus is the hypotenuse. Half of the length of each diagonal is the length of a leg of each right triangle. Therefore,

$$5^2 + b^2 = 10^2$$
$$25 + b^2 = 100$$
$$b^2 = 75$$
$$b = \sqrt{75} = \sqrt{25 \cdot 3} = \sqrt{25} \cdot \sqrt{3} = 5\sqrt{3}$$

Thus, the length of the whole diagonal is $2(5\sqrt{3})$ or $10\sqrt{3}$ cm ≈ 17.32 cm.

The solution for Example 5 also depends upon the Pythagorean Theorem, but it is slightly more demanding than the solution for Example 4. Indeed, it is one of those situations that may require some insight to solve. Note that the triangle described in Example 5 is *not* a right triangle because $4^2 + 5^2 \neq 6^2$.

EXAMPLE 5

A triangle has sides of lengths 4, 5, and 6, as shown in Figure 5.27. Find the length h of the altitude to the side of length 6.

SOLUTION The altitude to the side of length 6 separates that side into two parts whose lengths are given by x and $6 - x$. Using the two right triangles formed, we apply the Pythagorean Theorem twice.

$$x^2 + h^2 = 4^2 \quad \text{and} \quad (6 - x)^2 + h^2 = 5^2$$

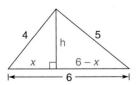

Figure 5.27

Subtracting the first equation from the second, we can calculate x.

$$\begin{array}{rl} 36 - 12x + x^2 + h^2 &= 25 \\ x^2 + h^2 &= 16 \\ \hline 36 - 12x \qquad\qquad\quad &= 9 \qquad \text{(subtraction)} \\ -12x &= -27 \\ x &= \dfrac{27}{12} = \dfrac{9}{4} \end{array}$$

Now we use $x = \frac{9}{4}$ to find h.

$$x^2 + h^2 = 4^2$$
$$\left(\frac{9}{4}\right)^2 + h^2 = 4^2$$
$$\frac{81}{16} + h^2 = 16$$
$$\frac{81}{16} + h^2 = \frac{256}{16}$$
$$h^2 = \frac{175}{16}, \text{ so } h = \sqrt{\frac{175}{16}} = \frac{\sqrt{175}}{\sqrt{16}}$$

Simplifying, $h = \dfrac{\sqrt{175}}{4} = \dfrac{\sqrt{25 \cdot 7}}{4} = \dfrac{\sqrt{25} \cdot \sqrt{7}}{4} = \dfrac{5\sqrt{7}}{4} \approx 3.31$

It is now possible to prove the HL method for proving the congruence of right triangles, a method that was introduced in Section 3.2.

THEOREM 5.4.6

If the hypotenuse and a leg of one right triangle are congruent to the hypotenuse and a leg of a second right triangle, then the triangles are congruent (HL).

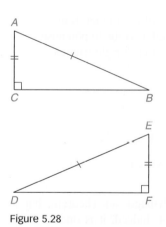

Figure 5.28

SSG **EXS. 7, 8**

GIVEN: Right $\triangle ABC$ with right $\angle C$ and right $\triangle DEF$ with right $\angle F$ as shown in Figure 5.28; $\overline{AB} \cong \overline{DE}$ and $\overline{AC} \cong \overline{EF}$

PROVE: $\triangle ABC \cong \triangle EDF$

PROOF: With right $\angle C$, the hypotenuse of $\triangle ABC$ is $\overline{AB}$; similarly, $\overline{DE}$ is the hypotenuse of right $\triangle EDF$. Because $\overline{AB} \cong \overline{DE}$, we denote the common length by c; that is, $AB = DE = c$. Because $\overline{AC} \cong \overline{EF}$, we also have $AC = EF = a$. Then

$$\begin{array}{llll} a^2 + (BC)^2 = c^2 & \text{and} & a^2 + (DF)^2 = c^2, \text{ which leads to} \\ BC = \sqrt{c^2 - a^2} & \text{and} & DF = \sqrt{c^2 - a^2} \end{array}$$

Then $BC = DF$ so that $\overline{BC} \cong \overline{DF}$. Hence, $\triangle ABC \cong \triangle EDF$ by SSS.

Our work with the Pythagorean Theorem would be incomplete if we did not address two issues. The first, Pythagorean triples, involves natural (or counting) numbers as possible choices of a, b, and c. The second leads to the classification of triangles according to the lengths of their sides, as found in Theorem 5.4.7 on page 239.

PYTHAGOREAN TRIPLES

DEFINITION

A **Pythagorean triple** is a set of three natural numbers (a, b, c) for which $a^2 + b^2 = c^2$.

Figure 5.29

Figure 5.30

Three sets of Pythagorean triples encountered in this section are (3, 4, 5), (5, 12, 13), and (8, 15, 17). These combinations of numbers, as lengths of sides, always lead to a right triangle.

Natural-number multiples of any of these triples also produce Pythagorean triples. For example, doubling (3, 4, 5) yields (6, 8, 10), which is also a Pythagorean triple; in Figure 5.29, the two triangles are similar by SSS ~.

The Pythagorean triple (3, 4, 5) also leads to the multiples (9, 12, 15), (12, 16, 20), and (15, 20, 25). The Pythagorean triple (5, 12, 13) leads to triples such as (10, 24, 26) and (15, 36, 39). Basic Pythagorean triples that are encountered less frequently include (7, 24, 25), (9, 40, 41), and (20, 21, 29).

Pythagorean triples can be generated by using select formulas. Where p and q are natural numbers and $p > q$, one formula uses $2pq$ for the length of one leg, $p^2 - q^2$ for the length of other leg, and $p^2 + q^2$ for the length of the hypotenuse (See Figure 5.30.).

Table 5.1 lists some Pythagorean triples corresponding to choices for p and q. The triples printed in boldface type are *basic triples,* also known as *primitive triples*; the entries a, b, and c of a primitive triple have only the common factor 1. In applications, knowledge of the primitive triples and their multiples will save you considerable time and effort. In the final column of Table 5.1, the resulting triple is provided in the order from a (small) to c (large).

TABLE 5.1
Pythagorean Triples

p	q	a (or b) $p^2 - q^2$	b (or a) $2pq$	c $p^2 + q^2$	(a, b, c)
2	1	3	4	5	**(3, 4, 5)**
3	1	8	6	10	(6, 8, 10)
3	2	5	12	13	**(5, 12, 13)**
4	1	15	8	17	**(8, 15, 17)**
4	3	7	24	25	**(7, 24, 25)**
5	1	24	10	26	(10, 24, 26)
5	2	21	20	29	**(20, 21, 29)**
5	3	16	30	34	(16, 30, 34)
5	4	9	40	41	**(9, 40, 41)**

THE CONVERSE OF THE PYTHAGOREAN THEOREM

The Converse of the Pythagorean Theorem (Theorem 5.4.5) allows us to recognize a right triangle by knowing the lengths of its sides. A variation on this converse allows us to determine whether a triangle is acute or obtuse. This theorem is stated without proof.

THEOREM 5.4.7

Let a, b, and c represent the lengths of the three sides of a triangle, with c the length of the longest side.

1. If $c^2 > a^2 + b^2$, then the triangle is obtuse and the obtuse angle lies opposite the side of length c.

2. If $c^2 < a^2 + b^2$, then the triangle is acute.

EXAMPLE 6

Determine the type of triangle represented if the lengths of its sides are as follows:

a) 4, 5, 7
b) 6, 7, 8
c) 9, 12, 15
d) 3, 4, 9

SOLUTION

a) Choosing $c = 7$, we have $7^2 > 4^2 + 5^2$, so the triangle is obtuse.
b) Choosing $c = 8$, we have $8^2 < 6^2 + 7^2$, so the triangle is acute.
c) Choosing $c = 15$, we have $15^2 = 9^2 + 12^2$, so the triangle is a right triangle.
d) Because $9 > 3 + 4$, no triangle is possible. (Remember that the sum of the lengths of two sides of a triangle must be greater than the length of the third side.)

SSG **EXS. 12, 13**

Exercises 5.4

1. By naming the vertices in order, state three different triangles that are similar to each other.

2. Use Theorem 5.4.2 to form a proportion in which SV is a geometric mean.

(HINT: $\triangle SVT \sim \triangle RVS$)

Exercises 1–6

3. Use Lemma 5.4.3 to form a proportion in which RS is a geometric mean.

(HINT: $\triangle RVS \sim \triangle RST$)

4. Use Lemma 5.4.3 to form a proportion in which TS is a geometric mean.

(HINT: $\triangle TVS \sim \triangle TSR$)

5. Use Theorem 5.4.2 to find RV if $SV = 6$ and $VT = 8$.

6. Use Lemma 5.4.3 to find RT if $RS = 6$ and $VR = 4$.

7. Find the length of $\overline{DF}$ if:
a) $DE = 8$ and $EF = 6$
b) $DE = 5$ and $EF = 3$

8. Find the length of $\overline{DE}$ if:
a) $DF = 13$ and $EF = 5$
b) $DF = 12$ and $EF = 6\sqrt{3}$

Exercises 7–10

9. Find EF if:
a) $DF = 17$ and $DE = 15$
b) $DF = 12$ and $DE = 8\sqrt{2}$

10. Find DF if:
a) $DE = 12$ and $EF = 5$
b) $DE = 12$ and $EF = 6$

11. Determine whether each triple (a, b, c) is a Pythagorean triple.
a) $(3, 4, 5)$ c) $(5, 12, 13)$
b) $(4, 5, 6)$ d) $(6, 13, 15)$

12. Determine whether each triple (a, b, c) is a Pythagorean triple.
a) $(8, 15, 17)$ c) $(6, 8, 10)$
b) $(10, 13, 19)$ d) $(11, 17, 20)$

13. Determine the type of triangle represented if the lengths of its sides are:
a) $a = 4$, $b = 3$, and $c = 5$
b) $a = 4$, $b = 5$, and $c = 6$
c) $a = 2$, $b = \sqrt{3}$, and $c = \sqrt{7}$
d) $a = 3$, $b = 8$, and $c = 15$

14. Determine the type of triangle represented if the lengths of its sides are:
a) $a = 1.5$, $b = 2$, and $c = 2.5$
b) $a = 20$, $b = 21$, and $c = 29$
c) $a = 10$, $b = 12$, and $c = 16$
d) $a = 5$, $b = 7$, and $c = 9$

15. A guy wire 25 ft long supports an antenna at a point that is 20 ft above the base of the antenna. How far from the base of the antenna is the guy wire secured?

16. A strong wind holds a kite 30 ft above the earth in a position 40 ft across the ground. How much string does the girl have out to the kite?

17. A boat is 6 m below the level of a pier and 12 m from the pier as measured across the water. How much rope is needed to reach the boat?

18. A hot-air balloon is held in place by the ground crew at a point that is 21 ft from a point directly beneath the basket of the balloon. If the rope is of length 29 ft, how far above ground level is the basket?

19. A drawbridge that is 104 ft in length is raised at its midpoint so that the uppermost points are 8 ft apart. How far has each of the midsections been raised?

20. A drawbridge that is 136 ft in length is raised at its midpoint so that the uppermost points are 16 ft apart. How far has each of the midsections been raised?

(HINT: Consider the drawing for Exercise 19.)

21. A rectangle has a width of 16 cm and a diagonal of length 20 cm. How long is the rectangle?

22. A right triangle has legs of lengths x and $2x + 2$ and a hypotenuse of length $2x + 3$. What are the lengths of its sides?

23. A rectangle has base length $x + 3$, altitude length $x + 1$, and diagonals of length $2x$ each. What are the lengths of its base, altitude, and diagonals?

24. The diagonals of a rhombus measure 6 m and 8 m. How long are each of the congruent sides?

25. Each side of a rhombus measures 12 in. If one diagonal is 18 in. long, how long is the other diagonal?

26. An isosceles right triangle has a hypotenuse of length 10 cm. How long is each leg?

27. Each leg of an isosceles right triangle has a length of $6\sqrt{2}$ in. What is the length of the hypotenuse?

28. In right $\triangle ABC$ with right $\angle C$, $AB = 10$ and $BC = 8$. Find the length of $\overline{MB}$ if M is the midpoint of $\overline{AC}$.

29. In right $\triangle ABC$ with right $\angle C$, $AB = 17$ and $BC = 15$. Find the length of $\overline{MN}$ if M and N are the midpoints of $\overline{AB}$ and $\overline{BC}$, respectively.

30. Find the length of the altitude to the 10-in. side of a triangle whose sides are 6, 8, and 10 inches in length.

31. Find the length of the altitude to the 26-in. side of a triangle whose sides are 10, 24, and 26 inches in length.

32. In quadrilateral $ABCD$, $\overline{BC} \perp \overline{AB}$ and $\overline{DC} \perp$ diagonal $\overline{AC}$. If $AB = 4$, $BC = 3$, and $DC = 12$, determine DA.

33. In quadrilateral $RSTU$, $\overline{RS} \perp \overline{ST}$ and $\overline{UT} \perp$ diagonal $\overline{RT}$. If $RS = 6$, $ST = 8$, and $RU = 15$, find UT.

34. *Given:* $\triangle ABC$ is not a right $\triangle$
Prove: $a^2 + b^2 \neq c^2$
(NOTE: $AB = c$, $AC = b$, and $CB = a$.)

*35. If $a = p^2 - q^2$, $b = 2pq$, and $c = p^2 + q^2$, show that $c^2 = a^2 + b^2$.

36. Given that the line segment shown has length 1, construct a line segment whose length is $\sqrt{2}$.

Exercises 36, 37

37. Using the line segment from Exercise 36, construct a line segment of length 2 and then a second line segment of length $\sqrt{5}$.

38. When the rectangle in the accompanying drawing (whose dimensions are 16 by 9) is cut into pieces and rearranged, a square can be formed. What is the perimeter of this square?

39. A, C, and F are three of the vertices of the cube shown in the accompanying figure. Given that each face of the cube is a square, what is the measure of angle ACF?

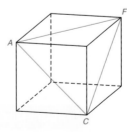

*40. Find the length of the altitude to the 8-in. side of a triangle whose sides are 4, 6, and 8 in. long.

(HINT: See Example 5.)

41. In the figure, square *RSTV* has its vertices on the sides of square *WXYZ* as shown. If $ZT = 5$ and $TY = 12$, find *TS*. Also find *RT*.

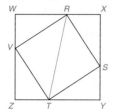

42. Prove that if (a, b, c) is a Pythagorean triple and n is a natural number, then (na, nb, nc) is also a Pythagorean triple.

43. Use Figure 5.19 to prove Theorem 5.4.2.

44. Use Figures 5.20 and 5.21 to prove Lemma 5.4.3.

45. For quadrilateral *ABCD*, $\overline{AC}$ and $\overline{BD}$ are diagonals. Also, $\overline{AB} \perp \overline{BD}$, $\overline{AC} \perp \overline{CD}$, and $\overline{CF} \perp \overline{BD}$. Give the reason why:
a) $\triangle ABE \sim \triangle CFE$
b) $\triangle CFE \sim \triangle DFC$
c) $\triangle ABE \sim \triangle DFC$

[HINT: For part (c), compare the results of (a) and (b)]

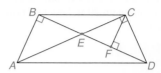

5.5 Special Right Triangles

KEY CONCEPTS	The 45°-45°-90° Triangle	The 30°-60°-90° Triangle

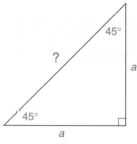

Figure 5.31

Many of the calculations that we do in this section involve square root radicals. To understand some of these calculations better, it may be necessary to review the Properties of Square Roots in Appendix A.5.

Certain right triangles occur so often that they deserve more attention than others. The two special right triangles that we consider in this section have angle measures of 45°, 45°, and 90° or of 30°, 60°, and 90°.

THE 45°-45°-90° RIGHT TRIANGLE

In the 45°-45°-90° triangle, the legs lie opposite the congruent angles and are also congruent. Rather than using a and b to represent the lengths of the legs, we use a for both lengths; see Figure 5.31. By the Pythagorean Theorem, it follows that

$$c^2 = a^2 + a^2$$
$$c^2 = 2a^2$$
$$c = \sqrt{2a^2}$$
$$c = \sqrt{2} \cdot \sqrt{a^2}$$
$$c = a\sqrt{2}$$

> **THEOREM 5.5.1 ■ 45-45-90 Theorem**
>
> In a right triangle whose angles measure 45°, 45°, and 90°, the legs are congruent and the hypotenuse has a length equal to the product of $\sqrt{2}$ and the length of either leg.

SSG **EXS. 1–3**

It is better to memorize the sketch in Figure 5.32 than to repeat the steps of the "proof" that precedes the 45-45-90 Theorem. The boxed information that follows summarizes the facts shown in Figure 5.32.

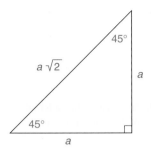

Figure 5.32

THE 45°-45°-90° TRIANGLE		
Angle Measure		**Length of Opposite Side**
45°	←——————→	a
45°	←——————→	a
90°	←——————→	$a\sqrt{2}$

EXAMPLE 1

Find the lengths of the missing sides in each triangle in Figure 5.33.

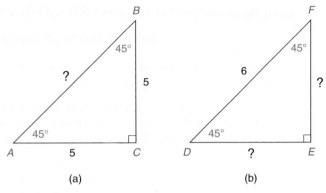

(a)   (b)

Figure 5.33

Reminder

If two angles of a triangle are congruent, then the sides opposite these angles are congruent.

SOLUTION

a) The length of hypotenuse $\overline{AB}$ is $5\sqrt{2}$, the product of $\sqrt{2}$ and the length of either of the equal legs.

b) Let a denote the length of $\overline{DE}$ and of $\overline{EF}$. The length of hypotenuse $\overline{DF}$ is $a\sqrt{2}$.

Then $a\sqrt{2} = 6$, so $a = \frac{6}{\sqrt{2}}$. Simplifying yields

$$a = \frac{6}{\sqrt{2}} \cdot \frac{\sqrt{2}}{\sqrt{2}}$$
$$= \frac{6\sqrt{2}}{2}$$
$$= 3\sqrt{2}$$

Therefore, $DE = EF = 3\sqrt{2} \approx 4.24$.

NOTE: If we use the Pythagorean Theorem to solve Example 1, the solution in part (a) can be found by solving the equation $5^2 + 5^2 = c^2$, and the solution in part (b) can be found by solving $a^2 + a^2 = 6^2$.

(a)

(b)

Figure 5.34

EXAMPLE 2

Each side of a square has a length of $\sqrt{5}$. Find the length of a diagonal.

SOLUTION The square shown in Figure 5.34(a) is separated by the diagonal into two 45°-45°-90° triangles. With each of the congruent legs represented by a in Figure 5.34(b), we see that $a = \sqrt{5}$ and the diagonal (hypotenuse) length is $a \cdot \sqrt{2} = \sqrt{5} \cdot \sqrt{2}$, so $a = \sqrt{10} \approx 3.16$.

SSG **EXS. 4–7**

(a)

(b)

(c)

Figure 5.35

THE 30°-60°-90° RIGHT TRIANGLE

The second special triangle is the 30°-60°-90° triangle.

> **THEOREM 5.5.2 ■ 30-60-90 Theorem**
>
> In a right triangle whose angles measure 30°, 60°, and 90°, the hypotenuse has a length equal to twice the length of the shorter leg, and the length of the longer leg is the product of $\sqrt{3}$ and the length of the shorter leg.

EXAMPLE 3

Study the picture proof of Theorem 5.5.2. See Figure 5.35(a).

PICTURE PROOF OF THEOREM 5.5.2

GIVEN: $\triangle ABC$ with $m\angle A = 30°$, $m\angle B = 60°$, $m\angle C = 90°$, and $BC = a$

PROVE: $AB = 2a$ and $AC = a\sqrt{3}$

PROOF: We reflect $\triangle ABC$ across $\overline{AC}$ to form an equiangular and therefore equilateral $\triangle ABD$. As shown in Figures 5.35(b) and 5.35(c), we have $AB = 2a$.
To find b in Figure 5.35(c), we apply the Pythagorean Theorem.

$$c^2 = a^2 + b^2$$
$$(2a)^2 = a^2 + b^2$$
$$4a^2 = a^2 + b^2$$
$$3a^2 = b^2$$
So
$$b^2 = 3a^2$$
$$b = \sqrt{3a^2}$$
$$b = \sqrt{3} \cdot \sqrt{a^2}$$
$$b = a\sqrt{3}$$

That is, $AC = a\sqrt{3}$.

It would be best to memorize the sketch in Figure 5.36. So that you will more easily recall which expression is used for each side, remember that the lengths of the sides follow the same order as the angles opposite them. See the summary below.

Figure 5.36

THE 30°-60°-90° TRIANGLE	
Angle Measure	**Length of Opposite Side**
30° ⟷	a
60° ⟷	$a\sqrt{3}$
90° ⟷	$2a$

SSG **EXS. 8–10**

EXAMPLE 4

Find the lengths of the missing sides of each triangle in Figure 5.37.

Figure 5.37

SOLUTION

a) $RT = 2 \cdot RS = 2 \cdot 5 = 10$
$ST = RS\sqrt{3} = 5\sqrt{3} \approx 8.66$

b) $UW = 2 \cdot VW \to 20 = 2 \cdot VW \to VW = 10$
$UV = VW\sqrt{3} = 10\sqrt{3} \approx 17.32$

c) $ZY = XY\sqrt{3} \to 9 = XY \cdot \sqrt{3} \to XY = \dfrac{9}{\sqrt{3}} = \dfrac{9}{\sqrt{3}} \cdot \dfrac{\sqrt{3}}{\sqrt{3}}$

$$= \dfrac{9\sqrt{3}}{3} = 3\sqrt{3} \approx 5.20$$

$XZ = 2 \cdot XY = 2 \cdot 3\sqrt{3} = 6\sqrt{3} \approx 10.39$

EXAMPLE 5

Each side of an equilateral triangle measures 6 in. Find the length of an altitude of the triangle.

(a)   (b)

Figure 5.38

SOLUTION The equilateral triangle shown in Figure 5.38(a) is separated into two 30°-60°-90° triangles by the altitude. In the 30°-60°-90° triangle in Figure 5.38(b), the side of the equilateral triangle becomes the hypotenuse, so $2a = 6$ and $a = 3$. The altitude lies opposite the 60° angle of the 30°-60°-90° triangle, so its length is $a\sqrt{3}$ or $3\sqrt{3}$ in. ≈ 5.20 in.

The converse of Theorem 5.5.1 is true and is described in the following theorem.

SSG **EXS. 11–13**

THEOREM 5.5.3

If the length of the hypotenuse of a right triangle equals the product of $\sqrt{2}$ and the length of either congruent leg, then the angles of the triangle measure 45°, 45°, and 90°.

GIVEN: The right triangle with lengths of sides a, a, and $a\sqrt{2}$ (See Figure 5.39).

PROVE: The triangle is a 45°-45°-90° triangle

PROOF

In Figure 5.39, the length of the hypotenuse is $a\sqrt{2}$, where a is the length of either leg. In a right triangle, the angles that lie opposite the congruent legs are also congruent. In a right triangle, the acute angles are complementary, so each of the congruent acute angles measures 45°.

Figure 5.39

EXAMPLE 6

In right $\triangle RST$, $RS = ST$ in Figure 5.40. What are the measures of the angles of the triangle? If $RT = 12\sqrt{2}$, what is the length of $\overline{RS}$ (or $\overline{ST}$)?

Figure 5.40

SOLUTION The longest side is the hypotenuse $\overline{RT}$, so the right angle is $\angle S$ and $m\angle S = 90°$. Because $\overline{RS} \cong \overline{ST}$, the congruent acute angles are $\angle$s R and T and $m\angle R = m\angle T = 45°$. Because $RT = 12\sqrt{2}$, $RS = ST = 12$.

The converse of Theorem 5.5.2 is also true and can be proved by the indirect method. Rather than construct the proof, we state and apply this theorem. See Figure 5.41.

Figure 5.41

THEOREM 5.5.4

If the length of the hypotenuse of a right triangle is twice the length of one leg of the triangle, then the angle of the triangle opposite that leg measures 30°.

An equivalent form of this theorem is stated as follows:

If one leg of a right triangle has a length equal to one-half the length of the hypotenuse, then the angle of the triangle opposite that leg measures 30° (see Figure 5.42).

Figure 5.42

EXAMPLE 7

In right $\triangle ABC$ with right $\angle C$, $AB = 24.6$ and $BC = 12.3$ (see Figure 5.43). What are the measures of the angles of the triangle? Also, what is the length of $\overline{AC}$?

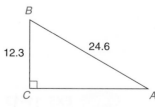

Figure 5.43

SOLUTION Because $\angle C$ is a right angle, $m\angle C = 90°$ and $\overline{AB}$ is the hypotenuse. Because $BC = \frac{1}{2}(AB)$, the angle opposite $\overline{BC}$ measures 30°. Thus, $m\angle A = 30°$ and $m\angle B = 60°$. Because $\overline{AC}$ lies opposite the 60° angle, $AC = (12.3)\sqrt{3} \approx 21.3$.

Exercises 5.5

1. For the 45°-45°-90° triangle shown, suppose that $AC = a$. Find:
   a) BC b) AB

2. For the 45°-45°-90° triangle shown, suppose that $AB = a\sqrt{2}$. Find:
   a) AC b) BC

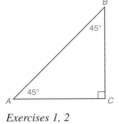

Exercises 1, 2

3. For the 30°-60°-90° triangle shown, suppose that $XZ = a$. Find:
   a) YZ b) XY

4. For the 30°-60°-90° triangle shown, suppose that $XY = 2a$. Find:
   a) XZ b) YZ

Exercises 3, 4

In Exercises 5 to 22, find the missing lengths. Give your answers in both simplest radical form and as approximations correct to two decimal places.

5. *Given:*   Right $\triangle XYZ$ with m$\angle X = 45°$ and $XZ = 8$
   Find: YZ and XY

Exercises 5–8

6. *Given:*   Right $\triangle XYZ$ with $\overline{XZ} \cong \overline{YZ}$ and $XY = 10$
   Find: XZ and YZ

7. *Given:*   Right $\triangle XYZ$ with $\overline{XZ} \cong \overline{YZ}$ and $XY = 10\sqrt{2}$
   Find: XZ and YZ

8. *Given:*   Right $\triangle XYZ$ with m$\angle X = 45°$ and $XY = 12\sqrt{2}$
   Find: XZ and YZ

9. *Given:*   Right $\triangle DEF$ with m$\angle E = 60°$ and $DE = 5$
   Find: DF and FE

Exercises 9–12

10. *Given:*   Right $\triangle DEF$ with m$\angle F = 30°$ and $FE = 12$
    Find: DF and DE

11. *Given:*   Right $\triangle DEF$ with m$\angle E = 60°$ and $FD = 12\sqrt{3}$
    Find: DE and FE

12. *Given:*   Right $\triangle DEF$ with m$\angle E = 2 \cdot$ m$\angle F$ and $EF = 12\sqrt{3}$
    Find: DE and DF

13. *Given:*   Rectangle $HJKL$ with diagonals $\overline{HK}$ and $\overline{JL}$ m$\angle HKL = 30°$ and $LK = 6\sqrt{3}$.
    Find: HL, HK, and MK

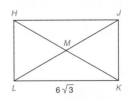

14. *Given:*   Right $\triangle RST$ with $RT = 6\sqrt{2}$ and m$\angle STV = 150°$
    Find: RS and ST

In Exercises 15–19, create drawings as needed.

15. *Given:* $\triangle ABC$ with m$\angle A =$ m$\angle B = 45°$ and $BC = 6$
    Find: AC and AB

16. *Given:*   Right $\triangle MNP$ with $MP = PN$ and $MN = 10\sqrt{2}$
    Find: PM and PN

17. *Given:* $\triangle RST$ with m$\angle T = 30°$, m$\angle S = 60°$, and $ST = 12$
    Find: RS and RT

18. *Given:* $\triangle XYZ$ with $\overline{XY} \cong \overline{XZ} \cong \overline{YZ}$ $\overline{ZW} \perp \overline{XY}$ with W on $\overline{XY}$ $YZ = 6$
    Find: ZW

19. *Given:*   Square $ABCD$ with diagonals $\overline{DB}$ and $\overline{AC}$ intersecting at E $DC = 5\sqrt{3}$
    Find: DB

20. *Given:* $\triangle NQM$ with angles as shown in the drawing $\overline{MP} \perp \overline{NQ}$
    Find: NM, MP, MQ, PQ, and NQ

21. *Given:* $\triangle XYZ$ with angles as shown in the drawing
    Find: XY

    (HINT: Compare this drawing to the one for Exercise 20.)

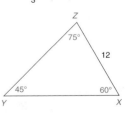

22. *Given:*   Rhombus $ABCD$ (not shown) in which diagonals $\overline{AC}$ and $\overline{DB}$ intersect at point E; $DB = AB = 8$
    Find: AC

23. A carpenter is working with a board that is $3\frac{3}{4}$ in. wide. After marking off a point down the side of length $3\frac{3}{4}$ in., the carpenter makes a cut along $\overline{BC}$ with a saw. What is the measure of the angle ($\angle ACB$) that is formed?

24. To unload groceries from a delivery truck at the Piggly Wiggly Market, an 8-ft ramp that rises 4 ft to the door of the trailer is used. What is the measure of the indicated angle ($\angle D$)?

25. A jogger runs along two sides of an open rectangular lot. If the first side of the lot is 200 ft long and the diagonal distance across the lot is 400 ft, what is the measure of the angle formed by the 200-ft and 400-ft dimensions? To the nearest foot, how much farther does the jogger run by traveling the two sides of the block rather than the diagonal distance across the lot?

26. Mara's boat leaves the dock at the same time that Meg's boat leaves the dock. Mara's boat travels due east at 12 mph. Meg's boat travels at 24 mph in the direction N 30° E. To the nearest tenth of a mile, how far apart will the boats be in half an hour?

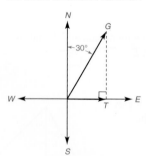

In Exercises 27 to 33, give both exact solutions and approximate solutions to two decimal places.

27. *Given:* In $\triangle ABC$, $\overrightarrow{AD}$ bisects $\angle BAC$
  $m\angle B = 30°$ and $AB = 12$
  Find: DC and DB

28. *Given:* In $\triangle ABC$, $\overrightarrow{AD}$ bisects $\angle BAC$
  $AB = 20$ and $AC = 10$
  Find: DC and DB

Exercises 27, 28

29. *Given:* $\triangle MNQ$ is equiangular and $NR = 6$
  $\overrightarrow{NR}$ bisects $\angle MNQ$
  $\overrightarrow{QR}$ bisects $\angle MQN$
  Find: NQ

30. *Given:* $\triangle STV$ is an isosceles right triangle
  M and N are midpoints of $\overline{ST}$ and $\overline{SV}$
  Find: MN

31. *Given:* Right $\triangle ABC$ with $m\angle C = 90°$ and $m\angle BAC = 60°$; point D on $\overline{BC}$; $\overrightarrow{AD}$ bisects $\angle BAC$ and $AB = 12$
  Find: BD

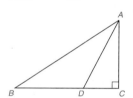

Exercises 31, 32

32. *Given:* Right $\triangle ABC$ with $m\angle C = 90°$ and $m\angle BAC = 60°$; point D on $\overline{BC}$; $\overrightarrow{AD}$ bisects $\angle BAC$ and $AC = 2\sqrt{3}$
  Find: BD

33. *Given:* $\triangle ABC$ with $m\angle A = 45°$, $m\angle B = 30°$, and $BC = 12$
  Find: AB

  (HINT: *Use altitude $\overline{CD}$ from C to $\overline{AB}$ as an auxiliary line.*)

*34. *Given:* Isosceles trapezoid $MNPQ$ with $QP = 12$ and $m\angle M = 120°$; the bisectors of $\angle$s MQP and NPQ meet at point T on $\overline{MN}$
  Find: The perimeter of $MNPQ$

35. In regular hexagon *ABCDEF*,
AB = 6 inches. Find the exact
length of
a) diagonal $\overline{BF}$.
b) diagonal $\overline{CF}$.

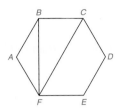

Exercises 35, 36

36. In regular hexagon *ABCDEF*, the
length of *AB* is *x* centimeters. In
terms of *x*, find the length of
a) diagonal $\overline{BF}$.
b) diagonal $\overline{CF}$.

*37. In right triangle *XYZ*, *XY* = 3 and *YZ* = 4. Where *V* is the
midpoint of $\overline{YZ}$ and m∠*VWZ* = 90°, find *VW*.

(HINT: Draw $\overline{XV}$.)

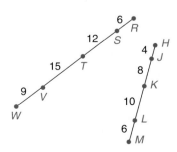

38. Diagonal $\overline{EC}$ separates pentagon *ABCDE* into square *ABCE*
and isosceles triangle *DEC*. If *AB* = 8 and *DC* = 5, find
the length of diagonal $\overline{DB}$.

(HINT: Draw $\overline{DF} \perp \overline{AB}$.)

5.6	**Segments Divided Proportionally**

KEY CONCEPTS	Segments Divided Proportionally	The Angle-Bisector Theorem	Ceva's Theorem

In this section, we begin with an informal description of the phrase *divided proportionally.* Suppose that three children have been provided with a joint savings account by their
parents. Equal monthly deposits have been made to the account for each child since birth.
If the ages of the children are 2, 4, and 6 (assume exactness of ages for simplicity) and the
total in the account is $7200, then the amount that each child should receive can be found
by solving the equation

$$2x + 4x + 6x = 7200$$

Solving this equation leads to the solution $1200 for the 2-year-old, $2400 for the 4-year-
old, and $3600 for the 6-year-old. We say that the amount has been divided proportion-
ally. Expressed as a proportion that compares the amount saved to age, we have

$$\frac{1200}{2} = \frac{2400}{4} = \frac{3600}{6}$$

Figure 5.44

In Figure 5.44, $\overline{AC}$ and $\overline{DF}$ are divided proportionally at points *B* and *E* if

$$\frac{AB}{DE} = \frac{BC}{EF} \qquad \text{or} \qquad \frac{AB}{BC} = \frac{DE}{EF}$$

Of course, a pair of segments may be divided proportionally by several points. In
Figure 5.45, $\overline{RW}$ and $\overline{HM}$ are divided proportionally when

Figure 5.45

$$\frac{RS}{HJ} = \frac{ST}{JK} = \frac{TV}{KL} = \frac{VW}{LM} \qquad \left(\text{notice that } \frac{6}{4} = \frac{12}{8} = \frac{15}{10} = \frac{9}{6} \right)$$

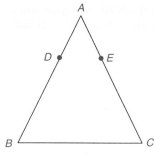

Figure 5.46

EXAMPLE 1

In Figure 5.46, points D and E divide $\overline{AB}$ and $\overline{AC}$ proportionally. If $AD = 4$, $DB = 7$, and $EC = 6$, find AE.

SOLUTION $\frac{AD}{AE} = \frac{DB}{EC}$. Where $AE = x$, $\frac{4}{x} = \frac{7}{6}$. Then $7x = 24$, so $x = AE = \frac{24}{7} = 3\frac{3}{7}$.

The following property will be proved in Exercise 31 of this section.

NUMERATOR-DENOMINATOR ADDITION PROPERTY

If $\frac{a}{b} = \frac{c}{d}$, then $\frac{a + c}{b + d} = \frac{a}{b} = \frac{c}{d}$

In words, we may restate this property as follows.

The fraction whose numerator and denominator are determined, respectively, by adding numerators and denominators of equal fractions is equal to each of those equal fractions.

Here is a numerical example of this claim:

$$\text{If } \frac{2}{3} = \frac{4}{6}, \qquad \text{then} \qquad \frac{2 + 4}{3 + 6} = \frac{2}{3} = \frac{4}{6}$$

In Example 2, the preceding property is necessary as a reason.

EXAMPLE 2

GIVEN: $\overline{RW}$ and $\overline{HM}$ are divided proportionally at the points shown in Figure 5.47.

PROVE: $\dfrac{RT}{HK} = \dfrac{TW}{KM}$

PROOF: $\overline{RW}$ and $\overline{HM}$ are divided proportionally so that

$$\frac{RS}{HJ} = \frac{ST}{JK} = \frac{TV}{KL} = \frac{VW}{LM}$$

Using the property, "If $\frac{a}{b} = \frac{c}{d}$, then $\frac{a + c}{b + d} = \frac{a}{b} = \frac{c}{d}$," we have

$$\frac{RS}{HJ} = \frac{RS + ST}{HJ + JK} = \frac{TV}{KL} = \frac{TV + VW}{KL + LM}$$

Because $RS + ST = RT$, $HJ + JK = HK$, $TV + VW = TW$, and $KL + LM = KM$, substitution leads to

$$\frac{RT}{HK} = \frac{TW}{KM}$$

Figure 5.47

SSG **EXS. 1, 2**

Two properties that were introduced earlier in Section 5.1 are now recalled.

$$\text{If } \frac{a}{b} = \frac{c}{d}, \text{ then } \frac{a \pm b}{b} = \frac{c \pm d}{d}$$

The subtraction operation of the property is needed for the proof of Theorem 5.6.1.

THEOREM 5.6.1

If a line is parallel to one side of a triangle and intersects the other two sides, then it divides these sides proportionally.

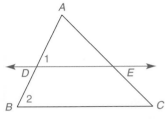

Figure 5.48

GIVEN: In Figure 5.48, $\triangle ABC$ with $\overleftrightarrow{DE} \parallel \overline{BC}$ so that $\overleftrightarrow{DE}$ intersects $\overline{AB}$ at D and $\overline{AC}$ at E

PROVE: $\dfrac{AD}{DB} = \dfrac{AE}{EC}$

PROOF: Because $\overleftrightarrow{DE} \parallel \overline{BC}$, $\angle 1 \cong \angle 2$. With $\angle A$ as a common angle for $\triangle ADE$ and $\triangle ABC$, $\angle A \cong \angle A$. It follows by AA that these triangles are similar. Now

$$\frac{AB}{AD} = \frac{AC}{AE} \quad \text{(by CSSTP)}$$

By the Difference Property of a Proportion,

$$\frac{AB - AD}{AD} = \frac{AC - AE}{AE}$$

Because $AB - AD = DB$ and $AC - AE = EC$, the proportion becomes

$$\frac{DB}{AD} = \frac{EC}{AE}$$

Using Property 2 of Section 5.1, we invert both fractions to obtain the desired conclusion:

$$\frac{AD}{DB} = \frac{AE}{EC}$$

SSG **EXS. 3–6**

COROLLARY 5.6.2

When three (or more) parallel lines are cut by a pair of transversals, the transversals are divided proportionally by the parallel lines.

GIVEN: $p_1 \parallel p_2 \parallel p_3$ in Figure 5.49.

PROVE: $\dfrac{AB}{BC} = \dfrac{DE}{EF}$

PICTURE PROOF OF COROLLARY 5.6.2

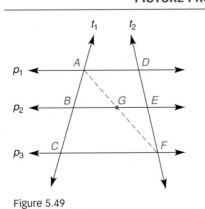

Figure 5.49

In Figure 5.49, draw $\overline{AF}$ as an auxiliary line segment.

On the basis of Theorem 5.6.1, we see that $\frac{AB}{BC} = \frac{AG}{GF}$ in $\triangle ACF$ and that $\frac{AG}{GF} = \frac{DE}{EF}$ in $\triangle ADF$.

By the Transitive Property of Equality, $\frac{AB}{BC} = \frac{DE}{EF}$.

NOTE: By interchanging the means, we can write the last proportion in the alternative form $\frac{AB}{DE} = \frac{BC}{EF}$.

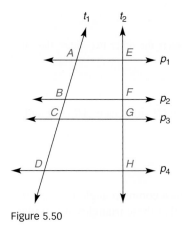

Figure 5.50

EXAMPLE 3

Given parallel lines p_1, p_2, p_3, and p_4 cut by t_1 and t_2 so that $AB = 4$, $EF = 3$, $BC = 2$, and $GH = 5$, find FG and CD. (See Figure 5.50.)

SOLUTION According to Corollary 5.6.2, the transversals are divided proportionally by the parallel lines.

Then
$$\frac{AB}{EF} = \frac{BC}{FG} = \frac{CD}{GH}$$

so
$$\frac{4}{3} = \frac{2}{FG} = \frac{CD}{5}$$

Then
$$4 \cdot FG = 6 \qquad \text{and} \qquad 3 \cdot CD = 20$$
$$FG = \frac{3}{2} = 1\frac{1}{2} \qquad \text{and} \qquad CD = \frac{20}{3} = 6\frac{2}{3}$$

SSG EXS. 7, 8

The following activity suggests the relationship described in Theorem 5.6.3.

DISCOVER

On a piece of paper, draw or construct $\triangle ABC$ whose sides measure $AB = 4$, $BC = 6$, and $AC = 5$. Then construct the angle bisector $\overrightarrow{BD}$ of $\angle B$. How does $\frac{AB}{AD}$ compare to $\frac{BC}{DC}$?

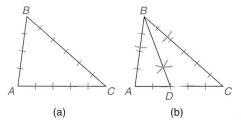

(a) (b)

ANSWER
Though not by chance, it may come as a surprise that $\frac{AB}{AD} = \frac{BC}{DC}$ (that is, $\frac{4}{2} = \frac{6}{3}$) and $\frac{AB}{AD} = \frac{BC}{DC}$ ($\frac{4}{2} = \frac{6}{3}$). It seems that the bisector of an angle included by two sides of a triangle separates the third side into segments whose lengths are proportional to the lengths of the two sides forming the angle.

The proof of Theorem 5.6.3 requires the use of Theorem 5.6.1.

(a)

THEOREM 5.6.3 ■ The Angle-Bisector Theorem

If a ray bisects one angle of a triangle, then it divides the opposite side into segments whose lengths are proportional to the lengths of the two sides that form the bisected angle.

GIVEN: $\triangle ABC$ in Figure 5.51(a), in which $\overrightarrow{CD}$ bisects $\angle ACB$

PROVE: $\dfrac{AD}{AC} = \dfrac{DB}{CB}$

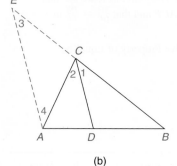

(b)

Figure 5.51

PROOF: We begin by extending $\overline{BC}$ beyond C (there is only one line through B and C) to meet the line drawn through A parallel to $\overrightarrow{DC}$. [See Figure 5.51(b).] Let E be the point of intersection. (These lines must intersect; otherwise, $\overline{AE}$ would have two parallels, $\overrightarrow{BC}$ and $\overrightarrow{CD}$, through point C.)

Because $\overrightarrow{CD} \parallel \overline{EA}$, we have

$$\frac{EC}{AD} = \frac{CB}{DB} \quad (*)$$

by Theorem 5.6.1. Now $\angle 1 \cong \angle 2$ because $\overrightarrow{CD}$ bisects $\angle ACB$, $\angle 1 \cong \angle 3$ (corresponding angles for parallel lines), and $\angle 2 \cong \angle 4$ (alternate interior angles for parallel lines). By the Transitive Property, $\angle 3 \cong \angle 4$, so $\triangle ACE$ is isosceles with $\overline{EC} \cong \overline{AC}$; then $EC = AC$. Using substitution, the starred (*) proportion becomes

$$\frac{AC}{AD} = \frac{CB}{DB} \qquad \text{or} \qquad \frac{AD}{AC} = \frac{DB}{CB} \qquad \textbf{(by inversion)}$$

The "Prove statement" of the preceding theorem indicates that one form of the proportion described is given by comparing lengths as shown:

$$\frac{\text{segment at left}}{\text{side at left}} = \frac{\text{segment at right}}{\text{side at right}}$$

Equivalently, the proportion could compare lengths like this:

$$\frac{\text{segment at left}}{\text{segment at right}} = \frac{\text{side at left}}{\text{side at right}}$$

Other forms of the proportion are also possible!

EXAMPLE 4

For $\triangle XYZ$ in Figure 5.52, $XY = 3$ and $YZ = 5$. If $\overrightarrow{YW}$ bisects $\angle XYZ$ and $XW = 2$, find WZ and XZ.

SOLUTION Let $WZ = x$. We know that $\frac{YX}{XW} = \frac{YZ}{WZ}$, so $\frac{3}{2} = \frac{5}{x}$. Therefore,

$$3x = 10$$
$$x = \frac{10}{3} = 3\tfrac{1}{3}$$

Then $WZ = 3\tfrac{1}{3}$.

Because $XZ = XW + WZ$, we have $XZ = 2 + 3\tfrac{1}{3} = 5\tfrac{1}{3}$.

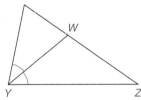

Figure 5.52

SSG EXS. 9–13

EXAMPLE 5

In Figure 5.52 (shown in Example 4), suppose that $\triangle XYZ$ has sides of lengths $XY = 3$, $YZ = 4$, and $XZ = 5$. If $\overrightarrow{YW}$ bisects $\angle XYZ$, find XW and WZ.

SOLUTION Let $XW = y$; then $WZ = 5 - y$, and $\frac{XY}{YZ} = \frac{XW}{WZ}$ becomes $\frac{3}{4} = \frac{y}{5 - y}$. From this proportion, we can find y as follows.

$$3(5 - y) = 4y$$
$$15 - 3y = 4y$$
$$15 = 7y$$
$$y = \frac{15}{7}$$

Then $XW = \frac{15}{7} = 2\tfrac{1}{7}$ and $WZ = 5 - 2\tfrac{1}{7} = 2\tfrac{6}{7}$.

In the following example, we provide an alternative method of solution to a problem of the type found in Example 5.

EXAMPLE 6

In Figure 5.52, $\overline{XZ} \cong \overline{YZ}$ and $\overline{YW}$ bisects $\angle XYZ$.
If $XY = 3$ and $YZ = 6$, find XW and WZ.

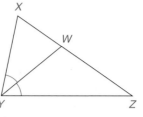

Figure 5.52

SOLUTION Because the ratio $XY{:}YZ$ is 3:6, or 1:2, the
ratio $XW{:}WZ$ is also 1:2. Thus, we can represent
these lengths by

$$XW = a \quad \text{and} \quad WZ = 2a$$

With $XZ = 6$ in triangle XYZ, the statement
$XW + WZ = XZ$ becomes $a + 2a = 6$, so $3a = 6$, and $a = 2$.
Now $XW = 2$ and $WZ = 4$.

CEVA'S THEOREM

You will find the proof of the following theorem in the Perspective on History section
at the end of this chapter. In Ceva's Theorem, point D is *any* point in the interior of the
triangle. See Figure 5.53(a). The auxiliary lines needed to complete the proof of Ceva's
Theorem are shown in Figure 5.53(b). In the figure, line ℓ is drawn through vertex C so
that it is parallel to $\overline{AB}$. Then $\overline{BE}$ and $\overline{AF}$ are extended to meet ℓ at R and S, respectively.

THEOREM 5.6.4 ■ Ceva's Theorem

Let point D be any point in the interior of $\triangle ABC$. Where E, F, and G lie on $\triangle ABC$, let
$\overline{BE}$, $\overline{AF}$, and $\overline{CG}$ be the line segments determined by D and vertices of $\triangle ABC$. Then the
product of the ratios of the lengths of the segments of each of the three sides (taken in
order from a given vertex of the triangle) equals 1; that is,

$$\frac{AG}{GB} \cdot \frac{BF}{FC} \cdot \frac{CE}{EA} = 1$$

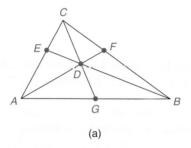

(a)   (b)

Figure 5.53

For Figure 5.53, Ceva's Theorem can be stated in many equivalent forms:

$$\frac{AE}{EC} \cdot \frac{CF}{FB} \cdot \frac{BG}{GA} = 1, \qquad \frac{CF}{FB} \cdot \frac{BG}{GA} \cdot \frac{AE}{EC} = 1, \qquad \text{etc.}$$

In each case, we select a vertex (such as A) and form the product of the ratios of the lengths
of segments of sides in a set order (clockwise or counterclockwise).
We will apply Ceva's Theorem in Example 7.

> **EXAMPLE 7**
>
> In $\triangle RST$ with interior point D, $RG = 6$, $GS = 4$, $SH = 4$, $HT = 3$, and $KR = 5$. Find TK. See Figure 5.54.
>
> **SOLUTION** Let $TK = a$. Applying Ceva's Theorem and following a counterclockwise path beginning at vertex R, we have $\frac{RG}{GS} \cdot \frac{SH}{HT} \cdot \frac{TK}{KR} = 1$. Then $\frac{6}{4} \cdot \frac{4}{3} \cdot \frac{a}{5} = 1$ and so $\frac{6^2}{4_1} \cdot \frac{4^1}{3_1} \cdot \frac{a}{5} = 1$ becomes $\frac{2a}{5} = 1$. Then $2a = 5$ and $a = 2.5$; thus, $TK = 2.5$.

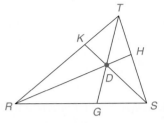

SSG **EX. 14**

Figure 5.54

Exercises 5.6

1. In preparing a certain recipe, a chef uses 5 oz of ingredient A, 4 oz of ingredient B, and 6 oz of ingredient C. If 90 oz of this dish are needed, how many ounces of each ingredient should be used?

2. In a chemical mixture, 2 g of chemical A are used for each gram of chemical B, and 3 g of chemical C are needed for each gram of B. If 72 g of the mixture are prepared, what amount (in grams) of each chemical is needed?

3. Given that $\frac{AB}{EF} = \frac{BC}{FG} = \frac{CD}{GH}$, are the following proportions true?

   a) $\dfrac{AC}{EG} = \dfrac{CD}{GH}$

   b) $\dfrac{AB}{EF} = \dfrac{BD}{FH}$

4. Given that $\overleftrightarrow{XY} \parallel \overline{TS}$, are the following proportions true?

   a) $\dfrac{TX}{XR} = \dfrac{RY}{YS}$

   b) $\dfrac{TR}{XR} = \dfrac{SR}{YR}$

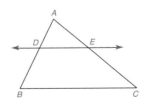

5. *Given:* $\ell_1 \parallel \ell_2 \parallel \ell_3 \parallel \ell_4$, $AB = 5$, $BC = 4$, $CD = 3$, $EH = 10$

   Find: EF, FG, GH
   (See the figure for Exercise 6).

6. *Given:* $\ell_1 \parallel \ell_2 \parallel \ell_3 \parallel \ell_4$, $AB = 7$, $BC = 5$, $CD = 4$, $EF = 6$

   Find: FG, GH, EH

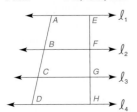

Exercises 5, 6

7. *Given:* $\ell_1 \parallel \ell_2 \parallel \ell_3$, $AB = 4$, $BC = 5$, $DE = x$, $EF = 12 - x$

   Find: x, DE, EF

Exercises 7, 8

8. *Given:* $\ell_1 \parallel \ell_2 \parallel \ell_3$, $AB = 5$, $BC = x$, $DE = x - 2$, $EF = 1$

   Find: x, BC, DE

9. *Given:* $\overleftrightarrow{DE} \parallel \overline{BC}$, $AD = 5$, $DB = 12$, $AE = 7$

   Find: EC

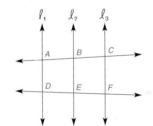

Exercises 9–12

10. *Given:* $\overleftrightarrow{DE} \parallel \overline{BC}$, $AD = 6$, $DB = 10$, $AC = 20$

    Find: EC

11. *Given:* $\overleftrightarrow{DE} \parallel \overline{BC}$, $AD = a - 1$, $DB = 2a + 2$, $AE = a$, $EC = 4a - 5$

    Find: a and AD

12. *Given:* $\overleftrightarrow{DE} \parallel \overline{BC}$, $AD = 5$, $DB = a + 3$, $AE = a + 1$, $EC = 3(a - 1)$

    Find: a and EC

13. *Given:* $\overrightarrow{RW}$ bisects $\angle SRT$
Do the following equalities hold?
a) $SW = WT$
b) $\dfrac{RS}{RT} = \dfrac{SW}{WT}$

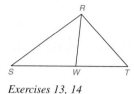

Exercises 13, 14

14. *Given:* $\overrightarrow{RW}$ bisects $\angle SRT$
Do the following equalities hold?
a) $\dfrac{RS}{SW} = \dfrac{RT}{WT}$
b) $m\angle S = m\angle T$

15. *Given:* $\overrightarrow{UT}$ bisects $\angle WUV$, $WU = 8$, $UV = 12$, $WT = 6$
Find: TV

Exercises 15, 16

16. *Given:* $\overrightarrow{UT}$ bisects $\angle WUV$, $WU = 9$, $UV = 12$, $WV = 9$
Find: WT

17. *Given:* $\overrightarrow{NQ}$ bisects $\angle MNP$, $NP = MQ$, $QP = 8$, $MN = 12$
Find: NP

Exercises 17–19

Exercises 18 and 19 are based on a theorem (not stated) that is the converse of Theorem 5.6.3. See the figure above.

18. *Given:* $NP = 4$, $MN = 8$, $PQ = 3$, and $MQ = 6$; $m\angle P = 63°$ and $m\angle M = 27°$
Find: $m\angle PNQ$
(HINT: $\dfrac{NP}{MN} = \dfrac{PQ}{MQ}$.)

19. *Given:* $NP = 6$, $MN = 9$, $PQ = 4$, and $MQ = 6$; $m\angle P = 62°$ and $m\angle M = 36°$
Find: $m\angle QNM$
(HINT: $\dfrac{NP}{MN} = \dfrac{PQ}{MQ}$.)

20. *Given:* In $\triangle ABC$, $\overrightarrow{AD}$ bisects $\angle BAC$ $AB = 20$ and $AC = 16$
Find: DC and DB

21. In $\triangle ABC$, $\angle ACB$ is trisected by $\overrightarrow{CD}$ and $\overrightarrow{CE}$ so that $\angle 1 \cong \angle 2 \cong \angle 3$. Write two different proportions that follow from this information.

22. In $\triangle ABC$, $m\angle CAB = 80°$, $m\angle ACB = 60°$, and $m\angle ABC = 40°$. With the angle bisectors as shown, which line segment is longer?
a) $\overline{AE}$ or $\overline{EC}$?   b) $\overline{CD}$ or $\overline{DB}$?   c) $\overline{AF}$ or $\overline{FB}$?

Exercises 22, 23

23. In $\triangle ABC$, $AC = 5.3$, $BC = 7.2$, and $BA = 6.7$. With angle bisectors as shown, which line segment is longer?
a) $\overline{AE}$ or $\overline{EC}$?
b) $\overline{CD}$ or $\overline{DB}$?
c) $\overline{AF}$ or $\overline{FB}$?

24. In right $\triangle RST$ (not shown) with right $\angle S$, $\overrightarrow{RV}$ bisects $\angle SRT$ so that V lies on side $\overline{ST}$. If $RS = 6$, $ST = 6\sqrt{3}$, and $RT = 12$, find SV and VT.

25. *Given:* $\overrightarrow{RV}$ bisects $\angle SRT$, $RS = x - 6$, $SV = 3$, $RT = 2 - x$, and $VT = x + 2$
Find: x
(HINT: You will need to apply the Quadratic Formula.)

26. *Given:* $\overrightarrow{MR}$ bisects $\angle NMP$, $MN = 2x$, $NR = x$, $RP = x + 1$, and $MP = 3x - 1$
Find: x

27. Given point D in the interior of $\triangle RST$, which statement(s) is (are) true?
a) $\dfrac{RK}{KT} \cdot \dfrac{TH}{HS} \cdot \dfrac{GS}{RG} = 1$
b) $\dfrac{TK}{KR} \cdot \dfrac{RG}{GS} \cdot \dfrac{SH}{HT} = 1$

Exercises 27–30

28. In $\triangle RST$, suppose that $\overline{RH}$, $\overline{TG}$, and $\overline{SK}$ are medians. Find the value of:
a) $\dfrac{RK}{KT}$ b) $\dfrac{TH}{HS}$

29. Given point D in the interior of $\triangle RST$, suppose that $RG = 3$, $GS = 4$, $SH = 4$, $HT = 5$, and $KT = 3$. Find RK.

30. Given point D in the interior of $\triangle RST$, suppose that $RG = 2$, $GS = 3$, $SH = 3$, and $HT = 4$. Find $\dfrac{KT}{KR}$.

31. Complete the proof (on page 257) of this property:
If $\dfrac{a}{b} = \dfrac{c}{d}$, then $\dfrac{a + c}{b + d} = \dfrac{a}{b}$ and $\dfrac{a + c}{b + d} = \dfrac{c}{d}$

PROOF

Statements	Reasons
1. $\frac{a}{b} = \frac{c}{d}$	1. ?
2. $b \cdot c = a \cdot d$	2. ?
3. $ab + bc = ab + ad$	3. ?
4. $b(a + c) = a(b + d)$	4. ?
5. $\frac{a + c}{b + d} = \frac{a}{b}$	5. Means-Extremes Property (symmetric form)
6. $\frac{a + c}{b + d} = \frac{c}{d}$	6. ?

32. *Given:* $\triangle RST$, with $\overleftrightarrow{XY} \parallel \overline{RT}, \overleftrightarrow{YZ} \parallel \overline{RS}$

Prove: $\dfrac{RX}{XS} = \dfrac{ZT}{RZ}$

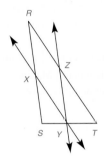

33. Use Theorem 5.6.1 and the drawing to complete the proof of this theorem: "If a line is parallel to one side of a triangle and passes through the midpoint of a second side, then it will pass through the midpoint of the third side."

Given: $\triangle RST$ with M the midpoint of $\overline{RS}; \overleftrightarrow{MN} \parallel \overline{ST}$
Prove: N is the midpoint of $\overline{RT}$

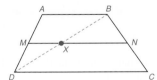

34. Use Exercise 33 and the following drawing to complete the proof of this theorem: "The length of the median of a trapezoid is one-half the sum of the lengths of the two bases."

Given: Trapezoid $ABCD$ with median $\overline{MN}$
Prove: $MN = \frac{1}{2}(AB + CD)$

35. Use Theorem 5.6.3 to complete the proof of this theorem: "If the bisector of an angle of a triangle also bisects the opposite side, then the triangle is an isosceles triangle."

Given: $\triangle XYZ; \overrightarrow{YW}$ bisects $\angle XYZ; \overline{WX} \cong \overline{WZ}$
Prove: $\triangle XYZ$ is isosceles

(HINT: Use a proportion to show that $YX = YZ$.)

★36. In right $\triangle ABC$ (not shown) with right $\angle C$, $\overrightarrow{AD}$ bisects $\angle BAC$ so that D lies on side $\overline{CB}$. If $AC = 6$ and $DC = 3$, find BD and AB.

(HINT: Let $BD = x$ and $AB = 2x$. Then use the Pythagorean Theorem.)

★37. *Given:* $\triangle ABC$ (not shown) is isosceles with $m\angle ABC = m\angle C = 72°$; $\overrightarrow{BD}$ bisects $\angle ABC$ and $AB = 1$
Find: BC

★38. *Given:* $\triangle RST$ with right $\angle RST$; $m\angle R = 30°$ and $ST = 6$; $\angle RST$ is trisected by $\overrightarrow{SM}$ and $\overrightarrow{SN}$
Find: $TN, NM,$ and MR

★39. In the figure, the angle bisectors of $\triangle ABC$ intersect at a point in the interior of the triangle. If $BC = 5$, $BA = 6$, and $CA = 4$, find:
a) CD and DB

(HINT: Use Theorem 5.6.3.)

b) CE and EA
c) BF and FA
d) Use results from parts (a), (b), and (c) to show that $\dfrac{BD}{DC} \cdot \dfrac{CE}{EA} \cdot \dfrac{AF}{FB} = 1$.

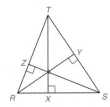

★40. In $\triangle RST$, the altitudes of the triangle intersect at a point in the interior of the triangle. The lengths of the sides of $\triangle RST$ are $RS = 14$, $ST = 15$, and $TR = 13$.
a) If $TX = 12$, find RX and XS.

(HINT: Use the Pythagorean Theorem)

b) If $RY = \frac{168}{15}$, find TY and YS.
c) If $SZ = \frac{168}{13}$, find ZR and TZ.
d) Use results from parts (a), (b), and (c) to show that $\dfrac{RX}{XS} \cdot \dfrac{SY}{YT} \cdot \dfrac{TZ}{ZR} = 1$.

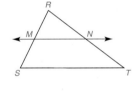

PERSPECTIVE ON HISTORY

CEVA'S PROOF

Giovanni Ceva (1647–1736) was the Italian mathematician for whom Ceva's Theorem is named. Although his theorem is difficult to believe, its proof is not lengthy. The proof follows.

THEOREM 5.6.4 ■ Ceva's Theorem

Let point D be any point in the interior of $\triangle ABC$. Where E, F, and G lie on $\triangle ABC$, let $\overline{BE}$, $\overline{AF}$, and $\overline{CG}$ be the line segments determined by D and vertices of $\triangle ABC$. Then the product of the ratios of the segments of each of the three sides (taken in order from a given vertex of the triangle) equals 1; that is, $\frac{AG}{GB} \cdot \frac{BF}{FC} \cdot \frac{CE}{EA} = 1$.

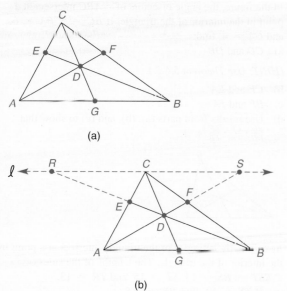

(a)

(b)

Figure 5.55

PROOF

Given $\triangle ABC$ with interior point D [see Figure 5.55(a)], draw a line ℓ through point C that is parallel to $\overline{AB}$. Now extend $\overline{BE}$ to meet ℓ at point R. Likewise, extend $\overline{AF}$ to meet ℓ at point S. See Figure 5.55(b). Using similar triangles, we will be able to substitute desired ratios into the statement $\frac{CS}{CR} \cdot \frac{AB}{CS} \cdot \frac{CR}{AB} = 1(*)$, obviously true because each numerator has a matching denominator. Because $\triangle AGD \sim \triangle SCD$ by AA, we have $\frac{AG}{CS} = \frac{GD}{CD}$. Also with $\triangle DGB \sim \triangle DCR$, we have $\frac{GD}{CD} = \frac{GB}{CR}$. By the Transitive Property of Equality, $\frac{AG}{CS} = \frac{GB}{CR}$, and by interchanging the means, we see that $\frac{AG}{BG} = \frac{CS}{CR}$. [The first ratio, $\frac{AG}{BG}$, of this proportion will replace the ratio $\frac{CS}{CR}$ in the starred (*) statement.]

From the fact that $\triangle CSF \sim \triangle BAF$, $\frac{AB}{SC} = \frac{BF}{FC}$. [The second ratio, $\frac{BF}{FC}$, of this proportion will replace the ratio $\frac{AB}{CS}$ in the starred (*) statement.]

With $\triangle RCE \sim \triangle BAE$, $\frac{CE}{EA} = \frac{CR}{AB}$. [The first ratio, $\frac{CE}{EA}$, of this proportion replaces $\frac{CR}{AB}$ in the starred (*) statement.] Making the indicated substitutions into the starred statement, we have

$$\frac{AG}{GB} \cdot \frac{BF}{FC} \cdot \frac{CE}{EA} = 1$$

PERSPECTIVE ON APPLICATIONS

AN UNUSUAL APPLICATION OF SIMILAR TRIANGLES

The following problem is one that can be solved in many ways. If methods of calculus are applied, the solution is found through many complicated and tedious calculations. The simplest solution, which follows, utilizes geometry and similar triangles.

Problem: A hiker is at a location 450 ft downstream from his campsite. He is 200 ft away from the straight stream, and his tent is 100 ft away, as shown in Figure 5.56(a). Across the flat field, he sees that a spark from his campfire has ignited the tent. Taking the empty bucket he is carrying, he runs to the river to get water and then on to the tent. To what point on the river should he run to minimize the distance he travels?

(a)

(b)

Figure 5.56

We wish to determine x in Figure 5.56(b) on page 258 so that the total distance $D = d_1 + d_2$ is as small as possible. Consider three possible choices of this point on the river. These are suggested by dashed, dotted, and solid lines in Figure 5.57(a). Also consider the reflections of the triangles across the river. [See Figure 5.57(b).]

Figure 5.58

(a)   (b)

Figure 5.57

The minimum distance D occurs where the segments of lengths d_1 and d_2 form a straight line. That is, the configuration with the solid line segments minimizes the distance. In that case, the triangle at left and the reflected triangle at right are similar, as shown in Figure 5.58.

Thus

$$\frac{200}{100} = \frac{450 - x}{x}$$

$$200x = 100(450 - x)$$
$$200x = 45,000 - 100x$$
$$300x = 45,000$$
$$x = 150$$

Accordingly, the desired point on the river is 300 ft (determined by $450 - x$) upstream from the hiker's location.

Summary

A Look Back at Chapter 5

One goal of this chapter has been to define similarity for two polygons. We postulated a method for proving triangles similar and showed that proportions are a consequence of similar triangles, a line parallel to one side of a triangle, and a ray bisecting one angle of a triangle. The Pythagorean Theorem and its converse were proved. We discussed the 30°-60°-90° triangle, the 45°-45°-90° triangle, and other special right triangles whose lengths of sides determine Pythagorean triples. The final section developed the concept known as "segments divided proportionally."

A Look Ahead to Chapter 6

In the next chapter, we will begin our work with the circle. Segments and lines of the circle will be defined, as will special angles in a circle. Several theorems dealing with the measurements of these angles and line segments will be proved. Our work with constructions will enable us to deal with the locus of points and the concurrence of lines that are found in Chapter 7.

Key Concepts

5.1
Ratio • Rate • Proportion • Extremes • Means • Means-Extremes Property • Geometric Mean • Extended Ratio • Extended Proportion

5.2
Similar Polygons • Congruent Polygons • Corresponding Vertices, Angles, and Sides

5.3
AAA • AA • CSSTP • CASTC • SAS~ • SSS~

5.4
Pythagorean Theorem • Converse of Pythagorean Theorem • Pythagorean Triple

5.5
The 45°-45°-90° Triangle • The 30°-60°-90° Triangle

5.6
Segments Divided Proportionally • The Angle-Bisector Theorem • Ceva's Theorem

Overview ■ Chapter 5

Methods of Proving Triangles Similar ($\triangle ABC \sim \triangle DEF$)

Figure (Note Marks)	Method	Steps Needed in Proof
	AA	$\angle A \cong \angle D;\ \angle C \cong \angle F$
	SSS~	$\dfrac{AB}{DE} = \dfrac{AC}{DF} = \dfrac{BC}{EF} = k$ (k is a constant.)
	SAS~	$\dfrac{AB}{DE} = \dfrac{BC}{EF} = k$ $\angle B \cong \angle E$

Pythagorean Theorem and Converse

Figure (Note Marks)	Relationship	Conclusion
	$m\angle C = 90°$	$c^2 = a^2 + b^2$
	$c^2 = a^2 + b^2$	$m\angle C = 90°$

continued

Special Relationships

Figure	Relationship	Conclusion(s)
	$45°\text{-}45°\text{-}90°\ \triangle$ Note: $BC = a$	$AC = a$ $AB = a\sqrt{2}$
	$30°\text{-}60°\text{-}90°\ \triangle$ Note: $BC = a$	$AC = a\sqrt{3}$ $AB = 2a$

Segments Divided Proportionally

Figure	Relationship	Conclusion(s)
	$\overleftrightarrow{DE} \parallel \overline{BC}$	$\dfrac{AD}{DB} = \dfrac{AE}{EC}$ $\dfrac{AD}{AE} = \dfrac{DB}{EC}$ or equivalent
	$\overleftrightarrow{AD} \parallel \overleftrightarrow{BE} \parallel \overleftrightarrow{CF}$	$\dfrac{AB}{BC} = \dfrac{DE}{EF}$ $\dfrac{AB}{DE} = \dfrac{BC}{EF}$ or equivalent
	$\overrightarrow{BD}$ bisects $\angle ABC$	$\dfrac{AB}{BC} = \dfrac{AD}{DC}$ $\dfrac{AB}{AD} = \dfrac{BC}{DC}$ or equivalent
	Ceva's Theorem (D is any point in the interior of $\triangle ABC$.)	$\dfrac{AG}{GB} \cdot \dfrac{BF}{FC} \cdot \dfrac{CE}{EA} = 1$ or equivalent

Chapter 5   Review Exercises

Answer true or false for Review Exercises 1 to 7.

1. The ratio of 12 hr to 1 day is 2 to 1.

2. If the numerator and the denominator of a ratio are multiplied by 4, the new ratio equals the given ratio.

3. The value of a ratio must be less than 1.

4. The three numbers 6, 14, and 22 are in a ratio of 3:7:11.

5. To express a ratio correctly, the terms must have the same unit of measure.

6. The ratio 3:4 is the same as the ratio 4:3.

7. If the second and third terms of a proportion are equal, then either is the geometric mean of the first and fourth terms.

8. Find the value(s) of x in each proportion:

a) $\dfrac{x}{6} = \dfrac{3}{x}$ e) $\dfrac{x-2}{x-5} = \dfrac{2x+1}{x-1}$

b) $\dfrac{x-5}{3} = \dfrac{2x-3}{7}$ f) $\dfrac{x(x+5)}{4x+4} = \dfrac{9}{5}$

c) $\dfrac{6}{x+4} = \dfrac{2}{x+2}$ g) $\dfrac{x-1}{x+2} = \dfrac{10}{3x-2}$

d) $\dfrac{x+3}{5} = \dfrac{x+5}{7}$ h) $\dfrac{x+7}{2} = \dfrac{x+2}{x-2}$

Use proportions to solve Review Exercises 9 to 11.

9. Four containers of fruit juice cost $3.52. How much do six containers cost?

10. Two packages of M&Ms cost $1.38. How many packages can you buy for $5?

11. A carpet measuring 20 square yards costs $350. How much would a 12 square-yard carpet of the same material cost?

12. The ratio of the measures of the sides of a quadrilateral is 2:3:5:7. If the perimeter is 68, find the length of each side.

13. The length and width of a rectangle are 18 and 12, respectively. A similar rectangle has length 27. What is its width?

14. The lengths of the sides of a triangle are 6, 8, and 9. The shortest side of a similar triangle has length 15. How long are its other sides?

15. The ratio of the measure of the supplement of an angle to that of the complement of the angle is 5:2. Find the measure of the supplement.

16. Name the method (AA, SSS~, or SAS~) that is used to show that the triangles are similar. Use the figure below.
a) $WU = 2 \cdot TR$, $WV = 2 \cdot TS$, and $UV = 2 \cdot RS$
b) $\angle T \cong \angle W$ and $\angle S \cong \angle V$
c) $\angle T \cong \angle W$ and $\dfrac{TR}{WU} = \dfrac{TS}{WV}$
d) $\dfrac{TR}{WU} = \dfrac{TS}{WV} = \dfrac{RS}{UV}$

17. *Given:* $ABCD$ is a parallelogram
$\overline{DB}$ intersects $\overline{AE}$ at point F
Prove: $\dfrac{AF}{EF} = \dfrac{AB}{DE}$

18. *Given:* $\angle 1 \cong \angle 2$
Prove: $\dfrac{AB}{AC} = \dfrac{BE}{CD}$

19. *Given:* $\triangle ABC \sim \triangle DEF$ (not shown)
$m\angle A = 50°$, $m\angle E = 33°$
$m\angle D = 2x + 40$
Find: x, $m\angle F$

20. *Given:*   In $\triangle ABC$ and $\triangle DEF$ (not shown)
$\angle B \cong \angle F$ and $\angle C \cong \angle E$
$AC = 9$, $DE = 3$, $DF = 2$, $FE = 4$
Find: AB, BC

For Review Exercises 21 to 23, $\overline{DE} \parallel \overline{AC}$.

Exercises 21–23

21. $BD = 6$, $BE = 8$, $EC = 4$, $AD = ?$

22. $AD = 4$, $BD = 8$, $DE = 3$, $AC = ?$

23. $AD = 2$, $AB = 10$, $BE = 5$, $BC = ?$

For Review Exercises 24 to 26, $\overrightarrow{GJ}$ bisects $\angle FGH$.

Exercises 24–26

24. *Given:* $FG = 10$, $GH = 8$, $FJ = 7$
Find: JH

25. *Given:* $GF{:}GH = 1{:}2$, $FJ = 5$
Find: JH

26. *Given:* $FG = 8$, $HG = 12$, $FH = 15$
Find: FJ

27. *Given:* $\overleftrightarrow{EF} \parallel \overleftrightarrow{GO} \parallel \overleftrightarrow{HM} \parallel \overleftrightarrow{JK}$, with transversals $\overline{FJ}$ and $\overline{EK}$; $FG = 2$, $GH = 8$, $HJ = 5$, $EM = 6$
Find: EO, EK

28. Prove that if a line bisects one side of a triangle and is parallel to a second side, then it bisects the third side.

29. Prove that the diagonals of a trapezoid divide themselves proportionally.

30. *Given:* $\triangle ABC$ with right $\angle BAC$
$\overline{AD} \perp \overline{BC}$
a) $BD = 3$, $AD = 5$, $DC = ?$
b) $AC = 10$, $DC = 4$, $BD = ?$
c) $BD = 2$, $BC = 6$, $BA = ?$
d) $BD = 3$, $AC = 3\sqrt{2}$, $DC = ?$

31. *Given:* $\triangle ABC$ with right $\angle ABC$
$\overline{BD} \perp \overline{AC}$
a) $BD = 12$, $AD = 9$, $DC = ?$
b) $DC = 5$, $BC = 15$, $AD = ?$
c) $AD = 2$, $DC = 8$, $AB = ?$
d) $AB = 2\sqrt{6}$, $DC = 2$, $AD = ?$

32. In the drawings shown, find x.

(a)   (b)

(c)   (d)

33. *Given:* $ABCD$ is a rectangle
E is the midpoint of $\overline{BC}$
$AB = 16$, $CF = 9$, $AD = 24$
Find: AE, EF, AF, $m\angle AEF$

34. Find the length of a diagonal of a square whose side is 4 in. long.

35. Find the length of a side of a square whose diagonal is 6 cm long.

36. Find the length of a side of a rhombus whose diagonals are 48 cm and 14 cm long.

37. Find the length of an altitude of an equilateral triangle if each side is 10 in. long.

38. Find the length of a side of an equilateral triangle if an altitude is 6 in. long.

39. The lengths of three sides of a triangle are 13 cm, 14 cm, and 15 cm. Find the length of the altitude to the 14-cm side.

40. In the drawings, find x and y.

(a)   (b)

(c)   (d)

41. An observation aircraft flying at a height of 12 km has detected a Brazilian ship at a distance of 20 km from the aircraft and in line with an American ship that is 13 km from the aircraft. How far apart are the U.S. and Brazilian ships?

42. Tell whether each set of numbers represents the lengths of the sides of an acute triangle, of an obtuse triangle, of a right triangle, or of no triangle:
a) 12, 13, 14   e) 8, 7, 16
b) 11, 5, 18   f) 8, 7, 6
c) 9, 15, 18   g) 9, 13, 8
d) 6, 8, 10   h) 4, 2, 3

Chapter 5 Test

1. Reduce to its simplest form:
   a) The ratio 12:20 _____
   b) The rate $\frac{200 \text{ miles}}{8 \text{ gallons}}$ _____

2. Solve each proportion for x. Show your work!
   a) $\frac{x}{5} = \frac{8}{13}$ _____   b) $\frac{x+1}{5} = \frac{16}{x-1}$ _____

3. The measures of two complementary angles are in the ratio 1:5. Find the measure of each angle.
   Smaller: _____; Larger: _____

4. $\triangle RTS \sim \triangle UWV$.
   a) Find $m\angle W$ if $m\angle R = 67°$ and $m\angle S = 21°$.
   _____
   b) Find WV if $RT = 4$, $UW = 6$, and $TS = 8$.
   _____

Exercises 4, 5

5. Give the reason (AA, SAS~, or SSS~) why $\triangle RTS \sim \triangle UTW$.
   a) $\angle R \cong \angle U$ and $\frac{TR}{WU} = \frac{RS}{UV}$ _____
   b) $\angle S \cong \angle V$; $\angle T$ and $\angle W$ are right angles _____

6. In right triangle ABC, $\overline{CD}$ is the altitude from C to hypotenuse $\overline{AB}$. Name three triangles that are similar to each other. _____

7. In $\triangle ABC$, $m\angle C = 90°$. Use a square root radical to represent:
   a) c, if $a = 5$ and $b = 4$
   _____
   b) a, if $b = 6$ and $c = 8$
   _____

8. Given its lengths of sides, is $\triangle RST$ a right triangle?
   a) $a = 15$, $b = 8$, and $c = 17$ _____ (Yes or No)
   b) $a = 11$, $b = 8$, and $c = 15$ _____ (Yes or No)

9. Given quadrilateral $ABCD$ with diagonal $\overline{AC}$. If $\overline{BC} \perp \overline{AB}$ and $\overline{AC} \perp \overline{DC}$, find DA if $AB = 4$, $BC = 3$, and $DC = 8$. Express the answer as a square root radical. _____

10. In $\triangle XYZ$, $\overline{XZ} \cong \overline{YZ}$ and $\angle Z$ is a right angle.
    a) Find XY if $XZ = 10$ in. _____
    b) Find XZ if $XY = 8\sqrt{2}$ cm. _____

11. In $\triangle DEF$, $\angle D$ is a right angle and $m\angle F = 30°$.
    a) Find DE if $EF = 10$ m.
    _____
    b) Find EF if $DF = 6\sqrt{3}$ ft.
    _____

12. In $\triangle ABC$, $\overleftrightarrow{DE} \parallel \overline{BC}$. If $AD = 6$, $DB = 8$, and $AE = 9$, find EC. _____

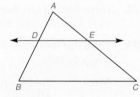

13. In $\triangle MNP$, $\overrightarrow{NQ}$ bisects $\angle MNP$. If $PN = 6$, $MN = 9$, and $MP = 10$, find PQ and QM.
    $PO =$ _____;
    $QM =$ _____

14. For $\triangle ABC$, the three angle bisectors are shown. Find the product $\frac{AE}{EC} \cdot \frac{CD}{DB} \cdot \frac{BF}{FA}$.
    _____

15. *Given:* $\angle 1 \cong \angle C$; M is the midpoint of $\overline{BC}$; $CM = MB = 6$ and $AD = 14$
    Find: x, the length of $\overline{DB}$

In Exercises 16 and 17, complete the statements and reasons in each proof.

16. *Given:* $\overline{MN} \parallel \overline{QR}$
    Prove: $\triangle MNP \sim \triangle QRP$

Statements	Reasons
1. _____	1. _____
2. $\angle N \cong \angle QRP$	2. If 2 $\parallel$ lines are cut by a trans., _____
3. _____	3. Identity
4. $\triangle MNP \sim \triangle QRP$	4. _____

17. *Given:*   In $\triangle ABC$, P is the
midpoint of $\overline{AC}$,
and R is the
midpoint of $\overline{CB}$.
Prove: $\angle PRC \cong \angle B$

Statements	Reasons
1. $\triangle ABC$	1. _____
2. $\angle C \cong \angle C$	2. _____
3. P is the midpoint of $\overline{AC}$, and R is the midpoint of $\overline{CB}$	3. _____
4. $\frac{PC}{AC} = \frac{1}{2}$ and $\frac{CR}{CB} = \frac{1}{2}$	4. Definition of midpoint
5. $\frac{PC}{AC} = \frac{CR}{CB}$	5. _____
6. $\triangle CPR \sim \triangle CAB$	6. _____
7. _____	7. CASTC

Kees Metselaar/Alamy

Chapter 6 — Circles

Towering! Displayed in the design of the Jardine House in Hong Kong are numerous windows that take the shape of circles. Circles appear everywhere in the real world, from the functional gear or pulley to the edible pancake. In this chapter, we will deal with the circle, related terminology, and properties. Based upon earlier principles, the theorems of this chapter follow logically from the properties found in previous chapters. There are numerous applications of the circle found in the examples and the exercises of this chapter. Another look at the Jardine House reveals that the circle has contemporary applications as well.

Additional video explanations of concepts, sample problems, and applications are available on DVD.

6.1 Circles and Related Segments and Angles

KEY CONCEPTS

Circle	Diameter	Minor Arc
Congruent Circles	Chord	Intercepted Arc
Concentric Circles	Semicircle	Congruent Arcs
Center	Arc	Central Angle
Radius	Major Arc	Inscribed Angle

> **Warning**
>
> If the phrase "in a plane" is omitted from the definition of a circle, the result is the definition of a sphere.

In this chapter, we will expand the terminology for circles that was found in Section 1.2. We also introduce some methods of measurement of arcs and angles and develop many properties of the circle.

> **DEFINITION**
>
> A **circle** is the set of all points in a plane that are at a fixed distance from a given point known as the *center* of the circle.

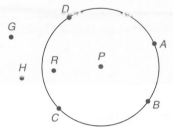

Figure 6.1

A circle is named by its center point. In Figure 6.1, point P is the center of the circle. The symbol for circle is $\odot$, so the circle in Figure 6.1 is $\odot P$. Points A, B, C, and D are points *of* (or *on*) the circle. Points P (the center) and R are in the *interior* of circle P; points G and H are in the *exterior* of the circle.

In $\odot Q$ of Figure 6.2, $\overline{SQ}$ is a radius of the circle. A **radius** is a segment that joins the center of the circle to a point on the circle. $\overline{SQ}$, $\overline{TQ}$, $\overline{VQ}$, and $\overline{WQ}$ are **radii** (plural of *radius*) of $\odot Q$. By definition, $SQ = TQ = VQ = WQ$; more generally, the following statement is a consequence of the definition of a circle.

> All radii of a circle are congruent.

A line segment (such as $\overline{TW}$ in Figure 6.2) that joins two points of a circle is a **chord** of the circle. A **diameter** of a circle is a chord that contains the center of the circle; if *T-Q-W* in Figure 6.2, then $\overline{TW}$ is a diameter of $\odot Q$.

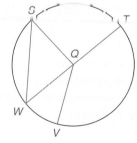

Figure 6.2

> **DEFINITION**
>
> **Congruent circles** are two or more circles that have congruent radii.

In Figure 6.3 on page 269, circles P and Q are congruent because their radii have equal lengths. We can slide $\odot P$ to the right to coincide with $\odot Q$.

Figure 6.3

Figure 6.4

SSG EXS. 1–3

Figure 6.5

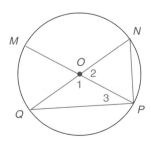

Figure 6.6

> ### DEFINITION
>
> **Concentric circles** are coplanar circles that have a common center.

The concentric circles in Figure 6.4 have the common center O.

In $\odot P$ of Figure 6.5, the part of the circle shown in red from point A to point B is **arc** AB, symbolized by $\overset{\frown}{AB}$. If $\overline{AC}$ is a diameter, then $\overset{\frown}{ABC}$ (three letters are used for clarity) is a **semicircle.** In Figure 6.5, a **minor arc** like $\overset{\frown}{AB}$ is part of a semicircle; a **major arc** such as $\overset{\frown}{ABCD}$ (also denoted by $\overset{\frown}{ABD}$ or $\overset{\frown}{ACD}$) is more than a semicircle but less than the entire circle.

> ### DEFINITION
>
> A **central angle** of a circle is an angle whose vertex is the center of the circle and whose sides are radii of the circle.

In Figure 6.6, $\angle NOP$ is a central angle of $\odot O$. The **intercepted arc** of $\angle NOP$ is $\overset{\frown}{NP}$. The intercepted arc of an angle is determined by the two points of intersection of the angle with the circle and all points of the arc in the interior of the angle.

In Example 1, we "check" the terminology just introduced.

> ### EXAMPLE 1
>
> In Figure 6.6, $\overline{MP}$ and $\overline{NQ}$ intersect at O, the center of the circle. Name:
>
> **a)** All four radii
> **b)** Both diameters
> **c)** All four chords
> **d)** One central angle
> **e)** One minor arc
>
> **f)** One semicircle
> **g)** One major arc
> **h)** Intercepted arc of $\angle MON$
> **i)** Central angle that intercepts $\overset{\frown}{NP}$

SOLUTION

a) $\overline{OM}, \overline{OQ}, \overline{OP},$ and $\overline{ON}$
b) $\overline{MP}$ and $\overline{QN}$
c) $\overline{MP}, \overline{QN}, \overline{QP},$ and $\overline{NP}$
d) $\angle QOP$ (other answers are possible)
e) $\overset{\frown}{NP}$ (other answers are possible)
f) $\overset{\frown}{MQP}$ (other answers are possible)
g) $\overset{\frown}{MQN}$ (can be named $\overset{\frown}{MQPN}$; other answers are possible)
h) $\overset{\frown}{MN}$ (lies in the interior of $\angle MON$ and includes points M and N)
i) $\angle NOP$ (also called $\angle 2$)

The following statement is a consequence of the Segment-Addition Postulate.

In a circle, the length of a diameter is twice the length of a radius; in symbols, $d = 2r$.

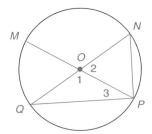

Figure 6.6

EXAMPLE 2

$\overline{QN}$ is a diameter of $\odot O$ in Figure 6.6 and $PN = ON = 12$. Find the length of chord $\overline{QP}$.

SOLUTION Because $PN = ON$ and $ON = OP$, $\triangle NOP$ is equilateral. Then $m\angle 2 = m\angle N = m\angle NPO = 60°$. Also, $OP = OQ$; so $\triangle POQ$ is isosceles with $m\angle 1 = 120°$, because this angle is supplementary to $\angle 2$. Now $m\angle Q = m\angle 3 = 30°$ because the sum of the measures of the angles of $\triangle POQ$ is $180°$. If $m\angle N = 60°$ and $m\angle Q = 30°$, then $\triangle NPQ$ is a right $\triangle$ whose angle measures are $30°$, $60°$, and $90°$. It follows that $QP = PN \cdot \sqrt{3} = 12\sqrt{3}$.

THEOREM 6.1.1

A radius that is perpendicular to a chord bisects the chord.

Figure 6.7

GIVEN: $\overline{OD} \perp \overline{AB}$ in $\odot O$ (See Figure 6.7.)

PROVE: $\overline{OD}$ bisects $\overline{AB}$

PROOF: $\overline{OD} \perp \overline{AB}$ in $\odot O$. Draw radii $\overline{OA}$ and $\overline{OB}$. Now $\overline{OA} \cong \overline{OB}$ because all radii of a circle are $\cong$. Because $\angle 1$ and $\angle 2$ are right $\angle$s and $\overline{OC} \cong \overline{OC}$, we see that $\triangle OCA \cong \triangle OCB$ by HL. Then $\overline{AC} \cong \overline{CB}$ by CPCTC, so $\overline{OD}$ bisects $\overline{AB}$.

ANGLE AND ARC RELATIONSHIPS IN THE CIRCLE

In Figure 6.8, the sum of the measures of the angles about point O (angles determined by perpendicular diameters $\overline{AC}$ and $\overline{BD}$) is $360°$. Similarly, the circle can be separated into 360 equal arcs, *each of which measures 1° of arc measure*; that is, each arc would be intercepted by a central angle measuring $1°$. Our description of arc measure leads to the following postulate.

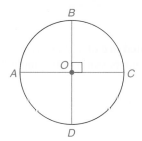

Figure 6.8

POSTULATE 16 ■ Central Angle Postulate

In a circle, the degree measure of a central angle is equal to the degree measure of its intercepted arc.

If $m\widehat{AB} = 90°$ in Figure 6.8, then $m\angle AOB = 90°$. The reflex angle that intercepts $\widehat{BCA}$ and that is composed of three right angles measures $270°$. In the same figure, $m\widehat{AB} = 90°$, $m\widehat{BCD} = 180°$, and $m\widehat{AD} = 90°$. It follows that $m\widehat{AB} + m\widehat{BCD} + m\widehat{AD} = 360°$. Consequently, we have the following generalization.

The sum of the measures of the consecutive arcs that form a circle is $360°$.

In ⊙Y [Figure 6.9(a)], if m∠XYZ = 76°, then m$\widehat{XZ}$ = 76° by the Central Angle Postulate. If two arcs have equal degree measures [Figures 6.9(b) and (c)] but are parts of two circles with unequal radii, then these arcs will not coincide. This observation leads to the following definition.

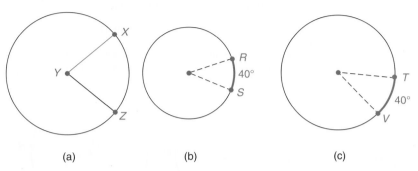

(a)   (b)   (c)

Figure 6.9

DEFINITION

In a circle or congruent circles, **congruent arcs** are arcs with equal measures.

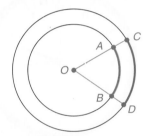

To further clarify the definition of congruent arcs, consider the concentric circles (having the same center) in Figure 6.10. Here the degree measure of ∠AOB of the smaller circle is the same as the degree measure of ∠COD of the larger circle. Even though m$\widehat{AB}$ = m$\widehat{CD}$, we conclude that $\widehat{AB}$ ≇ $\widehat{CD}$ because the arcs would not coincide.

(SSG) **EXS. 4–10**

Figure 6.10

EXAMPLE 3

In ⊙O of Figure 6.11, $\overrightarrow{OE}$ bisects ∠AOD. Using the measures indicated, find:

a) m$\widehat{AB}$ **b)** m$\widehat{BC}$ **c)** m$\widehat{BD}$ **d)** m∠AOD

e) m$\widehat{AE}$ **f)** m$\widehat{ACE}$ **g)** whether $\widehat{AE}$ ≅ $\widehat{ED}$

h) the measure of the reflex angle that intercepts $\widehat{ABCD}$

SOLUTION a) 105° **b)** 70° **c)** 105° **d)** 150°, from 360 − (105 + 70 + 35)
e) 75° because the corresponding central angle (∠AOE) is the result of bisecting ∠AOD, which was found to be 150° **f)** 285° (from 360 − 75, the measure of $\widehat{AE}$)
g) The arcs are congruent because both measure 75° and are arcs in the same circle. **h)** 210° (from 105° + 70° + 35°)

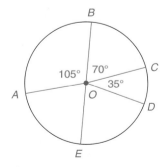

Figure 6.11

In Figure 6.11, note that m$\widehat{BC}$ + m$\widehat{CD}$ = m$\widehat{BD}$ (or m$\widehat{BCD}$). Because the union of $\widehat{BD}$ and $\widehat{DA}$ is the major arc $\widehat{BDA}$, we also see that m$\widehat{BD}$ + m$\widehat{DA}$ = m$\widehat{BDA}$. With the understanding that $\widehat{AB}$ and $\widehat{BC}$ do not overlap, we generalize the relationship as follows.

POSTULATE 17 ■ Arc-Addition Postulate

If $\widehat{AB}$ and $\widehat{BC}$ intersect only at point B, then m$\widehat{AB}$ + m$\widehat{BC}$ = m$\widehat{ABC}$.

When $\overset{\frown}{mAB} + \overset{\frown}{mBC} = \overset{\frown}{mABC}$ as in Figure 6.12(a), we say that point B is *between* A and C of circle O.

Given points A, B, and C on $\odot O$ as shown in Figure 6.12(a), suppose that radii $\overline{OA}$, $\overline{OB}$, and $\overline{OC}$ are drawn. Just as $m\angle AOB + m\angle BOC = m\angle AOC$ by the Angle-Addition Postulate, it follows that

$$\overset{\frown}{mAB} + \overset{\frown}{mBC} = \overset{\frown}{mABC}.$$

In the statement of the Arc-Addition Postulate, the reason for writing $\overset{\frown}{ABC}$ (rather than $\overset{\frown}{AC}$) is that the arc with endpoints at A and C *could be* a major arc.

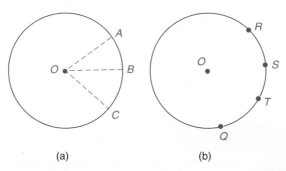

(a)   (b)

Figure 6.12

The Arc-Addition Postulate can easily be extended to include more than two arcs. In Figure 6.12(b), $\overset{\frown}{mRS} + \overset{\frown}{mST} + \overset{\frown}{mTQ} = \overset{\frown}{mRSTQ}$ (or $\overset{\frown}{mRQ}$ because $\overset{\frown}{RQ}$ is a minor arc). If $\overset{\frown}{mRS} = \overset{\frown}{mST}$, then point S is the **midpoint** of $\overset{\frown}{RT}$; alternately, $\overset{\frown}{RT}$ is **bisected** at point S.

In Example 4, we use the fact that the entire circle measures $360°$.

EXAMPLE 4

In Figure 6.13, determine the measure of the angle formed by the hands of a clock at 3:12 P.M.

SOLUTION The minute hand moves through 12 minutes, which is $\frac{12}{60}$ or $\frac{1}{5}$ of an hour. Thus, the minute hand points in a direction whose angle measure from the vertical is $\frac{1}{5}(360°)$ or $72°$. At exactly 3 P.M., the hour hand would form an angle of $90°$ with the vertical. However, gears inside the clock also turn the hour hand through $\frac{1}{5}$ of the $30°$ arc from the 3 toward the 4; that is, the hour hand moves another $\frac{1}{5}(30°)$ or $6°$ to form an angle of $96°$ with the vertical. At 3:12 P.M., the angle between the hands must measure $96° - 72°$ or $24°$.

The measure of an arc of a circle can be used to measure the corresponding central angle. The measure of an arc can also be used to measure an inscribed angle, which is defined as follows.

DEFINITION

An **inscribed angle** of a circle is an angle whose vertex is a point on the circle and whose sides are chords of the circle.

In Figure 6.14, $\angle B$ is an inscribed angle. Note that the word *inscribed* is often associated with the word *inside*. As suggested by the Discover activity at the left, the relationship between the measure of an inscribed angle and its intercepted arc leads to the following theorem.

Figure 6.13

Discover

In Figure 6.14, $\angle B$ is the inscribed angle whose sides are chords $\overline{BA}$ and $\overline{BC}$.

a. Use a protractor to find the measure of central $\angle AOC$.

b. Find the measure of $\overset{\frown}{AC}$.

c. Finally, measure inscribed $\angle B$.

d. How is the measure of inscribed $\angle B$ related to the measure of its intercepted arc $\overset{\frown}{AC}$?

ANSWERS

a) $58°$ b) $58°$ c) $29°$ d) $m\angle B = \frac{1}{2}m\overset{\frown}{AC}$

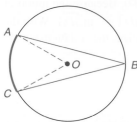

Figure 6.14

THEOREM 6.1.2

> The measure of an inscribed angle of a circle is one-half the measure of its intercepted arc.

The proof of Theorem 6.1.2 must be divided into three cases:

CASE 1. One side of the inscribed angle is a diameter. See Figure 6.16.

CASE 2. The diameter through the vertex of the inscribed angle lies in the interior of the angle. See Figure 6.15(a).

CASE 3. The diameter through the vertex of the inscribed angle lies in the exterior of the angle. See Figure 6.15(b).

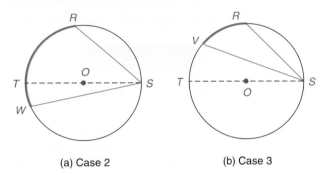

(a) Case 2   (b) Case 3

Figure 6.15

The proof of Case 1 follows, but proofs of the other cases are left as exercises.

GIVEN: $\odot O$ with inscribed $\angle RST$ and diameter $\overline{ST}$ (See Figure 6.16.)

PROVE: $m\angle S = \frac{1}{2}m\widehat{RT}$

PROOF OF CASE 1: We begin by drawing auxilary radius $\overline{RO}$. Then $m\angle ROT = m\widehat{RT}$ because the central angle has a measure equal to the measure of its intercepted arc. With $\overline{OR} \cong \overline{OS}$, $\triangle ROS$ is isosceles and $m\angle R = m\angle S$. Now the exterior angle of the triangle is $\angle ROT$, so $m\angle ROT = m\angle R + m\angle S$. Because $m\angle R = m\angle S$, $m\angle ROT = 2(m\angle S)$. Then $m\angle S = \frac{1}{2}m\angle ROT$. With $m\angle ROT = m\widehat{RT}$, we have $m\angle S = \frac{1}{2}m\widehat{RT}$ by substitution.

Although proofs in this chapter generally take the less formal paragraph form, it remains necessary to be able to justify each statement of the proof.

THEOREM 6.1.3

> In a circle (or in congruent circles), congruent minor arcs have congruent central angles.

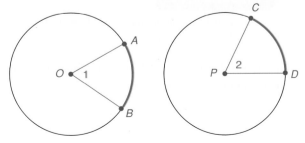

Figure 6.17

If $\widehat{AB} \cong \widehat{CD}$ in congruent circles O and P, then $\angle 1 \cong \angle 2$ by Theorem 6.1.3.

Technology Exploration

Use computer software if available.
1. Create circle O with inscribed angle RST; $\overline{TS}$ is a diameter.
2. Include radius $\overline{OR}$ in the figure. See Figure 6.16.
3. Measure $\widehat{RT}$, $\angle ROT$, and $\angle RST$.
4. Show that:
   $m\angle ROT = m\widehat{RT}$ and
   $m\angle RST = \frac{1}{2}m\widehat{RT}$

Figure 6.16

SSG EXS. 11–15

Reminder

The measure of an exterior angle of a triangle equals the sum of the measures of the two remote interior angles.

We suggest that the student make drawings to illustrate Theorems 6.1.4–6.1.6. Some of the proofs require the use of auxiliary radii.

THEOREM 6.1.4

In a circle (or in congruent circles), congruent central angles have congruent arcs.

In Theorems 6.1.5 and 6.1.6, the related chords and arcs share the same endpoints.

THEOREM 6.1.5

In a circle (or in congruent circles), congruent chords have congruent minor (major) arcs.

THEOREM 6.1.6

In a circle (or in congruent circles), congruent arcs have congruent chords.

On the basis of an earlier definition, we define the distance from the center of a circle to a chord to be the length of the perpendicular segment joining the center to that chord. Congruent triangles are used to prove Theorems 6.1.7 and 6.1.8.

THEOREM 6.1.7

Chords that are at the same distance from the center of a circle are congruent.

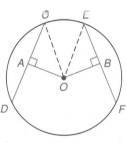

Figure 6.18

GIVEN: $\overline{OA} \perp \overline{CD}$ and $\overline{OB} \perp \overline{EF}$ in $\odot O$ (See Figure 6.18.)
$\overline{OA} \cong \overline{OB}$

PROVE: $\overline{CD} \cong \overline{EF}$

PROOF: Draw radii $\overline{OC}$ and $\overline{OE}$. With $\overline{OA} \perp \overline{CD}$ and $\overline{OB} \perp \overline{EF}$, $\angle OAC$ and $\angle OBE$ are right $\angle$s. $\overline{OA} \cong \overline{OB}$ is given, and $\overline{OC} \cong \overline{OE}$ because all radii of a circle are congruent. Thus, right triangles OAC and OBE are congruent by HL.

By CPCTC, $\overline{CA} \cong \overline{BE}$ so $CA = BE$. Then $2(CA) = 2(BE)$. But $2(CA) = CD$ because A is the midpoint of chord $\overline{CD}$. ($\overline{OA}$ bisects chord $\overline{CD}$ because $\overline{OA}$ is part of a radius. See Theorem 6.1.1). Likewise, $2(BE) = EF$, and it follows that $CD = EF$ and $\overline{CD} \cong \overline{EF}$

Proofs of Theorems 6.1.8 and 6.1.9 are left as exercises.

THEOREM 6.1.8

Congruent chords are located at the same distance from the center of a circle.

The student should make a drawing to illustrate Theorem 6.1.8. In a circle, draw two congruent chords that do not intersect.

> **THEOREM 6.1.9**
>
> An angle inscribed in a semicircle is a right angle.

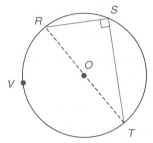

Figure 6.19

Theorem 6.1.9 is illustrated in Figure 6.19, where $\angle S$ is inscribed in the semicircle $\widehat{RST}$. Note that $\angle S$ also intercepts semicircle $\widehat{RVT}$; thus, an inscribed angle that intercepts a semicircle is a right angle.

> **THEOREM 6.1.10**
>
> If two inscribed angles intercept the same arc, then these angles are congruent.

(SSG) **EXS. 16, 17**

Theorem 6.1.10 is illustrated in Figure 6.20. Note that $\angle 1$ and $\angle 2$ both intercept $\widehat{XY}$. Because $m\angle 1 = \frac{1}{2}m\widehat{XY}$ and $m\angle 2 = \frac{1}{2}m\widehat{XY}$, $\angle 1 \cong \angle 2$.

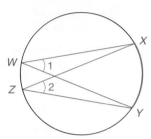

Figure 6.20

Exercises 6.1

For Exercises 1 to 8, use the figure provided.

Exercises 1–8

1. If $m\widehat{AC} = 58°$, find $m\angle B$.

2. If $m\widehat{DE} = 46°$, find $m\angle O$.

3. If $m\widehat{DE} = 47.6°$, find $m\angle O$.

4. If $m\widehat{AC} = 56.4°$, find $m\angle B$.

5. If $m\angle B = 28.3°$, find $m\widehat{AC}$.

6. If $m\angle O = 48.3°$, find $m\widehat{DE}$.

7. If $m\widehat{DE} = 47°$, find the measure of the reflex angle that intercepts $\widehat{DBACE}$.

8. If $m\widehat{ECABD} = 312°$, find $m\angle DOE$.

9. *Given:* $\overline{AO} \perp \overline{OB}$ and $\overline{OC}$ bisects $\widehat{ACB}$ in $\odot O$
   Find:
   a) $m\widehat{AB}$
   b) $m\widehat{ACB}$
   c) $m\widehat{BC}$
   d) $m\angle AOC$

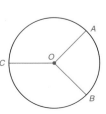

10. *Given:* $ST = \frac{1}{2}(SR)$ in $\odot Q$
    SR is a diameter
    Find:
    a) $m\widehat{ST}$
    b) $m\widehat{TR}$
    c) $m\widehat{STR}$
    d) $m\angle S$

    (HINT: Draw $\overline{QT}$.)

11. *Given:* $\odot Q$ in which $m\widehat{AB}=m\widehat{BC}=m\widehat{CA}$ ...
    Find:
    a) $m\widehat{AB}$
    b) $m\widehat{BC}$
    c) $m\widehat{CA}$
    d) $m\angle 1 (\angle AQB)$
    e) $m\angle 2 (\angle CQB)$
    f) $m\angle 3 (\angle CQA)$
    g) $m\angle 4 (\angle CAQ)$
    h) $m\angle 5 (\angle QAB)$
    i) $m\angle 6 (\angle QBC)$

    (HINT: Let $m\widehat{AB} = 2x$, $m\widehat{BC} = 3x$, and $m\widehat{CA} = 4x$.)

12. *Given:* $m\angle DOE = 76°$ and $m\angle EOG = 82°$ in $\odot O$
    $\overline{EF}$ is a diameter
    Find:
    a) $m\widehat{DE}$
    b) $m\widehat{DF}$
    c) $m\angle F$
    d) $m\angle DGE$
    e) $m\angle EHG$
    f) Whether $m\angle EHG = \frac{1}{2}(m\widehat{EG} + m\widehat{DF})$

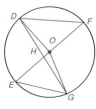

13. *Given:* ⊙O with $\overline{AB} \cong \overline{AC}$ and
m∠BOC = 72°

Find:
a) m$\widehat{BC}$
b) m$\widehat{AB}$
c) m∠A
d) m∠ABC
e) m∠ABO

14. In ⊙O (not shown), $\overline{OA}$ is a radius, $\overline{AB}$ is a diameter, and $\overline{AC}$ is a chord.
a) How does OA compare to AB?
b) How does AC compare to AB?
c) How does AC compare to OA?

15. *Given:* In⊙O, $\overline{OC} \perp \overline{AB}$ and
OC = 6

Find:
a) AB
b) BC

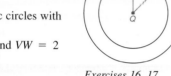

Exercise 15

16. *Given:* Concentric circles with center Q
SR = 3 and RQ = 4
$\overline{QS} \perp \overline{TV}$ at R

Find:
a) RV
b) TV

17. *Given:* Concentric circles with
center Q
TV = 8 and VW = 2
$\overline{RQ} \perp \overline{TV}$

Find: RQ

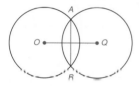

Exercises 16, 17

(HINT: Let RQ = x.)

18. $\overline{AB}$ is the **common chord** of ⊙O and ⊙Q. If AB = 12 and each circle has a radius of length 10, how long is $\overline{OQ}$?

Exercises 18, 19

19. Circles O and Q have the common chord $\overline{AB}$. If AB = 6, ⊙O has a radius of length 4, and ⊙Q has a radius of length 6, how long is $\overline{OQ}$?

20. Suppose that a circle is divided into three congruent arcs by points A, B, and C. What is the measure of each arc? What type of figure results when A, B, and C are joined by line segments?

21. Suppose that a circle is divided by points A, B, C, and D into four congruent arcs. What is the measure of each arc? If these points are joined in order, what type of quadrilateral results?

22. Following the pattern of Exercises 20 and 21, what type of figure results from dividing the circle equally by five points and joining those points in order? What type of polygon is formed by joining consecutively the n points that separate the circle into n congruent arcs?

23. Consider a circle or congruent circles, and explain why each statement is true:
a) Congruent arcs have congruent central angles.
b) Congruent central angles have congruent arcs.
c) Congruent chords have congruent arcs.
d) Congruent arcs have congruent chords.
e) Congruent central angles have congruent chords.
f) Congruent chords have congruent central angles.

24. State the measure of the angle formed by the minute hand and the hour hand of a clock when the time is
a) 1:30 P.M.
b) 2:20 A.M.

25. State the measure of the angle formed by the hands of the clock at
a) 6:30 P.M.
b) 5:40 A.M.

26. Five points are equally spaced on a circle. A five-pointed star (pentagram) is formed by joining nonconsecutive points two at a time. What is the degree measure of an arc determined by two consecutive points?

27. A cciling fan has equally spaced blades. What is the measure of the angle formed by two consecutive blades if there are
a) 5 blades?
b) 6 blades?

28. A wheel has equally spaced lug bolts. What is the measure of the central angle determined by two consecutive lug bolts if there are
a) 5 bolts?
b) 6 bolts?

29. An amusement park ride (the "Octopus") has eight support arms that are equally spaced about a circle. What is the measure of the central angle formed by two consecutive arms?

In Exercises 30 and 31, complete each proof.

30. *Given:* Diameters $\overline{AB}$ and $\overline{CD}$ in $\odot E$
Prove: $\overset{\frown}{AC} \cong \overset{\frown}{DB}$

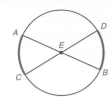

PROOF

Statements	Reasons
1. ?	1. Given
2. $\angle AEC \cong \angle DEB$	2. ?
3. $m\angle AEC = m\angle DEB$	3. ?
4. $m\angle AEC = m\overset{\frown}{AC}$ and $m\angle DEB = m\overset{\frown}{DB}$	4. ?
5. $m\overset{\frown}{AC} = m\overset{\frown}{DB}$	5. ?
6. ?	6. If two arcs of a circle have the same measure, they are $\cong$

31. *Given:* $\overline{MN} \parallel \overline{OP}$ in $\odot O$
Prove: $m\overset{\frown}{MQ} = 2(m\overset{\frown}{NP})$

PROOF

Statements	Reasons
1. ?	1. Given
2. $\angle 1 \cong \angle 2$	2. ?
3. $m\angle 1 = m\angle 2$	3. ?
4. $m\angle 1 = \frac{1}{2}(m\overset{\frown}{MQ})$	4. ?
5. $m\angle 2 = m\overset{\frown}{NP}$	5. ?
6. $\frac{1}{2}(m\overset{\frown}{MQ}) = m\overset{\frown}{NP}$	6. ?
7. $m\overset{\frown}{MQ} = 2(m\overset{\frown}{NP})$	7. Multiplication Prop. of Equality

In Exercises 32 to 37, write a paragraph proof.

32. *Given:* $\overline{RS}$ and $\overline{TV}$ are diameters of $\odot W$
Prove: $\triangle RST \cong \triangle VTS$

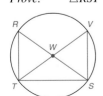

33. *Given:* Chords $\overline{AB}, \overline{BC}, \overline{CD}$, and $\overline{AD}$ in $\odot O$
Prove: $\triangle ABE \sim \triangle CDE$

34. Congruent chords are located at the same distance from the center of a circle.

35. A radius perpendicular to a chord bisects the arc of that chord.

36. An angle inscribed in a semicircle is a right angle.

37. If two inscribed angles intercept the same arc, then these angles are congruent.

38. If $\overleftrightarrow{MN} \parallel \overleftrightarrow{PQ}$ in $\odot O$, explain why $MNPQ$ is an isosceles trapezoid.

(HINT: Draw a diagonal.)

39. If $\overset{\frown}{ST} \cong \overset{\frown}{TV}$, explain why $\triangle STV$ is an isosceles triangle.

*40. Use a paragraph proof to complete this exercise.

Given: $\odot O$ with chords $\overline{AB}$ and $\overline{BC}$, radii $\overline{AO}$ and $\overline{OC}$
Prove: $m\angle ABC < m\angle AOC$

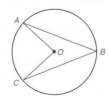

41. Prove Case 2 of Theorem 6.1.2.

42. Prove Case 3 of Theorem 6.1.2.

43. In $\odot O$, $OY = 5$ and $XZ = 6$. If $\overline{XW} \cong \overline{WY}$, find WZ.

| 6.2 | **More Angle Measures in the Circle** |

KEY CONCEPTS

Tangent
Point of Tangency
Secant
Polygon Inscribed
  in a Circle

Cyclic Polygon
Circumscribed Circle
Polygon Circumscribed
  about a Circle

Inscribed Circle
Interior and Exterior
  of a Circle

In this section, we consider lines, rays, and line segments that are related to the circle. We assume that the lines and circles are coplanar.

> **DEFINITION**
>
> A **tangent** is a line that intersects a circle at exactly one point; the point of intersection is the **point of contact,** or **point of tangency.**

The term *tangent* also applies to a line segment or ray that is part of a tangent line to a circle. In each case, the tangent touches the circle at one point.

> **DEFINITION**
>
> A **secant** is a line (or segment or ray) that intersects a circle at exactly two points.

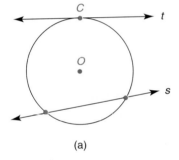

(a)

In Figure 6.21(a), line s is a secant to $\odot O$; also, line t is a tangent to $\odot O$, and point C is its point of contact. In Figure 6.21(b), $\overline{AB}$ is a tangent to $\odot Q$, and point T is its point of tangency; $\overrightarrow{CD}$ is a secant with points of intersection at E and F.

> **DEFINITION**
>
> A polygon is **inscribed in a circle** if its vertices are points on the circle and its sides are chords of the circle. Equivalently, the circle is said to be **circumscribed about the polygon.** The polygon inscribed in a circle is further described as a **cyclic polygon.**

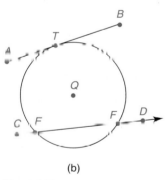

(b)

Figure 6.21

In Figure 6.22, $\triangle ABC$ is inscribed in $\odot O$ and quadrilateral $RSTV$ is inscribed in $\odot Q$. Conversely, $\odot O$ is circumscribed about $\triangle ABC$ and $\odot Q$ is circumscribed about quadrilateral $RSTV$. Note that $\overline{AB}$, $\overline{BC}$, and $\overline{AC}$ are chords of $\odot O$ and that $\overline{RS}$, $\overline{ST}$, $\overline{TV}$, and $\overline{RV}$ are chords of $\odot Q$. $\triangle ABC$ and quadrilateral $RSTV$ are cyclic polygons.

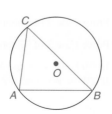

Figure 6.22

The preceding Discover activity prepares the way for the following theorem.

> **Reminder**
>
> A quadrilateral is said to be cyclic if its vertices lie on a circle.

THEOREM 6.2.1

If a quadrilateral is inscribed in a circle, the opposite angles are supplementary.
Alternative Form: The opposite angles of a cyclic quadrilateral are supplementary.

The proof of Theorem 6.2.1 follows. In the proof, we show that $\angle R$ and $\angle T$ are supplementary. In a similar proof, we could also show that $\angle S$ and $\angle V$ are supplementary as well.

GIVEN: *RSTV* is inscribed in ⊙Q (See Figure 6.23.)

PROVE: $\angle R$ and $\angle T$ are supplementary

PROOF: From Section 6.1, an inscribed angle is equal in measure to one-half the measure of its intercepted arc. Because $m\angle R = \frac{1}{2}m\widehat{STV}$ and $m\angle T = \frac{1}{2}m\widehat{SRV}$, it follows that

$$m\angle R + m\angle T = \frac{1}{2}m\widehat{STV} + \frac{1}{2}m\widehat{SRV}$$

$$= \frac{1}{2}(m\widehat{STV} + m\widehat{SRV})$$

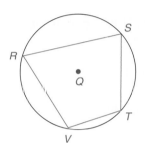

Figure 6.23

Because $\widehat{STV}$ and $\widehat{SRV}$ form the entire circle, $m\widehat{STV} + m\widehat{SRV} = 360°$.

By substitution, $m\angle R + m\angle T = \frac{1}{2}(360°) = 180°$

By definition, $\angle R$ and $\angle T$ are supplementary. ∎

The proof of Theorem 6.2.1 shows that $m\angle R + m\angle T = 180°$; see Figure 6.23. Because the sum of the interior angles of a quadrilateral is 360°, we know that

$$m\angle R + m\angle S + m\angle T + m\angle V = 360°.$$

Using substitution, it is easy to show that $m\angle S + m\angle V = 180°$; that is, $\angle S$ and $\angle V$ are also supplementary.

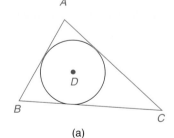

(a)

DEFINITION

A polygon is **circumscribed about a circle** if all sides of the polygon are line segments tangent to the circle; also, the circle is said to be **inscribed in the polygon.**

(b)

Figure 6.24

In Figure 6.24(a), $\triangle ABC$ is circumscribed about ⊙D. In Figure 6.24(b), square *MNPQ* is circumscribed about ⊙T. Furthermore, ⊙D is inscribed in $\triangle ABC$, and ⊙T is inscribed in square *MNPQ*. Note that $\overline{AB}$, $\overline{AC}$, and $\overline{BC}$ are tangents to ⊙D and that $\overline{MN}$, $\overline{NP}$, $\overline{PQ}$, and $\overline{MQ}$ are tangents to ⊙T.

We know that a central angle has a measure equal to the measure of its intercepted arc and that an inscribed angle has a measure equal to one-half the measure of its intercepted arc. Now we consider another type of angle found within the circle.

THEOREM 6.2.2

The measure of an angle formed by two chords that intersect within a circle is one-half the sum of the measures of the arcs intercepted by the angle and its vertical angle.

SSG **EXS. 1–6**

In Figure 6.25(a), $\angle 1$ and $\angle AEC$ are vertical angles; also $\angle 1$ intercepts $\overset{\frown}{DB}$ and $\angle AEC$ intercepts $\overset{\frown}{AC}$. According to Theorem 6.2.2,

$$m\angle 1 = \frac{1}{2}(m\overset{\frown}{AC} + m\overset{\frown}{DB})$$

To prove Theorem 6.2.2, we draw auxiliary line segment $\overline{CB}$ in Figure 6.25(b).

GIVEN: Chords $\overline{AB}$ and $\overline{CD}$ intersect at point E in $\odot O$

PROVE: $m\angle 1 = \frac{1}{2}(m\overset{\frown}{AC} + m\overset{\frown}{DB})$

PROOF: Draw $\overline{CB}$. Now $m\angle 1 = m\angle 2 + m\angle 3$ because $\angle 1$ is an exterior angle of $\triangle CBE$. Because $\angle 2$ and $\angle 3$ are inscribed angles of $\odot O$,

$$m\angle 2 = \frac{1}{2}m\overset{\frown}{DB} \quad \text{and} \quad m\angle 3 = \frac{1}{2}m\overset{\frown}{AC}$$

Substituting into the equation $m\angle 1 = m\angle 2 + m\angle 3$ leads to

$$m\angle 1 = \frac{1}{2}m\overset{\frown}{DB} + \frac{1}{2}m\overset{\frown}{AC}$$
$$= \frac{1}{2}(m\overset{\frown}{DB} + m\overset{\frown}{AC})$$

Equivalently,

$$m\angle 1 = \frac{1}{2}(m\overset{\frown}{AC} + m\overset{\frown}{DB})$$

In the preceding proof, we could have drawn auxiliary chord $\overline{AD}$. In that case, $\angle 1$ would be an exterior angle for $\triangle AED$ and a similar proof would follow.

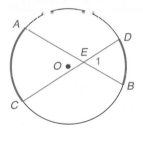

(a)

(b)

Figure 6.25

EXAMPLE 1

In Figure 6.25(a), $m\overset{\frown}{AC} = 84°$ and $m\overset{\frown}{DB} = 62°$. Find $m\angle 1$.

SOLUTION Applying Theorem 6.2.2,

$$m\angle 1 = \frac{1}{2}(m\overset{\frown}{AC} + m\overset{\frown}{DB})$$
$$= \frac{1}{2}(84° + 62°)$$
$$= \frac{1}{2}(146°) = 73°$$

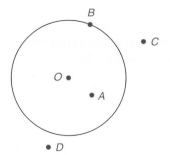

Figure 6.26

Recall that a circle separates points in the plane into three sets: points *in the interior* of the circle, points *on* the circle, and points *in the exterior* of the circle. In Figure 6.26, point A and center O are in the **interior** of ⊙O because their distances from center O are less than the length of the radius. Point B is on the circle, but points C and D are in the **exterior** of ⊙O because their distances from O are greater than the length of the radius. (See Exercise 46.) In the proof of Theorem 6.2.3, we use the fact that a tangent to a circle cannot contain an interior point of the circle.

> ### THEOREM 6.2.3
>
> The radius (or any other line through the center of a circle) drawn to a tangent at the point of tangency is perpendicular to the tangent at that point.

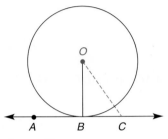

Figure 6.27

GIVEN: ⊙O with tangent $\overleftrightarrow{AB}$; point B is the point of tangency (See Figure 6.27.)

PROVE: $\overline{OB} \perp \overleftrightarrow{AB}$

PROOF: $\overleftrightarrow{AB}$ is tangent to ⊙O at point B. Let C name any point on $\overleftrightarrow{AB}$ except point B. Now $OC > OB$ because C lies in the exterior of the circle. It follows that $\overline{OB} \perp \overleftrightarrow{AB}$ because the shortest distance from a point to a line is determined by the perpendicular segment from that point to the line.

The following example illustrates an application of Theorem 6.2.3.

> ### EXAMPLE 2

© Marshall Space Flight Center/NASA

A manned NASA shuttle going to the moon has reached a position that is 5 mi above its surface. If the radius of the moon is 1080 mi, how far to the horizon can the crew members see? See Figure 6.28.

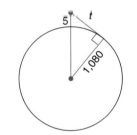

Figure 6.28

SOLUTION According to Theorem 6.2.3, the tangent determining the line of sight and the radius of the moon form a right angle. In the right triangle determined, let t represent the desired distance. Using the Pythagorean Theorem

$$1085^2 = t^2 + 1080^2$$
$$1,177,225 = t^2 + 1,166,400$$
$$t^2 = 10,825 \rightarrow t = \sqrt{10,825} \approx 104 \text{ mi}$$

A consequence of Theorem 6.2.3 is Corollary 6.2.4, which has three possible cases. In Case 1, we consider the measure of an angle formed by a tangent and the diameter drawn to the point of contact. Illustrated in Figure 6.29, only the first case is proved; proofs of the remaining two cases are left as exercises for the student in Exercises 44 and 45.

SSG EXS. 7–10

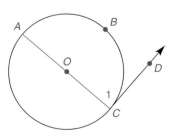

(a) Case 1
The chord is a diameter.

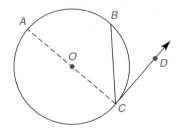

(b) Case 2
The diameter is in the
exterior of the angle.

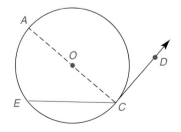

(c) Case 3
The diameter lies in the
interior of the angle.

Figure 6.29

COROLLARY 6.2.4

The measure of an angle formed by a tangent and a chord drawn to the point of tangency is one-half the measure of the intercepted arc. (See Figure 6.29 on page 281.)

GIVEN: Chord $\overrightarrow{CA}$ (which is a diameter) and tangent $\overrightarrow{CD}$ [See Figure 6.29(a).]

PROVE: $m\angle 1 = \frac{1}{2}m\widehat{ABC}$

PROOF: By Theorem 6.2.3, $\overline{AC} \perp \overrightarrow{CD}$. Then $\angle 1$ is a right angle and $m\angle 1 = 90°$. Because the intercepted arc $\widehat{ABC}$ is a semicircle, $m\widehat{ABC} = 180°$. Thus, it follows that $m\angle 1 = \frac{1}{2}m\widehat{ABC}$.

EXAMPLE 3

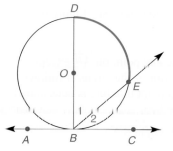

Figure 6.30

GIVEN: In Figure 6.30, $\odot O$ with diameter $\overline{DB}$, tangent $\overleftrightarrow{AC}$, and $m\widehat{DE} = 84°$

FIND: **a)** $m\angle 1$ **c)** $m\angle ABD$
       b) $m\angle 2$ **d)** $m\angle ABE$

SOLUTION

a) $\angle 1$ is an inscribed angle; $m\angle 1 = \frac{1}{2}m\widehat{DE} = \frac{1}{2}(84°) = 42°$.

b) With $m\widehat{DE} = 84°$ and $\widehat{DEB}$ a semicircle, $m\widehat{BE} = 180° - 84° = 96°$. By Corollary 6.2.4, $m\angle 2 = \frac{1}{2}m\widehat{BE} = \frac{1}{2}(96°) = 48°$.

c) Because $\overline{DB}$ is perpendicular to $\overleftrightarrow{AB}$, $m\angle ABD = 90°$.

d) $m\angle ABE = m\angle ABD + m\angle 1 = 90° + 42° = 132°$.

STRATEGY FOR PROOF ■ Proving Angle-Measure Theorems in the Circle

General Rule: With the help of an auxiliary line, Theorems 6.2.5, 6.2.6, and 6.2.7 can be proved by using Theorem 6.1.2 (measure of an inscribed angle).

Illustration: In the proof of Theorem 6.2.5, the auxiliary chord $\overline{BD}$ places $\angle 1$ in the position of an exterior angle of $\triangle BCD$.

THEOREM 6.2.5

The measure of an angle formed when two secants intersect at a point outside the circle is one-half the difference of the measures of the two intercepted arcs.

SSG EXS. 11, 12

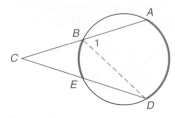

Figure 6.31

GIVEN: Secants $\overline{AC}$ and $\overline{DC}$ as shown in Figure 6.31

PROVE: $m\angle C = \frac{1}{2}(m\widehat{AD} - m\widehat{BE})$

PROOF: Draw $\overline{BD}$ to form $\triangle BCD$. Then the measure of the exterior angle of $\triangle BCD$ is given by

$$m\angle 1 = m\angle C + m\angle D, \text{ so } m\angle C = m\angle 1 - m\angle D.$$

Because $\angle 1$ and $\angle D$ are inscribed angles, $m\angle 1 = \frac{1}{2}m\widehat{AD}$ and $m\angle D = \frac{1}{2}m\widehat{BE}$. Then $m\angle C = \frac{1}{2}m\widehat{AD} - \frac{1}{2}m\widehat{BE}$ or

$$m\angle C = \frac{1}{2}(m\widehat{AD} - m\widehat{BE})$$

NOTE: In an application of Theorem 6.2.5, one subtracts the measure of the smaller arc from the measure of the larger arc.

EXAMPLE 4

GIVEN: In $\odot O$ of Figure 6.32, m$\angle AOB = 136°$ and m$\angle DOC = 46°$

FIND: m$\angle E$

SOLUTION If m$\angle AOB = 136°$, then m$\overset{\frown}{AB} = 136°$. If m$\angle DOC = 46°$, then m$\overset{\frown}{DC} = 46°$. By applying Theorem 6.2.5,

$$\text{m}\angle E = \frac{1}{2}(\text{m}\overset{\frown}{AB} - \text{m}\overset{\frown}{DC})$$

$$= \frac{1}{2}(136° - 46°)$$

$$= \frac{1}{2}(90°) = 45°$$

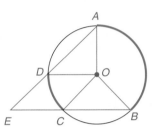

Figure 6.32

Theorems 6.2.5–6.2.7 show that any angle formed by two lines that intersect *outside* a circle has a measure equal to one-half of the difference of the measures of the two intercepted arcs. The auxiliary lines shown in Figures 6.33 and 6.34(a) will help us complete the proofs of Theorems 6.2.6 and 6.2.7.

THEOREM 6.2.6

If an angle is formed by a secant and a tangent that intersect in the exterior of a circle, then the measure of the angle is one-half the difference of the measures of its intercepted arcs.

In Figure 6.33, m$\angle L = \frac{1}{2}(\text{m}\overset{\frown}{HJ} - \text{m}\overset{\frown}{JK})$ according to Theorem 6.2.6. Again, we must subtract the measure of the smaller arc from the measure of the larger arc. A quick study of the figures that illustrate Theorems 6.2.5–6.2.7 shows that the smaller intercepted arc is "nearer" the vertex of the angle and that the larger intercepted arc is "farther from" the vertex of the angle.

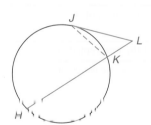

Figure 6.33

THEOREM 6.2.7

If an angle is formed by two intersecting tangents, then the measure of the angle is one-half the difference of the measures of the intercepted arcs.

In Figure 6.34(a), $\angle ABC$ intercepts the two arcs determined by points A and C. The small arc is a minor arc $(\overset{\frown}{AC})$, and the large arc is a major arc $(\overset{\frown}{ADC})$. According to Theorem 6.2.7,

$$\text{m}\angle ABC = \frac{1}{2}(\text{m}\overset{\frown}{ADC} - \text{m}\overset{\frown}{AC}).$$

(a)

Figure 6.34

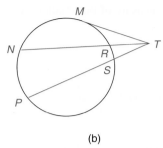

(b)

Figure 6.34

GIVEN: In Figure 6.34(b), $m\widehat{MN} = 70°$, $m\widehat{NP} = 88°$, $m\widehat{MR} = 46°$, and $m\widehat{RS} = 26°$

FIND: **a)** $m\angle MTN$
b) $m\angle NTP$
c) $m\angle MTP$

SOLUTION

a) $m\angle MTN = \frac{1}{2}(m\widehat{MN} - m\widehat{MR})$
$= \frac{1}{2}(70° - 46°)$
$= \frac{1}{2}(24°) = 12°$

b) $m\angle NTP = \frac{1}{2}(m\widehat{NP} - m\widehat{RS})$
$= \frac{1}{2}(88° - 26°)$
$= \frac{1}{2}(62°) = 31°$

c) $m\angle MTP = m\angle MTN + m\angle NTP$
Using results from (a) and (b), $m\angle MTP = 12° + 31° = 43°$

Before considering our final example, let's review the methods used to measure the different types of angles related to a circle. These are summarized in Table 6.1.

TABLE 6.1
Methods for Measuring Angles Related to a Circle

Location of the Vertex of the Angle	Rule for Measuring the Angle
Center of the circle (central angle)	The *measure* of the intercepted arc
In the *interior* of the circle (interior angle)	*One-half the sum* of the measures of the intercepted arcs
On the circle (inscribed angle)	*One-half the measure* of the intercepted arc
In the *exterior* of the circle (exterior angle)	*One-half the difference* of the measures of the two intercepted arcs

 EXS. 13–18

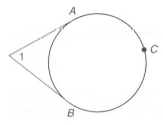

Figure 6.35

Given that $m\angle 1 = 46°$ in Figure 6.35, find the measures of $\widehat{AB}$ and $\widehat{ACB}$.

SOLUTION Let $m\widehat{AB} = x$ and $m\widehat{ACB} = y$. Now

$$m\angle 1 = \frac{1}{2}(m\widehat{ACB} - m\widehat{AB})$$

so

$$46 = \frac{1}{2}(y - x)$$

Multiplying by 2, we have $92 = y - x$.
Also, $y + x = 360$ because these two arcs form the entire circle. We add these equations as shown.

$$
\begin{aligned}
y + x &= 360 \\
y - x &= \underline{\quad 92} \\
2y &= 452 \\
y &= 226
\end{aligned}
$$

Because $x + y = 360$, we know that $x + 226 = 360$ and $x = 134$. Then $m\widehat{AB} = 134°$ and $m\widehat{ACB} = 226°$.

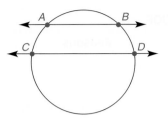

Figure 6.36

THEOREM 6.2.8

If two parallel lines intersect a circle, the intercepted arcs between these lines are congruent.

Where $\overleftrightarrow{AB} \parallel \overleftrightarrow{CD}$ in Figure 6.36, Theorem 6.2.8 states that $\overset{\frown}{AC} \cong \overset{\frown}{BD}$. Equivalently, $m\overset{\frown}{AC} = m\overset{\frown}{BD}$. The proof of Theorem 6.2.8 is left as Exercise 39.

Exercises 6.2

1. *Given:* $m\overset{\frown}{AB} = 92°$
$m\overset{\frown}{DA} = 114°$
$m\overset{\frown}{BC} = 138°$
Find: a) $m\angle 1$ ($\angle DAC$)
b) $m\angle 2$ ($\angle ADB$)
c) $m\angle 3$ ($\angle AFB$)
d) $m\angle 4$ ($\angle DEC$)
e) $m\angle 5$ ($\angle CEB$)

Exercises 1, 2

2. *Given:* $m\overset{\frown}{DC} = 30°$ and $\overset{\frown}{DABC}$ is trisected at points A and B
Find: a) $m\angle 1$ d) $m\angle 4$
b) $m\angle 2$ e) $m\angle 5$
c) $m\angle 3$

3. *Given:* Circle O with diameter $\overline{RS}$, tangent $\overleftrightarrow{SW}$, chord $\overline{TS}$, and $m\overset{\frown}{RT} = 26°$
Find: a) $m\angle WSR$
b) $m\angle RST$
c) $m\angle WST$

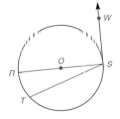

Exercises 3–5

4. Find $m\overset{\frown}{RT}$ if $m\angle RST:m\angle RSW = 1:5$.

5. Find $m\angle RST$ if $m\overset{\frown}{RT}:m\overset{\frown}{TS} = 1:4$

6. Is it possible for
a) a rectangle inscribed in a circle to have a diameter for a side? Explain.
b) a rectangle circumscribed about a circle to be a square? Explain.

7. *Given:* In $\odot Q$, $\overline{PR}$ contains Q, $\overline{MR}$ is a tangent, $m\overset{\frown}{MP} = 112°$, $m\overset{\frown}{MN} = 60°$, and $m\overset{\frown}{MT} = 46°$
Find: a) $m\angle MRP$
b) $m\angle 1$
c) $m\angle 2$

8. *Given:* $\overrightarrow{AB}$ and $\overrightarrow{AC}$ are tangent to $\odot O$, $m\overset{\frown}{BC} = 126°$
Find: a) $m\angle A$
b) $m\angle ABC$
c) $m\angle ACB$

9. *Given:* Tangents $\overrightarrow{AB}$ and $\overrightarrow{AC}$ to $\odot O$
$m\angle ACB = 68°$
Find: a) $m\overset{\frown}{BC}$
b) $m\overset{\frown}{BDC}$
c) $m\angle ABC$
d) $m\angle A$

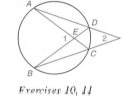

Exercises 8, 9

10. *Given:* $m\angle 1 = 72°$, $m\overset{\frown}{DC} = 34°$
Find: a) $m\overset{\frown}{AB}$
b) $m\angle 2$

11. *Given:* $m\angle 2 = 36°$
$m\overset{\frown}{AB} = 4 \cdot m\overset{\frown}{DC}$
Find: a) $m\overset{\frown}{AB}$
b) $m\angle 1$

Exercises 10, 11

(HINT: Let $m\overset{\frown}{DC} = x$ and $m\overset{\frown}{AB} = 4x$.)

In Exercises 12 and 13, R and T are points of tangency.

12. *Given:* $m\angle 3 = 42°$
Find: a) $m\overset{\frown}{RT}$
b) $m\overset{\frown}{RST}$

13. *Given:* $\overset{\frown}{RS} \cong \overset{\frown}{ST} \cong \overset{\frown}{RT}$
Find: a) $m\overset{\frown}{RT}$
b) $m\overset{\frown}{RST}$
c) $m\angle 3$

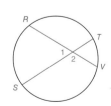

Exercises 12, 13

14. *Given:* $m\angle 1 = 63°$
$m\overset{\frown}{RS} = 3x + 6$
$m\overset{\frown}{VT} = x$
Find: $m\overset{\frown}{RS}$

15. *Given:* $m\angle 2 = 124°$
$m\overset{\frown}{TV} = x + 1$
$m\overset{\frown}{SR} = 3(x + 1)$
Find: $m\overset{\frown}{TV}$

Exercises 14, 15

16. *Given:*   m∠1 = 71°
          m∠2 = 33°
   Find: m$\widehat{CE}$ and m$\widehat{BD}$

17. *Given:*   m∠1 = 62°
          m∠2 = 26°
   Find: m$\widehat{CE}$ and m$\widehat{BD}$ *Exercises 16, 17*

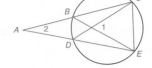

18. a) How are ∠R and ∠T related?
   b) Find m∠R if m∠T = 112°.

19. a) How are ∠S and ∠V related?
   b) Find m∠V if m∠S = 73°.

Exercises 18, 19

20. A quadrilateral *RSTV* is circumscribed about a circle so that its tangent sides are at the endpoints of two intersecting diameters.
   a) What type of quadrilateral is *RSTV*?
   b) If the diameters are also perpendicular, what type of quadrilateral is *RSTV*?

In Exercises 21 and 22, complete each proof.

21. *Given:* $\overline{AB}$ and $\overline{AC}$ are
          tangents to ⊙O
          from point A
   Prove: △ABC is isosceles

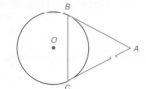

PROOF

Statements	Reasons
1. ?	1. Given
2. m∠B = $\frac{1}{2}$(m$\widehat{BC}$) and m∠C = $\frac{1}{2}$(m$\widehat{BC}$)	2. ?
3. m∠B = m∠C	3. ?
4. ∠B ≅ ∠C	4. ?
5. ?	5. If two ∠s of a △ are ≅, the sides opposite the ∠s are ≅
6. ?	6. If two sides of a △ are ≅, the △ is isosceles

---

22. *Given:* $\overline{RS} \parallel \overline{TQ}$
   Prove: $\widehat{RT} \cong \widehat{SQ}$

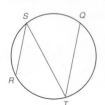

PROOF

Statements	Reasons
1. $\overline{RS} \parallel \overline{TQ}$	1. ?
2. ∠S ≅ ∠T	2. ?
3. ?	3. If two ∠s are ≅, the ∠s are = in measure
4. m∠S = $\frac{1}{2}$(m$\widehat{RT}$)	4. ?
5. m∠T = $\frac{1}{2}$(m$\widehat{SQ}$)	5. ?
6. $\frac{1}{2}$(m$\widehat{RT}$) = $\frac{1}{2}$(m$\widehat{SQ}$)	6. ?
7. m$\widehat{RT}$ = m$\widehat{SQ}$	7. Multiplication Property of Equality
8. ?	8. If two arcs of a ⊙ are = in measure, the arcs are ≅

In Exercises 23 to 25, complete a paragraph proof.

23. *Given:*   Tangent $\overline{AB}$ to ⊙O at point B
          m∠A = m∠B
   Prove: m$\widehat{BD}$ = 2 · m$\widehat{BC}$

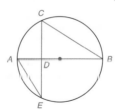

24. *Given:*   Diameter $\overline{AB} \perp \overline{CE}$ at D
   Prove: CD is the geometric mean of AD and DB

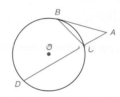

In Exercises 25 and 26, $\overline{CA}$ and $\overline{CD}$ are tangents.

25. *Given:*   m$\widehat{AB}$ = x
   Prove: m∠1 = 180° − x

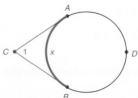

Exercises 25, 26

26. Use the result from Exercise 25 to find m∠1 if m$\widehat{AB}$ = 104°.

27. An airplane reaches an altitude of 3 mi above the earth. Assuming a clear day and that a passenger has binoculars, how far can the passenger see?

(HINT: The radius of the earth is approximately 4000 mi.)

28. From the veranda of a beachfront hotel, Manny is searching the seascape through his binoculars. A ship suddenly appears on the horizon. If Manny is 80 ft above the earth, how far is the ship out at sea?

(HINT: See Exercise 27 and note that 1 mi = 5280 ft.)

29. For the five-pointed star (a regular pentagram) inscribed in the circle, find the measures of ∠1 and ∠2.

30. For the six-pointed star (a regular hexagram) inscribed in the circle, find the measures of ∠1 and ∠2.

31. A satellite dish in the shape of a regular dodecagon (12 sides) is nearly "circular." Find:
a) $m\widehat{AB}$
b) $m\widehat{BC}$
c) $m\angle ABC$ (inscribed angle)

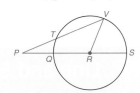

32. In the figure shown, $\triangle RST \sim \triangle WVT$ by the reason AA. Name two pairs of congruent angles in these similar triangles.

33. In the figure shown, $\triangle RXV \sim \triangle WXS$ by the reason AA. Name two pairs of congruent angles in these similar triangles.

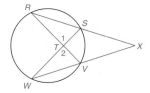

Exercises 32, 33

*34. On a fitting for a hex wrench, the distance from the center O to a vertex is 5 mm. The length of radius $\overline{OB}$ of the circle is 10 mm. If $\overline{OC} \perp \overline{DE}$ at F, how long is $\overline{FC}$?

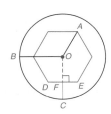

*35. *Given:* $\overline{AB}$ is a diameter of $\odot O$
M is the midpoint of chord $\overline{AC}$
N is the midpoint of chord $\overline{CB}$
$MB = \sqrt{73}$, $AN = 2\sqrt{13}$
Find: The length of diameter $\overline{AB}$

36. A surveyor sees a circular planetarium through an angle that measures 60°. If the surveyor is 45 ft from the door, what is the diameter of the planetarium?

*37. The larger circle is inscribed in a square with sides of length 4 cm. The smaller circle is tangent to the larger circle and to two sides of the square as shown. Find the length of the radius of the smaller circle.

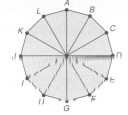

*38. In $\odot R$, $QS = 2(PT)$. Also, $m\angle P = 23°$. Find $m\angle VRS$.

In Exercises 39 to 47, provide a paragraph proof. Be sure to provide a drawing, Given, and Prove where needed.

39. If two parallel lines intersect a circle, then the intercepted arcs between these lines are congruent.

(HINT: See Figure 6.36 on page 285. Draw chord $\overline{AD}$.)

40. The line joining the centers of two circles that intersect at two points is the perpendicular bisector of the common chord.

41. If a trapezoid is inscribed in a circle, then it is an isosceles trapezoid.

42. If a parallelogram is inscribed in a circle, then it is a rectangle.

43. If one side of an inscribed triangle is a diameter, then the triangle is a right triangle.

44. Prove Case 2 of Corollary 6.2.4: The measure of an angle formed by a tangent and a chord drawn to the point of tangency is one-half the measure of the intercepted arc. (See Figure 6.29(b) on page 281.)

45. Prove Case 3 of Corollary 6.2.4.
(See Figure 6.29(c) on page 281.)

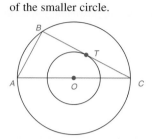

46. *Given:*   ⊙*O* with *P* in its
exterior; *O-Y-P*
Prove: *OP* > *OY*

47. *Given:*   Quadrilateral *RSTV*
inscribed in ⊙*Q*
Prove: m∠*R* + m∠*T* = m∠*V* + m∠*S*

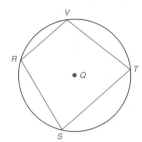

48. Given concentric circles with center *O*, △*ABC* is inscribed
in the larger circle as shown. If $\overline{BC}$ is tangent to the smaller
circle at point *T* and *AB* = 8, find the length of the radius
of the smaller circle.

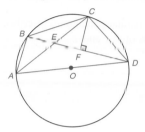

49. In the figure, quadrilateral *ABCD* is inscribed in ⊙*O*. Also,
$\overline{CF} \perp \overline{BD}$.
a) Explain why △*ABE* ~ △*DFC*.
b) Explain why ∠*DAB* and ∠*BCD* are supplementary.

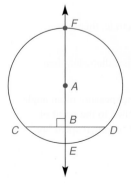

6.3	**Line and Segment Relationships in the Circle**

KEY CONCEPTS	Tangent Circles	Externally Tangent	Common External
	Internally Tangent	Circles	Tangents
	Circles	Line of Centers	Common Internal
		Common Tangent	Tangents

In this section, we consider additional line (and line segment) relationships for the circle.
Because Theorems 6.3.1–6.3.3 are so similar in wording, the student is strongly encour-
aged to make drawings and then compare the information that is given in each theorem to
the conclusion of that theorem.

> **THEOREM 6.3.1**
>
> If a line is drawn through the center of a circle perpendicular to a chord, then it bisects
> the chord and its arc.

NOTE: Note that the term *arc* generally refers to the minor arc, even though the major
arc is also bisected.

GIVEN: $\overleftrightarrow{AB} \perp$ chord $\overline{CD}$ in circle *A* (See Figure 6.37.)

PROVE: $\overline{CB} \cong \overline{BD}$ and $\overparen{CE} \cong \overparen{ED}$

Figure 6.37

The proof is left as an exercise for the student.

(HINT: Draw $\overline{AC}$ and $\overline{AD}$.)

Even though the Prove statement does not match the conclusion of Theorem 6.3.1, we know that $\overline{CD}$ is bisected by $\overleftrightarrow{AB}$ if $\overline{CB} \cong \overline{BD}$ and that $\overparen{CD}$ is bisected by $\overleftrightarrow{AE}$ if $\overparen{CE} \cong \overparen{ED}$. Similarly, $\overparen{CFD}$ is bisected because $\overparen{CF} \cong \overparen{DF}$.

THEOREM 6.3.2

If a line through the center of a circle bisects a chord other than a diameter, then it is perpendicular to the chord.

GIVEN: Circle O; $\overleftrightarrow{OM}$ is the bisector of chord $\overline{RS}$ (See Figure 6.38.)

PROVE: $\overleftrightarrow{OM} \perp \overline{RS}$

The proof of Theorem 6.3.2 is left as an exercise.

(HINT: Draw radii $\overline{OR}$ and $\overline{OS}$.)

Figure 6.39(a) illustrates the following theorem. However, Figure 6.39(b) is used in the proof.

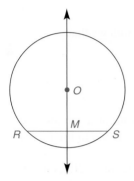

Figure 6.38

THEOREM 6.3.3

The perpendicular bisector of a chord contains the center of the circle.

(a)   (b)

Figure 6.39

GIVEN: In Figure 6.39(b), $\overleftrightarrow{QR}$ is the perpendicular bisector of chord $\overline{TV}$ in $\odot O$

PROVE: $\overleftrightarrow{QR}$ contains point O, as shown in Figure 6.39(a)

PROOF (BY INDIRECT METHOD): In Figure 6.39(b), suppose that O is not on $\overleftrightarrow{QR}$. Draw radii $\overline{OR}$, $\overline{OT}$, and $\overline{OV}$. Because $\overleftrightarrow{QR}$ is the perpendicular bisector of $\overline{TV}$, R must be the midpoint of $\overline{TV}$; then $\overline{TR} \cong \overline{RV}$. Also, $\overline{OT} \cong \overline{OV}$ (all radii of a $\odot$ are $\cong$). With $\overline{OR} \cong \overline{OR}$ by Identity, we have $\triangle ORT \cong \triangle ORV$ by SSS.

Now $\angle ORT \cong \angle ORV$ by CPCTC. It follows that $\overline{OR} \perp \overline{TV}$ because these line segments meet to form congruent adjacent angles.

Then $\overleftrightarrow{OR}$ is the perpendicular bisector of $\overline{TV}$. But $\overleftrightarrow{QR}$ is also the perpendicular bisector of $\overline{TV}$, which contradicts the uniqueness of the perpendicular bisector of a segment.

Thus, the supposition must be false, and it follows that center O is on $\overleftrightarrow{QR}$, the perpendicular bisector of chord $\overline{TV}$.

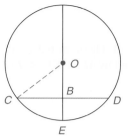

Figure 6.40

GIVEN: In Figure 6.40, $\odot O$ has a radius of length 5
$\overline{OE} \perp \overline{CD}$ at B and $OB = 3$

FIND: CD

SOLUTION Draw radius $\overline{OC}$. By the Pythagorean Theorem,

$$(OC)^2 = (OB)^2 + (BC)^2$$
$$5^2 = 3^2 + (BC)^2$$
$$25 = 9 + (BC)^2$$
$$(BC)^2 = 16$$
$$BC = 4$$

According to Theorem 6.3.1, we know that $CD = 2 \cdot BC$; then it follows that $CD = 2 \cdot 4 = 8$.

CIRCLES THAT ARE TANGENT

In this section, we assume that two circles are coplanar. Although concentric circles do not intersect, they do share a common center. For the concentric circles shown in Figure 6.41, the tangent of the smaller circle is a chord of the larger circle.

SSG **EXS. 1–4**

If two circles touch at one point, they are **tangent circles.** In Figure 6.42(a), circles P and Q are **internally tangent;** in Figure 6.42(b), circles O and R are **externally tangent.**

Figure 6.41

(a)   (b)

Figure 6.42

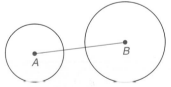

Figure 6.43

For two circles with different centers, the **line of centers** is the line (or line segment) containing the centers of both circles.

As the definition suggests, the line segment joining the centers of two circles is also commonly called the line of centers of the two circles. In Figure 6.43, $\overleftrightarrow{AB}$ or $\overline{AB}$ is the line of centers for circles A and B.

COMMON TANGENT LINES TO CIRCLES

A line, line segment, or ray that is tangent to each of two circles is a **common tangent** for these circles. If the common tangent *does not* intersect the line segment joining the centers, it is a **common external tangent.** In Figure 6.44(a), circles P and Q have one common external tangent, $\overleftrightarrow{ST}$; in Figure 6.44(b), circles A and B have two common external tangents, $\overleftrightarrow{WX}$ and $\overleftrightarrow{YZ}$.

Geometry in the Real World

Parts $\overline{AB}$ and $\overline{CD}$ of the chain belt represent common external tangents to the circular gears.

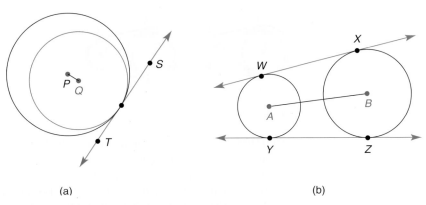

(a)   (b)

SSG   **EXS. 5–6**   Figure 6.44

If the common tangent for two circles *does* intersect the line of centers for these circles, it is a **common internal tangent** for the two circles. In Figure 6.45(a), $\overleftrightarrow{DE}$ is a common internal tangent for externally tangent circles O and R; in Figure 6.45(b), $\overleftrightarrow{AB}$ and $\overleftrightarrow{CD}$ are common internal tangents for $\odot M$ and $\odot N$.

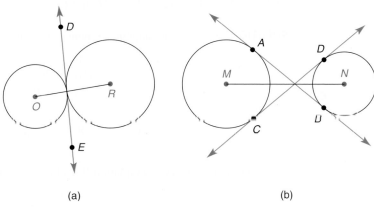

(a)   (b)

Figure 6.45

In the following theorem, a tangent segment with one endpoint outside the circle has its second endpoint at the point of tangency on the circle.

THEOREM 6.3.4

The tangent segments to a circle from an external point are congruent.

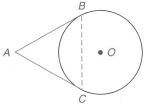

Figure 6.46

GIVEN: In Figure 6.46, $\overline{AB}$ and $\overline{AC}$ are tangents to $\odot O$ from point A

PROVE: $\overline{AB} \cong \overline{AC}$

PROOF: Draw $\overline{BC}$. Now $m\angle B = \frac{1}{2}m\widehat{BC}$ and $m\angle C = \frac{1}{2}m\widehat{BC}$. Then $\angle B \cong \angle C$ because these angles have equal measures. In turn, the sides opposite $\angle B$ and $\angle C$ of $\triangle ABC$ are congruent. That is, $\overline{AB} \cong \overline{AC}$. ∎

We apply Theorem 6.3.4 in Examples 2 and 3.

EXAMPLE 2

As shown in Figure 6.47 on page 292, a belt used in an automobile engine wraps around two pulleys with different lengths of radii. Explain why the straight pieces named $\overline{AB}$ and $\overline{CD}$ have the same length.

Figure 6.47

SOLUTION The pulley centered at O has the larger radius length, so we extend $\overline{AB}$ and $\overline{CD}$ to meet at point E. Because E is an external point to both $\odot O$ and $\odot P$, we know that $EB = ED$ and $EA = EC$ by Theorem 6.3.4. By subtracting equals from equals, $EA - EB = EC - ED$. Because $EA - EB = AB$ and $EC - ED = CD$, it follows that $AB = CD$.

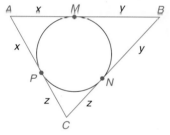

Figure 6.48

EXAMPLE 3

The circle shown in Figure 6.48 is inscribed in $\triangle ABC$; $AB = 9$, $BC = 8$, and $AC = 7$. Find the lengths AM, BM, and CN.

SOLUTION Because the tangent segments from an external point are $\cong$, we have

$$AM = AP = x$$
$$BM = BN = y$$
$$CN = CP = z$$

Now
$$x + y = 9 \quad \text{(from } AB = 9)$$
$$y + z = 8 \quad \text{(from } BC = 8)$$
$$x + z = 7 \quad \text{(from } AC = 7)$$

Subtracting the second equation from the first, we have

$$\begin{array}{r} x + y = 9 \\ y + z = 8 \\ \hline x - z = 1 \end{array}$$

Now we add this new equation to the equation, $x + z = 7$.

$$\begin{array}{r} x - z = 1 \\ x + z = 7 \\ \hline 2x = 8 \rightarrow x = 4 \rightarrow AM = 4 \end{array}$$

Because $x = 4$ and $x + y = 9$, $y = 5$. Then $BM = 5$. Because $x = 4$ and $x + z = 7$, $z = 3$, so $CN = 3$. Summarizing, $AM = 4$, $BM = 5$, and $CN = 3$.

SSG **EXS. 7–10**

LENGTHS OF LINE SEGMENTS IN A CIRCLE

To complete this section, we consider three relationships involving the lengths of chords, secants, or tangents. The first theorem is proved, but the proofs of the remaining theorems are left as exercises for the student.

STRATEGY FOR PROOF ■ Proving Segment-Length Theorems in the Circle

General Rule: With the help of auxiliary lines, Theorems 6.3.5, 6.3.6, and 6.3.7 can be proved by establishing similar triangles, followed by use of CSSTP and the Means-Extremes Property.

Illustration: In the proof of Theorem 6.3.5, the auxiliary chords drawn lead to similar triangles RTV and QSV.

Reminder

AA is the method used to prove triangles similar in this section.

THEOREM 6.3.5

If two chords intersect within a circle, then the product of the lengths of the segments (parts) of one chord is equal to the product of the lengths of the segments of the other chord.

GIVEN: Circle O with chords $\overline{RS}$ and $\overline{TQ}$ intersecting at point V (See Figure 6.49.)

PROVE: $RV \cdot VS = TV \cdot VQ$

PROOF: Draw $\overline{RT}$ and $\overline{QS}$. In $\triangle RTV$ and $\triangle QSV$, we have $\angle 1 \cong \angle 2$ (vertical $\angle$s). Also, $\angle R$ and $\angle Q$ are inscribed angles that intercept the same arc (namely $\overarc{TS}$), so $\angle R \cong \angle Q$. By AA, $\triangle RTV \sim \triangle QSV$.

Using CSSTP, we have $\frac{RV}{VQ} = \frac{TV}{VS}$ and so $RV \cdot VS = TV \cdot VQ$.

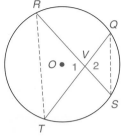

Figure 6.49

In the preceding proof, chords $\overline{RQ}$ and $\overline{TS}$ could also have been used as auxiliary lines.

Technology Exploration

Use computer software if available.

1. Draw a circle with chords $\overline{HJ}$ and $\overline{LM}$ intersecting at point P. (See Figure 6.50.)
2. Measure MP, PL, HP, and PJ.
3. Show that $MP \cdot PL = HP \cdot PJ$. (Answers are not "perfect.")

EXAMPLE 4

In Figure 6.50, $HP = 4$, $PJ = 5$, and $LP = 8$. Find PM.

SOLUTION Applying Theorem 6.3.5, we have

$$HP \cdot PJ = LP \cdot PM$$

Then
$$4 \cdot 5 = 8 \cdot PM$$
$$8 \cdot PM = 20$$
$$PM = 2.5$$

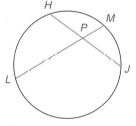

Figure 6.50

EXAMPLE 5

In Figure 6.50, $HP = 6$, $PJ = 4$, and $LM = 11$. Find LP and PM.

SOLUTION Because $LP + PM = LM$, it follows that $PM = LM - LP$. If $LM = 11$ and $LP = x$, then $PM = 11 - x$.

Now
$$HP \cdot PJ = LP \cdot PM \text{ becomes}$$
$$6 \cdot 4 = x(11 - x)$$
$$24 = 11x - x^2$$
$$x^2 - 11x + 24 = 0$$
$$(x - 3)(x - 8) = 0, \text{ so } x - 3 = 0 \text{ or } x - 8 = 0$$
$$x = 3 \quad \text{or} \quad x = 8$$

Therefore,
$$LP = 3 \quad \text{or} \quad LP = 8$$

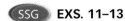 **EXS. 11–13**

If $LP = 3$, then $PM = 8$; conversely, if $LP = 8$, then $PM = 3$. That is, the segments of chord $\overline{LM}$ have lengths of 3 and 8.

In Figure 6.51 on page 294, we say that secant $\overline{AB}$ has internal segment (part) $\overline{RB}$ and external segment (part) $\overline{AR}$.

THEOREM 6.3.6

If two secant segments are drawn to a circle from an external point, then the products of the length of each secant with the length of its external segment are equal.

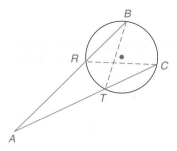

Figure 6.51

GIVEN: Secants $\overline{AB}$ and $\overline{AC}$ for the circle in Figure 6.51

PROVE: $AB \cdot RA = AC \cdot TA$

The proof is left as Exercise 46 for the student.

(HINT: First use the auxiliary lines shown to prove that $\triangle ABT \sim \triangle ACR$.)

EXAMPLE 6

GIVEN: In Figure 6.51, $AB = 14$, $BR = 5$, and $TC = 5$

FIND: AC and TA

SOLUTION Let $AC = x$. Because $TA + TC = AC$, we have $TA + 5 = x$, so $TA = x - 5$. If $AB = 14$ and $BR = 5$, then $AR = 9$. The statement

$$AB \cdot RA = AC \cdot TA \text{ becomes}$$
$$14 \cdot 9 = x(x - 5)$$
$$126 = x^2 - 5x$$
$$x^2 - 5x - 126 = 0$$
$$(x - 14)(x + 9) = 0, \text{ so } x - 14 = 0 \text{ or } x + 9 = 0$$
$$x = 14 \text{ or } x = -9 \quad (x = -9 \text{ is discarded because the length}$$
$$\text{of } \overline{AC} \text{ cannot be negative.})$$

Thus, $AC = 14$ and $TA = 9$.

THEOREM 6.3.7

If a tangent segment and a secant segment are drawn to a circle from an external point, then the square of the length of the tangent equals the product of the length of the secant with the length of its external segment.

GIVEN: Tangent $\overline{TV}$ with point of tangency V and secant $\overline{TW}$ in Figure 6.52

PROVE: $(TV)^2 = TW \cdot TX$

The proof is left as Exercise 47 for the student.

(HINT: Use the auxiliary lines shown to prove that $\triangle TVW \sim \triangle TXV$.)

Figure 6.52

EXAMPLE 7

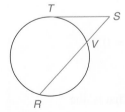

Figure 6.53

GIVEN: In Figure 6.53, $\overline{ST}$ is tangent to the circle at point T, and $\overline{SR}$ is a secant with $SV = 3$ and $VR = 9$

FIND: ST

SOLUTION If $SV = 3$ and $VR = 9$, then $SR = 12$. Using Theorem 6.3.7, we find that

$$(ST)^2 = SR \cdot SV$$
$$(ST)^2 = 12 \cdot 3$$
$$(ST)^2 = 36$$
$$ST = 6 \text{ or } -6$$

SSG **EXS. 14–17**   Because ST cannot be negative, $ST = 6$.

Exercises 6.3

1. *Given:* ⊙O with $\overline{OE} \perp \overline{CD}$
         $CD = OC$
    Find: m$\overset{\frown}{CF}$

2. *Given:* $OC = 8$ and $OE = 6$
         $\overline{OE} \perp \overline{CD}$ in ⊙O
    Find: CD

3. *Given:* $\overline{OV} \perp \overline{RS}$ in ⊙O
         $OV = 9$ and $OT = 6$
    Find: RS

Exercises 1, 2

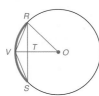

Exercises 3, 4

4. *Given:* V is the midpoint of $\overset{\frown}{RS}$ in ⊙O
         m∠S $= 15°$ and $OT = 6$
    Find: OR

5. Sketch two circles that have:
   a) No common tangents
   b) Exactly one common tangent
   c) Exactly two common tangents
   d) Exactly three common tangents
   e) Exactly four common tangents

6. Two congruent intersecting circles B and D (not shown) have a line (segment) of centers $\overline{BD}$ and a common chord $\overline{AC}$ that are congruent. Explain why quadrilateral $ABCD$ is a square.

In the figure for Exercises 7 to 16, O is the center of the circle. See Theorem 6.3.5.

7. *Given:* $AE = 6$, $EB = 4$, $DE = 8$
    Find: EC

8. *Given:* $DE = 12$, $EC = 5$, $AE = 8$
    Find: EB

9. *Given:* $AE = 8$, $EB = 6$, $DC = 16$
    Find: DE and EC

10. *Given:* $AE = 7$, $EB = 5$, $DC = 12$
    Find: DE and EC

11. *Given:* $AE = 6$, $EC = 3$, $AD = 8$
    Find: CB

12. *Given:* $AD = 10$, $BC = 4$, $AE = 7$
    Find: EC

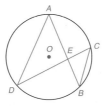

Exercises 7–16

13. *Given:* $AE = \frac{x}{2}$, $EB = 12$, $DE = \frac{x+6}{3}$, and
         $EC = 9$
    Find: x and AE

14. *Given:* $AE = \frac{x}{2}$, $EB = \frac{x}{3}$, $DE = \frac{5x}{6}$, and $EC = 6$
    Find: x and DE

15. *Given:* $AE = 9$ and $EB = 8$; $DE:EC = 2:1$
    Find: DE and EC

16. *Given:* $AE = 6$ and $EB = 4$; $DE:EC = 3:1$
    Find: DE and EC

For Exercises 17–20, see Theorem 6.3.6.

17. *Given:* $AB = 6$, $BC = 8$, $AE = 15$
    Find: DE

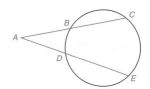

Exercises 17–20

18. *Given:* $AC = 12$, $AB = 6$, $AE = 14$
    Find: AD

19. *Given:* $AB = 4$, $BC = 5$, $AD = 3$
    Find: DE

20. *Given:* $AB = 5$, $BC = 6$, $AD = 6$
    Find: AE

In the figure for Exercises 21 to 24, $\overline{RS}$ is tangent to the circle at S. See Theorem 6.3.7.

21. *Given:* $RS = 8$ and $RV = 12$
    Find: RT

22. *Given:* $RT = 4$ and $TV = 6$
    Find: RS

23. *Given:* $\overline{RS} \cong \overline{TV}$ and $RT = 6$
    Find: RS

(HINT: Use the Quadratic Formula.)

Exercises 21–24

24. *Given:* $RT = \frac{1}{2} \cdot RS$ and $TV = 9$
    Find: RT

25. For the two circles in Figures (a), (b), and (c), find the total number of common tangents (internal and external).

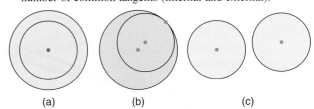

(a)   (b)   (c)

26. For the two circles in Figures (a), (b), and (c), find the total number of common tangents (internal and external).

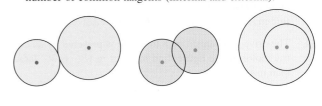

(a)   (b)   (c)

In Exercises 27 to 30, provide a paragraph proof.

27. *Given:* ⊙O and ⊙Q are tangent at point F
Secant $\overline{AC}$ to ⊙O
Secant $\overline{AE}$ to ⊙Q
Common internal tangent $\overline{AF}$
Prove: $AC \cdot AB = AE \cdot AD$

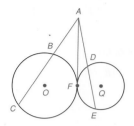

28. *Given:* ⊙O with $\overline{OM} \perp \overline{AB}$ and $\overline{ON} \perp \overline{BC}$
$\overline{OM} \cong \overline{ON}$
Prove: △ABC is isosceles

29. *Given:* Quadrilateral ABCD is circumscribed about ⊙O
Prove: $AB + CD = DA + BC$

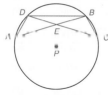

30. *Given:* $\overline{AB} \cong \overline{DC}$ in ⊙P
Prove: △ABD ≅ △CDB

31. Does it follow from Exercise 30 that △ADE is also congruent to △CBE? What can you conclude regarding $\overline{AE}$ and $\overline{CE}$ in the drawing? What can you conclude regarding $\overline{DE}$ and $\overline{EB}$?

Exercises 30, 31

32. In ⊙O (not shown), $\overline{RS}$ is a diameter and T is the midpoint of semicircle $\overset{\frown}{RTS}$. What is the value of the ratio $\frac{RT}{RS}$? The ratio $\frac{RT}{RO}$?

33. The cylindrical brush on a vacuum cleaner is powered by an electric motor. In the figure, the drive shaft is at point D. If $m\overset{\frown}{AC} = 160°$, find the measure of the angle formed by the drive belt at point D; that is, find m∠D.

Exercises 33, 34

34. The drive mechanism on a treadmill is powered by an electric motor. In the figure, find m∠D if $m\overset{\frown}{ABC}$ is 36° larger than $m\overset{\frown}{AC}$.

*35. *Given:* Tangents $\overline{AB}$, $\overline{BC}$, and $\overline{AC}$ to ⊙O at points M, N, and P, respectively
$AB = 14$, $BC = 16$, $AC = 12$
Find: AM, PC, and BN

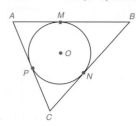

*36. *Given:* ⊙Q is inscribed in isosceles right △RST
The perimeter of △RST is $8 + 4\sqrt{2}$
Find: TM

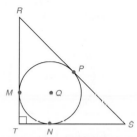

*37. *Given:* $\overline{AB}$ is an external tangent to ⊙O and ⊙Q at points A and B; radii lengths for ⊙O and ⊙Q are 4 and 9, respectively
Find: AB

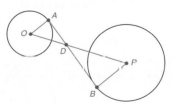

(HINT: The line of centers $\overline{OQ}$ contains point C, the point at which ⊙O and ⊙Q are tangent.)

38. The center of a circle of radius 3 inches is at a distance of 20 inches from the center of a circle of radius 9 inches. What is the exact length of common internal tangent $\overline{AB}$?

(HINT: Use similar triangles to find OD and DP. Then apply the Pythagorean Theorem twice.)

Exercises 38, 39

39. The center of a circle of radius 2 inches is at a distance of 10 inches from the center of a circle of radius length 3 inches. To the nearest tenth of an inch, what is the approximate length of a common internal tangent? Use the hint provided in Exercise 38.

40. Circles O, P, and Q are tangent (as shown) at points X, Y, and Z. Being as specific as possible, explain what type of triangle △PQO is if:
a) $OX = 2, PY = 3, QZ = 1$
b) $OX = 2, PY = 3, QZ = 2$

Exercises 40, 41

41. Circles O, P, and Q are tangent (as shown for Exercise 40 on page 296) at points X, Y, and Z. Being as specific as possible, explain what type of triangle $\triangle PQO$ is if:
a) $OX = 3$, $PY = 4$, $QZ = 1$
b) $OX = 2$, $PY = 2$, $QZ = 2$

*42. If the larger gear has 30 teeth and the smaller gear has 18, then the gear ratio (larger to smaller) is 5:3. When the larger gear rotates through an angle of 60°, through what angle measure does the smaller gear rotate?

Exercises 42, 43

*43. For the drawing in Exercise 42, suppose that the larger gear has 20 teeth and the smaller gear has 10 (the gear ratio is 2:1). If the smaller gear rotates through an angle of 90°, through what angle measure does the larger gear rotate?

In Exercises 44 to 47, prove the stated theorem.

44. If a line is drawn through the center of a circle perpendicular to a chord, then it bisects the chord and its minor arc. See Figure 6.37 on page 288.

(NOTE: The major arc is also bisected by the line.)

45. If a line through the center of a circle bisects a chord other than a diameter, then it is perpendicular to the chord. See Figure 6.38 on page 289.

46. If two secant segments are drawn to a circle from an external point, then the products of the length of each secant with the length of its external segment are equal. See Figure 6.51 on page 294.

47. If a tangent segment and a secant segment are drawn to a circle from an external point, then the square of the length of the tangent equals the product of the length of the secant with the length of its external segment. See Figure 6.52 on page 294.

48. The sides of $\angle ABC$ are tangent to $\odot D$ at A and C, respectively. Explain why quadrilateral $ABCD$ must be a kite.

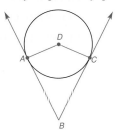

	Some Constructions and Inequalities
6.4	**for the Circle**

KEY CONCEPTS	Constructions of	Inequalities in
	Tangents to a Circle	the Circle

In Section 6.3, we proved that the radius drawn to a tangent at the point of contact is perpendicular to the tangent at that point. We now show, by using an indirect proof, that the converse of that theorem is also true. Recall that there is only one line perpendicular to a given line at a point on that line.

> **THEOREM 6.4.1**
>
> The line that is perpendicular to the radius of a circle at its endpoint on the circle is a tangent to the circle.

(a)

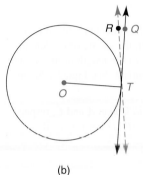

(b)

Figure 6.54

GIVEN: In Figure 6.54(a), $\odot O$ with radius $\overline{OT}$
$\overleftrightarrow{QT} \perp \overline{OT}$

PROVE: $\overleftrightarrow{QT}$ is a tangent to $\odot O$ at point T

PROOF: Suppose that $\overleftrightarrow{QT}$ is not a tangent to $\odot O$ at T. Then the tangent (call it $\overleftrightarrow{RT}$) can be drawn at T, the point of tangency. [See Figure 6.54(b).]

Now $\overline{OT}$ is the radius to tangent $\overleftrightarrow{RT}$ at T, and because a radius drawn to a tangent at the point of contact of the tangent is perpendicular to the tangent, $\overline{OT} \perp \overleftrightarrow{RT}$. But $\overline{OT} \perp \overleftrightarrow{QT}$ by hypothesis. Thus, two lines are perpendicular to $\overline{OT}$ at point T, contradicting the fact that there is only one line perpendicular to a line at a point on the line. Therefore, the supposition is false and $\overleftrightarrow{QT}$ must be the tangent to $\odot O$ at point T.

CONSTRUCTIONS OF TANGENTS TO CIRCLES

CONSTRUCTION 8 To construct a tangent to a circle at a point on the circle.

GIVEN: $\odot P$ with point X on the circle [See Figure 6.55(a).]

CONSTRUCT: A tangent $\overleftrightarrow{XW}$ to $\odot P$ at point X

PLAN: The strategy used in Construction 8 is based on Theorem 6.4.1.

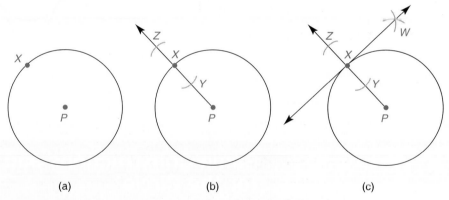

(a)   (b)   (c)

Figure 6.55

CONSTRUCTION: Figure 6.55(a): Consider $\odot P$ and point X on $\odot P$.
Figure 6.55(b): Draw radius $\overline{PX}$ and extend it to form $\overrightarrow{PX}$. Using X as the center and any radius length less than XP, draw two arcs to intersect $\overrightarrow{PX}$ at points Y and Z.

Figure 6.55(c): Complete the construction of the line perpendicular to $\overrightarrow{PX}$ at point X. From Y and Z, mark arcs with equal radii that have a length greater than XY. Calling the point of intersection W, draw $\overrightarrow{XW}$, the desired tangent to $\odot P$ at point X.

EXAMPLE 1

Make a drawing so that points A, B, C, and D are on $\odot O$ in that order. If tangents are constructed at points A, B, C, and D, what type of quadrilateral will be formed by the tangent segments if

a) $m\widehat{AB} = m\widehat{CD}$ and $m\widehat{BC} = m\widehat{AD}$?
b) all arcs $\widehat{AB}$, $\widehat{BC}$, $\widehat{CD}$, and $\widehat{DA}$ are congruent?

(a)

(b)

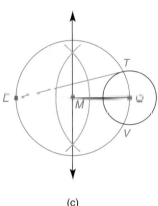

(c)

Figure 6.56

SOLUTION

a) A rhombus (all sides are congruent)
b) A square (all four ∠s are right ∠s; all sides ≅)

We now consider a more difficult construction.

> **CONSTRUCTION 9** To construct a tangent to a circle from an external point.

GIVEN: ⊙Q and external point E [See Figure 6.56(a).]

CONSTRUCT: A tangent $\overline{ET}$ to ⊙Q, with T as the point of tangency

CONSTRUCTION: Figure 6.56(a): Consider ⊙Q and external point E.
Figure 6.56(b): Draw $\overline{EQ}$. Construct the perpendicular bisector of $\overline{EQ}$ to intersect $\overline{EQ}$ at its midpoint M.

Figure 6.56(c): With M as center and MQ (or ME) as the length of radius, construct a circle. The points of intersection of circle M with circle Q are designated by T and V. Now draw $\overline{ET}$, the desired tangent.

NOTE: If drawn, $\overline{EV}$ would also be a tangent to ⊙Q and $\overline{EV} \cong \overline{ET}$.

In the preceding construction, $\overline{QT}$ (not shown) is a radius of the smaller circle Q. In the larger circle M, ∠ETQ is an inscribed angle that intercepts a semicircle. Thus, ∠ETQ is a right angle and $\overline{ET} \perp \overline{TQ}$. Because the line ($\overline{ET}$) drawn perpendicular to the radius ($\overline{TQ}$) of a circle at its endpoint on the circle is a tangent to the circle, $\overline{ET}$ is a tangent to circle Q.

INEQUALITIES IN THE CIRCLE

The remaining theorems in this section involve inequalities in the circle.

> **THEOREM 6.4.2**
>
> In a circle (or in congruent circles) containing two unequal central angles, the larger angle corresponds to the larger intercepted arc.

EXS. 1–3

GIVEN: ⊙O with central angles ∠1 and ∠2 in Figure 6.57; m∠1 > m∠2

PROVE: m$\widehat{AB}$ > m$\widehat{CD}$

PROOF: In ⊙O, m∠1 > m∠2. By the Central Angle Postulate, m∠1 = m$\widehat{AB}$ and m∠2 = m$\widehat{CD}$. By substitution, m$\widehat{AB}$ > m$\widehat{CD}$.

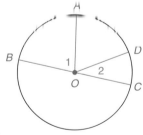

Figure 6.57

The converse of Theorem 6.4.2 follows, and it is also easily proved.

> **THEOREM 6.4.3**
>
> In a circle (or in congruent circles) containing two unequal arcs, the larger arc corresponds to the larger central angle.

GIVEN: In Figure 6.57, ⊙O with $\widehat{AB}$ and $\widehat{CD}$
m$\widehat{AB}$ > m$\widehat{CD}$

PROVE: m∠1 > m∠2

The proof is left as Exercise 35 for the student.

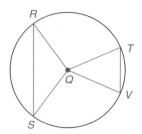

Figure 6.58

EXAMPLE 2

GIVEN: In Figure 6.58, $\odot Q$ with $m\overset{\frown}{RS} > m\overset{\frown}{TV}$.

a) Using Theorem 6.4.3, what conclusion can you draw regarding the measures of $\angle RQS$ and $\angle TQV$?

b) What does intuition suggest regarding RS and TV?

SOLUTION

a) $m\angle RQS > m\angle TQV$

b) $RS > TV$

Before we apply Theorem 6.4.4 and prove Theorem 6.4.5, consider the Discover activity at the left. The proof of Theorem 6.4.4 is not provided; however, the proof is similar to that of Theorem 6.4.5.

Discover

In Figure 6.59, *PT* measures the distance from center *P* to chord $\overline{EF}$. Likewise, *PR* measures the distance from *P* to chord $\overline{AB}$. Using a ruler, show that $PR > PT$. How do the lengths of chords $\overline{AB}$ and $\overline{EF}$ compare?

ANSWER
$AB > EF$

THEOREM 6.4.4

In a circle (or in congruent circles) containing two unequal chords, the shorter chord is at the greater distance from the center of the circle.

We apply Theorem 6.4.4 in Example 3.

EXAMPLE 3

In circle *P* of Figure 6.59, the radius length is 6 cm, and the chords have lengths $AB = 4$ cm, $DC = 6$ cm, and $EF = 10$ cm. Let $\overline{PR}$, $\overline{PS}$, and $\overline{PT}$ name perpendicular segments to these chords from center *P*.

a) Of $\overline{PR}$, $\overline{PS}$, and $\overline{PT}$, which is longest?

b) Of $\overline{PR}$, $\overline{PS}$, and $\overline{PT}$, which is shortest?

SOLUTION

a) $\overline{PR}$ is longest, because it is drawn to the shortest chord $\overline{AB}$.

b) $\overline{PT}$ is shortest, because it is drawn to the longest chord $\overline{EF}$.

In the proof of Theorem 6.4.5, the positive numbers *a* and *b* represent the lengths of line segments. If $a < b$, then $a^2 < b^2$; the converse is also true for positive numbers *a* and *b*.

THEOREM 6.4.5

In a circle (or in congruent circles) containing two unequal chords, the chord nearer the center of the circle has the greater length.

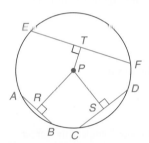

Figure 6.59

GIVEN: In Figure 6.60(a), $\odot Q$ with chords $\overline{AB}$ and $\overline{CD}$
   $\overline{QM} \perp \overline{AB}$ and $\overline{QN} \perp \overline{CD}$
   $QM < QN$

PROVE: $AB > CD$

PROOF: In Figure 6.60(b), we represent the lengths of $\overline{QM}$ and $\overline{QN}$ by *a* and *c*, respectively. Draw radii $\overline{QA}$, $\overline{QB}$, $\overline{QC}$, and $\overline{QD}$, and denote all lengths by *r*. $\overline{QM}$ is the perpendicular bisector of $\overline{AB}$, and $\overline{QN}$ is the perpendicular bisector of $\overline{CD}$, because a radius perpendicular to a chord bisects the chord and its arc. Let $MB = b$ and $NC = d$.

With right angles at *M* and *N*, we see that $\triangle QMB$ and $\triangle QNC$ are right triangles.

SSG EXS. 4–9

Figure 6.60

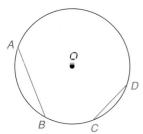

Figure 6.61

SSG **EXS. 10–16**

According to the Pythagorean Theorem, $r^2 = a^2 + b^2$ and $r^2 = c^2 + d^2$, so $b^2 = r^2 - a^2$ and $d^2 = r^2 - c^2$. If $QM < QN$, then $a < c$ and $a^2 < c^2$. Multiplication by -1 reverses the order of this inequality; therefore, $-a^2 > -c^2$. Adding r^2, we have $r^2 - a^2 > r^2 - c^2$ or $b^2 > d^2$, which implies that $b > d$. If $b > d$, then $2b > 2d$. But $AB = 2b$ and $CD = 2d$. Therefore, $AB > CD$. ∎

It is important that the phrase *minor arc* be used in Theorems 6.4.6 and 6.4.7. The proof of Theorem 6.4.6 is left to the student. In the statement for each theorem, the chord and related minor arc share common endpoints.

> **THEOREM 6.4.6**
>
> In a circle (or in congruent circles) containing two unequal chords, the longer chord corresponds to the greater minor arc.

According to Theorem 6.4.6, if $AB > CD$ in Figure 6.61, then $m\widehat{AB} > m\widehat{CD}$.

> **THEOREM 6.4.7**
>
> In a circle (or in congruent circles) containing two unequal minor arcs, the greater minor arc corresponds to the longer of the chords related to these arcs.

The proof of Theorem 6.4.7 is by the indirect method.

GIVEN: In Figure 6.61, $\odot O$ with $m\widehat{AB} > m\widehat{CD}$ and chords $\overline{AB}$ and $\overline{CD}$

PROVE: $AB > CD$

PROOF: Suppose that $AB \le CD$ in Figure 6.61. If $AB < CD$, then $m\widehat{AB} < m\widehat{CD}$ by Theorem 6.4.6; if $AB = CD$, then $m\widehat{AB} = m\widehat{CD}$ by Theorem 6.1.5. But it is given that $m\widehat{AB} > m\widehat{CD}$. Thus, the supposition must be false and it follows that $AB > CD$. ∎

NOTE: The negation of the statement $AB > CD$ is its denial, which is $AB \le CD$.

Exercises 6.4

In Exercises 1 to 8, use the figure provided.

1. If $m\widehat{CD} < m\widehat{AB}$, write an inequality that compares $m\angle CQD$ and $m\angle AQB$.

2. If $m\widehat{CD} < m\widehat{AB}$, write an inequality that compares CD and AB.

Exercises 1–8

3. If $m\widehat{CD} < m\widehat{AB}$, write an inequality that compares QM and QN.

4. If $m\widehat{CD} < m\widehat{AB}$, write an inequality that compares $m\angle A$ and $m\angle C$.

5. If $m\angle CQD < m\angle AQB$, write an inequality that compares CD to AB.

6. If $m\angle CQD < m\angle AQB$, write an inequality that compares QM to QN.

7. If $m\widehat{CD}:m\widehat{AB} = 3:2$, write an inequality that compares QM to QN.

8. If $QN:QM = 5:6$, write an inequality that compares $m\widehat{AB}$ to $m\widehat{CD}$.

9. Construct a circle O and choose some point D on the circle. Now construct the tangent to circle O at point D.

10. Construct a circle P and choose three points R, S, and T on the circle. Construct the triangle that has its sides tangent to the circle at R, S, and T.

11. X, Y, and Z are on circle O such that $m\widehat{XY} = 120°$, $m\widehat{YZ} = 130°$, and $m\widehat{XZ} = 110°$. Suppose that triangle XYZ is drawn and that the triangle ABC is constructed with its sides tangent to circle O at X, Y, and Z. Are $\triangle XYZ$ and $\triangle ABC$ similar triangles?

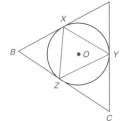

12. Construct the two tangent segments to circle P (not shown) from external point E.

13. Point V is in the exterior of circle Q (not shown) such that $\overline{VQ}$ is equal in length to the diameter of circle Q. Construct the two tangents to circle Q from point V. Then determine the measure of the angle that has vertex V and has the tangents as sides.

14. Given circle P and points R-P-T such that R and T are in the exterior of circle P, suppose that tangents are constructed from R and T to form a quadrilateral (as shown). Identify the type of quadrilateral formed
   a) when $RP > PT$.
   b) when $RP = PT$.

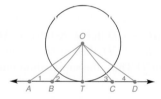

15. Given parallel chords $\overline{AB}$, $\overline{CD}$, $\overline{EF}$, and $\overline{GH}$ in circle O, which chord has the greatest length? Which has the least length? Why?

16. Given chords $\overline{MN}$, $\overline{RS}$, and $\overline{TV}$ in $\odot Q$ such that $QZ > QY > QX$, which chord has the greatest length? Which has the least length? Why?

17. Given circle O with radius $\overline{OT}$, tangent $\overleftrightarrow{AD}$, and line segments $\overline{OA}$, $\overline{OB}$, $\overline{OC}$, and $\overline{OD}$:
   a) Which line segment drawn from O has the smallest length?
   b) If $m\angle 1 = 40°$, $m\angle 2 = 50°$, $m\angle 3 = 45°$, and $m\angle 4 = 90°$, which line segment from point O has the greatest length?

18. a) If $m\widehat{RS} > m\widehat{TV}$, write an inequality that compares $m\angle 1$ with $m\angle 2$.
   b) If $m\angle 1 > m\angle 2$, write an inequality that compares $m\widehat{RS}$ with $m\widehat{TV}$.

19. a) If $MN > PQ$, write an inequality that compares the measures of minor arcs $\widehat{MN}$ and $\widehat{PQ}$.
   b) If $MN > PQ$, write an inequality that compares the measures of major arcs $\widehat{MPN}$ and $\widehat{PMQ}$.

20. a) If $m\widehat{XY} > m\widehat{YZ}$, write an inequality that compares the measures of inscribed angles 1 and 2.
   b) If $m\angle 1 < m\angle 2$, write an inequality that compares the measures of $\widehat{XY}$ and $\widehat{YZ}$.

21. Quadrilateral $ABCD$ is inscribed in circle P (not shown). If $\angle A$ is an acute angle, what type of angle is $\angle C$?

22. Quadrilateral $RSTV$ is inscribed in circle Q (not shown). If arcs $\widehat{RS}$, $\widehat{ST}$, and $\widehat{TV}$ are all congruent, what type of quadrilateral is $RSTV$?

23. In circle O, points A, B, and C are on the circle such that $m\widehat{AB} = 60°$ and $m\widehat{BC} = 40°$.
   a) How are $m\angle AOB$ and $m\angle BOC$ related?
   b) How are AB and BC related?

Exercises 23–25

24. In $\odot O$, $AB = 6$ cm and $BC = 4$ cm.
   a) How are $m\angle AOB$ and $m\angle BOC$ related?
   b) How are $m\widehat{AB}$ and $m\widehat{BC}$ related?

25. In $\odot O$, $m\angle AOB = 70°$ and $m\angle BOC = 30°$. See the figure above.
   a) How are $m\widehat{AB}$ and $m\widehat{BC}$ related?
   b) How are AB and BC related?

26. Triangle ABC is inscribed in circle O; $AB = 5$, $BC = 6$, and $AC = 7$.
   a) Which is the largest minor arc of $\odot O$: $\widehat{AB}$, $\widehat{BC}$, or $\widehat{AC}$?
   b) Which side of the triangle is nearest point O?

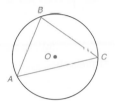

Exercises 26–29

27. Given circle O with $m\widehat{BC} = 120°$ and $m\widehat{AC} = 130°$:
   a) Which angle of triangle ABC is smallest?
   b) Which side of triangle ABC is nearest point O?

28. Given that $m\widehat{AC}:m\widehat{BC}:m\widehat{AB} = 4:3:2$ in circle O:
   a) Which arc is largest?
   b) Which chord is longest?

29. Given that $m\angle A:m\angle B:m\angle C = 2:4:3$ in circle O:
   a) Which angle is largest?
   b) Which chord is longest?
   (NOTE: See the figure for Exercise 26.)

30. Circle O has a diameter of length 20 cm. Chord $\overline{AB}$ has length 12 cm, and chord $\overline{CD}$ has length 10 cm. How much closer is $\overline{AB}$ than $\overline{CD}$ to point O?

31. Circle P has a radius of length 8 in. Points A, B, C, and D lie on circle P in such a way that $m\angle APB = 90°$ and $m\angle CPD = 60°$. How much closer to point P is chord $\overline{AB}$ than chord $\overline{CD}$?

32. A tangent $\overline{ET}$ is constructed to circle Q from external point E. Which angle and which side of triangle QTE are largest? Which angle and which side are smallest?

33. Two congruent circles, $\odot O$ and $\odot P$, do not intersect. Construct a common external tangent for $\odot O$ and $\odot P$.

34. Explain why the following statement is incorrect: "In a circle (or in congruent circles) containing two unequal chords, the longer chord corresponds to the greater major arc."

35. Prove: In a circle containing two unequal arcs, the larger arc corresponds to the larger central angle.

36. Prove: In a circle containing two unequal chords, the longer chord corresponds to the larger central angle.

(HINT: You may use any theorems stated in this section.)

*37. In ⊙O, chord $\overline{AB}$ ∥ chord $\overline{CD}$. Radius $\overline{OE}$ is perpendicular to $\overline{AB}$ and $\overline{CD}$ at points M and N, respectively. If $OE = 13$, $AB = 24$, and $CD = 10$, then the distance from O to $\overline{CD}$ is greater than the distance from O to $\overline{AB}$. Determine how much farther chord $\overline{CD}$ is from center O than chord $\overline{AB}$ is from center O; that is, find MN.

*38. In ⊙P, whose radius has length 8 in., $m\overarc{AB} = m\overarc{BC} = 60°$. Because $m\overarc{AC} = 120°$, chord $\overline{AC}$ is longer than either of the congruent chords $\overline{AB}$ and $\overline{BC}$. Determine how much longer $\overline{AC}$ is than $\overline{AB}$; that is, find the exact value and the approximate value of $AC - AB$.

39. Construct two externally tangent circles where the radius length of one circle is twice the radius length of the other circle.

PERSPECTIVE ON HISTORY

CIRCUMFERENCE OF THE EARTH

By traveling around the earth at the equator, one would traverse the circumference of the earth. Early mathematicians attempted to discover the numerical circumference of the earth. But the best approximation of the circumference was due to the work of the Greek mathematician Eratosthenes (276–194 B.C.). In his day, Eratosthenes held the highly regarded post as the head of the museum at the university in Alexandria.

What Eratosthenes did to calculate the earth's circumference was based upon several assumptions. With the sun at a great distance from the earth, its rays would be parallel as they struck the earth. Because of parallel lines, the alternate interior angles shown in the diagram would have the same measure (indicated by the Greek letter α). In Eratosthenes' plan, an angle measurement in Alexandria would be determined when the sun was *directly over the city of Syene*. While the angle suggested at the center of the earth could not be measured, the angle (in Alexandria) formed by the vertical and the related shadow could be measured; in fact, the measure was α ≈ 7.2°.

Eratosthenes' solution to the problem was based upon this fact: The ratio comparing angle measures is equivalent to the ratio comparing land distances. The distance between Syene and Alexandria was approximately 5,000 stadia

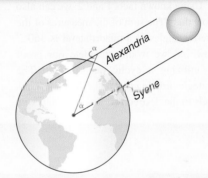

Figure 6.62

(1 stadium ≈ 516.73 ft). Where C is the circumference of the earth in stadia, this leads to the proportion

$$\frac{\alpha}{360°} = \frac{5000}{C} \quad \text{or} \quad \frac{7.2}{360} = \frac{5000}{C}$$

Solving the proportion and converting to miles, Eratosthenes' approximation of the earth's circumference was about 24,662 mi, which is about 245 mi less than the actual circumference.

Eratosthenes, a tireless student and teacher, lost his sight late in life. Unable to bear his loss of sight and lack of productivity, Eratosthenes committed suicide by refusing to eat.

PERSPECTIVE ON APPLICATIONS

SUM OF INTERIOR ANGLES OF A POLYGON

Suppose that we had studied the circle *before* studying polygons. Our methods of proof and justifications would be greatly affected. In particular, suppose that you do not know the sum of the measures of the interior angles of a triangle is 180° but you do know these facts:

1. The sum of the arc measures of a circle is 360°.
2. The measure of an inscribed angle of a circle is $\frac{1}{2}$ the measure of its intercepted arc.

Using these facts, we prove "The sum of the measures of the interior angles of a triangle is 180°."

Proof: In $\triangle ABC$, $m\angle A = \frac{1}{2}m\widehat{BC}$, $m\angle B = \frac{1}{2}m\widehat{AC}$, and $m\angle C = \frac{1}{2}m\widehat{AB}$. Then $m\angle A + m\angle B + m\angle C = \frac{1}{2}(m\widehat{BC} + m\widehat{AC} + m\widehat{AB}) = \frac{1}{2}(360°) = 180°$.

Figure 6.63

Using known facts 1 and 2, we can also show that "The sum of the measures of the interior angles of a quadrilateral is 360°." However, we would complete our proof by utilizing a cyclic quadrilateral. The strategic ordering and association of terms leads to the desired result.

Proof: For quadrilateral *HJKL* in Figure 6.64,

$m\angle H + m\angle J + m\angle K + m\angle L$
$= \frac{1}{2}m\widehat{LKJ} + \frac{1}{2}m\widehat{HLK} + \frac{1}{2}m\widehat{LHJ} + \frac{1}{2}m\widehat{HJK}$
$= \frac{1}{2}m\widehat{LKJ} + \frac{1}{2}m\widehat{LHJ} + \frac{1}{2}m\widehat{HLK} + \frac{1}{2}m\widehat{HJK}$
$= \frac{1}{2}(m\widehat{LKJ} + m\widehat{LHJ}) + \frac{1}{2}(m\widehat{HLK} + m\widehat{HJK})$
$= \frac{1}{2}(360°) + \frac{1}{2}(360°)$ or $360°$

By the Transitive Property,

$m\angle H + m\angle J + m\angle K + m\angle L = 360°$

Figure 6.64

We could continue in this manner to show that the sum of the measures of the five interior angles of a pentagon (using a cyclic pentagon) is 540° and that the sum of the measures of the *n* interior angles of a cyclic polygon of *n* sides is $(n - 2)180°$.

Summary

A Look Back at Chapter 6

One goal in this chapter has been to classify angles inside, on, and outside the circle. Formulas for finding the measures of these angles were developed. Line and line segments related to a circle were defined, and some ways of finding the measures of these segments were described. Theorems involving inequalities in a circle were proved. Some constructions that led to tangents of circles were considered.

A Look Ahead to Chapter 7

One goal of Chapter 7 is the study of loci (plural of locus), which has to do with point location. In fact, a locus of points is often nothing more than the description of some well-known geometric figure. Knowledge of locus leads to the determination of whether certain lines must be concurrent (meet at a common point). Finally, we will extend the notion of concurrence to develop further properties and terminology for regular polygons.

Key Concepts

6.1
Circle • Congruent Circles • Concentric Circles • Center • Radius • Diameter • Chord • Semicircle • Arc • Major Arc • Minor Arc • Intercepted Arc • Congruent Arcs • Central Angle • Inscribed Angle

6.2
Tangent • Point of Tangency • Secant • Polygon Inscribed in a Circle • Cyclic Polygon • Circumscribed Circle • Polygon Circumscribed about a Circle • Inscribed Circle • Interior and Exterior of a Circle

6.3
Tangent Circles • Internally Tangent Circles • Externally Tangent Circles • Line of Centers • Common Tangent • Common External Tangents • Common Internal Tangents

6.4
Constructions of Tangents to a Circle • Inequalities in the Circle

Overview ■ Chapter **6**

Selected Properties of Circles

Figure	Angle Measure	Segment Relationships
Central angle	$m\angle 1 = m\widehat{AB}$	$OA = OB$
Inscribed angle	$m\angle 2 = \frac{1}{2} m\widehat{HJ}$	Generally, $HK \neq KJ$
Angle formed by intersecting chords	$m\angle 3 = \frac{1}{2}(m\widehat{CE} + m\widehat{FD})$	$CG \cdot GD = EG \cdot GF$
Angle formed by intersecting secants	$m\angle 4 = \frac{1}{2}(m\widehat{PQ} - m\widehat{MN})$	$PL \cdot LM = QL \cdot LN$
Angle formed by intersecting tangents	$m\angle 5 = \frac{1}{2}(m\widehat{RVT} - m\widehat{RT})$	$SR = ST$
Angle formed by radius drawn to tangent	$m\angle 6 = 90°$	$\overline{OT} \perp \overline{TE}$

Chapter 6 | Review Exercises

1. The length of the radius of a circle is 15 mm. The length of a chord is 24 mm. Find the distance from the center of the circle to the chord.

2. Find the length of a chord that is 8 cm from the center of a circle that has a radius length of 17 cm.

3. Two circles intersect and have a common chord 10 in. long. The radius of one circle is 13 in. long and the centers of the circles are 16 in. apart. Find the radius of the other circle.

4. Two circles intersect and have a common chord 12 cm long. The measure of the angles formed by the common chord and a radius of each circle to the points of intersection of the circles is 45°. Find the length of the radius of each circle.

In Review Exercises 5 to 10, $\overrightarrow{BA}$ is tangent to the circle at point A in the figure shown.

5. $m\angle B = 25°$, $m\widehat{AD} = 140°$, $m\widehat{DC} = ?$

Exercises 5–10

6. $m\widehat{ADC} = 295°$, $m\widehat{AD} = 155°$, $m\angle B = ?$

7. $m\angle EAD = 70°$, $m\angle B = 30°$, $m\widehat{AC} = ?$

8. $m\angle D = 40°$, $m\widehat{DC} = 130°$, $m\angle B = ?$

9. *Given:* C is the midpoint of $\widehat{ACD}$ and $m\angle B = 40°$
   Find: $m\widehat{AD}$, $m\widehat{AC}$, $m\widehat{DC}$

10. *Given:* $m\angle B = 35°$ and $m\widehat{DC} = 70°$
    Find: $m\widehat{AD}$, $m\widehat{AC}$

11. *Given:* $\odot O$ with tangent ℓ and $m\angle 1 = 46°$
    Find: $m\angle 2$, $m\angle 3$, $m\angle 4$, $m\angle 5$

Exercises 11, 12

12. *Given:* $\odot O$ with tangent ℓ and $m\angle 5 = 40°$
    Find: $m\angle 1$, $m\angle 2$, $m\angle 3$, $m\angle 4$

13. Two circles are concentric. A chord of the larger circle is also tangent to the smaller circle. The lengths of the radii are 20 and 16, respectively. Find the length of the chord.

14. Two parallel chords of a circle each have length 16. The distance between these chords is 12. Find the length of the radius of the circle.

In Review Exercises 15 to 22, state whether the statements are always true (A), sometimes true (S), or never true (N).

15. In a circle, congruent chords are equidistant from the center.

16. If a triangle is inscribed in a circle and one of its sides is a diameter, then the triangle is an isosceles triangle.

17. If a central angle and an inscribed angle of a circle intercept the same arc, then they are congruent.

18. A trapezoid can be inscribed in a circle.

19. If a parallelogram is inscribed in a circle, then each of its diagonals must be a diameter.

20. If two chords of a circle are not congruent, then the shorter chord is nearer the center of the circle.

21. Tangents to a circle at the endpoints of a diameter are parallel.

22. Two concentric circles have at least one point in common.

23. a) $m\widehat{AB} = 80°$, $m\angle AEB = 75°$, $m\widehat{CD} = ?$
    b) $m\widehat{AC} = 62°$, $m\angle DEB = 45°$, $m\widehat{BD} = ?$
    c) $m\widehat{AB} = 88°$, $m\angle P = 24°$, $m\angle CED = ?$
    d) $m\angle CED = 41°$, $m\widehat{CD} = 20°$, $m\angle P = ?$
    e) $m\angle AEB = 65°$, $m\angle P = 25°$, $m\widehat{AB} = ?$, $m\widehat{CD} = ?$
    f) $m\angle CED = 50°$, $m\widehat{AC} + m\widehat{BD} = ?$

24. Given that $\overline{CF}$ is a tangent to the circle shown:
    a) $CF = 6$, $AC = 12$, $BC = ?$
    b) $AG = 3$, $BE = 10$, $BG = 4$, $DG = ?$
    c) $AC = 12$, $BC = 4$, $DC = 3$, $CE = ?$
    d) $AG = 8$, $GD = 5$, $BG = 10$, $GE = ?$
    e) $CF = 6$, $AB = 5$, $BC = ?$
    f) $EG = 4$, $GB = 2$, $AD = 9$, $GD = ?$
    g) $AC = 30$, $BC = 3$, $CD = ED$, $ED = ?$
    h) $AC = 9$, $BC = 5$, $ED = 12$, $CD = ?$
    i) $ED = 8$, $DC = 4$, $FC = ?$
    j) $FC = 6$, $ED = 9$, $CD = ?$

25. *Given:* $\overline{DF} \cong \overline{AC}$ in $\odot O$
    $OE = 5x + 4$
    $OB = 2x + 19$
    Find: OE

Exercises 25, 26

26. *Given:* $\overline{OE} \cong \overline{OB}$ in $\odot O$
$DF = x(x - 2)$
$AC = x + 28$
Find: DE and AC

In Review Exercises 27 to 29, give a proof for each statement.

27. *Given:* $\overrightarrow{DC}$ is tangent to circles B and A at points D and C, respectively
Prove: $AC \cdot ED = CE \cdot BD$

28. *Given:* $\odot O$ with $\overline{EO} \perp \overline{BC}$,
$\overline{DO} \perp \overline{BA}$, $\overline{EO} \cong \overline{OD}$
Prove: $\widehat{BC} \cong \widehat{BA}$

29. *Given:* $\overline{AP}$ and $\overline{BP}$ are tangent
to $\odot Q$ at A and B
C is the midpoint of $\widehat{AB}$
Prove: $\overrightarrow{PC}$ bisects $\angle APB$

30. *Given:* $\odot O$ with diameter $\overline{AC}$ and tangent $\overrightarrow{DE}$
$\widehat{mAD} = 136°$ and $\widehat{mBC} = 50°$
Find: The measures of $\angle 1$ through $\angle 10$

31. A square is inscribed in a circle with a radius of length 6 cm. Find the perimeter of the square.

32. A 30°-60°-90° triangle is inscribed in a circle with a radius of length 5 cm. Find the perimeter of the triangle.

33. A circle is inscribed in a right triangle. The length of the radius of the circle is 6 cm, and the length of the hypotenuse is 29 cm. Find the lengths of the two segments of the hypotenuse that are determined by the point of tangency.

34. *Given:* $\odot O$ is inscribed in $\triangle ABC$
$AB = 9, BC = 13, AC = 10$
Find: AD, BE, FC

35. In $\odot Q$ with $\triangle ABQ$ and $\triangle CDQ$, $\widehat{mAB} > \widehat{mCD}$. Also, $\overline{QP} \perp \overline{AB}$ and $\overline{QR} \perp \overline{CD}$.
a) How are AB and CD related?
b) How are QP and QR related?
c) How are $m\angle A$ and $m\angle C$ related?

36. In $\odot O$ (not shown), secant $\overleftrightarrow{AB}$ intersects the circle at A and B; C is a point on $\overleftrightarrow{AB}$ in the exterior of the circle.
a) Construct the tangent to $\odot O$ at point B.
b) Construct the tangents to $\odot O$ from point C.

In Review Exercises 37 and 38, use the figures shown.

37. Construct a right triangle so that one leg has length AB and the other has length twice AB.

Exercises 37, 38

38. Construct a rhombus with side $\overline{AB}$ and $\angle ABC$.

Chapter **6** | Test

1. a) If $\widehat{mAB} = 88°$, then
$\widehat{mACB} =$ _____.
b) If $\widehat{mAB} = 92°$ and C is the midpoint of major arc ACB, then $\widehat{mAC} =$ _____.

2. a) If $\widehat{mBC} = 69°$, then
$m\angle BOC =$ _____.
b) If $\widehat{mBC} = 64°$, then
$m\angle BAC =$ _____.

3. a) If $m\angle BAC = 24°$, then
$\widehat{mBC} =$ _____.
b) If $\widehat{AB} \cong \widehat{AC}$, then $\triangle ABC$ is a(n) _____ triangle.

Exercises 2, 3

4. Complete each theorem:
    a) An angle inscribed in a semicircle is a(n) _____ angle.
    b) The two tangent segments drawn to a circle from an external point are _____.

5. Given that m$\widehat{AB}$ = 106° and m$\widehat{DC}$ = 32°, find:
    a) m∠1 _____
    b) m∠2 _____

6. Given the tangents with m$\widehat{RT}$ = 146°, find:
    a) m$\widehat{RST}$ _____
    b) m∠3 _____

7. Given the tangents with m∠3 = 46°, find:
    a) m$\widehat{RST}$ _____
    b) m$\widehat{RT}$ _____

Exercises 6, 7

8. a) Because point Q is their common center, these circles are known as _____ circles.
    b) If RQ = 3 and QV = 5, find the length of chord $\overline{TV}$, a tangent to the inner circle. _____

9. In ⊙O with diameter $\overline{AC}$, OC = 5 and AB = 6. If M is the midpoint of $\overline{BC}$, find AM. _____

10. For the circles described and shown, how many common tangents do they possess?
    a) Internally tangent circles _____

    b) Circles that intersect in two points _____

11. a) If HP = 4, PJ = 5, and PM = 2, find LP. _____
    b) If HP = x + 1, PJ = x − 1, LP = 8, and PM = 3, find x. _____

12. In the figure, $\overline{TV}$ is a tangent and △TVW ~ △TXV. Find TV if TX = 3 and XW = 5. _____

13. Construct the tangent line to ⊙P at point X.

14. a) If m$\widehat{AB}$ > m$\widehat{CD}$, write an inequality that compares m∠AQB and m∠CQD. _____
    b) If OR > OP, write an inequality that compares AB and CD. _____

15. In ⊙P (not shown), the length of radius $\overline{PA}$ is 5. Also, chord $\overline{AB}$ ∥ chord $\overline{CD}$. If AB = 8 and CD = 6, find the distance between $\overline{AB}$ and $\overline{CD}$ if these chords
    a) lie on the same side of center P. _____
    b) lie on opposite sides of center P. _____

16. Provide the missing statements and reasons in the following proof.
    Given: In ⊙O, chords $\overline{AD}$ and $\overline{BC}$ intersect at E.
    Prove: $\frac{AE}{CE} = \frac{BE}{DE}$

PROOF	
Statements	**Reasons**
1. _____	1. _____
2. ∠AEB ≅ ∠DEC	2. _____
3. ∠B ≅ ∠D	3. If two inscribed angles intercept the same arc, these angles are congruent
4. △ABE ~ △CDE	4. _____
5. _____	5. CSSTP

Chapter 7

Locus and Concurrence

Gorgeous! Not only are the gardens at the Chateau de Villandry in France beautiful, but the layout of the garden also demonstrates the importance of location in this design. At the core of this chapter is the notion of *locus*, a Latin term that means "location." In the symmetry of the garden, each flower or shrub has a counterpart located on the opposite side of (and at the same distance from) the central path. The notion of locus provides the background necessary to develop properties for the concurrence of lines as well as further properties of regular polygons.

Additional video explanations of concepts, sample problems, and applications are available on DVD.

7.1 Locus of Points

KEY CONCEPTS   Locus of Points in a Plane   Locus of Points in Space

In some instances, we describe the set of points whose locations satisfy a given condition or set of conditions. The term used to describe the resulting geometric figure is *locus* (pronounced lō-kŭs), the plural of which is *loci* (pronounced lō-sī). The English word *location* is derived from the Latin word *locus*.

> **DEFINITION**
>
> A **locus** is the set of all points and only those points that satisfy a given condition (or set of conditions).

In this definition, the phrase "all points and only those points" has a dual meaning:

1. All points of the locus satisfy the given condition.
2. All points satisfying the given condition are included in the locus.

The set of points satisfying a given locus can be a well-known geometric figure such as a line or a circle. In Examples 1, 2, and 3, several points are located in a plane and then connected in order to form the locus.

Figure 7.1

EXAMPLE 1

Describe the locus of points in a plane that are at a fixed distance (*r*) from a given point (*P*).

SOLUTION See Figure 7.1. Each point shown in red is the same distance *r* from point *P*. Thus, the locus of points at fixed distance *r* from point *P* is the circle with center *P* and radius length *r*.

EXAMPLE 2

Describe the locus of points in a plane that are equidistant from two fixed points (*P* and *Q*).

SOLUTION See Figure 7.2. Each point shown in red is located the same distance from point *P* as it is from point *Q*. For instance, if $\overline{PX}$ and $\overline{QX}$ were drawn, then $PX = QX$. The locus of points in the plane that are equidistant from *P* and *Q* is the perpendicular bisector of $\overline{PQ}$.

Figure 7.2

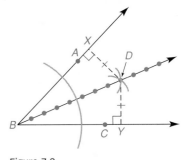

EXAMPLE 3

Describe the locus of points in a plane that are equidistant from the sides of an angle ($\angle ABC$) in that plane.

SOLUTION In Figure 7.3, each point shown in red is the same distance from $\overrightarrow{BA}$ as from $\overrightarrow{BC}$. For instance, $DX = DY$. Thus, the locus of points equidistant from the sides of $\angle ABC$ is the ray ($\overrightarrow{BD}$) that bisects $\angle ABC$.

Figure 7.3

Some definitions are given in a locus format; for example, the following is an alternative definition of the term **circle.**

Figure 7.4

DEFINITION

A **circle** is the locus of points in a plane that are at a fixed distance from a given point.

Each of the preceding examples includes the phrase "in a plane." If that phrase is omitted, the locus is found "in space." For instance, the locus of points that are at a fixed distance from a given point is actually a *sphere* (the three-dimensional object in Figure 7.4); the sphere has the fixed point as center, and the fixed distance determines the length of the radius. Thus, otherwise stated, we will consider the locus in the restricted in a plane.

EXS. 1–4

EXAMPLE 4

Describe the locus of points *in space* that are equidistant from two parallel planes (P and Q).

SOLUTION See Figure 7.5. Where plane $P \parallel$ plane Q, the points in red are the same distance from plane P as from plane Q. The locus is the plane parallel to each of the given planes and midway between them.

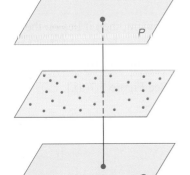

Figure 7.5

Following are two very important theorems involving the locus concept. The results of these two theorems will be used in Section 7.2. When we verify the locus theorems, we *must* establish two results:

1. If a point is in the locus, then it satisfies the condition.
2. If a point satisfies the condition, then it is a point of the locus.

THEOREM 7.1.1

The locus of points in a plane and equidistant from the sides of an angle is the angle bisector.

PROOF

(Note that *both* parts i and ii are necessary.)

  i) If a point is on the angle bisector, then it is equidistant from the sides of the angle.

(a)

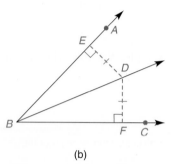

(b)

Figure 7.6

NOTE: In Figure 7.6(a), point D lies on the bisector of $\angle ABC$.

GIVEN: $\overrightarrow{BD}$ bisects $\angle ABC$
$\overline{DE} \perp \overrightarrow{BA}$ and $\overline{DF} \perp \overrightarrow{BC}$

PROVE: $\overline{DE} \cong \overline{DF}$

PROOF: In Figure 7.6(a), $\overrightarrow{BD}$ bisects $\angle ABC$; thus, $\angle ABD \cong \angle CBD$. $\overline{DE} \perp \overrightarrow{BA}$ and $\overline{DF} \perp \overrightarrow{BC}$, so $\angle DEB$ and $\angle DFB$ are $\cong$ right $\angle$s. By Identity, $\overline{BD} \cong \overline{BD}$. By AAS, $\triangle DEB \cong \triangle DFB$. Then $\overline{DE} \cong \overline{DF}$ by CPCTC.

ii) If a point is equidistant from the sides of an angle, then it is on the angle bisector.

NOTE: In Figure 7.6(b), point D is equidistant from the sides of $\angle ABC$.

GIVEN: $\angle ABC$ such that $\overline{DE} \perp \overrightarrow{BA}$ and $\overline{DF} \perp \overrightarrow{BC}$
$\overline{DE} \cong \overline{DF}$

PROVE: $\overrightarrow{BD}$ bisects $\angle ABC$; that is, D is on the bisector of $\angle ABC$

PROOF: In Figure 7.6(b), $\overline{DE} \perp \overrightarrow{BA}$ and $\overline{DF} \perp \overrightarrow{BC}$, so $\angle DEB$ and $\angle DFB$ are right angles. $\overline{DE} \cong \overline{DF}$ by hypothesis. Also, $\overline{BD} \cong \overline{BD}$ by Identity. Then $\triangle DEB \cong \triangle DFB$ by HL. With $\angle ABD \cong \angle CBD$ by CPCTC, $\overrightarrow{BD}$ bisects $\angle ABC$ by definition; of course, D lies on bisector $\overrightarrow{BD}$.

In locus problems, we must remember to demonstrate two relationships in order to validate results.

A second important theorem regarding a locus of points follows.

EXS. 5, 6

THEOREM 7.1.2

The locus of points in a plane that are equidistant from the endpoints of a line segment is the perpendicular bisector of that line segment.

PROOF

i) If a point is equidistant from the endpoints of a line segment, then it lies on the perpendicular bisector of the line segment.

NOTE: In Figure 7.7(a), point X is equidistant from the endpoints of $\overline{AB}$.

GIVEN: $\overline{AB}$ and point X not on $\overline{AB}$, so that $AX = BX$ [See Figure 7.7(a).]

PROVE: X lies on the perpendicular bisector of $\overline{AB}$

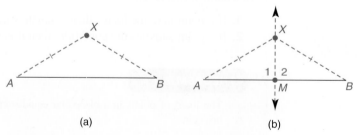

(a)   (b)

Figure 7.7

PROOF: Let M represent the midpoint of $\overline{AB}$. Then $\overline{AM} \cong \overline{MB}$. Draw $\overleftrightarrow{MX}$ as shown in Figure 7.7(b). Because $AX = BX$, we know that $\overline{AX} \cong \overline{BX}$. By Identity, $\overline{XM} \cong \overline{XM}$; thus, $\triangle AMX \cong \triangle BMX$ by SSS. By CPCTC, $\angle$s 1 and 2 are congruent so that $\overleftrightarrow{MX} \perp \overline{AB}$. By definition, $\overleftrightarrow{MX}$ is the perpendicular bisector of $\overline{AB}$, so X lies on the perpendicular bisector of $\overline{AB}$.

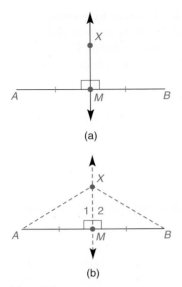

(a)

(b)

Figure 7.8

ii) If a point is on the perpendicular bisector of a line segment, then the point is equidistant from the endpoints of the line segment.

NOTE: In Figure 7.8(a), point X lies on the perpendicular bisector of $\overline{AB}$.

GIVEN: Point X lies on $\overleftrightarrow{MX}$, the perpendicular bisector of $\overline{AB}$ [See Figure 7.8(a).]

PROVE: X is equidistant from A and B ($AX = XB$) [See Figure 7.8(b).]

PROOF: X is on the perpendicular bisector of $\overline{AB}$, so $\angle$s 1 and 2 are congruent right angles and $\overline{AM} \cong \overline{MB}$. With $\overline{XM} \cong \overline{XM}$, $\triangle$s AMX and BMX are congruent by SAS; in turn, $\overline{XA} \cong \overline{XB}$ by CPCTC. Then $XA = XB$, and X is equidistant from A and B. ■

We now return to further considerations of a locus in a plane.

Suppose that a given line segment is to be used as the hypotenuse of a right triangle. How might one locate possible positions for the vertex of the right angle? One method might be to draw 30° and 60° angles at the endpoints so that the remaining angle formed must measure 90° [see Figure 7.9(a)]. This is only one possibility, but because of symmetry, it actually provides four permissible points, which are indicated in Figure 7.9(b). This problem is completed in Example 5.

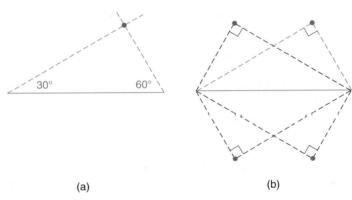

(a)

(b)

SSG **EXS. 7, 8**

Figure 7.9

EXAMPLE 5

Find the locus of the vertex of the right angle of a right triangle if the hypotenuse is AB in Figure 7.10(a).

Reminder

An angle inscribed in a semicircle is a right angle.

SOLUTION Rather than using a "hit or miss" approach for locating the possible vertices (as suggested in the paragraph preceding this example), recall that an angle inscribed in a semicircle is a right angle. Thus, we construct the circle whose center is the midpoint M of the hypotenuse and whose radius equals one-half the length of the hypotenuse.

Figure 7.10(b): First, the midpoint M of the hypotenuse $\overline{AB}$ is located.

(a)

Figure 7.10

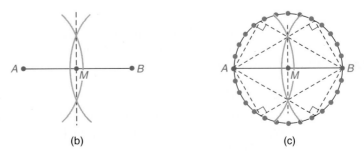

(b)

(c)

Figure 7.10(c): With the length of the radius of the circle equal to one-half the length of the hypotenuse (such as MB), the circle with center M is drawn.

The locus of the vertex of the right angle of a right triangle whose hypotenuse is given is the circle whose center is at the midpoint of the given segment and whose radius is equal in length to half the length of the given segment. Every point (except A and B) on $\odot M$ is the vertex of a right triangle with hypotenuse $\overline{AB}$; see Theorem 6.1.9.

In Example 5, the construction involves locating the midpoint M of $\overline{AB}$; the midpoint is found by constructing the perpendicular bisector of $\overline{AB}$. The compass is then opened to a radius whose length is MA or MB, and the circle is drawn.

When a construction is performed, it falls into one of two categories:

1. A basic construction method
2. A compound construction problem that may require several steps and may involve several basic construction methods

The next example falls into category 2.

Recall that the diagonals of a rhombus are perpendicular and also bisect each other. With this information, we can locate the vertices of the rhombus whose diagonals (lengths) are known.

EXAMPLE 6

Construct rhombus $ABCD$ given its diagonals $\overline{AC}$ and $\overline{BD}$. (See Figure 7.11(a).)

SOLUTION Figure 7.11(a): To begin, we construct the perpendicular bisector of $\overline{AC}$; we know that the remaining vertices B and D must lie on this line. As shown, M is the midpoint of $\overline{AC}$.

Figure 7.11(b): To locate the midpoint of $\overline{BD}$, we construct its perpendicular bisector as well. The midpoint of $\overline{BD}$ is also the midpoint of $\overline{AC}$.

Figure 7.11(c): Using an arc length equal to one-half the length of $\overline{BD}$ (such as MB), we mark off this distance both above and below $\overline{AC}$ on the perpendicular bisector determined in Figure 7.11(a).

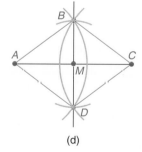

(a)   (b)   (c)   (d)

Figure 7.11

SSG EXS. 9, 10

Figure 7.11(d): Using the marked arcs to locate (determine) points B and D, we join A to B, B to C, C to D, and D to A. The completed rhombus is $ABCD$ as shown.

❯ Parabolas

A geometric figure that is conveniently defined by using the locus concept is the *parabola*. While *not* a parabola, the shape of the St. Louis Arch closely approximates its shape. Study the following definition.

DEFINITION

A **parabola** is the locus of points that are equidistant from a fixed line and a fixed point not on that line.

In this definition, the fixed point is known as the *focus* of the parabola while the fixed line is the *directrix* of the parabola. In the following description, we will indicate points that lie on the directrix by D_1, D_2, and so on. Using this *subscripted* notation, we also indicate points that are on the parabola by P_1, P_2, and so on. Subscripted notation will also be useful in the formulas of Chapters 8, 9, 10, and 11. For the moment, think of D_3 as "the third point located on directrix d."

Consider Figure 7.12(a), in which the indicated directrix d (a line) and the focus F (a point) are shown. In Figure 7.12(b), each point of the locus is equidistant from the line d (the directrix) and point F (the focus). In symbols, $D_1P_1 = P_1F$, $D_2P_2 = P_2F$, $D_3P_3 = P_3F$, and so on. Point P_1, the midpoint of the perpendicular line segment from focus F to directrix d, is the *vertex* of the parabola. Also, the line through F that is perpendicular to d is known as the *axis of symmetry* for the parabola; that is, each point on one side of the axis has a mirror image on the opposite side of the axis. Line segments that measure distances from P_1 to D_1 from P_2 to D_2 from P_3 to D_3, and so on, are always measured along a perpendicular line segment to the directrix d. The line segments that join mirror images of the parabola (such as P_2 and P_3) are perpendicular to the axis of symmetry. In Figure 7.12, the parabola is shown in black.

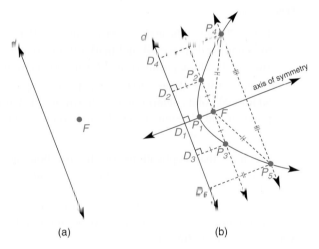

(a)   (b)

Figure 7.12

Many parabolas are drawn or constructed on a grid system (like the Cartesian system that we will study in Chapter 10). In such a system, it is fairly easy to draw lines parallel to the directrix and also easy to measure distances from the focus.

EXAMPLE 7

Sketch the parabola that has focus F and directrix d, as shown in Figure 7.13(a) on page 316.

SOLUTION

Figure 7.13(b): Draw $\overleftrightarrow{FA} \perp d$ at point A. Note that $\overleftrightarrow{FA}$ is the axis of symmetry of the parabola. Locate the midpoint of $\overline{FA}$, which is also the vertex V of the parabola.

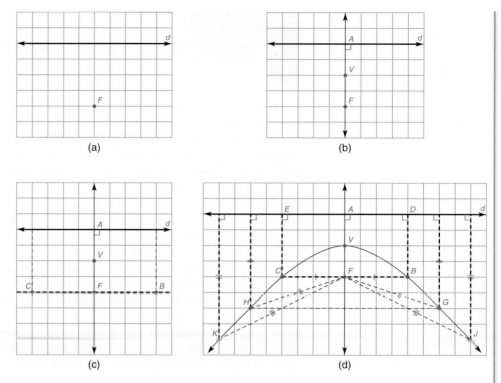

Figure 7.13a

Figure 7.13(c): Through F, draw a line parallel to d. *With a compass,* we measure the distance FA and mark off arcs on the parallel line (to d) on both sides of F. These points (B and C) are points on the parabola.

Figure 7.13(d): Where $\overleftrightarrow{BD} \perp d$ and $\overleftrightarrow{CE} \perp d$, $BF = BD$ and $CF = CE$. With additional points determined as shown, we sketch the parabola using vertex V, points B and C, points G and H, and points J and K.

We note that applications of the parabola are numerous. For one application, consider a satellite dish that reflects its incoming signals off the focus of a parabolic dish. Another application involves the automobile headlight; in this case, emitted light at the location of the focus of the parabola is reflected off the shiny parabolic surface behind the light source in such a way as to brighten and expand the illuminated path of the vehicle. See illustrations below.

Satellite dish with incoming signal

(a)

Automobile headlamp with filament at focus

(b)

Figure 7.13b

Exercises 7.1

1. In the figure, which of the points *A*, *B*, *C*, *D*, and *E* belong to "the locus of points in the plane that are at distance *r* from point *P*?"

2. In the figure, which of the points *F*, *G*, *H*, *J*, and *K* belong to "the locus of points in the plane that are at distance *r* from line *ℓ*?"

In Exercises 3 to 8, use the drawing provided.

3. *Given:* Obtuse △*ABC*
   Construct: The bisector of ∠*ABC*

Exercises 3–8

4. *Given:* Obtuse △*ABC*
   Construct: The bisector of ∠*BAC*

5. *Given:* Obtuse △*ABC*
   Construct: The perpendicular bisector of $\overline{AB}$

6. *Given:* Obtuse △*ABC*
   Construct: The perpendicular bisector of $\overline{AC}$

7. *Given:* Obtuse △*ABC*
   Construct: The altitude from *A* to $\overline{BC}$

   (HINT: Extend $\overline{BC}$.)

8. *Given:* Obtuse △*ABC*
   Construct: The altitude from *B* to $\overline{AC}$

9. *Given:* Right △*RST*
   Construct: The median from *S* to $\overline{RT}$

10. *Given:* Right △*RST*
    Construct: The median from *R* to $\overline{ST}$

Exercises 9–10

In Exercises 11 to 22, sketch and describe each locus in the plane.

11. Find the locus of points that are at a given distance from a fixed line.

12. Find the locus of points that are equidistant from two given parallel lines.

13. Find the locus of points that are at a distance of 3 in. from a fixed point *O*.

14. Find the locus of points that are equidistant from two fixed points *A* and *B*.

15. Find the locus of points that are equidistant from three noncollinear points *D*, *E*, and *F*.

16. Find the locus of the midpoints of the radii of a circle *O* that has a radius of length 8 cm.

17. Find the locus of the midpoints of all chords of circle *Q* that are parallel to diameter $\overline{PR}$.

18. Find the locus of points in the interior of a right triangle with sides of 6 in., 8 in., and 10 in. and at a distance of 1 in. from the triangle.

19. Find the locus of points that are equidistant from two given intersecting lines.

20. Given that congruent circles *O* and *P* have radii of length 4 in. and that the line of centers has length 6 in., find the locus of points that are 1 in. from each circle.

For Exercises 21 to 26, use the grid and your compass (as needed) to locate several points on the parabola having the given focus F and directrix d. Then sketch the parabola that is characterized by those points.

21.

22.

23.

24.

25.

26.

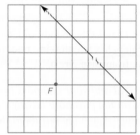

In Exercises 27 to 34, sketch and describe the locus of points in space.

27. Find the locus of points that are at a given distance from a fixed line.

28. Find the locus of points that are equidistant from two fixed points.

29. Find the locus of points that are at a distance of 2 cm from a sphere whose radius is 5 cm.

30. Find the locus of points that are at a given distance from a given plane.

31. Find the locus of points that are the midpoints of the radii of a sphere whose center is point *O* and whose radius has a length of 5 m.

*32. Find the locus of points that are equidistant from three non-collinear points *D*, *E*, and *F*.

33. In a room, find the locus of points that are equidistant from the parallel ceiling and floor, which are 8 ft apart.

34. Find the locus of points that are equidistant from all points on the surface of a sphere with center point *Q*.

In Exercises 35 and 36, use the method of proof of Theorem 7.1.1 to justify each construction method.

35. The perpendicular bisector method.

36. The construction of a perpendicular to a line from a point not on the line.

In Exercises 37 to 40, refer to the line segments shown.

37. Construct an isosceles right triangle that has hypotenuse $\overline{AB}$.

Exercises 37–40

38. Construct a rhombus whose sides are equal in length to *AB* and one diagonal of the rhombus has length *CD*.

39. Construct an isosceles triangle in which each leg has length *CD* and the altitude to the base has length *AB*.

40. Construct an equilateral triangle in which the altitude to any side has length *AB*.

41. Construct the three angle bisectors and then the inscribed circle for obtuse △*RST*.

Exercises 41, 42

42. Construct the three perpendicular bisectors of sides and then the circumscribed circle for obtuse △*RST*.

43. Use the following theorem to locate the center of the circle of which $\overset{\frown}{RT}$ is a part.
Theorem: The perpendicular bisector of a chord passes through the center of a circle.

*44. Use the following theorem to construct the geometric mean of the numerical lengths of the segments $\overline{WX}$ and $\overline{YZ}$.
Theorem: The length of the altitude to the hypotenuse of a right triangle is the geometric mean between the lengths of the segments of the hypotenuse.

45. Use the following theorem to construct a triangle similar to the given triangle but with sides that are twice the length of those of the given triangle.
Theorem: If the lengths of the three pairs of sides for two triangles are in proportion, then those triangles are similar (SSS ∼).

*46. Verify this locus theorem:
The locus of points equidistant from two fixed points is the perpendicular bisector of the line segment joining those points.

<table>
<tr><td>**7.2**</td><td>**Concurrence of Lines**</td></tr>
</table>

KEY CONCEPTS	Concurrent Lines Incenter Incircle	Circumcenter Circumcircle	Orthocenter Centroid

In this section, we consider coplanar lines that share a common point. However, any group of lines in space that have a single point in common are also known as *concurrent* lines.

> **DEFINITION**
>
> A number of lines are **concurrent** if they have exactly one point in common.

The three lines in Figure 7.14 are concurrent at point *A*. The three lines in Figure 7.15 are not concurrent even though any pair of lines (such as *r* and *s*) do intersect.

Parts of lines (rays or line segments) are concurrent if they are parts of concurrent lines and the parts share a common point.

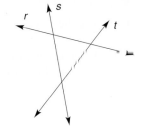

m, n, and *p are* concurrent   *r, s,* and *t* are *not* concurrent

Figure 7.14   Figure 7.15

SSG **EXS. 1, 2**

> **THEOREM 7.2.1**
>
> The three bisectors of the angles of a triangle are concurrent.

For the informal proofs of this section, no Given or Prove is stated. In more advanced courses, these parts of the proof are understood.

EXAMPLE 1

Give an informal proof of Theorem 7.2.1.

PROOF In Figure 7.16(a) on page 320, the bisectors of ∠*BAC* and ∠*ABC* intersect at point *E*. Because the bisector of ∠*BAC* is the locus of points equidistant from the sides of ∠*BAC*, we know that $\overline{EM} \cong \overline{EN}$ in Figure 7.16(b). Similarly, $\overline{EM} \cong \overline{EP}$ because *E* is on the bisector of ∠*ABC*.

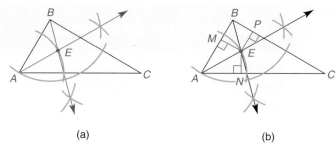

Figure 7.16

By the Transitive Property of Congruence, it follows that $\overline{EP} \cong \overline{EN}$. Because the bisector of an angle is the locus of points equidistant from the sides of the angle, E is also on the bisector of the third angle, $\angle ACB$. Thus, the three angle bisectors are concurrent at point E.

The point E at which the angle bisectors meet in Example 1 is the **incenter** of the triangle. In Example 1, we saw that $\overline{EN} \cong \overline{EM} \cong \overline{EP}$. As the following example shows, the term *incenter* is well deserved because this point is the *center* of the *in*scribed circle of the triangle.

Figure 7.17

EXAMPLE 2

Complete the construction of the inscribed circle for $\triangle ABC$ in Figure 7.16(b).

SOLUTION Having found the incenter E, we need the length of the radius. Because $\overline{EN} \perp \overline{AC}$ (as shown in Figure 7.17), the length of $\overline{EN}$ (or $\overline{EM}$ or $\overline{EP}$) is the desired radius; thus, the circle is completed.

NOTE: The sides of the triangle are tangents for the inscribed circle known as the **incircle** of the triangle. The *in*circle lies *in*side the triangle.

It is also possible to circumscribe a circle about a given triangle. The construction depends on the following theorem, the proof of which is sketched in Example 3.

THEOREM 7.2.2

 EXS. 3–7

The three perpendicular bisectors of the sides of a triangle are concurrent.

> **Reminder**
>
> A point on the perpendicular bisector of a line segment is equidistant from the endpoints of the line segment.

EXAMPLE 3

Give an informal proof of Theorem 7.2.2. See $\triangle ABC$ in Figure 7.18(a) on page 321.

PROOF Let $\overline{FS}$ and $\overline{FR}$ name the perpendicular bisectors of sides $\overline{BC}$ and $\overline{AC}$, respectively. Using Theorem 7.1.2, the point of concurrency F is equidistant from the endpoints of $\overline{BC}$; thus, $\overline{BF} \cong \overline{FC}$. In the same manner, $\overline{AF} \cong \overline{FC}$. By the Transitive Property, it follows that $\overline{AF} \cong \overline{BF}$; again citing Theorem 7.1.2, F must be on the perpendicular bisector of $\overline{AB}$ because this point is equidistant from the endpoints of $\overline{AB}$. Thus, F is the point of concurrency of the three perpendicular bisectors of the sides of $\triangle ABC$.

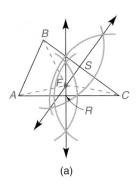

(a)　　　　　(b)

Figure 7.18

In Figure 7.18, the point at which the perpendicular bisectors of the sides of a triangle meet is the **circumcenter** F of the triangle. In Example 3, we see that $\overline{AF} \cong \overline{BF} \cong \overline{CF}$. The term *circumcenter* is easily remembered as the *center* of the *circum*scribed circle.

EXAMPLE 4

Complete the construction of the circumscribed circle for $\triangle ABC$ that was given in Figure 7.18(a).

SOLUTION We have already identified the center of the circle as point F. To complete the construction, we use F as the center and a radius of length equal to the distance from F to any one of the vertices A, B, or C. The circumscribed circle is shown above in Figure 7.18(b).

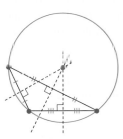

Figure 7.19

NOTE: The sides of the inscribed triangle are chords of the circumscribed circle, which is called the **circumcircle** of the triangle. The circumcircle of a polygon lies outside the polygon except where it contains the vertices of the polygon.

The incenter and the circumcenter of a triangle are generally distinct points. However, it is possible for the two centers to coincide in a special type of triangle. Although the incenter of a triangle always lies in the interior of the triangle, the circumcenter of an obtuse triangle will lie in the exterior of the triangle. See Figure 7.19.

To complete the discussion of concurrence, we include a theorem involving the altitudes of a triangle and a theorem involving the medians of a triangle.

SSG FXS. 8–12

THEOREM 7.2.3

The three altitudes of a triangle are concurrent.

The point of concurrence for the three altitudes of a triangle is the **orthocenter** of the triangle. In Figure 7.20(a), point N is the orthocenter of $\triangle DEF$. For the obtuse triangle in Figure 7.20(b), we see that orthocenter X lies in the exterior of $\triangle RST$.

Rather than proving Theorem 7.2.3, we sketch a part of that proof. In Figure 7.21(a) on page 322, $\triangle MNP$ is shown with its altitudes. To prove that the altitudes are concurrent requires

1. that we draw auxiliary lines through N parallel to $\overline{MP}$, through M parallel to $\overline{NP}$, and through P parallel to $\overline{NM}$. [See Figure 7.21(b).]
2. that we show that the altitudes of $\triangle MNP$ are perpendicular bisectors of the sides of the newly formed $\triangle RST$; thus, altitudes $\overline{PX}$, $\overline{MY}$, and $\overline{NZ}$ are concurrent (a consequence of Theorem 7.2.2).

(a)

(b)

Figure 7.20

Figure 7.21

(a)

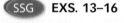

(b)

Figure 7.22

SKETCH OF PROOF THAT $\overline{PX}$ IS THE $\perp$ BISECTOR OF $\overline{RS}$:

Because $\overline{PX}$ is an altitude of $\triangle MNP$, $\overline{PX} \perp \overline{MN}$. But $\overline{RS} \parallel \overline{MN}$ by construction. Because a line perpendicular to one of two parallel lines must be perpendicular to the other, we have $\overline{PX} \perp \overline{RS}$. Now we need to show that $\overline{PX}$ bisects $\overline{RS}$. By construction, $\overline{MR} \parallel \overline{NP}$ and $\overline{RP} \parallel \overline{MN}$, so $MRPN$ is a parallelogram. Then $\overline{MN} \cong \overline{RP}$ because the opposite sides of a parallelogram are congruent. By construction, $MPSN$ is also a parallelogram and $\overline{MN} \cong \overline{PS}$. By the Transitive Property of Congruence, $\overline{RP} \cong \overline{PS}$. Thus, $\overline{RS}$ is bisected at point P, and $\overline{PX}$ is the $\perp$ bisector of $\overline{RS}$.

In Figure 7.21(b), similar arguments (leading to one long proof) could be used to show that $\overline{NZ}$ is the $\perp$ bisector of $\overline{TS}$ and also that $\overline{MY}$ is the $\perp$ bisector of $\overline{TR}$. Because the concurrent perpendicular bisectors of the sides of $\triangle RST$ are also the altitudes of $\triangle MNP$, these altitudes must be concurrent. ∎

The intersection of any two altitudes determines the orthocenter of a triangle. We use this fact in Example 5. If the third altitude were constructed, it would contain the same point of intersection (the orthocenter).

EXAMPLE 5

Construct the orthocenter of $\triangle ABC$ in Figure 7.22(a).

SOLUTION First construct the altitude from A to $\overline{BC}$; here, we draw an arc from A to intersect $\overline{BC}$ at X and Y. Now draw equal arcs from X and Y to intersect at Z. $\overline{AH}$ is the desired altitude. Repeat the process to construct altitude $\overline{CJ}$ from vertex C to side $\overline{AB}$. In Figure 7.22(b), the point of intersection O is the orthocenter of $\triangle ABC$.

Recall that a median of a triangle joins a vertex to the midpoint of the opposite side of the triangle. Through construction, we can show that the three medians of a triangle are concurrent. We will discuss the proof of the following theorem in Chapter 10.

SSG **EXS. 13–16**

THEOREM 7.2.4

The three medians of a triangle are concurrent at a point that is two-thirds the distance from any vertex to the midpoint of the opposite side.

The point of concurrence C for the three medians is the **centroid** of the triangle in Figure 7.23. M, N, and P are midpoints of the sides of $\triangle RST$; thus, $\overline{MR}$, $\overline{SN}$, and $\overline{TP}$ are medians of the triangle. According to Theorem 7.2.4, $RC = \frac{2}{3}(RM)$, $SC = \frac{2}{3}(SN)$, and $TC = \frac{2}{3}(TP)$.

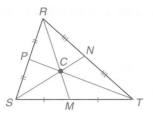

Figure 7.23

EXAMPLE 6

Suppose that the medians of $\triangle RST$ in Figure 7.23 have the lengths $RM = 12$, $SN = 15$, and $TP = 18$. The centroid of $\triangle RST$ is point C. Find the length of:

    a) RC **b)** CM **c)** SC

SOLUTION

    a) $RC = \frac{2}{3}(RM)$, so $RC = \frac{2}{3}(12) = 8$.

    b) $CM = RM - RC$, so $CM = 12 - 8 = 4$.

    c) $SC = \frac{2}{3}(SN)$, so $SC = \frac{2}{3}(15) = 10$.

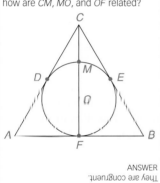
EXAMPLE 7

GIVEN: In Figure 7.24(a), isosceles $\triangle RST$ with $RS = RT = 15$, and $ST = 18$; medians $\overline{RZ}$, $\overline{TX}$, and $\overline{SY}$ meet at centroid Q.

FIND: RQ and QZ

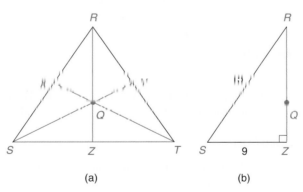

        (a)   (b)

Figure 7.24

SOLUTION Median $\overline{RZ}$ separates $\triangle RST$ into two congruent right triangles, $\triangle RZS$ and $\triangle RZT$; this follows from SSS. With Z the midpoint of $\overline{ST}$, $SZ = 9$. Using the Pythagorean Theorem with $\triangle RZS$ in Figure 7.24(b), we have

$$(RS)^2 = (RZ)^2 + (SZ)^2$$
$$15^2 = (RZ)^2 + 9^2$$
$$225 = (RZ)^2 + 81$$
$$(RZ)^2 = 144$$
$$RZ = 12$$

By Theorem 7.2.4, $RQ = \frac{2}{3}(RZ) = \frac{2}{3}(12) = 8$. Because $QZ = \frac{1}{2}(RQ)$, it follows that $QZ = 4$.

SSG **EXS. 17–22**

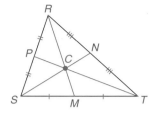

Based upon the figure at the left, the following table shows several interpretations of Theorem 7.2.4

TABLE 7.1 Centroid *C* in △*RST*

$RC = \frac{2}{3}(RM)$	$CM = \frac{1}{3}(RM)$	$RC = 2(CM)$	$CM = \frac{1}{2}(RC)$
$SC = \frac{2}{3}(SN)$	$CN = \frac{1}{3}(SN)$	$SC = 2(CN)$	$CN = \frac{1}{2}(SC)$
$TC = \frac{2}{3}(TP)$	$CP = \frac{1}{3}(TP)$	$TC = 2(CP)$	$CP = \frac{1}{2}(TC)$

The centroid of a triangular region is sometimes called its *center of mass* or *center of gravity*. This is because the region of uniform thickness "balances" upon the point known as its centroid. Consider the Discover activity at left.

It is *possible* for the angle bisectors of certain quadrilaterals to be concurrent. Likewise, the perpendicular bisectors of the sides of a quadrilateral *can* be concurrent. Of course, there are four angle bisectors and four perpendicular bisectors of sides to consider. In Example 8, we explore this situation.

Discover

Take a piece of cardboard or heavy poster paper. Draw a triangle on the paper and cut out the triangular shape. Now use a ruler to mark the midpoints of each side and draw the medians to locate the centroid. Place the triangle on the point of a pen or pencil at the centroid and see how well you can balance the triangular region.

EXAMPLE 8

Use intuition and Figure 7.25 to decide which of the following are concurrent.

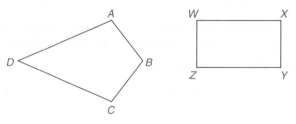

Figure 7.25

a) The angle bisectors of a kite

b) The perpendicular bisectors of the sides of a kite

c) The angle bisectors of a rectangle

d) The perpendicular bisectors of the sides of a rectangle

SOLUTION

a) The angle bisectors of the kite are concurrent at a point (the incenter of the kite).

b) The ⊥ bisectors of the sides of the kite are not concurrent (unless ∠*A* and ∠*C* are both right angles).

c) The angle bisectors of the rectangle are not concurrent (unless the rectangle is a square).

d) The ⊥ bisectors of the sides of the rectangle are concurrent (the circumcenter of the rectangle is also the point of intersection of the diagonals).

NOTE: The student should make drawings to verify the results in Example 8.

Exercises 7.2

1. In the figure, are lines *m, n,* and *p* concurrent?

2. If one exists, name the point of concurrence for lines *m, n,* and *p.*

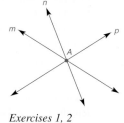

Exercises 1, 2

3. What is the general name of the point of concurrence for the three angle bisectors of a triangle?

4. What is the general name of the point of concurrence for the three altitudes of a triangle?

5. What is the general name of the point of concurrence for the three perpendicular bisectors of the sides of a triangle?

6. What is the general name of the point of concurrence for the three medians of a triangle?

7. Which lines or line segments or rays must be drawn or constructed in a triangle to locate its
a) incenter?
b) circumcenter?
c) orthocenter?
d) centroid?

8. Is it really necessary to construct all three bisectors of the angles of a triangle to locate its incenter?

9. Is it really necessary to construct all three perpendicular bisectors of the sides of a triangle to locate its circumcenter?

10. a) To locate the orthocenter, is it necessary to construct all three altitudes of a right triangle?
b) What point is the orthocenter of any right triangle?

11. To locate the centroid of a triangle, is it necessary to construct all three medians?

12. For what type of triangle are the angle bisectors, the medians, the perpendicular bisectors of sides, and the altitudes all the same?

13. What point on a right triangle is the circumcenter of the right triangle?

14. Must the centroid of an isosceles triangle lie on the altitude to the base?

15. Draw a triangle and, by construction, find its incenter.

16. Draw an acute triangle and, by construction, find its circumcenter.

17. Draw an obtuse triangle and, by construction, find its circumcenter.

18. Draw an acute triangle and, by construction, find its orthocenter.

19. Draw an obtuse triangle and, by construction, find its orthocenter.

(HINT: You will have to extend the sides opposite the acute angles.)

20. Draw an acute triangle and, by construction, find the centroid of the triangle.

(HINT: Begin by constructing the perpendicular bisectors of the sides.)

21. Draw an obtuse triangle and, by construction, find the centroid of the triangle.

(HINT: Begin by constructing the perpendicular bisectors of the sides.)

22. Is the incenter always located in the interior of the triangle?

23. Is the circumcenter always located in the interior of the triangle?

24. Find the length of the radius of the inscribed circle for a right triangle whose legs measure 6 and 8.

25. Find the distance from the circumcenter to each vertex of an equilateral triangle whose sides have the length 10.

26. A triangle has angles measuring $30°$, $30°$, and $120°$. If the congruent sides measure 6 units each, find the length of the radius of the circumscribed circle.

27. *Given:* Isosceles $\triangle RST$
$RS = RT = 17$ and $ST = 16$
Medians $\overline{RZ}, \overline{TX},$ and $\overline{SY}$ meet at centroid Q
Find: RQ and SQ

28. *Given:* Isosceles $\triangle RST$
$RS = RT = 10$
and $ST = 16$
Medians $\overline{RZ}, \overline{TX},$
and $\overline{SY}$ meet at Q
Find: RQ and QT

Exercises 27, 28

29. In $\triangle MNP$, medians $\overline{MB}, \overline{NA},$ and $\overline{PC}$ intersect at centroid Q.
a) If $MQ = 8$, find QB.
b) If $QC = 3$, find PQ.
c) If $AQ = 3.5$, find AN.

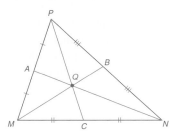

Exercises 29, 30

30. In △*MNP* for Exercise 29, medians $\overline{MB}$, $\overline{NA}$, and $\overline{PC}$ intersect at centroid *Q*.
    a) Find *QB* if *MQ* = 8.2.
    b) Find *PQ* if $QC = \frac{7}{2}$.
    c) Find *AN* if *AQ* = 4.6.

31. Draw a triangle. Construct its inscribed circle.

32. Draw a triangle. Construct its circumscribed circle.

33. For what type of triangle will the incenter and the circumcenter be the same?

34. Does a rectangle have (a) an incenter? (b) a circumcenter?

35. Does a square have (a) an incenter? (b) a circumcenter?

36. Does a regular pentagon have (a) an incenter? (b) a circumcenter?

37. Does a rhombus have (a) an incenter? (b) a circumcenter?

38. Does an isosceles trapezoid have (a) an incenter? (b) a circumcenter?

39. A distributing company plans an Illinois location that would be the same distance from each of its principal delivery sites at Chicago, St. Louis, and Indianapolis. Use a construction method to locate the approximate position of the distributing company.

   (Note: Trace the outline of the two states on your own paper.)

40. There are plans to locate a disaster response agency in an area that is prone to tornadic activity. The agency is to be located at equal distances from Wichita, Tulsa, and Oklahoma City. Use a construction method to locate the approximate position of the agency.

   (Note: Trace the outline of the two states on your own paper.)

*41. A circle is inscribed in an isosceles triangle with legs of length 10 in. and a base of length 12 in. Find the length of the radius for the circle.

For Exercises 42 to 44, ⊙O is the inscribed circle for △ABC. In Exercises 43 and 44, use the result from Exercise 42.

42. Explain why △*AOF* ≅ △*AOE*.

*43. Explain why ∠*FOA* and ∠*BOC* are supplementary.

*44. Where $s = \frac{1}{2}(a + b + c)$, show that *AF* = *s* − *a*.

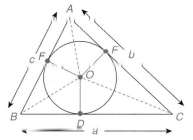

7.3 More About Regular Polygons

KEY CONCEPTS	Regular Polygon	Center and Central Angle of a Regular Polygon	Radius and Apothem of a Regular Polygon

Several interesting properties of regular polygons are developed in this section. For instance, every regular polygon has both an inscribed circle and a circumscribed circle; furthermore, these two circles are concentric. In Example 1, we use bisectors of the angles of a square to locate the center of the inscribed circle. The center, which is found by using the bisectors of any two *consecutive* angles, is equidistant from the sides of the square.

EXAMPLE 1

Given square *ABCD* in Figure 7.26(a), construct inscribed ⊙*O*.

Reminder

A regular polygon is both equilateral and equiangular.

(a)

(b)

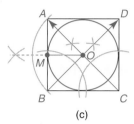
(c)

Figure 7.26

SOLUTION Figure 7.26(b): The center of an inscribed circle must lie at the same distance from each side. Center *O* is the point of concurrency of the angle bisectors of the square. Thus, we construct the angle bisectors of ∠*B* and ∠*C* to identify point *O*.
Figure 7.26(c): Constructing $\overline{OM} \perp \overline{AB}$, *OM* is the distance from *O* to $\overline{AB}$ and the length of the radius of the inscribed circle. Finally we construct inscribed ⊙*O* with radius $\overline{OM}$ as shown.

In Example 2, we use the perpendicular bisectors of two consecutive sides of a regular hexagon to locate the center of the circumscribed circle. The center determines a point that is equidistant from the vertices of the hexagon.

EXAMPLE 2

Given regular hexagon *MNPQRS* in Figure 7.27(a), construct circumscribed ⊙*X*.

(a)

(b)

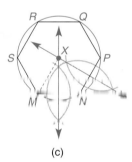
(c)

Figure 7.27

SOLUTION Figure 7.27(b): The center of a circumscribed circle must lie at the same distance from each vertex of the hexagon. In Figure 7.27(b), we construct the perpendicular bisectors of $\overline{MN}$ and $\overline{NP}$ to locate point *X*. Center *X* is the point of concurrency of the perpendicular bisectors of two consecutive sides of the hexagon.
Figure 7.27(c): Where *XM* is the distance from *X* to vertex *M*, we use radius $\overline{XM}$ to construct circumscribed ⊙*X*.

Figure 7.28

Figure 7.29

For a rectangle, which is not a regular polygon, we can only circumscribe a circle (see Figure 7.28). For a rhombus (also not a regular polygon), we can only inscribe a circle (see Figure 7.29).
As we shall see, we can construct both inscribed and circumscribed circles for regular polygons because they are both equilateral and equiangular. A few of the regular polygons are shown in Figure 7.30 on the next page.

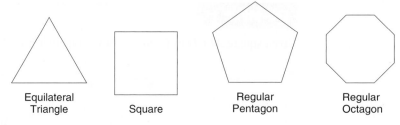

Equilateral Triangle   Square   Regular Pentagon   Regular Octagon

SSG   EXS. 1–6   Figure 7.30

In the following table, we recall these facts from Chapter 2.

TABLE 7.2

Regular Polygons (n sides)

	Interior Angles	**Exterior Angles**
Sum	$(n - 2) \cdot 180°$	$360°$
Each Angle	$\dfrac{(n - 2) \cdot 180°}{n}$	$\dfrac{360°}{n}$

The number of diagonals is $D = \dfrac{n(n - 3)}{2}$.

EXAMPLE 3

a) Find the measure of each interior angle of a regular polygon with 15 sides.
b) Find the number of sides of a regular polygon if each interior angle measures 144°.
c) Find the number of diagonals for the polygon in part(a).

SOLUTION

a) Because all of the n angles have equal measures, the formula for the measure of each interior angle,

$$I = \frac{(n - 2)180}{n}$$

becomes

$$I = \frac{(15 - 2)180}{15}$$

which simplifies to 156°.

b) Because $I = 144°$, we can determine the number of sides by solving the equation

$$\frac{(n - 2)180}{n} = 144$$

Then

$$(n - 2)180 = 144n$$
$$180n - 360 = 144n$$
$$36n = 360$$
$$n = 10$$

c) With $n = 15$, $D = \dfrac{n(n - 3)}{2}$ becomes

$$D = \frac{15(15 - 3)}{2} = \frac{15 \cdot 12}{2} = \frac{180}{2} = 90$$

The polygon has 90 diagonals.

NOTE: In Example 3(a), we could have found the measure of each exterior angle and then used the fact that the interior angle is its supplement. With $n = 15$, $E = \frac{360^\circ}{n}$ leads to $E = 24^\circ$. It follows that $I = 180^\circ - 24^\circ$ or 156°. In Example 3(b), the fact that $I = 144^\circ$ leads to $E = 36^\circ$. In turn, $E = \frac{360^\circ}{n}$ becomes $36^\circ = \frac{360^\circ}{n}$, which leads to $n = 10$.

SSG **EXS. 7, 8**

Regular polygons allow us to inscribe and to circumscribe a circle. The proof of the following theorem will establish the following relationships:

1. The centers of the inscribed and circumscribed circles of a regular polygon are the same.
2. The angle bisectors of two consecutive angles or the perpendicular bisectors of two consecutive sides can be used to locate the common center of the inscribed circle and the circumscribed circle.
3. The inscribed circle's radius is any line segment from the center drawn perpendicular to a side of the regular polygon; also, the radius of the circumscribed circle joins the center to any vertex of the regular polygon.

THEOREM 7.3.1

A circle can be circumscribed about (or inscribed in) any regular polygon.

GIVEN: Regular polygon *ABCDEF* [See Figure 7.31(a).]

PROVE: A circle *O* can be circumscribed about *ABCDEF* and a circle with center *O* can be inscribed in *ABCDEF*

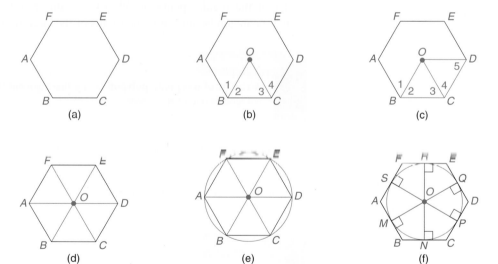

Figure 7.31

PROOF: By the definition of a regular polygon, $\angle ABC \cong \angle BCD$. Let point *O* be the point at which the angle bisectors for $\angle ABC$ and $\angle BCD$ meet. [See Figure 7.31(b).] Applying Theorem 1.7.8, $\angle 1 \cong \angle 2 \cong \angle 3 \cong \angle 4$.

With $\angle 2 \cong \angle 3$, it follows that $\overline{OB} \cong \overline{OC}$ (sides opposite $\cong \angle$s of a $\triangle$ are also $\cong$).

From the facts that $\angle 3 \cong \angle 4$, $\overline{OC} \cong \overline{OC}$, and $\overline{BC} \cong \overline{CD}$, it follows that $\triangle OCB \cong \triangle OCD$ by SAS. [See Figure 7.31(c).] In turn, $\overline{OC} \cong \overline{OD}$ by CPCTC, so $\angle 4 \cong \angle 5$ because these lie opposite $\overline{OC}$ and $\overline{OD}$. Because $\angle 5 \cong \angle 4$ and $m\angle 4 = \frac{1}{2}m\angle BCD$, it follows that $m\angle 5 = \frac{1}{2}m\angle BCD$. But $\angle BCD \cong \angle CDE$ because these are angles of a regular polygon. Thus, $m\angle 5 = \frac{1}{2}m\angle CDE$, and $\overline{OD}$ bisects $\angle CDE$.

By continuing this procedure, we can show that $\overline{OE}$ bisects $\angle DEF$, $\overline{OF}$ bisects $\angle EFA$, and $\overline{OA}$ bisects $\angle FAB$. The resulting triangles, $\triangle AOB$, $\triangle BOC$, $\triangle COD$, $\triangle DOE$, $\triangle EOF$, and $\triangle FOA$, are congruent by ASA. [See Figure 7.31(d).] By CPCTC, $\overline{OA} \cong \overline{OB} \cong \overline{OC} \cong \overline{OD} \cong \overline{OE} \cong \overline{OF}$. With O as center and $\overline{OA}$ as radius, circle O can be circumscribed about $ABCDEF$, as shown in Figure 7.31(e).

Because $\triangle AOB$, $\triangle BOC$, $\triangle COD$, $\triangle DOE$, $\triangle EOF$, and $\triangle FOA$ are congruent, we see that $\overline{OM} \cong \overline{ON} \cong \overline{OP} \cong \overline{OQ} \cong \overline{OR} \cong \overline{OS}$ because these are the corresponding altitudes to the bases of the congruent triangles.

Again with O as center, but now with a radius equal in length to OM, we complete the inscribed circle in $ABCDEF$. [See Figure 7.31(f).]

In the proof of Theorem 7.3.1, a regular hexagon was drawn. The method of proof would not change, regardless of the number of sides of the polygon chosen. In the proof, point O was the common center of the circumscribed and inscribed circles for $ABCDEF$. Because any regular polygon can be inscribed in a circle, any regular polygon is cyclic.

DEFINITION

The **center of a regular polygon** is the common center for the inscribed and circumscribed circles of the polygon.

The preceding definition does not tell us how to locate the center of a regular polygon. The center is the intersection of the bisectors of two consecutive angles; alternatively, the intersection of the perpendicular bisectors of two consecutive sides can be used to locate the center of the regular polygon. Note that a regular polygon has a center, whether or not either of the related circles is shown. In Figure 7.32, point O is the center of the regular pentagon $RSTVW$. In this figure, $\overline{OR}$ is called a "radius" of the regular pentagon because it is the radius of the circumscribed circle.

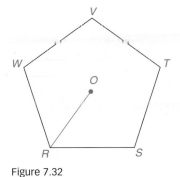

Figure 7.32

DEFINITION

A **radius of a regular polygon** is any line segment that joins the center of the regular polygon to one of its vertices.

Based upon the proof of Theorem 7.3.1, we state the following relationship.

All radii of a regular polygon are congruent.

DEFINITION

An **apothem** of a regular polygon is any line segment drawn from the center of that polygon perpendicular to one of the sides.

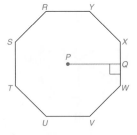

Figure 7.33

In regular octagon $RSTUVWXY$ with center P in Figure 7.33, the segment $\overline{PQ}$ is an apothem. An apothem of a regular polygon is a radius of the inscribed circle. Any regular polygon of n sides has n apothems and n radii. The proof of Theorem 7.3.1 also establishes the following relationship.

All apothems of a regular polygon are congruent.

Figure 7.34

DEFINITION

A **central angle of a regular polygon** is an angle formed by two consecutive radii of the regular polygon.

In regular hexagon *ABCDEF* with center *Q* (in Figure 7.34), angle *EQD* is a central angle. Due to the congruences of the triangles in the proof of Theorem 7.3.1, we also see that Theorem 7.3.2 is a consequence of the following relationship.

All central angles of a regular polygon are congruent.

THEOREM 7.3.2

The measure of any central angle of a regular polygon of *n* sides is given by $c = \frac{360}{n}$.

We apply Theorem 7.3.2 in parts (a) and (b) of Example 4.

Discover

The base of a garden gazebo is to have the shape of a regular hexagon. The carpenter building the gazebo measures the lengths of diagonals $\overline{AD}$, $\overline{BE}$, and $\overline{CF}$. What does he hope to find?

ANSWER
$\overline{AD} \cong \overline{BE} \cong \overline{CF}$

EXAMPLE 4

a) Find the measure of the central angle of a regular polygon of 9 sides.
b) Find the number of sides of a regular polygon whose central angle measures 72°.
c) Is there a regular polygon with 20 diagonals? If so, how many sides does the polygon have?

SOLUTION

a) $c = \frac{360}{9} = 40°$

b) $72 = \frac{360}{n} \rightarrow 72n = 360 \rightarrow n = 5$ sides

c) $D = \frac{n(n - 3)}{2}$, so $20 = \frac{n(n - 3)}{2}$

$$40 = n(n - 3)$$
$$40 = n^2 - 3n$$
$$0 = n^2 - 3n - 40$$
$$0 = (n - 8)(n - 5)$$
$$n - 8 = 0 \quad \text{or} \quad n + 5 = 0$$
$$n = 8 \quad \text{or} \quad n = -5$$

Yes, the polygon exists and has 8 sides.

The final theorems also follow from the proof of Theorem 7.3.1.

THEOREM 7.3.3

Any radius of a regular polygon bisects the angle at the vertex to which it is drawn.

THEOREM 7.3.4

Any apothem of a regular polygon bisects the side of the polygon to which it is drawn.

EXAMPLE 5

Given that each side of regular hexagon *ABCDEF* has the length 4 in., find the length of

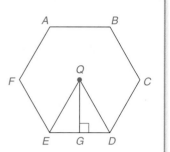

a) radius $\overline{QE}$.
b) apothem $\overline{QG}$.

SOLUTION

a) By Theorem 7.3.2, the measure of $\angle EQD$ is $\frac{360°}{6}$, or 60°. With $\overline{QE} \cong \overline{QD}$, $\triangle QED$ is equiangular and equilateral. Then $QE = 4$ in.

b) With apothem $\overline{QG}$ as shown, $\triangle QEG$ is a 30°-60°-90° triangle in which $m\angle EQG = 30°$. Using Theorem 7.3.4 or the 30°-60°-90° relationship, $EG = 2$ in. With $\overline{QG}$ opposite the 60° angle of $\triangle QEG$, it follows that $QG = 2\sqrt{3}$ in.

SSG **EXS. 9–20**

Exercises 7.3

1. Describe, if possible, how you would inscribe a circle within kite *ABCD*.

2. What condition must be satisfied for it to be possible to circumscribe a circle about kite *ABCD*?

Exercises 1, 2

3. Describe, if possible, how you would inscribe a circle in rhombus *JKLM*.

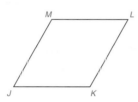

4. What condition must be satisfied for it to be possible to circumscribe a circle about trapezoid *RSTV*?

In Exercises 5 to 8, perform constructions.

5. Inscribe a regular octagon within a circle.

6. Inscribe an equilateral triangle within a circle.

7. Circumscribe a square about a circle.

8. Circumscribe an equilateral triangle about a circle.

9. Find the perimeter of a regular octagon if the length of each side is 3.4 in.

10. In a regular polygon with each side of length 6.5 cm, the perimeter is 130 cm. How many sides does the regular polygon have?

11. If the perimeter of a regular dodecagon (12 sides) is 99.6 cm, how long is each side?

12. If the apothem of a square measures 5 cm, find the perimeter of the square.

13. Find the lengths of the apothem and the radius of a square whose sides have length 10 in.

14. Find the lengths of the apothem and the radius of a regular hexagon whose sides have length 6 cm.

15. Find the lengths of the side and the radius of an equilateral triangle whose apothem has the length 8 ft.

16. Find the lengths of the side and the radius of a regular hexagon whose apothem has the length 10 m.

17. Find the measure of a central angle of a regular polygon of
a) 3 sides.   c) 5 sides.
b) 4 sides.   d) 6 sides.

18. Find the measure of a central angle of a regular polygon of
a) 8 sides.   c) 9 sides.
b) 10 sides.   d) 12 sides.

19. Find the number of sides of a regular polygon that has a central angle measuring
a) 90°.   c) 60°.
b) 45°.   d) 24°.

20. Find the number of sides of a regular polygon that has a central angle measuring
a) 30°.   c) 36°.
b) 72°.   d) 20°.

21. Find the measure of each interior angle of a regular polygon whose central angle measures
a) 40°.   b) 45°.

22. Find the measure of each interior angle of a regular polygon whose central angle measures
a) 60°   b) 90°

23. Find the measure of each exterior angle of a regular polygon whose central angle measures
a) 30°.   b) 40°.

24. Find the measure of each exterior angle of a regular polygon whose central angle measures
a) 45°   b) 120°

25. Find the number of sides for a regular polygon in which the measure of each interior angle is 60° greater than the measure of each central angle.

26. Find the number of sides for a regular polygon in which the measure of each interior angle is 90° greater than the measure of each central angle.

27. Is there a regular polygon for which each central angle measures
a) 40°?   c) 60°?
b) 50°?   d) 70°?

28. Given regular hexagon $ABCDEF$ with each side of length 6, find the length of diagonal $\overline{AC}$.

(HINT: With G on $\overline{AC}$, draw $\overline{BG} \perp \overline{AC}$.)

29. Given regular octagon $RSTUVWXY$ with each side of length 4, find the length of diagonal RU.

(HINT: Extended sides, as shown, form a square.)

30. Given that $RSTVQ$ is a regular pentagon and $\triangle PQR$ is equilateral in the figure shown, determine
a) the *type* of triangle represented by $\triangle VPQ$.
b) the *type* of quadrilateral represented by $TVPS$.

31. *Given:*   Regular pentagon
$RSTVQ$ with
equilateral $\triangle PQR$
Find: m∠VPS

Exercises 30, 31

32. *Given:*   Regular pentagon $JKLMN$ (not shown) with diagonals $\overline{LN}$ and $\overline{KN}$
Find: m∠LNK

33. Is there a regular polygon with 8 diagonals? If so, how many sides does it have?

34. Is there a regular polygon with 12 diagonals? If so, how many sides does it have?

35. Find the measure of a central angle of a regular polygon that has 54 diagonals.

36. Find the measure of a central angle of a regular polygon that has 35 diagonals.

▴37. *Prove:*   If a circle is divided into n congruent arcs $(n \geq 3)$, the chords determined by joining consecutive endpoints of these arcs form a regular polygon.

★38. *Prove:*   If a circle is divided into n congruent arcs $(n \geq 3)$, the tangents drawn at the endpoints of these arcs form a regular polygon.

PERSPECTIVE ON HISTORY

THE VALUE OF π

In geometry, any two figures that have the same shape are described as similar. Because all circles have the same shape, we say that all circles are similar to each other. Just as a proportionality exists among the corresponding sides of similar triangles, we can demonstrate a proportionality among the circumferences (distances around) and diameters (distances across) of circles. By representing the circumferences of the circles in Figure 7.35 by C_1, C_2, and C_3 and their corresponding lengths of diameters by d_1, d_2, and d_3, we claim that

$$\frac{C_1}{d_1} = \frac{C_2}{d_2} = \frac{C_3}{d_3} = k$$

for some constant of proportionality k.

Figure 7.35

We denote the constant k described above by the Greek letter π. Thus, $\pi = \frac{C}{d}$ in any circle. It follows that $C = \pi d$ or $C = 2\pi r$ (because $d = 2r$ in any circle). In applying these formulas for the circumference of a circle, we often leave π in the answer so that the result is exact. When an approximation for the circumference (and later for the area) of a circle is needed, several common substitutions are used for π. Among these are $\pi \approx \frac{22}{7}$ and $\pi \approx 3.14$. A calculator may display the value $\pi \approx 3.1415926535$.

Because π is needed in many applications involving the circumference or area of a circle, its approximation is often necessary; but finding an accurate approximation of π was not quickly or easily done. The formula for circumference can be expressed as $C = 2\pi r$, but the formula for the area of the circle is $A = \pi r^2$. This and other area formulas will be given more attention in Chapter 8.

Several references to the value of π are made in literature. One of the earliest comes from the Bible; the passage from I Kings, Chapter 7, verse 23, describes the distance around a vat as three times the distance across the vat (which suggests that π equals 3, a very rough approximation). Perhaps no greater accuracy was needed in some applications of that time.

In the content of the Rhind papyrus (a document over 3000 years old), the Egyptian scribe Ahmes gives the formula for the area of a circle as $\left(d - \frac{1}{9}d\right)^2$. To determine the Egyptian approximation of π, we need to expand this expression as follows:

$$\left(d - \frac{1}{9}d\right)^2 = \left(\frac{8}{9}d\right)^2 = \left(\frac{8}{9} \cdot 2r\right)^2 = \left(\frac{16}{9}r\right)^2 = \frac{256}{81}r^2$$

In the formula for the area of the circle, the value of π is the multiplier (coefficient) of r^2. Because this coefficient is $\frac{256}{81}$ (which has the decimal equivalent of 3.1604), the Egyptians found a better approximation of π than was given in the book of I Kings.

Archimedes, the brilliant Greek geometer, knew that the formula for the area of a circle was $A = \frac{1}{2}Cr$ (with C the circumference and r the length of radius). His formula was equivalent to the one we use today and is developed as follows.

$$A = \frac{1}{2}Cr = \frac{1}{2}(2\pi r)r = \pi r^2$$

The second proposition of Archimedes' work *Measure of the Circle* develops a relationship between the area of a circle and the area of the square in which it is inscribed. (See Figure 7.36.) Specifically, Archimedes claimed that the ratio of the area of the circle to that of the square was 11:14. This leads to the following set of equations and to an approximation of the value of π.

$$\frac{\pi r^2}{(2r)^2} \approx \frac{11}{14}$$

$$\frac{\pi r^2}{4r^2} \approx \frac{11}{14}$$

$$\frac{\pi}{4} \approx \frac{11}{14}$$

$$\pi \approx 4 \cdot \frac{11}{14} \approx \frac{22}{7}$$

Figure 7.36

Archimedes later improved his approximation of π by showing that

$$3\frac{10}{71} < \pi < 3\frac{1}{7}$$

Today's calculators provide excellent approximations for the irrational number π. We should recall, however, that π is an irrational number that can be expressed as an exact value only by the symbol π.

PERSPECTIVE ON APPLICATIONS

THE NINE-POINT CIRCLE

In the study of geometry, there is a curiosity known as the Nine-Point Circle—a curiosity because its practical value consists of the reasoning needed to verify its plausibility.

In △ABC, in Figure 7.37 we locate these points:

M, N, and *P,* the midpoints of the sides of △ABC, *D, E,* and *F,* points on △ABC determined by its altitudes, and *X, Y,* and *Z,* the midpoints of the line segments determined by orthocenter *O* and the vertices of △ABC.

Through these nine points, it is possible to draw or construct the circle shown in Figure 7.37.

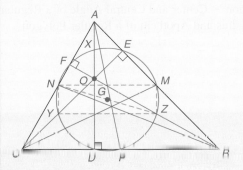

Figure 7.37

To understand why the Nine-Point Circle can be drawn, we show that the quadrilateral *NMZY* is both a parallelogram and a rectangle. Because $\overline{NM}$ joins the midpoints of $\overline{AC}$ and $\overline{AB}$, we know that $\overline{NM} \parallel \overline{CB}$ and $NM = \frac{1}{2}(CB)$. Likewise, *Y* and *Z* are midpoints of the sides of △OBC, so $\overline{YZ} \parallel \overline{CB}$ and $YZ = \frac{1}{2}(CB)$. By Theorem 4.2.1, *NMZY* is a parallelogram. Then $\overline{NY}$ must be parallel to $\overline{MZ}$. With $\overline{CB} \perp \overline{AD}$, it follows that $\overline{NM}$ must be perpendicular to $\overline{AD}$ as well. In turn, $\overline{MZ} \perp \overline{NM}$, and *NMZY* is a rectangle in Figure 7.37. It is possible to circumscribe a circle about any rectangle; in fact, the length of the radius of the circumscribed circle is one-half the length of a diagonal of the rectangle, so we choose $r = \frac{1}{2}(NZ) = NG$. This circle certainly contains the points *N, M, Z,* and *Y.*

Figure 7.38

Although we do not provide the details, it can be shown that quadrilateral *XZPN* of Figure 7.38 is a rectangle as well. Further, $\overline{NZ}$ is also a diagonal of rectangle *XZPN.* Then we can choose the radius of the circumscribed circle for rectangle *XZPN* to have the length $r = \frac{1}{2}(NZ) = NG$. Because it has the same center *G* and the same length of radius *r* as the circle that was circumscribed about rectangle *NMZY,* we see that the same circle must contain points *N, X, M, Z, P,* and *Y.*

Finally, we need to show that the circle in Figure 7.39 with center *G* and radius $r = \frac{1}{2}(NZ)$ will contain the points *D, E,* and *F.* This can be done by an indirect argument. If we suppose that these points do *not* lie on the circle, then we contradict the fact that an angle inscribed in a semicircle must be a right angle. Of course, $\overline{AD}, \overline{BF},$ and $\overline{CE}$ were altitudes of △ABC, so inscribed angles at *D, E,* and *F* must measure 90°; in turn, these angles must lie inside semicircles. In Figure 7.37, ∠NFZ intercepts an arc (a semicircle) determined by diameter $\overline{NZ}$. So *D, E,* and *F* are on the same circle that has center *G* and radius *r.* Thus, the circle described in the preceding paragraphs is the anticipated Nine-Point Circle!

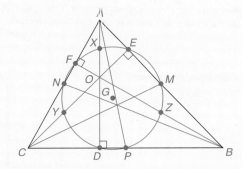

Figure 7.39

Summary

A Look Back at Chapter 7

In Chapter 7, we used the locus of points concept to establish the concurrence of lines relationships in Section 7.2. In turn, these concepts of locus and concurrence allowed us to show that a regular polygon has both an inscribed and a circumscribed circle; in particular, these two circles have a common center. Several new properties of regular polygons were discovered.

A Look Ahead to Chapter 8

One goal of the next chapter is to deal with the areas of triangles, certain quadrilaterals, and regular polygons. We will consider perimeters of polygons and the circumference of a circle. The area of a circle and the area of a sector of a circle will be discussed. Special right triangles will play an important role in determining the areas of some of these plane figures.

Key Concepts

7.1

Locus of Points in a Plane • Locus of Points in Space

7.2

Concurrent Lines • Incenter • Incircle • Circumcenter • Circumcircle • Orthocenter • Centroid

7.3

Regular Polygon • Center and Central Angle of a Regular Polygon • Radius and Apothem of a Regular Polygon

Overview ■ Chapter 7

Selected Locus Problems (in a plane)

Locus	Figure	Description
Locus of points that are at a fixed distance r from fixed point P		The circle with center P and radius r
Locus of points that are equidistant from the sides of an angle		The bisector $\overrightarrow{BD}$ of $\angle ABC$
Locus of points that are equidistant from the endpoints of a line segment		The perpendicular bisector ℓ of $\overline{RS}$
Locus of points that are equidistant from a fixed line and a point not on that line		The parabola with directrix d, focus F, and vertex V

Concurrence of Lines (in a triangle)

Type of Lines	Figure	Point of Concurrence
Angle bisectors		Incenter D of $\triangle ABC$ D is the center of the inscribed circle of $\triangle ABC$.
Perpendicular bisectors of the sides		Circumcenter T of $\triangle XYZ$ T is the center of the circumscribed circle of $\triangle XYZ$.
Altitudes		Orthocenter N of $\triangle DEF$
Medians		Centroid C of $\triangle RST$ (See Table 7.1 on page 324.)

Properties of Regular Polygons

Regular Polygon	Figure	Description
Point O in the center of regular pentagon $ABCDE$.		$\overline{OA}$ is a radius of $ABCDE$; $\overline{OA}$ bisects $\angle BAE$. $\overline{OP}$ is an apothem of $ABCDE$; $\overline{OP}$ is the perpendicular bisector of side ED.

Chapter 7 Review Exercises

In Review Exercises 1 to 6, use the figure shown.

1. Construct a right triangle so that one leg has length *AB* and the other leg has length twice *AB*.

2. Construct a right triangle so that one leg has length *AB* and the hypotenuse has length twice *AB*.

3. Construct an isosceles triangle with vertex angle *B* and legs the length of $\overline{AB}$ (from the line segment shown).

Exercises 1–6

4. Construct an isosceles triangle with vertex angle *B* and an altitude with the length of $\overline{AB}$ from vertex *B* to the base.

5. Construct a square with sides of length *AB*.

6. Construct a rhombus with side $\overline{AB}$ and ∠*ABC*.

In Review Exercises 7 to 13, sketch and describe the locus in a plane.

7. Find the locus of points equidistant from the sides of ∠*ABC*.

8. Find the locus of points that are 1 in. from a given point *B*.

Exercises 7, 8

9. Find the locus of points equidistant from points *D* and *E*.

Exercises 9, 10

10. Find the locus of points that are $\frac{1}{2}$ in. from $\overleftrightarrow{DE}$.

11. Find the locus of the midpoints of the radii of a circle.

12. Find the locus of the centers of all circles passing through two given points.

13. What is the locus of the center of a penny that rolls around and remains tangent to a half-dollar?

In Exercises 14 to 17, sketch and describe the locus in space.

14. Find the locus of points 2 cm from a given point *A*.

15. Find the locus of points 1 cm from a given plane *P*.

16. Find the locus of points less than 3 units from a given point.

17. Find the locus of points equidistant from two parallel planes.

In Review Exercises 18 to 23, use construction methods with the accompanying figure.

18. *Given:* △*ABC*
    Find: The incenter

19. *Given:* △*ABC*
    Find: The circumcenter

Exercises 18–23

For Exercises 20 to 23, see the figure for Exercise 18.

20. *Given:* △*ABC*
    Find: The orthocenter

21. *Given:* △*ABC*
    Find: The centroid

22. Use the result from Exercise 18 to inscribe a circle in △*ABC*.

23. Use the result from Exercise 19 to circumscribe a circle about △*ABC*.

24. *Given:* △*ABC* with medians $\overline{AE}, \overline{DC}, \overline{BF}$
    Find:
    a) *BG* if *BF* = 18
    b) *GE* if *AG* = 4
    c) *DG* if *CG* = $4\sqrt{3}$

Exercises 24, 25

25. *Given:* △*ABC* with medians $\overline{AE}, \overline{DC}, \overline{BF}$
    $AG = 2x + 2y, GE = 2x - y$
    $BG = 3y + 1, GF = x$
    Find: *BF* and *AE*

26. For a regular pentagon, find the measure of each
    a) central angle.
    b) interior angle.
    c) exterior angle.

27. For a regular decagon (10 sides), find the measure of each
    a) central angle.
    b) interior angle.
    c) exterior angle.

28. In a regular polygon, each central angle measures 45°.
    a) How many sides does the regular polygon have?
    b) How many diagonals does this regular polygon have?
    c) If each side measures 5 cm and each apothem is approximately 6 cm in length, what is the perimeter of the polygon?

29. In a regular polygon, the apothem measures 3 in. Each side of the same regular polygon measures 6 in.
    a) Find the perimeter of the regular polygon.
    b) Find the length of radius for this polygon.

30. Can a circle be circumscribed about each of the following figures?
    a) Parallelogram   c) Rectangle
    b) Rhombus   d) Square

31. Can a circle be inscribed in each of the following figures?
    a) Parallelogram   c) Rectangle
    b) Rhombus   d) Square

32. The length of the radius of a circle inscribed in an equilateral triangle is 7 in. Find the length of the radius of the triangle.

33. The length of the radius of a circle inscribed in a regular hexagon is 10 cm. Find the perimeter of the hexagon.

34. Sketch the parabola that has directrix *d* and focus *F*.

Chapter **7** | Test

1. Draw and describe the locus of points in the plane that are equidistant from parallel lines ℓ and *m*.

_____

_____

2. Draw and describe the locus of points in the plane that are equidistant from the sides of ∠*ABC*.

_____

_____

3. Draw and describe the locus of points in the plane that are equidistant from the endpoints of $\overline{DE}$.

_____

_____

4. Describe the locus of points in a plane that are at a distance of 3 cm from point *P*.

• *P*

_____

_____

Exercises 4, 5

5. Describe the locus of points *in space* that are at a distance of 3 cm from point *P*.

_____

_____

6. For a given triangle (such as △*ABC*), what *word* describes the point of concurrency for
   a) the three angle bisectors?

   _____

   b) the three medians?

   _____

Exercises 6, 7

7. For a given triangle (such as △*ABC*), what *word* describes the point of concurrency for
   a) the three perpendicular bisectors of sides? _____
   b) the three altitudes? _____

8. In what type of triangle are the angle bisectors, perpendicular bisectors of sides, altitudes, and medians the same? _____

9. Which of the following *must be* concurrent at an interior point of any triangle?
   angle bisectors   perpendicular bisectors of sides
   altitudes   medians

   _____

10. Classify as true/false:
   a) A circle can be inscribed in any regular polygon.

   _____

   b) A regular polygon can be circumscribed about any circle. _____
   c) A circle can be inscribed in any rectangle. _____
   d) A circle can be circumscribed about any rhombus.

   _____

11. An equilateral triangle has a radius of length 3 in. Find the length of
   a) an apothem. _____
   b) a side. _____

12. For a regular pentagon, find the measure of each
   a) central angle.
   b) interior angle. _____

13. The measure of each central angle of a regular polygon is 36°.
   a) How many sides does this regular polygon have?

   _____

   b) How many diagonals does this regular polygon have?

   _____

14. For a regular octagon, the length of the apothem is approximately 12 cm and the length of the radius is approximately 13 cm. To the nearest centimeter, find the perimeter of the regular octagon.

   _____

15. For regular hexagon *ABCDEF*, the length of side $\overline{AB}$ is 4 in. Find the exact length of
   a) diagonal $\overline{AC}$. _____
   b) diagonal $\overline{AD}$. _____

16. For rectangle *MNPQ*, points *A*,*B*,*C*, and *D* are the midpoints of the sides.
a) What type of quadrilateral (not shown) is *ABCD*?
   _____
b) Are the inscribed circle for *ABCD* and the circumscribed circle for *MNPQ* concentric circles?
   _____

17. Sketch the parabola that has directrix *d* and focus *F*.

Chapter 8

Areas of Polygons and Circles

Powerful! The unique shape and the massive size of the Pentagon in Washington, D.C., manifest the notion of strength. In this chapter, we introduce the concept of area. The area of an enclosed plane region is a measure of size that has applications in construction, farming, real estate, and more. Some of the units that are used to measure area include the square inch and the square centimeter. While the areas of square and rectangular regions are generally easily calculated, we will also develop formulas for the areas of less common polygonal regions. In particular, Section 8.3 is devoted to calculating the areas of regular polygons, such as the Pentagon shown in the photograph. Many real-world applications of the area concept are found in the examples and exercise sets of this chapter.

Additional video explanations of concepts, sample problems, and applications are available on DVD.

8.1 Area and Initial Postulates

KEY CONCEPTS

Plane Region	Area of a Rectangle,	Altitude and Base
Square Unit	a Parallelogram,	of a Parallelogram
Area Postulates	and a Triangle	and a Triangle

(a)

1 in.

1 in.

(b)

Figure 8.1

Lines are said to be *one-dimensional* because we can measure only the length of a line segment. Of course, such measures are only good approximations. A line segment is measured in linear units such as inches, centimeters, or yards. When a line segment measures 5 centimeters, we write $AB = 5$ cm (or $AB = 5$ if units are not stated). The instrument for measuring length is the ruler.

A plane is an infinite *two-dimensional* surface. A closed or bounded portion of the plane is called a **region.** When a region such as R in plane M [see Figure 8.1(a)] is measured, we call this measure the "area of the plane region." The unit used to measure area is known as a **square unit** because it is a square with each side of length 1 [see Figure 8.1(b)]. The measure of the area of region R is the number of nonoverlapping square units that can be placed adjacent to each other in the region.

Square units (not linear units) are used to measure area. Using an exponent, we write square inches as in^2. The unit represented by Figure 8.1(b) is 1 square inch or 1 in^2.

One application of area involves measuring the floor area to be covered by carpeting, which could be measured in square feet (ft^2) or square yards (yd^2). Another application of area involves calculating the number of squares of shingles needed to cover a roof; in this situation, a "square" is the number of shingles needed to cover a 100-ft^2 section of the roof.

In Figure 8.2, the regions have measurable areas and are bounded by figures encountered in earlier chapters. A region is **bounded** if we can distinguish between its interior and its exterior; in calculating area, we measure the interior of the region.

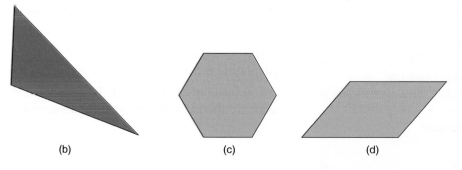

(a)　　　　(b)　　　　(c)　　　　(d)

Figure 8.2

We can measure the area of the region within a triangle [see Figure 8.2(b)]. However, we cannot actually measure the area of the triangle itself (three line segments do not have area). Nonetheless, the area of the region within a triangle is commonly referred to as the *area of the triangle.*

The preceding discussion does not formally define a region or its area. These are accepted as the undefined terms in the following postulate.

POSTULATE 18 ■ Area Postulate

Corresponding to every bounded region is a unique positive number A, known as the area of that region.

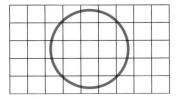

Figure 8.3

As with the length of a line segment, the area measure of an enclosed plane region is also an approximation. One way to estimate the area of a region is to place it in a grid, as shown in Figure 8.3. Counting only the number of whole squares inside the region gives an approximation that is less than the actual area. On the other hand, counting squares that are inside or partially inside provides an approximation that is greater than the actual area. A fair estimate of the area of a region is often given by the average of the smaller and larger approximations just described. If the area of the circle shown in Figure 8.3 is between 9 and 21 square units, we might estimate its area to be $\frac{9 + 21}{2}$ or 15 square units.

To explore another property of area, we consider $\triangle ABC$ and $\triangle DEF$ (which are congruent) in Figure 8.4. One triangle can be placed over the other so that they coincide. How are the areas of the two triangles related? The answer is found in the following postulate.

Figure 8.4

POSTULATE 19

If two closed plane figures are congruent, then their areas are equal.

Discover

Complete this analogy: An inch is to the length of a line segment as a ?   ? is to the area of a plane region.

ANSWER
square inch

EXAMPLE 1

In Figure 8.5, points B and C trisect $\overline{AD}$; $\overline{EC} \perp \overline{AD}$. Name two triangles with equal areas.

SOLUTION $\triangle ECD \cong \triangle ECB$ by SAS. Then $\triangle ECB$ and $\triangle ECD$ have equal areas according to Postulate 19.

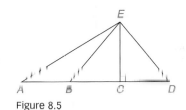

Figure 8.5

NOTE: $\triangle EBA$ is also equal in area to $\triangle ECB$ and $\triangle ECD$, but this relationship cannot be established until we consider Theorem 8.1.3.

Consider Figure 8.6. The entire region is bounded by a curve and then subdivided by a line segment into smaller regions R and S. These regions have a common boundary and do not overlap. Because a numerical area can be associated with each region R and S, the area of $R \cup S$ (read as "R union S" and meaning region R joined to region S) is equal to the sum of the areas of R and S. This leads to Postulate 20, in which A_R represents the "area of region R," A_S represents the "area of region S," and $A_{R \cup S}$ represents the "area of region $R \cup S$."

Figure 8.6

POSTULATE 20 ■ Area-Addition Postulate

Let R and S be two enclosed regions that do not overlap. Then

$$A_{R \cup S} = A_R + A_S$$

Figure 8.7

Figure 8.8

EXAMPLE 2

In Figure 8.7, the pentagon *ABCDE* is composed of square *ABCD* and △*ADE*. If the area of the square is 36 in² and that of △*ADE* is 12 in², find the area of pentagon *ABCDE*.

SOLUTION Square *ABCD* and △*ADE* do not overlap and have a common boundary $\overline{AD}$. By the Area-Addition Postulate,

$$\text{Area (pentagon } ABCDE) = \text{ area (square } ABCD) + \text{ area (}\triangle ADE)$$
$$\text{Area (pentagon } ABCDE) = 36 \text{ in}^2 + 12 \text{ in}^2 = 48 \text{ in}^2$$

It is convenient to provide a subscript for *A* (area) that names the figure whose area is indicated; in the subscript *ABCDE*, the letter *A* also names a vertex of the pentagon shown in Figure 8.7. The symbol A_{ABCDE} represents the area *A* of the pentagon *ABCDE*. The principle used in Example 2 is conveniently and compactly stated in the form

$$A_{ABCDE} = A_{ABCD} + A_{ADE}$$

AREA OF A RECTANGLE

Discover

Study rectangle *MNPQ* in Figure 8.8, and note that it has dimensions of 3 cm and 4 cm. The number of squares, 1 cm on a side, in the rectangle is 12. Rather than counting the number of squares in the figure, how can you calculate the area?

ANSWER
Multiply 3 × 4 = 12.

Warning
Although 1 ft = 12 in., 1 ft² = 144 in². See Figure 8.9.

In the preceding Discover activity, the unit of area is cm². Multiplication of dimensions is handled like algebraic multiplication. Compare

$$3x \cdot 4x = 12x^2 \qquad \text{and} \qquad 3 \text{ cm} \cdot 4 \text{ cm} = 12 \text{ cm}^2$$

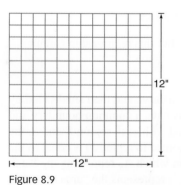

Figure 8.9

If the units used to measure the dimensions of a region are *not* the same, then they must be converted into like units in order to calculate area. For instance, if we need to multiply 2 ft by 6 in., we note that 2 ft = 2(12 in.) = 24 in., so *A* = 2 ft · 6 in. = 24 in. · 6 in., and *A* = 144 in². Alternatively, 6 in. = $6\left(\frac{1}{12} \text{ ft}\right)$ = $\frac{1}{2}$ ft, so *A* = 2 ft · $\frac{1}{2}$ ft = 1 ft². Because the area is unique, we know that 1 ft² = 144 in²; in Figure 8.9 we see the same result.

Recall that one side of a rectangle is called its *base* and that any line segment between sides and perpendicular to the base is called an *altitude* of the rectangle. The length of the altitude is known as the *height* of the rectangle. In the statement of Postulate 21, we assume that *b* and *h* are measured in like units.

POSTULATE 21

The area *A* of a rectangle whose base has length *b* and whose altitude has length *h* is given by *A* = *bh*.

It is also common to describe the dimensions of a rectangle as length ℓ and width *w*. The formula for the area of the rectangle is then written *A* = ℓ*w*.

Figure 8.10

EXAMPLE 3

Find the area of rectangle *ABCD* in Figure 8.10 if $AB = 12$ cm and $AD = 7$ cm.

SOLUTION Because it makes no difference which dimension is chosen as base b and which as altitude h, we arbitrarily choose $AB = b = 12$ cm and $AD = h = 7$ cm. Then

$$
\begin{aligned}
A &= bh \\
&= 12 \text{ cm} \cdot 7 \text{ cm} \\
&= 84 \text{ cm}^2
\end{aligned}
$$

If units are not provided for the dimensions of a region, we assume that they are alike. In such a case, we simply give the area as a number of square units.

THEOREM 8.1.1

The area A of a square whose sides are each of length s is given by $A = s^2$.

No proof is given for Theorem 8.1.1, which follows immediately from Postulate 21.

AREA OF A PARALLELOGRAM

A rectangle's altitude can be one of its sides, but that is not true of a parallelogram. An **altitude** of a parallelogram is a perpendicular line segment drawn from one side to the opposite side, known as the **base**. A side may have to be extended in order to show this altitude-base relationship in a drawing. In Figure 8.11(a), if $\overline{RS}$ is designated as the base, then any of the segments $\overline{ZR}$, $\overline{VX}$, or $\overline{YS}$ is an altitude corresponding to that base (or, for that matter, to base $\overline{VT}$).

(a)   (b)

Figure 8.11

Another look at $\square RSTV$ [in Figure 8.11(b)] shows that $\overline{ST}$ (or $\overline{VR}$) could just as well have been chosen as the base. Possible choices for the corresponding altitude in this case include $\overline{VH}$ and $\overline{SG}$. In the theorem that follows, it is necessary to select a base and an altitude drawn to that base!

THEOREM 8.1.2

The area A of a parallelogram with a base of length b and with corresponding altitude of length h is given by
$$A = bh$$

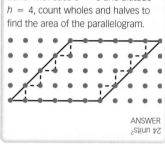
GIVEN: In Figure 8.12(a), $\square RSTV$ with $\overline{VX} \perp \overline{RS}$, $RS = b$, and $VX = h$

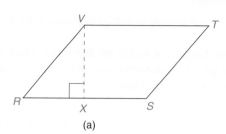

(a)   (b)

Figure 8.12

PROVE: $A_{RSTV} = bh$

PROOF: From vertex S, construct $\overline{SY} \perp \overline{VT}$. Also, from vertex R, construct $\overline{RZ} \perp \overline{VT}$, where Z lies on the extension of $\overline{VT}$. See Figure 8.12(b). Right $\angle Z$ and right $\angle SYT$ are $\cong$. Also, $\overline{ZR} \cong \overline{SY}$ because parallel lines are everywhere equidistant. Because $\angle 1$ and $\angle 2$ are $\cong$ corresponding angles for parallel segments $\overline{VR}$ and $\overline{TS}$, $\triangle RZV \cong \triangle SYT$ by AAS. Then $A_{RZV} = A_{SYT}$ because congruent $\triangle$s have equal areas. Because $A_{RSTV} = A_{RSYV} + A_{SYT}$, it follows that $A_{RSTV} = A_{RSYV} + A_{RZV}$. But $RSYV \cup RZV$ is rectangle $RSYZ$, which has the area $RS \cdot SY = RS \cdot VX = bh$. Therefore, $A_{RSTV} = A_{RSYZ} = bh$.

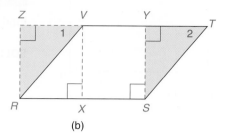

Figure 8.13

EXAMPLE 4

Given that all dimensions in Figure 8.13 are in inches, find the area of $\square MNPQ$ by using base

a) MN.   **b)** PN.

SOLUTION

a) $MN = QP = b = 8$, and the corresponding altitude is of length $QT = h = 5$. Then

$$A = 8 \cdot 5$$
$$= 40 \text{ in}^2$$

b) $PN = b = 6$, so the corresponding altitude length is $MR = h = 6\frac{2}{3}$. Then

$$A = 6 \cdot 6\frac{2}{3}$$
$$= 6 \cdot \frac{20}{3}$$
$$= 40 \text{ in}^2$$

In Example 4, the area of $\square MNPQ$ was not changed when a different base (length) and the length of its corresponding altitude were used to calculate its area. See Postulate 18. The uniqueness of the area of a polygon is also utilized in the solution of Example 5.

Figure 8.14

EXAMPLE 5

GIVEN: In Figure 8.14, $\square MNPQ$ with $PN = 8$ and $QP = 10$
Altitude $\overline{QR}$ to base $\overline{MN}$ has length $QR = 6$

FIND: SN, the length of the altitude between $\overline{QM}$ and $\overline{PN}$

SOLUTION Choosing $MN = b = 10$ and $QR = h = 6$, we see that

$$A = bh = 10 \cdot 6 = 60$$

Now we choose $PN = b = 8$ and $SN = h$, so $A = 8h$. Because the area of the parallelogram is unique and must equal 60, it follows that

$$8h = 60$$
$$h = \frac{60}{8} = 7.5$$

That is, $SN = 7.5$.

 EXS. 11–14

AREA OF A TRIANGLE

The formula used to calculate the area of a triangle follows easily from the formula for the area of a parallelogram. In the formula, any side of the triangle can be chosen as its base; however, we must use the length of the corresponding altitude for that base.

> **THEOREM 8.1.0**
>
> The area A of a triangle whose base has length b and whose corresponding altitude has length h is given by
>
> $$A = \frac{1}{2}bh$$

Following is a picture proof of Theorem 8.1.3.

PICTURE PROOF OF THEOREM 8.1.3

(a)

Figure 8.15

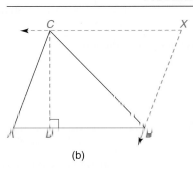

(b)

GIVEN: In Figure 8.15(a), $\triangle ABC$ with $\overline{CD} \perp \overline{AB}$
$AB = b$ and $CD = h$

PROVE: $A = \frac{1}{2}bh$

PROOF: Let lines through C parallel to $\overline{AB}$ and through B parallel to $\overline{AC}$ meet at point X [see Figure 8.15(b)]. With $\square ABXC$ and congruent triangles ABC and XCB, we see that $A_{ABC} = \frac{1}{2} \cdot A_{ABXC} = \frac{1}{2}bh$.

> **EXAMPLE 6**

In the figure, find the area of $\triangle ABC$ if $AB = 10$ cm and $CD = 7$ cm.

SOLUTION With $\overline{AB}$ as base, $b = 10$ cm.
The corresponding altitude for base $\overline{AB}$ is $\overline{CD}$, so $h = 7$ cm. Now

$$A = \frac{1}{2}bh$$

becomes

$$A = \frac{1}{2} \cdot 10 \text{ cm} \cdot 7 \text{ cm}$$

$$A = 35 \text{ cm}^2$$

> **W**arning
>
> The phrase *area of a polygon* really means the area of the region enclosed by the polygon.

The following theorem is a corollary of Theorem 8.1.3.

> **COROLLARY 8.1.4**
>
> The area of a right triangle with legs of lengths a and b is given by $A = \frac{1}{2}ab$.

In the proof of Corollary 8.1.4, the length of either leg can be chosen as the base; in turn, the length of the altitude to that base is the length of the remaining leg. This follows from the fact "The legs of a right triangle are perpendicular."

Discover

On the pegboard shown, each vertical (and horizontal) space between consecutive pegs measures one unit. By counting or applying area formulas, find the area of each polygonal region.

(a)   (b)

ANSWER
(a) 8.5 units² (b) 14 units²

SSG **EXS. 15–20**

EXAMPLE 7

GIVEN: In Figure 8.16, right $\triangle MPN$ with $PN = 8$ and $MN = 17$

FIND: A_{MNP}

SOLUTION With $\overline{PN}$ as one leg of $\triangle MPN$, we need the length of the second leg $\overline{PM}$. By the Pythagorean Theorem,

$$17^2 = (PM)^2 + 8^2$$
$$289 = (PM)^2 + 64$$

Then $(PM)^2 = 225$, so $PM = 15$.
With $PN = a = 8$ and $PM = b = 15$, we apply Corollary 8.1.4.

$$A = \frac{1}{2}ab$$

becomes

$$A = \frac{1}{2} \cdot 8 \cdot 15 = 60 \text{ units}^2$$

Figure 8.16

See the Discover activity at the left.

Exercises 8.1

1. Suppose that two triangles have equal areas. Are the triangles congruent? Why or why not? Are two squares with equal areas necessarily congruent? Why or why not?

2. The area of the square is 12, and the area of the circle is 30. Does the area of the entire shaded region equal 42? Why or why not?

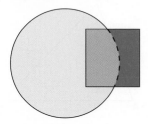

Exercises 2, 3

3. Consider the information in Exercise 2, but suppose you know that the area of the region defined by the intersection of the square and the circle measures 5. What is the area of the entire colored region?

4. If $MNPQ$ is a rhombus, which formula from this section should be used to calculate its area?

Exercises 4–6

5. In rhombus $MNPQ$, how does the length of the altitude from Q to $\overline{PN}$ compare to the length of the altitude from Q to $\overline{MN}$? Explain.

6. When the diagonals of rhombus $MNPQ$ are drawn, how do the areas of the four resulting smaller triangles compare to each other and to the area of the given rhombus?

7. △*ABC* is an obtuse triangle with obtuse angle *A*. △*DEF* is an acute triangle. How do the areas of △*ABC* and △*DEF* compare?

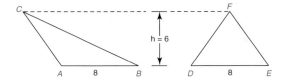

Exercises 7, 8

8. Are △*ABC* and △*DEF* congruent?

In Exercises 9 to 18, find the areas of the figures shown or described.

9. A rectangle's length is 6 cm and its width is 9 cm.

10. A right triangle has one leg measuring 20 in. and a hypotenuse measuring 29 in.

11. A 45-45-90 triangle has a leg measuring 6 m.

12. A triangle's altitude to the 15-in. side measures 8 in.

13.

14.

15.

16.

17.

18.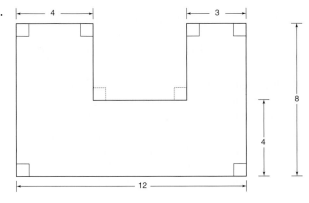

In Exercises 19 to 22, find the area of the shaded region.

19.

20.

21.

   ▱ *PQST*

22.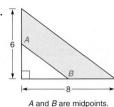

   A and *B* are midpoints.

23. A triangular corner of a store has been roped off to be used as an area for displaying Christmas ornaments. Find the area of the display section.

24. Carpeting is to be purchased for the family room and hallway shown. What is the area to be covered?

25. The exterior wall (the gabled end of the house shown) remains to be painted.
a) What is the area of the outside wall?
b) If each gallon of paint covers approximately 105 ft², how many gallons of paint must be purchased?
c) If each gallon of paint is on sale for $22.50, what is the total cost of the paint?

26. The roof of the house shown needs to be reshingled.
a) Considering that the front and back sections of the roof have equal areas, find the total area to be reshingled.
b) If roofing is sold in squares (each covering 100 ft²), how many squares are needed to complete the work?
c) To remove old shingles and replace with new shingles costs $97.50 per square. What is the cost of reroofing?

27. A beach tent is designed so that one side is open. Find the number of square feet of canvas needed to make the tent.

28. Gary and Carolyn plan to build the deck shown.
a) Find the total floor space (area) of the deck.
b) Find the approximate cost of building the deck if the estimated cost is $6.40 per ft².

29. A *square yard* is a square with sides 1 yard in length.
a) How many *square feet* are in 1 square yard?
b) How many *square inches* are in 1 square yard?

30. The following problem is based on this theorem: "A median of a triangle separates it into two triangles of equal area." In the figure, △RST has median $\overline{RV}$.
a) Explain why $A_{RSV} = A_{RVT}$.
b) If $A_{RST} = 40.8$ cm², find A_{RSV}.

For Exercises 31 and 32, X is the midpoint of $\overline{VT}$ and Y is the midpoint of $\overline{TS}$.

31. If $A_{RSTV} = 48$ cm², find A_{RYTX}.

32. If $A_{RYTX} = 13.5$ in², find A_{RSTV}.

Exercises 31, 32

33. Given △ABC with midpoints M, N, and P of the sides, explain why $A_{ABC} = 4 \cdot A_{MNP}$.

In Exercises 34 to 36, provide paragraph proofs.

34. *Given:* Right △ABC
Prove: $h = \dfrac{ab}{c}$

35. *Given:* Square HJKL with $LJ = d$
Prove: $A_{HJKL} = \dfrac{d^2}{2}$

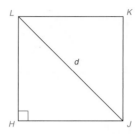

36. *Given:* ▱RSTV with $\overline{VW} \cong \overline{VT}$
Prove: $A_{RSTV} = (RS)^2$

37. *Given:* The area of right △ABC (not shown) is 40 in².
$m\angle C = 90°$
$AC = x$
$BC = x + 2$
Find: x

38. The lengths of the legs of a right triangle are consecutive even integers. The numerical value of the area is three times that of the longer leg. Find the lengths of the legs of the triangle.

***39.** *Given:* △ABC, whose sides are 13 in., 14 in., and 15 in.
Find: a) BD, the length of the altitude to the 14-in. side

(*HINT: Use the Pythagorean Theorem twice.*)

b) The area of △ABC, using the result from part (a)

Exercises 39, 40

***40.** *Given:* △ABC, whose sides are 10 cm, 17 cm, and 21 cm
Find: a) BD, the length of the altitude to the 21-cm side
b) The area of △ABC, using the result from part (a)

41. If the length of the base of a rectangle is increased by 20 percent and the length of the altitude is increased by 30 percent, by what percentage is the area increased?

42. If the length of the base of a rectangle is increased by 20 percent but the length of the altitude is decreased by 30 percent, by what percentage is the area changed? Is this an increase or a decrease in area?

43. Given region $R \cup S$, explain why $A_{R \cup S} > A_R$.

44. Given region $R \cup S \cup T$, explain why $A_{R \cup S \cup T} = A_R + A_S + A_T$.

45. The algebra method of FOIL multiplication is illustrated geometrically in the drawing. Use the drawing with rectangular regions to complete the following rule:
$(a + b)(c + d) = $ _____

46. Use the square configuration to complete the following algebra rule: $(a + b)^2 = $ _____

(**NOTE**: Simplify where possible.)

In Exercises 47 to 50, use the fact that the area of the polygon is unique.

47. In the right triangle, find the length of the altitude drawn to the hypotenuse.

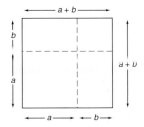

48. In the triangle whose sides are 13, 20, and 21 cm long, the length of the altitude drawn to the 21-cm side is 12 cm. Find the lengths of the remaining altitudes of the triangle.

49. In $\square MNPQ$, $QP = 12$ and $QM = 9$. The length of altitude $\overline{QR}$ (to side $\overline{MN}$) is 6. Find the length of altitude $\overline{QS}$ from Q to $\overline{PN}$.

50. In $\square ABCD$, $AB = 7$ and $BC = 12$. The length of altitude $\overline{AF}$ (to side $\overline{BC}$) is 5. Find the length of altitude $\overline{AE}$ from A to $\overline{DC}$.

*51. The area of a rectangle is 48 in.² Where x is the width and y is the length, express the perimeter P of the rectangle in terms only of x.

*52. The perimeter of a rectangle is 32 cm. Where x is the width and y is the length, express the area A of the rectangle in terms only of x.

*53. Square $DEFG$ is inscribed in right $\triangle ABC$ as shown. If $AD = 6$ and $EB = 8$, find the area of square $DEFG$.

*54. $\overline{TV}$ bisects $\angle STR$ of $\triangle STR$. $ST = 6$ and $TR = 9$. If the area of $\triangle RST$ is 25 m², find the area of $\triangle SVT$.

55. a) Find a lower estimate of the area of the figure by counting whole squares within the figure.
   b) Find an upper estimate of the area of the figure by counting whole and partial squares within the figure.
   c) Use the average of the results in parts (a) and (b) to provide a better estimate of the area of the figure.
   d) Does intuition suggest that the area estimate of part (c) is the exact answer?

56. a) Find a lower estimate of the area of the figure by counting whole squares within the figure.
   b) Find an upper estimate of the area of the figure by counting whole and partial squares within the figure.
   c) Use the average of the results in parts (a) and (b) to provide a better estimate of the area of the figure.
   d) Does intuition suggest that the area estimate of part (c) is the exact answer?

8.2 Perimeter and Area of Polygons

KEY CONCEPTS	Perimeter of a Polygon
	Semiperimeter of a
	  Triangle
	Heron's Formula
		Brahmagupta's Formula
		Area of a Trapezoid, a
		  Rhombus, and a Kite
			Areas of Similar
			  Polygons

Geometry in the Real World

The outside boundary of an enclosure is called its perimeter or its periphery.

We begin this section with a reminder of the meaning of perimeter.

DEFINITION

The **perimeter** of a polygon is the sum of the lengths of all sides of the polygon.

Table 8.1 summarizes perimeter formulas for selected types of triangles, and Table 8.2 summarizes formulas for the perimeters of selected types of quadrilaterals. However, it is more important to understand the concept of perimeter than to memorize formulas. Study each figure so that you can explain its corresponding formula.

TABLE 8.1
Perimeter of a Triangle

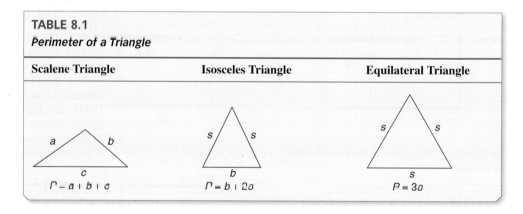

Scalene Triangle	Isosceles Triangle	Equilateral Triangle
$P = a + b + c$	$P = b + 2a$	$P = 3a$

TABLE 8.2
Perimeter of a Quadrilateral

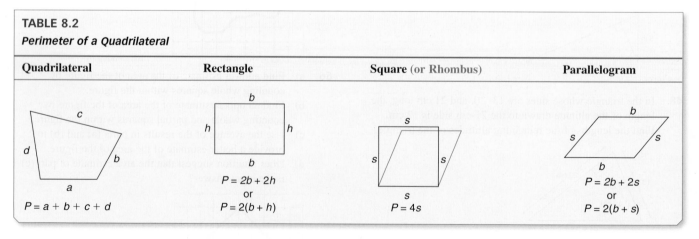

Quadrilateral	Rectangle	Square (or Rhombus)	Parallelogram
$P = a + b + c + d$	$P = 2b + 2h$ or $P = 2(b + h)$	$P = 4s$	$P = 2b + 2s$ or $P = 2(b + s)$

Figure 8.17

EXAMPLE 1

Find the perimeter of $\triangle ABC$ shown in Figure 8.17 if:

    a) $AB = 5$ in., $AC = 6$ in., and $BC = 7$ in.
    b) $AD = 8$ cm, $BC = 6$ cm, and $\overline{AB} \cong \overline{AC}$

SOLUTION

    a) $P_{ABC} = AB + AC + BC$
                $= 5 + 6 + 7$
                $= 18$ in.

    b) With $\overline{AB} \cong \overline{AC}$, $\triangle ABC$ is isosceles. Then $\overline{AD}$ is the $\perp$ bisector of $\overline{BC}$. If $BC = 6$, it follows that $DC = 3$. Using the Pythagorean Theorem, we have

$$(AD)^2 + (DC)^2 = (AC)^2$$
$$8^2 + 3^2 = (AC)^2$$
$$64 + 9 = (AC)^2$$
$$AC = \sqrt{73}$$

Now $P_{ABC} = 6 + \sqrt{73} + \sqrt{73} = 6 + 2\sqrt{73} \approx 23.1$ cm

NOTE: Because $x + x = 2x$, we have $\sqrt{73} + \sqrt{73} = 2\sqrt{73}$.

We apply the perimeter concept in a more general manner in Example 2.

EXAMPLE 2

While remodeling, the Gibsons have decided to replace the old woodwork with Colonial-style oak woodwork.

    a) Using the floor plan provided in Figure 8.18, find the amount of baseboard (in linear feet) needed for the room. Do *not* make any allowances for doors!
    b) The baseboard to be used is sold in 8-foot lengths. How many 8-foot pieces are needed?
    c) Find the cost of the baseboard used if the price is $8.75 per piece.

Figure 8.18

SOLUTION

    a) Dimensions not shown measure $20 - 12$ or 8 ft and $18 - 12$ or 6 ft. Starting with the upper left vertex of Figure 8.18, the perimeter, or "distance around" the room is

$$12 + 6 + 8 + 12 + 20 + 18 = 76 \text{ linear ft}$$

    b) The Gibsons need 10 pieces of the 8-foot sections.
    c) The cost of the baseboard is 10($8.75) = $87.50.

SSG   **EXS. 1–4**

HERON'S FORMULA

If the lengths of the sides of a triangle are known, the formula generally used to calculate the area is **Heron's Formula** (named in honor of Heron of Alexandria, circa A.D. 75). One of the numbers found in this formula is the *semiperimeter* of a triangle, which is defined as one-half the perimeter. For the triangle that has sides of lengths a, b, and c, the semiperimeter is $s = \frac{1}{2}(a + b + c)$. We apply Heron's Formula in Example 3. The proof of Heron's Formula can be found at the website for this textbook.

> **THEOREM 8.2.1 ■ Heron's Formula**
>
> If the three sides of a triangle have lengths a, b, and c, then the area A of the triangle is given by
>
> $$A = \sqrt{s(s - a)(s - b)(s - c)},$$
>
> where the semiperimeter of the triangle is
>
> $$s = \frac{1}{2}(a + b + c)$$

EXAMPLE 3

Find the area of a triangle that has sides of lengths 4, 13, and 15. (See Figure 8.19.)

SOLUTION If we designate the sides as $a = 4$, $b = 13$, and $c = 15$, the semiperimeter of the triangle is given by $s = \frac{1}{2}(4 + 13 + 15) = \frac{1}{2}(32) = 16$. Therefore,

$$
\begin{aligned}
A &= \sqrt{s(s - a)(s - b)(s - c)} \\
&= \sqrt{16(16 - 4)(16 - 13)(16 - 15)} \\
&= \sqrt{16(12)(3)(1)} = \sqrt{576} = 24 \text{ units}^2
\end{aligned}
$$

Figure 8.19

When the lengths of the sides of a quadrilateral are known, we can apply Heron's Formula to find its area if the length of either diagonal is also known.

EXAMPLE 4

In quadrilateral $ABCD$ (Figure 8.20), the lengths of the sides and of diagonal $\overline{BD}$ are shown. Find the area of $ABCD$.

SOLUTION
For $ABCD$, we see that $A_{ABCD} = A_{ABD} + A_{BCD}$. For $\triangle ABD$, $(8, 15, 17)$ is a Pythagorean Triple in which $a = 8$, $b = 15$, and $c = 17$. Then

$$A_{ABD} = \frac{1}{2}ab = \frac{1}{2} \cdot 8 \cdot 15 = 60.$$

Using Heron's Formula with $\triangle BCD$, we know

$$s = \frac{1}{2}(13 + 14 + 15) = \frac{1}{2}(42) = 21.$$

Then

$$
\begin{aligned}
A_{BCD} &= \sqrt{21(21 - 13)(21 - 14)(21 - 15)} \\
&= \sqrt{21(8)(7)(6)} \\
&= \sqrt{7056} = 84
\end{aligned}
$$

Now, $A_{ABCD} = A_{ABD} + A_{BCD} = 60 + 84 = 144 \text{ units}^2$.

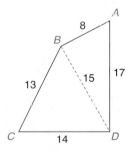

Figure 8.20

The following theorem is named in honor of Brahmagupta, a Hindu mathematician born in A.D. 598. We include the theorem without its rather lengthy proof. As it happens, Heron's Formula for the area of any triangle is actually a special case of Brahmagupta's Formula, which is used to determine the area of a cyclic quadrilateral. In Brahmagupta's Formula, as in Heron's Formula, the letter s represents the numerical value of the semiperimeter. The formula is applied in essentially the same manner as Heron's Formula. See Exercises 11, 12, 43, and 44 of this section.

> **THEOREM 8.2.2 ■ Brahmagupta's Formula**
>
> For a cyclic quadrilateral with sides of lengths a, b, c, and d, the area is given by
>
> $$A = \sqrt{(s - a)(s - b)(s - c)(s - d)},$$
>
> where $\quad s = \frac{1}{2}(a + b + c + d)$

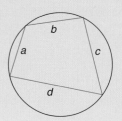

Brahmagupta's Formula becomes Heron's Formula when the length d of the fourth side shrinks (the length d approaches 0) so that the quadrilateral becomes a triangle with sides of lengths a, b, and c.

The remaining theorems of this section contain *numerical subscripts*. In practice, subscripts enable us to distinguish quantities. For instance, the lengths of the two unequal bases of a trapezoid are written b_1 (read "b sub 1") and b_2. The following chart emphasizes the use of numerical subscripts and their interpretations.

SSG EXS. 5–8

Theorem	Subscripted Symbol	Meaning
Theorem 8.2.3	b_1	Length of the *first* base of a trapezoid
Corollary 8.2.5	d_2	Length of the *second* diagonal of a rhombus
Theorem 8.2.7	A_1	Area of the *first* triangle

AREA OF A TRAPEZOID

Recall that the two parallel sides of a trapezoid are its *bases*. The *altitude* is any line segment that is drawn perpendicular from one base to the other. In Figure 8.21, $\overline{AB} \parallel \overline{DC}$ so $\overline{AB}$ and $\overline{DC}$ are bases and $\overline{AE}$ is an altitude for the trapezoid.

We use the more common formula for the area of a triangle (namely, $A = \frac{1}{2}bh$) to develop our remaining theorems.

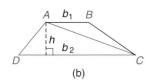

Figure 8.21

> **STRATEGY FOR PROOF ■ Proving Area Relationships**
>
> *General Rule:* Many area relationships depend upon the use of the Area-Addition Postulate.
>
> *Illustration:* In the proof of Theorem 8.2.3, the area of the trapezoid is developed as the sum of the areas of two triangles.

(a)

> **THEOREM 8.2.3**
>
> The area A of a trapezoid whose bases have lengths b_1 and b_2 and whose altitude has length h is given by
>
> $$A = \frac{1}{2}h(b_1 + b_2)$$

(b)

(c)

Figure 8.22

GIVEN: Trapezoid $ABCD$ with $\overline{AB} \parallel \overline{DC}$; $AB = b_1$ and $DC = b_2$.

PROVE: $A_{ABCD} = \frac{1}{2}h(b_1 + b_2)$

PROOF: Draw $\overline{AC}$ as shown in Figure 8.22(a). Now $\triangle ADC$ has an altitude of length h and a base of length b_2. As shown in Figure 8.22(b),

$$A_{ADC} = \frac{1}{2}hb_2$$

Also, $\triangle ABC$ has an altitude of length h and a base of length b_1. [See Figure 8.22(c).] Then

$$A_{ABC} = \frac{1}{2}hb_1$$

Thus,

$$A_{ABCD} = A_{ABC} + A_{ADC}$$
$$= \frac{1}{2}hb_1 + \frac{1}{2}hb_2$$
$$= \frac{1}{2}h(b_1 + b_2)$$

$\overline{RS} \parallel \overline{VT}$

Figure 8.23

EXAMPLE 5

Given that $\overline{RS} \parallel \overline{VT}$, find the area of the trapezoid in Figure 8.23. Note that $RS = 5$, $TV = 13$, and $RW = 6$.

SOLUTION Let $RS = 5 = b_1$ and $TV = 13 = b_2$. Also, $RW = h = 6$.

Now,
$$A = \frac{1}{2}h(b_1 + b_2)$$

becomes
$$A = \frac{1}{2} \cdot 6(5 + 13)$$
$$= \frac{1}{2} \cdot 6 \cdot 18$$
$$= 3 \cdot 18 = 54 \text{ units}^2$$

The following activity reinforces the formula for the area of a trapezoid.

Discover

Cut out two trapezoids that are copies of each other and place one next to the other to form a parallelogram.

a) How long is the base of the parallelogram?

b) What is the area of the parallelogram?

c) What is the area of the trapezoid?

ANSWERS

a) $b_1 + b_2$ b) $h(b_1 + b_2)$ c) $\frac{1}{2}h(b_1 + b_2)$

QUADRILATERALS WITH PERPENDICULAR DIAGONALS

 EXS. 9–12

The formula in Theorem 8.2.4 will also be used to find the area of a rhombus (Corollary 8.2.5) and the area of a kite (Corollary 8.2.6).

(a)

(b)

Figure 8.24

Figure 8.25

Figure 8.26

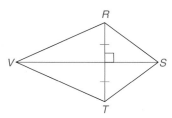

Figure 8.27

THEOREM 8.2.4

The area of any quadrilateral with perpendicular diagonals of lengths d_1 and d_2 is given by

$$A = \frac{1}{2}d_1 d_2$$

GIVEN: Quadrilateral $ABCD$ with $\overline{AC} \perp \overline{BD}$ [See Figure 8.24(a).]

PROVE: $A_{ABCD} = \frac{1}{2}d_1 d_2$

PROOF: Through points A and C, draw lines parallel to $\overline{DB}$. Likewise, draw lines through B and D parallel to $\overline{AC}$. Let the points of intersection of these lines be R, S, T, and V, as shown in Figure 8.24(b). Because each of the quadrilaterals $ARDE$, $ASBE$, $BECT$, and $CEDV$ is a parallelogram containing a right angle, each is a rectangle.

Furthermore, $A_{\triangle ADE} = \frac{1}{2} \cdot A_{ARDE}$, $A_{\triangle ABE} = \frac{1}{2} \cdot A_{ASBE}$, $A_{\triangle BEC} = \frac{1}{2} \cdot A_{BECT}$, and $A_{\triangle DEC} = \frac{1}{2} \cdot A_{CEDV}$.

Then $A_{ABCD} = \frac{1}{2} \cdot A_{RSTV}$. But $RSTV$ is a rectangle because it is a parallelogram containing a right angle. Because $RSTV$ has dimensions d_1 and d_2, its area is $d_1 d_2$. By substitution, $A_{ABCD} = \frac{1}{2}d_1 d_2$. ■

AREA OF A RHOMBUS

Recall that a rhombus is a parallelogram with two congruent adjacent sides. Among the properties of the rhombus, we proved "The diagonals of a rhombus are perpendicular." Thus, we have the following corollary of Theorem 8.2.4. See Figure 8.25.

COROLLARY 8.2.5

The area A of a rhombus whose diagonals have lengths d_1 and d_2 is given by

$$A = \frac{1}{2}d_1 d_2$$

Example 6 illustrates Corollary 8.2.5.

EXAMPLE 6

Find the area of the rhombus $MNPQ$ in Figure 8.26 if $MP = 12$ and $NQ = 16$.

SOLUTION Applying Corollary 8.2.5,

$$A_{MNPQ} = \frac{1}{2}d_1 d_2 = \frac{1}{2} \cdot 12 \cdot 16 = 96 \text{ units}^2$$

In problems involving the rhombus, we often utilize the fact that its diagonals are perpendicular bisectors of each other. If the length of a side and the length of either diagonal are known, the length of the other diagonal can be found by applying the Pythagorean Theorem.

AREA OF A KITE

For a kite, we proved in Exercise 27 of Section 4.2 that one diagonal is the perpendicular bisector of the other. See Figure 8.27.

> **COROLLARY 8.2.6**
>
> The area A of a kite whose diagonals have lengths d_1 and d_2 is given by
>
> $$A = \frac{1}{2}d_1 d_2$$

We apply Corollary 8.2.6 in Example 7.

EXAMPLE 7

Find the length of $\overline{RT}$ in Figure 8.28 if the area of the kite $RSTV$ is 360 in^2 and $SV = 30$ in.

SOLUTION $A = \frac{1}{2}d_1 d_2$ becomes $360 = \frac{1}{2}(30)d$, in which d is the length of the remaining diagonal $\overline{RT}$. Then $360 = 15d$, which means that $d = 24$. Then $RT = 24$ in.

Figure 8.28

SSG EXS. 13–17

AREAS OF SIMILAR POLYGONS

The following theorem compares the areas of similar triangles. In Figure 8.29, we refer to the areas of the similar triangles as A_1 and A_2. The triangle with area A_1 has sides of lengths a_1, b_1, and c_1, and the triangle with area A_2 has sides of lengths a_2, b_2, and c_2. Where a_1 corresponds to a_2, b_1 to b_2, and c_1 to c_2, Theorem 8.2.7 implies that

$$\frac{A_1}{A_2} = \left(\frac{a_1}{a_2}\right)^2 \quad \text{or} \quad \frac{A_1}{A_2} = \left(\frac{b_1}{b_2}\right)^2 \quad \text{or} \quad \frac{A_1}{A_2} = \left(\frac{c_1}{c_2}\right)^2$$

We prove only the first relationship; the other proofs are analogous.

> **THEOREM 8.2.7**
>
> The ratio of the areas of two similar triangles equals the square of the ratio of the lengths of any two corresponding sides; that is,
>
> $$\frac{A_1}{A_2} = \left(\frac{a_1}{a_2}\right)^2$$

Figure 8.29

GIVEN: Two similar triangles, as shown in Figure 8.29

PROVE: $\dfrac{A_1}{A_2} = \left(\dfrac{a_1}{a_2}\right)^2$

PROOF: For the similar triangles, h_1 and h_2 are the respective lengths of altitudes to the corresponding sides of lengths b_1 and b_2. Now $A_1 = \frac{1}{2}b_1 h_1$ and $A_2 = \frac{1}{2}b_2 h_2$, so

$$\frac{A_1}{A_2} = \frac{\frac{1}{2}b_1 h_1}{\frac{1}{2}b_2 h_2} \quad \text{or} \quad \frac{A_1}{A_2} = \frac{\frac{1}{2}}{\frac{1}{2}} \cdot \frac{b_1}{b_2} \cdot \frac{h_1}{h_2}$$

Simplifying, we have

$$\frac{A_1}{A_2} = \frac{b_1}{b_2} \cdot \frac{h_1}{h_2}$$

Because the triangles are similar, we know that $\frac{b_1}{b_2} = \frac{a_1}{a_2}$. Because the lengths of corresponding altitudes of similar triangles have the same ratio as the lengths of a pair of corresponding sides (Theorem 5.3.2), we also know that $\frac{h_1}{h_2} = \frac{a_1}{a_2}$. Through substitution, $\frac{A_1}{A_2} = \frac{b_1}{b_2} \cdot \frac{h_1}{h_2}$ becomes $\frac{A_1}{A_2} = \frac{a_1}{a_2} \cdot \frac{a_1}{a_2}$. Then $\frac{A_1}{A_2} = \left(\frac{a_1}{a_2}\right)^2$. ■

Because Theorem 8.2.7 can be extended to any pair of similar polygons, we could also prove that the ratio of the areas of two squares equals the square of the ratio of the lengths of any two sides. We apply this relationship in Example 8.

s_1

s_2

Figure 8.30

EXAMPLE 8

Use the ratio $\frac{A_1}{A_2}$ to compare the areas of

a) two similar triangles in which the sides of the first triangle are $\frac{1}{2}$ as long as the sides of the second triangle.

b) two squares in which each side of the first square is 3 times as long as each side of the second square.

SOLUTION

a) $s_1 = \frac{1}{2}s_2$, so $\frac{s_1}{s_2} = \frac{1}{2}$. (See Figure 8.30.)

Now $\frac{A_1}{A_2} = \left(\frac{s_1}{s_2}\right)^2$, so that $\frac{A_1}{A_2} = \left(\frac{1}{2}\right)^2$ or $\frac{A_1}{A_2} = \frac{1}{4}$. That is, the area of the first triangle is $\frac{1}{4}$ the area of the second triangle.

b) $s_1 = 3s_2$, so $\frac{s_1}{s_2} = 3$. (See Figure 8.31.)

$\frac{A_1}{A_2} = \left(\frac{s_1}{s_2}\right)^2$, so that $\frac{A_1}{A_2} = (3)^2$ or $\frac{A_1}{A_2} = 9$. That is, the area of the first square is 9 times the area of the second square.

NOTE: For Example 8, Figures 8.30 and 8.31 provide visual evidence of the relationship described in Theorem 8.2.7.

s_1

s_2

Figure 8.31

SSG **EXS. 18–21**

Exercises 8.2

In Exercises 1 to 8, find the perimeter of each polygon.

1.

5 in.

12 in.

2.

←13 in.→

B   C

8 in.

A ←7 in.→ D

▱ ABCD

3.

B   C

d_2

d_1

A   D

▱ ABCD with $\overline{AB} \cong \overline{BC}$

$d_1 = 4$ m
$d_2 = 10$ m

4.
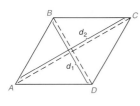
B   C

5

O

4

A   D

▱ ABCD in ⊙O

5.

A   7 ft   D

4 ft

B

13 ft

C

Trapezoid ABCD with $\overline{AB} \cong \overline{DC}$

6.

$3\sqrt{5}$

x

2x

7.

D

13   13

$12 - \sqrt{11}$

B

$\sqrt{11}$

A   C

10

$\overline{AB} \cong \overline{BC}$ in concave quadrilateral ABCD

8.

20 cm

16 cm

5 cm

In Exercises 9 and 10, use Heron's Formula.

9. Find the area of a triangle whose sides measure 13 in., 14 in., and 15 in.

10. Find the area of a triangle whose sides measure 10 cm, 17 cm, and 21 cm.

For Exercises 11 and 12, use Brahmagupta's Formula.

11. For cyclic quadrilateral *ABCD*, find the area if $AB = 39$ mm, $BC = 52$ mm, $CD = 25$ mm, and $DA = 60$ mm.

12. For cyclic quadrilateral *ABCD*, find the area if $AB = 6$ cm, $BC = 7$ cm, $CD = 2$ cm, and $DA = 9$ cm.

In Exercises 13 to 18, find the area of the given polygon.

13.

Trapezoid *ABCD* with $\overline{AB} \cong \overline{DC}$

14.

15.

▱ *ABCD*

16.

▱ *ABCD* with $\overline{BC} \cong \overline{CD}$

17.

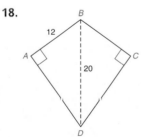

Kite *ABCD* with $BD = 12$
$m\angle BAC = 45°$, $m\angle BCA = 30°$

18.

Kite *ABCD*

19. In a triangle of perimeter 76 in., the length of the first side is twice the length of the second side, and the length of the third side is 12 in. more than the length of the second side. Find the lengths of the three sides.

20. In a triangle whose area is 72 in², the base has a length of 8 in. Find the length of the corresponding altitude.

21. A trapezoid has an area of 96 cm². If the altitude has a length of 8 cm and one base has a length of 9 cm, find the length of the other base.

22. The numerical difference between the area of a square and the perimeter of that square is 32. Find the length of a side of the square.

23. Find the ratio $\frac{A_1}{A_2}$ of the areas of two similar triangles if

   a) the ratio of the lengths of the corresponding sides is $\frac{s_1}{s_2} = \frac{3}{2}$.

   b) the lengths of the sides of the first triangle are 6 in., 8 in., and 10 in., and those of the second triangle are 3 in., 4 in., and 5 in.

24. Find the ratio $\frac{A_1}{A_2}$ of the areas of two similar rectangles if

   a) the ratio of the lengths of the corresponding sides is $\frac{s_1}{s_2} = \frac{2}{5}$.

   b) the length of the first rectangle is 6 m, and the length of the second rectangle is 4 m.

In Exercises 25 and 26, give a paragraph form of proof. Provide drawings as needed.

25. Given: Equilateral $\triangle ABC$ with each side of length s
   Prove: $A_{ABC} = \frac{s^2}{4}\sqrt{3}$

   (HINT: Use Heron's Formula.)

26. Given: Isosceles $\triangle MNQ$ with $QM = QN = s$ and $MN = 2a$
   Prove: $A_{MNQ} = a\sqrt{s^2 - a^2}$
   (NOTE: $s > a$.)

In Exercises 27 to 30, find the area of the figure shown.

27. Given: In $\odot O$, $OA = 5$,
   $BC = 6$, and
   $CD = 4$
   Find: A_{ABCD}

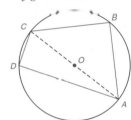

28. Given: Hexagon *RSTVWX*
   with $\overline{WV} \parallel \overline{XT} \parallel \overline{RS}$
   $RS = 10$
   $ST = 8$
   $TV = 5$
   $WV = 16$
   $\overline{WX} \cong \overline{VT}$
   Find: A_{RSTVWX}

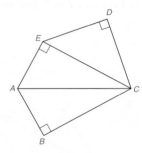

29. Given: Pentagon *ABCDE*
   with $\overline{DC} \cong \overline{DE}$
   $AE = AB = 5$
   $BC = 12$
   Find: A_{ABCDE}

30. Given: Pentagon *RSTVW*
   with $m\angle VRS = m\angle VSR = 60°$,
   $RS = 8\sqrt{2}$, and
   $\overline{RW} \cong \overline{WV} \cong \overline{VT} \cong \overline{TS}$
   Find: A_{RSTVW}

31. Mary Frances has a rectangular garden plot that encloses an area of 48 yd². If 28 yd of fencing are purchased to enclose the garden, what are the dimensions of the rectangular plot?

32. The perimeter of a right triangle is 12 m. If the hypotenuse has a length of 5 m, find the lengths of the two legs.

33. Farmer Watson wishes to fence a rectangular plot of ground measuring 245 ft by 140 ft.
   a) What amount of fencing is needed?
   b) What is the total cost of the fencing if it costs $1.59 per foot?

34. The farmer in Exercise 33 has decided to take the fencing purchased and use it to enclose the subdivided plots shown.
   a) What are the overall dimensions of the rectangular enclosure shown?
   b) What is the total area of the enclosure shown?

35. Find the area of the room whose floor plan is shown.

Exercises 35, 36

36. Find the perimeter of the room in Exercise 35.

37. Examine several rectangles, each with a perimeter of 40 in., and find the dimensions of the rectangle that has the largest area. What type of figure has the largest area?

38. Examine several rectangles, each with an area of 36 in²., and find the dimensions of the rectangle that has the smallest perimeter. What type of figure has the smallest perimeter?

39. Square RSTV is inscribed in square WXYZ as shown. If ZT = 5 and TY = 12, find
   a) the perimeter of RSTV.
   b) the area of RSTV.

Exercises 39, 40

40. Square RSTV is inscribed in square WXYZ as shown. If ZT = 8 and TY = 15, find
   a) the perimeter of RSTV.
   b) the area of RSTV.

For Exercises 41 and 42, the sides of square ABCD are trisected at the indicated points.

41. Find the ratio:
   a) $\dfrac{P_{EGIK}}{P_{ABCD}}$
   b) $\dfrac{A_{EGIK}}{A_{ABCD}}$

Exercises 41, 42

42. Find the ratio:
   a) $\dfrac{P_{EHIL}}{P_{ABCD}}$
   b) $\dfrac{A_{EHIL}}{A_{ABCD}}$

43. Although not all kites are cyclic, one with sides of lengths 5 in., 1 ft, 1 ft, and 5 in. would be cyclic. Find the area of this kite. Give the resulting area in *square inches*.

44. Although not all trapezoids are cyclic, one with bases of lengths 12 cm and 28 cm and both legs of length 10 cm would be cyclic. Find the area of this isosceles trapezoid.

For Exercises 45 and 46, use this information: Let a, b, and c be the integer lengths of the sides of a triangle. If the area of the triangle is also an integer, then (a, b, c) is known as a Heron triple.

45. Which of these are Heron triples?
   a) (5, 6, 7)
   b) (13, 14, 15)

46. Which of these are Heron triples?
   a) (9, 10, 17)
   b) (8, 10, 12)

47. Prove that the area of a trapezoid whose altitude has length *h* and whose median has length *m* is A = *hm*.

For Exercises 48 and 49, use the formula found in Exercise 47.

48. Find the area of a trapezoid with an altitude of length 4.2 m and a median of length 6.5 m.

49. Find the area of a trapezoid with an altitude of length $5\frac{1}{3}$ ft and a median of length $2\frac{1}{4}$ ft.

50. Prove that the area of a square whose diagonal has length *d* is $A = \frac{1}{2}d^2$.

For Exercises 51 and 52, use the formula found in Exercise 50.

51. Find the area of a square whose diagonal has length $\sqrt{10}$ in.

52. Find the area of a square whose diagonal has length 14.5 cm.

*53. The shaded region is that of a trapezoid. Determine the height of the trapezoid.

A and *B* are midpoints.

54. Trapezoid *ABCD* (not shown) is inscribed in ⊙*O* so that side $\overline{DC}$ is a diameter of ⊙*O*. If *DC* = 10 and *AB* = 6, find the exact area of trapezoid *ABCD*.

55. Each side of square *RSTV* has length 8. Point *W* lies on $\overline{VR}$ and point *Y* lies on $\overline{TS}$ in such a way to form parallelogram *VWSY*, which has an area of 16 units². Find the length of $\overline{VW}$.

56. For the cyclic quadrilateral *MNPQ*, the sides have lengths *a*, *b*, *c*, and *d*. If $a^2 + b^2 = c^2 + d^2$, explain why the area of the quadrilateral is $A = \dfrac{ab + cd}{2}$.

8.3 Regular Polygons and Area

KEY CONCEPTS

Regular Polygon
Center and Central Angle
of a Regular Polygon

Radius and Apothem
of a Regular Polygon

Area of a Regular
Polygon

Regular polygons are, of course, both equilateral and equiangular. As we saw in Section 7.3, we can inscribe a circle within any regular polygon and we can circumscribe a circle about any regular polygon. For regular hexagon *ABCDEF* shown in Figure 8.32, suppose that $\overline{QE}$ and $\overline{QD}$ bisect the interior angles of *ABCDEF* as shown. In terms of hexagon *ABCDEF*, recall these terms and theorems.

Figure 8.32

1. Point *Q*, the *center* of regular hexagon *ABCDEF*, is the common center of both the inscribed and circumscribed circles for regular hexagon *ABCDEF*.
2. $\overline{QE}$ is a *radius* of regular hexagon *ABCDEF* because it joins the center of the regular polygon to a vertex.
3. $\overline{QG}$ is an *apothem* of regular hexagon *ABCDEF* because it is drawn from the center of the regular polygon perpendicular to a side of the polygon.
4. ∠*EQD* is a *central angle* of regular hexagon *ABCDEF* because center *Q* is the vertex of the central angle, while the sides are consecutive radii of the polygon. The measure of a central angle of a regular polygon of *n* sides is $c = \dfrac{360°}{n}$.
5. Any radius of a regular polygon bisects the interior angle to which it is drawn.
6. Any apothem of a regular polygon bisects the side to which it is drawn.

Among regular polygons are the square and the equilateral triangle. As we saw in Section 8.1, the area of a square whose sides have length *s* is given by $A = s^2$.

EXAMPLE 1

Find the area of the square whose apothem length is $a = 2$ in.

SOLUTION The apothem is the perpendicular distance from the center to a side of length s. For the square, $s = 2a$; that is, $s = 4$ in. Then $A = s^2$ becomes $A = 4^2$ and $A = 16$ in².

Figure 8.33

SSG **EXS. 1–4**

In Exercise 25 of Section 8.2, we showed that the area of an equilateral triangle whose sides are of length s is given by

$$A = \frac{s^2}{4}\sqrt{3}$$

Following is a picture proof of this area relationship.

PICTURE PROOF

Figure 8.34

GIVEN: The equilateral triangle with sides of length s

PROVE: $A = \dfrac{s^2}{4}\sqrt{3}$

PROOF: Based upon the 30°-60°-90° triangle in Figure 8.34, $A = \frac{1}{2}bh$ becomes

$$A = \frac{1}{2} \cdot s \cdot \frac{s}{2}\sqrt{3}$$

so

$$A = \frac{s^2}{4}\sqrt{3}.$$

EXAMPLE 2

Find the area of an equilateral triangle (not shown) in which each side measures 4 inches.

SOLUTION $A = \dfrac{s^2}{4}\sqrt{3}$ becomes $A = \dfrac{4^2}{4}\sqrt{3}$ or $A = 4\sqrt{3}$ in².

EXAMPLE 3

Figure 8.35

Find the area of equilateral triangle ABC in which apothem $\overline{OD}$ has a length of 6 cm.

SOLUTION See Figure 8.35. If $OD = 6$ cm, then $AD = 6\sqrt{3}$ cm in the indicated 30°-60°-90° triangle AOD. In turn, $AB = 12\sqrt{3}$ cm. Now $A = \frac{s^2}{4}\sqrt{3}$ becomes

$$A = \frac{(12\sqrt{3})^2}{4}\sqrt{3} = \frac{432}{4}\sqrt{3} = 108\sqrt{3} \text{ cm}^2$$

SSG **EXS. 5–8**

AREA OF A REGULAR POLYGON

Before we consider the area of a regular polygon, we consider its perimeter. The following fact is a consequence of the definition of perimeter.

For a regular polygon of n sides, each side of length s, its perimeter is $P = n \cdot s$.

We now have the framework necessary to develop a formula for the area of a regular polygon. In the proof of Theorem 8.3.1, the figure chosen is a regular pentagon; however, the proof is similiar for any regular polygon.

THEOREM 8.3.1

The area A of a regular polygon whose apothem has length a and whose perimeter is P is given by

$$A = \frac{1}{2}aP$$

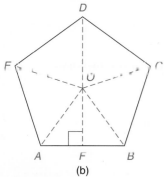

Figure 8.36

GIVEN: Regular polygon $ABCDE$ in Figure 8.36(a) so that $OF = a$ and the perimeter of $ABCDE$ is P

PROVE: $A_{ABCDE} = \frac{1}{2}aP$

PROOF: From center O, draw radii $\overline{OA}$, $\overline{OB}$, $\overline{OC}$, $\overline{OD}$, and $\overline{OE}$. [See Figure 8.36(b).] Now $\triangle AOB$, $\triangle BOC$, $\triangle COD$, $\triangle DOE$, and $\triangle EOA$ are all $\cong$ by SSS. Where s represents the length of each of the congruent sides of the regular polygon and a is the length of an apothem, the area of each triangle is $\frac{1}{2}sa$ (from $A = \frac{1}{2}bh$). Therefore, the area of the regular polygon is

$$A_{ABCDE} = \left(\frac{1}{2}sa\right) + \left(\frac{1}{2}sa\right) + \left(\frac{1}{2}sa\right) + \left(\frac{1}{2}sa\right) + \left(\frac{1}{2}sa\right)$$

$$= \frac{1}{2}a(s + s + s + s + s)$$

$$= \frac{1}{2}a(n \cdot s)$$

$$= \frac{1}{2}aP$$

$$A_{ABCDE} = \frac{1}{2}aP$$

EXAMPLE 4

Use $A = \frac{1}{2}aP$ to find the area of the square whose apothem length is $a = 2$ in.

SOLUTION For this repeat of Example 1, see Figure 8.33 as needed. When the length of apothem of a square is $a = 2$, the length of side is $s = 4$. In turn, the perimeter is $P = 16$ in.
Now $A = \frac{1}{2}aP$ becomes $A = \frac{1}{2} \cdot 2 \cdot 16$, so $A = 16$ in^2.
NOTE: As expected, the answer from Example 1 is repeated in Example 4.

EXAMPLE 5

Use $A = \frac{1}{2}aP$ to find the area of the equilateral triangle whose apothem has the length 6 cm.

SOLUTION For this repeat of Example 3, refer to Figure 8.35 on page 363. Because the length of apothem $\overline{OD}$ is 6 cm, the length of $\overline{AD}$ is $6\sqrt{3}$ cm. In turn, the length of side $\overline{AB}$ is $12\sqrt{3}$ cm. For the equilateral triangle, the perimeter is $P = 3(12\sqrt{3}$ cm) or $36\sqrt{3}$ cm.

Now $A = \frac{1}{2}aP$ becomes $A = \frac{1}{2} \cdot 6 \cdot 36\sqrt{3}$, so $A = 108\sqrt{3}$ cm².

NOTE: The answer found in Example 5 must repeat that of Example 3.

For Examples 6 and 7, the measures of the line segments that represent the length of the apothem, the radius, or the side of a regular polygon depend upon relationships that are developed in the study of trigonometry. The methods used to find related measures will be developed in Chapter 11 but are not given attention at this time. Using those methods, many of the measures provided in the following examples are actually only *good approximations*.

EXAMPLE 6

In Figure 8.36(a) on page 364, find the area of the regular pentagon *ABCDE* with center *O* if $OF = 4$ and $AB = 5.9$.

SOLUTION $OF = a = 4$ and $AB = 5.9$. Therefore, $P = 5(5.9)$ or $P = 29.5$. Consequently,

$$A = \frac{1}{2}aP \text{ becomes}$$

$$A_{ABCDE} = \frac{1}{2} \cdot 4(29.5)$$

$$= 59 \text{ units}^2$$

EXAMPLE 7

Find the area of the regular octagon shown in Figure 8.37. The center of *PQRSTUVW* is point *O*. The length of apothem $\overline{OX}$ is 12.1 cm, and the length of side $\overline{QR}$ is 10 cm.

Figure 8.37

SOLUTION If $QR = 10$ cm, then the perimeter of regular octagon *PQRSTUVW* is $8 \cdot 10$ cm or 80 cm. With the length of apothem being $OX = 12.1$ cm, the area formula $A = \frac{1}{2}aP$ becomes $A = \frac{1}{2} \cdot 12.1 \cdot 80$, so $A = 484$ cm².

SSG   **EXS. 12–15**

EXAMPLE 8

Find the exact area of equilateral triangle *ABC* in Figure 8.38 on page 366 if each side measures 12 in. Use the formula $A = \frac{1}{2}aP$.

SOLUTION In $\triangle ABC$, the perimeter is $P = 3 \cdot 12$ or 36 in.

To find the length a of an apothem, we draw the radius $\overline{OA}$ from center *O* to point *A* and the apothem $\overline{OM}$ from *O* to side $\overline{AB}$. Because the radius bisects $\angle BAC$, m$\angle OAB = 30°$. Because apothem $\overline{OM} \perp \overline{AB}$, m$\angle OMA = 90°$. $\overline{OM}$ also bisects $\overline{AB}$. Using the 30°-60°-90° relationship in $\triangle OMA$, we see that $a\sqrt{3} = 6$. Thus

$$a = \frac{6}{\sqrt{3}} = \frac{6}{\sqrt{3}} \cdot \frac{\sqrt{3}}{\sqrt{3}} = \frac{6\sqrt{3}}{3} = 2\sqrt{3}$$

Then $a = 2\sqrt{3}$ while $P = 36$.

Figure 8.38

Now $A = \frac{1}{2}aP$ becomes $A = \frac{1}{2} \cdot 2\sqrt{3} \cdot 36 = 36\sqrt{3}$ in^2.

NOTE: Using the calculator's value for $\sqrt{3}$ leads to an approximation of the area rather than to an exact area.

Discover

TESSELLATIONS

Tessellations are patterns composed strictly of interlocking and nonoverlapping regular polygons. All of the regular polygons of a given number of sides will be congruent. Tessellations are commonly used in design, but especially in flooring (tiles and vinyl sheets). A *pure tessellation* is one formed by using only one regular polygon in the pattern. An *impure tessellation* is one formed by using two different regular polygons.

In the accompanying pure tessellation, only the regular hexagon appears. In nature, the beehive has compartments that are regular hexagons. The sum of the measures of the adjacent angles must equal 360°; in this case, 120° + 120° + 120° = 360°. It would also be possible to form a pure tessellation of congruent squares because the sum of the adjacent angles' measures would be 90° + 90° + 90° + 90° = 360°.

In the impure tessellation shown, the regular octagon and the square are used; of course, the sides of the square and the octagon must be congruent. In Champaign-Urbana, sidewalks found on the University of Illinois campus use this tessellation pattern. Again it is necessary that the sum of the adjacent angles' measures be 360°; for this impure tessellation, 135° + 135° + 90° = 360°.

a) Can congruent equilateral triangles be used to form a pure tessellation?

b) Can two regular hexagons and a square be used to build an impure tessellation?

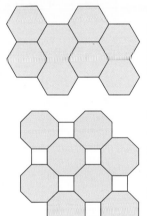

ANSWERS

a) Yes, because 6 × 60° = 360° b) No, because 120° + 120° + 90° ≠ 360°

Exercises 8.3

1. Find the area of a square with
   a) sides of length 3.5 cm each.
   b) an apothem of length 4.7 in.

2. Find the area of a square with
   a) a perimeter of 14.8 cm.
   b) a radius of length $4\sqrt{2}$ in.

3. Find the area of an equilateral triangle with
   a) sides of length 2.5 m each.
   b) an apothem of length 3 in.

4. Find the area of an equiangular triangle with
   a) a perimeter of 24.6 cm.
   b) a radius of length 4 in.

5. In a regular polygon, each central angle measures 30°. If each side of the regular polygon measures 5.7 in., find the perimeter of the polygon.

6. In a regular polygon, each interior angle measures 135°. If each side of the regular polygon measures 4.2 cm, find the perimeter of the polygon.

7. For a regular hexagon, the length of the apothem is 10 cm. Find the length of the radius for the circumscribed circle for this hexagon.

8. For a regular hexagon, the length of the radius is 12 in. Find the length of the radius for the inscribed circle for this hexagon.

9. In a particular type of regular polygon, the length of the radius is exactly the same as the length of a side of the polygon. What type of regular polygon is it?

10. In a particular type of regular polygon, the length of the apothem is exactly one-half the length of a side. What type of regular polygon is it?

11. In one type of regular polygon, the measure of each interior angle $\left(I = \frac{(n - 2)180°}{n}\right)$ is equal to the measure of each central angle. What type of regular polygon is it?

12. If the area $\left(A = \frac{1}{2}aP\right)$ and the perimeter of a regular polygon are numerically equal, find the length of the apothem of the regular polygon.

13. Find the area of a square with apothem $a = 3.2$ cm and perimeter $P = 25.6$ cm.

14. Find the area of an equilateral triangle with apothem $a = 3.2$ cm and perimeter $P = 19.2\sqrt{3}$ cm.

15. Find the area of an equiangular triangle with apothem $a = 4.6$ in. and perimeter $P = 27.6\sqrt{3}$ in.

16. Find the area of a square with apothem $a = 8.2$ ft and perimeter $P = 65.6$ ft.

In Exercises 17 to 30, use the formula $A = \frac{1}{2}aP$ to find the area of the regular polygon described.

17. Find the area of a regular pentagon with an apothem of length $a = 5.2$ cm and each side of length $s = 7.5$ cm.

18. Find the area of a regular pentagon with an apothem of length $a = 6.5$ in. and each side of length $s = 9.4$ in.

19. Find the area of a regular octagon with an apothem of length $a = 9.8$ in. and each side of length $s = 8.1$ in.

20. Find the area of a regular octagon with an apothem of length $a = 7.9$ ft and each side of length $s = 6.5$ ft.

21. Find the area of a regular hexagon whose sides have length 6 cm.

22. Find the area of a square whose apothem measures 5 cm.

23. Find the area of an equilateral triangle whose radius measures 10 in.

24. Find the approximate area of a regular pentagon whose apothem measures 6 in. and each of whose sides measures approximately 8.9 in.

25. In a regular octagon, the approximate ratio of the length of an apothem to the length of a side is 6:5. For a regular octagon with an apothem of length 15 cm, find the approximate area.

26. In a regular dodecagon (12 sides), the approximate ratio of the length of an apothem to the length of a side is 15:8. For a regular dodecagon with a side of length 12 ft, find the approximate area.

27. In a regular dodecagon (12 sides), the approximate ratio of the length of an apothem to the length of a side is 15:8. For a regular dodecagon with an apothem of length 12 ft, find the approximate area.

28. In a regular octagon, the approximate ratio of the length of an apothem to the length of a side is 6:5. For a regular octagon with a side of length 15 ft, find the approximate area.

29. In a regular polygon of 12 sides, the measure of each side is 2 in., and the measure of an apothem is exactly $(2 + \sqrt{3})$ in. Find the exact area of this regular polygon.

30. In a regular octagon, the measure of each apothem is 4 cm, and each side measures exactly $8(\sqrt{2} - 1)$ cm. Find the exact area of this regular polygon.

31. Find the ratio of the area of a square circumscribed about a circle to the area of a square inscribed in the circle.

*32. Given regular hexagon $ABCDEF$ with each side of length 6 and diagonal $\overline{AC}$, find the area of pentagon $ACDEF$.

*33. Given regular octagon $RSTUVWXY$ with each side of length 4 and diagonal $\overline{RU}$, find the area of hexagon $RYXWVU$.

34. Regular octagon $ABCDEFGH$ is inscribed in a circle with radius $r = \frac{7}{2}\sqrt{2}$ cm. Considering that the area of the octagon is less than the area of the circle and greater than the area of the square $ACEG$, find the two integers between which the area of the octagon must lie.

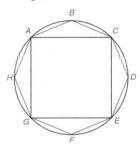

(Note: For the circle, use $A = \pi r^2$ with $\pi \approx \frac{22}{7}$.)

*35. Given regular pentagon $RSTVQ$ and equilateral triangle PQR, the length of an apothem (not shown) of $RSTVQ$ is 12, while the length of each side of the equilateral triangle is 10. If $PV \approx 8.2$, find the approximate area of kite $VPST$.

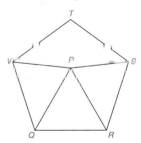

*36. Consider regular pentagon $RSTVQ$ (not shown). Given that diagonals $\overline{QT}$ and $\overline{VR}$ intersect at point F, show that $VF \cdot FR = TF \cdot FQ$.

*37. Consider a regular hexagon $ABCDEF$ (not shown). By joining midpoints of consecutive sides, a smaller regular hexagon $MNPQRS$ is formed. Show that the ratio of areas is
$$\frac{A_{MNPQRS}}{A_{ABCDEF}} = \frac{3}{4}.$$

*38. The length of each side of a regular hexagon measures 6 in. Find the area of the inscribed regular hexagram shaded in the figure.

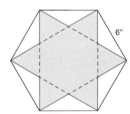

8.4 Circumference and Area of a Circle

KEY CONCEPTS   Circumference of π (Pi)   Limit
                   a Circle   Length of an Arc   Area of a Circle

Figure 8.39

In geometry, any two figures that have the same shape are similar; for this reason, all circles are similar to each other. Just as a proportionality exists among the lengths of the sides of similar triangles, experimentation shows that there is a proportionality among the circumferences (distances around) and the lengths of diameters (distances across) of circles; see the Discover activity on this page. Representing the circumferences of the circles in Figure 8.39 by C_1, C_2, and C_3 and their respective lengths of diameters by d_1, d_2, and d_3, we claim that

$$\frac{C_1}{d_1} = \frac{C_2}{d_2} = \frac{C_3}{d_3} = k$$

where k is the constant of proportionality.

POSTULATE 22

The ratio of the circumference of a circle to the length of its diameter is a unique positive constant.

The constant of proportionality k (described in the opening paragraph of this section, in Postulate 22, and in the Discover activity) is represented by the Greek letter $\pi\,(pi)$.

DEFINITION

π is the ratio between the circumference C and the diameter length d of any circle; thus, $\pi = \frac{C}{d}$ in any circle.

In the following theorem, the lengths of the diameter and radius of the circle are represented by d and r, respectively; of course, $d = 2r$.

THEOREM 8.4.1

The circumference of a circle is given by the formula

$$C = \pi d \qquad \text{or} \qquad C = 2\pi r$$

Discover

Find an object of circular shape, such as the lid of a jar. Using a flexible tape measure (such as a seamstress or carpenter might use), measure both the distance around (circumference) and the distance across (length of diameter) the circle. Now divide the circumference C by the diameter length d. What is your result?

ANSWER
The ratio should be slightly larger than 3.

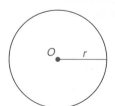

Figure 8.40

GIVEN: Circle O with length of diameter d and length of radius r. (See Figure 8.40.)

PROVE: $C = 2\pi r$

PROOF: By definition, $\pi = \frac{C}{d}$. Multiplying each side of the equation by d, we have $C = \pi d$. Because $d = 2r$, the formula for the circumference can be written $C = \pi(2r)$ or $C = 2\pi r$. ∎

VALUE OF π

In calculating the circumference of a circle, we generally leave the symbol π in the answer in order to state an *exact* result. However, the value of π is irrational and cannot be represented exactly by a common fraction or by a terminating decimal.

SSG **EXS. 1–2**

> *Commonly used Approximations of* π
>
> $$\pi \approx \frac{22}{7} \qquad \pi \approx 3.14 \qquad \pi \approx 3.1416$$

When a calculator is used to determine π with greater accuracy, we see an approximation such as $\pi = 3.141592654$.

EXAMPLE 1

In $\odot O$ in Figure 8.41, $OA = 7$ cm. Using $\pi \approx \frac{22}{7}$,

   a) find the approximate circumference C of $\odot O$.
   b) find the approximate length of the minor arc $\overset{\frown}{AB}$.

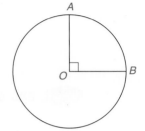

Figure 8.41

SOLUTION

a) $C = 2\pi r$
$\quad = 2 \cdot \frac{22}{7} \cdot 7$
$\quad = 44$ cm

b) Because the degree of measure of $\overset{\frown}{AB}$ is $90°$, the arc length is $\frac{90}{360}$ or $\frac{1}{4}$ of the circumference, 44 cm.

Thus, the length of $\overset{\frown}{AB} = \frac{90}{360} \cdot 44 = \frac{1}{4} \cdot 44 = 11$ cm.

EXAMPLE 2

The exact circumference of a circle is 17π in.

   a) Find the length of the radius.   b) Find the length of the diameter.

SOLUTION

a) $\quad C = 2\pi r$
$\quad 17\pi = 2\pi r$
$\quad \dfrac{17\pi}{2\pi} = \dfrac{2\pi r}{2\pi}$
$\quad\quad r = \dfrac{17}{2} = 8.5$ in.

b) Because $d = 2r$, $d = 2(8.5)$ or $d = 17$ in.

EXAMPLE 3

2.37

Figure 8.42

A thin circular rubber gasket is used as a seal to prevent oil from leaking from a tank (see Figure 8.42). If the gasket has a radius of 2.37 in., use the value of π provided by your calculator to find the circumference of the gasket to the nearest hundredth of an inch.

SOLUTION Using the calculator with $C = 2\pi r$, we have $C = 2 \cdot \pi \cdot 2.37$ or $C \approx 14.89114918$. Rounding to the nearest hundredth of an inch, $C \approx 14.89$ in.

SSG **EXS. 3–5**

LENGTH OF AN ARC

In Example 1(b), we used the phrase *length of arc* without a definition. Informally, the length of an arc is the distance between the endpoints of the arc as though it were measured along a straight line. If we measured one-third of the circumference of the rubber gasket (a $120°$ arc) in Example 3, we would expect the length to be slightly less than 5 in. This measurement could be accomplished by holding that part of the gasket taut in a straight line, but not so tightly that it would be stretched.

Two further considerations regarding the measurement of arc length follow.

1. The ratio of the degree measure m of the arc to 360 (the degree measure of the entire circle) is the same as the ratio of the length ℓ of the arc to the circumference; that is, $\frac{m}{360} = \frac{\ell}{C}$.

2. Just as $m\overparen{AB}$ denotes the degree measure of an arc, $\ell\overparen{AB}$ denotes the *length* of the arc. Whereas $m\overparen{AB}$ is measured in degrees, $\ell\overparen{AB}$ is measured in linear units such as inches, feet, or centimeters.

THEOREM 8.4.2

In a circle whose circumference is C, the length ℓ of an arc whose degree measure is m is given by

$$\ell = \frac{m}{360} \cdot C$$

NOTE: For arc AB, $\ell\overparen{AB} = \frac{m\overparen{AB}}{360} \cdot C$.

 EXS. 6–8

Figure 8.43

EXAMPLE 4

Find the approximate length of major arc ABC in a circle of radius 7 in. if $m\overparen{AC} = 45°$. See Figure 8.43. Use $\pi \approx \frac{22}{7}$.

SOLUTION $m\overparen{ABC} = 360° - 45° = 315°$. According to Theorem 8.4.2,

$\ell\overparen{ABC} = \frac{m\overparen{ABC}}{360} \cdot C$, or $\ell\overparen{ABC} = \frac{315}{360} \cdot 2 \cdot \frac{22}{7} \cdot 7$, which can be simplified to $\ell\overparen{ABC} = 38\frac{1}{2}$ in.

LIMITS

In the discussion that follows, we use the undefined term *limit*; in practice, a limit represents a numerical measure. In some situations, we seek an upper limit, a lower limit, or both. The following example illustrates this notion.

EXAMPLE 5

Find the upper limit (largest possible number) for the length of a chord in a circle whose length of radius is 5 cm.

SOLUTION By considering several chords in the circle in Figure 8.44, we see that the greatest possible length of a chord is that of a diameter. Thus, the limit of the length of a chord is 10 cm.

NOTE: Although the length of a chord must be a positive number, the lower limit of this length is 0.

Figure 8.44

AREA OF A CIRCLE

Now consider the problem of finding the area of a circle. To do so, let a regular polygon of n sides be inscribed in the circle. As we consider larger and larger values of n (often written as $n \to \infty$ and read "n approaches infinity"), two observations can be made:

1. The length of an apothem of the regular polygon approaches the length of a radius of the circle as its limit ($a \to r$).

2. The perimeter of the regular polygon approaches the circumference of the circle as its limit ($P \to C$).

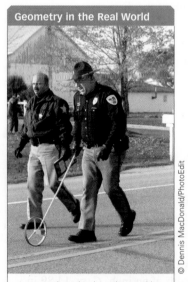

Geometry in the Real World

A measuring wheel can be used by a police officer to find the length of skid marks or by a cross-country coach to determine the length of a running course.

 EXS. 9–11

In Figure 8.45, the area of an inscribed regular polygon with n sides approaches the area of the circle as its limit as n increases. Using observations 1 and 2, we make the following claim. Because the formula for the area of a regular polygon is

$$A = \frac{1}{2}aP \quad \text{and} \quad a \to r \text{ while } P \to C,$$

the area of the circumscribed circle is given by the limit

$$A = \frac{1}{2}rC$$

Because $C = 2\pi r$, this formula becomes $A = \frac{1}{2}r(2\pi r)$.

Figure 8.45

Simplifying, $\qquad A = \frac{1}{2}r(\cancel{2}\pi r) \qquad \text{or} \qquad A = \pi r^2$

Based upon the preceding discussion, we state Theorem 8.4.3.

THEOREM 8.4.3

The area A of a circle whose radius has length r is given by $A = \pi r^2$.

Another justification of the formula $A = \pi r^2$ is found in the following Discover activity.

Discover

AREA OF A CIRCLE

Use a protractor to divide a circle into several congruent "sectors." For instance, 15° central angles will divide the circle into $\frac{360}{15} = 24$ sectors. If these sectors are alternated as shown, the resulting figure approximates a parallelogram. This parallelogram has a base of length πr (half the circumference of the circle) and an altitude of length r (radius of the circle). Based upon the formula $A = bh$, the area of the parallelogram (and of the circle) can be seen to be $A = (\pi r) r$, or $A = \pi r^2$.

EXAMPLE 6

Find the approximate area of a circle whose radius has a length of 10 in. Use $\pi \approx 3.14$.

SOLUTION $A = \pi r^2$ becomes $A = 3.14(10)^2$. Then

$$A = 3.14(100) = 314 \text{ in}^2$$

EXAMPLE 7

The approximate area of a circle is 38.5 cm². Find the length of the radius of the circle. Use $\pi \approx \frac{22}{7}$.

SOLUTION By substituting known values, the formula $A = \pi r^2$ becomes $38.5 = \frac{22}{7} \cdot r^2$, or $\frac{77}{2} = \frac{22}{7} \cdot r^2$. Multiplying each side of the equation by $\frac{7}{22}$, we have

$$\frac{7}{\cancel{22}} \cdot \frac{\cancel{77}^{7}}{2} = \frac{\cancel{7}}{\cancel{22}} \cdot \frac{\cancel{22}}{\cancel{7}} \cdot r^2$$

or

$$r^2 = \frac{49}{4}$$

Taking the positive square root for the approximate length of radius,

$$r = \sqrt{\frac{49}{4}} = \frac{\sqrt{49}}{\sqrt{4}} = \frac{7}{2} = 3.5 \text{ cm}$$

Figure 8.46

A plane figure bounded by concentric circles is known as a *ring* or *annulus* (see Figure 8.46). The piece of hardware known as a *washer* has the shape of an annulus.

 EXS. 12–17

EXAMPLE 8

A machine cuts washers from a flat piece of metal. The radius of the inside circular boundary of the washer is 0.3 in., and the radius of the outer circular boundary is 0.5 in. What is the area of the annulus? Give both an exact answer and an approximate answer rounded to tenths of a square inch. Using the approximate answer, determine the number of square inches of material used to produce 1000 washers.

SOLUTION Where R is the larger radius and r is the smaller radius, $A = \pi R^2 - \pi r^2$. Then $A = \pi(0.5)^2 - \pi(0.3)^2$, or $A = 0.16\pi$. The exact number of square inches used in producing a washer is 0.16π in^2, or approximately 0.5 in^2. When 1000 washers are produced, approximately 500 in^2 of metal is used.

Many students have a difficult time remembering which expression ($2\pi r$ or πr^2) is used in the formula for the circumference or for the area of a circle. This is understandable because each expression contains a 2, a radius length r, and the factor π. To remember that $C = 2\pi r$ gives the circumference and $A = \pi r^2$ gives the area, *think about the units involved*. Considering a circle of radius 3 in., $C = 2\pi r$ becomes $C = 2 \times 3.14 \times 3$ in., or Circumference equals 18.84 *inches*. (We measure the *distance around* a circle in *linear* units such as inches.) For the circle of radius 3 in., $A = \pi r^2$ becomes $A = 3.14 \times 3$ in. $\times 3$ in., or Area equals 28.26 in^2.

Exercises 8.4

1. Find the exact circumference and area of a circle whose radius has length 8 cm.

2. Find the exact circumference and area of a circle whose diameter has length 10 in.

3. Find the approximate circumference and area of a circle whose radius has length $10\frac{1}{2}$ in. Use $\pi \approx \frac{22}{7}$.

4. Find the approximate circumference and area of a circle whose diameter has length 20 cm. Use $\pi \approx 3.14$.

5. Find the exact lengths of the radius and the diameter of a circle whose circumference is:
   a) 44π in.
   b) 60π ft

6. Find the approximate lengths of the radius and the diameter of a circle whose circumference is:
   a) 88 in. (Use $\pi \approx \frac{22}{7}$.)
   b) 157 m (Use $\pi \approx 3.14$.)

7. Find the exact lengths of the radius and the diameter of a circle whose area is:
   a) 25π in^2
   b) 2.25π cm^2

8. Find the exact length of the radius and the exact circumference of a circle whose area is:
   a) 36π m^2
   b) 6.25π ft^2

9. Find exactly $\ell\widehat{AB}$, where $\widehat{AB}$ refers to the minor arc of the circle.

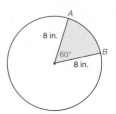

10. Find exactly $\ell \widehat{CD}$ for the minor arc shown.

11. Use your calculator value of π to find the approximate circumference of a circle with radius length 12.38 in.

12. Use your calculator value of π to find the approximate area of a circle with radius length 12.38 in.

13. A metal circular disk whose area is 143 cm² is used as a knockout on an electrical service in a factory. Use your calculator value of π to find the length of the radius of the disk to the nearest tenth of a cm.

14. A circular lock washer whose outside circumference measures 5.48 cm is used in an electric box to hold an electrical cable in place. Use your calculator value of π to find the length of the radius to the nearest tenth of a cm.

15. The central angle corresponding to a circular brake shoe measures 60°. Approximately how long is the curved surface of the brake shoe if the length of radius is 7 in.?

16. Use your calculator to find the approximate lengths of the radius and the diameter of a circle with area 56.35 in².

17. A rectangle has a perimeter of 16 in. What is the limit (largest possible value) of the area of the rectangle?

18. A rectangle has an area of 36 in². What is the limit (smallest possible value) of the perimeter of the rectangle?

19. The legs of an isosceles triangle each measure 10 cm. What are the limits of the length of the base?

20. Two sides of a triangle measure 5 in. and 7 in. What are the limits of the length of the third side?

21. Let N be any point on side $\overline{BC}$ of the right triangle ABC. Find the upper and lower limits for the length of $\overline{AN}$.

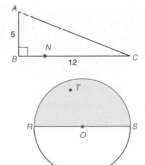

22. What is the limit of $m\angle RTS$ if T lies in the interior of the shaded region?

In Exercises 23 to 26, find the exact areas of the shaded regions.

23.

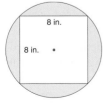

Square inscribed in a circle

24.

25.

$d_1 = 30$ ft
$d_2 = 40$ ft
Rhombus

26.

Regular hexagon inscribed in a circle

In Exercises 27 and 28, use your calculator value of π to solve each problem. Round answers to the nearest integer.

27. Find the length of the radius of a circle whose area is 154 cm².

28. Find the length of the diameter of a circle whose circumference is 157 in.

29. Assuming that a 90° arc has an exact length of 4π in., find the length of the radius of the circle.

30. The ratio of the circumferences of two circles is 2:1. What is the ratio of their areas?

31. Given concentric circles with radii of lengths R and r, where $R > r$, explain why $A_{ring} = \pi(R + r)(R - r)$.

32. Given a circle with diameter of length d, explain why $A_{circle} = \frac{1}{4}\pi d^2$.

33. The radii of two concentric circles differ in length by exactly 1 in. If their areas differ by exactly 7π in², find the lengths of the radii of the two circles.

In Exercises 34 to 44, use your calculator value of π unless otherwise stated. Round answers to two decimal places.

34. The carpet in the circular entryway of a church needs to be replaced. The diameter of the circular region to be carpeted is 18 ft.
 a) What length (in feet) of a metal protective strip is needed to bind the circumference of the carpet?
 b) If the metal strips are sold in lengths of 6 ft, how many will be needed?
 (Note: Assume that these can be bent to follow the circle and that they can be placed end to end.)
 c) If the cost of the metal strip is $2.59 per linear foot, find the cost of the metal strips needed.

35. At center court on a gymnasium floor, a large circular emblem is to be painted. The circular design has a radius length of 8 ft.
 a) What is the area to be painted?
 b) If a quart of paint covers 70 ft², how many quarts of paint are needed to complete the job?
 c) If each quart of paint costs $15.89, find the cost of the paint needed.

36. A track is to be constructed around the football field at a junior high school. If the straightaways are 100 yd in length, what length of radius is needed for each of the semicircles shown if the total length around the track is to be 440 yd?

100 yd

100 yd

37. A circular grass courtyard at a shopping mall has a 40-ft diameter. This area needs to be reseeded.
   a) What is the total area to be reseeded? (Use $\pi \approx 3.14$.)
   b) If 1 lb of seed is to be used to cover a 60-ft² region, how many pounds of seed will be needed?
   c) If the cost of 1 lb of seed is $1.65, what is the total cost of the grass seed needed?

38. Find the approximate area of a regular polygon that has 20 sides if the length of its radius is 7 cm.

39. Find the approximate perimeter of a regular polygon that has 20 sides if the length of its radius is 7 cm.

40. In a two-pulley system, the centers of the pulleys are 20 in. apart. If the radius of each pulley measures 6 in., how long is the belt used in the pulley system?

6"   6"
20"

41. If two gears, each of radius 4 in., are used in a chain drive system with a chain of length 54 in., what is the distance between the centers of the gears?

4"   4"

42. A pizza with a 12-in. diameter costs $6.95. A 16-in. diameter pizza with the same ingredients costs $9.95. Which pizza is the better buy?

43. A communications satellite forms a circular orbit 375 mi above the earth. If the earth's radius is approximately 4000 mi, what distance is traveled by the satellite in one complete orbit?

44. The radius of the Ferris wheel's circular path is 40 ft. If a "ride" of 12 revolutions is made in 3 minutes, at what rate in *feet per second* is the passenger in a cart moving during the ride?

45. The diameter of a carousel (merry-go-round) is 30 ft. At full speed, it makes a complete revolution in 6 s. At what rate, in *feet per second*, is a horse on the outer edge moving?

46. A tabletop is semicircular when its three congruent drop-leaves are used. By how much has the table's area increased when the drop leaves are raised? Give the answer to the nearest *whole* percent.

*47. Given that the length of each side of a rhombus is 8 cm and that an interior angle (shown) measures 60°, find the area of the inscribed circle.

8 cm
60°

8.5   More Area Relationships in the Circle

KEY CONCEPTS	Sector Area and Perimeter of a Sector	Segment of a Circle Area and Perimeter of a Segment	Area of a Triangle with an Inscribed Circle

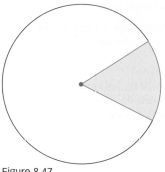

> **DEFINITION**
>
> A **sector** of a circle is a region bounded by two radii of the circle and an arc intercepted by those radii. (See Figure 8.47.)

A sector will generally be shaded to avoid confusion about whether the related arc is a major arc or a minor arc. As shown in Figure 8.47, the sector of a circle generally has the shape of a center cut piece of pie.

Figure 8.47

AREA OF A SECTOR

Just as the length of an arc is part of the circle's circumference, the area of a sector is part of the area of this circle. When fractions are illustrated by using circles, $\frac{1}{4}$ is represented by shading a 90° sector, and $\frac{1}{3}$ is represented by shading a 120° sector (see Figure 8.48). Thus, we make the following assumption about the measure of the area of a sector.

$$\frac{1}{4} = \frac{90°}{360°} \qquad \frac{1}{3} = \frac{120°}{360°}$$

Figure 8.48

POSTULATE 23

The ratio of the degree measure m of the arc (or central angle) of a sector to 360° is the same as the ratio of the area of the sector to the area of the circle; that is,

$$\frac{\text{area of sector}}{\text{area of circle}} = \frac{m}{360}.$$

THEOREM 8.5.1

In a circle of radius length r, the area A of a sector whose arc has degree measure m is

$$A = \frac{m}{360}\pi r^2.$$

Theorem 8.5.1 follows directly from Postulate 23.

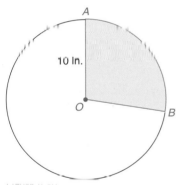

Figure 8.49

EXAMPLE 1

If $m\angle O = 100°$, find the area of the 100° sector shown in Figure 8.49. Use your calculator and round the answer to the nearest hundredth of a square inch.

SOLUTION

$$A = \frac{m}{360}\pi r^2$$

becomes

$$A = \frac{100}{360} \cdot \pi \cdot 10^2 \approx 87.27 \text{ in}^2$$

In applications with circles, we often seek exact answers for circumference and area; in such cases, we simply leave π in the result. For instance, in a circle of radius length 5 in., the exact circumference is 10π in. and the exact area is expressed as 25π in².

Because a sector is bounded by two radii and an arc, the perimeter of a sector is the sum of the lengths of the two radii and the length of its arc. In Example 2, we use the intercepted arc $\overarc{AB}$ and apply the formula, $P_{\text{sector}} = 2r + \ell\overarc{AB}$.

EXAMPLE 2

Find the perimeter of the sector shown in Figure 8.49. Use the calculator value of π and round your answer to the nearest hundredth of an inch.

SOLUTION Because $r = 10$ and $m\angle O = 100°$, $\ell\overarc{AB} = \frac{100}{360} \cdot 2 \cdot \pi \cdot 10 \approx 17.45$ in.
Now $P_{\text{sector}} = 2r + \ell\overarc{AB}$ becomes $P_{\text{sector}} = 2(10) + 17.45 \approx 37.45$ in.

Because a semicircle is one-half of a circle, a semicircular region corresponds to a central angle of 180°. As stated in the following corollary to Theorem 8.5.1, the area of the semicircular region is $\frac{180}{360}$ (or one-half) the area of the entire circle.

SSG EXS. 1–6

COROLLARY 8.5.2

The area of a semicircular region of radius length r is $A = \frac{1}{2}\pi r^2$.

Figure 8.50

EXAMPLE 3

In Figure 8.50, a square of side 8 in. is shown with semicircles cut away. Find the exact shaded area leaving π in the answer.

SOLUTION To find the shaded area A, we see that $A + 2 \cdot A_{\text{semicircle}} = A_{\text{square}}$.
It follows that $A = A_{\text{square}} - 2 \cdot A_{\text{semicircle}}$.
If the side of the square is 8 in., then the radius of each semicircle is 4 in. Now $A = 8^2 - 2\left(\frac{1}{2}\pi \cdot 4^2\right)$, or $A = 64 - 2(8\pi)$, so $A = (64 - 16\pi)$ in².

Discover

In statistics, a pie chart can be used to represent the breakdown of a budget. In the pie chart shown, a 90° sector (one-fourth the area of the circle) is shaded to show that 25% of a person's income (one-fourth of the income) is devoted to rent payment. What degree measure of sector must be shaded if a sector indicates that 20% of the person's income is used for a car payment?

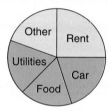

ANSWER
72° (from 20% of 360°)

AREA OF A SEGMENT

DEFINITION

A **segment** of a circle is a region bounded by a chord and its minor (or major) arc.

Figure 8.51

In Figure 8.51, the segment of the circle is bounded by chord $\overline{AB}$ and its minor arc $\overparen{AB}$. Again, we avoid confusion by shading the segment whose area or perimeter we seek.

EXAMPLE 4

Find the exact area of the segment bounded by the chord and the arc whose measure is 90°. The radius has length 12 in., as shown in Figure 8.52.

SOLUTION Let $A_{\triangle}$ represent the area of the triangle shown. Because $A_{\triangle} + A_{\text{segment}} = A_{\text{sector}}$, we see that $A_{\text{segment}} = A_{\text{sector}} - A_{\triangle}$

$$A_{\text{segment}} = \frac{90}{360} \cdot \pi \cdot 12^2 - \frac{1}{2} \cdot 12 \cdot 12$$

$$= \frac{1}{4} \cdot 144\pi - \frac{1}{2} \cdot 144$$

$$= (36\pi - 72) \text{ in}^2$$

Figure 8.52

In Example 4, the boundaries of the segment shown are chord $\overline{AB}$ and minor arc $\overparen{AB}$. Therefore, the perimeter of the segment is given by $P_{\text{segment}} = AB + \ell\overparen{AB}$. We use this formula in Example 5.

EXAMPLE 5

Find the exact perimeter of the segment shown in Figure 8.53. Then use your calculator to approximate this answer to the nearest hundredth of an inch.

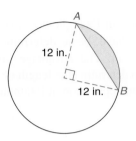

Figure 8.53

SOLUTION Because $\ell\overparen{AB} = \frac{90}{360} \cdot 2 \cdot \pi \cdot r$, we have $\ell\overparen{AB} = \frac{1}{4} \cdot 2 \cdot \pi \cdot 12 = 6\pi$ in. Using either the Pythagorean Theorem or the 45°-45°-90° relationship, $AB = 12\sqrt{2}$. Now $P_{\text{segment}} = AB + \ell\overparen{AB}$ becomes $P_{\text{segment}} = (12\sqrt{2} + 6\pi)$ in. Using a calculator, we find that the approximate perimeter is 35.82 in.

GGG EXS 7-11

AREA OF A TRIANGLE WITH AN INSCRIBED CIRCLE

THEOREM 8.5.3

Where P represents the perimeter of a triangle and r represents the length of the radius of its inscribed circle, the area of the triangle is given by

$$A = \frac{1}{2}rP$$

PICTURE PROOF OF THEOREM 8.5.3

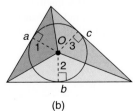

(a)

(b)

Figure 8.54

GIVEN: A triangle with perimeter P, whose sides measure a, b, and c; the radius of the inscribed circle measures r. See Figure 8.54(a).

PROVE: $A = \frac{1}{2}rP$

PROOF: In Figure 8.54(b), the triangle has been separated into three smaller triangles (each with altitude r). Hence

$$A = A_1 + A_2 + A_3$$
$$A = \frac{1}{2}r \cdot a + \frac{1}{2}r \cdot b + \frac{1}{2}r \cdot c$$
$$A = \frac{1}{2}r(a + b + c)$$
$$A = \frac{1}{2}rP$$

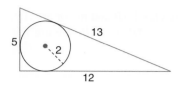

Figure 8.55

EXAMPLE 6

Find the area of a triangle whose sides measure 5 cm, 12 cm, and 13 cm, if the radius length of the inscribed circle is 2 cm. See Figure 8.55.

SOLUTION With the given lengths of sides, the perimeter of the triangle is
$P = 5 + 12 + 13 = 30$ cm. Using $A = \frac{1}{2}rP$, we have $A = \frac{1}{2} \cdot 2 \cdot 30$, or $A = 30$ cm^2.

Because the triangle shown in Example 6 is a right triangle ($5^2 + 12^2 = 13^2$), the area of the triangle could have been determined by using either $A = \frac{1}{2}ab$ or $A = \sqrt{s(s - a)(s - b)(s - c)}$. The advantage provided by Theorem 8.5.3 lies in applications where we need to determine the length of the radius of the inscribed circle of a triangle. We consider such an application in Example 7.

EXAMPLE 7

In an attic, wooden braces supporting the roof form a triangle whose sides measure 4 ft, 6 ft, and 6 ft; see Figure 8.56. To the nearest inch, find the radius of the largest circular cold-air duct that can be run through the opening formed by the braces.

SOLUTION Where s is the semiperimeter of the triangle, Heron's Formula states that $A = \sqrt{s(s - a)(s - b)(s - c)}$. With $s = \frac{1}{2}(a + b + c) = \frac{1}{2}(4 + 6 + 6) = 8$, we have $A = \sqrt{8(8 - 4)(8 - 6)(8 - 6)} = \sqrt{8(4)(2)(2)} = \sqrt{128}$. We can simplify the area expression to $\sqrt{64} \cdot \sqrt{2}$, so $A = 8\sqrt{2}$ ft^2.

Recalling Theorem 8.5.3, we know that $A = \frac{1}{2}rP$. Substitution leads to $8\sqrt{2} = \frac{1}{2}r(4 + 6 + 6)$, or $8\sqrt{2} = 8r$. Then $r = \sqrt{2} \approx 1.414$ ft. Converting to inches, it follows that $r \approx 1.414(12$ in.$) \approx 17$ in.

NOTE: If the ductwork is a flexible plastic tubing, the duct having radius 17 in. can probably be used. If the ductwork were a rigid metal or heavy plastic, the radius might need to be restricted to perhaps 16 in.

Figure 8.56

 EXS. 12–15

Exercises 8.5

1. Given ⊙O with radii $\overline{OA}$ and $\overline{OB}$ and chord $\overline{AB}$.
  a) What type of figure (sector or segment) is bounded by $\overline{OA}$, $\overline{OB}$, and $\widehat{AB}$?
  b) If $OA = 7$ cm and $\ell\widehat{AB} = 11$ cm, find the perimeter of the figure in (a).

2. Given ⊙O with radii $\overline{OA}$ and $\overline{OB}$ and chord $\overline{AB}$.
  a) What type of figure (sector or segment) is bounded by $\overline{AB}$ and $\widehat{AB}$?
  b) If $AB = 9.7$ cm and $\ell\widehat{AB} = 11.4$ cm, find the perimeter of the figure in (a).

3. In the semicircular region shaded, $DQ = 6''$.
  a) Find the exact perimeter of the region.
  b) Find the exact area of the region.

Exercises 1, 2

Exercises 3, 4

4. For the semicircular region of Exercise 3, the length of the radius is *r*.
   a) Find an expression for the perimeter of the region.
   b) Find an expression for the area of the region.

5. In the circle, the radius length is 10 in. and the length of $\overset{\frown}{AB}$ is 14 in. What is the perimeter of the shaded sector?

Exercises 5, 6

6. If the area of the circle is 360 in², what is the area of the sector if its central angle measures 90°?

7. If the area of the 120° sector is 50 cm², what is the area of the entire circle?

8. If the area of the 120° sector is 40 cm² and the area of △*MON* is 16 cm², what is the area of the segment bounded by chord $\overline{MN}$ and $\overset{\frown}{MN}$?

Exercises 7, 8

9. Suppose that a circle of radius *r* is inscribed in an equilateral triangle whose sides have length *s*. Find an expression for the area of the triangle in terms of *r* and *s*.

   (HINT: Use Theorem 8.5.3.)

10. Suppose that a circle of radius *r* is inscribed in a rhombus each of whose sides has length *s*. Find an expression for the area of the rhombus in terms of *r* and *s*.

11. Find the perimeter of a segment of a circle whose boundaries are a chord measuring 24 mm (millimeters) and an arc of length 30 mm.

12. A sector with perimeter 30 in. has a bounding arc of length 12 in. Find the length of the radius of the circle.

13. A circle is inscribed in a triangle having sides of lengths 6 in., 8 in., and 10 in. If the length of the radius of the inscribed circle is 2 in., find the area of the triangle.

14. A circle is inscribed in a triangle having sides of lengths 5 in., 12 in., and 13 in. If the length of the radius of the inscribed circle is 2 in., find the area of the triangle.

15. A triangle with sides of lengths 3 in., 4 in., and 5 in. has an area of 6 in². What is the length of the radius of the inscribed circle?

16. The approximate area of a triangle with sides of lengths 3 in., 5 in., and 6 in. is 7.48 in². What is the approximate length of the radius of the inscribed circle?

17. Find the exact perimeter and area of the sector shown.

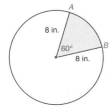

18. Find the exact perimeter and area of the sector shown.

19. Find the approximate perimeter of the sector shown. Answer to the nearest hundredth of an inch.

20. Find the approximate area of the sector shown. Answer to the nearest hundredth of a square inch.

Exercises 19, 20

21. Find the exact perimeter and area of the segment shown, given that m∠*O* = 60° and *OA* = 12 in.

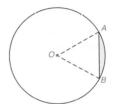

Exercises 21, 22

22. Find the exact perimeter and area of the segment shown, given that m∠*O* = 120° and *AB* = 10 in.

In Exercises 23 and 24, find the exact areas of the shaded regions.

23.

24.

Square *ABCD*

25. Assuming that the exact area of a sector determined by a 40° arc is $\frac{9}{4}\pi$ cm², find the length of the radius of the circle.

26. For concentric circles with radii of lengths 3 in. and 6 in., find the area of the smaller segment determined by a chord of the larger circle that is also a tangent of the smaller circle.

*27. A circle can be inscribed in the trapezoid shown. Find the area of that circle.

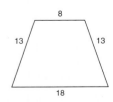

*28. A circle can be inscribed in an equilateral triangle, each of whose sides has length 10 cm. Find the area of that circle.

29. In a circle whose radius has length 12 m, the length of an arc is 6π m. What is the degree measure of that arc?

30. At the Pizza Dude restaurant, a 12-in. pizza costs $5.40 to make, and the manager wants to make at least $4.80 from the sale of each pizza. If the pizza will be sold by the slice and each pizza is cut into 6 pieces, what is the minimum charge per slice?

31. At the Pizza Dude restaurant, pizza is sold by the slice. If the pizza is cut into 6 pieces, then the selling price is $1.95 per slice. If the pizza is cut into 8 pieces, then each slice is sold for $1.50. In which way will the Pizza Dude restaurant clear more money from sales?

32. Determine a formula for the area of the shaded region determined by the square and its inscribed circle.

33. Determine a formula for the area of the shaded region determined by the circle and its inscribed square.

34. Find a formula for the area of the shaded region, which represents one-fourth of an annulus (ring).

35. A company logo on the side of a building shows an isosceles triangle with an inscribed circle. If the sides of the triangle measure 10 ft, 13 ft, and 13 ft, find the length of the radius of the inscribed circle.

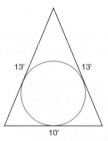

36. In a right triangle with sides of lengths a, b, and c (where c is the length of the hypotenuse), show that the length of the radius of the inscribed circle is $r = \frac{ab}{a + b + c}$.

37. In a triangle with sides of lengths a, b, and c and semiperimeter s, show that the length of the radius of the inscribed circle is

$$r = \frac{2\sqrt{s(s - a)(s - b)(s - c)}}{a + b + c}$$

38. Use the results from Exercises 36 and 37 to find the length of the radius of the inscribed circle for a triangle with sides of lengths
   a) 8, 15, and 17.   b) 7, 9, and 12.

39. Use the results from Exercises 36 and 37 to find the length of the radius of the inscribed circle for a triangle with sides of lengths
   a) 7, 24, and 25.   b) 9, 10, and 17.

40. Three pipes, each of radius length 4 in., are stacked as shown. What is the height of the stack?

41. A windshield wiper rotates through a 120° angle as it cleans a windshield. From the point of rotation, the wiper blade begins at a distance of 4 in. and ends at a distance of 18 in. (The wiper blade is 14 inches in length.) Find the area cleaned by the wiper blade.

42. A goat is tethered to a barn by a 12-ft chain. If the chain is connected to the barn at a point 6 ft from one end of the barn, what is the area of the pasture that the goat is able to graze?

43. An exit ramp from one freeway onto another freeway forms a 90° arc of a circle. The ramp is scheduled for resurfacing. As shown, its inside radius is 370 ft, and its outside radius is 380 ft. What is the area of the ramp?

*44. In $\triangle ABC$, m$\angle C = 90°$ and m$\angle B = 60°$. If $AB = 12$ in., find the radius of the inscribed circle. Give the answer to the nearest tenth of an inch.

*45. A triangle has sides of lengths 6 cm, 8 cm, and 10 cm. Find the distance between the center of the inscribed circle and the center of the circumscribed circle for this triangle. Give the answer to the nearest tenth of a centimeter.

PERSPECTIVE ON HISTORY

SKETCH OF PYTHAGORAS

Pythagoras (circa 580–500 B.C.) was a Greek philosopher and mathematician. Having studied under some of the great minds of the day, he formed his own school around 529 B.C. in Crotona, Italy.

Students of his school fell into two classes, the listeners and the elite Pythagoreans. Included in the Pythagoreans were brilliant students, including 28 women, and all were faithful followers of Pythagoras. The Pythagoreans, who adhered to a rigid set of beliefs, were guided by the principle "Knowledge is the greatest purification."

The apparent areas of study for the Pythagoreans included arithmetic, music, geometry, and astronomy, but underlying principles that led to a cult-like existence included self-discipline, temperance, purity, and obedience. The Pythagoreans recognized fellow members by using the pentagram (five-pointed star) as their symbol. With their focus on virtue, politics, and religion, the members of the group saw themselves as above others. Because of their belief in *transmigration* (movement of the soul after death to another human or animal), the Pythagoreans refused to eat meat or fish. On one occasion, it is said that Pythagoras came upon a person beating a dog. Approaching that person, Pythagoras said, "Stop beating the dog, for in this dog lives the soul of my friend; I recognize him by his voice."

In time, the secrecy, clannishness, and supremacy of the Pythagoreans led to suspicion and fear on the part of other factions of society. Around 500 B.C., the revolution against the Pythagoreans led to the burning of their primary meeting house. Although many of the Pythagoreans died in the ensuing inferno, it is unclear whether Pythagoras himself died or escaped.

PERSPECTIVE ON APPLICATIONS

ANOTHER LOOK AT THE PYTHAGOREAN THEOREM

Some of the many proofs of the Pythagorean Theorem depend on area relationships. One such proof was devised by President James A. Garfield (1831–1881), twentieth president of the United States.

In his proof, the right triangle with legs of lengths a and b and a hypotenuse of length c is transformed into a trapezoid. See Figures 8.57(a) and (b).

In Figure 8.57(b), the points A, B, and C are collinear. With $\angle 1$ and $\angle 2$ being complementary and the sum of the angles' measures about point B being 180°, it follows that $\angle 3$ is a right angle.

If the drawing is perceived as a trapezoid (as shown in Figure 8.58), the area is given by

$$
\begin{aligned}
A &= \frac{1}{2}h(b_1 + b_2) \\
&= \frac{1}{2}(a + b)(a + b) \\
&= \frac{1}{2}(a + b)^2 \\
&= \frac{1}{2}(a^2 + 2ab + b^2) \\
&= \frac{1}{2}a^2 + ab + \frac{1}{2}b^2
\end{aligned}
$$

Figure 8.58

Now we treat the trapezoid as a composite of three triangles as shown in Figure 8.59.

Figure 8.59

(a)

(b)

Figure 8.57

The total area of regions (triangles) I, II, and III is given by

$$A = A_I + A_{II} + A_{III}$$
$$= \frac{1}{2}ab + \frac{1}{2}ab + \left(\frac{1}{2}c \cdot c\right)$$
$$= ab + \frac{1}{2}c^2$$

Equating the areas of the trapezoid in Figure 8.58 and the composite in Figure 8.59, we find that

$$\frac{1}{2}a^2 + ab + \frac{1}{2}b^2 = ab + \frac{1}{2}c^2$$
$$\frac{1}{2}a^2 + \frac{1}{2}b^2 = \frac{1}{2}c^2$$

Multiplying by 2, we have

$$a^2 + b^2 = c^2$$

An earlier proof (over 2000 years earlier!) of this theorem by the Greek mathematician Pythagoras is found in many historical works on geometry. It is not difficult to see the relationship between the two proofs.

In the proof credited to Pythagoras, a right triangle with legs of lengths a and b and hypotenuse of length c is reproduced several times to form a square. Again, points A, B, C (and C, D, E, and so on) must be collinear. [See Figure 8.60(c).]

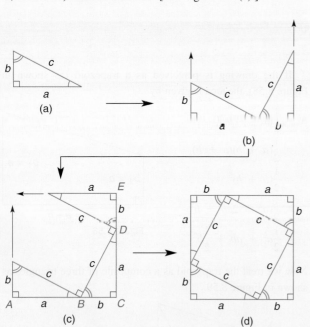

Figure 8.60

The area of the large square in Figure 8.61(a) is given by

$$A = (a + b)^2$$
$$= a^2 + 2ab + b^2$$

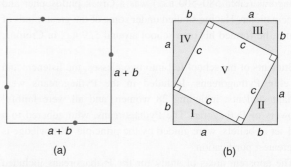

Figure 8.61

Considering the composite in Figure 8.61(b), we find that

$$A = A_I + A_{II} + A_{III} + A_{IV} + A_V$$
$$= 4 \cdot A_I + A_V$$

because the four right triangles are congruent. Then

$$A = 4\left(\frac{1}{2}ab\right) + c^2$$
$$= 2ab + c^2$$

Again, because of the uniqueness of area, the results (area of square and area of composite) must be equal. Then

$$a^2 + 2ab + b^2 = 2ab + c^2$$
$$a^2 + b^2 = c^2$$

Another look at the proofs by President Garfield and by Pythagoras makes it clear that the results must be consistent. In Figure 8.62, observe that Garfield's trapezoid must have one-half the area of Pythagoras' square, while maintaining the relationship that

$$c^2 = a^2 + b^2$$

Figure 8.62

Summary

A Look Back at Chapter 8

One goal of this chapter was to determine the areas of triangles, certain quadrilaterals, and regular polygons. We also explored the circumference and area of a circle and the area of a sector of a circle. The area of a circle is sometimes approximated by using $\pi \approx 3.14$ or $\pi \approx \frac{22}{7}$. At other times, the exact area is given by leaving π in the answer.

A Look Ahead to Chapter 9

Our goal in the next chapter is to deal with a type of geometry known as solid geometry. We will find the surface areas of solids with polygonal or circular bases. We will also find the volumes of these solid figures. Select polyhedra will be discussed.

Key Concepts

8.1

Plane Region • Square Unit • Area Postulates • Area of a Rectangle, a Parallelogram, and a Triangle • Altitude and Base of a Parallelogram and a Triangle

8.2

Perimeter of a Polygon • Semiperimeter of a Triangle • Heron's Formula • Brahmagupta's Formula • Area of a Trapezoid, a Rhombus, and a Kite • Areas of Similar Polygons

8.3

Regular Polygon • Center and Central Angle of a Regular Polygon • Radius and Apothem of a Regular Polygon • Area of a Regular Polygon

8.4

Circumference of a Circle • π (Pi) • Length of an Arc • Limit • Area of a Circle

8.5

Sector • Area and Perimeter of a Sector • Segment of a Circle • Area and Perimeter of a Segment • Area of a Triangle with an Inscribed Circle

Overview ■ Chapter 8

Area and Perimeter Relationships

Figure	Drawing	Area	Perimeter or Circumference
Rectangle		$A = \ell w$ (or $A = bh$)	$P = 2\ell + 2w$ (or $P = 2b + 2h$)
Square		$A = s^2$	$P = 4s$
Parallelogram		$A = bh$	$P = 2b + 2s$
Triangle		$A = \frac{1}{2}bh$ $A = \sqrt{s(s-a)(s-b)(s-c)}$, where $s = \frac{1}{2}(a + b + c)$	$P = a + b + c$
Right triangle		$A = \frac{1}{2}ab$	$P = a + b + c$

(continued)

Figure	Drawing	Area	Perimeter or Circumference
Trapezoid (base of lengths b_1 and b_2)		$A = \frac{1}{2}h(b_1 + b_2)$	$P = s_1 + s_2 + b_1 + b_2$
Rhombus (diagonals of lengths d_1 and d_2)		$A = \frac{1}{2}d_1 d_2$	$P = 4s$
Kite (diagonals of lengths d_1 and d_2)		$A = \frac{1}{2}d_1 d_2$	$P = 2b + 2s$
Regular polygon (n sides; s is the length of a side; a is the length of an apothem)		$A = \frac{1}{2}aP$ ($P =$ perimeter)	$P = ns$
Circle with radius length r		$A = \pi r^2$	$C = 2\pi r$
Sector ($\text{m}\widehat{AB}$ is the degree measure of $\widehat{AB}$ and of central angle AOB)		$A = \frac{\text{m}\widehat{AB}}{360}\pi r^2$	$P = 2r + \ell\widehat{AB}$, where $\ell\widehat{AB} = \frac{\text{m}\widehat{AB}}{360} \cdot 2\pi r$
Triangle with inscribed circle of radius length r		$A = \frac{1}{2}rP$ ($P =$ perimeter)	$P = a + b + c$

Chapter **8** Review Exercises

In Review Exercises 1 to 3, draw a figure that enables you to solve each problem.

1. *Given:* □$ABCD$ with $BD = 34$ and $BC = 30$
  m∠$C = 90°$
  Find: A_{ABCD}

2. *Given:* □$ABCD$ with $AB = 8$ and $AD = 10$
  Find: A_{ABCD} if:
  a) m∠$A = 30°$
  b) m∠$A = 60°$
  c) m∠$A = 45°$

3. *Given:* □$ABCD$ with $\overline{AB} \cong \overline{BD}$ and $AD = 10$
  $\overline{BD} \perp \overline{DC}$
  Find: A_{ABCD}

In Review Exercises 4 and 5, draw △ABC, if necessary, to solve each problem.

4. *Given:* $AB = 26$, $BC = 25$, and $AC = 17$
  Find: A_{ABC}

5. *Given:* $AB = 30$, $BC = 26$, and $AC = 28$
  Find: A_{ABC}

6. *Given:* Trapezoid $ABCD$, with $AD \parallel CD$, $DC = 6$,
  $AD = 12$, and $AB = 5$
  Find: A_{ABCD}

Exercises 6, 7

7. *Given:* Trapezoid $ABCD$, with $AB = 6$ and $BC = 8$,
  $\overline{AB} \cong \overline{CD}$
  Find: A_{ABCD} if:
  a) m∠$A = 45°$
  b) m∠$A = 30°$
  c) m∠$A = 60°$

8. Find the area and the perimeter of a rhombus whose diagonals have lengths 18 in. and 24 in.

9. Tom Morrow wants to buy some fertilizer for his yard. The lot size is 140 ft by 160 ft. The outside measurements of his house are 80 ft by 35 ft. The driveway measures 30 ft by 20 ft. All shapes are rectangular.
  a) What is the square footage of his yard that needs to be fertilized?
  b) If each bag of fertilizer covers 5000 ft², how many bags should Tom buy?
  c) If the fertilizer costs $18 per bag, what is his total cost?

10. Alice's mother wants to wallpaper two adjacent walls in Alice's bedroom. She also wants to put a border along the top of all four walls. The bedroom is 9 ft by 12 ft by 8 ft high.
  a) If each double roll covers approximately 60 ft² and the wallpaper is sold in double rolls only, how many double rolls are needed?
  b) If the border is sold in rolls of 5 yd each, how many rolls of the border are needed?

11. *Given:* Isosceles trapezoid
  $ABCD$
  Equilateral △FBC
  Right △AED
  $BC = 12$, $AB = 5$,
  and $ED = 16$
  Find: a) A_{EAFD}
  b) P_{EAFD}

12. *Given:* Kite $ABCD$
  with $AB = 10$,
  $BC = 17$, and
  $BD = 16$
  Find: A_{ABCD}

13. One side of a rectangle is 2 cm longer than a second side. If the area is 35 cm², find the dimensions of the rectangle.

14. One side of a triangle is 10 cm longer than a second side, and the third side is 5 cm longer than the second side. The perimeter of the triangle is 60 cm.
  a) Find the lengths of the three sides.
  b) Find the area of the triangle.

15. Find the area of △ABD as shown.

16. Find the area of an equilateral triangle if each of its sides has length 12 cm.

17. If $\overline{AC}$ is a diameter of ⊙O, find the area of the shaded triangle.

18. For a regular pentagon, find the measure of each:
  a) central angle
  b) interior angle
  c) exterior angle

Exercise 17

19. Find the area of a regular hexagon, each of whose sides has length 8 ft.

20. The area of an equilateral triangle is $108\sqrt{3}$ in². If the length of each side of the triangle is $12\sqrt{3}$ in, find the length of an apothem of the triangle.

21. Find the area of a regular hexagon whose apothem has length 9 in.

22. In a regular polygon, each central angle measures 45°.
    a) How many sides does the regular polygon have?
    b) If each side measures 5 cm and the length of each apothem is approximately 6 cm, what is the approximate area of the polygon?

23. Can a circle be circumscribed about each of the following figures? Why or why not?
    a) Parallelogram   c) Rectangle
    b) Rhombus   d) Square

24. Can a circle be inscribed in each of the following figures? Why or why not?
    a) Parallelogram   c) Rectangle
    b) Rhombus   d) Square

25. The length of the radius of a circle inscribed in an equilateral triangle is 7 in. Find the area of the triangle.

26. The Turners want to install outdoor carpet around their rectangular pool. The dimensions for the rectangular area formed by the pool and its walkway are 20 ft by 30 ft. The pool is 12 ft by 24 ft.
    a) How many square feet need to be covered?
    b) Approximately how many square yards does the area in part (a) represent?
    c) If the carpet costs $12.95 per square yard, what will be the cost of the carpet?

Find the exact areas of the shaded regions in Exercises 27 to 31.

27.

28.

Square

29.

30.
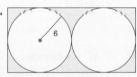

Two ≅ tangent circles, inscribed in a rectangle

31.

Equilateral triangle

32. The arc of a sector measures 40°. Find the exact length of the arc and the exact area of the sector if the radius measures $3\sqrt{5}$ cm.

33. The circumference of a circle is 66 ft.
    a) Find the diameter of the circle using $\pi \approx \frac{22}{7}$.
    b) Find the area of the circle using $\pi \approx \frac{22}{7}$.

34. A circle has an exact area of 27π ft^2.
    a) What is the area of a sector of this circle if the arc of the sector measures 80°?
    b) What is the exact perimeter of the sector in part (a)?

35. An isosceles right triangle is inscribed in a circle that has a diameter of 12 in. Find the exact area between one of the legs of the triangle and its corresponding arc.

36. *Given:*   Concentric circles with radii of lengths R and r, with $R > r$
    Prove: $A_{\text{ring}} = \pi(BC)^2$

37. Prove that the area of a circle circumscribed about a square is twice the area of the circle inscribed within the square.

38. Prove that if semicircles are constructed on each of the sides of a right triangle, then the area of the semicircle on the hypotenuse is equal to the sum of the areas of the semicircles on the two legs.

39. Jeff and Helen want to carpet their family room, except for the entranceway and the semicircle in front of the fireplace, both of which they want to tile.
    a) How many square yards of carpeting are needed?
    b) How many square feet are to be tiled?

40. Sue and Dave's semicircular driveway is to be resealed, and then flowers are to be planted on either side.
   a) What is the number of square feet to be resealed?
   b) If Dave can reseal the driveway at a cost of $0.45 per square foot, what is the cost of resealing the driveway?
   c) If individual flowers are to be planted 1 foot from the edge of the driveway at intervals of approximately 1 foot on both sides of the driveway, how many flowers are needed?

12 ft   30 ft

Chapter **8**   Test

1. Complete each statement.
   a) Given that the length and the width of a rectangle are measured in inches, its area is measured in
   _____.
   b) If two closed plane figures are congruent, then their areas are _____.

2. Give each formula.
   a) The formula for the area of a square whose sides are of length s is _____.
   b) The formula for the circumference of a circle with radius length r is _____.

3. Determine whether the statement is True or False.
   a) The area of a circle with radius length r is given by $A = \pi r^2$. _____
   b) With lengths of the corresponding sides of similar polygons having the ratio $\frac{s_1}{s_2} = \frac{1}{2}$, the ratio of their areas is $\frac{A_1}{A_2} = \frac{1}{2}$. _____

4. If the area of rectangle $ABCD$ is 46 cm², find the area of $\triangle ABE$. _____

5. In *square feet*, find the area of $\square EFGH$. _____

10 ft

2 yd   3 yd

6. Find the area of rhombus $MNPQ$ given that $QN = 8$ ft and $PM = 6$ ft. _____

7. Use Heron's Formula, $A = \sqrt{s(s-a)(s-b)(s-c)}$, to find the exact area of a triangle that has lengths of sides 4 cm, 13 cm, and 15 cm. _____

8. In trapezoid $ABCD$, $AB = 7$ ft and $DC = 13$ ft. If the area of trapezoid $ABCD$ is 60 ft², find the length of altitude AE. _____

9. A regular pentagon has an apothem of length 4.0 in. and each side is of length $s = 5.8$ in. For the regular pentagon, find its:
   a) Perimeter _____
   b) Area _____

10. For the circle shown below, the length of the radius is $r = 5$ in. Find the exact:
   a) Circumference _____
   b) Area _____

   (*HINT: Leave π in the answer in order to achieve exactness.*)

11. Where $\pi \approx \frac{22}{7}$, find the approximate length of $\overparen{AC}$.
$\ell\overparen{AC} \approx$ _____

12. Where $\pi \approx 3.14$, find the approximate area of a circle (not shown) whose diameter measures 20 cm. _____

13. In the figure, a square is inscribed in a circle. If each side of the square measures $4\sqrt{2}$ in., find an expression for the exact area of the shaded region. _____

Square inscribed
in a circle

14. Find the exact area of the 135° sector shown.
_____

15. Find the exact area of the shaded segment.
_____

16. The area of a right triangle whose sides have lengths 5 in., 12 in., and 13 in. is exactly 30 in². Use the formula $A = \frac{1}{2}rP$ to find the length of the radius of the circle that can be inscribed in this triangle. _____

17. Pascual is remodeling a house and plans a complete drywall for one 8-foot high room. Dimensions for the room (the floor) are 12 ft. by 16 ft.
   a) To replace the drywall on the 4 walls and ceiling, how many 4 ft by 8 ft sheets of drywall are needed?
   b) If the cost of the drywall and other materials runs $0.40 per ft², what does Pascual spend for the drywall and materials?

Chapter 9

Surfaces and Solids

Colossal! Located near Cairo, Egypt, the Great Pyramids illustrate one of the types of solids that we study in Chapter 9. The architectural designs of buildings often illustrate other solid shapes that we study in this chapter. The real world is three-dimensional; that is, solids and space figures can be characterized by contrasting three defining measures of length, width, and depth. Each solid determines a bounded region of space that has a measure known as *volume.* Some units that are used to measure volume include the cubic foot and the cubic meter. The same technique that is used to measure the volume of the pyramid in Section 9.2 could be used to measure the volumes of the Great Pyramids.

Additional video explanations of concepts, sample problems, and applications are available on DVD.

9.1 Prisms, Area, and Volume

KEY CONCEPTS

Prisms (Right and Oblique)	Edges	Regular Prism
	Faces	Cube
Bases	Lateral Area	Cubic Unit
Altitude	Total (Surface) Area	
Vertices	Volume	

PRISMS

Suppose that two congruent polygons lie in parallel planes in such a way that their corresponding sides are parallel. If the corresponding vertices of these polygons [such as A and A' in Figure 9.1(a)] are joined by line segments, then the "solid" or "space figure" that results is known as a **prism**. The congruent figures that lie in the parallel planes are the **bases** of the prism. The parallel planes need not be shown in the drawings of prisms. Suggested by an empty box, the prism is like a shell that encloses a portion of space by the parts of planes that form the prism; thus, a prism does not contain interior points. In practice, it is sometimes convenient to call a prism such as a brick a *solid*; of course, this interpretation of a prism contains its interior points.

$\triangle ABC \cong \triangle A'B'C'$

(a)

Square $DEFG \cong$ Square $D'E'F'G'$

(b)

Figure 9.1

In Figure 9.1(a), $\overline{AB}$, $\overline{AC}$, $\overline{BC}$, $\overline{A'B'}$, $\overline{A'C'}$, and $\overline{B'C'}$ are **base edges** of the prism, while $\overline{AA'}$, $\overline{BB'}$, and $\overline{CC'}$ are **lateral edges** of the prism. Because the lateral edges of this prism are perpendicular to its base edges, the **lateral faces** (such as quadrilateral $ACC'A'$) are rectangles. The bases and lateral faces are known collectively as the **faces** of the prism. Any point at which three faces are concurrent is a **vertex** of the prism. Points A, B, C, A', B', and C' are the **vertices** of the prism.

In Figure 9.1(b), the lateral edges of the prism are not perpendicular to its base edges; with respect to the base edges, the lateral edges are often described as **oblique** (slanted). For the oblique prism, the lateral faces are parallelograms. Considering the prisms in Figure 9.1, we are led to the following definitions.

Part of the description used to classify a prism depends on its base. For instance, the prism in Figure 9.1(a) is a *right triangular prism*; in this case, the word *right* describes the prism, whereas the word *triangular* refers to the triangular base. Similarly, the prism in Figure 9.1(b) is an *oblique square prism*. Both prisms in Figure 9.1 have an **altitude** (a perpendicular segment between the planes that contain the bases) of length *h*, also known as the *height* of the prism.

> **EXAMPLE 1**
>
> Name each type of prism in Figure 9.2.
>
>
>
> Bases are equilateral triangles
>
> (a)   (b)   (c)
>
> Figure 9.2
>
> **SOLUTION**
>
> a) The lateral edges are perpendicular to the base edges of the hexagonal base. The prism is a *right hexagonal prism*.
> b) The lateral edges are oblique to the base edges of the pentagonal base. The prism is an *oblique pentagonal prism*.
> c) The lateral edges are perpendicular to the base edges of the triangular base. Because the base is equilateral, the prism is a *right equilateral triangular prism*.

SSG   EXS. 1, 2   **AREA OF A PRISM**

In the right triangular prism of Figure 9.3, *a, b,* and *c* are the lengths of the sides of either base. These dimensions are used along with the length of the altitude (denoted by *h*) to calculate the lateral area, the sum of the areas of rectangles $ACC'A'$, $ABB'A'$, and $BCC'B'$. The lateral area *L* of the right triangular prism can be found as follows:

$$L = ah + bh + ch$$
$$= h(a + b + c)$$
$$= hP$$

where *P* is the perimeter of a base of the prism. This formula, $L = hP$, is valid for finding the lateral area of any *right* prism. Although lateral faces of an oblique prism are parallelograms, it is easy to show that the formula $L = hP$ can be used to find its lateral area as well.

Figure 9.3

> ### THEOREM 9.1.1
>
> The lateral area L of any prism whose altitude has measure h and whose base has perimeter P is given by $L = hP$.

Many students (and teachers) find it easier to calculate the lateral area of a prism without using the formula $L = hP$. We illustrate this in Example 2.

EXAMPLE 2

The bases of the right prism shown in Figure 9.4 are equilateral pentagons with sides of length 3 in. each. If the altitude measures 5 in., find the lateral area of the prism.

SOLUTION Each lateral face is a rectangle with dimensions 3 in. by 5 in. The area of each rectangular face is 3 in. $\times$ 5 in. $= 15$ in^2. Because there are five congruent lateral faces, the lateral area of the pentagonal prism is 5×15 in$^2 = 75$ in^2.

NOTE: When applied in Example 2, the formula $L = hP$ leads to $L = 5$ in. $\times$ 15 in. $= 75$ in^2.

Figure 9.4

> ### DEFINITION
>
> For any prism, the **total area** T is the sum of the lateral area and the areas of the bases.

NOTE: The total area of the prism is also known as its surface area.

Recall that the bases and the lateral faces are known as *faces* of a prism. Thus, the total area T of the prism is the sum of the areas of all its faces.

> ### THEOREM 9.1.2
>
> The total area T of any prism with lateral area L and base area B is given by $T = L + 2B$.

PICTURE PROOF OF THEOREM 9.1.2

GIVEN: The pentagonal prism of Figure 9.5(a)

PROVE: $T = L + 2B$

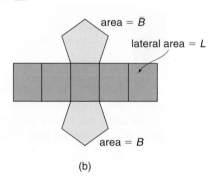

(a)

(b)

Figure 9.5

PROOF: When the prism is "taken apart" and laid flat, as shown in Figure 9.5(b), we see that the total area depends upon the lateral area (shaded darker) and the areas of the two bases (shaded lighter); that is,

$$T = L + 2B$$

Recalling Heron's Formula, we know that the base area B of the right triangular prism in Figure 9.6 can be found by the formula

$$B = \sqrt{s(s - a)(s - b)(s - c)}$$

in which s is the semiperimeter of the triangular base. We will apply Heron's Formula in Example 3.

Figure 9.6

EXAMPLE 3

Find the total area of the right triangular prism with an altitude of length 8 in. if the sides of the triangular bases have lengths of 13 in., 14 in., and 15 in. See Figure 9.6.

SOLUTION The lateral area is found by adding the areas of the three rectangular lateral faces. That is,

$$L = 8 \text{ in.} \cdot 13 \text{ in.} + 8 \text{ in.} \cdot 14 \text{ in.} + 8 \text{ in.} \cdot 15 \text{ in.}$$
$$= 104 \text{ in}^2 + 112 \text{ in}^2 + 120 \text{ in}^2 = 336 \text{ in}^2$$

Using Heron's Formula to find the area of each base, $s = \frac{1}{2}(13 + 14 + 15)$
$= 21$, $B = \sqrt{21(21 - 13)(21 - 14)(21 - 15)} = \sqrt{21(8)(7)(6)} = \sqrt{7056} = 84$.
Calculating the total area (or surface area) of the triangular prism,

$$T = L + 2B \quad \text{becomes} \quad T = 336 + 2(84) \quad \text{or} \quad T = 504 \text{ in}^2 \quad ■$$

DEFINITION

A regular prism is a right prism whose bases are regular polygons.

Consider this definition, the prism in Figure 9.2(c) on page 391 could be called a regular triangular prism.

In the following example, each base of the prism is a regular hexagon. Because this regular hexagonal prism is a right prism, the lateral faces are congruent rectangles.

EXAMPLE 4

Find the lateral area L and the surface area T of the regular hexagonal prism in Figure 9.7(a)

(a)

(b)

Figure 9.7

SOLUTION In Figure 9.7(a) on page 393, there are six congruent lateral faces, each rectangular with dimensions of 4 in. by 10 in. Then

$$L = 6(4 \cdot 10)$$
$$= 240 \text{ in}^2$$

For the regular hexagonal base [see Figure 9.7(b)], the apothem measures $a = 2\sqrt{3}$ in., and the perimeter is $P = 6 \cdot 4 = 24$ in. Then the area B of each base is given by the formula for the area of a regular polygon.

$$B = \frac{1}{2}aP$$
$$= \frac{1}{2} \cdot 2\sqrt{3} \cdot 24$$
$$= 24\sqrt{3} \text{ in}^2 \approx 41.57 \text{ in}^2$$

Now
$$T = L + 2B$$
$$= (240 + 48\sqrt{3}) \text{ in}^2 \approx 323.14 \text{ in}^2$$

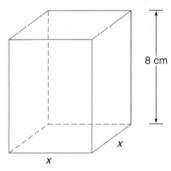

Figure 9.8

EXAMPLE 5

The total area of the right square prism in Figure 9.8 is 210 cm². Find the length of a side of the square base if the altitude of the prism is 8 cm.

SOLUTION Let x be the length in cm of a side of the square. Then the area of the base is $B = x^2$ and the area of each of the four lateral faces is $8x$. Therefore,

$$\underset{\text{2 bases}}{2(x^2)} + \underset{\substack{\text{4 lateral} \\ \text{faces}}}{4(8x)} = 210$$

$$2x^2 + 32x - 210$$
$$2x^2 + 32x - 210 = 0$$
$$x^2 + 16x - 105 = 0 \qquad \text{(dividing by 2)}$$
$$(x + 21)(x - 5) = 0 \qquad \text{(factoring)}$$
$$x + 21 = 0 \qquad \text{or} \qquad x - 5 = 0$$
$$x = -21 \qquad \text{or} \qquad x = 5 \qquad \text{(reject } -21 \text{ as a solution)}$$

Then each side of the square base measures 5 cm.

SSG EXS. 3–7

DEFINITION

A **cube** is a right square prism whose edges are congruent.

As we shall see, the cube is very important in determining the volume of a solid.

VOLUME OF A PRISM

To introduce the notion of *volume*, we recognize that a prism encloses a portion of space. Without a formal definition, we say that the **volume** of the solid is a number that measures the amount of enclosed space. To begin, we need a unit for measuring volume. Just as the meter can be used to measure length and the square yard can be used to measure area, a **cubic unit** is used to measure the amount of space enclosed within a bounded region of space. One such unit is described in the following paragraph.

The volume enclosed by the cube shown in Figure 9.9 is 1 cubic inch or 1 in³. The volume of a solid is the number of cubic units within the solid. Thus, we assume that the volume of any solid is a positive number of cubic units.

Figure 9.9

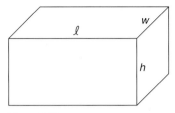

Figure 9.10

> **POSTULATE 24 ■ Volume Postulate**
>
> Corresponding to every solid is a unique positive number V known as the volume of that solid.

The simplest space figure for which we can determine volume is the **right rectangular prism**. Such a solid might be described as a **parallelpiped** or as a "box." Because boxes are used as containers for storage and shipping (such as a boxcar), it is important to calculate volume as a measure of capacity. A right rectangular prism is shown in Figure 9.10; its dimensions are length ℓ, width w, and height h.

The volume of a right rectangular prism of length 4 in., width 3 in., and height 2 in. is easily shown to be 24 in³. The volume is the product of the three dimensions of the given solid. We see not only that $4 \cdot 3 \cdot 2 = 24$ but also that the units of volume are in. $\cdot$ in. $\cdot$ in. $=$ in³. Figures 9.11(a) and (b) illustrate that the 4 by 3 by 2 box must have the volume 24 in³. We see that there are four layers of blocks, each of which is a 2 by 3 configuration of 6 in³. Figure 9.11 provides the insight that leads us to our following postulate.

(a)   (b)

Figure 9.11

> **POSTULATE 25**
>
> The volume of a right rectangular prism is given by
>
> $$V = \ell w h$$
>
> where ℓ measures the length, w the width, and h the altitude of the prism.

In order to apply the formula found in Postulate 25, the units used for dimensions ℓ, w, and h must be alike, as illustrated in Example 6.

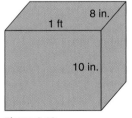

Figure 9.12

EXAMPLE 6

Find the volume of a box whose dimensions are 1 ft, 8 in., and 10 in. (See Figure 9.12.)

SOLUTION Although it makes no difference which dimension is chosen for ℓ or w or h, it is most important that the units of measure be the same. Thus, 1 ft is replaced by 12 in. in the formula for volume:

$$V = \ell w h$$
$$= 12 \text{ in.} \cdot 8 \text{ in.} \cdot 10 \text{ in.}$$
$$= 960 \text{ in}^3$$

Note that the formula for the volume of the right rectangular prism, $V = \ell w h$, could be replaced by the formula $V = Bh$, where B is the area of the base of the prism; for a rectangular prism, $B = \ell w$. As stated in the following postulate, this volume relationship is true for right prisms in general.

> **POSTULATE 26**
>
> The volume of a right prism is given by
>
> $$V = Bh$$
>
> where B is the area of a base and h is the length of the altitude of the prism.

In real-world applications, the formula $V = Bh$ is valid for calculating the volumes of oblique prisms as well as right prisms.

SSG **EXS. 8–13**

Technology Exploration

On your calculator, determine the method of "cubing." That is, find a value such as 2.1^3. On many calculators, we enter 2.1, a caret $\wedge$, and 3. Some calculators have an x^3 function.

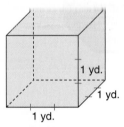

1 yd.

1 yd.

1 yd.

Figure 9.13

EXAMPLE 7

Find the volume of the regular hexagonal prism found in Figure 9.7 on page 393.

SOLUTION In Example 4, we found that the area of the hexagonal base was $24\sqrt{3}$ in^2. Because the altitude of the hexagonal prism is 10 in., the volume is $V = Bh = (24\sqrt{3}$ in$^2)(10$ in.$)$. Then $V = 240\sqrt{3}$ in$^3 \approx 415.69$ in^3.

NOTE: Just as $x^2 \cdot x = x^3$, the units in Example 7 are in$^2 \cdot$ in. $=$ in^3.

In the final example of this section, we use the fact that 1 yd$^3 = 27$ ft^3. In the cube shown in Figure 9.13, each dimension measures 1 yd, or 3 ft. The cube's volume is given by 1 yd $\cdot$ 1 yd $\cdot$ 1 yd $= 1$ yd^3 or 3 ft $\cdot$ 3 ft $\cdot$ 3 ft $= 27$ ft^3. It follows that 1 yd$^3 = 27$ ft^3 or 1 ft$^3 = \frac{1}{27}$ yd^3.

EXAMPLE 8

Sarah is having a concrete driveway poured at her house. The section to be poured is rectangular, measuring 12 ft by 40 ft by 4 in. deep. How many cubic yards of concrete are needed?

SOLUTION Using $V = \ell wh$, we must be consistent with units. Thus, $\ell = 12$ ft, $w = 40$ ft, and $h = \frac{1}{3}$ ft (from 4 in.).

$$V = 12 \text{ ft} \cdot 40 \text{ ft} \cdot \frac{1}{3} \text{ ft}$$

$$V = 160 \text{ ft}^3$$

Now 160 ft$^3 = 160(\frac{1}{27}$ yd$^3) = \frac{160}{27}$ yd^3 or $5\frac{25}{27}$ yd^3.

SSG **EXS. 14, 15**   **NOTE:** Sarah will be charged for 6 yd^3 of concrete, the result of rounding upward.

Exercises 9.1

1. Consider the solid shown.
   a) Does it appear to be a prism?
   b) Is it right or oblique?
   c) What type of base(s) does the solid have?
   d) Name the type of solid.
   e) What type of figure is each lateral face?

Exercises 1, 3, 5, 7, 9

2. Consider the solid shown.
   a) Does it appear to be a prism?
   b) Is it right or oblique?
   c) What type of base(s) does the solid have?
   d) Name the type of solid.
   e) What type of figure is each lateral face?

Exercises 2, 4, 6, 8, 10

3. Consider the hexagonal prism shown in Exercise 1.
   a) How many vertices does it have?
   b) How many edges (lateral edges plus base edges) does it have?
   c) How many faces (lateral faces plus bases) does it have?

4. Consider the triangular prism shown in Exercise 2.
   a) How many vertices does it have?
   b) How many edges (lateral edges plus base edges) does it have?
   c) How many faces (lateral faces plus bases) does it have?

5. If each edge of the hexagonal prism in Exercise 1 is measured in centimeters, what unit is used to measure its (a) surface area? (b) volume?

6. If each edge of the triangular prism in Exercise 2 is measured in inches, what unit is used to measure its (a) lateral area? (b) volume?

7. Suppose that each of the bases of the hexagonal prism in Exercise 1 has an area of 12 cm² and that each lateral face has an area of 18 cm². Find the total (surface) area of the prism.

8. Suppose that each of the bases of the triangular prism in Exercise 2 has an area of 3.4 in² and that each lateral face has an area of 4.6 in². Find the total (surface) area of the prism.

9. Suppose that each of the bases of the hexagonal prism in Exercise 1 has an area of 12 cm² and that the altitude of the prism measures 10 cm. Find the volume of the prism.

10. Suppose that each of the bases of the triangular prism in Exercise 2 has an area of 3.4 cm² and that the altitude of the prism measures 1.2 cm. Find the volume of the prism.

11. A solid is an octagonal prism.
    a) How many vertices does it have?
    b) How many lateral edges does it have?
    c) How many base edges are there in all?

12. A solid is a pentagonal prism.
    a) How many vertices does it have?
    b) How many lateral edges does it have?
    c) How many base edges are there in all?

13. Generalize the results found in Exercises 11 and 12 by answering each of the following questions. Assume that the number of sides in each base of the prism is *n*. For the prism, what is the
    a) number of vertices?
    b) number of lateral edges?
    c) number of base edges?
    d) total number of edges?
    e) number of lateral faces?
    f) number of bases?
    g) total number of faces?

14. In the accompanying regular pentagonal prism, suppose that each base edge measures 6 in. and that the apothem of the base measures 4.1 in. The altitude of the prism measures 10 in.
    a) Find the lateral area of the prism.
    b) Find the total area of the prism.
    c) Find the volume of the prism.

Base

Exercises 14, 15

15. In the regular pentagonal prism shown above, suppose that each base edge measures 9.2 cm and that the apothem of the base measures 6.3 cm. The altitude of the prism measures 14.6 cm.
    a) Find the lateral area of the prism.
    b) Find the total area of the prism.
    c) Find the volume of the prism.

16. For the right triangular prism, suppose that the sides of the triangular base measure 4 m, 5 m, and 6 m. The altitude is 7 m.
    a) Find the lateral area of the prism.
    b) Find the total area of the prism.
    c) Find the volume of the prism.

Exercises 16, 17

17. For the right triangular prism found in Exercise 16, suppose that the sides of the triangular base measure 3 ft, 4 ft, and 5 ft. The altitude is 6 ft in length.
    a) Find the lateral area of the prism.
    b) Find the total area of the prism.
    c) Find the volume of the prism.

18. Given that 100 cm = 1 m, find the number of cubic centimeters in 1 cubic meter.

19. Given that 12 in. = 1 ft, find the number of cubic inches in 1 cubic foot.

20. Find the volume and the surface area of a "closed box" that has dimensions of 9 in., 10 in., and 1 ft.

21. Find the volume and the surface area of a "closed box" that has dimensions of 15 cm, 20 cm, and 0.25 m.
    (Hint: 1 m = 100 cm.)

22. A cereal box measures 2 in. by 8 in. by 10 in. What is the volume of the box? How many square inches of cardboard make up its surface? (Disregard any hidden flaps.)

23. The measures of the sides of the square base of a box are twice the measure of the height of the box. If the volume of the box is 108 in³, find the dimensions of the box.

24. For a given box, the height measures 4 m. If the length of the rectangular base is 2 m greater than the width of the base and the lateral area L is 96 m^2, find the dimensions of the box.

25. For the box shown, the total area is 94 cm^2. Determine the value of x.

Exercises 25, 26

26. If the volume of the box is 252 in^3, find the value of x. (See the figure for Exercise 25.)

27. The box with dimensions indicated is to be constructed of materials that cost 1 cent per square inch for the lateral surface and 2 cents per square inch for the bases. What is the total cost of constructing the box?

28. A hollow steel door is 32 in. wide by 80 in. tall by $1\frac{3}{8}$ in. thick. How many cubic inches of foam insulation are needed to fill the door?

29. A storage shed is in the shape of a pentagonal prism. The front represents one of its two pentagonal bases. What is the storage capacity (volume) of its interior?

30. A storage shed is in the shape of a trapezoidal prism. Each trapezoid represents one of its bases. With dimensions as shown, what is the storage capacity (volume) of its interior?

31. A cube is a right square prism in which all edges have the same length. For the cube with edge e,
    a) show that the total area is $T = 6e^2$.
    b) find the total area if $e = 4$ cm.
    c) show that the volume is $V = e^3$.
    d) find the volume if $e = 4$ cm.

Exercises 31–33

32. Use the formulas and drawing in Exercise 31 to find (a) the total area T and (b) the volume V of a cube with edges of length 5.3 ft each.

33. When the length of each edge of a cube is increased by 1 cm, the volume is increased by 61 cm^3. What is the length of each edge of the original cube?

34. The numerical value of the volume of a cube equals the numerical value of its total surface area. What is the length of each edge of the cube?

35. The sum of the lengths of all edges of a cube is 60 cm. Find the volume V and the surface area T of the cube.

36. A concrete pad 4 in. thick is to have a length of 36 ft and a width of 30 ft. How many cubic yards of concrete must be poured?

37. Zaidah plans a raised flower bed 2 ft high by 12 ft wide by 15 ft long. The mulch, soil, and peat mixture used to fill the raised bed costs $15.75 per cubic yard. What is the total cost of the ingredients used to fill the raised garden?

38. In excavating for a new house, a contractor digs a hole in the shape of a right rectangular prism. The dimensions of the hole are 54 ft long by 36 ft wide by 9 ft deep. How many cubic yards of dirt were removed?

39. Kristine creates an open box by cutting congruent squares from the four corners of a square piece of cardboard that has a length of 24 in. per side. If the congruent squares that are removed have sides that measure 6 in. each, what is the volume of the box formed by folding and sealing the flaps?

40. As in Exercise 39, find the volume of the box if four congruent squares with sides of length 6 in. are cut from the corners of a rectangular piece of poster board that is 20 in. wide by 30 in. long.

41. Kianna's aquarium is "box shaped" with dimensions of 2 ft by 1 ft by 8 in. If 1 ft^3 corresponds to 7.5 gal of water, what is the water capacity of her aquarium in *gallons*?

42. The gasoline tank on an automobile is "box-shaped" with dimensions of 24 in. by 20 in. by 9 in. If 1 ft^3 corresponds to 7.5 gal of gasoline, what is the capacity of the automobile's fuel tank in *gallons*?

For Exercises 43 to 45, consider the oblique regular pentagonal prism shown. Each side of the base measures 12 cm, and the altitude measures 12 cm.

Exercises 43–45

43. Find the lateral area of the prism.

(*HINT: Each lateral face is a parallelogram.*)

44. Find the total area of the prism.

45. Find the volume of the prism.

46. It can be shown that the length of a diagonal of a right rectangular prism with dimensions ℓ, w, and h is given by $d = \sqrt{\ell^2 + w^2 + h^2}$. Use this formula to find the length of the diagonal when $\ell = 12$ in., $w = 4$ in., and $h = 3$ in.

*47. A diagonal of a cube joins two vertices so that the remaining points of the diagonal lie in the interior of the cube. Show the diagonal of the cube having edges of length e is $e\sqrt{3}$ units long.

9.2 Pyramids, Area, and Volume

KEY CONCEPTS	Pyramid	Faces	Slant Height of a
	Base	Vertex (Apex) of	Regular Pyramid
	Altitude	a Pyramid	Lateral Area
	Vertices	Regular Pyramid	Total (Surface) Area
	Edges		Volume

The solids (space figures) shown in Figure 9.14 below are **pyramids**. In Figure 9.14(a), point *A* is noncoplanar with square *base BCDE*. In Figure 9.14(b), point *F* is noncoplanar with its base, $\triangle GHJ$. In each space pyramid, the noncoplanar point is joined to each vertex as well as each point of the base. A solid pyramid results when the noncoplanar point is joined both to points on the polygon as well as to points in its interior. Point *A* is known as the **vertex** or **apex** of the **square pyramid**; likewise, point *F* is the vertex or apex of the **triangular pyramid**. The pyramid of Figure 9.14(b) has *four* triangular faces; for this reason, it is called a **tetrahedron**.

Figure 9.14

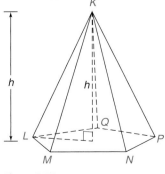

Figure 9.15

The pyramid in Figure 9.15 is a **pentagonal pyramid**. It has vertex *K*, pentagon *LMNPQ* for its **base**, and **lateral edges** $\overline{KL}$, $\overline{KM}$, $\overline{KN}$, $\overline{KP}$, and $\overline{KQ}$. Although *K* is called *the* **vertex of the pyramid**, there are actually six vertices: *K, L, M, N, P,* and *Q*. The sides of the base $\overline{LM}$, $\overline{MN}$, $\overline{NP}$, $\overline{PQ}$, and $\overline{QL}$ are **base edges**. All **lateral faces** of a pyramid are triangles; $\triangle KLM$ is one of the five lateral faces of the pentagonal pyramid. Including base *LMNPQ*, this pyramid has a total of six faces. The **altitude** of the pyramid, of length *h*, is the line segment from the vertex *K* perpendicular to the plane of the base. In every pyramid, the lateral edges are concurrent at the vertex (apex) of the pyramid. Likewise, the lateral faces are concurrent at the apex of the pyramid.

> **DEFINITION**
>
> A **regular pyramid** is a pyramid whose base is a regular polygon and whose lateral edges are all congruent.

Suppose that the pyramid in Figure 9.15 on page 399, is a regular pentagonal pyramid. Then the lateral faces are necessarily congruent to each other; by SSS, $\triangle KLM \cong \triangle KMN \cong \triangle KNP \cong \triangle KPQ \cong \triangle KQL$. Each lateral face is an isosceles triangle. In a regular pyramid, the altitude joins the apex of the pyramid to the center of the regular polygon that is the base of the pyramid. The length of the altitude is height h.

> **DEFINITION**
>
> The **slant height** of a regular pyramid is the altitude from the vertex (apex) of the pyramid to the base of any of the congruent lateral faces of the regular pyramid.

SSG ▸ **EXS. 1, 2**

NOTE: Among pyramids, only a regular pyramid has a slant height.

In our formulas and explanations, we use ℓ to represent the length of the slant height of a regular pyramid. See Figure 9.16(c) in Example 1.

> **EXAMPLE 1**
>
> For a regular square pyramid with height 4 in. and base edges of length 6 in. each, find the length of the slant height ℓ.
>
> **SOLUTION** In Figure 9.16, it can be shown that the apothem to any side has length 3 in. (one-half the length of the side of the square base). Also, the slant height is the hypotenuse of a right triangle with legs equal to the lengths of the altitude of the pyramid and the apothem of the base. See Figure 9.16(c). Applying the Pythagorean Theorem,
>
> $$\ell^2 = a^2 + h^2$$
> $$\ell^2 = 3^2 + 4^2$$
> $$\ell^2 = 9 + 16$$
> $$\ell^2 = 25$$
> $$\ell = 5 \text{ in.}$$

(a)

(b)

(c)

Figure 9.16

The principle found in the following theorem was used in the solution of Example 1. We accept Theorem 9.2.1 on the basis of the visual proof that Figure 9.16(c) provides.

> **THEOREM 9.2.1**
>
> In a regular pyramid, the lengths of the apothem a of the base, the altitude h, and the slant height ℓ satisfy the Pythagorean Theorem; that is, $\ell^2 = a^2 + h^2$.

SSG ▸ **EXS. 3, 4**

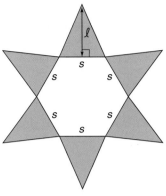

Figure 9.17

SURFACE AREA OF A PYRAMID

To lay the groundwork for the next theorem, we justify the result by "taking apart" one of the regular pyramids and laying it out flat. Although we use a regular hexagonal pyramid for this purpose, the argument is similar if the base is any regular polygon.

When the lateral faces of the regular pyramid are folded down into the plane, as shown in Figure 9.17, the shaded lateral area is the sum of the areas of the congruent triangular lateral faces. Using $A = \frac{1}{2}bh$, we find that the area of each triangular face is $\frac{1}{2} \cdot s \cdot \ell$ (each side of the base of the pyramid has length s, and the slant height has length ℓ). The combined areas of the triangles give the lateral area. Because there are n triangles,

$$L = n \cdot \frac{1}{2} \cdot s \cdot \ell$$
$$= \frac{1}{2} \cdot \ell(n \cdot s)$$
$$= \frac{1}{2}\ell P$$

where P is the perimeter of the base.

THEOREM 9.2.2

The lateral area L of a regular pyramid with slant height of length ℓ and perimeter P of the base is given by

$$L = \frac{1}{2}\ell P$$

We will illustrate the use of Theorem 9.2.2 in Example 2.

EXAMPLE 2

Find the lateral area of the regular pentagonal pyramid in Figure 9.18(a) if the sides of the base measure 8 cm and the lateral edges measure 10 cm each.

SOLUTION For the triangular lateral face in Figure 9.18(b), the slant height bisects the base edge as indicated. Applying the Pythagorean Theorem, we have

$$4^2 + \ell^2 = 10^2, \text{ so } 16 + \ell^2 = 100$$
$$\ell^2 = 84$$
$$\ell = \sqrt{84} = \sqrt{4 \cdot 21} = \sqrt{4} \cdot \sqrt{21} = 2\sqrt{21}$$

(a)   (b)

Figure 9.18

Now $L = \frac{1}{2}\ell P$ becomes $L = \frac{1}{2} \cdot 2\sqrt{21} \cdot (5 \cdot 8) = \frac{1}{2} \cdot 2\sqrt{21} \cdot 40 = 40\sqrt{21}$ cm^2 ≈ 183.30 cm^2.

It may be easier to find the lateral area of a regular pyramid without using the formula of Theorem 9.2.2; simply find the area of one lateral face and multiply by the number of faces. In Example 2, the area of each triangular face is $\frac{1}{2} \cdot 8 \cdot 2\sqrt{21}$ or $8\sqrt{21}$; thus, the lateral area of the regular pentagonal pyramid is $5 \cdot 8\sqrt{21} = 40\sqrt{21}$ cm².

THEOREM 9.2.3

The total area (surface area) T of a pyramid with lateral area L and base area B is given by $T = L + B$.

The formula for the total area T of the pyramid can be written $T = \frac{1}{2}\ell P + B$.

EXAMPLE 3

Find the total area of the regular square pyramid in Figure 9.19(a) that has base edges of length 4 ft and lateral edges of length 6 ft

(a)   (b)

Figure 9.19

SOLUTION To determine the lateral area, we need the length of the slant height. In Figure 9.19(b),

$$\ell^2 + 2^2 = 6^2$$
$$\ell^2 + 4 = 36$$
$$\ell^2 = 32$$
$$\ell = \sqrt{32} = \sqrt{16 \cdot 2} = \sqrt{16} \cdot \sqrt{2} = 4\sqrt{2}$$

The lateral area is $L = \frac{1}{2}\ell P$. For the square, $P = 16$ ft, so

$$L = \frac{1}{2} \cdot 4\sqrt{2}(16) = 32\sqrt{2} \text{ ft}^2$$

Because the area of the square base is $B = 4^2$ or 16 ft², the total area is

$$T = 32\sqrt{2} + 16 \approx 61.25 \text{ ft}^2$$

SSG EXS. 5–7

The pyramid in Figure 9.20(a) on the following page is a regular square pyramid rather than just a square pyramid. It has congruent lateral edges and congruent lateral faces. The pyramid shown in Figure 9.20(b) is oblique. The oblique square pyramid has neither congruent lateral edges nor congruent lateral faces.

Regular square pyramid
(a)

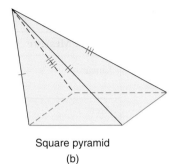

Square pyramid
(b)

Figure 9.20

VOLUME OF A PYRAMID

The final theorem in this section is presented without any attempt to construct the proof. In an advanced course such as calculus, the statement can be proved. The factor "one-third" in the formula for the volume of a pyramid provides exact results. This formula can be applied to any pyramid, even one that is not regular; in Figure 9.20(b), the length of the altitude is the perpendicular distance from the vertex to the plane of the square base. Read the Discover activity in the margin at the left before considering Theorem 9.2.4 and its applications.

Discover

There are kits that contain a hollow pyramid and a hollow prism that have congruent bases and the same altitude. Using a kit, fill the pyramid with water and then empty the water into the prism.

a) How many times did you have to empty the pyramid in order to fill the prism?

b) As a fraction, the volume of the pyramid is what part of the volume of the prism?

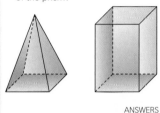

ANSWERS
(a) Three times (b) $\frac{1}{3}$

> **THEOREM 9.2.4**
>
> The volume V of a pyramid having a base area B and an altitude of length h is given by
>
> $$V = \frac{1}{3}Bh$$

EXAMPLE 4

Find the volume of the regular square pyramid with height $h = 4$ in. and base edges of length $s = 6$ in. (This was the pyramid in Example 1.)

SOLUTION The area of the square base is $B = (6 \text{ in.})^2$ or 36 in^2. Because $h = 4$ in., the formula $V = \frac{1}{3}Bh$ becomes

$$V = \frac{1}{3}(36 \text{ in}^2)(4 \text{ in.}) = 48 \text{ in}^3$$

To find the volume of a pyramid by using the formula $V = \frac{1}{3}Bh$, it is often necessary to determine B or h from other information that has been provided. In Example 5, calculating the length of the altitude h is a challenge! In Example 6, the difficulty lies in finding the area of the base. Before we consider either problem, Table 9.1 reminds us of the types of units necessary in different types of applications requiring measure.

TABLE 9.1

Type of Measure	Geometric Measure	Type of Unit
Linear	Length of segment, such as length of slant height	in., cm, etc.
Area	Amount of plane region enclosed, such as area of lateral face	in^2, cm^2, etc.
Volume	Amount of space enclosed, such as volume of a pyramid	in^3, cm^3, etc.

In Example 5, we apply Theorem 9.2.5. This application of the Pythagorean Theorem relates the lengths of the lateral edge, the radius of the base, and the altitude of a *regular* pyramid. Figure 9.21(c) provides a visual interpretation of the theorem.

THEOREM 9.2.5

In a regular pyramid, the lengths of altitude h, radius r of the base, and lateral edge e satisfy the Pythagorean Theorem; that is, $e^2 = h^2 + r^2$.

EXAMPLE 5

Find the volume of the regular square pyramid in Figure 9.21(a).

Figure 9.21

SOLUTION The length of the altitude of the pyramid is represented by h, which is determined as follows.

The altitude meets the diagonals of the square base at their common midpoint [see Figure 9.21(b)]. Each diagonal of the base has the length $4\sqrt{2}$ ft by the 45°-45°-90° relationship. Thus, the radius of the square base measures $r = 2\sqrt{2}$. In Figure 9.21(c), we have a right triangle whose legs are of lengths $2\sqrt{2}$ ft and h, and the hypotenuse has length 6 ft (the length of the lateral edge). That is, $r = 2\sqrt{2}$ and $e = 6$.

Using the formula $h^2 + r^2 = e^2$ from Theorem 9.2.5, we have

$$h^2 + (2\sqrt{2})^2 = 6^2$$
$$h^2 + 8 = 36$$
$$h^2 = 28$$
$$h = \sqrt{28} = \sqrt{4 \cdot 7} = \sqrt{4} \cdot \sqrt{7} = 2\sqrt{7}$$

The area of the square base is $B = 4^2$, or $B = 16$ ft^2. Now we have

$$V = \frac{1}{3}Bh$$

$$= \frac{1}{3}(16)(2\sqrt{7})$$

$$= \frac{32}{3}\sqrt{7} \text{ ft}^3 \approx 28.22 \text{ ft}^3$$

EXAMPLE 6

Find the volume of a regular hexagonal pyramid whose base edges have length 4 in. and whose altitude measures 12 in. [See Figure 9.22(a).]

4"
(a)

60°
2
|←4"→|
$2\sqrt{3}$
(b)

Figure 9.22

SOLUTION In the formula $V = \frac{1}{3}Bh$, the altitude is $h = 12$. To find the area of
the base, we use the formula $B = \frac{1}{2}aP$ (this was written $A = \frac{1}{2}aP$ in Chapter 8).
In Figure 9.22(b), the 30°-60°-90° triangle formed by the apothem, the radius,
and a side of the regular hexagon has an apothem length $a = 2\sqrt{3}$ in.
Now $B = \frac{1}{2} \cdot 2\sqrt{3} \cdot (6 \cdot 4)$, or $B = 24\sqrt{3}$ in^2.
In turn, $V = \frac{1}{3}Bh$ becomes $V = \frac{1}{3}(24\sqrt{3})(12)$, so $V = 96\sqrt{3}$ in$^3 \approx 166.28$ in^3.

EXAMPLE 7

A church steeple has the shape of a regular square pyramid. Measurements taken
show that the base edges measure 10 ft and that the length of a lateral edge is 13 ft.
To determine the amount of roof needing to be reshingled, find the lateral area of
the pyramid. (See Figure 9.23.)

13'
ℓ
5'
|←— 10' —→|

Figure 9.23

SOLUTION The slant height ℓ of each triangular face is determined by solving
the equation

$$5^2 + \ell^2 = 13^2$$
$$25 + \ell^2 = 169$$
$$\ell^2 = 144$$
$$\ell = 12$$

Using the formula $A = \frac{1}{2}bh$ with $b = 10$ and $h = 12$, the area of a lateral face
is $A = \frac{1}{2} \cdot 10 \cdot 12 = 60$ ft^2.
Considering the four lateral faces, the area to be reshingled measures

$$L = 4 \cdot 60 \text{ ft}^2 \qquad \text{or} \qquad L = 240 \text{ ft}^2$$

Reminder

It is sometimes easier to find the lateral area without memorizing and using another new formula.

SSG EXS. 8–11

Plane and solid figures may have line symmetry and point symmetry. However, solid figures may also have **plane symmetry.** To have this type of symmetry, a plane can be drawn for which each point of the space figure has a corresponding point on the opposite side of the plane at the same distance from the plane.

Each solid in Figure 9.24 has more than one plane of symmetry. In Figure 9.24(a), the plane of symmetry shown is determined by the midpoints of the indicated edges of the "box." Note that the box also has both line symmetry and point symmetry. In Figure 9.24(b), the plane determined by the apex and the midpoints of opposite sides of the square base leads to plane symmetry for the pyramid. The pyramid also has line symmetry, but it does *not* have point symmetry.

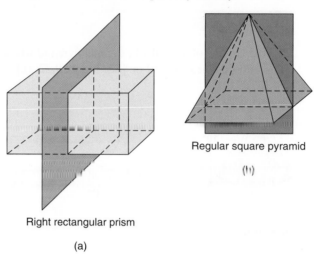

Regular square pyramid

(b)

Right rectangular prism

(a)

Figure 9.24

Exercises 9.2

In Exercises 1 to 4, name the solid that is shown. Answers are based on Sections 9.1 and 9.2.

1. a)

Bases are not regular.

b)

Bases are not regular.

2. a)

Bases are regular.

b)

Bases are not regular.

3. a)

Lateral faces are congruent; base is a square.

b)

Base is a square.

4. a)

Lateral faces are congruent; base is a regular polygon.

b)

Lateral faces are not congruent.

5. In the solid shown, base *ABCD* is a square.
   a) Is the solid a prism or a pyramid?
   b) Name the vertex (apex) of the pyramid.
   c) Name the lateral edges.
   d) Name the lateral faces.
   e) Is the solid a regular square pyramid?

Exercises 5, 7, 9, 11

Exercises 6, 8, 10, 12

6. In the solid on page 406, the base is a regular hexagon.
   a) Name the vertex (apex) of the pyramid.
   b) Name the base edges of the pyramid.
   c) Assuming that lateral edges are congruent, are the lateral faces also congruent?
   d) Assuming that lateral edges are congruent, is the solid a regular hexagonal pyramid?

7. Consider the square pyramid in Exercise 5.
   a) How many vertices does it have?
   b) How many edges (lateral edges plus base edges) does it have?
   c) How many faces (lateral faces plus bases) does it have?
   d) At which point are the lateral faces concurrent?

8. Consider the hexagonal pyramid in Exercise 6.
   a) How many vertices does it have?
   b) How many edges (lateral edges plus base edges) does it have?
   c) How many faces (lateral faces plus bases) does it have?
   d) At which point are the lateral edges concurrent?

9. Suppose that the lateral faces of the pyramid in Exercise 5 have $A_{ABE} = 12$ in^2, $A_{BCE} = 16$ in^2, $A_{CED} = 12$ in^2, and $A_{ADE} = 10$ in^2. If each side of the square base measures 4 in., find the total surface area of the pyramid.

10. Suppose that the base of the hexagonal pyramid in Exercise 6 has an area of 41.6 cm^2 and that each lateral face has an area of 20 cm^2. Find the total (surface) area of the pyramid.

11. Suppose that the base of the square pyramid in Exercise 5 has an area of 16 cm^2 and that the altitude of the pyramid measures 6 cm. Find the volume of the square pyramid.

12. Suppose that the base of the hexagonal pyramid in Exercise 6 has an area of 41.6 cm^2 and that the altitude of the pyramid measures 3.7 cm. Find the volume of the hexagonal pyramid.

13. Assume that the number of sides in the base of a pyramid is n. Generalize the results found in earlier exercises by answering each of the following questions.
    a) What is the number of vertices?
    b) What is the number of lateral edges?
    c) What is the number of base edges?
    d) What is the total number of edges?
    e) What is the number of lateral faces?
    f) What is the total number of faces?

    (Note: Lateral faces and base = faces.)

14. Refer to the prisms of Exercises 1 and 2. Which of these have symmetry with respect to one (or more) plane(s)?

15. Refer to the pyramids of Exercises 3 and 4. Which of these have symmetry with respect to one (or more) plane(s)?

16. Refer to the prisms of Exercises 1 and 2. Which of these prisms have symmetry with respect to a point?

17. Refer to the pyramids of Exercises 3 and 4. Which of these pyramids have symmetry with respect to a line?

18. Consider any regular pyramid. Indicate which line segment has the greater length:
    a) Slant height or altitude?
    b) Lateral edge or radius of the base?

19. Consider any regular pyramid. Indicate which line segment has the greater length:
    a) Slant height or apothem of base?
    b) Lateral edge or slant height?

In Exercises 20 and 21, use Theorem 9.2.1 in which the lengths of apothem a, altitude h, and slant height ℓ of a regular pyramid are related by the equation $\ell^2 = a^2 + h^2$.

20. In a regular square pyramid whose base edges measure 8 in., the apothem of the base measures 4 in. If the height of the pyramid is 8 in., find the length of its slant height.

21. In a regular hexagonal pyramid whose base edges measure $2\sqrt{3}$ in., the apothem of the base measures 3 in. If the slant height of the pyramid is 5 in., find the length of its altitude.

22. In the regular pentagonal pyramid, each lateral edge measures 8 in., and each base edge measures 6 in. The apothem of the base measures 4.1 in.
    a) Find the lateral area of the pyramid.
    b) Find the total area of the pyramid.

Base

Exercises 22, 23

23. In the regular pentagonal pyramid, each base edge measures 9.2 cm and the apothem of the base measures 6.3 cm. The altitude of the pyramid measures 14.6 cm.
    a) Find the base area of the pyramid.
    b) Find the volume of the pyramid.

24. For the regular square pyramid shown, suppose that the sides of the square base measure 10 m each and that the lateral edges measure 13 m each.
    a) Find the lateral area of the pyramid.
    b) Find the total area of the pyramid.
    c) Find the volume of the pyramid.

Exercises 24, 25

25. For the regular square pyramid shown in Exercise 24, suppose that the sides of the square base measure 6 ft each and that the altitude is 4 ft in length.
    a) Find the lateral area L of the pyramid.
    b) Find the total area T of the pyramid.
    c) Find the volume V of the pyramid.

26. The figure below is a regular hexagonal pyramid.
  a) Find the lateral area L of the pyramid.
  b) Find the total area T of the pyramid.
  c) Find the volume V of the pyramid.

27. For a regular square pyramid, suppose that the altitude has a measure equal to that of each edge of the base. If the volume of the pyramid is 72 in³, find the total area of the pyramid.

Exercises 27, 28

28. For a regular square pyramid, the slant height of each lateral face has a measure equal to that of each edge of the base. If the lateral area is 200 in², find the volume of the pyramid.

29. A church steeple in the shape of a regular square pyramid needs to be reshingled. The part to be covered corresponds to the lateral area of the square pyramid. If each lateral edge measures 17 ft and each base edge measures 16 ft, how many square feet of shingles need to be replaced?

Exercises 29, 30

30. Before the shingles of the steeple (see Exercise 29) are replaced, an exhaust fan is to be installed in the steeple. To determine what size exhaust fan should be installed, it is necessary to know the volume of air in the attic (steeple). Find the volume of the regular square pyramid described in Exercise 29.

31. A teepee is constructed by using 12 poles. The construction leads to a regular pyramid with a dodecagon (12 sides) for the base. With the base as shown, and knowing that the height of the teepee is 15 ft, find its volume.

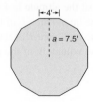

Exercises 31, 32

32. For its occupants to be protected from the elements, it was necessary that the teepee in Exercise 31 be enclosed. Find the amount of area to be covered; that is, determine the lateral area of the regular dodecagonal pyramid. Recall that its altitude measures 15 ft.

33. The street department's storage building, which is used to store the rock, gravel, and salt used on the city's roadways, is built in the shape of a regular hexagonal pyramid. The altitude of the pyramid has the same length as any side of the base. If the volume of the interior is 11,972 ft³, find the length of the altitude and of each side of the base to the nearest foot.

34. The foyer planned as an addition to an existing church is designed as a regular octagonal pyramid. Each side of the octagonal floor has a length of 10 ft, and its apothem measures 12 ft. If 800 ft² of plywood is needed to cover the exterior of the foyer (that is, the lateral area of the pyramid is 800 ft²), what is the height of the foyer?

35. The exhaust chute on a wood chipper has a shape like the part of a pyramid known as the *frustrum of the pyramid*. With dimensions as indicated, find the volume (capacity) of the chipper's exhaust chute.

36. A popcorn container at a movie theater has the shape of a frustrum of a pyramid (see Exercise 35). With dimensions as indicated, find the volume (capacity) of the container.

37. A regular tetrahedron is a regular triangular pyramid in which all faces (lateral faces and base) are congruent. If each edge has length e,
  a) show that the area of each face is $A = \frac{e^2\sqrt{3}}{4}$.
  b) show that the total area of the tetrahedron is $T = e^2\sqrt{3}$.
  c) find the total area if each side measures $e = 4$ in.

Exercises 37, 38

*38. Each edge of a regular tetrahedron (see Exercise 37) has length e.

a) Show that the altitude of the tetrahedron measures $h = \frac{\sqrt{2}}{\sqrt{3}}e$.

b) Show that the volume of the tetrahedron is $V = \frac{\sqrt{2}}{12}e^3$.

c) Find the volume of the tetrahedron if each side measures $e = 4$ in.

Exercises 39 and 40 are based upon the "uniqueness of volume."

39. A tetrahedron (not rectangular) has vertices at A, B, C, and D. The length of the altitude from A to the base ($\triangle BCD$) measures 6 in. It is given that $m\angle BCD = 90°$, $BC = 4$ in., and $CD = 8$ in.

a) Find the volume of the pyramid.

b) Find the length of the altitude from vertex D to the base ($\triangle ABC$); note that $A_{ABC} = 12$ in^2.

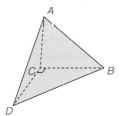

Exercises 39, 40

40. The volume of the pyramid shown is 32 in^3.

a) Find the area of base $\triangle ACD$ if the length of the altitude from vertex B to base $\triangle ACD$ is 4 in.

b) To the nearest tenth of an inch, find the length of the altitude from vertex C to base $\triangle ABD$ if $A_{ABD} \approx 31.23$ in^2.

41. Consider the accompanying figure. When the four congruent isosceles triangles are folded upward, a regular square pyramid is formed. What is the surface area (total area) of the pyramid?

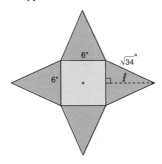

Exercises 41, 42

42. Find the volume of the regular square pyramid that was formed in Exercise 41.

43. Where e_1 and e_2 are the lengths of two corresponding edges of similar prisms or pyramids, the ratio of their volumes is $\frac{V_1}{V_2} = \left(\frac{e_1}{e_2}\right)^3$. Write a ratio to compare volumes for two similar regular square pyramids in which $e_1 = 4$ in. and $e_2 = 2$ in.

44. Use the information from Exercise 43 to find the ratio of volumes $\frac{V_1}{V_2}$ for two cubes in which $e_1 = 2$ cm and $e_2 = 6$ cm.

(Note: $\frac{V_1}{V_2}$ can be found by determining the actual volumes of the cubes.)

45. A hexagonal pyramid (not regular) with base $ABCDEF$ has plane symmetry with respect to a plane determined by apex G and vertices A and D of its base. If the volume of the pyramid with apex G and base $ABCD$ is 19.7 in^3, find the volume of the given hexagonal pyramid.

9.3 | Cylinders and Cones

KEY CONCEPTS	Cylinders (Right and Oblique)	Base and Altitude of a Cone	Total (Surface) Area
	Bases and Altitude of a Cylinder	Volume	
		Vertex (Apex) and Slant Height of a Cone	Solid of Revolution
	Axis of a Cylinder	Axis of a Cone	Axis of a Solid of Revolution
	Cones (Right and Oblique)	Lateral Area	

CYLINDERS

Consider the solids in Figure 9.25 on page 410, in which congruent circles lie in parallel planes. For the circles on the left, suppose that centers O and O' are joined to form $\overline{OO'}$; similarly, suppose that $\overline{QQ'}$ joins the centers of the circles on the right. Let segments such as $\overline{XX'}$ join two points of the circles on the left, so that $\overline{XX'} \parallel \overline{OO'}$. If all such segments (such as $\overline{XX'}$, $\overline{YY'}$, and $\overline{ZZ'}$) are parallel to each other, then a **cylinder** is generated.

Because $\overline{OO'}$ is not perpendicular to planes P and P', the solid on the left is an **oblique circular cylinder**. With $\overline{QQ'}$ perpendicular to planes P and P', the solid on the right is a **right circular cylinder**. For both cylinders, the distance h between the planes P and P' is the length of the **altitude** of the cylinder; h is also called the *height* of the cylinder. The congruent circles are known as the **bases** of each cylinder.

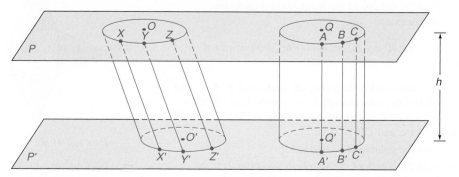

Figure 9.25

A right circular cylinder is shown in Figure 9.26; however, the parallel planes (such as P and P' in Figure 9.25) are not pictured. The line segment joining the centers of the two circular bases is known as the **axis** of the cylinder. For a right circular cylinder, it is necessary that the axis be perpendicular to the planes of the circular bases; in such a case, the length of the altitude h is the length of the axis.

Figure 9.26

SURFACE AREA OF A CYLINDER

Discover

Think of the labeled part of the aluminum can as a right circular cylinder. For the cylinder, the lateral surface is the "label" of the can. If the label were sliced downward by a perpendicular line between the planes, removed, and rolled out flat, it would be rectangular in shape. As shown below, that rectangle would have a length equal to the circumference of the circular base and a width equal to the height of the cylinder. Thus, the lateral area is given by $A = bh$, which becomes $L = Ch$, or $L = 2\pi rh$.

SSG   **EXS. 1, 2**

The formula for the lateral area of a right circular cylinder (found in the following theorem) should be compared to the formula $L = hP$, the lateral area of a right prism whose base has perimeter P.

THEOREM 9.3.1

The lateral area L of a right circular cylinder with altitude of length h and circumference C of the base is given by $L = hC$.
Alternative Form: The lateral area of the right circular cylinder can be expressed in the form $L = 2\pi rh$, where r is the length of the radius of the circular base.

Rather than constructing a formal proof of Theorem 9.3.1, consider the Discover activity shown on page 410.

THEOREM 9.3.2

The total area T of a right circular cylinder with base area B and lateral area L is given by $T = L + 2B$.
Alternative Form: Where r is the length of the radius of the base and h is the length of the altitude of the cylinder, the total area can be expressed in the form $T = 2\pi rh + 2\pi r^2$.

EXAMPLE 1

For the right circular cylinder shown in Figure 9.27, find the
a) exact lateral area L.
b) exact surface area T.

Figure 9.27

SOLUTION

a) $L = 2\pi rh$
   $= 2 \cdot \pi \cdot 5 \cdot 12$
   $= 120\pi \text{ in}^2$
b) $T = L + 2B$
   $= 2\pi rh + 2\pi r^2$
   $= 120\pi + 2 \cdot \pi \cdot 5^2$
   $= 120\pi + 50\pi$
   $= 170\pi \text{ in}^2$

 EXS. 3–5

VOLUME OF A CYLINDER

In considering the volume of a right circular cylinder, recall that the volume of a prism is given by $V = Bh$, where B is the area of the base. In Figure 9.28, we inscribe a prism in the cylinder as shown. Suppose that the prism is regular and that the number of sides in the inscribed polygon's base increases without limit; thus, the base approaches a circle in this limiting process. The area of the polygonal base also approaches the area of the circle, and the volume of the prism approaches that of the right circular cylinder. Our conclusion is stated without proof in the following theorem.

Figure 9.28

> **THEOREM 9.3.3**
>
> The volume V of a right circular cylinder with base area B and altitude of length h is given by $V = Bh$.
> *Alternative Form:* Where r is the length of the radius of the base, the volume for the right circular cylinder can be written $V = \pi r^2 h$.

Figure 9.29

EXAMPLE 2

If $d = 4$ cm and $h = 3.5$ cm, use a calculator to find the approximate volume of the right circular cylinder shown in Figure 9.29. Give the answer correct to two decimal places.

SOLUTION $d = 4$, so $r = 2$. Thus, $V = Bh$ or $V = \pi r^2 h$ becomes

$$V = \pi \cdot 2^2(3.5)$$
$$= \pi \cdot 4(3.5) = 14\pi \approx 43.98 \text{ cm}^3$$

EXAMPLE 3

In the right circular cylinder shown in Figure 9.29, suppose that the height equals the diameter of the circular base. If the exact volume is 128π in^3, find the exact lateral area L of the cylinder.

SOLUTION

$$h = 2r$$
so $$V = \pi r^2 h$$
becomes $$V = \pi r^2(2r)$$
$$V = 2\pi r^3$$

Thus, $$2\pi r^3 = 128\pi,$$
Dividing by 2π, $$r^3 = 64$$
$$r = 4$$
$$h = 8 \quad \text{(from } h = 2r\text{)}$$
Now $$L = 2\pi r h$$
$$= 2 \cdot \pi \cdot 4 \cdot 8$$
$$= 64\pi \text{ in}^2$$

Table 9.2 should help us recall and compare the area and volume formulas found in Sections 9.1 and 9.3.

TABLE 9.2

	Lateral Area	Total Area	Volume
Prism	$L = hP$	$T = L + 2B$	$V = Bh$
Cylinder	$L = hC$	$T = L + 2B$	$V = Bh$

 EXS. 6, 7

CONES

In Figure 9.30 on page 413, point P lies outside the plane containing circle O. A surface known as a **cone** results when line segments are drawn from P to points on the circle. However, if P is joined to all possible points on the circle as well as to points in the

Figure 9.30

interior of the circle, a solid is formed. If $\overline{PO}$ is not perpendicular to the plane of circle O in Figure 9.30, the cone is an **oblique circular cone**.

In Figures 9.30 and 9.31, point P is the **vertex** (or **apex**) of the cone, and circle O is the **base**. The segment $\overline{PO}$, which joins the vertex to the center of the circular base, is the **axis** of the cone. If the axis is perpendicular to the plane containing the base, as in Figure 9.31, the cone is a **right circular cone**. In any cone, the perpendicular segment from the vertex to the plane of the base is the **altitude** of the cone. In a right circular cone, the length h of the altitude equals the length of the axis. For a right circular cone, and only for this type of cone, any line segment that joins the vertex to a point on the circle is a **slant height** of the cone; we will denote the length of the slant height by ℓ as shown in Figure 9.31.

SURFACE AREA OF A CONE

SSG **EXS. 8, 9**

Recall now that the lateral area for a regular pyramid is given by $L = \frac{1}{2}\ell P$. For a right circular cone, consider an inscribed regular pyramid as in Figure 9.32. As the number of sides of the inscribed polygon's base grows larger, the perimeter of the inscribed polygon approaches the circumference of the circle as a limit. In addition, the slant height of the congruent triangular faces approaches that of the slant height of the cone. Thus, the lateral area of the right circular cone can be compared to $L = \frac{1}{2}\ell P$; for the cone, we have

Figure 9.31

$$L = \frac{1}{2}\ell C$$

in which C is the circumference of the base. The fact that $C = 2\pi r$ leads to

$$L = \frac{1}{2}\ell(2\pi r)$$

so

$$L = \pi r \ell$$

Figure 9.32

> ### THEOREM 9.3.4
>
> The lateral area L of a right circular cone with slant height of length ℓ and circumference C of the base is given by $L = \frac{1}{2}\ell C$.
> *Alternative Form:* Where r is the length of the radius of the base, $L = \pi r \ell$.

The following theorem follows easily from Theorem 9.3.4 and is given without proof.

> ### THEOREM 9.3.5
>
> The total area T of a right circular cone with base area B and lateral area L is given by $T = B + L$.
> *Alternative Form:* Where r is the length of the radius of the base and ℓ is the length of the slant height, $T = \pi r^2 + \pi r \ell$.

EXAMPLE 4

For the right circular cone in which $r = 3$ cm and $h = 6$ cm (see Figure 9.33), find the

a) exact and approximate lateral area L.
b) exact and approximate total area T.

Figure 9.33

SOLUTION

a) We need the length of the slant height ℓ for each problem part, so we apply the Pythagorean Theorem:

$$\ell^2 = r^2 + h^2$$
$$\ell^2 = 3^2 + 6^2$$
$$\ell^2 = 9 + 36 = 45$$

Then
$$\ell = \sqrt{45} = \sqrt{9 \cdot 5}$$
$$\ell = \sqrt{9} \cdot \sqrt{5} = 3\sqrt{5}$$

Using $L = \pi r \ell$, we have

$$L = \pi \cdot 3 \cdot 3\sqrt{5}$$
$$L = 9\pi\sqrt{5} \text{ cm}^2 \approx 63.22 \text{ cm}^2$$

b) We also have

$$T = B + L$$
$$T = \pi r^2 + \pi r \ell$$
$$T = \pi \cdot 3^2 + 9\pi\sqrt{5}$$
$$T = (9\pi + 9\pi\sqrt{5}) \text{ cm}^2 \approx 91.50 \text{ cm}^2$$

The following theorem was illustrated in the solution of Example 4. Consider Figure 9.31 on page 413 as you read Theorem 9.3.6.

EXS. 10, 11

THEOREM 9.3.6

In a right circular cone, the lengths of the radius r of the base, the altitude h, and the slant height ℓ satisfy the Pythagorean Theorem; that is, $\ell^2 = r^2 + h^2$ in every right circular cone.

VOLUME OF A CONE

Recall that the volume of a pyramid is given by the formula $V = \frac{1}{3}Bh$. Consider a regular pyramid inscribed in a right circular cone. If its number of sides increases indefinitely, the volume of the pyramid approaches that of the right circular cone (see Figure 9.34). Then the volume of the right circular cone is $V = \frac{1}{3}Bh$. Because the area of the base of the cone is $B = \pi r^2$, an alternative formula for the volume of the cone is

$$V = \frac{1}{3}\pi r^2 h$$

Figure 9.34

We state this result as a theorem.

THEOREM 9.3.7

The volume V of a right circular cone with base area B and altitude of length h is given by $V = \frac{1}{3}Bh$.
Alternative Form: Where r is the length of the radius of the base, the formula for the volume of the cone is usually written $V = \frac{1}{3}\pi r^2 h$.

Table 9.3 should help us to recall and compare the area and volume formulas found in Sections 9.2 and 9.3.

TABLE 9.3

	Lateral Area	**Total Area**	**Volume**	**Slant Height**
Pyramid	$L = \frac{1}{2}\ell P$	$T = B + L$	$V = \frac{1}{3}Bh$	$\ell^2 = a^2 + h^2$
Cone	$L = \frac{1}{2}\ell C$	$T = B + L$	$V = \frac{1}{3}Bh$	$\ell^2 = r^2 + h^2$

NOTE: The formulas that contain the slant height ℓ are used only with the regular pyramid and the right circular cone.

SSG **EXS. 12, 13**

SOLIDS OF REVOLUTION

Suppose that part of the boundary for a plane region is a line segment. When the plane region is revolved about this line segment, the locus of points generated in space is called a **solid of revolution**. The complete 360° rotation moves the region about the edge until the region returns to its original position. The side (edge) used is called the **axis** of the resulting solid of revolution. Consider Example 5.

EXAMPLE 5

Describe the solid of revolution that results when

   a) a rectangular region with dimensions 2 ft by 5 ft is revolved about the 5-ft side as shown in Figure 9.35(a).

   b) a semicircular region with radius of length 3 cm is revolved about the diameter shown in Figure 9.35(b).

(a)

(b)

Figure 9.35

SOLUTION

   a) In Figure 9.35(a), the rectangle on the left is revolved about the 5-ft side to form the solid on the right. The solid of revolution generated is a right circular cylinder that has a base radius length of 2 ft and an altitude measuring 5 ft.

   b) In Figure 9.35(b), the semicircle on the left is revolved about its diameter to form the solid on the right. The solid of revolution generated is a *sphere* with a radius of length 3 cm.

NOTE: We will study the sphere in greater detail in Section 9.4.

EXAMPLE 6

Determine the exact volume of the solid of revolution formed when the region bounded by a right triangle with legs of lengths 4 in. and 6 in. is revolved about the 6-in. side. The triangular region is shown in Figure 9.36(a) on page 416.

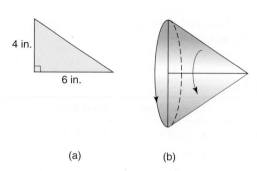

4 in.

6 in.

(a)   (b)

Figure 9.36

SOLUTION As shown in Figure 9.36(b), the resulting solid is a cone whose altitude measures 6 in. and whose radius of the base measures 4 in. Using $V = \frac{1}{3}Bh$, we have

$$V = \frac{1}{3}\pi r^2 h$$

$$= \frac{1}{3} \cdot \pi \cdot 4^2 \cdot 6 = 32\pi \text{ in}^3$$

It may come as a surprise that the formulas that are used to calculate the volumes of an oblique circular cylinder and a right circular cylinder are identical. To see why the formula $V = Bh$ or $V = \pi r^2 h$ can be used to calculate the volume of an oblique circular cylinder, consider the stacks of pancakes shown in Figures 9.37(a) and 9.37(b). With each stack h units high, the volume is the same regardless of whether the stack is vertical or oblique.

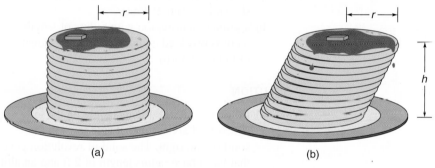

(a)   (b)

Figure 9.37

It is also true that the formula for the volume of an oblique circular cone is the same as the formula for the volume of the right circular cone. In fact, the motivating argument preceding Theorem 9.3.7 could be repeated, with the exception that the inscribed pyramid is oblique. For both the right circular cone and the oblique circular cone, $V = \frac{1}{3}Bh$ or $V = \frac{1}{3}\pi r^2 h$.

SSG **EXS. 14, 15**

Exercises 9.3

1. Does a right circular cylinder such as an aluminum can have
   a) symmetry with respect to at least one plane?
   b) symmetry with respect to at least one line?
   c) symmetry with respect to a point?

2. Does a right circular cone such as a wizard's cap have
   a) symmetry with respect to at least one plane?
   b) symmetry with respect to at least one line?
   c) symmetry with respect to a point?

3. For the right circular cylinder shown on page 417, $r = 3.2$ cm and $h = 5.1$ cm. Find the approximate volume of the cylinder.

4. For the right circular cylinder shown on page 417, $r = 1.75$ in. and $h = 4.23$ in. Find the approximate volume of the cylinder.

5. For the right circular cylinder, suppose that $r = 5$ in. and $h = 6$ in. Find the exact and approximate
   a) lateral area.
   b) total area.
   c) volume.

Exercises 3–6

6. Suppose that $r = 12$ cm and $h = 15$ cm in the right circular cylinder. Find the exact and approximate
   a) lateral area.
   b) total area.
   c) volume.

7. The tin can shown at the right has the indicated dimensions. Estimate the number of square inches of tin required for its construction.

   (HINT: Include the lid and the base in the result.)

Exercises 7, 8

8. What is the volume of the tin can? If it contains 16 oz of green beans, what is the volume of the can used for 20 oz of green beans? Assume a proportionality between weight and volume.

9. If the exact volume of a right circular cylinder is 200π cm³ and its altitude measures 8 cm, what is the measure of the radius of the circular base?

10. Suppose that the volume of an aluminum can is to be 9π in³. Find the dimensions of the can if the diameter length of the base is three-fourths the length of the altitude.

11. For an aluminum can, the lateral surface area is 12π in². If the length of the altitude is 1 in. greater than the length of the radius of the circular base, find the dimensions of the can.

12. Find the height of a storage tank in the shape of a right circular cylinder that has a circumference measuring 6π m and a volume measuring 81π m³.

13. Find the volume of the oblique circular cylinder. The axis meets the plane of the base to form a 45° angle.

14. A cylindrical orange juice container has metal bases of radius length 1 in. and a cardboard lateral surface 3 in. high. If the cost of the metal used is 0.5 cent per square inch and the cost of the cardboard is 0.2 cent per square inch, what is the approximate cost of constructing one container? Use $\pi \approx 3.14$.

In Exercises 15 to 20, use the fact that $r^2 + h^2 = \ell^2$ in a right circular cone (Theorem 9.3.6).

15. Find the length of the slant height ℓ of a right circular cone with $r = 4$ cm and $h = 6$ cm.

16. Find the length of the slant height ℓ of a right circular cone with $r = 5.2$ ft and $h = 3.9$ ft.

17. Find the height h of a right circular cone in which the diameter of the base measures $d = 9.6$ m and $\ell = 5.2$ m.

18. Find the length of the radius r of a right circular cone in which $h = 6$ yd and $\ell = 8$ yd.

19. Find the length of the slant height ℓ of a right circular cone with $r = 6$ in., length of altitude h, and $\ell = 2h$ in.

20. Find the length of the radius r of a right circular cone with $\ell = 12$ in. and $h = 3r$ in.

21. The oblique circular cone has an altitude and a diameter of base that are each of length 6 cm. The line segment joining the vertex to the center of the base is the axis of the cone. What is the length of the axis?

22. For the accompanying right circular cone, $h = 6$ m and $r = 4$ m. Find the exact and approximate
   a) lateral area.
   b) total area.
   c) volume.

Exercises 22, 23

23. For the right circular cone shown in Exercise 22, suppose that $h = 7$ in. and $r = 6$ in. Find the exact and approximate
   a) lateral area.
   b) total area.
   c) volume.

24. Rukia discovers that the teepee with a circular floor has a radius length of 6 ft and a height of 15 ft. Find the volume of the enclosure.

25. A rectangle has dimensions of 6 in. by 3 in. Find the exact volume of the solid of revolution formed when the rectangle is rotated about its 6-in. side.

26. A rectangle has dimensions of 6 in. by 3 in. Find the exact volume of the solid of revolution formed when the rectangle is rotated about its 3-in. side.

27. A triangle has sides that measure 15 cm, 20 cm, and 25 cm. Find the exact volume of the solid of revolution formed when the triangle is revolved about the side of length 15 cm.

28. A triangle has sides that measure 15 cm, 20 cm, and 25 cm. Find the exact volume of the solid of revolution formed when the triangle is revolved about the side of length 20 cm.

29. A triangle has sides that measure 15 cm, 20 cm, and 25 cm. Find the exact volume of the solid of revolution formed when the triangle is revolved about the side of length 25 cm.

(HINT: The altitude to the 25-cm side has length 12 cm.)

30. Where r is the length of the radius of a sphere, the volume of the sphere is given by $V = \frac{4}{3}\pi r^3$. Find the exact volume of the sphere that was formed in Example 5(b).

31. If a right circular cone has a circular base with a diameter of length 10 cm and a volume of 100π cm³, find its lateral area.

32. A right circular cone has a slant height of 12 ft and a lateral area of 96π ft². Find its volume.

33. A solid is formed by cutting a conical section away from a right circular cylinder. If the radius measures 6 in. and the altitude measures 8 in., what is the volume of the resulting solid?

In Exercises 34 and 35, give a paragraph proof for each claim.

34. The total area T of a right circular cylinder whose altitude is of length h and whose circular base has a radius of length r is given by $T = 2\pi r(r + h)$.

35. The volume V of a washer that has an inside radius of length r, an outside radius of length R, and an altitude of measure h is given by $V = \pi h(R + r)(R - r)$.

36. For a right circular cone, the slant height has a measure equal to twice that of the radius length of the base. If the total area of the cone is 48π in², what are the dimensions of the cone?

37. For a right circular cone, the ratio of the slant height to the length of the radius is 5:3. If the volume of the cone is 96π in³, find the lateral area of the cone.

38. If the length of the radius and the height of a right circular cylinder are both doubled to form a larger cylinder, what is the ratio of the volume of the larger cylinder to the volume of the smaller cylinder?

(Note: The two cylinders are said to be "similar.")

39. For the two similar cylinders in Exercise 38, what is the ratio of the lateral area of the larger cylinder to that of the smaller cylinder?

40. For a right circular cone, the dimensions are $r = 6$ cm and $h = 8$ cm. If the length of the radius is doubled while the height is made half as large in forming a new cone, will the volumes of the two cones be equal?

41. A cylindrical storage tank has a depth of 5 ft and a radius measuring 2 ft. If each cubic foot can hold 7.5 gal of gasoline, what is the total storage capacity of the tank measured in gallons?

42. If the tank in Exercise 41 needs to be painted and 1 pt of paint covers 50 ft², how many pints are needed to paint the exterior of the storage tank?

43. A frustrum of a cone is the portion of the cone bounded between the circular base and a plane parallel to the base. With dimensions as indicated, show that the volume of the frustrum of the cone is

$$V = \frac{1}{3}\pi R^2 H - \frac{1}{3}\pi r^2 h$$

In Exercises 44 and 45, use the formula from Exercise 43. Similar triangles were used to find h and H.

44. A margarine tub has the shape of the frustrum of a cone. With the lower base having diameter length 11 cm and the upper base having diameter length 14 cm, the volume of such a container $6\frac{2}{3}$ cm tall can be determined by using $H = 32\frac{2}{3}$ cm, and $h = 26$ cm. Find its volume.

45. A container of yogurt has the shape of the frustrum of a cone. With the lower base having diameter length 6 cm and the upper base having diameter length 8 cm, the volume of such a container 7.5 cm tall can be determined by using $H = 30$ cm and $h = 22.5$ cm. Find its volume.

46. An oil refinery has storage tanks in the shape of right circular cylinders. Each tank has a height of 16 ft and a radius length of 10 ft for its circular base. If 1 ft³ of volume contains 7.5 gal of oil, what is the capacity of the fuel tank in *gallons*? Round the result to the nearest hundred of gallons.

47. A farmer has a fuel tank in the shape of a right circular cylinder. The tank has a height of 6 ft and a radius of length 1.5 ft for its circular base. If 1 ft³ of volume contains 7.5 gal of gasoline, what is the approximate capacity of the fuel tank in *gallons*?

48. When radii $\overline{OA}$ and $\overline{OB}$ are placed so that they coincide, a 240° sector of a circle is sealed to form a right circular cone. If the length of the radius of the circle is 6.4 cm, what is the length of the approximate lateral area of the cone that is formed? Use a calculator and round the answer to the nearest *tenth* of a square inch.

49. A lawn roller in the shape of a right circular cylinder has a radius of length 18 in. and a length (height) of 4 ft. Find the area rolled during one complete revolution of the roller. Use the calculator value of π, and give the answer to the nearest square foot.

| **9.4** | **Polyhedrons and Spheres** |

KEY CONCEPTS	Dihedral Angle	Regular Polyhedrons	Sphere (Center, Radius,
	Polyhedron (Convex	(Tetrahedron,	Diameter, Great
	and Concave)	Hexahedron,	Circle, Hemisphere)
	Vertices	Octahedron,	Surface Area and
	Edges and Faces	Dodecahedron,	Volume of a Sphere
	Euler's Equation	Icosahedron)	

POLYHEDRONS

When two planes intersect, the angle formed by two half-planes with a common edge (the line of intersection) is a **dihedral angle**. The angle shown in Figure 9.38 is such an angle; in that figure, the measure of the dihedral angle is the same as that of the angle determined by two rays that

1. have a vertex (the common endpoint) on the edge.
2. lie in the planes so that they are perpendicular to the edge.

Recall that three (or more) planes can intersect in a point; in this situation, the planes are said to be concurrent at that point. A **polyhedron** (plural *polyhedrons* or *polyhedra*) is a solid bounded by four or more plane regions. Polygons form the **faces** of the solid, and the line segments common to these polygons are the **edges** of the polyhedron. Endpoints of the edges are the **vertices** of the polyhedron. When a polyhedron is **convex**, each face determines a plane for which all remaining faces lie on the same side of that plane. Figure 9.39(a) illustrates a convex polyhedron, and Figure 9.39(b) illustrates a **concave** polyhedron. In the concave polyhedron, at least one diagonal lies in the exterior of the polyhedron.

Figure 9.38

Convex polyhedron

(a)

Concave polyhedron

(b)

Figure 9.39

The prisms and pyramids discussed in Sections 9.1 and 9.2 were special types of polyhedrons. For instance, a pentagonal pyramid can be described as a hexahedron because it has six faces. Because some of their surfaces do not lie in planes, the cylinders and cones of Section 9.3 are not polyhedrons.

Leonhard Euler (Swiss, 1707–1763) found that the number of vertices, edges, and faces of any polyhedron are related by **Euler's equation**. This equation is given in the following theorem, which is stated without proof.

(a)

(b)

Figure 9.40

SSG EXS. 1–5

THEOREM 9.4.1 ■ Euler's Equation

The number of vertices V, the number of edges E, and the number of faces F of a polyhedron are related by the equation

$$V + F = E + 2$$

EXAMPLE 1

Verify Euler's equation for the (a) tetrahedron and (b) square pyramid shown in Figure 9.40(a) and (b), respectively.

SOLUTION

a) The tetrahedron has four vertices ($V = 4$), six edges ($E = 6$), and four faces ($F = 4$). So Euler's equation becomes $4 + 4 = 6 + 2$, which is true.

b) The pyramid has five vertices (apex + vertices from the base), eight edges (4 base edges + 4 lateral edges), and five faces (4 triangular faces + 1 square base). Now $V + F = E + 2$ becomes $5 + 5 = 8 + 2$, which is also true.

A polyhedron must have at least four faces. Special names are given to polyhedra having a specific number of faces. See Table 9.4.

TABLE 9.4
Naming Polyhedra

Polyhedron name	Number of Faces
Tetrahedron	4
Pentahedron	5
Hexahedron	6
Heptahedron	7
Octahedron	8
Nonahedron	9
Decahedron	10
Dodecahedron	12
Icosahedron	20

REGULAR POLYHEDRONS

DEFINITION

A **regular polyhedron** is a convex polyhedron whose faces are congruent regular polygons, all of the same type.

There are exactly five regular polyhedrons, named as follows:

1. Regular **tetrahedron**: with 4 faces that are congruent equilateral triangles
2. Regular **hexahedron** (or **cube**): with 6 faces that are congruent squares
3. Regular **octahedron**: with 8 faces that are congruent equilateral triangles
4. Regular **dodecahedron**: with 12 faces that are congruent regular pentagons
5. Regular **icosahedron**: with 20 faces that are congruent equilateral triangles

Four of the regular polyhedrons are shown in Figure 9.41.

Regular Polyhedrons

Tetrahedron

Hexahedron

Octahedron

Dodecahedron

Figure 9.41

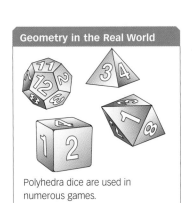

Polyhedra dice are used in numerous games.

Because each regular polyhedron has a central point, each solid is said to have a center. Except for the tetrahedron, these polyhedrons have *point symmetry* at the center. All regular polyhedra have line symmetry and plane symmetry as well.

EXAMPLE 2

Consider a die that is a regular tetrahedron with faces numbered 1, 2, 3, and 4. Assuming that each face has an equal chance of being rolled, what is the likelihood (probability) that one roll produces (a) a "1"? (b) a result larger than "1"?

SOLUTION

a) With four equally likely results (1, 2, 3, and 4), the probability of a "1" is $\frac{1}{4}$.

b) With four equally likely results (1, 2, 3, and 4) and three "favorable" outcomes (2, 3, and 4), the probability of rolling a number larger than a "1" is $\frac{3}{4}$.

 SSG EXS. 6, 7

SPHERES

Another type of solid with which you are familiar is the sphere. Although the surface of a basketball correctly depicts the sphere, we often use the term *sphere* to refer to a solid like a baseball as well. A sphere can be inscribed in or circumscribed about any regular polyhedron because it has point symmetry about its center.

Reminder

The sphere was described as a locus of points in Chapter 7.

In space, the sphere is characterized in three ways:

1. A **sphere** is the locus of points at a fixed distance r from a given point O. Point O is known as the **center** of the sphere, even though it is not a part of the spherical surface.
2. A **sphere** is the surface determined when a circle (or semicircle) is rotated about any of its diameters.
3. A **sphere** is the surface that represents the theoretical limit of an "inscribed" regular polyhedron whose number of faces increases without limit.

NOTE: In characterization 3, suppose that the number of faces of the regular polyhedron could grow without limit. In theory, the resulting regular polyhedra would appear more "spherical" as the number of faces increases without limit. In reality, a regular polyhedron can have no more than 20 faces (the regular icosahedron). It will be necessary to use this third characterization of the sphere when we determine the formula for its volume.

Each characterization of the sphere has its advantages.

▶ Characterization 1

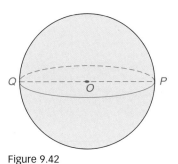

Figure 9.42

In Figure 9.42, a sphere was generated as the locus of points in space at a distance r from point O. The line segment $\overline{OP}$ is a **radius** of sphere O, and $\overline{QP}$ is a **diameter** of the sphere. The intersection of a sphere and a plane that contains its center is a **great circle** of the sphere.

For the earth, the equator is a great circle that separates the earth into two **hemispheres**. The sphere has line symmetry about any line containing the center of the sphere. Similarly, the sphere has symmetry about any plane that contains the center of the sphere.

Discover

Suppose that you use scissors to cut out each pattern. (You may want to copy and enlarge this page.) Then glue or tape the indicated tabs (shaded) to form regular polyhedra. Which regular polyhedron is formed in each pattern?

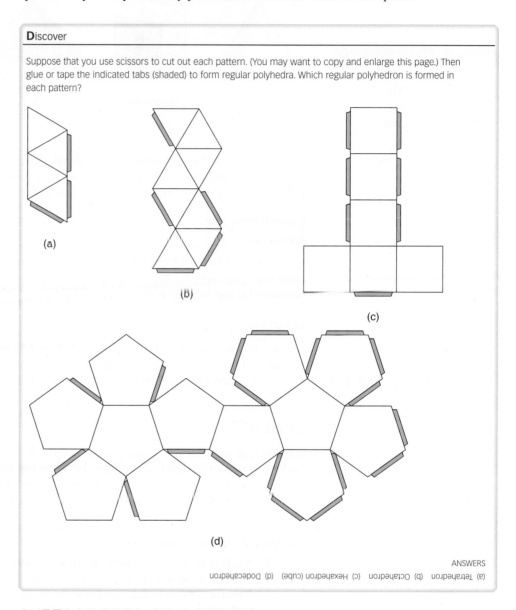

(a)

(b)

(c)

(d)

ANSWERS

(a) Tetrahedron   (b) Octahedron   (c) Hexahedron (cube)   (d) Dodecahedron

SURFACE AREA OF A SPHERE

▶ Characterization 2

The following theorem claims that the surface area of a sphere equals four times the area of a great circle of that sphere. This theorem, which is proved in many calculus textbooks, treats the sphere as a surface of revolution.

THEOREM 9.4.2

The surface area S of a sphere whose radius has length r is given by $S = 4\pi r^2$.

Geometry in the Real World

Fruits such as oranges have the shape of a sphere.

EXAMPLE 3

Find the surface area of a sphere whose radius length is $r = 7$ in. Use your calculator to approximate the result.

SOLUTION

$$S = 4\pi r^2 \rightarrow S = 4\pi \cdot 7^2 = 196\pi \text{ in}^2$$

SSG **EXS. 8–10**

Then $S \approx 615.75 \text{ in}^2$.

Although half of a circle is called a *semicircle*, remember that half of a sphere is generally called a *hemisphere*.

VOLUME OF A SPHERE

▶ Characterization 3

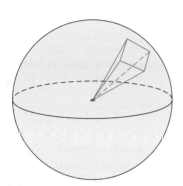

Figure 9.43

The third description of the sphere enables us to find its volume. To accomplish this, we treat the sphere as the theoretical limit of an inscribed regular polyhedron whose number of faces n increases without limit. The polyhedron can be separated into n congruent pyramids; the center of the sphere is the vertex of each pyramid. As n increases, the length of the altitude of each pyramid approaches the length of the radius of the sphere. Next we find the sum of the volumes of these pyramids, the limit of which is the volume of the sphere.

In Figure 9.43, one of the pyramids described in the preceding paragraph is shown. We designate the height of each and every pyramid by h. Where the areas of the bases of the pyramids are written B_1, B_2, B_3, and so on, the sum of the volumes of the n pyramids forming the polyhedron is

$$\frac{1}{3}B_1h + \frac{1}{3}B_2h + \frac{1}{3}B_3h + \cdots + \frac{1}{3}B_nh$$

Next we write the volume of the polyhedron in the form

$$\frac{1}{3}h(B_1 + B_2 + B_3 + \cdots + B_n)$$

As n increases, $h \rightarrow r$ and $B_1 + B_2 + B_3 + \cdots + B_n \rightarrow S$, the surface area of the sphere. That is, $\frac{1}{3}h(B_1 + B_2 + B_3 + \cdots + B_n) \rightarrow \frac{1}{3}rS$. Because the surface area of the sphere is $S = 4\pi r^2$, the sum approaches the following limit as the volume of the sphere:

$$V = \frac{1}{3}r \cdot 4\pi r^2 = \frac{4}{3}\pi r^3$$

The preceding discussion suggests the following theorem.

Discover

A farmer's silo is a composite shape. That is, it is actually composed of two solids. What are they?

© James Horning/Shutterstock.com

ANSWER
Cylinder and hemisphere

THEOREM 9.4.3

The volume V of a sphere with a radius of length r is given by $V = \frac{4}{3}\pi r^3$.

EXAMPLE 4

Find the exact volume of a sphere whose length of radius is 1.5 in.

SOLUTION This calculation can be done more easily if we replace 1.5 by $\frac{3}{2}$.

$$V = \frac{4}{3}\pi r^3$$

$$= \frac{4}{3} \cdot \pi \cdot \frac{3}{2} \cdot \frac{3}{2} \cdot \frac{3}{2}$$

$$= \frac{9\pi}{2} \text{ in}^3$$

EXAMPLE 5

A spherical propane gas storage tank has a volume of $\frac{792}{7}$ ft³. Using $\pi \approx \frac{22}{7}$, find the radius of the sphere.

SOLUTION $V = \frac{4}{3}\pi r^3$ becomes $\frac{792}{7} = \frac{4}{3} \cdot \frac{22}{7} \cdot r^3$. Then $\frac{88}{21}r^3 = \frac{792}{7}$. In turn,

$$\frac{\cancel{21}}{\cancel{88}} \cdot \frac{\cancel{88}}{\cancel{21}}r^3 = \frac{\cancel{21}}{\cancel{88}_1} \cdot \frac{\overset{9}{\cancel{792}}}{\cancel{7}_1} \to r^3 = 27 \to r = \sqrt[3]{27} = 3$$

The radius of the tank is 3 ft.

Just as two concentric circles have the same center but different lengths of radii, two spheres can also be concentric. This fact is the basis for the solution of the problem in the following example.

EXAMPLE 6

A child's hollow plastic ball has an inside diameter length of 10 in. and is approximately $\frac{1}{8}$ in. thick (see the cross-section of the ball in Figure 9.44). Approximately how many cubic inches of plastic were needed to construct the ball?

SOLUTION The volume of plastic used is the difference between the outside volume and the inside volume. Where R denotes the length of the outside radius and r denotes the length of the inside radius, $R \approx 5.125$ and $r = 5$.

$$V = \frac{4}{3}\pi R^3 - \frac{4}{3}\pi r^3, \qquad \text{so} \qquad V = \frac{4}{3}\pi(5.125)^3 - \frac{4}{3}\pi \cdot 5^3$$

Then $V \approx 563.86 - 523.60 \approx 40.26$

The volume of plastic used was approximately 40.26 in³.

5 in.

5.125 in.

Figure 9.44

Like circles, spheres may have tangent lines and secant lines as illustrated in Figure 9.45(a). However, spheres also have tangent planes as shown in Figure 9.45(b). In Figure 9.45(c), spheres T and V are externally tangent; although no such drawing has been provided, two spheres may be internally tangent as well.

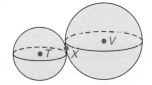

Line t is tangent to sphere O at point P; line s is a secant.

(a)

Plane R is tangent to sphere Q at point S.

(b)

Spheres T and V are externally tangent at point X.

(c)

SSG **EXS. 11–13**   **Figure 9.45**

In Section 9.3, each solid of revolution was generated by revolving a plane region about a horizontal line segment. It is also possible to form a solid of revolution by rotating a region about a vertical or oblique line segment. Example 7 illustrates the solid formed when a region is rotated about a vertical line segment.

EXAMPLE 7

Describe the solid of revolution that is formed when a semicircular region having a vertical diameter of length 12 cm [see Figure 9.46(a)] is revolved about that diameter. Then find the exact volume of the solid formed [see Figure 9.46(b)].

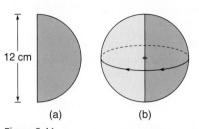

12 cm

(a)   (b)

Figure 9.46

SOLUTION The solid that is formed is a sphere with length of radius $r = 6$ cm. The formula we use to find the volume is $V = \frac{4}{3}\pi r^3$. Then $V = \frac{4}{3}\pi \cdot 6^3$, which simplifies to $V = 288\pi \text{ cm}^3$.

When a circular region is revolved about a line in the circle's exterior, a doughnut-shaped solid results. The formal name of the resulting solid of revolution, shown in Figure 9.47, is the *torus*. Methods of calculus are necessary to calculate both the surface area and the volume of the torus.

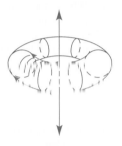

SSG **EXS. 14–16**   Figure 9.47

Exercises 9.4

1. Which of these two polyhedrons is concave? Note that the interior dihedral angle formed by the planes containing $\triangle EJF$ and $\triangle KJF$ is larger than 180°.

(a)

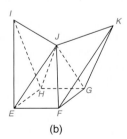

(b)

2. For Figure (a) of Exercise 1, find the number of faces, vertices, and edges in the polyhedron. Then verify Euler's equation for that polyhedron.

3. For Figure (b) of Exercise 1, find the number of faces, vertices, and edges in the polyhedron. Then verify Euler's equation for that polyhedron.

4. For a regular tetrahedron, find the number of faces, vertices, and edges in the polyhedron. Then verify Euler's equation for that polyhedron.

5. For a regular hexahedron, find the number of faces, vertices, and edges in the polyhedron. Then verify Euler's equation for that polyhedron.

6. A regular polyhedron has 12 edges and 8 vertices.
   a) Use Euler's equation to find the number of faces.
   b) Use the result from part (a) to name the regular polyhedron.

7. A regular polyhedron has 12 edges and 6 vertices.
   a) Use Euler's equation to find the number of faces.
   b) Use the result from part (a) to name the regular polyhedron.

8. A polyhedron (not regular) has 10 vertices and 7 faces. How many edges does it have?

9. A polyhedron (not regular) has 14 vertices and 21 edges. How many faces must it have?

In Exercises 10 to 12, the probability is the ratio
$\frac{number\ of\ favorable\ outcomes}{number\ of\ possible\ outcomes}$. *Use Example 2 of this section as a guide.*

10. Assume that a die of the most common shape, a regular hexahedron, is rolled. What is the likelihood that
   a) a "2" results?
   b) an even number results?
   c) the result is larger than 2?

11. Assume that a die in the shape of a regular dodecahedron is rolled. What is the probability that
   a) an even number results?
   b) a prime number (2, 3, 5, 7, or 11) results?
   c) the result is larger than 2?

12. Assume that a die in the shape of a regular icosahedron is rolled. What is the likelihood that
   a) an odd number results?
   b) a prime number (2, 3, 5, 7, 11, 13, 17, or 19) results?
   c) the result is larger than 2?

13. In sphere O, the length of radius $\overline{OP}$ is 6 in. Find the length of the chord:
   a) $\overline{QR}$ if m$\angle QOR = 90°$
   b) $\overline{QS}$ if m$\angle SOP = 60°$

Exercises 13, 14

14. Find the approximate surface area and volume of the sphere if $OP = 6$ in. Use your calculator.

15. Find the total area (surface area) of a regular octahedron if the area of each face is 5.5 in^2.

16. Find the total area (surface area) of a regular dodecahedron (12 faces) if the area of each face is 6.4 cm^2.

17. Find the total area (surface area) of a regular hexahedron if each edge has a length of 4.2 cm.

18. Find the total area (surface area) of a regular tetrahedron if each edge has a length of 6 in.

19. The total area (surface area) of a regular hexahedron is 105.84 m^2. Find the
   a) area of each face.
   b) length of each edge.

20. The total area (surface area) of a regular octahedron is $32\sqrt{3}$ ft^2. Find the
   a) area of each face.
   b) length of each edge.

21. Find the approximate volume of a sphere with radius length $r = 2.7$ cm.

22. Find the approximate volume of a sphere with radius length $r = 33.5$ mm.

23. The surface of a soccer ball is composed of 12 regular pentagons and 20 regular hexagons. With each side of each regular polygon measuring 4.5 cm, the area of each regular pentagon is 34.9 cm^2 and the area of each regular hexagon is 52.5 cm^2.
   a) What is the surface area of the soccer ball?
   b) If the material used to construct the ball costs 0.8 of a cent per square centimeter, what is the cost of the materials used in construction?

24. A calendar is determined by using each of the 12 faces of a regular dodecahedron for one month of the year. With each side of the regular pentagonal face measuring 4 cm, the area of each face is approximately 27.5 cm^2.
   a) What is the total surface area of the calendar?
   b) If the material used to construct the calendar costs 0.8 of a cent per square centimeter, what is the cost of the materials used in construction?

25. A sphere is inscribed within a right circular cylinder whose altitude and diameter have equal measures.
   a) Find the ratio of the surface area of the cylinder to that of the sphere.
   b) Find the ratio of the volume of the cylinder to that of the sphere.

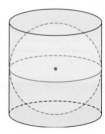

26. Given that a right circular cylinder is inscribed within a sphere, what is the least possible volume of the cylinder?

(HINT: Consider various lengths for radius and altitude.)

27. In calculus, it can be shown that the largest possible volume for the inscribed right circular cylinder in Exercise 26 occurs when its altitude has a length equal to the diameter length of the circular base. Find the length of the radius and the altitude of the cylinder of greatest volume if the radius length of the sphere is 6 in.

28. Given that a *regular* polyhedron of n faces is inscribed in a sphere of radius length 6 in., find the maximum (largest) possible volume for the polyhedron.

29. A right circular cone is inscribed in a sphere. If the slant height of the cone has a length equal to that of its diameter, find the length of the
a) radius of the base of the cone.
b) altitude of the cone.

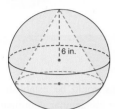

The radius of the sphere has
a length of 6 in.

30. A sphere is inscribed in a right circular cone whose slant height has a length equal to that of the diameter of its base. What is the length of the radius of the sphere if the slant height and the diameter of the cone both measure 12 cm?

In Exercises 31 and 32, use the calculator value of π.

31. For a sphere whose radius has length 3 m, find the approximate
a) surface area.
b) volume.

32. For a sphere whose radius has length 7 cm, find the approximate
a) surface area.
b) volume.

33. A sphere has a volume equal to $\frac{99}{7}$ in³. Determine the length of the radius of the sphere. $\left(\text{Use } \pi \approx \frac{22}{7}.\right)$

34. A sphere has a surface area equal to 154 in². Determine the length of the radius of the sphere. $\left(\text{Use } \pi \approx \frac{22}{7}.\right)$

35. The spherical storage tank described in Example 5 had a length of radius of 3 ft. Because the tank needs to be painted, we need to find its surface area. Also determine the number of pints of rust-proofing paint needed to paint the tank if 1 pt covers approximately 40 ft². Use your calculator.

36. An observatory has the shape of a right circular cylinder surmounted by a hemisphere. If the radius of the cylinder is 14 ft and its altitude measures 30 ft, what is the surface area of the observatory? If 1 gal of paint covers 300 ft², how many gallons are needed to paint the surface if it requires two coats? Use your calculator.

30 ft

|←14 ft→|

37. A leather soccer ball has an inside diameter length of 8.5 in. and a thickness of 0.1 in. Find the volume of leather needed for its construction. Use your calculator.

38. An ice cream cone is filled with ice cream as shown. What is the volume of the ice cream? Use your calculator.

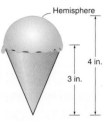

Hemisphere
4 in.
3 in.

For Exercises 39 to 44, make drawings as needed.

39. Can two spheres
a) be internally tangent?
b) have no points in common?

40. If two spheres intersect at more than one point, what type of geometric figure is determined by their intersection?

41. Two planes are tangent to a sphere at the endpoints of a diameter. How are the planes related?

42. Plane R is tangent to sphere O at point T. How are radius $\overline{OT}$ and plane R related?

43. Two tangent segments are drawn to sphere O from external point E. Where A and B are the points of tangency on sphere O, how are $\overline{EA}$ and $\overline{EB}$ related?

44. How many common tangent planes do two externally tangent spheres have?

45. Suppose that a semicircular region with a vertical diameter of length 6 is rotated about that diameter. Determine the exact surface area and the exact volume of the resulting solid of revolution.

46. Suppose that a semicircular region with a vertical diameter of length 4 is rotated about that diameter. Determine the exact surface area and the exact volume of the resulting solid of revolution.

47. Sketch the torus that results when the given circle of radius length 1 unit is revolved about the horizontal line that lies 4 units below the center of that circle.

48. Sketch the solid that results when the given circle of radius length 1 unit is revolved about the horizontal line that lies 1 unit below the center of that circle.

49. Explain how the following formula used in Example 6 was obtained:

$$V = \frac{4}{3}\pi R^3 - \frac{4}{3}\pi r^3$$

50. Derive a formula for the total surface area of the hollow-core sphere.

(Note: Include both interior and exterior surface areas.)

PERSPECTIVE ON HISTORY

SKETCH OF RENÉ DESCARTES

René Descartes was born in Tours, France, on March 31, 1596, and died in Stockholm, Sweden, on February 11, 1650. He was a contemporary of Galileo, the Italian scientist responsible for many discoveries in the science of dynamics. Descartes was also a friend of the French mathematicians Marin Mersenne (Mersenne Numbers) and Blaise Pascal (Pascal's Triangle).

As a small child, René Descartes was in poor health much of the time. Because he spent so much time reading in bed during his illnesses, he became a very well-educated young man. When Descartes was older and stronger, he joined the French army. It was during his time as a soldier that Descartes had three dreams that vastly influenced his future. The dreams, dated to November 10, 1619, shaped his philosophy and laid the framework for his discoveries in mathematics.

Descartes resigned his commission with the army in 1621 so that he could devote his life to studying philosophy, science, and mathematics. In the ensuing years, Descartes came to be highly regarded as a philosopher and mathematician and was invited to the learning centers of France, Holland, and Sweden.

Descartes's work in mathematics, in which he used an oblique coordinate system as a means of representing points, led to the birth of analytical geometry. His convention for locating points was eventually replaced by a coordinate system with perpendicular axes. In this system, algebraic equations could be represented by geometric figures; subsequently, many conjectured properties of these figures could be established through algebraic (analytic) proof. The rectangular coordinate system (which is called the Cartesian system in honor of Descartes) can also be used to locate the points of intersection of geometric figures such as lines and circles. Much of the material in Chapter 10 depends on his work.

Generally, the phrase *conic sections* refers to four geometric figures: the **circle**, the **parabola**, the **ellipse**, and the **hyperbola**. These figures are shown in Figure 9.48 both individually and also in relation to the upper and lower nappes of a cone. The conic sections are formed when a plane intersects the nappe(s) of a cone.

Other mathematical works of Descartes were devoted to the study of tangent lines to curves. The notion of a tangent to a curve is illustrated in Figure 9.49; this concept is the basis for the branch of mathematics known as **differential calculus**.

Descartes's final contributions to mathematics involved his standardizing the use of many symbols. To mention a few of these, Descartes used (1) a^2 rather than aa and a^3 rather than aaa; (2) ab to indicate multiplication; and (3) a, b, and c as constants and x, y, and z as variables.

Figure 9.49

Figure 9.48

PERSPECTIVE ON APPLICATIONS

BIRDS IN FLIGHT

The following application of geometry is not so much practical as classical.

Figure 9.50

Two birds have been attracted to bird feeders that rest atop vertical poles. The bases of these poles where their perches are located are 20 ft apart. The poles are themselves 10 ft and 16 ft tall. See Figure 9.50. Each bird eyes birdseed that has fallen on the ground at the base of the other pole. Leaving their perches, the birds fly in a straight-line path toward their goal. Avoiding a collision in flight, the birds take paths that have them pass at a common point X. How far is the point above ground level?

The solution to the problem follows. However, we redraw the figure to indicate that the 20-ft distance between poles is separated along the ground into line segments with lengths of a and $20 - a$ as shown. See Figures 9.51(a), 9.51(b), and 9.51(c).

(a)

(b)

(c)

Figure 9.51

From Figures 9.51(b) and 9.51(c), we form the following equations based upon the similarity of the right triangles:

$$\frac{10}{20} = \frac{h}{20 - a} \quad \text{and} \quad \frac{16}{20} = \frac{h}{a}$$

By the Means-Extremes Property of Proportions, $10(20 - a) = 20h$ and $16a = 20h$. By substitution,

$$10(20 - a) = 16a$$
$$200 - 10a = 16a$$
$$26a = 200$$
$$a = \frac{200}{26} \quad \text{or} \quad a = \frac{100}{13}$$

From the fact that $16a = 20h$, we see that

$$16 \cdot \frac{100}{13} = 20h$$

so

$$h = \frac{1}{\cancel{20}} \cdot 16 \cdot \frac{\overset{5}{\cancel{100}}}{13} = \frac{80}{13} = 6\frac{2}{13} \text{ ft}$$

The point at which the birds flew past each other was $6\frac{2}{13}$ feet above the ground.

Summary

A Look Back at Chapter 9

Our goal in this chapter was to deal with a type of geometry known as solid geometry. We found formulas for the lateral area, the total area (surface area), and the volume of prisms, pyramids, cylinders, cones, and spheres. Some of the formulas used in this chapter were developed using the concept of "limit." Regular polyhedra were introduced.

A Look Ahead to Chapter 10

Our focus in the next chapter is analytic (or coordinate) geometry. This type of geometry relates algebra and geometry. Formulas for the midpoint of a line segment, the length of a line segment, and the slope of a line will be developed. We will not only graph the equations of lines but also determine equations for given lines. We will see that proofs of many geometric theorems can be completed by using analytic geometry.

Key Concepts

9.1

Prisms (Right and Oblique) • Bases • Altitude • Vertices • Edges • Faces • Lateral Area • Total (Surface) Area • Volume • Regular Prism • Cube • Cubic Unit

9.2

Pyramid • Base • Altitude • Vertices • Edges • Faces • Vertex (Apex) of a Pyramid • Regular Pyramid • Slant Height of a Regular Pyramid • Lateral Area • Total (Surface) Area • Volume

9.3

Cylinders (Right and Oblique) • Bases and Altitude of a Cylinder • Axis of a Cylinder • Cones (Right and Oblique) • Base and Altitude of a Cone • Vertex (Apex) and Slant Height of a Cone • Axis of a Cone • Lateral Area • Total (Surface) Area • Volume • Solid of Revolution • Axis of a Solid of Revolution

9.4

Dihedral Angle • Polyhedron (Convex and Concave) • Vertices • Edges and Faces • Euler's Equation • Regular Polyhedrons (Tetrahedron, Hexahedron, Octahedron, Dodecahedron, and Icosahedron) • Sphere (Center, Radius, Diameter, Great Circle, Hemisphere) • Surface Area and Volume of a Sphere

Overview ■ Chapter 9

Volume and Area Relationships for Solids

Solid	Figure	Volume	Area
Rectangular prism (box)		$V = \ell wh$	$T = 2\ell w + 2\ell h + 2wh$
Cube		$V = e^3$	$T = 6e^2$
Prism (right prism shown)		$V = Bh$ (B = area of base)	$L = hP$ (P = perimeter of base) $T = L + 2B$
Regular pyramid (with slant height ℓ)		$V = \frac{1}{3}Bh$ (B = area of base)	$L = \frac{1}{2}\ell P$ (P = perimeter of base) $T = L + B$ **NOTE:** $\ell^2 = a^2 + h^2$
Right circular cylinder		$V = Bh$ or $V = \pi r^2 h$	$L = 2\pi rh$ $T = L + 2B$ or $T = 2\pi rh + 2\pi r^2$
Right circular cone (with slant height ℓ)		$V = \frac{1}{3}Bh$ or $V = \frac{1}{3}\pi r^2 h$	$L = \pi r\ell$ $T = L + B$ or $T = \pi r\ell + \pi r^2$ **NOTE:** $\ell^2 = r^2 + h^2$
Sphere		$V = \frac{4}{3}\pi r^3$	$S = 4\pi r^2$

Chapter 9 Review Exercises

1. Each side of the base of a right octagonal prism is 7 in. long. The altitude of the prism measures 12 in. Find the lateral area.

2. The base of a right prism is a triangle whose sides measure 7 cm, 8 cm, and 12 cm. The altitude of the prism measures 11 cm. Calculate the lateral area of the right prism.

3. The height of a square box is 2 in. more than three times the length of a side of the base. If the lateral area is 480 in², find the dimensions of the box and the volume of the box.

4. The base of a right prism is a rectangle whose length is 3 cm more than its width. If the altitude of the prism is 12 cm and the lateral area is 360 cm², find the total area and the volume of the prism

5. The base of a right prism is a triangle whose sides have lengths of 9 in., 15 in., and 12 in. The height of the prism is 10 in. Find the
   a) lateral area.   b) total area.   c) volume.

6. The base of a right prism is a regular hexagon whose sides are 8 cm in length. The height of the prism is 13 cm. Find the
   a) lateral area.   b) total area.   c) volume.

7. A regular square pyramid has a base whose sides are of length 10 cm each. The altitude of the pyramid measures 8 cm. Find the length of the slant height.

8. A regular hexagonal pyramid has a base whose sides are of length $6\sqrt{3}$ in. each. If the slant height is 12 in., find the length of the altitude of the pyramid.

9. The radius of the base of a right circular cone measures 5 in. If the altitude of the cone measures 7 in., what is the length of the slant height?

10. The diameter of the base of a right circular cone is equal in length to the slant height. If the height of the cone is 6 cm, find the length of the radius of the base.

11. The slant height of a regular square pyramid measures 15 in. One side of the base measures 18 in. Find the
    a) lateral area.   b) total area.   c) volume.

12. The base of a regular pyramid is an equilateral triangle each of whose sides measure 12 cm. The height of the pyramid is 8 cm. Find the exact and approximate
    a) lateral area.   b) total area.   c) volume.

13. The radius length of the base of a right circular cylinder is 6 in. The height of the cylinder is 10 in. Find the exact
    a) lateral area.   b) total area.   c) volume.

14. a) For the trough in the shape of a half-cylinder, find the volume of water it will hold. (Use $\pi \approx 3.14$ and disregard the thickness.)
    b) If the trough is to be painted inside and out, find the number of square feet to be painted. (Use $\pi \approx 3.14$.)

15. The slant height of a right circular cone is 12 cm. The angle formed by the slant height and the altitude is 30°. Find the exact and approximate
    a) lateral area.   b) total area.   c) volume.

16. The volume of a right circular cone is 96π in³. If the radius of the base measures 6 in., find the length of the slant height.

17. Find the surface area of a sphere if the radius has the length 7 in. Use $\pi \approx \frac{22}{7}$.

18. Find the volume of a sphere if the diameter has the length 12 cm. Use $\pi \approx 3.14$.

19. The solid shown consists of a hemisphere (half of a sphere), a cylinder, and a cone. Find the exact volume of the solid.

20. If the radius length of one sphere is three times as long as the radius length of another sphere, how do the surface areas of the spheres compare? How do the volumes compare?

21. Find the volume of the solid of revolution that results when a right triangle with legs of lengths 5 in. and 7 in. is rotated about the 7-in. leg. Use $\pi \approx \frac{22}{7}$.

22. Find the exact volume of the solid of revolution that results when a rectangular region with dimensions of 6 cm and 8 cm is rotated about a side of length 8 cm.

23. Find the exact volume of the solid of revolution that results when a semicircular region with diameter of length 4 in. is rotated about that diameter.

24. A plastic pipe is 3 ft long and has an inside radius length of 4 in. and an outside radius length of 5 in. How many cubic inches of plastic are in the pipe? (Use $\pi \approx 3.14$.)

25. A sphere with a diameter length of 14 in. is inscribed in a regular hexahedron. Find the exact volume of the space inside the hexahedron but outside the sphere.

26. a) An octahedron has _____ faces that are _____.
    b) A tetrahedron has _____ faces that are _____.
    c) A dodecahedron has _____ faces that are _____.

27. A drug manufacturing company wants to manufacture a capsule that contains a spherical pill inside. The diameter of the pill measures 4 mm, and the capsule is cylindrical with hemispheres on either end. The length of the capsule between the two hemispheres is 10 mm. What is the exact volume that the capsule will hold, excluding the volume of the pill?

28. For each of the following solids, verify Euler's equation by determining V, the number of vertices; E, the number of edges; and F, the number of faces.
    a) Right octagonal prism
    b) Tetrahedron
    c) Octahedron

29. Find the volume of cement used in the block shown.

30. Given a die in the shape of a regular octahedron, find the probability that one roll produces
a) an even-number result.
b) a result of 4 or more.

31. Find the total surface area of
a) a regular dodecahedron with each face having an area of 6.5 in.^{2}
b) a regular tetrahedron with each edge measuring 4 cm.

32. Three spheres are tangent to each other in pairs. Their radii have lengths of 1 in., 2 in., and 3 in., respectively. What type of triangle is formed by the lines of centers?

Chapter 9   Test

1. For the regular pentagonal prism shown below, find the total number of
a) edges. _____
b) faces. _____

Exercises 1, 2

2. For the regular pentagonal base, each edge measures 3.2 cm and the apothem measures 2 cm.
a) Find the area of the base $\left(\text{use } A = \frac{1}{2}aP\right)$. _____
b) Find the total area of the regular pentagonal prism if its altitude measures 5 cm. _____
c) Find the volume of the prism. _____

3. For the regular square pyramid shown, find the total number of
a) vertices. _____
b) lateral faces. _____

Exercises 3, 4

4. For the regular square pyramid shown above, find
a) the lateral area. _____
b) the total area. _____

5. For the regular square pyramid shown, find the length of the slant height. _____

6 Find the height of a regular square pyramid (not shown) if each edge of the base measures 8 in. and the length of the slant height is 5 in. _____

7. Find the volume of the regular square pyramid shown if each edge of the base measures 5 ft and the altitude measures 6 ft.

_____

8. Determine whether the statement is true or false.
a) A right circular cone has exactly two bases. _____
b) The lateral area L of a right circular cylinder with radius length r and height h is given by $L = 2\pi rh$. _____

9. Determine whether the statement is true or false.
a) The volume of a right circular cone is given by $V = \frac{1}{3}Bh$, which can also be expressed in the form $V = \frac{1}{3}\pi r^2 h$. _____
b) A regular dodecahedron has exactly 12 faces. _____

10. Recall Euler's equation, $V + F = E + 2$. For a certain polyhedron, there are eight faces and six vertices. How many edges does it have? _____

11. Find the slant height of the right circular cone below. Leave the answer in simplified radical form. _____

12. For the right circular cylinder shown, $r = 4$ cm and $h = 6$ cm. Find the exact
a) lateral area. _____
b) volume. _____

13. The exact volume of a right circular cone (not shown) is 32π in^3. If the length of the base radius is 4 in., find the length of the altitude of the cone. _____

14. Assume that a die used for gaming is in the shape of a regular octahedron. The faces are numbered 1, 2, 3, 4, . . ., and 8. When this die is rolled once, what is the probability that the roll produces
    a) an even number result? _____
    b) a result greater than or equal to 6? _____

15. A spherical storage tank filled with water has a radius length of 10 ft. Use the calculator value of π to find to the nearest tenth of a unit the approximate
    a) surface area of the sphere. _____
    b) volume of the sphere. _____

16. A pump moves water at a rate of 8π ft^3 per minute. How long will it take the pump to empty the tank in Exercise 15? (Answer to the nearest whole minute.) _____

Chapter 10 | Analytic Geometry

Guidance! The French mathematician René Descartes is considered the father of analytic geometry. His inspiration relating algebra and geometry, the Cartesian coordinate system (or Cartesian plane), was a major breakthrough in the development of much of mathematics. The photograph illustrates the use of a GPS (global positioning system). The system allows one to pinpoint locations such as that of a moving vehicle or that of a destination. On the map, locations identified by latitude and longitude are comparable to points whose x and y coordinates locate a position in the Cartesian coordinate system. Cartesian space extends the relationship between algebra and geometry to three dimensions; in this system, each point has three coordinates.

Additional video explanations of concepts, sample problems, and applications are available on DVD.

<table>
<tr><td>10.1</td><td colspan="3"># The Rectangular Coordinate System</td></tr>
</table>

10.1 The Rectangular Coordinate System

KEY CONCEPTS

Analytic Geometry	*y* Axis	Ordered Pair
Cartesian (Rectangular)	Quadrants	Distance Formula
Coordinate System	Origin	Linear Equation
Cartesian Plane	*x* Coordinate	Midpoint Formula
x Axis	*y* Coordinate	

Figure 10.1

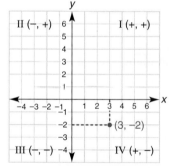

Figure 10.2

Graphing the solution sets for $3x - 2 = 7$ and $3x - 2 > 7$ requires a single number line to indicate the value of x. (See Appendices A.2 and A.3 for further information.) In this chapter, we deal with equations containing two variables; to relate such algebraic statements to plane geometry, we will need two number lines.

The study of the relationships between number pairs and points is known as **analytic geometry.** The **Cartesian coordinate system** or **rectangular coordinate system** is the plane that results when two number lines intersect perpendicularly at the origin (the point corresponding to the number 0 of each line). This two-dimensional coordinate plane is often called the Cartesian plane, named in honor of René Descartes (see Perspective on History for Chapter 9). The horizontal number line is known as the *x* **axis**, and its numerical coordinates increase from left to right. On the vertical number line, the *y* **axis**, values increase from bottom to top; see Figure 10.1. The scale (unit size) is generally the same for the *x* axis and the *y* axis. The two axes separate the plane into four **quadrants** which are numbered counterclockwise I, II, III, and IV, as shown. The point that marks the common origin of the two number lines is the **origin** of the rectangular coordinate system. It is convenient to identify the origin as (0, 0); this notation indicates that the *x* **coordinate** (listed first) is 0 and also that the *y* **coordinate** (listed second) is 0.

In the coordinate system, each point has the order (x, y) and is called an **ordered pair.** In Figure 10.2, the point $(3, -2)$ for which $x = 3$ and $y = -2$ is shown. The point $(3, -2)$ is located by moving 3 units to the right of the origin and then 2 units down from the *x* axis. The dashed lines shown emphasize the reason why the grid is called the *rectangular coordinate system.* The point $(3, -2)$ is located in Quadrant IV. In Figure 10.2, ordered pairs of positive and negative signs characterize the signs of the coordinates of a point located in each quadrant.

EXAMPLE 1

Plot points $A (-3, 4)$ and $B (2, 4)$, and find the distance between them.

SOLUTION Point A is located by moving 3 units to the left of the origin and then 4 units up from the *x* axis. Point B is located by moving 2 units to the right of the origin and then 4 units up from the *x* axis. In Figure 10.3, $\overline{AB}$ is a horizontal segment.

In the rectangular coordinate system, $ABCD$ is a rectangle in which $DC = 5$; $\overline{DC}$ is easily measured because it lies on the *x* axis. Because the opposite sides of a rectangle are congruent, it follows that $AB = 5$.

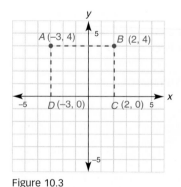

Figure 10.3

SSG EXS. 1–4

In Example 1, the points $(-3, 4)$ and $(2, 4)$ have the same y coordinates. In this case, the distance between the points on a horizontal line is merely the positive difference in the x coordinates; thus, the distance between A and B is $2 - (-3)$, or 5. It is also easy to find the distance between two points on a vertical line; if the x coordinates of two points are the same, the distance between points is the positive difference in the y coordinates. In Figure 10.3, where C is $(2, 0)$ and B is $(2, 4)$, the distance between the points is $4 - 0$ or 4.

In the following definition, the repeated y coordinates for the endpoints of the line segment indicate that the segment is horizontal.

DEFINITION

Given points $A(x_1, y_1)$ and $B(x_2, y_1)$ on a horizontal line segment $\overline{AB}$, the distance between these points is

$$AB = x_2 - x_1 \text{ if } x_2 > x_1 \qquad \text{or} \qquad AB = x_1 - x_2 \text{ if } x_1 > x_2$$

In the following definition, repeated x coordinates for the endpoints of a line segment determine a vertical line segment. In both definitions, the distance between points is found by subtracting the smaller from the larger of the two unequal coordinates.

DEFINITION

Given points $C(x_1, y_1)$ and $D(x_1, y_2)$ on a vertical line segment $\overline{CD}$, the distance between these points is

$$CD = y_2 - y_1 \text{ if } y_2 > y_1 \qquad \text{or} \qquad CD = y_1 - y_2 \text{ if } y_1 > y_2$$

EXAMPLE 2

In Figure 10.4, name the coordinates of points C and D, and find the distance between them. The x coordinates of C and D are identical because the points lie on a vertical line.

SOLUTION C is the point $(0, 1)$ because C is 1 unit above the origin; similarly, D is the point $(0, 5)$. We designate the coordinates of point C by $x_1 = 0$ and $y_1 = 1$ and the coordinates of point D by $x_1 = 0$ and $y_2 = 5$. Using the preceding definition,

$$CD = y_2 - y_1 = 5 - 1 = 4$$

We now turn our attention to the more general problem of finding the distance between any two points. In this situation, the line segment described is neither vertical nor horizontal.

Figure 10.4

THE DISTANCE FORMULA

The following formula enables us to find the distance between two points that lie on a "slanted" line.

Discover

Plot the points $A(0, 0)$ and $B(4, 3)$. Now find AB by using the Pythagorean Theorem. To accomplish this, you will need to form a path from A to B along horizontal and vertical line segments.

ANSWER
5

THEOREM 10.1.1 ■ Distance Formula

The distance between two points (x_1, y_1) and (x_2, y_2) is given by the formula

$$d = \sqrt{(x_2 - x_1)^2 + (y_2 - y_1)^2}$$

PROOF

In the coordinate system in Figure 10.5 are points $P_1\,(x_1, y_1)$ and $P_2\,(x_2, y_2)$. In addition to drawing the segment joining these points, we draw an auxiliary horizontal segment through P_1 and an auxiliary vertical segment through P_2; these meet at point $C\,(x_2, y_1)$ in Figure 10.5(a). Using Figure 10.5(b) and the definitions for lengths of horizontal and vertical segments,

$$P_1C = x_2 - x_1 \quad \text{and} \quad P_2C = y_2 - y_1$$

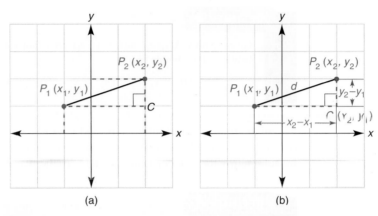

(a)   (b)

Figure 10.5

In right triangle P_1P_2C of Figure 10.5(b), let $d = P_1P_2$. By the Pythagorean Theorem,

$$d^2 = (x_2 - x_1)^2 + (y_2 - y_1)^2$$

Taking the positive square root for length d yields

$$d = \sqrt{(x_2 - x_1)^2 + (y_2 - y_1)^2}$$

Figure 10.6

EXAMPLE 3

In Figure 10.6, find the distance between points $A(5, -1)$ and $B(-1, 7)$.

SOLUTION Using the Distance Formula and choosing $x_1 = 5$ and $y_1 = -1$ (from point A) and $x_2 = -1$ and $y_2 = 7$ (from point B), we obtain

$$\begin{aligned} d &= \sqrt{(-1 - 5)^2 + [7 - (-1)]^2} \\ &= \sqrt{(-6)^2 + (8)^2} = \sqrt{36 + 64} \\ &= \sqrt{100} = 10 \end{aligned}$$

NOTE: If the coordinates of point A were designated as $x_2 = 5$ and $y_2 = -1$ and those of point B were designated as $x_1 = -1$ and $y_1 = 7$, the distance would remain the same.

Figure 10.7

If we look back at the proof of the Distance Formula, Figure 10.5 shows only one of several possible placements of points. If the placement had been as shown in Figure 10.7, then $AC = x_2 - x_1$ because $x_2 > x_1$, and $BC = y_1 - y_2$ because $y_1 > y_2$. The Pythagorean Theorem leads to what looks like a different result:

$$d^2 = (x_2 - x_1)^2 + (y_1 - y_2)^2$$

But this can be converted to the earlier formula by using the fact that

$$(y_1 - y_2)^2 = (y_2 - y_1)^2.$$

This follows from the fact that $(-a)^2 = a^2$ for any real number a.

The following example reminds us of the form of a **linear equation,** an equation whose graph is a straight line. In general, the form of a linear equation is $Ax + By = C$ for constants A, B, and C (where A and B do not both equal 0). We will consider the graphing of linear equations in Section 10.2.

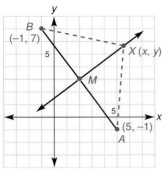

Figure 10.8

EXAMPLE 4

Find the equation that describes all points (x, y) that are equidistant from $A(5, -1)$ and $B(-1, 7)$. See Figure 10.8.

SOLUTION In Chapter 7, we saw that the locus of points equidistant from two fixed points is a line. This line ($\overleftrightarrow{MX}$ in Figure 10.8) is the perpendicular bisector of $\overline{AB}$.

If X is on the locus, then $AX = BX$. By the Distance Formula, we have

$$\sqrt{(x - 5)^2 + [y - (-1)]^2} = \sqrt{[x - (-1)]^2 + (y - 7)^2}$$

or

$$(x - 5)^2 + (y + 1)^2 = (x + 1)^2 + (y - 7)^2$$

after simplifying and squaring. Then

$$x^2 - 10x + 25 + y^2 + 2y + 1 = x^2 + 2x + 1 + y^2 - 14y + 49$$

Eliminating x^2 and y^2 terms by subtraction leads to the equation

$$-12x + 16y = 24$$

When we divide by 4, the equation of the line becomes

$$-3x + 4y = 6$$

If we divide the equation $-12x + 16y = 24$ by -4, an equivalent solution is

$$3x - 4y = -6$$

NOTE: The equations $-3x + 4y = 6$ and $3x - 4y = -6$ are said to be **equivalent** because their solutions are the same. For instance, $(-2, 0)$, $(2, 3)$, and $(0, 0)$ are all solutions for *both* equations.

Discover

On a number line, x_2 lies to the right of x_1. Then $x_2 > x_1$ and the distance between points is $(x_2 - x_1)$. To find the number a that is midway between x_1 and x_2, we add one-half the distance $(x_2 - x_1)$ to x_1. That is,

$$a = x_1 + \frac{1}{2}(x_2 - x_1).$$

Complete the simplification of a.

ANSWER

$a = \frac{1}{2}(x_1 + x_2)$ or $a = \frac{x_1 + x_2}{2}$

THE MIDPOINT FORMULA

In Figure 10.8, point M is the midpoint of $\overline{AB}$. It will be shown in Example 5(a) that M is the point $(2, 3)$. The Distance Formula can be used to verify that $MB = MA$.

A generalized midpoint formula is given in Theorem 10.1.2. The result shows that the coordinates of the midpoint M of a line segment are the averages of the coordinates of the endpoints. See the Discover activity at the left.

THEOREM 10.1.2 ■ Midpoint Formula

The midpoint M of the line segment joining $A(x_1, y_1)$ and $B(x_2, y_2)$ has coordinates x_M and y_M, where

$$(x_M, y_M) = \left(\frac{x_1 + x_2}{2}, \frac{y_1 + y_2}{2}\right)$$

That is,

$$M = \left(\frac{x_1 + x_2}{2}, \frac{y_1 + y_2}{2}\right)$$

We apply the Midpoint Formula in Example 5.

EXAMPLE 5

Use the Midpoint Formula to find the midpoint of the line segment joining:

a) $(5, -1)$ and $(-1, 7)$ **b)** (a, b) and (c, d)

SOLUTION

a) Using the Midpoint Formula and setting $x_1 = 5$, $y_1 = -1$, $x_2 = -1$, and $y_2 = 7$, we have

$$M = \left(\frac{5 + (-1)}{2}, \frac{-1 + 7}{2} \right) = \left(\frac{4}{2}, \frac{6}{2} \right), \quad \text{so} \quad M = (2, 3)$$

b) Using the Midpoint Formula and setting $x_1 = a$, $y_1 = b$, $x_2 = c$, and $y_2 = d$, we have

$$M = \left(\frac{a + c}{2}, \frac{b + d}{2} \right)$$

For Example 5(a), the line segment described is shown in Figure 10.8 on page 439; from appearances, the solution seems reasonable! In Example 5(b), we are generalizing the coordinates in preparation for the analytic geometry proofs that appear later in the chapter. For those proofs, we choose the x and y values for each point in such a way as to represent the general case.

 EXS. 9–12

PROOF OF THE MIDPOINT FORMULA (OPTIONAL)

For the segment joining P_1 and P_2, we name the midpoint M, as shown in Figure 10.9(a). Let the coordinates of M be designated by (x_M, y_M). Now construct horizontal segments through P_1 and M and vertical segments through M and P_2 to intersect at points A and B, as shown in Figure 10.9(b). Because $\angle A$ and $\angle B$ are right angles, $\angle A \cong \angle B$.

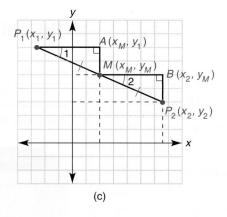

(a)   (b)   (c)

Figure 10.9

Because $\overline{P_1A}$ and $\overline{MB}$ are both horizontal, these segments are parallel in Figure 10.9(c). Then $\angle 1 \cong \angle 2$ because these are corresponding angles. With $\overline{P_1M} \cong \overline{MP_2}$ by the definition of a midpoint, it follows that $\triangle P_1AM \cong \triangle MBP_2$ by AAS. Because A is the point (x_M, y_1), we have $P_1A = x_M - x_1$. Likewise, the coordinates of B are (x_2, y_M), so $MB = x_2 - x_M$. Because $\overline{P_1A} \cong \overline{MB}$ by CPCTC, we represent the common length of the segments $\overline{P_1A}$ and $\overline{MB}$ by a. From the first equation, $x_M - x_1 = a$, so $x_M = x_1 + a$. From the second equation, $x_2 - x_M = a$, so $x_2 = x_M + a$. Substituting $x_1 + a$ for x_M into the second equation, we have

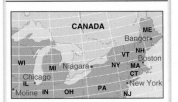
$$(x_1 + a) + a = x_2$$
$$x_1 + 2a = x_2$$

Then

$$2a = x_2 - x_1$$

so

$$a = \frac{x_2 - x_1}{2}$$

It follows that

$$x_M = x_1 + a$$
$$= x_1 + \frac{x_2 - x_1}{2}$$
$$= \frac{2x_1}{2} + \frac{x_2 - x_1}{2}$$
$$= \frac{x_1 + x_2}{2}$$

Similarly, it can be shown that the y coordinate of the midpoint is $y_M = \frac{y_1 + y_2}{2}$.

Thus,

$$M = \left(\frac{x_1 + x_2}{2}, \frac{y_1 + y_2}{2} \right)$$

The following example is based upon the definitions of symmetry with respect to a line and symmetry with respect to a point. Make drawings, if necessary, to verify each solution.

EXAMPLE 6

In the rectangular coordinate system, consider the point $A(2, -3)$. Then find point B if points A and B have symmetry with respect to

    a) the y axis   **b)** the x axis   **c)** the origin

SOLUTION

 EXS. 13–15

    a) $(-2, -3)$ **b)** $(2, 3)$ **c)** $(-2, 3)$

Exercises 10.1

1. Plot and then label the points $A(0, -3)$, $B(3, -4)$, $C(5, 6)$, $D(-2, -5)$, and $E(-3, 5)$.

2. Give the coordinates of each point A, B, C, D, and E. Also name the quadrant in which each point lies.

3. Find the distance between each pair of points:
    a) $(5, -3)$ and $(5, 1)$ c) $(0, 2)$ and $(0, -3)$
    b) $(-3, 4)$ and $(5, 4)$ d) $(-2, 0)$ and $(7, 0)$

4. If the distance between $(-2, 3)$ and $(-2, a)$ is 5 units, find all possible values of a.

5. If the distance between $(b, 3)$ and $(7, 3)$ is 3.5 units, find all possible values of b.

6. Find an expression for the distance between (a, b) and (a, c) if $b > c$.

7. Find the distance between each pair of points:
    a) $(0, -3)$ and $(4, 0)$ c) $(3, 2)$ and $(5, -2)$
    b) $(-2, 5)$ and $(4, -3)$ d) $(a, 0)$ and $(0, b)$

8. Find the distance between each pair of points:
    a) $(-3, -7)$ and $(2, 5)$ c) $(-a, -b)$ and (a, b)
    b) $(0, 0)$ and $(-2, 6)$ d) $(2a, 2b)$ and $(2c, 2d)$

9. Find the midpoint of the line segment that joins each pair of points:
    a) $(0, -3)$ and $(4, 0)$ c) $(3, 2)$ and $(5, -2)$
    b) $(-2, 5)$ and $(4, -3)$ d) $(a, 0)$ and $(0, b)$

10. Find the midpoint of the line segment that joins each pair of points:
    a) $(-3, -7)$ and $(2, 5)$ c) $(-a, -b)$ and (a, b)
    b) $(0, 0)$ and $(-2, 6)$ d) $(2a, 2b)$ and $(2c, 2d)$

11. Points A and B have *symmetry with respect to the origin O*. Find the coordinates of B if A is the point:
a) $(3, -4)$ c) $(a, 0)$
b) $(0, 2)$ d) (b, c)

12. Points A and B have *symmetry with respect to point C(2, 3)*. Find the coordinates of B if A is the point:
a) $(3, -4)$ c) $(5, 0)$
b) $(0, 2)$ d) (a, b)

13. Points A and B have *symmetry with respect to point C*. Find the coordinates of C given the points:
a) $A(3, -4)$ and $B(5, -1)$ c) $A(5, -3)$ and $B(2, 1)$
b) $A(0, 2)$ and $B(0, 6)$ d) $A(2a, 0)$ and $B(0, 2b)$

14. Points A and B have *symmetry with respect to the x axis*. Find the coordinates of B if A is the point:
a) $(3, -4)$ c) $(0, a)$
b) $(0, 2)$ d) (b, c)

15. Points A and B have *symmetry with respect to the x axis*. Find the coordinates of A if B is the point:
a) $(3, 1)$ c) $(2, 0)$
b) $(0, 5)$ d) (b, c)

16. Points A and B have *symmetry with respect to the vertical line where $x = 2$*. Find the coordinates of A if B is the point:
a) $(5, 1)$ c) $(-6, a)$
b) $(0, 5)$ d) (b, c)

17. Points A and B have *symmetry with respect to the y axis*. Find the coordinates of A if B is the point:
a) $(3, -4)$ c) $(a, 0)$
b) $(2, 0)$ d) (b, c)

18. Points A and B have *symmetry with respect to either the x axis or the y axis*. Name the axis of symmetry for:
a) $A(3, -4)$ and $B(3, 4)$ c) $A(3, -4)$ and $B(-3, -4)$
b) $A(2, 0)$ and $B(-2, 0)$ d) $A(a, b)$ and $B(a, -b)$

19. Points A and B have *symmetry with respect to* a vertical line $(x = a)$ or a horizontal line $(y = b)$. Give an equation such as $x = 3$ for the axis of symmetry for:
a) $A(3, -4)$ and $B(5, -4)$
b) $A(a, 0)$ and $B(a, -b)$
c) $A(7, -4)$ and $B(-3, -4)$
d) $A(a, 7)$ and $B(a, -1)$

In Exercises 20 to 22, apply the Midpoint Formula.

20. $M(3, -4)$ is the midpoint of $\overline{AB}$, in which A is the point $(-5, 7)$. Find the coordinates of B.

21. $M(2.1, -5.7)$ is the midpoint of $\overline{AB}$, in which A is the point $(1.7, 2.3)$. Find the coordinates of B.

22. A circle has its center at the point $(-2, 3)$. If one endpoint of a diameter is at $(3, -5)$, find the other endpoint of the diameter.

23. A rectangle $ABCD$ has three of its vertices at $A(2, -1)$, $B(6, -1)$, and $C(6, 3)$. Find the fourth vertex D and the area of rectangle $ABCD$.

24. A rectangle $MNPQ$ has three of its vertices at $M(0, 0)$, $N(a, 0)$, and $Q(0, b)$. Find the fourth vertex P and the area of the rectangle $MNPQ$.

25. Use the Distance Formula to determine the type of triangle that has these vertices:
a) $A(0, 0)$, $B(4, 0)$, and $C(2, 5)$
b) $D(0, 0)$, $E(4, 0)$, and $F(2, 2\sqrt{3})$
c) $G(-5, 2)$, $H(-2, 6)$, and $K(2, 3)$

26. Use the method of Example 4 to find the equation of the line that describes all points equidistant from the points $(-3, 4)$ and $(3, 2)$.

27. Use the method of Example 4 to find the equation of the line that describes all points equidistant from the points $(1, 2)$ and $(4, 5)$.

28. For coplanar points A, B, and C, suppose that you have used the Distance Formula to show that $AB = 5$, $BC = 10$, and $AC = 15$. What can you conclude regarding points A, B, and C?

29. If two vertices of an equilateral triangle are at $(0, 0)$ and $(2a, 0)$, what point is the third vertex?

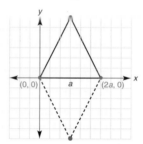

30. The rectangle whose vertices are $A(0, 0)$, $B(a, 0)$, $C(a, b)$, and $D(0, b)$ is shown. Use the Distance Formula to draw a conclusion concerning the lengths of the diagonals $\overline{AC}$ and $\overline{BD}$.

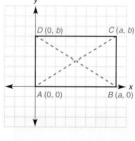

★31. There are two points on the y axis that are located a distance of 6 units from the point $(3, 1)$. Determine the coordinates of each point.

★32. There are two points on the x axis that are located a distance of 6 units from the point $(3, 1)$. Determine the coordinates of each point.

33. The triangle that has vertices at $M(-4, 0)$, $N(3, -1)$, and $Q(2, 4)$ has been boxed in as shown. Find the area of $\triangle MNQ$.

Exercises 33, 34

34. Use the method suggested in Exercise 33 to find the area of $\triangle RST$, with $R(-2, 4)$, $S(-1, -2)$, and $T(6, 5)$.

35. Determine the area of $\triangle ABC$ if $A = (2, 1)$, $B = (5, 3)$, and C is the reflection of B across the x axis.

36. Find the area of $\triangle ABC$ in Exercise 35, but assume that C is the reflection of B across the y axis.

For Exercises 37 to 42, refer to formulas for Chapter 9.

37. Find the exact volume of the solid that results when the triangular region with vertices at $(0, 0)$, $(5, 0)$, and $(0, 9)$ is rotated about the
a) x axis.   b) y axis.

38. Find the exact volume of the solid that results when the triangular region with vertices at $(0, 0)$, $(6, 0)$, and $(6, 4)$ is rotated about the
a) x axis.   b) y axis.

39. Find the exact volume of the solid formed when the rectangular region with vertices at $(0, 0)$, $(6, 0)$, $(6, 4)$, and $(0, 4)$ is revolved about the
a) x axis.   b) y axis.

40. Find the exact volume of the solid formed when the region bounded in Quadrant I by the axes and the lines $x = 9$ and $y = 5$ is revolved about the
a) x axis.   b) y axis.

41. Find the exact lateral area of each solid in Exercise 40.

*42. Find the volume of the solid formed when the triangular region having vertices at $(2, 0)$, $(4, 0)$, and $(2, 4)$ is rotated about the y axis.

*43. By definition, an *ellipse* is the locus of points whose sum of distances from two fixed points F_1 and F_2 (called foci) is constant. In the grid provided, find points whose sum of distances from points $F_1(3, 0)$ and $F_2(-3, 0)$ is 10. That is, locate some points for which $PF_1 + PF_2 = 10$; point $P(5, 0)$ is one such point. Then sketch the ellipse.

*44. By definition, a *hyperbola* is the locus of points whose positive difference of distances from two fixed points F_1 and F_2 (called foci) is constant. In the grid provided, find points whose difference of distances from points $F_1(5, 0)$ and $F_2(-5, 0)$ is 6. That is, locate some points for which either $PF_1 - PF_2 = 6$ or $PF_2 - PF_1 = 6$; point $P(3, 0)$ is one such point, Then sketch the hyperbola.

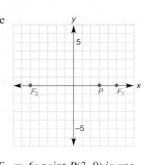

*45. Use the Distance Formula to show that the equation of the parabola with focus $F(0, 1)$ and directrix $y = -1$ is
$$y = \frac{1}{4}x^2.$$

*46. Use the Distance Formula to show that the equation of the parabola with focus $F(0, 2)$ and directrix $y = -2$ is
$$y = \frac{1}{8}x^2.$$

47. Following a 90° counterclockwise rotation about the origin, the image of $A(3, 1)$ is point $B(-1, 3)$. What is the image of point A following a counterclockwise rotation of
a) 180° about the origin?
b) 270° about the origin?
c) 360° about the origin?

48. Consider the point $C(a, b)$. What is the image of C after a counterclockwise rotation of
a) 90° about the origin?
b) 180° about the origin?
c) 360° about the origin?

49. Given the point $D(3, 2)$, find the image of D after a counterclockwise rotation of
a) 90° about the point $E(3, 4)$.
b) 180° about the point $F(4, 5)$.
c) 360° about the point $G(a, b)$.

10.2   Graphs of Linear Equations and Slope

KEY CONCEPTS	Graphs of Equations	*y* Intercept	Slope Formula
	x Intercept	Slope	Negative Reciprocal

In Section 10.1, we were reminded that the general form of the equation of a line is $Ax + By = C$ (where A and B do not both equal 0). Some examples of *linear* equations are $2x + 3y = 12$, $3x - 4y = 12$, and $3x = -6$; as we shall see, the graph of each of these equations is a line.

THE GRAPH OF AN EQUATION

> **DEFINITION**
>
> In the rectangular coordinate system, the **graph of an equation** is the set of all points (x, y) whose ordered pairs satisfy the equation.

EXAMPLE 1

Draw the graph of the equation $2x + 3y = 12$.

SOLUTION We begin by completing a table. It is convenient to use one point for which $x = 0$, a second point for which $y = 0$, and a third point as a check for collinearity.

$$2x + 3y = 12$$

$$x = 0 \rightarrow 2(0) + 3y = 12 \rightarrow y = 4$$
$$y = 0 \rightarrow 2x + 3(0) = 12 \rightarrow x = 6$$
$$x = 3 \rightarrow 2(3) + 3y = 12 \rightarrow y = 2$$

x	y	(x, y)
0	4	$(0, 4)$
6	0	$(6, 0)$
3	2	$(3, 2)$

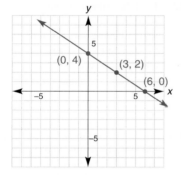

Figure 10.10

Upon plotting the third point, we see that the three points are collinear. The graph of a linear equation must be a straight line, as shown in Figure 10.10.

SSG   **EXS. 1–3**

For the graph and equation in Example 1, every point on the line must also satisfy the given equation. Notice that the point $(-3, 6)$ lies on the line shown in Figure 10.10. This ordered pair also satisfies the equation $2x + 3y = 12$; that is, $2(-3) + 3(6) = 12$ because $-6 + 18 = 12$.

EXAMPLE 2

Which point(s) lie on the graph of the equation $2x - 3y = 12$?

a) $(3, -2)$ **b)** $(9, 2)$

Technology Exploration

Use a graphing calculator if one is available.

1. To graph $2x + 3y = 12$, solve for y.

2. Enter your result from part (1) as the value of Y_1.

$$[Y_1 = -\left(\frac{2}{3}\right)x + 4]$$

3. Now GRAPH to see the line of Figure 10.10.

SOLUTION:
a) $x = 3$ and $y = -2$;
  Testing $2x - 3y = 12$, we see that $2(3) - 3(-2) = 6 + 6 = 12$, which is true; $(3, -2)$ is on the graph.
b) $x = 9$ and $y = 2$;
  Testing $2x - 3y = 12$, we see that $2(9) - 3(2) = 18 - 6 = 12$, which is true; $(9, 2)$ is also on the graph.

For the equation $2x + 3y = 12$, the number 6 is known as the **x intercept** because $(6, 0)$ is the point at which the graph crosses the x axis; similarly, the number 4 is known as the **y intercept.** See Figure 10.10 on page 444. Most linear equations have two intercepts; these are generally represented by a (the x intercept) and b (the y intercept).

For the equation $Ax + By = C$, we determine the

a) x intercept by choosing $y = 0$. Solve the resulting equation for x.
b) y intercept by choosing $x = 0$. Solve the resulting equation for y.

EXAMPLE 3

Find the x and y intercepts of the equation $3x - 4y = -12$, and use the intercepts to graph the equation.

SOLUTION The x intercept is found when $y = 0$, $3x - 4(0) = -12$, so $x = -4$. The x intercept is $a = -4$, so $(-4, 0)$ is on the graph. The y intercept results when $x = 0$; $3(0) - 4y = -12$, so $y = 3$. The y intercept is $b = 3$, so $(0, 3)$ is on the graph. Once the points $(-4, 0)$ and $(0, 3)$ have been plotted, the graph can be completed by drawing the line through these points. See Figure 10.11.

As we shall see in Example 4, a linear equation may have only one intercept. It is impossible for a linear equation to have no intercepts whatsoever.

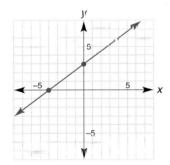

Figure 10.11

EXAMPLE 4

Draw the graphs of the following equations:
a) $x = -2$ b) $y = 3$

SOLUTION First note that each equation is a linear equation and can be written in the form $Ax + By = C$.

$$x = -2 \text{ is equivalent to } (1 \cdot x) + (0 \cdot y) = -2$$
$$y = 3 \text{ is equivalent to } (0 \cdot x) + (1 \cdot y) = 3$$

a) The equation $x = -2$ claims that the value of x is -2, regardless of the value of y; this leads to the following table:

x	y	$\rightarrow$	(x, y)
-2	-2	$\rightarrow$	$(-2, -2)$
-2	0	$\rightarrow$	$(-2, 0)$
-2	5	$\rightarrow$	$(-2, 5)$

Figure 10.12

 EXS. 4–8

b) The equation $y = 3$ claims that the value of y is 3, regardless of the value of x; this leads to the following table:

x	y	→	(x, y)
-4	3	→	$(-4, 3)$
0	3	→	$(0, 3)$
5	3	→	$(5, 3)$

The graphs of the equations are shown in Figure 10.12.

NOTE: When an equation can be written in the form $x = a$ (for constant a), its graph is the vertical line containing the point $(a, 0)$. When an equation can be written in the form $y = b$ (for constant b), its graph is the horizontal line containing the point $(0, b)$.

THE SLOPE OF A LINE

Most lines are oblique; that is, the line is neither horizontal nor vertical. Especially for oblique lines, it is convenient to describe the amount of "slant" by a number called the *slope* of the line.

> **DEFINITION ■ Slope Formula**
>
> The slope of the line that contains the points (x_1, y_1) and (x_2, y_2) is given by
> $$m = \frac{y_2 - y_1}{x_2 - x_1} \text{ for } x_1 \neq x_2$$

NOTE: When $x_1 = x_2$, the denominator of the Slope Formula becomes 0 and we say that the slope of the line is undefined or that the slope of the line does not exist.

Whereas the uppercase italic M means midpoint, we use the lowercase italic m to represent the slope of a line. Other terms that are used to describe the slope of a line include *pitch* and *grade*. A carpenter may say that a roofline has a $\frac{5}{12}$ pitch. [See Figure 10.13(a).] In constructing a stretch of roadway, an engineer may say that this part of the roadway has a grade of $\frac{3}{100}$, or 3 percent. [See Figure 10.13(b).]

(a)

(b)

Figure 10.13

Whether in geometry, carpentry, or engineering, the **slope** of a line is a number. The slope of the line is the ratio of the vertical change (rise) to the horizontal change (run). For any two points on the line in question, a line that "rises" from left to right has a *positive* slope, and a line that "falls" from left to right has a *negative* slope. The lines shown in Figures 10.14(a) and (b) on page 447 confirm these claims.

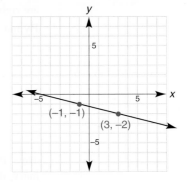

$$m = \frac{y_2 - y_1}{x_2 - x_1}$$

$$m = \frac{5 - 3}{6 - 0} = \frac{2}{6} = \frac{1}{3}$$

(a)

$$m = \frac{y_2 - y_1}{x_2 - x_1}$$

$$m = \frac{-2 - (-1)}{3 - (-1)} = -\frac{1}{4}$$

(b)

Figure 10.14

Any horizontal line has slope 0; any vertical line has an undefined slope. Figure 10.15 shows an example of each of these types of lines.

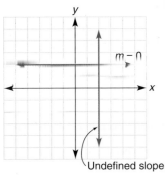

Figure 10.15

EXAMPLE 5

Without graphing, find the slope of the line that contains:

a) (2, 2) and (5, 3) b) (1, −1) and (1, 3)

SOLUTION

a) Using the Slope Formula and choosing $x_1 = 2$, $y_1 = 2$, $x_2 = 5$, and $y_2 = 3$, we have

$$m = \frac{3 - 2}{5 - 2} = \frac{1}{3},$$

NOTE: If drawn, the line in part (a) will slant upward from left to right.

b) Let $x_1 = 1$, $y_1 = -1$, $x_2 = 1$, and $y_2 = 3$. Then we calculate

$$m = \frac{3 - (-1)}{1 - 1} = \frac{4}{0},$$

which is undefined.

NOTE: If drawn, the line in part (b) will be vertical because the x coordinates of the points are the same.

SSG **EXS. 9–12**

The slope of a line is unique; that is, the slope does not change when:

1. The order of the two points is reversed in the Slope Formula.
2. Different points on the line are selected.

The first situation is true because $\frac{-a}{-b} = \frac{a}{b}$. The second situation is more difficult to explain because it depends on similar triangles.

For an explanation of point 2, consider Figure 10.16, in which points P_1, P_2, and P_3 are collinear. We wish to show that the slope of the line is the same whether P_1 and P_2 or P_2 and P_3 are used in the Slope Formula. If horizontal and vertical segments are drawn as shown in Figure 10.16, we can show that triangles P_1P_2A and P_2P_3B are similar. The similarity follows from the facts that $\angle 1 \cong \angle 2$ (because $\overline{P_1A} \parallel \overline{P_2B}$) and that $\angle A$ and $\angle B$

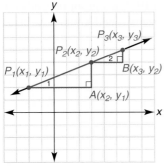

Figure 10.16

are right angles. Then $\frac{P_2A}{P_3B} = \frac{P_1A}{P_2B}$ because these are corresponding sides of similar triangles. By interchanging the means, we have $\frac{P_2A}{P_1A} = \frac{P_3B}{P_2B}$. But

$$\frac{P_2A}{P_1A} = \frac{y_2 - y_1}{x_2 - x_1} \qquad \text{and} \qquad \frac{P_3B}{P_2B} = \frac{y_3 - y_2}{x_3 - x_2}$$

so

$$\frac{y_2 - y_1}{x_2 - x_1} = \frac{y_3 - y_2}{x_3 - x_2}$$

Thus, the slope of the line is not changed by our having used either pair of points. In summary, the slopes agree because of similar triangles.

If points P_1, P_2, and P_3 are collinear, then the slopes of $\overline{P_1P_2}$, $\overline{P_1P_3}$, and $\overline{P_2P_3}$ are the same. The converse of this statement is also true. If the slopes of $\overline{P_1P_2}$, $\overline{P_1P_3}$, and $\overline{P_2P_3}$ are equal, then P_1, P_2, and P_3 are collinear. See Example 6 for an application.

EXAMPLE 6

Are the points $A(2, -3)$, $B(5, 1)$, and $C(-4, -11)$ collinear?

SOLUTION Let m_{AB} and m_{BC} represent the slopes of $\overline{AB}$ and $\overline{BC}$, respectively. By the Slope Formula, we have

$$m_{\overline{AB}} = \frac{1 - (-3)}{5 - 2} = \frac{4}{3} \qquad \text{and} \qquad m_{\overline{BC}} = \frac{11 - 1}{-4 - 5} = \frac{-12}{-9} = \frac{4}{3}$$

Because $m_{\overline{AB}} = m_{\overline{BC}}$, it follows that A, B, and C are collinear.

As we trace a line from one point to a second point, the Slope Formula tells us that

$$m = \frac{\text{change in } y}{\text{change in } x} = \frac{\text{vertical change}}{\text{horizontal change}} = \frac{\text{rise}}{\text{run}}$$

SSG **EXS. 13–15**

This interpretation of slope is used in Example 7.

EXAMPLE 7

Draw the line through $(-1, 5)$ so that it has slope $m = -\frac{2}{3}$.

SOLUTION First we plot the point $(-1, 5)$. The slope can be written as $m = \frac{-2}{3}$. Thus, we let the change in y from the first to the second point be -2 while the change in x is 3. From the first point $(-1, 5)$, we locate the second point by moving 2 units down and 3 units to the right. The line is then drawn as shown in Figure 10.17.

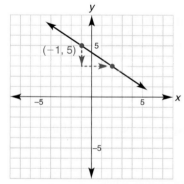

Figure 10.17

Theorems 10.2.1 and 10.2.2 are stated without proof. However, drawings needed for the proofs of the theorems are found in Figures 10.18 and 10.19. Each proof depends on similar triangles created through the use of the auxiliary segments found in the drawings. See Exercises 43 and 44 in this section.

THEOREM 10.2.1

If two nonvertical lines are parallel, then their slopes are equal.
Alternative Form: If $\ell_1 \parallel \ell_2$, then $m_1 = m_2$.

Figure 10.18

Figure 10.19

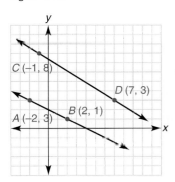

Figure 10.20

SSG   **EXS. 16–18**

In Figure 10.18, note that $\overline{AC} \parallel \overline{DF}$. Also, $\overline{AB}$ and $\overline{DE}$ are horizontal, and $\overline{BC}$ and $\overline{EF}$ are auxiliary vertical segments. In the proof of Theorem 10.2.1, the goal is to show that $m_{\overline{AC}} = m_{\overline{DF}}$. The converse of Theorem 10.2.1 is also true; that is, if $m_1 = m_2$, then $\ell_1 \parallel \ell_2$.

> ### THEOREM 10.2.2
>
> If two lines (neither horizontal nor vertical) are perpendicular, then the product of their slopes is -1.
> *Alternative Form:* If $\ell_1 \perp \ell_2$, then $m_1 \cdot m_2 = -1$.

In Figure 10.19, auxiliary segments have been included. To prove Theorem 10.2.2, we need to show that $m_{\overline{AC}} \cdot m_{\overline{CE}} = -1$. Because the product of the slopes is -1, the slopes are **negative reciprocals.** In general, negative reciprocals take the forms $\frac{a}{b}$ and $\frac{-b}{a}$. The converse of Theorem 10.2.2 is also true; if $m_1 \cdot m_2 = -1$, then $\ell_1 \perp \ell_2$.

EXAMPLE 8

Given the points $A(-2, 3)$, $B(2, 1)$, $C(-1, 8)$, and $D(7, 3)$, are $\overline{AB}$ and $\overline{CD}$ parallel, perpendicular, or neither? See Figure 10.20.

SOLUTION

$$m_{\overline{AB}} = \frac{1 - 3}{2 - (-2)} = \frac{-2}{4} = -\frac{1}{2}$$

$$m_{\overline{CD}} = \frac{3 - 8}{7 - (-1)} = \frac{-5}{8} \text{ or } -\frac{5}{8}$$

Because $m_{\overline{AB}} \neq m_{\overline{CD}}$, $\overline{AB} \nparallel \overline{CD}$. The slopes are not negative reciprocals, so $\overline{AB}$ is not perpendicular to $\overline{CD}$. Neither relationship holds for $\overline{AB}$ and $\overline{CD}$.

In Example 8, it was worthwhile to sketch the lines described. It is apparent from Figure 10.20 that the lines are not perpendicular. A sketch may help to show that a relationship does *not* exist, but sketching is not a precise method for showing that lines are parallel or perpendicular.

EXAMPLE 9

Are the lines that are the graphs of $2x + 3y = 6$ and $3x - 2y = 12$ parallel, perpendicular, or neither?

SOLUTION Because $2x + 3y = 6$ contains the points $(3, 0)$ and $(0, 2)$, its slope is $m_1 = \frac{2 - 0}{0 - 3} = -\frac{2}{3}$. The line $3x - 2y = 12$ contains $(0, -6)$ and $(4, 0)$; thus, its slope is $m_2 = \frac{0 - (-6)}{4 - 0} = \frac{6}{4}$ or $\frac{3}{2}$. Because the product of the slopes is $m_1 \cdot m_2 = -\frac{2}{3} \cdot \frac{3}{2} = -1$, the lines described are perpendicular.

EXAMPLE 10

Determine the value of a for which the line through $(2, -3)$ and $(5, a)$ is perpendicular to the line $3x + 4y = 12$.

SOLUTION The line $3x + 4y = 12$ contains the points $(4, 0)$ and $(0, 3)$; this line has the slope

$$m = \frac{3 - 0}{0 - 4} = -\frac{3}{4}$$

For the two lines to be perpendicular, the second line must have slope $\frac{4}{3}$. Using the Slope Formula, we find that the second line has the slope

$$\frac{a - (-3)}{5 - 2}$$

so

$$\frac{a + 3}{3} = \frac{4}{3}$$

Multiplying by 3, we obtain $a + 3 = 4$. It follows that $a = 1$.

EXAMPLE 11

In Figure 10.21, show that the quadrilateral with vertices $A(0, 0)$, $B(a, 0)$, $C(a, b)$, and $D(0, b)$ is a rectangle.

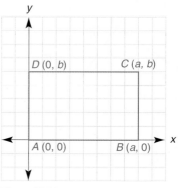

Figure 10.21

SOLUTION By applying the Slope Formula, we see that

$$m_{\overline{AB}} = \frac{0 - 0}{a - 0} = 0$$

and

$$m_{\overline{DC}} = \frac{b - b}{a - 0} = 0.$$

Then $\overline{AB}$ and $\overline{DC}$ are both horizontal and, therefore, parallel to each other.

For $\overline{DA}$ and $\overline{CB}$, the slopes are undefined because the denominators in the Slope Formula both equal 0. Then $\overline{DA}$ and $\overline{CB}$ are both vertical and, therefore, parallel to each other.

Thus, $ABCD$ is a parallelogram. With $\overline{AB}$ being horizontal and $\overline{DA}$ vertical, it follows that $\overline{DA} \perp \overline{AB}$. Therefore, $ABCD$ is a rectangle by definition.

SSG **EX. 19**

Exercises 10.2

In Exercises 1 to 8, draw the graph of each equation. Name any intercepts.

1. $3x + 4y = 12$

2. $3x + 5y = 15$

3. $x - 2y = 5$

4. $x - 3y = 4$

5. $2x + 6 = 0$

6. $3y - 9 = 0$

7. $\frac{1}{2}x + y = 3$

8. $\frac{2}{3}x - y = 1$

9. Which point(s) lie on the graph of $5x - 4y = 32$?
   a) $(4, -3)$ b) $(8, 1)$

10. Which point(s) lie on the graph of $3x - 2y = 12$?
   a) $(6, 2)$ b) $(-2, 3)$

11. Find the slopes of the lines containing:
   a) $(2, -3)$ and $(4, 5)$ d) $(-2.7, 5)$ and $(-1.3, 5)$
   b) $(3, -2)$ and $(3, 7)$ e) (a, b) and (c, d)
   c) $(1, -1)$ and $(2, -2)$ f) $(a, 0)$ and $(0, b)$

12. Find the slopes of the lines containing:
   a) $(3, -5)$ and $(-1, 2)$
   b) $(-2, -3)$ and $(-5, -7)$
   c) $(2\sqrt{2}, -3\sqrt{6})$ and $(3\sqrt{2}, 5\sqrt{6})$
   d) $(\sqrt{2}, \sqrt{7})$ and $(\sqrt{2}, \sqrt{3})$
   e) $(a, 0)$ and $(a + b, c)$
   f) (a, b) and $(-b, -a)$

13. Find x so that $\overline{AB}$ has slope m, where:
   a) A is $(2, -3)$, B is $(x, 5)$, and $m = 1$
   b) A is $(x, -1)$, B is $(3, 5)$, and $m = -0.5$

14. Find y so that $\overline{CD}$ has slope m, where:
   a) C is $(2, -3)$, D is $(4, y)$, and $m = \frac{3}{2}$
   b) C is $(-1, -4)$, D is $(3, y)$, and $m = -\frac{2}{3}$

15. Are these points collinear?
   a) $A(-2, 5)$, $B(0, 2)$, and $C(4, -4)$
   b) $D(-1, -1)$, $E(2, -2)$, and $F(5, -5)$

16. Are these points collinear?
   a) $A(-1, -2)$, $B(3, 2)$, and $C(5, 5)$
   b) $D(a, c - d)$, $E(b, c)$, and $F(2b - a, c + d)$

17. Parallel lines ℓ_1 and ℓ_2 have slopes m_1 and m_2, respectively. Find m_2 if m_1 equals:
   a) $\frac{3}{4}$ b) $-\frac{5}{3}$ c) -2 d) $\frac{a - b}{c}$

18. Parallel lines ℓ_1 and ℓ_2 have slopes m_1 and m_2, respectively. Find m_2 if m_1 equals:
   a) $\frac{4}{5}$ b) $-\frac{1}{5}$ c) 3 d) $\frac{f + g}{h + j}$

19. Perpendicular lines ℓ_1 and ℓ_2 have slopes m_1 and m_2, respectively. Find m_2 if m_1 equals:
   a) $-\frac{1}{2}$ b) $\frac{3}{4}$ c) 3 d) $\frac{f + g}{h + j}$

20. Perpendicular lines ℓ_1 and ℓ_2 have slopes m_1 and m_2, respectively. Find m_2 if m_1 equals:
   a) 5 b) $-\frac{5}{7}$ c) $-\frac{1}{9}$ d) $\frac{a - b}{c}$

In Exercises 21 to 24, state whether the lines are parallel, perpendicular, the same (coincident), or none of these.

21. $2x + 3y = 6$ and $2x - 3y = 12$

22. $2x + 3y = 6$ and $4x + 6y = -12$

23. $2x + 3y = 6$ and $3x - 2y = 12$

24. $2x + 3y = 6$ and $4x + 6y = 12$

25. Find x such that the points $A(x, 5)$, $B(2, 3)$, and $C(4, -5)$ are collinear.

26. Find a such that the points $A(1, 3)$, $B(4, 5)$, and $C(a, a)$ are collinear.

27. Find x such that the line through $(2, -3)$ and $(3, 2)$ is perpendicular to the line through $(-2, 4)$ and $(x, -1)$.

28. Find x such that the line through $(2, -3)$ and $(3, 2)$ is parallel to the line through $(-2, 4)$ and $(x, -1)$.

In Exercises 29 to 34, draw the line described.

29. Through $(3, -2)$ and with $m = 2$

30. Through $(-2, -5)$ and with $m = \frac{5}{7}$

31. With y intercept 5 and with $m = -\frac{3}{4}$

32. With x intercept -3 and with $m = 0.25$

33. Through $(-2, 1)$ and parallel to the line $2x - y = 6$

34. Through $(-2, 1)$ and perpendicular to the line that has intercepts $a = -2$ and $b = 3$

35. Use slopes to decide whether the triangle with vertices at $(6, 5)$, $(-3, 0)$, and $(4, -2)$ is a right triangle.

36. If $A(2, 2)$, $B(7, 3)$, and $C(4, x)$ are the vertices of a right triangle with right angle C, find the value of x.

*37. If $(2, 3)$, $(5, -2)$, and $(7, 2)$ are three vertices (not necessarily consecutive) of a parallelogram, find the possible locations of the fourth vertex.

38. Three vertices of rectangle $ABCD$ are $A(-5, 1)$, $B(-2, -3)$, and $C(6, y)$. Find the value of y and also the fourth vertex.

39. Show that quadrilateral $RSTV$ is an isosceles trapezoid.

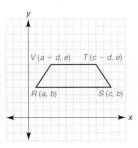

40. Show that quadrilateral $ABCD$ is a parallelogram.

41. Quadrilateral $EFGH$ has the vertices $E(0, 0)$, $F(a, 0)$, $G(a + b, c)$, and $H(2b, 2c)$. Verify that $EFGH$ is a trapezoid by showing that the slopes of two sides are equal.

42. Find an equation involving a, b, c, d, and e if $\overleftrightarrow{AC} \perp \overleftrightarrow{BC}$.
   (HINT: Use slopes.)

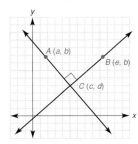

43. Prove that if two nonvertical lines are parallel, then their slopes are equal.
   (HINT: See Figure 10.18.)

*44. Prove that if two lines (neither horizontal nor vertical) are perpendicular, then the product of their slopes is -1.

(*HINT: See Figure 10.19. You need to show and use the fact that $\triangle ABC \sim \triangle EDC$.*)

*45. Where $m < 0$ and $b > 0$, the graph of $y = mx + b$ (along with the x and y axes) determines a triangular region in Quadrant I. Find an expression for the area of the triangle in terms of m and b.

*46. Where $m > 0$, $a > 0$, and $b > 0$, the graph of $y = mx + b$, the axes, and the vertical line through $(a, 0)$ determines a trapezoidal region in Quadrant I. Find an expression for the area of this trapezoid in terms of a, b, and m.

10.3 Preparing to Do Analytic Proofs

KEY CONCEPTS   Formulas and Relationships   Placement of Figure

In this section, we lay the groundwork for constructing analytic proofs of geometric theorems. An analytic proof requires the use of the coordinate system and the application of the formulas found in earlier sections of this chapter. Because of the need for these formulas, a summary follows in Table 10.1. Be sure that you have committed to memory these formulas and know when and how to use them.

TABLE 10.1
Formulas of Analytic Geometry

Distance	$d = \sqrt{(x_2 - x_1)^2 + (y_2 - y_1)^2}$
Midpoint	$M = \left(\frac{x_1 + x_2}{2}, \frac{y_1 + y_2}{2}\right)$
Slope	$m = \frac{y_2 - y_1}{x_2 - x_1}$ where $x_1 \neq x_2$
Special relationships for lines	$\ell_1 \parallel \ell_2 \leftrightarrow m_1 = m_2$
	$\ell_1 \perp \ell_2 \leftrightarrow m_1 \cdot m_2 = -1$, where neither ℓ_1 nor ℓ_2 is a vertical line or a horizontal line.

SSG **EXS. 1–6**

EXAMPLE 1

Suppose that you are to prove the following relationships:

  a) Two lines are parallel.
  b) Two lines are perpendicular.
  c) Two line segments are congruent.

Which formula(s) from Table 10.1 would you need to use? How would you complete your proof?

SOLUTION
  a) Use the Slope Formula to find the slope of each line. Then show that the slopes are equal.
  b) Use the Slope Formula to find the slope of each line. Then show that $m_1 \cdot m_2 = -1$.
  c) Use the Distance Formula to find the length of each line segment. Then show that the resulting lengths are equal.

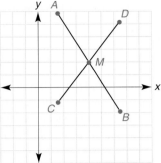

Figure 10.22

The following example has a proof that is subtle. Figure 10.22 is provided to help you understand the concept.

EXAMPLE 2

How can the Midpoint Formula be used to verify that the two line segments shown in Figure 10.22 bisect each other?

SOLUTION If $\overline{AB}$ bisects $\overline{CD}$, and conversely, then M is the common midpoint of the two line segments. The Midpoint Formula is used to find the midpoint of each line segment, and the results are then shown to be the same point. This establishes that each line segment has been bisected at a point that is found on each line segment.

EXAMPLE 3

Suppose that line ℓ_1 has slope $\frac{c}{d}$. Use this fact to identify the slopes of the following lines:

    a) ℓ_2 if $\ell_1 \parallel \ell_2$ **b)** ℓ_3 if $\ell_1 \perp \ell_3$

SOLUTION
    a) $m_2 = \frac{c}{d}$ $(m_1 = m_2$ when $\ell_1 \parallel \ell_2.)$
    b) $m_3 = -\frac{d}{c}$ $(m_1 \cdot m_3 = -1$ when $\ell_1 \perp \ell_3.)$

EXAMPLE 4

What can you conclude if you know that the point (p, q) lies on the line $y = mx + b$?

SOLUTION Because (p, q) is on the line, it is also a solution for the equation $y = mx + b$. Therefore, $q = mp + b$.

EX3. 7–9

To construct proofs of geometric theorems by using analytic methods, we must use the hypothesis to determine the drawing. Unlike the drawings in Chapters 1–9, the figure must be placed in the coordinate system. Making the drawing requires careful placement of the figure and strategically naming the coordinates of the vertices in that figure. The following guidelines should prove helpful in positioning the figure and in naming its vertices.

STRATEGY FOR PROOF ■ The Drawing for an Analytic Proof

Some considerations for preparing the drawing:

1. Coordinates of the vertices must be general; for instance, you may use (a, b) as a vertex, but do *not* use a specific point such as $(2, 3)$.
2. Make the drawing satisfy the hypothesis without providing any additional qualities; if the theorem describes a rectangle, draw and label a rectangle but *not* a square.
3. For simplicity in your calculations, drop the figure into the rectangular coordinate system in such a manner that
    a) as many 0 coordinates as possible are used.
    b) the remaining coordinates represent positive numbers due to the positioning of the remaining vertices in Quadrant I.

NOTE: In some cases, it is convenient to place a figure so that it has symmetry with respect to the y axis, in which case some negative coordinates are present.

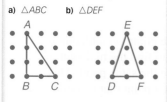
4. When possible, use horizontal and vertical line segments because you know their parallel and perpendicular relationships.
5. Use as few variable names in the coordinates as possible.

Now consider Example 5, which clarifies the list of suggestions found in the "Strategy for Proof." As you observe the drawing in each part of the example, imagine that △ABC has been cut out of a piece of cardboard and dropped into the coordinate system in the position indicated. Because we have freedom of placement, we choose the positioning that allows the simplest solution for a proof or problem.

EXAMPLE 5

Suppose that you need to make a drawing for the following theorem, which is to be proved analytically: "The midpoint of the hypotenuse of a right triangle is equidistant from the three vertices of the triangle." Explain why the placement of right △ABC in each part of Figure 10.23 could be improved.

(a)

(d)

(b)

(e)

(c)

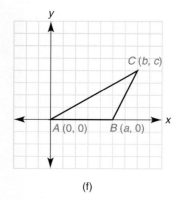

(f)

Figure 10.23

SOLUTION Refer to Figure 10.23.

a) The choice of vertices causes $AB = BC$, so the triangle is also an isosceles triangle. This contradicts point 2 of the list of suggestions.

b) The coordinates are too specific! This contradicts point 1 of the list. A proof with these coordinates would *not* establish the general case.

c) The drawing does not make use of horizontal and vertical lines to obtain the right angle. This violates point 4 of the list.

d) This placement fails point 3 of the list because b is a negative number. The length of $\overline{AB}$ would be $-b$, which could be confusing.

e) This placement fails point 3 because we have not used as many 0 coordinates as we could have used. As we shall see, it also fails point 5.

f) This placement fails point 2. The triangle is not a right triangle unless $a = b$. ▪

In Example 5, we wanted to place $\triangle ABC$ in the coordinate system so that we met as many of the conditions found in the "Strategy" on pages 453 and 454 as possible. Two convenient placements are given in Figure 10.24. The triangle in Figure 10.24(b) is slightly better than the one in 10.24(a) in that it uses four 0 coordinates rather than three. Another advantage of Figure 10.24(b) is that the placement forces angle B to be a right angle, because the x and y axes are perpendicular.

We now turn our attention to the role of the conclusion of the theorem for the proof. A second list examines some considerations for proving statements analytically.

(a)

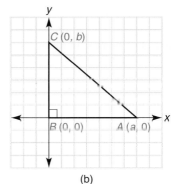

(b)

Figure 10.24

> **STRATEGY FOR PROOF ▪ The Conclusion for an Analytic Proof**
>
> Three considerations for using the conclusion as a guide:
>
> 1. If the conclusion is a conjunction "P and Q," be sure to verify both parts of the conclusion.
>
> 2. The following pairings indicate the formulas from Table 10.1 used to prove statements of the type shown in the left column.
>
> | To prove the conclusion: | Use the: |
> |---|---|
> | a) Segments are congruent or have equal lengths (such as $AB = CD$). | Distance Formula |
> | b) Segments are parallel (such as $\overline{AB} \parallel \overline{CD}$). | Slope Formula (need $m_{AB} = m_{CD}$) |
> | c) Segments are perpendicular (such as $\overline{AB} \perp \overline{CD}$). | Slope Formula (need $m_{\overline{AB}} \cdot m_{\overline{CD}} = -1$) |
> | d) A segment is bisected. | Distance Formula |
> | e) Segments bisect each other. | Midpoint Formula |
>
> 3. Anticipate the proof by thinking of the steps of the proof in reverse order; that is, reason backward from the conclusion.

SSG EXS. 10–15

EXAMPLE 6

a) Provide an ideal drawing for the following theorem: "The midpoint of the hypotenuse of a right triangle is equidistant from the three vertices of the triangle."

b) By studying the theorem, name at least two of the formulas that will be used to complete the proof. Explain your choices.

SOLUTION

a) We improve Figure 10.24(b) by giving the value $2a$ to the x coordinate of A and the value $2b$ to the y coordinate of C. (A factor of 2 makes it easier to calculate and represent the midpoint M of $\overline{AC}$. See Figure 10.25 on page 456.)

Figure 10.25

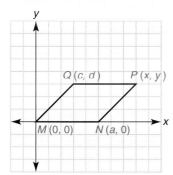

b) The Midpoint Formula is applied to describe the midpoint of $\overline{AC}$. Using the formula, we find that

$$M = \left(\frac{x_1 + x_2}{2}, \frac{y_1 + y_2}{2} \right) = \left(\frac{2a + 0}{2}, \frac{0 + 2b}{2} \right) = \left(\frac{2a}{2}, \frac{2b}{2} \right) = (a, b).$$

So the midpoint of the hypotenuse is (a, b). The Distance Formula will also be needed because the theorem states that the distances from *M* to *A*, from *M* to *B*, and from *M* to *C* should all be equal.

The purpose of our next example is to demonstrate efficiency in the labeling of vertices. Our goal is to use fewer variables in characterizing the vertices of the parallelogram found in Figure 10.26.

EXAMPLE 7

If *MNPQ* is a parallelogram in Figure 10.26, find the coordinates of point *P* in terms of *a*, *c*, and *d*.

SOLUTION Consider $\square MNPQ$ in which we refer to point *P* as (x, y).
    Because $\overline{MN} \parallel \overline{QP}$, we have $m_{\overline{MN}} = m_{\overline{QP}}$. But $m_{\overline{MN}} = \frac{0 - 0}{a - 0} = 0$ and $m_{\overline{QP}} = \frac{y - d}{x - c}$, so we are led to the equation

$$\frac{y - d}{x - c} = 0 \rightarrow y - d = 0 \rightarrow y = d$$

Now *P* is described by (x, d). Because $\overline{MQ} \parallel \overline{NP}$, we are also led to equal slopes for these segments. But

$$m_{\overline{MQ}} = \frac{d - 0}{c - 0} = \frac{d}{c} \qquad \text{and} \qquad m_{\overline{NP}} = \frac{d - 0}{x - a} = \frac{d}{x - a}$$

Then
$$\frac{d}{c} = \frac{d}{x - a}$$

By using the Means-Extremes Property, we have

$$
\begin{aligned}
d(x - a) &= d \cdot c && \text{(with } d \neq 0\text{)}\\
x - a &= c && \text{(dividing by } d\text{)}\\
x &= a + c && \text{(adding } a\text{)}
\end{aligned}
$$

Therefore, *P* is the point $(a + c, d)$.

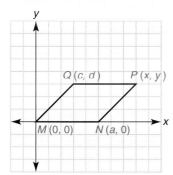

Figure 10.26

In retrospect, Example 7 shows that $\square MNPQ$ is characterized by vertices $M(0, 0)$, $N(a, 0)$, $P(a + c, d)$, and $Q(c, d)$. Because $\overline{MN}$ and $\overline{QP}$ are horizontal segments, it is obvious that $\overline{MN} \parallel \overline{QP}$. Both $\overline{MQ}$ (starting at M) and $\overline{NP}$ (starting at N) trace paths that move along each segment c *units to the right* and d *units upward*. Thus, the slopes of $\overline{MQ}$ and $\overline{NP}$ are both $\frac{d}{c}$, and it follows that $\overline{MQ} \parallel \overline{NP}$.

In Example 7, we named the coordinates of the vertices of a parallelogram with the fewest possible letters. We now extend our result in Example 7 to allow for a rhombus—a parallelogram with two congruent adjacent sides.

Figure 10.27

EXAMPLE 8

In Figure 10.27, find an equation that relates a, c, and d if $\square ABCD$ is a rhombus.

SOLUTION As we saw in Example 7, the coordinates of the vertices of $ABCD$ define a parallelogram. For emphasis, we note that $\overline{AB} \parallel \overline{DC}$ and $\overline{AD} \parallel \overline{BC}$ because

$$m_{\overline{AB}} = m_{\overline{DC}} = 0 \qquad \text{and} \qquad m_{\overline{AD}} = m_{\overline{BC}} = \frac{d}{c}$$

For Figure 10.27 to represent a rhombus, it is necessary that $AB = AD$. Now $AB = a - 0 = a$ because $\overline{AB}$ is a horizontal segment. To find an expression for the length of $\overline{AD}$, we need to use the Distance Formula.

$$\begin{aligned} AD &= \sqrt{(x_2 - x_1)^2 + (y_2 - y_1)^2} \\ &= \sqrt{(c - 0)^2 + (d - 0)^2} \\ &= \sqrt{c^2 + d^2} \end{aligned}$$

Because $AB = AD$, we are led to $a = \sqrt{c^2 + d^2}$. Squaring, we have the desired equation, $a^2 = c^2 + d^2$.

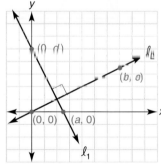

Figure 10.28

EXAMPLE 9

If $\ell_1 \perp \ell_2$ in Figure 10.28, find a relationship among the variables a, b, c, and d.

SOLUTION First we find the slopes of lines ℓ_1 and ℓ_2. For ℓ_1, we have

$$m_1 = \frac{0 - d}{a - 0} = -\frac{d}{a}$$

For ℓ_2, we have

$$m_2 = \frac{c - 0}{b - 0} = \frac{c}{b}$$

With $\ell_1 \perp \ell_2$, it follows that $m_1 \cdot m_2 = -1$. Substituting the slopes found above into the equation $m_1 \cdot m_2 = -1$, we have

$$-\frac{d}{a} \cdot \frac{c}{b} = -1 \qquad \text{so} \qquad -\frac{dc}{ab} = -1$$

Equivalently, $\dfrac{dc}{ab} = 1$ and it follows that $dc = ab$.

Exercises 10.3

1. Find an expression for
   a) The distance between $(a, 0)$ and $(0, a)$.
   b) The slope of the segment joining (a, b) and (c, d).

2. Find the coordinates of the midpoint of the segment that joins the points
   a) $(a, 0)$ and $(0, b)$.   b) $(2a, 0)$ and $(0, 2b)$.

3. Find the slope of the line containing the points
   a) $(a, 0)$ and $(0, a)$.   b) $(a, 0)$ and $(0, b)$.

4. Find the slope of the line that is
   a) parallel to the line containing $(a, 0)$ and $(0, b)$.
   b) perpendicular to the line through $(a, 0)$ and $(0, b)$.

In Exercises 5 to 10, the real numbers a, b, c, and d are positive.

5. Consider the triangle with vertices at $A(0, 0)$, $B(a, 0)$, and $C(a, b)$. Explain why $\triangle ABC$ is a right triangle.

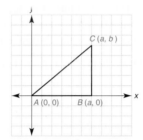

6. Consider the triangle with vertices at $R(-a, 0)$, $S(a, 0)$, and $T(0, b)$. Explain why $\triangle RST$ is an isosceles triangle.

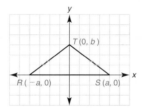

7. Consider the quadrilateral with vertices at $M(0, 0)$, $N(a, 0)$, $P(a + b, c)$, and $Q(b, c)$. Explain why $MNPQ$ is a parallelogram.

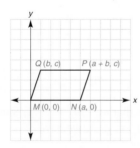

8. Consider the quadrilateral with vertices at $A(0, 0)$, $B(a, 0)$, $C(b, c)$, and $D(d, c)$. Explain why $ABCD$ is a trapezoid.

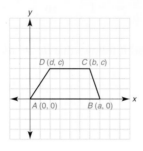

9. Consider the quadrilateral with vertices at $M(0, 0)$, $N(a, 0)$, $P(a, b)$, and $Q(0, b)$. Explain why $MNPQ$ is a rectangle.

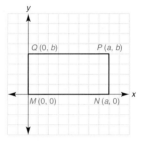

10. Consider the quadrilateral with vertices at $R(0, 0)$, $S(a, 0)$, $T(a, a)$, and $V(0, a)$. Explain why $RSTV$ is a square.

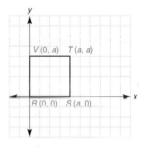

In Exercises 11 to 16, supply the missing coordinates for the vertices, using as few variables as possible.

11.

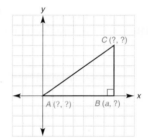

ABC is a right triangle.

12.

DEF is an isosceles triangle with $\overline{DF} \cong \overline{FE}$.

13.

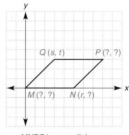

$MNPQ$ is a parallelogram.

14.

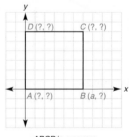

$ABCD$ is a square.

15.

$ABCD$ is an isosceles trapezoid; $\overline{AB} \parallel \overline{DC}$ and $\overline{AD} \cong \overline{BC}$.

16.

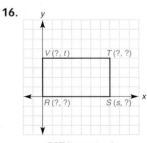

$RSTV$ is a rectangle.

In Exercises 17 to 22, draw an ideally placed figure in the coordinate system; then name the coordinates of each vertex of the figure.

17. a) A square
b) A square (midpoints of sides are needed)

18. a) A rectangle
b) A rectangle (midpoints of sides are needed)

19. a) A parallelogram
b) A parallelogram (midpoints of sides are needed)

20. a) A triangle
b) A triangle (midpoints of sides are needed)

21. a) An isosceles triangle
b) An isosceles triangle (midpoints of sides are needed)

22. a) A trapezoid
b) A trapezoid (midpoints of sides are needed)

In Exercises 23 to 28, find the equation (relationship) requested. Then eliminate fractions and square root radicals from the equation.

23. If □*MNPQ* is a rhombus, state an equation that relates *r*, *s*, and *t*.

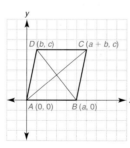

24. For □*RSTV*, suppose that *RT* = *VS*. State an equation that relates *s*, *t*, and *v*.

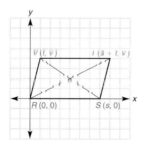

25. For □*ABCD*, suppose that diagonals $\overline{AC}$ and $\overline{DB}$ are perpendicular. State an equation that relates *a*, *b*, and *c*.

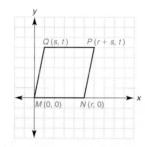

26. For quadrilateral *RSTV*, suppose that $\overline{RV} \parallel \overline{ST}$. State an equation that relates *m*, *n*, *p*, *q*, and *r*.

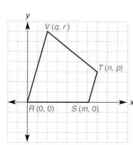

27. Suppose that △*ABC* is an equilateral triangle. State an equation that relates variables *a* and *b*.

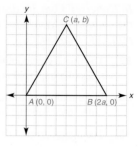

28. Suppose that △*RST* is an isosceles triangle, with $\overline{RS} \cong \overline{RT}$. State an equation that relates *s*, *t*, and *v*.

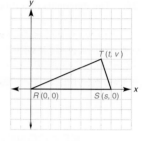

29. The drawing shows isosceles △*ABC* with $\overline{AC} \cong \overline{BC}$.
a) What type of number is *a*?
b) What type of number is −*a*?
c) Find an expression for the length of $\overline{AB}$.

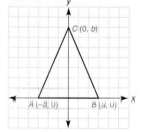

30. The drawing shows parallelogram *RSTV*.
a) What type of number is *r*?
b) Find an expression for *RS*.
c) Describe the coordinate *t* in terms of the other variables shown.

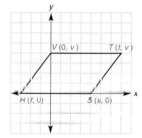

31. Which formula would you use to establish each of the following claims?
a) $\overline{AC} \perp \overline{DB}$
b) $\overline{AC} = \overline{DB}$
c) $\overline{DB}$ and $\overline{AC}$ bisect each other
d) $\overline{AD} \parallel \overline{BC}$

ABCD is a parallelogram.

32. Which formula would you use to establish each of the following claims?
a) The coordinates of *X* are (*d*, *c*).
b) $m_{\overline{VT}} = 0$
c) $\overline{VT} \parallel \overline{RS}$
d) The length of $\overline{RV}$ is $2\sqrt{d^2 + c^2}$.

Trapezoid *RSTV*; *X* is the midpoint of $\overline{RV}$.

In Exercises 33 to 36, draw and label a well-placed figure in the coordinate system for each theorem. Do not attempt to prove the theorem!

33. The line segment joining the midpoints of the two nonparallel sides of a trapezoid is parallel to each base of the trapezoid.

34. If the midpoints of the sides of a quadrilateral are joined in order, the resulting quadrilateral is a parallelogram.

35. The diagonals of a rectangle are equal in length.

36. The diagonals of a rhombus are perpendicular to each other.

*37. In $\triangle RST$, $\overline{SV}$ bisects $\angle RST$. Find the coordinates of point T in terms of a.

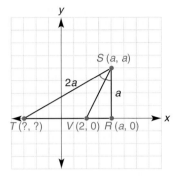

10.4 Analytic Proofs

KEY CONCEPTS   Analytic Proof   Synthetic Proof

When we use algebra along with the rectangular coordinate system to prove a geometric theorem, the method of proof is **analytic.** The analytic (algebraic) approach relies heavily on the placement of the figure in the coordinate system and on the application of the formulas in Table 10.1. In order to contrast analytic proof with **synthetic** proof (the two-column or paragraph proofs used in earlier chapters), we repeat in this section some earlier theorems and prove these analytically.

In Section 10.3, we saw how to place triangles having special qualities in the coordinate system. We review this information in Table 10.2; in Example 1, we consider the

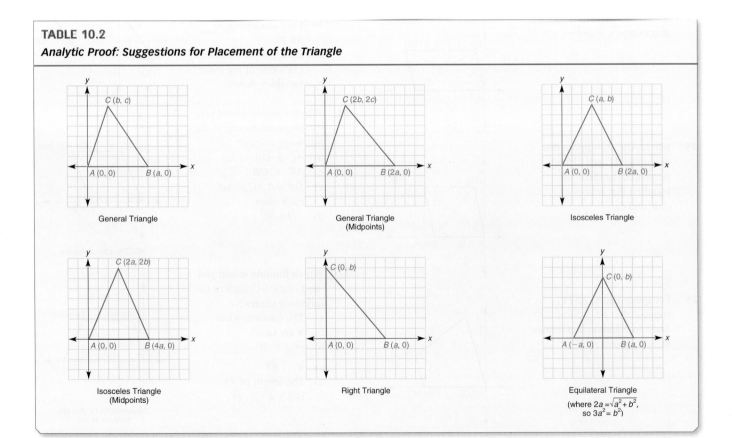

TABLE 10.2

Analytic Proof: Suggestions for Placement of the Triangle

General Triangle

General Triangle (Midpoints)

Isosceles Triangle

Isosceles Triangle (Midpoints)

Right Triangle

Equilateral Triangle (where $2a = \sqrt{a^2 + b^2}$, so $3a^2 = b^2$)

proof of a theorem involving triangles. In Table 10.2, you will find that the figure determined by any positive numerical choices of *a*, *b*, and *c* matches the type of triangle described. When midpoints are involved, we use coordinates such as 2*a* or 2*b*.

EXAMPLE 1

SSG **EXS. 1–4**

Prove the following theorem by the analytic method (see Figure 10.29).

THEOREM 10.4.1

The line segment determined by the midpoints of two sides of a triangle is parallel to the third side.

PLAN: Use the Slope Formula; if $m_{\overline{MN}} = m_{\overline{AC}}$, then $\overline{MN} \parallel \overline{AC}$.

PROOF: As shown in Figure 10.29, $\triangle ABC$ has vertices at $A(0, 0)$, $B(2a, 0)$, and $C(2b, 2c)$. With M the midpoint of $\overline{BC}$, and N the midpoint of $\overline{AB}$,

$$M = \left(\frac{2a + 2b}{2}, \frac{0 + 2c}{2} \right) = (a + b, c)$$

$$N = \left(\frac{0 + 2a}{2}, \frac{0 + 0}{2} \right) = (a, 0)$$

Next we apply the Slope Formula to determine $m_{\overline{MN}}$ and $m_{\overline{AC}}$. Now

$m_{\overline{MN}} = \frac{c - 0}{(a + b) - a} = \frac{c}{b}$; also, $m_{\overline{AC}} = \frac{2c - 0}{2b - 0} = \frac{2c}{2b} = \frac{c}{b}$. Because $m_{\overline{MN}} = m_{\overline{AC}}$, we see that $\overline{MN} \parallel \overline{AC}$.

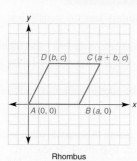

Figure 10.29

In Table 10.3, we review convenient placements for types of quadrilaterals.

TABLE 10.3
Analytic Proof: Suggestions for Placement of the Quadrilateral

General Quadrilateral

General Quadrilateral
(Midpoints)

Parallelogram

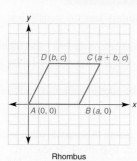

Rhombus

(where $a = \sqrt{b^2 + c^2}$, so $a^2 = b^2 + c^2$)

Rectangle

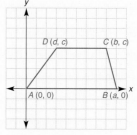

Trapezoid

As we did in Example 1, we include a "plan" for Example 2. Although no plan is shown for Example 3 or Example 4, one is necessary before the proof can be written.

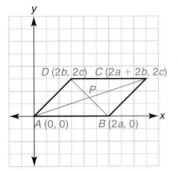

Figure 10.30

EXAMPLE 2

Prove the following theorem by the analytic method. See Figure 10.30.

THEOREM 10.4.2

The diagonals of a parallelogram bisect each other.

PLAN: Use the Midpoint Formula to show that the two diagonals have a common midpoint. Use a factor of 2 in the coordinates of the vertices.

PROOF: With coordinates as shown in Figure 10.30, quadrilateral $ABCD$ is a parallelogram. The diagonals intersect at point P. By the Midpoint Formula, we have

$$M_{\overline{AC}} = \left(\frac{0 + (2a + 2b)}{2}, \frac{0 + 2c}{2} \right)$$
$$= (a + b, c)$$

Also, the midpoint of $\overline{DB}$ is

$$M_{\overline{DB}} = \left(\frac{2a + 2b}{2}, \frac{0 + 2c}{2} \right)$$
$$= (a + b, c)$$

Thus, $(a + b, c)$ is the common midpoint of the two diagonals and must be the point of intersection of $\overline{AC}$ and $\overline{DB}$. Then $\overline{AC}$ and $\overline{DB}$ bisect each other at point $P(a + b, c)$.

EXS. 5–9

The proof of Theorem 10.4.2 is not unique! In Section 10.5, we could prove Theorem 10.4.2 by using a three-step proof:

1. Find the equations of the two diagonals.
2. Determine the point of intersection of these lines.
3. Show that this point of intersection is the common midpoint.

But the phrase *bisect each other* in Theorem 10.4.2 implied the use of the Midpoint Formula. Our approach to Example 2 was far easier and just as valid as the three steps described above. The use of the Midpoint Formula is generally the best approach when the phrase *bisect each other* appears in the statement of a theorem.

We now outline the method of analytic proof.

STRATEGY FOR PROOF ■ Completing an Analytic Proof

1. Read the theorem carefully to distinguish the hypothesis and the conclusion. The hypothesis characterizes the figure to use.
2. Use the hypothesis (and nothing more) to determine a convenient placement of the figure in the rectangular coordinate system. Then label the figure. See Tables 10.1 and 10.2.
3. If any special quality is provided by the hypothesis, be sure to state this early in the proof. (For example, a rhombus should be described as a parallelogram that has two congruent adjacent sides.)
4. Study the conclusion, and devise a plan to prove this claim; this may involve reasoning backward from the conclusion step by step until the hypothesis is reached.
5. Write the proof, being careful to order the statements properly and to justify each statement.

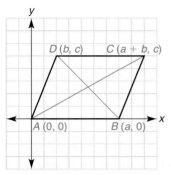

Figure 10.31

SSG **EXS. 10, 11**

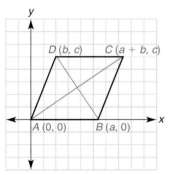

Figure 10.32

EXAMPLE 3

Prove Theorem 10.4.3 by the analytic method. (See Figure 10.31.)

THEOREM 10.4.3

The diagonals of a rhombus are perpendicular.

SOLUTION In Figure 10.31, $ABCD$ has the coordinates of a parallelogram. Because $\square ABCD$ is a rhombus, $AB = AD$. Then $a = \sqrt{b^2 + c^2}$ by the Distance Formula, and squaring gives $a^2 = b^2 + c^2$. The Slope Formula leads to

$$m_{\overline{AC}} = \frac{c - 0}{(a + b) - 0} \quad \text{and} \quad m_{\overline{DB}} = \frac{0 - c}{a - b}.$$

Simplifying, $\quad m_{\overline{AC}} = \dfrac{c}{a + b} \quad$ and $\quad m_{\overline{DB}} = \dfrac{-c}{a - b}.$

Then the product of the slopes of the diagonals is

$$m_{\overline{AC}} \cdot m_{\overline{DB}} = \frac{c}{a + b} \cdot \frac{-c}{a - b}$$

$$= \frac{-c^2}{a^2 - b^2}$$

$$\frac{-c^2}{(b^2 + c^2) - b^2} \qquad \text{(replaced } a^2 \text{ by } b^2 + c^2\text{)}$$

$$= \frac{-c^2}{c^2} = -1$$

Thus, $\overline{AC} \perp \overline{DB}$ because the product of their slopes equals -1.

In Example 3, we had to use the condition that two adjacent sides of the rhombus were congruent to complete the proof. Had that condition been omitted, the product of slopes could not have been shown to equal -1. In general, the diagonals of a parallelogram are not perpendicular.

In our next example, we consider the proof of the converse of an earlier theorem. Although it is easy to complete an analytic proof of the statement "The diagonals of a rectangle are equal in length," the proof of the converse is not as straightforward.

EXAMPLE 4

Prove Theorem 10.4.4 by the analytic method. See Figure 10.32.

THEOREM 10.4.4

If the diagonals of a parallelogram are equal in length, then the parallelogram is a rectangle.

SOLUTION In parallelogram $ABCD$ of Figure 10.32, $AC = DB$. Applying the Distance Formula, we have

$$AC = \sqrt{[(a + b) - 0]^2 + (c - 0)^2}$$

and

$$DB = \sqrt{(a - b)^2 + (0 - c)^2}$$

Because the diagonals have the same length,

$$\sqrt{(a + b)^2 + c^2} = \sqrt{(a - b)^2 + (-c)^2}$$
$$(a + b)^2 + c^2 = (a - b)^2 + (-c)^2 \qquad \text{(squaring)}$$
$$a^2 + 2ab + b^2 + c^2 = a^2 - 2ab + b^2 + c^2 \qquad \text{(simplifying)}$$
$$4ab = 0$$
$$a \cdot b = 0 \qquad \text{(dividing by 4)}$$

Thus, $a = 0$ or $b = 0$

Because $a \neq 0$ (otherwise, points A and B would coincide), it is necessary that $b = 0$; so point D is on the y axis. With $b = 0$, the coordinates of the figure are $A(0, 0)$, $B(a, 0)$, $C(a, c)$, and $D(0, c)$. Because $\overline{AB}$ is horizontal and $\overline{AD}$ is vertical, $ABCD$ must be a rectangle with a right angle at A.

SSG **EXS. 12, 13**

Exercises 10.4

In Exercises 1 to 17, complete an analytic proof for each theorem.

1. The diagonals of a rectangle are equal in length.

2. The opposite sides of a parallelogram are equal in length.

3. The diagonals of a square are perpendicular bisectors of each other.

4. The diagonals of an isosceles trapezoid are equal in length.

5. The median from the vertex of an isosceles triangle to the base is perpendicular to the base.

6. The medians to the congruent sides of an isosceles triangle are equal in length.

7. The line segments that join the midpoints of the consecutive sides of a quadrilateral form a parallelogram.

8. The line segments that join the midpoints of the opposite sides of a quadrilateral bisect each other.

9. The line segments that join the midpoints of the consecutive sides of a rectangle form a rhombus.

10. The line segments that join the midpoints of the consecutive sides of a rhombus form a rectangle.

11. The midpoint of the hypotenuse of a right triangle is equidistant from the three vertices of the triangle.

12. The median of a trapezoid is parallel to the bases of the trapezoid and has a length equal to one-half the sum of the lengths of the two bases.

13. The line segment that joins the midpoints of two sides of a triangle is parallel to the third side and has a length equal to one-half the length of the third side.

14. The perpendicular bisector of the base of an isosceles triangle contains the vertex of the triangle.

15. If the midpoint of one side of a rectangle is joined to the endpoints of the opposite side, then an isosceles triangle is formed.

*16. If the median to one side of a triangle is also an altitude of the triangle, then the triangle is isosceles.

*17. If the diagonals of a parallelogram are perpendicular, then the parallelogram is a rhombus.

18. Use the analytic method to decide what type of quadrilateral is formed when the midpoints of the consecutive sides of a parallelogram are joined by line segments.

19. Use the analytic method to decide what type of triangle is formed when the midpoints of the sides of an isosceles triangle are joined by line segments.

20. Use slopes to verify that the graphs of the equations

$$Ax + By = C \qquad \text{and} \qquad Ax + By = D$$

are parallel.

(Note: $A \neq 0$, $B \neq 0$ and $C \neq D$.)

21. Use slopes to verify that the graphs of the equations

$$Ax + By = C \qquad \text{and} \qquad Bx - Ay = D$$

are perpendicular.

(Note: $A \neq 0$ and $B \neq 0$.)

22. Use the result in Exercise 20 to find the equation of the line that contains $(4, 5)$ and is parallel to the graph of $2x + 3y = 6$.

23. Use the result in Exercise 21 to find the equation of the line that contains $(4, 5)$ and is perpendicular to the graph of $2x + 3y = 6$.

24. Use the Distance Formula to show that the circle with center $(0, 0)$ and radius length r has the equation $x^2 + y^2 = r^2$.

25. Use the result in Exercise 24 to find the equation of the circle with center $(0, 0)$ and radius length $r = 3$.

26. Use the result in Exercise 24 to find the equation of the circle that has center $(0, 0)$ and contains the point $(3, 4)$.

27. Suppose that the circle with center (0, 0) and radius length r contains the point (a, b). Find the slope of the tangent line to the circle at the point (a, b).

28. Consider the circle with center (h, k) and radius length r. If the circle contains the point (c, d), find the slope of the tangent line to the circle at the point (c, d).

29. Would the theorem of Exercise 7 remain true for a concave quadrilateral like the one shown?

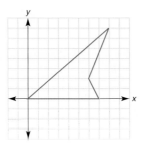

Exercise 29

*30. Complete an analytic proof of the following theorem: In a triangle that has sides of lengths a, b, and c, if $c^2 = a^2 + b^2$, then the triangle is a right triangle.

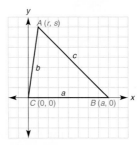

Exercise 30

10.5 Equations of Lines

KEY CONCEPTS Slope-Intercept Form of a Line   Point-Slope Form of a Line   Systems of Equations

In Section 10.2, we saw that equations such as $2x + 3y = 6$ and $4x - 12y = 60$ have graphs that are lines. To graph an equation of the general form $Ax + By = C$, that equation is often replaced with an equivalent equation of the form $y = mx + b$. For instance, $2x + 3y = 6$ can be transformed into $y = -\frac{2}{3}x + 2$; equations such as these are known as *equivalent* because their ordered-pair solutions (and graphs) are identical. In particular, we must express a linear equation in the form $y = mx + b$ in order to plot it on a graphing calculator.

EXAMPLE 1

Write the equation $4x - 12y = 60$ in the form $y = mx + b$.

SOLUTION Given $4x - 12y = 60$, we subtract $4x$ from each side of the equation to obtain $-12y = -4x + 60$. Dividing by -12,

$$\frac{-12y}{-12} = \frac{-4x}{-12} + \frac{60}{-12}$$

SSG **EXS. 1–3**

Then $y = \frac{1}{3}x - 5$.

SLOPE-INTERCEPT FORM OF A LINE

We now turn our attention to a method for finding the equation of a line. In the following technique, the equation can be found if the slope m and the y intercept b of the line are known. The form $y = mx + b$ is known as the Slope-Intercept Form of a line.

> ### THEOREM 10.5.1 ■ Slope-intercept form of a Line
>
> The line whose slope is m and whose y intercept is b has the equation $y = mx + b$.

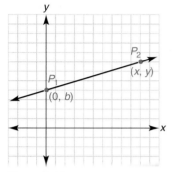

Figure 10.33

PROOF

Consider the line whose slope is m (see Figure 10.33). Using the Slope Formula

$$m = \frac{y_2 - y_1}{x_2 - x_1},$$

we use (x, y) for P_2 and $(0, b)$ for P_1. Then

$$m = \frac{y - b}{x - 0} \qquad \text{or} \qquad m = \frac{y - b}{x}$$

Multiplying by x, we have $mx = y - b$. Then $mx + b = y$, or $y = mx + b$. ■

Discover

Use a graphing calculator to graph $Y_1 = x$, $Y_2 = x^2$, and $Y_3 = x^3$. Which of these is (are) a line(s)?

ANSWER

$y = x$

EXAMPLE 2

Find the general equation $Ax + By = C$ for the line with slope $m = -\frac{2}{3}$ and y intercept -2.

SOLUTION In the form $y = mx + b$, we have

$$y = -\frac{2}{3}x - 2$$

Multiplying by 3, we obtain

$$3y = -2x - 6 \qquad \text{so} \qquad 2x + 3y = -6$$

NOTE: An equivalent and correct solution is $-2x - 3y = 6$.

It is often easier to graph an equation if it is in the form $y = mx + b$. When an equation has this form, we know that its graph is a line that has slope m and contains $(0, b)$.

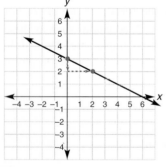

Figure 10.34

EXAMPLE 3

Draw the graph of $\frac{1}{2}x + y = 3$.

SOLUTION Solving for y, we have $y = -\frac{1}{2}x + 3$. Then the slope is $m = -\frac{1}{2}$, and the y intercept is $b = 3$.

We first plot the point $(0, 3)$ in Figure 10.34. Because $m = -\frac{1}{2}$ or $\frac{-1}{2}$, the vertical change of -1 corresponds to a horizontal change of $+2$. Thus, a second point is located 1 unit down from and 2 units to the right of the first point. The line is drawn in Figure 10.34.

Another look at Figure 10.34 shows that the graph contains the points $(2, 2)$ and $(6, 0)$. Both ordered pairs are easily shown to be solutions for the equation $\frac{1}{2}x + y = 3$ of Example 3.

POINT-SLOPE FORM OF A LINE

SSG **EXS. 4–8**

If slope m and a point other than the y intercept of a line are known, we generally do not use the Slope-Intercept Form to find the equation of the line. Instead, the Point-Slope Form of the equation of a line is used. This form is also used when the coordinates of two points of the line are known; in that case, the value of m is found by the Slope Formula. The form $y - y_1 = m(x - x_1)$ is known as the Point-Slope Form of the line with slope m and containing the point (x_1, y_1).

THEOREM 10.5.2 ■ Point-Slope Form of a Line

The line that has slope m and contains the point (x_1, y_1) has the equation

$$y - y_1 = m(x - x_1)$$

PROOF

Let P_1 be the given point (x_1, y_1) on the line, and let $P_2(x, y)$ represent any other point on the line. See Figure 10.35. Using the Slope Formula, we have

$$m = \frac{y - y_1}{x - x_1}$$

Multiplying the equation by $(x - x_1)$ yields

$$m(x - x_1) = y - y_1$$

It follows that

$$y - y_1 = m(x - x_1)$$

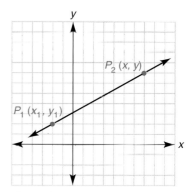

Figure 10.35

EXAMPLE 4

Find an equation $(Ax + By = C)$ for the line that has the slope $m = 2$ and contains the point $(-1, 3)$.

SOLUTION We have $m = 2, x_1 = -1$, and $y_1 = 3$. Applying the Point-Slope Form, we find that the line in Figure 10.36 has the equation

$$
\begin{aligned}
y - y_1 &= m(x - x_1), \text{ or} \\
y - 3 &= 2[x - (-1)] \\
y - 3 &= 2(x + 1) \\
y - 3 &= 2x + 2 \\
-2x + y &= 5
\end{aligned}
$$

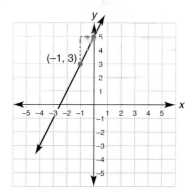

Figure 10.36

An equivalent answer for Example 4 is the equation $2x - y = -5$. The form $y = 2x + 5$ emphasizes that the slope is $m = 2$ and that the y intercept is $(0, 5)$. With $m = 2 \left(\text{or } \frac{2}{1}\right)$, the vertical change of 2 corresponds to a horizontal change of 1 as shown in Figure 10.36.

EXAMPLE 5

Find an equation for the line containing the points $(-1, 2)$ and $(4, 1)$.

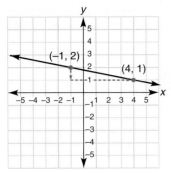

Figure 10.37

SOLUTION To use the Point-Slope Form, we need to know the slope of the line (see Figure 10.37). When we choose $P_1(-1, 2)$ and $P_2(4, 1)$, the Slope Formula reads

$$m = \frac{1-2}{4-(-1)} = \frac{-1}{5} = -\frac{1}{5}$$

Then $y - y_1 = m(x - x_1)$ becomes

$$y - 2 = -\frac{1}{5}[x - (-1)]$$

$$y - 2 = -\frac{1}{5}[x + 1]$$

$$y - 2 = -\frac{1}{5}x - \frac{1}{5}$$

Multiplying the equation by 5, we obtain

$$5y - 10 = -1x - 1 \qquad \text{so} \qquad x + 5y = 9$$

NOTE: Other forms of the answer are $-x - 5y = -9$ and $y = -\frac{1}{5}x + \frac{9}{5}$. In any correct form of the solution, the coordinates of the given points P_1 and P_2 must satisfy the equation.

In Example 6, we use the Point-Slope Form to find an equation for a median of a triangle.

SSG **EXS. 9–12**

EXAMPLE 6

For $\triangle ABC$, the vertices are $A(0, 0)$, $B(2a, 0)$, and $C(2b, 2c)$. Find the equation of median $\overline{CM}$ in the form $y = mx + b$. See Figure 10.38.

SOLUTION For $\overline{CM}$ to be a median of $\triangle ABC$, M must be the midpoint of $\overline{AB}$. Then

$$M = \left(\frac{0 + 2a}{2}, \frac{0 + 0}{2}\right) - (a, 0)$$

To determine an equation for $\overline{CM}$, we also need to know its slope. With $M(a, 0)$ and $C(2b, 2c)$ on $\overline{CM}$, the slope is $m_{\overline{CM}} = \frac{2c - 0}{2b - a}$ or $\frac{2c}{2b - a}$. With $M = (a, 0)$ as the point on the line, $y - y_1 = m(x - x_1)$ becomes

$$y - 0 = \frac{2c}{2b - a}(x - a) \qquad \text{or} \qquad y = \frac{2c}{2b - a}x - \frac{2ac}{2b - a}$$

Figure 10.38

SOLVING SYSTEMS OF EQUATIONS

In earlier chapters, we solved systems of equations such as

$$x + 2y = 6$$
$$2x - y = 7$$

by using the Addition Property or the Subtraction Property of Equality. We review the method in Example 7. The solution for the system is an ordered pair; in fact, the solution is the point of intersection of the graphs of the given equations.

EXAMPLE 7

Solve the following system by using algebra:

$$\begin{cases} x + 2y = 6 \\ 2x - y = 7 \end{cases}$$

SOLUTION When we multiply the second equation by 2, the system becomes

$$\begin{cases} x + 2y = 6 \\ 4x - 2y = 14 \end{cases}$$

Adding these equations yields $5x = 20$, so $x = 4$. Substituting $x = 4$ into the first equation, we find that $4 + 2y = 6$, so $2y = 2$. Then $y = 1$. The solution is the ordered pair (4, 1).

Another method for solving a system of equations is geometric and requires graphing. Solving by graphing amounts to finding the point of intersection of the linear graphs. That point is the ordered pair that is the common solution (when one exists) for the two equations. Notice that Example 8 repeats the system of Example 7. The graphs of the Technology Exploration should also have the appearance of Figure 10.39.

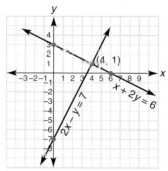

Figure 10.39

EXAMPLE 8

Solve the following system by graphing:

$$\begin{cases} x + 2y = 6 \\ 2x - y = 7 \end{cases}$$

SOLUTION Each equation is changed to the form $y = mx + b$ so that the slope m and the y intercept b can be used in graphing these equations:

$$x + 2y = 6 \rightarrow 2y = -1x + 6 \rightarrow y = -\frac{1}{2}x + 3$$
$$2x - y = 7 \rightarrow -y = -2x + 7 \rightarrow y = 2x - 7$$

The graph of $y = -\frac{1}{2}x + 3$ is a line with y intercept $b = 3$ and slope $m = -\frac{1}{2}$. The graph of $y = 2x - 7$ is a line with y intercept $b = -7$ and slope $m = 2$.

The graphs are drawn in the same coordinate system. See Figure 10.39. The point of intersection (4, 1) is the common solution for each of the given equations and thus is the solution of the system.

NOTE: To verify the result (solution) found in Examples 7 and 8, we show that (4, 1) satisfies both of the given equations:

$$x + 2y = 6 \rightarrow 4 + 2(1) = 6 \text{ is } true.$$
$$2x - y = 7 \rightarrow 2(4) - 1 = 7 \text{ is } true.$$

The solution is verified in that both statements are true.

Advantages of the method of solving a system of equations by graphing include the following:

1. It is easy to understand why a system such as

$$\begin{cases} x + 2y = 6 \\ 2x - y = 7 \end{cases} \text{ can be replaced by } \begin{cases} x + 2y = 6 \\ 4x - 2y = 14 \end{cases}$$

when we are solving by addition or subtraction. We know that the graphs of $2x - y = 7$ and $4x - 2y = 14$ are coincident (the same line) because each equation can be changed to the form $y = 2x - 7$.

2. It is easy to understand why a system such as

$$\begin{cases} x + 2y = 6 \\ 2x + 4y = -4 \end{cases}$$

has no solution. In Figure 10.40, the graphs of these equations are parallel lines.

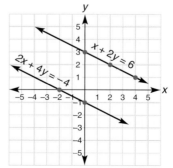

Figure 10.40

The first equation is equivalent to $y = -\frac{1}{2}x + 3$, and the second equation can be changed to $y = -\frac{1}{2}x - 1$. Both lines have slope $m = -\frac{1}{2}$, but have different y intercepts. Therefore, the lines are parallel and distinct.

Two lines with the same slope and the same y intercept are *coincident*; otherwise, lines having the same slope but different y intercepts are *parallel*.

Algebraic substitution can also be used to solve a system of equations. In our approach, we write each equation in the form $y = mx + b$ and then equate the expressions for y. Once the x coordinate of the solution is known, we substitute this value of x into either equation to find the value of y.

EXAMPLE 9

Use substitution to solve $\begin{cases} x + 2y = 6 \\ 2x - y = 7 \end{cases}$

SOLUTION Solving for y, we have

$$x + 2y = 6 \rightarrow 2y = -1x + 6 \rightarrow y = -\frac{1}{2}x + 3$$
$$2x - y = 7 \rightarrow -1y = -2x + 7 \rightarrow y = 2x - 7$$

Equating values for y, $-\frac{1}{2}x + 3 = 2x - 7$. Then $-2\frac{1}{2}x = -10$, or $-2.5x = -10$. Dividing by -2.5, $x = 4$. Substitution of 4 for x in the equation $y = 2x - 7$ leads to $y = 2(4) - 7$, so $y = 1$. The solution is the ordered pair $(4, 1)$.

NOTE: Substitution of $x = 4$ into the equation $y = -\frac{1}{2}x + 3$ would lead to the same value of y, namely $y = 1$. Thus, one can substitute into either equation.

SSG **EXS. 13–16**

The method illustrated in Example 9 is also used in our final example. In the proof of Theorem 10.5.3, we use equations of lines to determine the centroid of a triangle.

EXAMPLE 10

Formulate a plan to complete the proof of Theorem 10.5.3.

THEOREM 10.5.3

The three medians of a triangle are concurrent at a point that is two-thirds the distance from any vertex to the midpoint of the opposite side.

(a)

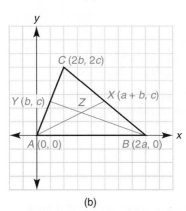

(b)

Figure 10.41

SOLUTION The proof can be completed as follows:

1. Find the coordinates of the two midpoints X and Y. See Figures 10.41(a) and 10.41(b) at the left. Note that

$$X = (a + b, c) \qquad \text{and} \qquad Y = (b, c)$$

2. Find the equations of the lines containing $\overline{AX}$ and $\overline{BY}$. The equations for $\overline{AX}$ and $\overline{BY}$ are $y = \frac{c}{a + b}x$ and $y = \frac{-c}{2a - b}x + \frac{2ac}{2a - b}$, respectively.

3. Find the point of intersection Z of $\overline{AX}$ and $\overline{BY}$, as shown in Figure 10.41(b). Solving the system provides the solution

$$Z = \left(\frac{2}{3}(a + b), \frac{2}{3}c \right)$$

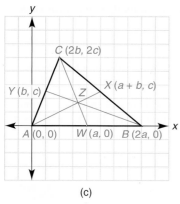

(c)

Figure 10.41

4. It can now be shown that $AZ = \frac{2}{3} \cdot AX$ and $BZ = \frac{2}{3} \cdot BY$. See Figure 10.41(b), in which we can show that

$$AZ = \frac{2}{3}\sqrt{(a + b)^2 + c^2} \qquad \text{and} \qquad AX = \sqrt{(a + b)^2 + c^2}$$

5. It can also be shown that point Z lies on the third median $\overline{CW}$, whose equation is $y = \frac{2c}{2b - a}(x - a)$. See Figure 10.41(c).

6. We can also show that $CZ = \frac{2}{3} \cdot CW$, which would complete the proof.

Exercises 10.5

In Exercises 1 to 4, use division to write an equation of the form $Ax + By = C$ that is equivalent to the one provided. Then write the given equation in the form $y = mx + b$.

1. $8x + 16y = 48$

2. $15x - 35y = 105$

3. $-6x + 18y = -240$

4. $27x - 36y = 108$

In Exercises 5 to 8, draw the graph of each equation by using the method of Example 3.

5. $y - 2x = 3$

6. $y = -2x + 5$

7. $\frac{9}{5}x + y = 6$

8. $3x - 2y = 12$

In Exercises 9 to 24, find an equation of the line described. Leave the solution in the form $Ax + By = C$.

9. The line has slope $m = -\frac{2}{3}$ and contains $(0, 5)$.

10. The line has slope $m = -3$ and contains $(0, -2)$.

11. The line contains $(2, 4)$ and $(0, 6)$.

12. The line contains $(-2, 5)$ and $(2, -1)$.

13. The line contains $(0, -1)$ and $(3, 1)$.

14. The line contains $(-2, 0)$ and $(4, 3)$.

15. The line contains $(0, b)$ and $(a, 0)$.

16. The line contains (b, c) and has slope d.

17. The line has intercepts $a = 2$ and $b = -2$.

18. The line has intercepts $a = -3$ and $b = 5$.

19. The line contains $(-1, 5)$ and is parallel to the line $5x + 2y = 10$.

20. The line contains $(0, 3)$ and is parallel to the line $3x + y = 7$.

21. The line contains $(0, -4)$ and is perpendicular to the line $y = \frac{3}{4}x - 5$.

22. The line contains $(2, -3)$ and is perpendicular to the line $2x - 3y = 6$.

23. The line is the perpendicular bisector of the line segment that joins $(3, 5)$ and $(5, -1)$.

24. The line is the perpendicular bisector of the line segment that joins $(-4, 5)$ and $(1, 1)$.

In Exercises 25 and 26, find the equation of the line in the form $y = mx + b$.

25. The line contains (g, h) and is *perpendicular* to the line $y = \frac{a}{b}x + c$.

26. The line contains (g, h) and is *parallel* to the line $y = \frac{a}{b}x + c$.

In Exercises 27 to 32, use graphing to find the point of intersection of the two lines. Use Example 8 as a guide.

27. $y = \frac{1}{2}x - 3$ and $y = \frac{1}{3}x - 2$

28. $y = 2x + 3$ and $y = 3x$

29. $2x + y = 6$ and $3x - y = 19$

30. $\frac{1}{2}x + y = -3$ and $\frac{3}{4}x - y = 8$

31. $4x + 3y = 18$ and $x - 2y = 10$

32. $2x + 3y = 3$ and $3x - 2y = 24$

In Exercises 33 to 38, use algebra to find the point of intersection of the two lines whose equations are provided. Use Example 7 as a guide.

33. $2x + y = 8$ and $3x - y = 7$

34. $2x + 3y = 7$ and $x + 3y = 2$

35. $2x + y = 11$ and $3x + 2y = 16$

36. $x + y = 1$ and $4x - 2y = 1$

37. $2x + 3y = 4$ and $3x - 4y = 23$

38. $5x - 2y = -13$ and $3x + 5y = 17$

In Exercises 39 to 42, use substitution to solve the system. Use Example 9 as a guide.

39. $y = \frac{1}{2}x - 3$ and $y = \frac{1}{3}x - 2$

40. $y = 2x + 3$ and $y = 3x$

41. $y = a$ and $y = bx + c$

42. $x = d$ and $y = fx + g$

43. For $\triangle ABC$, the vertices are $A(0, 0)$, $B(a, 0)$, and $C(b, c)$. In terms of a, b, and c, find the coordinates of the orthocenter of $\triangle ABC$. (The orthocenter is the point of concurrence for the altitudes of a triangle.)

44. For isosceles $\triangle PNQ$, the vertices are $P(-2a, 0)$, $N(2a, 0)$, and $Q(0, 2b)$. In terms of a and b, find the coordinates of the circumcenter of $\triangle PNQ$. (The circumcenter is the point of concurrence for the perpendicular bisectors of the sides of a triangle.)

In Exercises 45 and 46, complete an analytic proof for each theorem.

45. The altitudes of a triangle are concurrent.

46. The perpendicular bisectors of the sides of a triangle are concurrent.

47. Describe the steps of the procedure that enables us to find the distance from a point $P(a, b)$ to the line $Ax + By = C$.

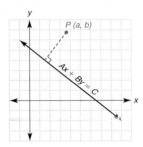

*48. Where C is a real number, the lines $3x + 2y = C$, $2x + y = 5$, and $x - y = 4$ are concurrent. Determine the value of C.

*49. Where a and b are real numbers, $(a + b, a - 2b)$ is the point of intersection for the lines $3x - 2y = 19$ and $2x - 5y = 9$. Find the values of a and b.

10.6 The Three-Dimensional Coordinate System

KEY CONCEPTS Cartesian Space   Distance Formula   Equation of a Sphere
Direction Vector   Midpoint Formula
Equation of a Line   Equation of a Plane

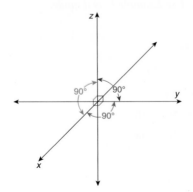

Figure 10.42

In three-dimensional space (the real world in which we live), an object can be located by its latitude, longitude, and altitude. In mathematics, we can extend the coordinate system to include three axes; in this extension, the third axis (the z axis) is perpendicular to the xy plane (the Cartesian plane) at the point that is the common origin of all three number lines (axes). In Figure 10.42, the three axes are mutually perpendicular, meaning that any two axes are perpendicular. The three-dimensional coordinate system is known as *Cartesian space*; at times, the system will be called the *xyz* system.

Figure 10.43

POINTS

Points in the three-dimensional Cartesian system are characterized by ordered triples of the form (x, y, z). While the *origin* of this system is the point $(0, 0, 0)$, the point $(4, 5, 3)$ has the *x coordinate* 4, the *y coordinate* 5, and the *z coordinate* 3. To plot the point $(4, 5, 3)$, we need to know the orientation of the axes. In space, the *x* axis moves forward and back, the *y* axis moves right and left, and the *z* axis moves up and down. Table 10.2 indicates the positive and negative directions of the three axes. As shown in Figure 10.43, we use the same scale on each of the three axes.

TABLE 10.4			
Point Location			
	x axis	*y* axis	*z* axis
Positive direction	Forward	Right	Upward
Negative direction	Back	Left	Down

EXAMPLE 1

Plot each point in the three-dimensional coordinate system:

a) $A(4, 5, 3)$ **b)** $B(5, -6, -3)$

SOLUTION

a) Beginning at the origin, the point is located by moving 4 units forward, 5 units to the right, and up 3 units.
b) Beginning at the origin, this located by moving 5 units forward, 6 units to the left, and down 3 units.

(a)

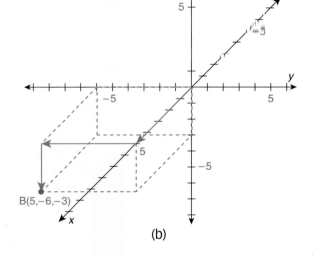

(b)

SSG **EXS. 1, 2** Figure 10.44

LINES

A line is determined by exactly two points in any coordinate system. Because the slope concept of the two-dimensional Cartesian system is a ratio of two numbers, there is no slope concept for a line in three dimensions. In Cartesian space, we give direction to a line by using a *direction vector*. A direction vector of the form (a, b, c) provides changes in x, y, and z (respectively) as we trace movement along the line from one point to another point.

DEFINITION

The line through the points $P_1(x_1, y_1, z_1)$ and $P_2(x_2, y_2, z_2)$ has the **direction vector** $(x_2 - x_1, y_2 - y_1, z_2 - z_1)$.

EXAMPLE 2

Find a direction vector for the line through $(4, 5, 3)$ and $(2, -3, -1)$.

SOLUTION Using the definition above, a direction vector is $(2 - 4, (-3) - 5, (-1) - 3)$ or $(-2, -8, -4)$. Another choice of direction vector is $(2, 8, 4)$, found by negating the signs of the entries.

NOTE: Any nonzero multiple of a direction vector of the line is also a direction vector. For instance, multiplying the direction vectors named above by $\frac{1}{2}$ leads to $(-1, -4, -2)$ and $(1, 4, 2)$ as direction vectors for the line.

The equation for a line in three dimensions is actually a sum determined by a fixed point on the line and any nonzero multiple of the chosen direction vector. Before we consider the definition of the vector form of a line, consider the following operations.

DEFINITION

The real number **multiple** of a vector is $n(a, b, c) = (na, nb, nc)$, where n is any real number. Also, the **sum** of two vectors (points) is $(a, b, c) + (d, e, f) = (a + d, b + e, c + f)$.

From the definition above, $-2(3, -4, 7) = (-6, 8, -14)$ determines a multiple while $(3, -4, 7) + (1, 2, -3) = (4, -2, 4)$ provides a sum.

DEFINITION

Where (x, y, z) represents a point on the line, the **vector form** of the line through $P_1(x_1, y_1, z_1)$ and $P_2(x_2, y_2, z_2)$ is the equation

$$(x, y, z) = (x_1, y_1, z_1) + n(a, b, c);$$

n is any real number and $(a, b, c) = (x_2 - x_1, y_2 - y_1, z_2 - z_1)$ is the direction vector.

The *vector form* of the line, $(x, y, z) = (x_1, y_1, z_1) + n(a, b, c)$ can be simplified and written in the form $(x, y, z) = (x_1 + na, y_1 + nb, z_1 + nc)$.

EXAMPLE 3

Find an equation for the line containing the points $(4, 5, 3)$ and $(2, -3, -1)$.

SOLUTION: From Example 2, a direction vector is $(1, 4, 2)$; thus, one equation for the line is $(x, y, z) = (4, 5, 3) + n(1, 4, 2)$. In Figure 10.45, the point $(4, 5, 3)$ is located by arrows shown in blue while $(2, -3, -1)$ is located by arrows shown in red and the line containing these points in green.

NOTE: For Example 3, each solution names a known point on the line and a direction vector. Where r is any real number, another correct form of the solution is $(x, y, z) = (2, -3, -1) + r(-1, -4, -2)$.

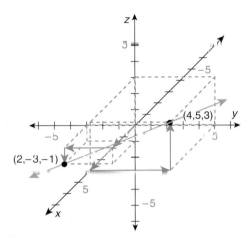

SSG   **EXS. 3–5**   Figure 10.45

THE DISTANCE FORMULA

Some parallels between the two-dimensional coordinate system and the three-dimensional coordinate system are found in the Distance Formula and the Midpoint Formula. We state these formulas in the following theorems. The proof of Theorem 10.6.1 is based upon Exercise 46 of Section 9.1.

THEOREM 10.6.1 ■ The Distance Formula

In the xyz coordinate system, the distance d between the points $P_1(x_1, y_1, z_1)$ and $P_2(x_2, y_2, z_2)$ is given by

$$d = P_1P_2 = \sqrt{(x_2 - x_1)^2 + (y_2 - y_1)^2 + (z_2 - z_1)^2}.$$

EXAMPLE 4

Find the distance d between the points $(5, -7, 2)$ and $(2, 5, 6)$.

SOLUTION When applying the formula from Theorem 10.6.1, we choose $P_1 = (5, -7, 2)$ and $P_2(2, 5, 6)$.

$$d = P_1P_2 = \sqrt{(x_2 - x_1)^2 + (y_2 - y_1)^2 + (z_2 - z_1)^2} \text{ becomes}$$
$$d = \sqrt{(2 - 5)^2 + (5 - [-7])^2 + (6 - 2)^2}$$
$$d = \sqrt{(-3)^2 + 12^2 + 4^2}$$
$$d = \sqrt{9 + 144 + 16} \text{ or } \sqrt{169} \text{ or } 13.$$

That is, $d = P_1P_2 = 13$.

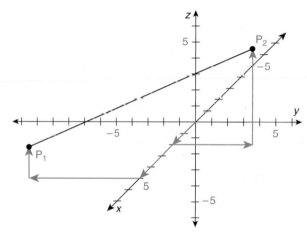

Figure 10.46

THE MIDPOINT FORMULA

Reminder

In the Cartesian plane, the Midpoint Formula is

$$M = \left(\frac{x_1 + x_2}{2}, \frac{y_1 + y_2}{2}\right).$$

The formula found in Theorem 10.6.2 is also merely an extension of the Midpoint Formula of Section 10.1. In the theorem, we characterize the midpoint by the letter M, where coordinates of the designated midpoint are $M = (x_M, y_M, z_M)$.

THEOREM 10.6.2 ■ The Midpoint Formula

In the xyz system, the midpoint of the line segment joining the points $P_1(x_1, y_1, z_1)$ and $P_2(x_2, y_2, z_2)$ is given by

$$M = (x_M, y_M, z_M) = \left(\frac{x_1 + x_2}{2}, \frac{y_1 + y_2}{2}, \frac{z_1 + z_2}{2}\right).$$

SSG   **EXS. 6–9**

EXAMPLE 5

Find midpoint M of $\overline{P_1P_2}$, which joins $P_1(5, -7, 2)$ and $P_2(2, 5, 6)$ as shown in Figure 10.46.

SOLUTION Applying Theorem 10.6.2 with $\overline{P_1P_2}$ of Figure 10.46, we have

$$M = \left(\frac{5 + 2}{2}, \frac{-7 + 5}{2}, \frac{2 + 6}{2}\right) = \left(\frac{7}{2}, \frac{-2}{2}, \frac{8}{2}\right), \text{ so } M = (3.5, -1, 4)$$

NOTE: The Distance Formula could be used to verify that $P_1M = MP_2$.

The proof of Theorem 10.6.2 is best accomplished by using points with coordinates such as $P_1(2a, 2b, 2c)$ and $P_2(2d, 2e, 2f)$. Once we show that the midpoint of $\overline{P_1P_2}$ is the point $M(a + d, b + e, c + f)$, we apply the Distance Formula to show that $P_1M = MP_2$.

PLANES

In space, a plane is determined by three *noncollinear* points, two *intersecting* lines, or two *parallel* lines. In space, we first consider planes that are determined by two intersecting axes. The xy plane that is determined by the intersection of the x and y axes has the equation $z = 0$. Determined by two intersecting axes, there is an xz plane (where $y = 0$) and a yz plane (where $x = 0$). Where a, b, and c are constants, the planes that are most easily described are those of the form $x = a$, $y = b$, or $z = c$. For instance, $z = 2$ is a plane that is parallel to and 2 units above the xy plane ($z = 0$). Likewise, $x = -4$ is a plane that is parallel to the yz plane ($x = 0$) and 4 units behind it.

There are numerous similarities in the two-dimensional and three-dimensional coordinate systems; however, the graph of the equation for the linear equation $Ax + By = C$ in two dimensions is a line, while the graph of the linear equation $Ax + By + Cz = D$ is actually a plane. It is still convenient to graph this plane by using *intercepts*; the x intercept has the form $(a, 0, 0)$ and is the point where the graph intersects the x axis. Similarly, the y intercept has the form $(0, b, 0)$ while the z intercept has the form $(0, 0, c)$.

EXAMPLE 6

Sketch the graph of the equation $x + 2y + 3z = 12$ in the xyz system.

SOLUTION The x intercept is the point for which $y = 0$ and $z = 0$; thus, $(12, 0, 0)$ is the x intercept. Remaining intercepts are $(0, 6, 0)$ and $(0, 0, 4)$. The graph is shown in Figure 10.47.

NOTE: Because $(2, 2, 2)$ is a solution for the equation $x + 2y + 3z = 12$, the point $(2, 2, 2)$ must lie on the plane shown.

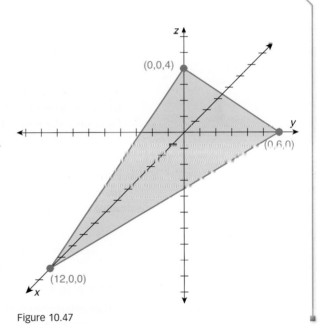

Figure 10.47

SSG **EX. 10**

LINE AND PLANE RELATIONSHIPS

Reminder

Two lines that do not intersect and are not parallel are known as skew lines.

There are four possible relationships between two lines drawn in Cartesian space. First, two lines with different direction vectors will intersect in a point if the x, y, and z coordinates of the *intersecting lines* are respectively equal. A second possibility occurs when two lines neither move in the same direction nor intersect at a point; such lines are called *skew lines*. Third, two lines that have the same direction vectors may be *parallel lines*; for example $\ell_1 = (1, 2, 3) + n(2, 5, -6)$ and $\ell_2 = (5, 2, 7) + r(-2, -5, 6)$ are parallel because their direction vectors are multiples and the lines do *not* have a point in common. Finally, two lines with direction vectors that are multiples coincide if they contain a common point; for instance, $\ell_1 = (1, 2, 3) + n(2, 5, -6)$ and $\ell_2 = (1, 2, 3) + r(-2, -5, 6)$ are known as *coincident lines*. Next, we illustrate a method used to determine whether two lines with different direction vectors intersect or else are skew.

EXAMPLE 7

Decide if $\ell_1 = (1, 2, 3) + n(2, 5, -6)$ and $\ell_2 = (5, 7, -7) + r(1, 5, -4)$ intersect or are skew. If ℓ_1 and ℓ_2 intersect, what is the point of intersection?

SOLUTION If the lines intersect, then there are values of n and r for which a point on the lines is identical. Point forms of the lines are
$\ell_1 = (1 + 2n, 2 + 5n, 3 - 6n)$ and $\ell_2 = (5 + r, 7 + 5r, -7 - 4r)$.
If there is a point of intersection for the two lines, then

$$1 + 2n = 5 + r, \; 2 + 5n = 7 + 5r, \text{ and } 3 - 6n = -7 - 4r.$$

From the first equation, $r = -4 + 2n$. Substituting into the second equation,

$$2 + 5n = 7 + 5(-4 + 2n)$$
or $\qquad 2 + 5n = 7 - 20 + 10n,$
so $\qquad 15 = 5n \text{ and } n = 3.$

If $n = 3$, then $r = 2$ (from $r = -4 + 2n$); in turn, the points on the lines are

$$\ell_1 = (1 + 2n, 2 + 5n, 3 - 6n)$$
$$= (1 + 2 \cdot 3, 2 + 5 \cdot 3, 3 - 6 \cdot 3) \text{ or } (7, 17, -15), \text{ and}$$

$$\ell_2 = (5 + r, 7 + 5r, -7 - 4r)$$
$$= (5 + 2, 7 + 5 \cdot 2, -7 - 4 \cdot 2) \text{ or } (7, 17, -15).$$

Yes, the lines intersect in that the point having the common values for x and y also has a common value for z.
Of course, the point of intersection is $(7, 17, -15)$.

NOTE 1: Equating x and y produced values of n and r that could lead to different z coordinates for the two lines; in that case, the two lines are necessarily skew lines.

NOTE 2: The system of equations in n and r can be solved by the addition-subtraction method. If the first and third equations are used to form a system, then a common y value is sought; using the second and third equations, then a common x value is found.

The four relationships for two lines in Cartesian space are summarized in Table 10.5.

TABLE 10.5

Line Relationships in Cartesian Space

ℓ_1 and ℓ_2	Direction vectors are multiples	Direction vectors are *not* multiples
have a common point	Coincide	Intersect
have no common point	Parallel	Skew

While two planes generally intersect in a line, two planes can be parallel. If two planes are parallel, the coefficients for their equations will be multiples; however, the resulting constants will not be multiples. For example, the planes represented by $x + 2y - 3z = 6$ and $2x + 4y - 6z = -18$ are parallel; the graphs will have different intercepts and yet the same orientation in Cartesian space. When two planes are parallel, they have (or simplify to) equations that have forms such as $Ax + By + Cz = D$ and $Ax + By + Cz = E$, where $D \neq E$.

As with lines, two equations for planes could lead to coincident planes (the same plane). For example, the graphs of $x + 2y - 3z = 6$ and $2x + 4y - 6z = 12$ are the same; notice that the second equation is a *multiple* of the first equation. When one equation is a multiple of the other, as with $x + 2y - 3z = 6$ and $2x + 4y - 6z = 12$, the equations are said to be *equivalent equations*; when graphed, the planes having equivalent equations will have identical x, y, and z intercepts so that the planes coincide.

As we saw in Chapter 7, three or more lines that intersect at a common point are said to be concurrent. Similarly, three or more planes that intersect at a single point are also known as *concurrent planes*.

INTERSECTION OF PLANES

In Cartesian space, finding the line of intersection of two intersecting planes by a visual (geometric) approach is virtually impossible. Furthermore, the algebraic technique used to determine the vector equation of the line of intersection is a real challenge. To be complete, we examine the intersection of two nonparallel planes in Example 8.

EXAMPLE 8

The intersection of the planes $x + 2y + 3z = 12$ and $2x + 3y + z = 18$ is the line given in vector form by $(x, y, z) = (0, 6, 0) + n(7, -5, 1)$.

    a) Name a point on the line of intersection.
    b) State a direction vector for the line of intersection.
    c) Using $n = -1$, name a second point on the line of intersection.
    d) Show that the point named in the solution for part (a) lies in both planes.
    e) Show that the point named in the solution for part (c) lies in both planes.

SOLUTION
    a) By its vector equation form, the line of intersection must contain the point $(0, 6, 0)$.
    b) A direction vector of the line is $(7, -5, 1)$.
    c) With $n = -1$, a second point is $(0, 6, 0) + -1(7, -5, 1)$ or
        $(0, 6, 0) + (-7, 5, -1)$ or $(-7, 11, -1)$.
    d) For $(0, 6, 0)$ to be on both planes, it must be a solution for the equation of each plane:
        $x + 2y + 3z = 12$ becomes $0 + 2(6) + 3(0)$ or 12, so $(0, 6, 0)$ is a solution;
        $2x + 3y + z = 18$ becomes $2(0) + 3(6) + 0$ or 18, so $(0, 6, 0)$ is a solution.
    e) Similarly, $(-7, 11, -1)$ must be a solution for the equation of each plane:
        $x + 2y + 3z = 12$ becomes $-7 + 2(11) + 3(-1)$ or 12, so $(-7, 11, -1)$ is a solution;
        $2x + 3y + z = 18$ becomes $2(-7) + 3(11) + (-1)$ or 18, so $(-7, 11, -1)$ is a solution.
    NOTE: If $(0, 6, 0)$ and $(-7, 11, -1)$ lie on both planes then these two points determine the line of intersection for the two planes.

As we saw earlier in this section, there is an alternative form for the equation of a line in three variables. Found by simplification of a real number multiple and addition, it provides a form that makes it easier to recognize points on the line.

> **DEFINITION**
>
> Where (x, y, z) is any point on the line through the point $P_1(x_1, y_1, z_1)$ and having a direction vector (a, b, c), the *point form* of the line is given by
>
> $$(x, y, z) = (x_1 + na, y_1 + nb, z_1 + nc),$$
>
> where n is any real number.

SSG EXS. 11–13

Using this definition and the line in Example 8, the point form of the line of intersection is $(x, y, z) = (7n, 6 - 5n, n)$. Where $n = -2$, another point that lies on the line of intersection is the point $(x, y, z) = (7[-2], 6 - 5[-2], -2)$ or $(-14, 16, -2)$.

EXAMPLE 9

For the two planes found in Example 8, the line of intersection has the point form $(x, y, z) = (7n, 6 - 5n, n)$. Show that for any n, all points on this line are solutions for the equations $x + 2y + 3z = 12$ and $2x + 3y + z = 18$.

SOLUTION Substitution of $7n$ (for x), $6 - 5n$ (for y), and n (for z) into the equation $x + 2y + 3z = 12$ leads to $7n + 2(6 - 5n) + 3n$ or $7n + 12 - 10n + 3n$, which equals 12; similarly, substitution of $(7n, 6 - 5n, n)$ into the equation $2x + 3y + z = 18$ leads to $2(7n) + 3(6 - 5n) + n$ or $14n + 18 - 15n + n$, which equals 18.

SPHERES

In Cartesian space, the counterpart of the circle is the sphere. To find the equation for a sphere, we apply the Distance Formula. Where (h, k, ℓ) is the center of the sphere of radius length r, the points on the sphere (x, y, z) must lie at distance r from the center. This relationship leads to the equation $\sqrt{(x - h)^2 + (y - k)^2 + (z - \ell)^2} = r$. Squaring each side of the equation, we have the following theorem.

> **THEOREM 10.6.3**
>
> The equation for the sphere with center (h, k, ℓ) and radius length r is given by the equation $(x - h)^2 + (y - k)^2 + (z - \ell)^2 = r^2$.

The *general form* for the equation of the sphere can be found by expanding the equation found in Theorem 10.6.3; thus, the general form for the equation of the sphere is $x^2 + y^2 + z^2 + Dx + Ey + Fz + G = 0$.

The following corollary shows the equation for the sphere with center at the origin.

> **COROLLARY 10.6.4**
>
> The equation for the sphere with center $(0, 0, 0)$ and radius length r is given by the equation $x^2 + y^2 + z^2 = r^2$.

EXAMPLE 10

Find an equation for the sphere with

a) center at the origin and radius length $r = 5$.

b) center $(2, -3, 4)$ and radius length $r = 4$.

SOLUTION

a) Using Corollary 10.6.4, $x^2 + y^2 + z^2 = 5^2$, so $x^2 + y^2 + z^2 = 25$.

b) Substituting into the equation in Theorem 10.6.3,

$$(x - h)^2 + (y - k)^2 + (z - \ell)^2 = r^2, \text{ we have}$$
$$(x - 2)^2 + (y - [-3])^2 + (z - 4)^2 = 4^2 \text{ or}$$
$$(x - 2)^2 + (y + 3)^2 + (z - 4)^2 = 16.$$

Expanding the equation,

$$x^2 - 4x + 4 + y^2 + 6y + 9 + z^2 - 8z + 16 = 16.$$

In general form, the equation of the sphere is

SSG **EXS. 14–17**

$$x^2 + y^2 + z^2 - 4x + 6y - 8z + 13 = 0.$$

The graph of the equation for the sphere described in Example 10(a) is shown in Figure 10.48. The sphere has symmetry about its center, each axis (as well as any line that contains the center of the sphere), and with respect to planes such as the *xy* plane, the *xz* plane, and the *yz* plane (as well as any plane that contains the center of the sphere).

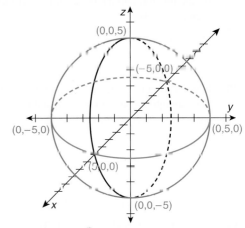

Figure 10.48

Exercises 10.6

1. In the Cartesian system *below*, name the ordered triple (x, y, z) represented by point A. Also, plot the point $B(5, 6, 4)$.

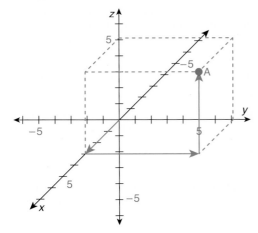

Exercise 1

2. In the Cartesian system below, name the ordered triple (x, y, z) represented by point C. Also, plot the point $D(5, -4, 6)$.

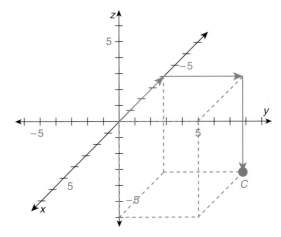

Exercise 2

3. Find a direction vector for the line containing the origin $(0, 0, 0)$ and the point $(2, 3, -1)$.

4. Find a direction vector for the line containing the points $(-1, 5, 2)$ and the point $(2, 3, -1)$.

5. For the line ℓ: $(x, y, z) = (2, 3, 4) + n(3, -2, 5)$, find
   a) a point of the line.
   b) a direction vector for the line.

6. For the line ℓ: $(x, y, z) = (5, -3, 2) + n(1, 2, -2)$, find
   a) a point of the line.
   b) a direction vector for the line.

7. Line m is expressed in point form $(x, y, z) = (2 + 3n, 4 - 5n, 5 + 2n)$. Find
   a) a point of the line.
   b) a direction vector for the line.

8. In vector form (as in Exercises 5 and 6), find an equation for the line through the point $(4, 1, -3)$ and with direction vector $(1, 2, 3)$.

9. In vector form (as in Exercises 5 and 6), find an equation for the line through the point $(4, -3, 7)$ and with direction vector $(4, 2, -3)$.

10. In point form (as in Exercise 7), find an equation for the line through the point $(4, 1, -3)$ and parallel to a line that has the direction vector $(-1, 2, 5)$.

In Exercises 11 to 14, find the distance between the two points P_1 and P_2.

11. $P_1 = (0, 0, 0)$ and $P_2 = (1, 2, 4)$

12. $P_1 = (-1, 2, 3)$ and $P_2 = (2, -2, 9)$

13. $P_1 = (1, -1, 1)$ and $P_2 = (2, 1, 3)$

14. $P_1 = (1, 0, 2)$ and $P_2 = (3, 4, -1)$

In Exercises 15 to 18, find the midpoint of the line segment $\overline{P_1P_2}$.

15. $P_1 = (0, 0, 0)$ and $P_2 = (-6, 4, 14)$

16. $P_1 = (-1, 2, 3)$ and $P_2 = (3, 6, 6)$

17. $P_1 = (2, -1, 1)$ and $P_2 = (5, 7, -7)$

18. $P_1 = (1, 0, 2)$ and $P_2 = (0, 5, 9)$

In Exercises 19 and 20, use the x, y, and z intercepts to sketch the plane for each equation.

19. $x + 2y + z = 6$

20. $2x - y + 4z = 8$

21. Which point(s) lie in the plane $2x + 3y + 4z = 24$?
    a) $(0, 0, 6)$ b) $(0, 3, 4)$
    c) $(4, 4, 1)$ d) $(-2, 6, 3)$

22. Which point(s) lie in the plane $2x + y - z = 10$?
    a) $(0, 0, 10)$ b) $(5, 3, -3)$
    c) $(2, 4, -2)$ d) $(-3, 6, -10)$

In Exercises 23 to 26, find an equation for each sphere.

23. With center $(0, 0, 0)$ and radius length $r = 5$.

24. With center $(0, 0, 0)$ and containing the point $(3, 12, -5)$.

25. With center $(1, 2, 3)$ and containing the point $(5, 5, 3)$.

26. With center $(-1, 2, 4)$ and radius length $r = 7$.

27. Which point(s) lie on the line $(x, y, z) = (2, -1, 5) + n(1, 2, -1)$?
    a) $(1, -3, 4)$ b) $(5, 5, 2)$

28. Which point(s) lie on the line $(x, y, z) = (2, 1, -3) + r(3, 2, -4)$?
    a) $(2, 1, -3)$ b) $(8, 5, -11)$

29. Explain why the lines below are parallel.
    ℓ_1: $(x, y, z) = (2, 3, 4) + n(1, 2, -3)$ and
    ℓ_2: $(x, y, z) = (3, 7, -2) + r(-1, -2, 3)$

30. Explain why the lines below intersect.
    ℓ_1: $(x, y, z) = (2, 3, 4) + n(1, 2, -3)$ and
    ℓ_2: $(x, y, z) = (2, 3, 4) + r(2, 3, 5)$

31. Explain why the lines below are coincident.
    ℓ_1: $(x, y, z) = (0, 0, 0) + n(1, 2, -3)$ and
    ℓ_2: $(x, y, z) = (1, 2, -3) + r(-1, -2, 3)$

32. Explain why the lines below are skew.
    ℓ_1: $(x, y, z) = (-2, -2, -2) + n(1, 2, 3)$ and
    ℓ_2: $(x, y, z) = (1, 1, 1) + r(1, -3, 5)$

For Exercises 33 and 34, apply the following theorem (stated without proof).

"Two intersecting lines are perpendicular if their direction vectors $v_1 = (a, b, c)$ and $v_2 = (d, e, f)$ satisfy the condition that $ad + be + cf = 0$."

33. Are these lines perpendicular?
    ℓ_1: $(x, y, z) = (2, 3, 4) + n(1, 1, 2)$ and
    ℓ_2: $(x, y, z) = (2, 3, 4) + r(-2, -4, 3)$

34. Are these lines perpendicular?
    ℓ_1: $(x, y, z) = (1, -2, 5) + n(4, 1, -3)$ and
    ℓ_2: $(x, y, z) = (1, -2, 5) + r(-3, 6, -2)$

35. The planes $x + 2y + 3z = 12$ and $2x + 3y + z = 18$ intersect in the line ℓ whose equation is $(x, y, z) = (0, 6, 0) + n(7, -5, 1)$. Find the point in both planes for which:
    a) $x = 7$ b) $y = 16$

36. The planes $x + 2y + z = 9$ and $3x + y - 2z = 2$ intersect in the line ℓ whose equation is $(x, y, z) = (-1, 5, 0) + n(1, -1, 1)$. Find the point in both planes for which:
    a) $x = 7$ b) $y = 16$

37. The line $(x, y, z) = (3, 4, 5) + n(3, 4, -5)$ intersects the sphere $x^2 + y^2 + z^2 = 100$ in two points. Find each point.

38. The line $(x, y, z) = (0, 0, 0) + n(1, 1, 2)$ intersects the sphere $x^2 + y^2 + z^2 = 54$ in two points. Find each point.

39. Find the surface area and volume of the sphere $x^2 + y^2 + z^2 = 100$.

40. For the spheres $(x - 1)^2 + (y + 2)^2 + (z - 4)^2 = 36$ and $x^2 + y^2 + z^2 = 64$, find the ratio of their
    a) surface areas.
    b) volumes.

41. Does the sphere $x^2 + y^2 + z^2 = 100$ have symmetry with respect to the
    a) x axis?
    b) xy plane?

42. Does the sphere $x^2 + y^2 + z^2 = 100$ have symmetry with respect to
    a) the line through the points $(0, 0, 0)$ and $(0, 5, 5\sqrt{5})$?
    b) the plane with the equation $y = 5$?

*43. Lines $\ell_1: (x, y, z) = (2, 3, -1) + n(1, 1, 2)$ and $\ell_2: (x, y, z) = (7, 7, 2) + r(-2, -1, 3)$ intersect at point P. Find the coordinates (x, y, z) of point P.

*44. Lines $\ell_1: (x, y, z) = (2, 0, 3) + n(2, -3, 5)$ and $\ell_2: (x, y, z) = (4, 1, -4) + r(-1, 2, -4)$ intersect at point P. Find the coordinates (x, y, z) of point P.

*45. The planes with the equations $x + 2z = 12$ and $y - 3z = 6$ intersect in a line. Find the equation for the line in the form $(x, y, z) = (x_1, y_1, z_1) + n(a, b, c)$.

*46. The planes with the equations $x + 3z = 12$ and $y - 5z = 10$ intersect in a line. Find the equation for the line in the form $(x, y, z) = (x_1, y_1, z_1) + n(a, b, c)$.

*47. Does the sphere $(x - 1)^2 + (y - 2)^2 + (z + 5)^2 = 49$ have symmetry with respect to
    a) the point $P = (1, 2, -5)$?
    b) the line $\ell: (x, y, z) = (-5, 4, -13) + n(3, -1, 4)$?

*48. Does the sphere described in Exercise 47 have symmetry with respect to the plane whose equation is $3x - 4y + 5z = -30$?

*49. Determine the point of intersection, if such a point exists, for the line $\ell: (x, y, z) = (-1, -6, 5) + n(1, 3, -2)$ and the plane $2x + 3y + z = 48$.

*50. Determine the point of intersection, if such a point exists, for the line $\ell: (x, y, z) = (3, -1, 7) + r(-5, 2, 1)$ and the plane $2x + 3y + 4z = 24$.

PERSPECTIVE ON HISTORY

THE BANACH-TARSKI PARADOX

In the 1920s, two Polish mathematicians proposed a mathematical dilemma to their colleagues. Known as the Banach-Tarski paradox, their proposal has puzzled students of geometry for decades. What was most baffling was that the proposal suggested that matter could be created through rearrangement of the pieces of a figure! The following steps outline the Banach-Tarski paradox.

First consider the square whose sides are each of length 8. [See Figure 10.49(a).] By counting squares or by applying a formula, it is clear that the 8-by-8 square must have an area of 64 square units. We now subdivide the square (as shown) to form two right triangles and two trapezoids. Note the dimensions indicated on each piece of the square in Figure 10.49(b).

(a)

(b)

Figure 10.50

The puzzle is real, but you may also sense that something is wrong. This paradox can be explained by considering the slopes of lines. The triangles, which have legs of lengths 3 and 8, determine a hypotenuse whose slope is $-\frac{3}{8}$. Although the side of the trapezoid appears to be collinear with the hypotenuse, it actually has a slope of $-\frac{2}{5}$. It was easy to accept that the segments were collinear because the slopes are nearly equal; in fact, $-\frac{3}{8} = -0.375$ and $\frac{2}{5} = -0.400$. In Figure 10.51 (which is somewhat exaggerated), a very thin parallelogram appears in the space between the original segments of the cut-up square. One may quickly conclude that the area of that parallelogram is 1 square unit, and the paradox has been resolved once more!

(a)

(b)

Figure 10.49

The parts of the square are now rearranged to form a rectangle [see Figure 10.50(a)] whose dimensions are 13 and 5. The rectangle redrawn in Figure 10.50(b) clearly has an area that measures 65 square units, 1 square unit more than the given square! How is it possible that the second figure has an area greater than that of the first?

Figure 10.51

PERSPECTIVE ON APPLICATIONS

THE POINT-OF-DIVISION FORMULAS

The subject of this feature is a generalization of the formulas that led to the Midpoint Formula. Recall that the midpoint of the line segment that joins $A(x_1, y_1)$ to $B(x_2, y_2)$ is given by $M = \left(\frac{x_1 + x_2}{2}, \frac{y_1 + y_2}{2}\right)$, which is derived from the formulas $x = x_1 + \frac{1}{2}(x_2 - x_1)$ and $y = y_1 + \frac{1}{2}(y_2 - y_1)$. The formulas for a more general location of a point between A and B follow; to better understand how these formulas can be applied, we note that r represents the fractional part of the distance from point A toward point B on $\overline{AB}$; in the Midpoint Formula, $r = \frac{1}{2}$.

Point-of-Division Formulas: Let $A(x_1, y_1)$ and $B(x_2, y_2)$ represent the endpoints of $\overline{AB}$. Where r represents a common fraction ($0 < r < 1$), the coordinates of the point P that lies part r of the distance from A to B are given by

$$x = x_1 + r(x_2 - x_1) \quad \text{and} \quad y = y_1 + r(y_2 - y_1)$$

Table 10.6 clarifies the use of the formulas above.

TABLE 10.6

Value of r	Location of Point P on $\overline{AB}$
$\frac{1}{3}$	Point P lies $\frac{1}{3}$ of the distance from A to B.
$\frac{3}{4}$	Point P lies $\frac{3}{4}$ of the distance from A to B.

EXAMPLE 1

Find the point P on $\overline{AB}$ that is one-third of the distance from $A(-1, 2)$ to $B(8, 5)$.

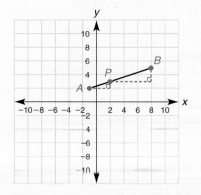

SOLUTION

$$x = x_1 + r(x_2 - x_1) \quad \text{and} \quad y = y_1 + r(y_2 - y_1)$$
$$x = -1 + \tfrac{1}{3}(8 - [-1]) \quad \text{and} \quad y = 2 + \tfrac{1}{3}(5 - 2)$$

Then $\quad x = -1 + \tfrac{1}{3}(9) \quad$ so $\quad x = -1 + 3$ or 2

Also, $\quad y = 2 + \tfrac{1}{3}(3) \quad$ so $\quad y = 2 + 1$ or 3

The desired point is $P(2, 3)$.

NOTE: See the figure above, in which similar triangles can be used to explain why point P is the desired point.

In some higher-level courses, the value of r is not restricted to values between 0 and 1. For instance, we could choose $r = 2$ or $r = -1$. For such values of r, the point P produced by the Point of-Division Formulas remains collinear with A and B; however, the point P that is produced does not lie between A and B.

The Point-of-Division Formulas extend to Cartesian space. That is, the point (x, y, z) that is part r of the distance from $A(x_1, y_1, z_1)$ to $B(x_2, y_2, z_2)$ is given by

$$x = x_1 + r(x_2 - x_1),$$
$$y = y_1 + r(y_2 - y_1), \text{ and}$$
$$z = z_1 + r(z_2 - z_1).$$

Summary

A Look Back at Chapter 10

Our goal in this chapter was to relate algebra and geometry. This relationship is called *analytic geometry* or *coordinate geometry*. Formulas for the distance between points, the midpoint of a line segment, and the slope of a line were developed. We found the equation for a line and used it for graphing. Analytic proofs were provided for a number of theorems of geometry. In the final section, we extended the Midpoint Formula and the Distance Formula to three dimensions; also, we considered the equations of lines, planes, and spheres.

A Look Ahead to Chapter 11

In the next chapter, we will again deal with the right triangle. Three trigonometric ratios (sine, cosine, and tangent) will be defined for an acute angle of the right triangle in terms of its sides. An area formula for triangles will be derived using the sine ratio. We will also prove the Law of Sines and the Law of Cosines for acute triangles

Key Concepts

10.1

Analytic Geometry • Cartesian (Rectangular) Coordinate System • Cartesian Plane • *x* Axis • *y* Axis • Quadrants

• Origin • *x* Coordinate • *y* Coordinate • Ordered Pair • Distance Formula • Linear Equation • Midpoint Formula

10.2

Graphs of Equations • *x* Intercept • *y* Intercept • Slope • Slope Formula • Negative Reciprocal

10.3

Formulas and Relationships • Placement of Figure

10.4

Analytic Proof • Synthetic Proof

10.5

Slope-Intercept Form of a Line • Point-Slope Form of a Line • Systems of Equations

10.6

Cartesian Space • Direction Vector • Equation of a Line • Distance Formula • Midpoint Formula • Equation of a Plane • Equation of a Sphere

Chapter 10 Review Exercises

1. Find the distance between each pair of points:
   a) $(6, 4)$ and $(6, -3)$ c) $(-5, 2)$ and $(7, -3)$
   b) $(1, 4)$ and $(-5, 4)$ d) $(x - 3, y + 2)$ and $(x, y - 2)$

2. Find the distance between each pair of points:
   a) $(2, -3)$ and $(2, 5)$ c) $(-4, 1)$ and $(4, 5)$
   b) $(3, -2)$ and $(-7, -2)$ d) $(x - 2, y - 3)$ and $(x + 4, y + 5)$

3. Find the midpoint of the line segment that joins each pair of points in Exercise 1.

4. Find the midpoint of the line segment that joins each pair of points in Exercise 2.

5. Find the slope of the line containing each pair of points in Exercise 1.

6. Find the slope of the line containing each pair of points in Exercise 2.

7. $(2, 1)$ is the midpoint of $\overline{AB}$, in which A has coordinates $(8, 10)$. Find the coordinates of B.

8. The y axis is the perpendicular bisector of $\overline{RS}$. Find the coordinates of R if S is the point $(-3, 7)$.

9. If A has coordinates $(2, 1)$ and B has coordinates $(x, 3)$, find x so that the slope of $\overleftrightarrow{AB}$ is -3.

10. If R has coordinates $(-5, 2)$ and S has coordinates $(2, y)$, find y so that the slope of $\overleftrightarrow{RS}$ is $\frac{-6}{7}$.

11. Without graphing, determine whether the pairs of lines are parallel, perpendicular, the same, or none of these:
    a) $x + 3y = 6$ and $3x - y = -7$
    b) $2x - y = -3$ and $y = 2x - 14$
    c) $y + 2 = -3(x - 5)$ and $2y = 6x + 11$
    d) $0.5x + y = 0$ and $2x - y = 10$

12. Determine whether the points $(-6, 5)$, $(1, 7)$, and $(16, 10)$ are collinear.

13. Find x so that $(-2, 3)$, $(x, 6)$, and $(8, 8)$ are collinear.

14. Draw the graph of $3x + 7y = 21$, and name the x intercept a and the y intercept b.

15. Draw the graph of $4x - 3y = 9$ by first changing the equation to Slope-Intercept Form.

16. Draw the graph of $y + 2 = \frac{-2}{3}(x - 1)$.

17. Write an equation for
   a) the line through $(2, 3)$ and $(-3, 6)$.
   b) the line through $(-2, -1)$ and parallel to the line through $(6, -3)$ and $(8, -9)$.
   c) the line through $(3, -2)$ and perpendicular to the line $x + 2y = 4$.
   d) the line through $(-3, 5)$ and parallel to the x axis.

18. Show that the triangle whose vertices are $A(-2, -3)$, $B(4, 5)$, and $C(-4, 1)$ is a right triangle.

19. Show that the triangle whose vertices are $A(3, 6)$, $B(-6, 4)$, and $C(1, -2)$ is an isosceles triangle.

20. Show that quadrilateral $RSTV$ with vertices $R(-5, -3)$, $S(1, -11)$, $T(7, -6)$, and $V(1, 2)$ is a parallelogram.

In Exercises 21 and 22, find the intersection of the graphs of the two equations by graphing.

21. $4x - 3y = -3$ **22.** $y = x + 3$
       $x + 2y = 13$ $y = 4x$

In Exercises 23 and 24, solve the systems of equations in Exercises 21 and 22 by using algebraic methods.

23. Refer to Exercise 21.

24. Refer to Exercise 22.

25. Three of the four vertices of a parallelogram are $(0, -2)$, $(6, 8)$, and $(10, 1)$. Find the possibilities for the coordinates of the remaining vertex.

26. $A(3, 1)$, $B(5, 9)$, and $C(11, 3)$ are the vertices of $\triangle ABC$.
   a) Find the length of the median from B to $\overline{AC}$.
   b) Find the slope of the altitude from B to $\overline{AC}$.
   c) Find the slope of a line through B parallel to $\overline{AC}$.

In Exercises 27 to 30, supply the missing coordinates for the vertices, using as few variables as possible.

27.

Isosceles $\triangle ABC$
with base $\overline{AC}$

28.

Rectangle $DEFG$ with
$DG = 2 \cdot DE$

29.

Isosceles trapezoid $RSTU$
with $\overline{RV} \cong \overline{RU}$

30.

Parallelogram $MPQN$

31. $A(2a, 2b)$, $B(2c, 2d)$, and $C(0, 2e)$ are the vertices of $\triangle ABC$.
   a) Find the length of the median from C to $\overline{AB}$.
   b) Find the slope of the altitude from B to $\overline{AC}$.
   c) Find the equation of the altitude from B to $\overline{AC}$.

Prove the statements in Exercises 32 to 36 by using analytic geometry.

32. The line segments that join the midpoints of consecutive sides of a parallelogram form another parallelogram.

33. If the diagonals of a rectangle are perpendicular, then the rectangle is a square.

34. If the diagonals of a trapezoid are equal in length, then the trapezoid is an isosceles trapezoid.

35. If two medians of a triangle are equal in length, then the triangle is isosceles.

36. The line segments joining the midpoints of consecutive sides of an isosceles trapezoid form a rhombus.

37. Determine whether $\triangle ABC$, with vertices $A(0, 0, 0)$, $B(1, 2, 4)$, and $C(0, 0, 8)$, is an isosceles triangle.

38. For the line that contains the points $P_1(-1, 2, 4)$ and $P_2(3, 4, 7)$, find
   a) a direction vector.
   b) an equation for the line.

39. Consider the line that contains the point $(2, -3, 5)$ and that has the direction vector $(1, 2, 4)$.
   a) Write the point form of the line.
   b) Determine the point at which the line in part (a) intersects the plane $2x - y + 5z = -8$.

40. For the sphere with equation $(x - 1)^2 + (y + 2)^2 + (z + 4)^2 = 36$, find
   a) the center.
   b) the length of the radius.

Chapter 10 Test

1. In the coordinate system provided, give the coordinates of
   a) point A in the form (x, y). _____
   b) point B in the form (x, y). _____

2. In the coordinate system for Exercise 1, plot and label each point: $C(-6, 1)$ and $D(0, 9)$

   Exercises 1–4

3. Use $d = \sqrt{(x_2 - x_1)^2 + (y_2 - y_1)^2}$ to find the length of $\overline{CD}$ as described in Exercise 2. _____

4. In the form (x, y), determine the midpoint of $\overline{CD}$ as described in Exercise 2. _____

5. Complete the following table of x and y coordinates of points on the graph of the equation $2x + 3y = 12$.

x	0	3		9
y		4		

6. Using the table from Exercise 5, sketch the graph of $2x + 3y = 12$.

   Exercises 5–6

7. Find the slope m of a line containing these points:
   a) $(-1, 3)$ and $(2, -6)$ _____
   b) (a, b) and (c, d) _____

8. Line ℓ has slope $m = \frac{1}{3}$. Find the slope of any line that is
   a) parallel to ℓ. _____
   b) perpendicular to ℓ. _____

9. What type of quadrilateral $ABCD$ is represented if its vertices are $A(0, 0)$, $B(a, 0)$, $C(a + b, c)$, and $D(b, c)$? _____

10. For quadrilateral $ABCD$ of Exercise 9 to be a rhombus, it would be necessary that $AB = AD$. Using a, b, and c (as in Exercise 9), write the equation stating that $AB = AD$.

    _____

11. Being as specific as possible, describe the type of polygon shown in each figure.

   a) _____   b) _____

12. What formula (by name) is used to establish that
    a) two lines are parallel? _____
    b) two line segments are congruent? _____

13. Using as few variables as possible, state the coordinates of each point if $\triangle DEF$ is isosceles with $\overline{DF} \cong \overline{FE}$.

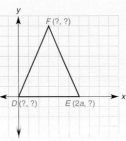

    DEF is an isosceles triangle with $\overline{DF} \cong \overline{FE}$

    $D(\underline{}, \underline{})$, $E(2a, \underline{})$, $F(\underline{}, \underline{})$.

14. For proving the theorem "The midpoint of the hypotenuse of a right triangle is equidistant from all three vertices," which drawing is best? _____

15. In the figure, we see that $m_{\overline{RS}} = m_{\overline{VT}} = 0$. Find the equation that relates r, s, and t if it is known that $RSTV$ is a parallelogram. _____

16. In the form $y = mx + b$, find the equation of the line that
    a) contains the points $(0, 4)$ and $(2, 6)$. _____
    b) contains $(0, -3)$ and is parallel to the line $y = \frac{3}{4}x - 5$.

    _____

17. Use $y - y_1 = m(x - x_1)$ to find the equation of the line that contains (a, b) and is perpendicular to the line $y = -\frac{1}{c}x + d$. Leave the answer (equation) in the form $y = mx + b$. _____

18. Use the graphs provided to solve the system consisting of the equations $x + 2y = 6$ and $2x - y = 7$. _____

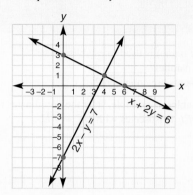

19. Use algebra to solve the system consisting of the equations $5x - 2y = -13$ and $3x + 5y = 17$.

_____

20. Use the drawing provided to complete the proof of the theorem "The line segment that joins the midpoints of two sides of a triangle is parallel to the third side of the triangle."

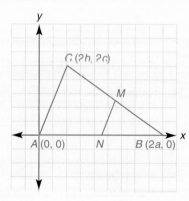

Proof: Given $\triangle ABC$ with vertices as shown, let M and N name the midpoints of sides $\overline{CB}$ and $\overline{AB}$, respectively. Then

_____

_____

_____

_____

21. In Cartesian space, line segment $\overline{P_1P_2}$ has endpoints $P_1(-2, 3, 5)$ and $P_2(2, -1, -7)$.
   a) Find midpoint M of $\overline{P_1P_2}$. _____
   b) Find the length d of $\overline{P_1P_2}$. _____

22. In Cartesian space, the point form of a line is
$\ell: (x, y, z) = (2 + r, -3 + 2r, 5 - 5r)$.
   a) Find the point on the line for which $r = -3$. _____
   b) Does the point $(6, 5, -12)$ lie on line ℓ? _____

23. Does the sphere $x^2 + y^2 + z^2 = 25$ have symmetry with respect to
   a) the xy plane? _____
   b) the line $\ell: (x, y, z) = (-1, 2, 3) + n(1, -2, -3)$?

   _____

Chapter 11

Introduction to Trigonometry

Surreal! The Pontusval Lighthouse is located on the rugged shoreline of the Bretagne (Brittany) peninsula in northwest France. As with any lighthouse, it sends a "Welcome" message as well as a "Caution" message to the people on board an approaching vessel. Methods of trigonometry enable the ship captain to determine the distance from his or her ship to the rocky shoreline beneath the lighthouse. The word *trigonometry*, which means "the measure of a triangle," provides methods for the measurement of parts (sides and angles) of a triangle. Found in Chapter 11 are some techniques that enable you to find measures of one part of a right triangle when the measures of other parts are known. These methods can be expanded to include techniques for measuring parts of acute triangles as well.

For the applications of this chapter, you will find it necessary to use a scientific or graphing calculator.

Additional video explanations of concepts, sample problems, and applications are available on DVD.

11.1 | The Sine Ratio and Applications

KEY CONCEPTS

Greek Letters: α, β, γ, θ
Opposite Side (Leg)
Hypotenuse

Sine Ratio:
$\sin \theta = \frac{\text{opposite}}{\text{hypotenuse}}$

Angle of Elevation
Angle of Depression

In this section, we deal exclusively with similar right triangles. In Figure 11.1, $\triangle ABC \sim \triangle DEF$ and $\angle C$ and $\angle F$ are right angles. Consider corresponding angles A and D; if we compare the length of the side opposite each angle to the length of the hypotenuse in each triangle, we obtain this result by the reason CSSTP:

$$\frac{BC}{AB} = \frac{EF}{DE} \qquad \text{or} \qquad \frac{3}{5} = \frac{6}{10}$$

In the two similar right triangles, the ratio of each pair of corresponding sides depends on the measure of acute $\angle A$ (or $\angle D$, because $m\angle A = m\angle D$); for each angle, the numerical value of the ratio

$$\frac{\text{length of side opposite the acute angle}}{\text{length of hypotenuse}}$$

is unique. This ratio becomes smaller for smaller measures of $\angle A$ and larger for larger measures of $\angle A$. This ratio is unique for each measure of an acute angle even though the lengths of the sides of the two similar right triangles containing the angle are different.

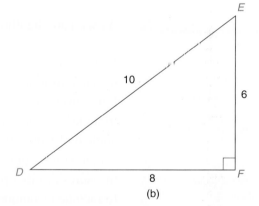

Figure 11.1

In Figure 11.2, we name the measures of the angles of the right triangle by the Greek letters α (alpha) at vertex A, β (beta) at vertex B, and γ (gamma) at vertex C. The lengths of the sides opposite vertices A, B, and C are a, b, and c, respectively. Relative to the acute angle, the lengths of the sides of the right triangle in the following definition are described as "opposite" and "hypotenuse." The word **opposite** is used to mean the length of the side opposite the angle named; the word **hypotenuse** is used to mean the length of the hypotenuse.

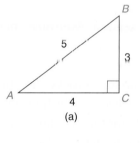

Figure 11.2

> **Geometry in the Real World**
>
> A surveyor uses trigonometry to find both angle measurements and distances.

DEFINITION

In a right triangle, the **sine ratio** for an acute angle is the ratio $\frac{\text{opposite}}{\text{hypotenuse}}$.

NOTE: In right $\triangle ABC$ in Figure 11.2, we say that $\sin \alpha = \frac{a}{c}$ and $\sin \beta = \frac{b}{c}$, where "sin" is an abbreviation of the word *sine* (pronounced like *sign*). It is also common to say that $\sin A = \frac{a}{c}$ and $\sin B = \frac{b}{c}$.

EXAMPLE 1

In Figure 11.3, find $\sin \alpha$ and $\sin \beta$ for right $\triangle ABC$.

SOLUTION $a = 3$, $b = 4$, and $c = 5$. Therefore,

$$\sin \alpha = \frac{a}{c} = \frac{3}{5}$$

and

$$\sin \beta = \frac{b}{c} = \frac{4}{5}$$

Figure 11.3

NOTE: In Example 1, it can also be stated that $\sin A = \frac{3}{5}$ and $\sin B = \frac{4}{5}$.

EXAMPLE 2

Figure 11.4

In Figure 11.4, find $\sin \alpha$ and $\sin \beta$ for right $\triangle ABC$.

SOLUTION Where $a = 5$ and $c = 13$, we know that $b = 12$ because (5, 12, 13) is a Pythagorean triple. We verify this result using the Pythagorean Theorem.

$$c^2 = a^2 + b^2$$
$$13^2 = 5^2 + b^2$$
$$169 = 25 + b^2$$
$$b^2 = 144$$
$$b = 12$$

Therefore, $\sin \alpha = \frac{a}{c} = \frac{5}{13}$ and $\sin \beta = \frac{b}{c} = \frac{12}{13}$

Where α is the measure of an acute angle of a right triangle, the value of $\sin \alpha$ is unique. The following Discover activity is designed to give you a better understanding of the meaning of an expression such as $\sin 53°$ as well as its uniqueness.

Discover

Given that an acute angle of a right triangle measures 53°, find the approximate value of sin 53°. We can estimate the value of sin 53° as follows (refer to the triangle at the left).

1. Draw right $\triangle ABC$ so that $\alpha = 53°$ and $\gamma = 90°$.
2. For convenience, mark off the length of the hypotenuse as 4 cm.
3. Using a ruler, measure the length of the leg opposite the angle measuring 53°. It is approximately 3.2 cm long.
4. Now divide $\frac{\text{opposite}}{\text{hypotenuse}}$ or $\frac{3.2}{4}$ to find that sin 53° ≈ 0.8.

NOTE: A calculator provides greater accuracy than the geometric approach found in the Discover activity; in particular, sin 53° ≈ 0.7986.

SSG **EXS. 1–5**

Repeat the procedure in the preceding Discover activity and use it to find an approximation for $\sin 37°$. You will need to use the Pythagorean Theorem to find AC. You should find that $\sin 37° \approx 0.6$.

Although the sine ratios for angle measures are readily available on a calculator, we can justify several of the calculator's results by using special triangles. For certain angles, we can find *exact* results, whereas the calculator provides approximations.

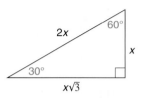

Figure 11.5

Recall the 30°-60°-90° relationship, in which the side opposite the 30° angle has a length equal to one-half that of the hypotenuse; the remaining leg has a length equal to the product of the length of the shorter leg and $\sqrt{3}$.

EXAMPLE 3

Find exact and approximate values for sin 30° and sin 60°. See Figure 11.5.

SOLUTION Using the 30°-60°-90° triangle shown in Figure 11.5, $\sin 30° = \dfrac{x}{2x} = \dfrac{1}{2}$

while $\sin 60° = \dfrac{x\sqrt{3}}{2x} = \dfrac{\sqrt{3}}{2} \approx 0.866$

Although the exact value of sin 30° is 0.5 and the exact value of sin 60° is $\frac{\sqrt{3}}{2}$, a calculator would give an approximate value for sin 60° such as 0.8660254. If we round the ratio for sin 60° to four decimal places, then $\sin 60° \approx 0.8660$. Use your calculator to show that $\frac{\sqrt{3}}{2} \approx 0.8660$ as well.

In Example 4, we apply the 45°-45°-90° triangle. In this triangle, the legs of the right triangle are congruent; also, the length of the hypotenuse is $\sqrt{2}$ times the length of either leg.

EXAMPLE 4

Find exact and approximate values for sin 45°.

SOLUTION Using the 45°-45°-90° triangle in Figure 11.6, we
see that $\sin 45° = \dfrac{x}{x\sqrt{2}} = \dfrac{1}{\sqrt{2}}$. Equivalently, $\sin 45° = \dfrac{\sqrt{2}}{2}$.
In turn, $\sin 45° \approx 0.7071$.

Figure 11.6

Figure 11.7

We will now use the Angle-Bisector Theorem (from Section 5.6) to determine the sine ratios for angles that measure 15° and 75°. Recall that the bisector of one angle of a triangle divides the opposite side into two segments whose lengths are proportional to those of the sides forming the bisected angle. In Figure 11.5, we bisect the 30° angle and choose $x = 1$. Using the resulting triangle shown in Figure 11.7, we are led to the proportion

$$\frac{y}{1 - y} = \frac{\sqrt{3}}{2}$$

Applying the Means-Extremes Property, we have

$$2y = \sqrt{3} - y\sqrt{3}$$
$$2y + y\sqrt{3} = \sqrt{3}$$
$$(2 + \sqrt{3})y = \sqrt{3}$$
$$y = \frac{\sqrt{3}}{2 + \sqrt{3}} \approx 0.4641$$

The number 0.4641 is the length of the side that lies opposite the 15° angle of the 15°-75°-90° triangle in Figure 11.8. Using the Pythagorean Theorem, we can show that the length of the hypotenuse is approximately 1.79315. In turn, $\sin 15° = \frac{0.46410}{1.79315} \approx 0.2588$. Using the same triangle, $\sin 75° = \frac{1.73205}{1.79315} \approx 0.9659$.

We now begin to formulate a small table of values of sine ratios. In Table 11.1, the Greek letter θ (theta) designates the angle measure in degrees. The second column has the heading sin θ and provides the ratio for the corresponding angle; this ratio is generally given to four decimal places of accuracy. Note that the values of sin θ increase as θ increases in measure.

Figure 11.8

TABLE 11.1
Sine Ratios

θ	sin θ
15°	0.2588
30°	0.5000
45°	0.7071
60°	0.8660
75°	0.9659

Warning

Note that $\sin\left(\frac{1}{2}\theta\right) \neq \frac{1}{2}\sin\theta$. See Table 11.1. If θ 60°, sin 60° 1/2 sin 60° 0.5000 (1/2)(0.8660)

NOTE: Most of the sine ratios that are found in tables or that are displayed on a calculator are approximations. Although we use the equality symbol (=) when reading values from a table (or calculator), the solutions to the problems that follow are generally approximations.

$$\sin\theta = \frac{\text{opposite}}{\text{hypotenuse}} \quad \text{in each right triangle}$$

Figure 11.9

In each right triangle shown in Figure 11.9, $\angle\theta$ is an acute angle with a side that is horizontal. In the figure, note that the length of the hypotenuse is constant—it is always equal to the length of the radius of the circle. However, the length of the side opposite $\angle\theta$ gets larger as θ increases in measure. In fact, as θ approaches 90° ($\theta \to 90°$), the length of the leg opposite $\angle\theta$ approaches the length of the hypotenuse. As $\theta \to 90°$, $\sin\theta \to 1$. As θ decreases, $\sin\theta$ also decreases. As θ decreases ($\theta \to 0°$), the length of the side opposite $\angle\theta$ approaches 0. As $\theta \to 0°$, $\sin\theta \to 0$. These observations lead to the following definition.

DEFINITION

$\sin 0° = 0$ and $\sin 90° = 1$

 EXS. 6–10

NOTE: Use your calculator to verify the results found in the definition.

EXAMPLE 5

Using Table 11.1, find the length of a in Figure 11.10 to the nearest tenth of an inch.

SOLUTION

$$\sin 15° = \frac{\text{opposite}}{\text{hypotenuse}} = \frac{a}{10}$$

From Table 11.1, we have $\sin 15° = 0.2588$.

$$\frac{a}{10} = 0.2588 \quad \text{(by substitution)}$$
$$a = 2.588$$

Therefore, $a \approx 2.6$ in. when rounded to tenths.

In an application problem, the sine ratio can be used to find the measure of either a side or an angle of a right triangle. To find the sine ratio of the angle involved, you may

Figure 11.10

use a table of ratios or a calculator. Table 11.2 provides an expanded list of sine ratios; for each angle measure θ, the sine ratio is found to its immediate right. As with calculators, the sine ratios found in tables are only approximations.

Technology Exploration

If you have a graphing calculator, draw the graph of $y = \sin x$ subject to these conditions:

i. Calculator in degree mode.

ii. Window has $0 \le x \le 90$ and $0 \le y \le 1$.

Show by your graph that $y = \sin x$ increases as x increases.

TABLE 11.2

Sine Ratios

θ	$\sin \theta$	θ	$\sin \theta$	θ	$\sin \theta$	θ	$\sin \theta$
0°	0.0000	23°	0.3907	46°	0.7193	69°	0.9336
1°	0.0175	24°	0.4067	47°	0.7314	70°	0.9397
2°	0.0349	25°	0.4226	48°	0.7431	71°	0.9455
3°	0.0523	26°	0.4384	49°	0.7547	72°	0.9511
4°	0.0698	27°	0.4540	50°	0.7660	73°	0.9563
5°	0.0872	28°	0.4695	51°	0.7771	74°	0.9613
6°	0.1045	29°	0.4848	52°	0.7880	75°	0.9659
7°	0.1219	30°	0.5000	53°	0.7986	76°	0.9703
8°	0.1392	31°	0.5150	54°	0.8090	77°	0.9744
9°	0.1564	32°	0.5299	55°	0.8192	78°	0.9781
10°	0.1736	33°	0.5446	56°	0.8290	79°	0.9816
11°	0.1908	34°	0.5592	57°	0.8387	80°	0.9848
12°	0.2079	35°	0.5736	58°	0.8480	81°	0.9877
13°	0.2250	36°	0.5878	59°	0.8572	82°	0.9903
14°	0.2419	37°	0.6018	60°	0.8660	83°	0.9925
15°	0.2588	38°	0.6157	61°	0.8746	84°	0.9945
16°	0.2756	39°	0.6293	62°	0.8829	85°	0.9962
17°	0.2924	40°	0.6428	63°	0.8910	86°	0.9976
18°	0.3090	41°	0.6561	64°	0.8988	87°	0.9986
19°	0.3256	42°	0.6691	65°	0.9063	88°	0.9994
20°	0.3420	43°	0.6820	66°	0.9135	89°	0.9998
21°	0.3584	44°	0.6947	67°	0.9205	90°	1.0000
22°	0.3746	45°	0.7071	68°	0.9272		

NOTE: In later sections, we will use the calculator rather than tables to find values of trigonometric ratios such as sin 36°.

EXAMPLE 6

Find sin 36°, using

    a) Table 11.2.
    b) a scientific or graphing calculator.

SOLUTION

    a) Find 36° under the heading θ. Now read the number under the sin θ heading: sin 36° = 0.5878
    b) On a *scientific calculator* that is in degree mode, use the following key sequence:

$$\boxed{3} \rightarrow \boxed{6} \rightarrow \boxed{\sin} \rightarrow \boxed{\textbf{0.5878}}$$

The result is sin 36° = 0.5878, correct to four decimal places.

NOTE 1: The boldfaced number in the box represents the final answer.

NOTE 2: The key sequence for a *graphing calculator* follows. Here, the calculator is in degree mode and the answer is rounded to four decimal places.

$$\boxed{\sin} \rightarrow \boxed{(} \rightarrow \boxed{3} \rightarrow \boxed{6} \rightarrow \boxed{)} \rightarrow \boxed{\text{Enter}} \rightarrow \boxed{\textbf{0.5878}}$$

The table or a calculator can also be used to find the measure of an angle. This is possible when the sine of the angle is known. Because $\sin 30° = \frac{1}{2}$, the following statement is true. "The acute angle whose sine ratio is $\frac{1}{2}$ measures 30°." To find the missing angle measure on a scientific calculator, we would input "inverse sine of $\frac{1}{2}$." On the graphing calculator, we find the angle's measure by the input " $\sin^{-1}$ of $\frac{1}{2}$." The method is illustrated in Example 7.

EXAMPLE 7

If $\sin \theta = 0.7986$, find θ to the nearest degree by using

  a) Table 11.2.   b) a calculator.

SOLUTION

  a) Find 0.7986 under the heading $\sin \theta$. Now look to the left to find the degree measure of the angle in the θ column:

$$\sin \theta = 0.7986 \rightarrow \theta = 53°$$

  b) On a scientific calculator that is in degree mode, you can use the following key sequence to find θ:

$$\boxed{\cdot} \rightarrow \boxed{7} \rightarrow \boxed{9} \rightarrow \boxed{8} \rightarrow \boxed{6} \rightarrow \boxed{\text{inv}} \rightarrow \boxed{\text{sin}} \rightarrow \boxed{\mathbf{53}}$$

The combination "inv" and "sin" yields the angle whose sine ratio is known, so $\theta = 53°$.

NOTE: On a graphing calculator that is in degree mode, use this sequence:

$$\boxed{\sin^{-1}} \rightarrow \boxed{(} \rightarrow \boxed{\cdot} \rightarrow \boxed{7} \rightarrow \boxed{9} \rightarrow \boxed{8} \rightarrow \boxed{6} \rightarrow \boxed{)} \rightarrow \boxed{\text{ENTER}} \rightarrow \boxed{\mathbf{53}}$$

(SSG) **EXS. 11–15**

The calculator function $\boxed{\sin^{-1}}$ is found by pressing $\boxed{\text{2nd}}$ followed by $\boxed{\text{sin}}$.

In most application problems, a drawing provides a good deal of information and affords some insight into the method of solution. For some drawings and applications, the phrases *angle of elevation* and *angle of depression* are used. These angles are measured from the horizontal as illustrated in Figures 11.11(a) and 11.11(b). In Figure 11.11(a), the angle α measured upward from the horizontal ray is the **angle of elevation**. In Figure 11.11(b), the angle β measured downward from the horizontal ray is the **angle of depression**.

(a)   (b)

Figure 11.11

EXAMPLE 8

The tower for a radio station stands 200 ft tall. A guy wire 250 ft. long helps to support the antenna, as shown in Figure 11.12. Find the measure of α, the angle of elevation, to the nearest degree.

250 ft   200 ft

α

Figure 11.12

SOLUTION

$$\sin \alpha = \frac{\text{opposite}}{\text{hypotenuse}} = \frac{200}{250} = 0.8$$

From Table 11.2 (or from a calculator), we find that the angle whose sine ratio is 0.8 is $\alpha \approx 53°$.

SSG **EXS. 16, 17**

Exercises 11.1

In Exercises 1 to 6, find sin α and sin β for the triangle shown.

1.

2.

3.

4.

5.

6.
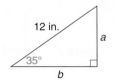

In Exercises 7 to 14, use either Table 11.2 or a calculator to find the sine of the indicated angle to four decimal places.

7. sin 90°

8. sin 0°

9. sin 17°

10. sin 23°

11. sin 82°

12. sin 46°

13. sin 72°

14. sin 57°

In Exercises 15 to 20, find the lengths of the sides named by the variables. Use either Table 11.2 or a calculator, and round answers to the nearest tenth of a unit.

15.

16.

17.

18.

19.

20.

In Exercises 21 to 26, find the measures of the angles named to the nearest degree.

21.

22.

23.

24.

25.

26.

In Exercises 27 to 34, use the drawings where provided to solve each problem. Angle measures should be given to the nearest degree; distances should be given to the nearest tenth of a unit.

27. The pitch or slope of a roofline is 5 to 12. Find the measure of angle α.
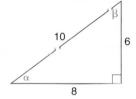

28. Zaidah is flying a kite at an angle of elevation of 67° from a point on the ground. If 100 ft of kite string is out, how far is the kite above the ground?

100 ft

67°

29. Richard sees a balloon that is 100 ft above the ground. If the angle of elevation from Richard to the balloon is 75°, how far from Richard is the balloon?

100 ft

75°

30. Over a 2000-ft span of highway through a hillside, there is a 100-ft rise in the roadway. What is the measure of the angle formed by the road and the horizontal?

2000 ft

100 ft

α

31. From a cliff, Phillip observes an automobile through an angle of depression of 23°. If the cliff is 50 ft high, how far is the automobile from Phillip?

23°

50 ft

32. A 12-ft rope secures a rowboat to a pier that is 4 ft above the water. Assume that the lower end of the rope is at "water level." What is the angle formed by the rope and the water? Assume that the rope is taut.

12 ft

4 ft

33. A 10-ft ladder is leaning against a vertical wall so that the bottom of the ladder is 4 ft away from the base of the wall. How large is the angle formed by the ladder and the wall?

34. An airplane flying at the rate of 350 feet per second begins to climb at an angle of 10°. What is the increase in altitude over the next 15 seconds?

10°

350 ft/s during climb

For Exercises 35 to 38, make drawings as needed.

35. In parallelogram *ABCD*, *AB* = 6 ft and *AD* = 10 ft. If m∠*A* = 65° and $\overline{BE}$ is the altitude to $\overline{AD}$, find
 a) *BE* correct to tenths.
 b) the area of ▱*ABCD*.

36. In right △*ABC*, γ = 90° and β = 55°. If *AB* = 20 in., find
 a) *a* (the length of $\overline{BC}$) correct to tenths.
 b) *b* (the length of $\overline{AC}$) correct to tenths.
 c) the area of right △*ABC*.

37. In a right circular cone, the slant height is 13 cm and the height is 10 cm. To the nearest degree, find the measure of the angle θ that is formed by the radius and slant height.

38. In a right circular cone, the slant height is 13 cm. Where θ is the angle formed by the radius and the slant height, θ = 48°. Find the length of the altitude of the cone correct to tenths.

*39. In regular pentagon *ABCDE*, sides $\overline{AB}$ and $\overline{BC}$ along with diagonal $\overline{AC}$ form isosceles △*ABC*. Let *AB* = *BC* = *s*. In terms of *s*, find an expression for
 a) *h*, the length of the altitude of △*ABC* from vertex *B* to side $\overline{AC}$
 b) *d*, the length of diagonal *AC* of regular pentagon *ABCDE*.

11.2 The Cosine Ratio and Applications

Adjacent Side (Leg)   Cosine Ratio:
$$\cos\theta = \frac{\text{adjacent}}{\text{hypotenuse}}$$
  Identity:
$$\sin^2\theta + \cos^2\theta = 1$$

In this section, we again deal exclusively with similar right triangles. In Figure 11.13, $\angle A \cong \angle D$ and $\angle C \cong \angle F$; thus, $\triangle ABC \sim \triangle DEF$ by AA. In $\triangle ABC$, $\overline{BC}$ is the leg opposite angle A, while $\overline{AC}$ is the leg *adjacent* to angle A. In the two triangles, the ratios of the form

$$\frac{\text{length of adjacent leg}}{\text{length of hypotenuse}}$$

are equal; that is,

$$\frac{AC}{AB} = \frac{DF}{DE} \qquad \text{or} \qquad \frac{4}{5} = \frac{8}{10}$$

This relationship follows from the fact that corresponding sides of similar triangles are proportional (CSSTP).

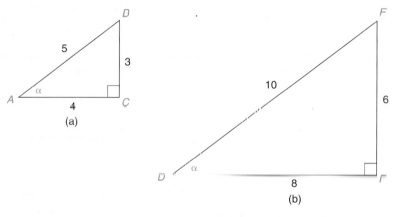

Figure 11.13

As with the sine ratio, the *cosine ratio* depends on the measure of acute angle A (or D) in Figure 11.13. In the following definition, the term *adjacent* refers to the length of the leg that is adjacent to the angle named.

DEFINITION

In a right triangle, the **cosine ratio** for an acute angle is the ratio $\frac{\text{adjacent}}{\text{hypotenuse}}$.

Figure 11.14

NOTE: For right $\triangle ABC$ in Figure 11.14, we have $\cos\alpha = \frac{b}{c}$ and $\cos\beta = \frac{a}{c}$; in each statement, "cos" is an abbreviated form of the word *cosine*. These statements can also be expressed in the forms $\cos A = \frac{b}{c}$ and $\cos B = \frac{a}{c}$.

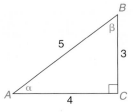

Figure 11.15

Find cos α and cos β for right $\triangle ABC$ in Figure 11.15.

SOLUTION $a = 3$, $b = 4$, and $c = 5$ for the triangle shown in Figure 11.15. Because b is the length of the leg adjacent to α and a is the length of the leg adjacent to β,

$$\cos \alpha = \frac{b}{c} = \frac{4}{5} \qquad \text{and} \qquad \cos \beta = \frac{a}{c} = \frac{3}{5}$$

EXAMPLE 2

Find cos α and cos β for right $\triangle ABC$ in Figure 11.16.

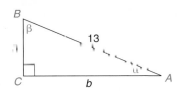

SSG **EXS. 1–5**   Figure 11.16

SOLUTION $a = 5$ and $c = 13$. Then $b = 12$ from the Pythagorean triple $(5, 12, 13)$. Consequently,

$$\cos \alpha = \frac{b}{c} = \frac{12}{13} \qquad \text{and} \qquad \cos \beta = \frac{a}{c} = \frac{5}{13}$$

Just as the sine ratio of any angle is unique, the cosine ratio of any angle is also unique. Using the 30°-60°-90° and 45°-45°-90° triangles of Figure 11.17, we see that

$$\cos 30° = \frac{x\sqrt{3}}{2x} = \frac{\sqrt{3}}{2} \approx 0.8660$$

$$\cos 45° = \frac{x}{x\sqrt{2}} = \frac{1}{\sqrt{2}} = \frac{\sqrt{2}}{2} \approx 0.7071$$

$$\cos 60° = \frac{x}{2x} = \frac{1}{2} = 0.5$$

Figure 11.17

Now we use the 15°-75°-90° triangle shown in Figure 11.18 to find cos 75° and cos 15°. From Section 11.1, sin 15° = $\frac{a}{c}$ and sin 15° = 0.2588 (see page 492). But cos 75° = $\frac{a}{c}$, so cos 75° = 0.2588. Similarly, because sin 75° = $\frac{b}{c}$ = 0.9659, we see that cos 15° = $\frac{b}{c}$ = 0.9659.

In Figure 11.19 on page 500, the cosine ratios become larger as θ decreases and become smaller as θ increases. To understand why, consider the definition

$$\cos \theta = \frac{\text{length of adjacent leg}}{\text{length of hypotenuse}}$$

Figure 11.18

and Figure 11.19. Recall that the symbol $\rightarrow$ is read "approaches." As $\theta \rightarrow 0°$, length of adjacent leg $\rightarrow$ length of hypotenuse, and therefore cos 0° $\rightarrow$ 1. Similarly, cos 90° $\rightarrow$ 0 because the length of the adjacent leg grows smaller as $\theta \rightarrow 90°$. Consequently, we have the following definition.

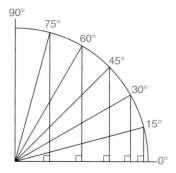

Figure 11.19

Technology Exploration

If you have a graphing calculator, draw the graph of $y = \cos x$ subject to these conditions:

i. Calculator in degree mode.

ii. Window has $0 \leq x \leq 90$ and $0 \leq y \leq 1$.

Show by your graph that $y = \cos x$ decreases as x increases.

> **DEFINITION**
>
> $\cos 0° = 1$ and $\cos 90° = 0$.

Use your calculator to verify the results found in Table 11.3.

TABLE 11.3
Cosine Ratios

θ	$\cos \theta$
0°	1.0000
15°	0.9659
30°	0.8660
45°	0.7071
60°	0.5000
75°	0.2588
90°	0.0000

This textbook does not provide an expanded table of cosine ratios comparable to Table 11.2 for sine ratios. Nonetheless, we can use Table 11.3 or a calculator to find the cosine ratio of an angle. We illustrate the application of such a table in Example 3.

SSG **EXS. 6–10**

EXAMPLE 3

Using Table 11.3, find the length of b in Figure 11.20 correct to the nearest tenth.

Figure 11.20

SOLUTION $\cos 15° = \frac{\text{adjacent}}{\text{hypotenuse}} = \frac{b}{10}$ from the triangle. Also, $\cos 15° = 0.9659$ from the table. Then

$$\frac{b}{10} = 0.9659 \quad \text{(because both equal } \cos 15°\text{)}$$

$$b = 9.659$$

Therefore, $\qquad b \approx 9.7 \text{ in.}$

when rounded to the nearest tenth of an inch.

Reminder

$\sin \theta = \frac{\text{opposite}}{\text{hypotenuse}}$

$\cos \theta = \frac{\text{adjacent}}{\text{hypotenuse}}$

In a right triangle, the cosine ratio can often be used to find either an unknown length or an unknown angle measure. Whereas the sine ratio requires that we use *opposite* and *hypotenuse*, the cosine ratio requires that we use *adjacent* and *hypotenuse*.

An equation of the form $\sin \alpha = \frac{a}{c}$ or $\cos \alpha = \frac{b}{c}$ contains three variables; for the equation $\cos \alpha = \frac{b}{c}$, the variables are α, b, and c. When the values of two of the variables are known, the value of the third variable can be determined; however, we must decide which trigonometric ratio is needed to solve the problem.

In Example 4, we deal with the question, "Should we write an equation containing the sine or the cosine of the angle measure?"

EXAMPLE 4

In Figure 11.21 on page 501, which trigonometric ratio would you use to find

a) α, if $a = 3$ and $c = 5$?

b) b, if $\alpha = 32°$ and $c = 5$?

c) c, if $a = 4$ and $\alpha = 35°$?

d) β, if $a = 3.5$ and $c = 5$?

Figure 11.21

SOLUTION

a) sine, because $\sin \alpha = \frac{a}{c}$ becomes $\sin \alpha = \frac{3}{5}$, with α the remaining variable.

b) cosine, because $\cos \alpha = \frac{b}{c}$ becomes $\cos 32° = \frac{b}{5}$, with b the remaining variable.

c) sine, because $\sin \alpha = \frac{a}{c}$ becomes $\sin 35° = \frac{4}{c}$, with c the remaining variable.

d) cosine, because $\cos \beta = \frac{a}{c}$, become $\cos \beta = \frac{3.5}{5}$, with β the remaining variable.

To solve application problems, we generally use a calculator. As we saw in Example 4, we will need to determine numbers such as cos 32° and sin 35°.

EXAMPLE 5

Find cos 67° correct to four decimal places by using a calculator.

SOLUTION On a scientific calculator that is in degree mode, use the following key sequence:

$\boxed{6} \rightarrow \boxed{7} \rightarrow \boxed{\text{cos}} \rightarrow \boxed{\textbf{0.3907}}$

Using a graphing calculator (in degree mode), follow this key sequence:

$\boxed{\text{cos}} \rightarrow \boxed{(} \rightarrow \boxed{6} \rightarrow \boxed{7} \rightarrow \boxed{)} \rightarrow \boxed{\text{ENTER}} \rightarrow \boxed{\textbf{0.3907}}$

That is, $\cos 67° \approx 0.3907$.

In Example 4, we found equations such as $\sin \alpha = \frac{3}{5}$ and $\cos \beta = \frac{3.5}{5}$. In Example 6, we illustrate the use of the calculator to find the measure of the angle named in the given equation.

EXAMPLE 6

Use a calculator to find the measure of angle θ to the nearest degree if $\cos \theta = 0.5878$.

SOLUTION Using a scientific calculator in degree mode, follow this key sequence:

$\boxed{.} \rightarrow \boxed{5} \rightarrow \boxed{8} \rightarrow \boxed{7} \rightarrow \boxed{8} \rightarrow \boxed{\text{inv}} \rightarrow \boxed{\text{cos}} \rightarrow \boxed{\textbf{54}}$

Using a graphing calculator in degree mode, follow this key sequence:

$\boxed{\text{cos}^{-1}} \rightarrow \boxed{(} \rightarrow \boxed{.} \rightarrow \boxed{5} \rightarrow \boxed{8} \rightarrow \boxed{7} \rightarrow \boxed{8} \rightarrow \boxed{)} \rightarrow \boxed{\text{ENTER}} \rightarrow \boxed{\textbf{54}}$

Thus, $\theta = 54°$.

SSG EXS. 11–15   **NOTE:** By pressing $\boxed{\text{2nd}}$ and $\boxed{\text{cos}}$ on a graphing calculator, you obtain $\boxed{\text{cos}^{-1}}$.

EXAMPLE 7

For a regular pentagon, the length of the apothem is 12 in. Find the length of the pentagon's radius to the nearest tenth of an inch.

SOLUTION The central angle of the regular pentagon measures $\frac{360}{5}$, or 72°. An apothem bisects this angle, so the angle formed by the apothem and the radius measures 36°. In Figure 11.22,

$$\cos 36° = \frac{\text{adjacent}}{\text{hypotenuse}} = \frac{12}{r}$$

Figure 11.22

Thus, $\cos 36° = \frac{12}{r}$.

Using a calculator, $\cos 36° = 0.8090$. Then $\frac{12}{r} = 0.8090$ and $0.8090r = 12$. By division, $r \approx 14.8$ in.

NOTE: If $\cos 36° = \frac{12}{r}$, then $r \cdot \cos 36° = 12$. Thus, the solution in Example 7 can also be calculated as $r = \frac{12}{\cos 36°}$.

We now consider the proof of a statement that is called an **identity** because it is true for all angles; we refer to this statement as a theorem. As you will see, the proof of this statement is based entirely on the Pythagorean Theorem.

THEOREM 11.2.1 ■ The Pythagorean Identity

In any right triangle in which α is the measure of an acute angle,

$$\sin^2 \alpha + \cos^2 \alpha = 1$$

NOTE: $\sin^2 \alpha$ means $(\sin \alpha)^2$ and $\cos^2 \alpha$ means $(\cos \alpha)^2$.

PROOF

In Figure 11.23, $\sin \alpha = \frac{a}{c}$ and $\cos \alpha = \frac{b}{c}$. Then

$$\sin^2 \alpha + \cos^2 \alpha = \left(\frac{a}{c}\right)^2 + \left(\frac{b}{c}\right)^2 = \frac{a^2}{c^2} + \frac{b^2}{c^2} = \frac{a^2 + b^2}{c^2}$$

In the right triangle in Figure 11.23, $a^2 + b^2 = c^2$ by the Pythagorean Theorem. Substituting c^2 for $a^2 + b^2$ into the equation above, we have

$$\sin^2 \alpha + \cos^2 \alpha = \frac{c^2}{c^2} = 1$$

It follows that $\sin^2 \alpha + \cos^2 \alpha = 1$ for any angle α.

NOTE: Use your calculator to show that $(\sin 67°)^2 + (\cos 67°)^2 = 1$. Theorem 11.2.1 is also true for $\alpha = 0°$ and for $\alpha = 90°$.

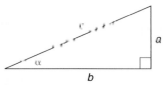

Figure 11.23

EXAMPLE 8

In right triangle ABC (not shown), $\sin \alpha = \frac{2}{3}$. Find $\cos \alpha$.

SOLUTION Applying the Pythagorean Identity,

$$\sin^2 \alpha + \cos^2 \alpha = 1$$
$$\left(\frac{2}{3}\right)^2 + \cos^2 \alpha = 1$$
$$\frac{4}{9} + \cos^2 \alpha = 1$$
$$\cos^2 \alpha = \frac{5}{9}$$

SSG **EXS. 16, 17**

Therefore, $\cos \alpha = \sqrt{\frac{5}{9}} = \frac{\sqrt{5}}{\sqrt{9}} = \frac{\sqrt{5}}{3}$.

NOTE: Because $\cos \alpha > 0$, $\cos \alpha = \frac{\sqrt{5}}{3}$ rather than $-\frac{\sqrt{5}}{3}$.

Theorem 11.2.1 represents one of many trigonometric identities. In Exercises 33–36 of Section 11.3, we will discover further trigonometric identities.

Exercises 11.2

In Exercises 1 to 6, find cos α and cos β.

1.

2.

3.

4.

5.

6.

7. For the triangles in Exercises 1 to 6,
 a) why does sin α = cos β?
 b) why does cos α = sin β?

8. Using the right triangle from Exercise 1, show that
 $\sin^2 \alpha + \cos^2 \alpha = 1$.

In Exercises 9 to 16, use a calculator to find the indicated cosine ratio to four decimal places.

9. cos 23°   **10.** cos 0°   **11.** cos 17°   **12.** cos 73°

13. cos 90°   **14.** cos 42°   **15.** cos 82°   **16.** cos 7°

In Exercises 17 to 22, use either the sine ratio or the cosine ratio to find the lengths of the indicated sides of the triangle correct to the nearest tenth of a unit.

17.

18.

19.

20.

21.

22.
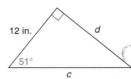

In Exercises 23 to 28, use the sine ratio or the cosine ratio as needed to find the measure of each indicated angle to the nearest degree.

23.

24.

25.

26.

27.

28.

Rectangle *ABCD*

In Exercises 29 to 37, angle measures should be given to the nearest degree; distances should be given to the nearest tenth of a unit.

29. In building a garage onto his house, Gene wants to use a sloped 12-ft roof to cover an expanse that is 10 ft wide. Find the measure of angle θ.

30. Gene redesigned the garage from Exercise 29 so that the 12-ft roof would rise 2 ft as shown. Find the measure of angle θ.

31. When an airplane is descending to land, the angle of depression is 5°. When the plane has a reading of 100 ft on the altimeter, what is its distance *x* from touchdown?

32. At a point 200 ft from the base of a cliff, Journey sees the top of the cliff through an angle of elevation of 37°. How tall is the cliff?

33. Find the length of each apothem in a regular pentagon whose radii measure 10 in. each.

34. Dale looks up to see his friend Lisa waving from her apartment window 30 ft from him. If Dale is standing 10 ft from the building, what is the angle of elevation as Dale looks up at Lisa?

35. Find the length of the radius in a regular decagon for which each apothem has a length of 12.5 cm.

36. In searching for survivors of a boating accident, a helicopter moves horizontally across the ocean at an altitude of 200 ft above the water. If a man clinging to a life raft is seen through an angle of depression of 12°, what is the distance from the helicopter to the man in the water?

37. What is the size of the angle α formed by a diagonal of a cube and one of its edges?

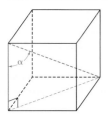

38. In the right circular cone,
   a) find r correct to tenths.
   b) use $L = \pi r \ell$ to find the lateral area of the cone.

39. In parallelogram $ABCD$, find, to the nearest degree:
   a) $m\angle A$ b) $m\angle B$

40. A ladder of length AB is carried horizontally through an L-shaped turn in a hallway. Show that the ladder has the length $L = \frac{6}{\sin \theta} + \frac{6}{\cos \theta}$.

41. Use the drawing provided to show that the area of the isosceles triangle is $A = s^2 \sin \theta \cos \theta$.

For Exercises 42 and 43, use the drawing and the formula from Exercise 41.

42. Find the area of an isosceles triangle for which $s = 10.6$ cm and the measure of the *vertex* angle is 46°.

43. Find the area of an isosceles triangle for which $s = 4.8$ inches and the measure of a *base* angle is 72°.

44. In regular pentagon $ABCDE$, each radius has length r. In terms of r, find an expression for the perimeter of $ABCDE$.

45. Consider regular pentagon $ABCDE$ of Exercise 44. In terms of radius length r, find an expression for the area of $ABCDE$.

11.3 The Tangent Ratio and Other Ratios

KEY CONCEPTS	Tangent Ratio: $\tan \theta = \frac{\text{opposite}}{\text{adjacent}}$	Cotangent Secant	Cosecant Reciprocal Ratios

As in Sections 11.1 and 11.2, we again deal exclusively with right triangles in this section. The next trigonometric ratio that we consider is the **tangent** ratio, which is defined for an acute angle of the right triangle by

$$\frac{\text{length of leg } opposite \text{ acute angle}}{\text{length of leg } adjacent \text{ to acute angle}}$$

Like the sine ratio, the tangent ratio increases as the measure of the acute angle increases. Unlike the sine and cosine ratios, whose values range from 0 to 1, the value of the tangent ratio is from 0 upward; that is, there is no greatest value for the tangent.

Figure 11.24

In a right triangle, the **tangent ratio** for an acute angle is the ratio $\frac{opposite}{adjacent}$.

NOTE: In right $\triangle ABC$ in Figure 11.24, $\tan \alpha = \frac{a}{b}$ and $\tan \beta = \frac{b}{a}$, in which "tan" is an abbreviated form of the word tangent.

Technology Exploration

If you have a graphing calculator, draw the graph of $y = \tan x$ subject to these conditions:

i. Calculator in degree mode.
ii. Window has $0 \le x \le 90$ and $0 \le y \le 4$

Show that $y = \tan x$ increases as x increases.

EXAMPLE 1

Find the values of $\tan \alpha$ and $\tan \beta$ for the triangle in Figure 11.25.

SOLUTION Using the fact that the tangent ratio is $\frac{opposite}{adjacent}$, we find that

$$\tan \alpha = \frac{a}{b} = \frac{8}{15}$$

and

$$\tan \beta = \frac{b}{a} = \frac{15}{8}$$

Figure 11.25

SSG **EXS. 1–4**

The value of $\tan \theta$ increases from 0 for a $0°$ angle to an immeasurably large value as the measure of the acute angle approaches $90°$. That the tangent ratio $\frac{opposite}{adjacent}$ becomes infinitely large as $\theta \to 90°$ follows from the fact that the denominator becomes smaller (approaching 0) while the numerator increases.

Study Figure 11.26 to see why the value of the tangent of an angle grows immeasurably large as the angle approaches $90°$ in size. As the angle grows larger, consider the value of the ratio $\frac{opposite}{adjacent}$. We often express this relationship by writing: As $\theta \to 90°$, $\tan \theta \to \infty$. The symbol ∞ is read "infinity" and implies that $\tan 90°$ is not measurable; thus, $\tan 90°$ is *undefined*.

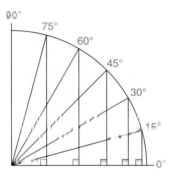

Figure 11.26

$\tan 0° = 0$ and $\tan 90°$ is undefined.

Use your calculator to verify that $\tan 0° = 0$. When you use your calculator to find $\tan 90°$, you will find an "Error" message.

Certain tangent ratios are found by using special right triangles. By observing the triangles in Figure 11.27 and using the fact that $\tan \theta = \frac{opposite}{adjacent}$, we have

$$\tan 30° = \frac{x}{x\sqrt{3}} = \frac{1}{\sqrt{3}} = \frac{\sqrt{3}}{3} \approx 0.5774$$

$$\tan 45° = \frac{x}{x} = 1$$

$$\tan 60° = \frac{x\sqrt{3}}{x} = \sqrt{3} \approx 1.7321$$

We will use a calculator to find the value of the tangent ratio or the angle measure in Examples 2 and 3.

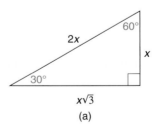

(a)

(b)

Figure 11.27

EXAMPLE 2

A ski lift moves each chair through an angle of elevation of $25°$, as shown in Figure 11.28 on page 506. What vertical change (rise) accompanies a horizontal change (run) of 845 ft?

Figure 11.28

SOLUTION In the triangle, $\tan 25° = \frac{\text{opposite}}{\text{adjacent}} = \frac{a}{845}$. With $\tan 25° = \frac{a}{845}$, we multiply by 845 to obtain $a = 845 \cdot \tan 25°$. Using a calculator, we find that $a \approx 394$ ft.

The tangent ratio can also be used to find the measure of an angle if the lengths of the legs of a right triangle are known. This is illustrated in Example 3.

Geometry in the Real World

In the coordinate system shown, we see that the slope of the line is
$$m = \tan \alpha$$

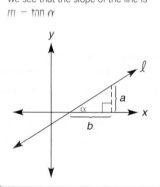

EXAMPLE 3

Mary Katherine sees a small airplane flying over Mission Rock, which is 1 mi away. If Mission Rock is known to be 335 ft high and the airplane is 520 ft above it, then what is the angle of elevation through which Mary Katherine sees the plane?

SOLUTION In Figure 11.29, the altitude of the airplane is $335 + 520$ or 855 feet. Using the fact that $1 \text{ mi} = 5280 \text{ ft}$,

$$\tan \theta = \frac{\text{opposite}}{\text{adjacent}} = \frac{855}{5280}$$

Then $\tan \theta \approx 0.1619$, so $\theta = 9°$ to the nearest degree.

Figure 11.29

NOTE: Determining the measure of θ in Example 3 required the use of a calculator. When we use a scientific calculator in degree mode, the typical key sequence is

$$\boxed{\cdot} \to \boxed{1} \to \boxed{6} \to \boxed{1} \to \boxed{9} \to \boxed{\text{inv}} \to \boxed{\tan} \to \boxed{9}$$

When we use a graphing calculator in degree mode, the typical key sequence is

$$\boxed{\tan^{-1}} \to \boxed{(} \to \boxed{\cdot} \to \boxed{1} \to \boxed{6} \to \boxed{1} \to \boxed{9} \to \boxed{)} \to \boxed{\text{ENTER}} \to \boxed{9}$$

For the right triangle in Figure 11.30, we now have three ratios that can be used in problem solving. These are summarized as follows:

Figure 11.30

$$\sin \alpha = \frac{\text{opposite}}{\text{hypotenuse}}$$
$$\cos \alpha = \frac{\text{adjacent}}{\text{hypotenuse}}$$
$$\tan \alpha = \frac{\text{opposite}}{\text{adjacent}}$$

The equation $\tan \alpha = \frac{a}{b}$ contains three variables: α, a, and b. If the values of two of the variables are known, the value of the third variable can be found.

EXAMPLE 4

In Figure 11.31, name the ratio that should be used to find:

a) a, if $\alpha = 57°$ and $c = 12$
b) α, if $a = 7.2$ and $b = 4.5$
c) β, if $a = 6.9$ and $c = 9.2$
d) b, if $a = 6.7$ and $\beta = 36°$

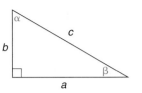

Figure 11.31

SOLUTION

a) sine, because $\sin \alpha = \dfrac{a}{c}$ leads to $\sin 57° = \dfrac{a}{12}$

b) tangent, because $\tan \alpha = \dfrac{a}{b}$ leads to $\tan \alpha = \dfrac{7.2}{4.5}$

c) cosine, because $\cos \beta = \dfrac{a}{c}$ leads to $\cos \beta = \dfrac{6.9}{9.2}$

d) tangent, because $\tan \beta = \dfrac{b}{a}$ leads to $\tan 36° = \dfrac{b}{6.7}$

EXAMPLE 5

Figure 11.32

Two apartment buildings are 40 ft apart. From a window in her apartment, Izzi can see the top of the other apartment building through an angle of elevation of 47°. She can also see the base of the other building through an angle of depression of 33°. Approximately how tall is the other building?

SOLUTION In Figure 11.32, the height of the building is the sum $x + y$. Using the upper and lower right triangles, we have

$$\tan 47° = \frac{x}{40} \qquad \text{and} \qquad \tan 33° = \frac{y}{40}$$

Now $x = 40 \cdot \tan 47°$ and $y = 40 \cdot \tan 33°$

Then $x \approx 43$ and $y \approx 26$, so $x + y \approx 43 + 26 = 69$. The building is approximately 69 ft tall.

NOTE. In Example 5, you can determine the height of the building $(x + y)$ by entering the expression $40 \cdot \tan 47° + 40 \cdot \tan 33°$ on your calculator.

Technology Exploration

If you have a graphing calculator, show that $\tan 23°$ equals $\frac{\sin 23°}{\cos 23°}$. The identity $\tan \alpha = \frac{\sin \alpha}{\cos \alpha}$ is true as long as $\cos \alpha \neq 0$.

There are a total of six trigonometric ratios. We define the remaining ratios for completeness; however, we will be able to solve all application problems in this chapter by using only the sine, cosine, and tangent ratios. The remaining ratios are the **cotangent** (abbreviated "cot"), **secant** (abbreviated "sec"), and **cosecant** (abbreviated "csc"). These are defined in terms of the right triangle shown in Figure 11.33.

Figure 11.33

$$\cot \alpha = \frac{\text{adjacent}}{\text{opposite}}$$

$$\sec \alpha = \frac{\text{hypotenuse}}{\text{adjacent}}$$

$$\csc \alpha = \frac{\text{hypotenuse}}{\text{opposite}}$$

For the fraction $\frac{a}{b}$ (where $b \neq 0$), the reciprocal is $\frac{b}{a}$ ($a \neq 0$). It is easy to see that $\cot \alpha$ is the reciprocal of $\tan \alpha$, $\sec \alpha$ is the reciprocal of $\cos \alpha$, and $\csc \alpha$ is the reciprocal of $\sin \alpha$.

EXAMPLE 6

Use the given information to find the missing ratio.

a) $\csc \alpha$, if $\sin \alpha = \dfrac{2}{3}$

b) $\cot \alpha$, if $\tan \alpha = 5$

c) $\cos \alpha$, if $\sec \alpha = \dfrac{7}{3}$

SOLUTION

a) $\csc \alpha = \dfrac{3}{2}$, the reciprocal of $\sin \alpha$

b) $\cot \alpha = \dfrac{1}{5}$, the reciprocal of $\tan \alpha$

c) $\cos \alpha = \dfrac{3}{7}$, the reciprocal of $\sec \alpha$

In Table 11.4, we invert the trigonometric ratio on the left to obtain the reciprocal ratio named to its right.

TABLE 11.4

The Six Trigonometric Ratios

$\text{sine } \alpha = \dfrac{\text{opposite}}{\text{hypotenuse}}$	$\text{cosecant } \alpha = \dfrac{\text{hypotenuse}}{\text{opposite}}$
$\text{cosine } \alpha = \dfrac{\text{adjacent}}{\text{hypotenuse}}$	$\text{secant } \alpha = \dfrac{\text{hypotenuse}}{\text{adjacent}}$
$\text{tangent } \alpha = \dfrac{\text{opposite}}{\text{adjacent}}$	$\text{cotangent } \alpha = \dfrac{\text{adjacent}}{\text{opposite}}$

Calculators display only the sine, cosine, and tangent ratios. By using the reciprocal key, $\boxed{1/x}$ or $\boxed{x^{-1}}$, you can obtain the values for the remaining ratios. See Example 7 for details.

EXAMPLE 7

Use a calculator to evaluate

a) $\csc 37°$ **b)** $\cot 51°$ **c)** $\sec 84°$

SOLUTION

a) First we use the calculator to find $\sin 37° \approx 0.6081$. Now use the $\boxed{1/x}$ or the $\boxed{x^{-1}}$ key to show that $\csc 37° \approx 1.6616$.

b) First we use the calculator to find $\tan 51° \approx 1.2349$. Now use the $\boxed{1/x}$ or the $\boxed{x^{-1}}$ key to show that $\cot 51° \approx 0.8098$.

c) First we use the calculator to find $\cos 84° \approx 0.1045$. Now use the $\boxed{1/x}$ or the $\boxed{x^{-1}}$ key to show that $\sec 84° \approx 9.5668$.

NOTE: In part (a), the value of $\csc 37°$ can also be determined by using the display $1 \div (\sin 37)$ on the scientific calculator or $(\sin 37)^{-1}$ on the graphing calculator. Similar displays can be used in parts (b) and (c).

SSG **EXS. 12–16**

In Example 8, a calculator is not needed to determine exact results.

EXAMPLE 8

For the triangle in Figure 11.34, find the exact values of all six trigonometric ratios for angle θ.

Figure 11.34

SOLUTION We need to know the length c of the hypotenuse, which we find by applying the Pythagorean Theorem.

$$c^2 = 5^2 + 6^2$$
$$c^2 = 25 + 36$$
$$c^2 = 61$$
$$c = \sqrt{61}$$

Therefore,

$$\sin \theta = \frac{\text{opposite}}{\text{hypotenuse}} = \frac{6}{\sqrt{61}} = \frac{6}{\sqrt{61}} \cdot \frac{\sqrt{61}}{\sqrt{61}} = \frac{6\sqrt{61}}{61}$$

$$\cos \theta = \frac{\text{adjacent}}{\text{hypotenuse}} = \frac{5}{\sqrt{61}} = \frac{5}{\sqrt{61}} \cdot \frac{\sqrt{61}}{\sqrt{61}} = \frac{5\sqrt{61}}{61}$$

$$\tan \theta = \frac{\text{opposite}}{\text{adjacent}} = \frac{6}{5}$$

$$\cot \theta = \frac{\text{adjacent}}{\text{opposite}} = \frac{5}{6}$$

$$\sec \theta = \frac{\text{hypotenuse}}{\text{adjacent}} = \frac{\sqrt{61}}{5}$$

$$\csc \theta = \frac{\text{hypotenuse}}{\text{opposite}} = \frac{\sqrt{61}}{6}$$

NOTE: The arrows in Example 8 indicate which pairs of ratios are reciprocals.

EXAMPLE 9

Evaluate the ratio named by using the given ratio:

a) $\tan \theta$, if $\cot \theta = \frac{2}{3}$
b) $\sin \alpha$, if $\csc \alpha = 1.25$
c) $\sec \beta$, if $\cos \beta = \frac{\sqrt{3}}{2}$
d) $\csc \gamma$, if $\sin \gamma = 1$

SOLUTION

a) If $\cot \theta = \frac{2}{3}$, then $\tan \theta = \frac{3}{2}$ (the reciprocal of $\cot \theta$).

b) If $\csc \alpha = 1.25$ or $\frac{5}{4}$, then $\sin \alpha = \frac{4}{5}$ (the reciprocal of $\csc \alpha$).

c) If $\cos \beta = \frac{\sqrt{3}}{2}$, then $\sec \beta = \frac{2}{\sqrt{3}}$ or $\frac{2\sqrt{3}}{3}$ (the reciprocal of $\cos \beta$).

d) If $\sin \gamma = 1$, then $\csc \gamma = 1$ (the reciprocal of $\sin \gamma$).

EXAMPLE 10

To the nearest degree, what is the measure of θ in the triangle in Figure 11.35 if $\cot\theta = \frac{8}{5}$?

SOLUTION Because $\cot\theta = \frac{8}{5}$, we can use its reciprocal to find θ. That is,

$$\tan\theta = \frac{5}{8}$$

With a scientific calculator, we determine θ by using the key sequence

$$\boxed{5} \rightarrow \boxed{\div} \rightarrow \boxed{8} \rightarrow \boxed{=} \rightarrow \boxed{\text{inv}} \rightarrow \boxed{\tan} \rightarrow \boxed{\mathbf{32}}$$

When we use a graphing calculator, the key sequence is

$$\boxed{\tan^{-1}} \rightarrow \boxed{(} \rightarrow \boxed{5} \rightarrow \boxed{\div} \rightarrow \boxed{8} \rightarrow \boxed{)} \rightarrow \boxed{\text{ENTER}} \rightarrow \boxed{\mathbf{32}}$$

Thus, $\theta \approx 32°$.

Figure 11.35

In Example 11 and the application exercises that follow this section, you will generally have to decide which trigonometric ratio enables you to solve the problem.

EXAMPLE 11

As his fishing vessel moves into the bay, the captain notes that the angle of elevation to the top of the lighthouse is 11°. If the lighthouse is 200 ft tall, how far is the vessel from the base of the lighthouse? See Figure 11.36.

SOLUTION Again we use the tangent ratio; in Figure 11.36,

$$\tan 11° = \frac{200}{x}$$
$$x \cdot \tan 11° = 200$$
$$x = \frac{200}{\tan 11°} \approx 1028.91$$

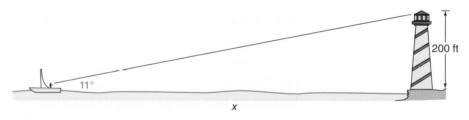

Figure 11.36

SSG **EXS. 17, 18**   The vessel is approximately 1029 feet from the base of the lighthouse.

Exercises 11.3

In Exercises 1 to 4, find tan α and tan β for each triangle.

1.

2.

3.

4.

Rectangle *ABCD*

In Exercises 5 to 10, find the value (or expression) for each of the six trigonometric ratios of angle α. Use the Pythagorean Theorem as needed.

5.

6.

7.

8.

9.

10.

In Exercises 11 to 14, use a calculator to find the indicated tangent ratio correct to four decimal places.

11. tan 15°

12. tan 45°

13. tan 57°

14. tan 78°

In Exercises 15 to 20, use the sine, cosine, or tangent ratio to find the lengths of the indicated sides to the nearest tenth of a unit.

15.

16.

17.

18.

19.

20.

Rectangle *ABCD*

In Exercises 21 to 26, use the sine, cosine, or tangent ratio to find the indicated angle measures to the nearest degree.

21.

22.

23.

24.

25.

26.

In Exercises 27 to 32, use a calculator and reciprocal relationships to find each ratio correct to four decimal places.

27. cot 34°

28. sec 15°

29. csc 30°

30. cot 67°

31. sec 42°

32. csc 72°

In Exercises 33 to 36, we expand the list of trigonometric identities. As you may recall, an identity is a statement that is true for all permissible choices of the variable (see page 502).

33. a) For $\alpha \neq 90°$, prove the identity

$$\tan \alpha = \frac{\sin \alpha}{\cos \alpha}$$

(*HINT:* $\sin \alpha = \frac{a}{c}$ and $\cos \alpha = \frac{b}{c}$.)

b) Use your calculator to show that $\tan 23°$ and $\frac{\sin 23°}{\cos 23°}$ are equivalent.

Exercises 33–36

34. a) For $\alpha \neq 0°$, prove the identity $\cot \alpha = \frac{\cos \alpha}{\sin \alpha}$.
    b) Use your calculator to determine $\cot 57°$ by dividing $\cos 57°$ by $\sin 57°$.

35. a) For $\alpha \neq 90°$, prove the identity $\sec \alpha = \frac{1}{\cos \alpha}$.
    b) Use your calculator to determine $\sec 82°$.

36. a) For $\alpha \neq 0°$, prove the identity $\csc \alpha = \frac{1}{\sin \alpha}$.
    b) Use your calculator to determine $\csc 12.3°$.

In Exercises 37 to 43, angle measures should be given to the nearest degree; distances should be given to the nearest tenth of a unit.

37. When her airplane is descending to land, the pilot notes an angle of depression of 5°. If the altimeter shows an altitude reading of 120 ft, what is the distance x from the plane to touchdown?

38. Kristine observes the top of a lookout tower from a point 270 ft from its base. If the indicated angle of elevation measures 37°, how tall is the tower?

39. Find the length of the apothem drawn to of the 6-in. sides of a regular pentagon.

*40. What is the measure of the angle between the diagonal of a cube and the diagonal of the face of the cube?

41. Upon approaching a house, Mary hears Lynn shout to her. Mary, who is standing 10 ft from the house, looks up to see Lynn in the third-story window approximately 32 ft away. What is the measure of the angle of elevation as Mary looks up at Lynn?

*42. While a helicopter hovers 1000 ft above the water, its pilot spies a man in a lifeboat through an angle of depression of 28°. Along a straight line, a rescue boat can also be seen through an angle of depression of 14°. How far is the rescue boat from the lifeboat?

*43. From atop a 200-ft lookout tower, a fire is spotted due north through an angle of depression of 12°. Firefighters located 1000 ft due east of the tower must work their way through heavy foliage to the fire. By their compasses, through what angle (measured from the north toward the west) must the firefighters travel?

44. In the triangle shown, find each measure to the nearest tenth of a unit.
    a) r b) y
    c) A, the area of the triangle

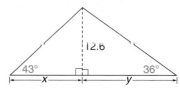

45. At an altitude of 12,000 ft, a pilot sees two towns through angles of depression of 37° and 48° as shown. To the nearest 10 ft, how far apart are the towns?

46. Consider the regular square pyramid shown.
    a) Find the length of the slant height ℓ correct to tenths.
    b) Use ℓ from part (a) to find the lateral area L of the pyramid.

47. Consider the regular square pyramid shown.
    a) Find the height h correct to the nearest tenth of a unit.
    b) Use h from part (a) to find the volume of the pyramid.

Exercises 46, 47

11.4 Applications with Acute Triangles

KEY CONCEPTS

Area of a Triangle:
$$A = \tfrac{1}{2}bc \sin \alpha$$
$$A = \tfrac{1}{2}ac \sin \beta$$
$$A = \tfrac{1}{2}ab \sin \gamma$$

Law of Sines:
$$\frac{\sin \alpha}{a} = \frac{\sin \beta}{b} = \frac{\sin \gamma}{c}$$

Law of Cosines:
$$c^2 = a^2 + b^2 - 2ab \cos \gamma$$
$$b^2 = a^2 + c^2 - 2ac \cos \beta$$
$$a^2 = b^2 + c^2 - 2bc \cos \alpha$$

or

$$\cos \alpha = \frac{b^2 + c^2 - a^2}{2bc}$$
$$\cos \beta = \frac{a^2 + c^2 - b^2}{2ac}$$
$$\cos \gamma = \frac{a^2 + b^2 - c^2}{2ab}$$

In Sections 11.1 through 11.3, our focus was strictly upon right triangles. We now turn our attention to some relationships that we will prove for (and apply with) acute triangles. The first relationship provides a formula for the area of a triangle in which α, β, and γ are all acute angles.

AREA OF A TRIANGLE

THEOREM 11.4.1

The area of an acute triangle equals one-half the product of the lengths of two sides and the sine of the included angle.

(a)

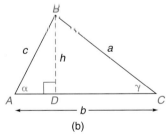

(b)

Figure 11.37

GIVEN: Acute $\triangle ABC$, as shown in Figure 11.37(a).

PROVE: $A = \tfrac{1}{2}bc \sin \alpha$

PROOF: The area of the triangle is given by $A = \tfrac{1}{2}bh$. With the altitude BD of length h [see Figure 11.37(b)], we see that $\sin \alpha = \frac{h}{c}$ in right $\triangle ABD$. Then $h = c \sin \alpha$. Consequently, $A = \tfrac{1}{2}bh$ becomes

$$A = \frac{1}{2}b(c \sin \alpha), \qquad \text{so} \qquad A = \frac{1}{2}bc \sin \alpha.$$

Theorem 11.4.1 has three equivalent forms, as shown in the following box. Proofs of the counterparts are similar to the one above.

AREA OF A TRIANGLE
$$A = \frac{1}{2}bc \sin \alpha$$

Equivalently, we can prove that

$$A = \frac{1}{2}ac \sin \beta$$

$$A = \frac{1}{2}ab \sin \gamma$$

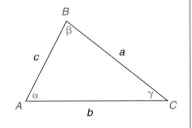

In a more advanced study of trigonometry, the area formula found in Theorem 11.4.1 can also be proved for obtuse triangles. In a right triangle with $\gamma = 90°$, the formula $A = \tfrac{1}{2}ab \sin \gamma$ reduces to $A = \tfrac{1}{2}ab$ since $\sin \gamma = 1$. Recall Corollary 8.1.4.

EXAMPLE 1

In Figure 11.38, find the area of $\triangle ABC$.

Figure 11.38

SOLUTION We use the form $A = \frac{1}{2}bc \sin \alpha$, because α, b, and c are known.

$$A = \frac{1}{2} \cdot 6 \cdot 10 \cdot \sin 33°$$
$$= 30 \cdot \sin 33°$$
$$\approx 16.3 \text{ in}^2$$

EXAMPLE 2

In $\triangle ABC$ in Figure 11.39, $a = 7.6$ and $c = 10.2$. If the area of $\triangle ABC$ is approximately 38.3 square units, find β to the nearest degree. Note that $\angle B$ is acute.

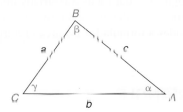

Figure 11.39

SSG **EXS. 1–4**

SOLUTION Using the form $A = \frac{1}{2}ac \sin \beta$, we have

$38.3 = (0.5)(7.6)(10.2) \sin \beta$, or $38.3 = 38.76 \sin \beta$.

Thus, $\sin \beta = \frac{38.3}{38.76}$, so $\beta = \sin^{-1}\left(\frac{38.3}{38.76}\right)$.

Then $\beta \approx 81°$ (rounded from 81.16).

LAW OF SINES

Because the area of a triangle is unique, we can equate the three area expressions characterized by Theorem 11.4.1 as follows:

$$\frac{1}{2}bc \sin \alpha = \frac{1}{2}ac \sin \beta = \frac{1}{2}ab \sin \gamma$$

Multiplying by 2, $\qquad bc \sin \alpha = ac \sin \beta = ab \sin \gamma$

Dividing each part of this equality by abc, we find

$$\frac{bc \sin \alpha}{abc} = \frac{ac \sin \beta}{abc} = \frac{ab \sin \gamma}{abc}$$

So $\qquad \dfrac{\sin \alpha}{a} = \dfrac{\sin \beta}{b} = \dfrac{\sin \gamma}{c}$

This relationship between the lengths of the sides of an acute triangle and the sines of their opposite angles is known as the Law of Sines. In trigonometry, it is shown that the Law of Sines is true for right triangles and obtuse triangles as well.

> **THEOREM 11.4.2 ■ Law of Sines**
>
> In any acute triangle, the three ratios between the sines of the angles and the lengths of the opposite sides are equal. That is,
>
> $$\frac{\sin \alpha}{a} = \frac{\sin \beta}{b} = \frac{\sin \gamma}{c} \qquad \text{or} \qquad \frac{a}{\sin \alpha} = \frac{b}{\sin \beta} = \frac{c}{\sin \gamma}$$

When solving a problem, we equate only two of the equal ratios described in Theorem 11.4.2. For instance, we could use

$$\frac{\sin \alpha}{a} = \frac{\sin \beta}{b} \qquad \text{or} \qquad \frac{\sin \alpha}{a} = \frac{\sin \gamma}{c} \qquad \text{or} \qquad \frac{\sin \beta}{b} = \frac{\sin \gamma}{c}$$

In Example 3, we will use exact values for sin 45° and sin 60° in order to determine an exact length.

EXAMPLE 3

Use the Law of Sines to find the exact length *ST* in Figure 11.40

SOLUTION Because we know *RT* and the measures of angles *S* and *R*, we use $\frac{\sin S}{RT} = \frac{\sin R}{ST}$. Substitution of known values leads to

$$\frac{\sin 45°}{10} = \frac{\sin 60°}{x}$$

Because $\sin 45° = \frac{\sqrt{2}}{2}$ and $\sin 60° = \frac{\sqrt{3}}{2}$, we have

$$\frac{\frac{\sqrt{2}}{2}}{10} = \frac{\frac{\sqrt{3}}{2}}{x}$$

By the Means-Extremes Property of a Proportion,

$$\frac{\sqrt{2}}{2} \cdot x = \frac{\sqrt{3}}{2} \cdot 10$$

Multiplying by 2, we have

$$\sqrt{2} \cdot x = 10\sqrt{3}$$

Then

$$x = \frac{10\sqrt{3}}{\sqrt{2}} = \frac{10\sqrt{3}}{\sqrt{2}} \cdot \frac{\sqrt{2}}{\sqrt{2}} = \frac{10\sqrt{6}}{2} = 5\sqrt{6}$$

Then $ST = 5\sqrt{6}$ m.

EXAMPLE 4

In △*ABC* shown in Figure 11.41, *b* = 12, *c* = 10, and β = 83°. Find γ to the nearest degree.

SOLUTION Knowing values of *b*, *c*, and β, we use the following form of the Law of Sines to find γ:

$$\frac{\sin \beta}{b} = \frac{\sin \gamma}{c}.$$

$$\frac{\sin 83°}{12} = \frac{\sin \gamma}{10}, \qquad \text{so} \qquad 12 \sin \gamma = 10 \sin 83°$$

Then $\sin \gamma = \frac{10 \sin 83°}{12} \approx 0.8271$, so $\gamma = \sin^{-1}(0.8271) \approx 56°$.

Figure 11.41

(a)

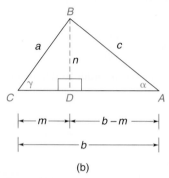

(b)

Figure 11.42

LAW OF COSINES

The final relationship that we consider is again proved only for an acute triangle. Like the Law of Sines, this relationship (known as the Law of Cosines) can be used to find unknown measures in a triangle. The Law of Cosines (which can also be established for obtuse triangles in a more advanced course) can be stated in words, "The square of the length of one side of a triangle equals the sum of the squares of the lengths of the two remaining sides decreased by twice the product of the lengths of those two sides and the cosine of their included angle." See Figure 11.42(a) as you read Theorem 11.4.3.

THEOREM 11.4.3 ■ Law of Cosines

In acute $\triangle ABC$,

$$c^2 = a^2 + b^2 - 2ab \cos \gamma$$
$$b^2 = a^2 + c^2 - 2ac \cos \beta$$
$$a^2 = b^2 + c^2 - 2bc \cos \alpha$$

The proof of the first form of the Law of Cosines follows.

GIVEN: Acute $\triangle ABC$ in Figure 11.42(a)

PROVE: $c^2 = a^2 + b^2 - 2ab \cos \gamma$

PROOF: In Figure 11.42(a), draw the altitude $\overline{BD}$ from B to $\overline{AC}$. We designate lengths of the line segments as shown in Figure 11.42(b). Now

$$(b - m)^2 + n^2 = c^2 \qquad \text{and} \qquad m^2 + n^2 = a^2$$

by applying the Pythagorean Theorem twice.
The second equation is equivalent to $m^2 = a^2 - n^2$. After we expand $(b - m)^2$, the first equation becomes

$$b^2 - 2bm + m^2 + n^2 = c^2$$

Then we replace m^2 by $(a^2 - n^2)$ to obtain

$$b^2 - 2bm + (a^2 - n^2) + n^2 = c^2$$

Simplifying yields

$$c^2 = a^2 + b^2 - 2bm$$

In right $\triangle CDB$,

$$\cos \gamma = \frac{m}{a} \qquad \text{so} \qquad m = a \cos \gamma$$

Hence $\quad c^2 = a^2 + b^2 - 2bm$ becomes
$$c^2 = a^2 + b^2 - 2b(a \cos \gamma)$$
$$c^2 = a^2 + b^2 - 2ab \cos \gamma$$

Arguments similar to the preceding proof can be provided for both remaining forms of the Law of Cosines. The statement $c^2 = a^2 + b^2 - 2ab \cos \gamma$ reduces to the Pythagorean Theorem when $\gamma = 90°$ because $\cos 90° = 0$; thus, $c^2 = a^2 + b^2$ when $\gamma = 90°$.

EXAMPLE 5

Find the length of $\overline{AB}$ in the triangle in Figure 11.43.
Then find m$\angle B$ and m$\angle A$.

Figure 11.43

SOLUTION Referring to the 30° angle as γ, we use
the following form of the Law of Cosines:

$$c^2 = a^2 + b^2 - 2ab \cos \gamma$$
$$c^2 = (4\sqrt{3})^2 + 4^2 - 2 \cdot 4\sqrt{3} \cdot 4 \cdot \cos 30°$$
$$c^2 = 48 + 16 - 2 \cdot 4\sqrt{3} \cdot 4 \cdot \frac{\sqrt{3}}{2}$$
$$c^2 = 48 + 16 - 48$$
$$c^2 = 16$$
$$c = 4$$

Therefore, $AB = 4$ in.

Now $\triangle ABC$ is isosceles because $\overline{AB} \cong \overline{AC}$. Therefore, $\angle B \cong \angle C$. It follows
that m$\angle B = 30°$ and m$\angle A = 120°$.

The Law of Cosines can also be used to find the measure of an angle of a triangle
when the lengths of its three sides are known. In such applications, it is convenient to apply
the following alternative form of Theorem 11.4.3.

THEOREM 11.4.3 ■ Law of Cosines-Alternative Form

In acute $\triangle ABC$,

$$\cos \alpha = \frac{b^2 + c^2 - a^2}{2bc}$$
$$\cos \beta = \frac{a^2 + c^2 - b^2}{2ac}$$
$$\cos \gamma = \frac{a^2 + b^2 - c^2}{2ab}$$

PROOF OF THE THIRD FORM

If $c^2 = a^2 + b^2 - 2ab \cos \gamma$, then

$$2ab \cos \gamma = a^2 + b^2 - c^2.$$

Dividing each side of the equation by $2ab$, we have

$$\cos \gamma = \frac{a^2 + b^2 - c^2}{2ab}.$$

Arguments for the remaining alternative forms are similar.

We use an alternative form of the Law of Cosines in Example 6.

EXAMPLE 6

In acute $\triangle ABC$ in Figure 11.44, find β to the nearest degree.

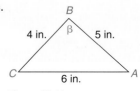

Figure 11.44

SOLUTION The alternative form of the Law of Cosines
involving β is

$$\cos \beta = \frac{a^2 + c^2 - b^2}{2ac}$$

With $a = 4$, $b = 6$, and $c = 5$, we have

$$\cos \beta = \frac{4^2 + 5^2 - 6^2}{2 \cdot 4 \cdot 5}$$

so

$$\cos \beta = \frac{16 + 25 - 36}{40} = \frac{5}{40} = \frac{1}{8} = 0.1250$$

With $\beta = \cos^{-1}(0.1250)$, we find that $\beta \approx 83°$.

To find the measure of a side or of an angle of an acute triangle, we often have to decide which form of the Law of Sines or of the Law of Cosines should be applied. Table 11.5 deals with that question and is based on the acute triangle shown in Figure 11.45. Note that a, b, and c represent the lengths of the sides and that α, β, and γ represent the measures of the opposite angles, respectively.

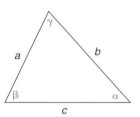

Figure 11.45

Warning

If we know *only* the measures of the three angles of the triangle, then no length of side can be determined.

TABLE 11.5

When to Use the Law of Sines/Law of Cosines

1. *Three sides are known:* Use the Law of Cosines to find *any* angle.
   Known measures: a, b, and c
   Desired measure: α
   ∴ Use $a^2 = b^2 + c^2 - 2bc \cos \alpha$
   or $\cos \alpha = \dfrac{b^2 + c^2 - a^2}{2bc}$

2. *Two sides and a nonincluded angle are known:* Use the Law of Sines to find the remaining nonincluded angle.
   Known measures: a, b, and α
   Desired measure: β
   ∴ Use $\dfrac{\sin \alpha}{a} = \dfrac{\sin \beta}{b}$

3. *Two sides and an included angle are known:* Use the Law of Cosines to find the remaining side.
   Known measures: a, b, and γ
   Desired measure: c
   ∴ Use $c^2 = a^2 + b^2 - 2ab \cos \gamma$

4. *Two angles and a nonincluded side are known:* Use the Law of Sines to find the other nonincluded side.
   Known measures: a, α, and β
   Desired measure: b
   ∴ Use $\dfrac{\sin \alpha}{a} = \dfrac{\sin \beta}{b}$

EXAMPLE 7

In the design of a child's swing set, each of the two metal posts that support the top bar measures 8 ft. At ground level, the posts are to be 6 ft apart (see Figure 11.46 on page 519). At what angle should the two metal posts be secured? Give the answer to the nearest degree.

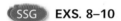

Figure 11.46

SOLUTION Call the desired angle measure α. Because the three sides of the triangle are known, we use the following form of the Law of Cosines:

$$\cos \alpha = \frac{b^2 + c^2 - a^2}{2bc}$$

Because a represents the length of the side opposite the angle α, $a = 6$ while $b = 8$ and $c = 8$. Consequently, we have

$$\cos \alpha = \frac{8^2 + 8^2 - 6^2}{2 \cdot 8 \cdot 8}$$

$$\cos \alpha = \frac{64 + 64 - 36}{128}$$

$$\cos \alpha = \frac{92}{128}, \text{ so } \cos \alpha = \frac{23}{32}.$$

SSG **EXS. 8–10**   Then $\alpha = \cos^{-1}\left(\frac{23}{32}\right)$ and $\alpha \approx 44^\circ$.

Exercises 11.4

In Exercises 1 and 2, use the given information to find an expression for the area of $\triangle ABC$. Give the answer in a form such as $A = \frac{1}{2}(3)(4) \sin 32^\circ$. See the figure for Exercises 1–8.

1. a) $a = 5$, $b = 6$, and $\gamma = 78^\circ$
  b) $a = 5$, $b = 7$, $\alpha = 36^\circ$, and $\beta = 88^\circ$

2. a) $b = 7.3$, $c = 8.6$, and $\alpha = 38^\circ$
  b) $a = 5.3$, $c = 8.4$, $\alpha = 36^\circ$, and $\gamma = 87^\circ$

In Exercises 3 and 4, state the form of the Law of Sines used to solve the problem. Give the answer in a form such as $\frac{\sin 72^\circ}{6.3} = \frac{\sin 55^\circ}{a}$.

3. a) Find β if it is
    known that $a = 5$,
    $b = 8$, and
    $\alpha = 40^\circ$.
  b) Find c if it is
    known that
    $a = 5.3$, $\alpha = 41^\circ$,
    and $\gamma = 87^\circ$.

Exercises 1–8

4. a) Find β if it is known that $b = 8.1$, $c = 8.4$, and $\gamma = 86^\circ$.
  b) Find c if it is known that $a = 5.3$, $\alpha = 40^\circ$, and $\beta = 80^\circ$.

In Exercises 5 and 6, state the form of the Law of Cosines used to solve the problem. Using the values provided, give the answer in a form such as $a^2 = b^2 + c^2 - 2bc \cos \alpha$. See the figure for Exercises 1–8.

5. a) Find c if it is known that $a = 5.2$, $b = 7.9$, and $\gamma = 83^\circ$.
  b) Find α if it is known that $a = 6$, $b = 9$, and $c = 10$.

6. a) Find b if it is known that $a = 5.7$, $c = 8.2$, and $\beta = 79^\circ$.
  b) Find β if it is known that $a = 6$, $b = 8$, and $c = 9$.

In Exercises 7 and 8, state the form of the Law of Sines or the Law of Cosines that you would use to solve the problem. See the figure for Exercises 1–8.

7. a) Find α if you know the values of a, b, and β.
  b) Find α if you know the values of a, b, and c.

8. a) Find b if you know the values of a, c, and β.
  b) Find b if you know the values of a, α, and β.

9. For $\triangle ABC$ (not shown), suppose you know that $a = 3$, $b = 4$, and $c = 5$.
  a) Explain why you do *not* need to apply the Law of Sines or the Law of Cosines to find the measure of γ.
  b) Find γ.

10. For △*ABC* (not shown), suppose you know that *a* = 3, *α* = 57°, and *β* = 84°.
  a) Explain why you do not need to apply the Law of Sines or the Law of Cosines to find the measure of *γ*.
  b) Find *γ*.

In Exercises 11 to 14, find the area of each triangle shown. Give the answer to the nearest tenth of a square unit.

11.

4 in.
30°
8 in.

12.

12 cm
60°
8 cm

13.

40°
6 ft
70°

14.

4 m
64°
10 m

In Exercises 15 and 16, find the area of the given figure. Give the answer to the nearest tenth of a square unit.

15.

Q   *P*
6 ft
25°
M   *N*
Rhombus *MNPQ*

16.

6 in.
5 in.   7 in.
70°   42°
Trapezoid

In Exercises 17 to 22, use a form of the Law of Sines to find the measure of the indicated side or angle. Angle measures should be found to the nearest degree and lengths of sides to the nearest tenth of a unit.

17.
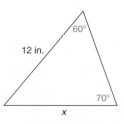
60°
12 in.
70°
x

18.

6 m   10 m
65°   *α*

19.

75°
y
35°
15 m

20.

7 cm
70°   *β*
7 cm

21.

80°   10 in.
γ
12 in.

22.
65°
5 ft   *x*
70°

In Exercises 23 to 28, use a form of the Law of Cosines to find the measure of the indicated side or angle. Angle measures should be found to the nearest degree and lengths of sides to the nearest tenth of a unit.

23.

5   7
α
9

24.
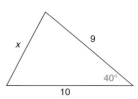
β
6   8
10

25.
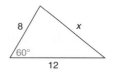
8   *x*
60°
12

26.

x   9
40°
10

27.

D   *C*
x   6
60°
A   4   *B*
Parallelogram *ABCD*

28.
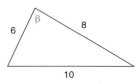
P
6
5   *Q*
M   *α*   7   *N*
$\overrightarrow{MQ}$ bisects ∠*PMN*

In Exercises 29 to 34, use the Law of Sines or the Law of Cosines to solve each problem. Angle measures should be found to the nearest degree and areas and distances to the nearest tenth of a unit.

29 A triangular lot has street dimensions of 150 ft and 180 ft and an included angle of 80° for these two sides.
  a) Find the length of the remaining side of the lot.
  b) Find the area of the lot in square feet.

30. Phil and Matt observe a balloon. They are 500 ft apart, and their angles of observation are 47° and 65°, as shown. Find the distance *x* from Matt to the balloon.

x
47°   65°
500 ft

31. A surveillance aircraft at point *C* sights an ammunition warehouse at *A* and enemy headquarters at *B* through the angles indicated. If points *A* and *B* are 10,000 m apart, what is the distance from the aircraft to enemy headquarters?
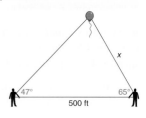
C
53°   62°
A   *B*
├─ 10,000 m ─┤

32. Above one room of a house the rafters meet as shown. What is the measure of the angle *α* at which they meet?

24 ft   *α*   10 ft
26 ft

33. In an A-frame house, a bullet is found embedded at a point 8 ft up the sloped wall. If it was fired at a 30° angle with the horizontal, how far from the base of the wall was the gun fired?

34. Clay pigeons are released at an angle of 30° with the horizontal. Massimo hits one of the clay pigeons when shooting through an angle of elevation of 70°. If the point of release is 120 m from Massimo, how far (x) is he from the target when it is hit?

35. For the triangle shown, the area is exactly $18\sqrt{3}$ units^2. Determine the length x.

36. For the triangle shown, use the Law of Cosines to determine b.

37. In the support structure for the Ferris wheel, $m\angle CAB = 30°$. If $AB = AC = 27$ ft, find BC.

38. Show that the form of the Law of Cosines written $c^2 = a^2 + b^2 - 2ab \cos \gamma$ reduces to the Pythagorean Theorem when $\gamma = 90°$.

39. Explain why the area of the parallelogram shown is given by the formula $A = ab \sin \gamma$.

(*HINT:* You will need to use $\overline{QN}$.)

Exercises 39–42

40. Find the area of $\square MNPQ$ if $a = 8$ cm, $b = 12$ cm, and $\gamma = 70°$. Answer to the nearest tenth of a square centimeter. (See Exercise 39.)

41. Find the area of $\square MNPQ$ if $a = 6.3$ cm, $b = 8.9$ cm, and $\gamma = 67.5°$. Answer to the nearest *tenth* of a square centimeter. (See Exercise 39.)

42. The sides of a rhombus have length a. Two adjacent sides meet to form acute angle θ. Use the formula from Exercise 39 to show that the area of the rhombus is given by $A = a^2 \sin \theta$.

43. Two sides of a triangle have measures a inches and b inches, respectively. In terms of a and b, what is the largest (maximum) possible area for the triangle?

44. Use Theorem 11.4.1 to show that the area of an equilateral triangle with sides of length s is given by $A = \frac{s^2}{4}\sqrt{3}$

PERSPECTIVE ON HISTORY

SKETCH OF PLATO

Plato (428–348 B.C.) was a Greek philosopher who studied under Socrates in Athens until the time of Socrates' death. Because his master had been forced to drink poison, Plato feared for his own life and left Greece to travel. His journey began around 400 B.C. and lasted for 12 years, taking Plato to Egypt, Sicily, and Italy, where he became familiar with the Pythagoreans (see page 381.)

Plato eventually returned to Athens where he formed his own school, the Academy. Though primarily a philosopher, Plato held that the study of mathematical reasoning provided the most perfect training for the mind. So insistent was Plato that his students have some background in geometry that he placed a sign above the door to the Academy that read, "Let no man ignorant of geometry enter here."

Plato was the first to insist that all constructions be performed by using only two instruments, the compass and the straightedge. Given a line segment of length 1, Plato constructed line segments of lengths $\sqrt{2}$, $\sqrt{3}$, and so on. Unlike Archimedes (see page 161), Plato had no interest in applied mathematics. In fact, Plato's methodology was quite strict and required accurate definitions, precise hypotheses, and logical reasoning. Without doubt, his methods paved the way for the compilation of geometric knowledge in the form of *The Elements* by Euclid (see page 112).

Commenting on the life of Plato, Proclus stated that Plato caused mathematics (and geometry in particular) to make great advances. At that time, many of the discoveries in mathematics were made by Plato's students and by those who studied at the Academy after the death of Plato. It is ironic that although Plato was not himself a truly great mathematician, yet he was largely responsible for its development in his time.

PERSPECTIVE ON APPLICATIONS

RADIAN MEASURE OF ANGLES

In much of this textbook, we have considered angle measures from 0° to 180°. As you study mathematics, you will find that two things are true:

1. Angle measures used in applications do not have to be limited to measures from 0° to 180°.
2. The degree is not the only unit used in measuring angles.

We will address the first of these issues in Examples 1, 2, and 3.

EXAMPLE 1

As the time changes from 1 P.M. to 1:45 P.M., through what angle does the minute hand rotate? See Figure 11.47.

(a)   (b)

Figure 11.47

SOLUTION Because the rotation is $\frac{3}{4}$ of a complete circle (360°), the result is $\frac{3}{4}$ (360°) or 270°.

EXAMPLE 2

An airplane pilot is instructed to circle the control tower twice during a holding pattern before receiving clearance to land. Through what angle does the airplane move? See Figure 11.48.

Figure 11.48

SOLUTION Two circular rotations give 2(360°) or 720°.

In trigonometry, negative measures for angles are used to distinguish the direction of rotation. A counterclockwise rotation is measured as positive, a clockwise rotation as negative. The arcs with arrows in Figure 11.49 are used to indicate the direction of rotation.

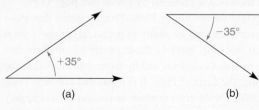

(a)   (b)

Figure 11.49

EXAMPLE 3

To tighten a hex bolt, a mechanic applies clockwise rotations of 60° several times. What is the measure of each rotation? See Figure 11.50.

Figure 11.50

SOLUTION Tightening occurs if the angle is −60°.

NOTE: If the angle of rotation is 60° (that is, +60°), the bolt is loosened.

Our second concern is with an alternative unit for measuring angles, a unit often used in the study of trigonometry and calculus.

DEFINITION

In a circle, a **radian** (rad) is the measure of a central angle that intercepts an arc whose length is equal to the radius of the circle.

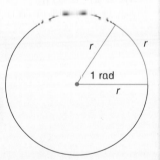

Figure 11.51

In Figure 11.51, the length of each radius and the intercepted arc are all equal to r. Thus, the central angle shown measures 1 radian. A complete rotation about the circle corresponds to 360° and to $2\pi r$. Thus, the arc length of 1 radius corresponds to the central angle measure of 1 rad, and the circumference of 2π radii corresponds to the complete rotation of 2π rad.

The angle relationship found in the preceding paragraph allows us to equate 360° and 2π radians. As suggested by Figure 11.52, there are approximately 6.28 rad (or exactly 2π radians) about the circle. The exact result leads to an important relationship.

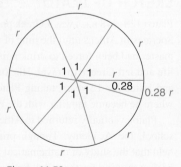

Figure 11.52

$$2\pi \text{ rad} = 360°$$

or

$$360° = 2\pi \text{ rad}$$

Through division by 2, the preceding relationship is often restated as follows:

$$\pi \, \text{rad} = 180°$$

or

$$180° = \pi \, \text{rad}$$

With $\pi \, \text{rad} = 180°$, we divide each side of this equation by π to obtain the following relationship:

$$1 \, \text{rad} = \frac{180°}{\pi} \approx 57.3°$$

To compare angle measures, we can also divide each side of the equation $180° = \pi \, \text{rad}$ by 180 to get the following relationship:

$$1° = \frac{\pi}{180} \, \text{rad}$$

EXAMPLE 4

Using the fact that $1° = \frac{\pi}{180}$ rad, find the radian equivalencies for:

a) 30° **b)** 45° **c)** 60° **d)** $-90°$

SOLUTION

a) $30° = 30(1°) = 30\left(\frac{\pi}{180} \, \text{rad}\right) = \frac{\pi}{6} \, \text{rad}$

b) $45° = 45(1°) = 45\left(\frac{\pi}{180} \, \text{rad}\right) = \frac{\pi}{4} \, \text{rad}$

c) $60° = 60(1°) = 60\left(\frac{\pi}{180} \, \text{rad}\right) = \frac{\pi}{3} \, \text{rad}$

d) $-90° = -90(1°) = -90\left(\frac{\pi}{180}\right) \text{rad} = -\frac{\pi}{2} \, \text{rad}$

EXAMPLE 5

Using the fact that $\pi \, \text{rad} = 180°$, find the degree equivalencies for the following angles measured in radians:

a) $\frac{\pi}{6}$ **b)** $\frac{2\pi}{5}$ **c)** $\frac{-3\pi}{4}$ **d)** $\frac{\pi}{2}$

SOLUTION

a) $\frac{\pi}{6} = \frac{180°}{6} = 30°$

b) $\frac{2\pi}{5} = \frac{2}{5} \cdot \pi = \frac{2}{5} \cdot 180° = 72°$

c) $\frac{-3\pi}{4} = -\frac{3}{4} \cdot \pi = -\frac{3}{4} \cdot 180° = -135°$

d) $\frac{\pi}{2} = \frac{180°}{2} = 90°$

Although we did not use this method of measuring angles in the earlier part of this textbook, you may need to use this method of angle measurement in a more advanced course.

Summary

A Look Back at Chapter 11

One goal of this chapter was to define the sine, cosine, and tangent ratios in terms of the sides of a right triangle. We derived a formula for finding the area of a triangle, given two sides and the included angle. We also proved the Law of Sines and the Law of Cosines for acute triangles. Another unit for measuring angles, the radian, was introduced in the Perspective on Applications section.

Key Concepts

11.1

Greek Letters: α, β, γ, θ • Opposite Side (Leg) • Hypotenuse • Sine Ratio: $\sin \theta = \frac{\text{opposite}}{\text{hypotenuse}}$ • Angle of Elevation • Angle of Depression

11.2

Adjacent Side (Leg) • Cosine Ratio: $\cos \theta = \frac{\text{adjacent}}{\text{hypotenuse}}$ • Identity: $\sin^2 \theta + \cos^2 \theta = 1$

11.3

Tangent Ratio: $\tan \theta = \frac{\text{opposite}}{\text{adjacent}}$ • Cotangent • Secant • Cosecant • Reciprocal Ratios

11.4

Area of a Triangle:
$$A = \frac{1}{2}bc \sin \alpha$$
$$A = \frac{1}{2}ac \sin \beta$$
$$A = \frac{1}{2}ab \sin \gamma \text{ .}$$

Law of Sines: $\dfrac{\sin \alpha}{a} = \dfrac{\sin \beta}{b} = \dfrac{\sin \gamma}{c}$.

Law of Cosines:
$$c^2 = a^2 + b^2 - 2ab \cos \gamma$$
$$b^2 = a^2 + c^2 - 2ac \cos \beta$$
$$a^2 = b^2 + c^2 - 2bc \cos \alpha$$

or
$$\cos \gamma = \frac{a^2 + b^2 - c^2}{2ab}$$
$$\cos \beta = \frac{a^2 + c^2 - b^2}{2ac}$$
$$\cos \alpha = \frac{b^2 + c^2 - a^2}{2bc}$$

Chapter 11   Review Exercises

In Exercises 1 to 4, state the ratio needed, and use it to find the measure of the indicated line segment to the nearest tenth of a unit.

1.

2.

3.

⊘ *ABCD*

4.

Regular pentagon with radius = 5 ft

In Exercises 5 to 8, state the ratio needed, and use it to find the measure of the indicated angle to the nearest degree.

5.

6.

Isosceles trapezoid *ABCD*

7.

Rhombus *ABCD*

8.

Circle *O*

In Exercises 9 to 12, use the Law of Sines or the Law of Cosines to find the indicated length of side or angle measure. Angle measures should be found to the nearest degree; distances should be found to the nearest tenth of a unit.

9.

10.

11.

12.

In Exercises 13 to 17, use the Law of Sines or the Law of Cosines to solve each problem. Angle measures should be found to the nearest degree; distances should be found to the nearest tenth of a unit.

13. A building 50 ft tall is on a hillside. A surveyor at a point on the hill observes that the angle of elevation to the top of the building measures 43° and the angle of depression to the base of the building measures 16°. How far is the surveyor from the base of the building?

14. Two sides of a parallelogram are 50 cm and 70 cm long. Find the length of the shorter diagonal if a larger angle of the parallelogram measures 105°.

15. The length of the sides of a rhombus are 6 in. each, and the longer diagonal measures 11 in. Find the measure of each of the acute angles of the rhombus.

16. The area of $\triangle ABC$ is 9.7 in.² If $a = 6$ in. and $c = 4$ in., find the measure of angle B.

17. Find the area of the rhombus in Exercise 15.

In Exercises 18 to 20, prove each statement without using a table or a calculator. Draw an appropriate right triangle.

18. If $m\angle R = 45°$, then $\tan R = 1$.

19. If $m\angle S = 30°$, then $\sin S = \frac{1}{2}$.

20. If $m\angle T = 60°$, then $\sin T = \frac{\sqrt{3}}{2}$.

In Exercises 21 to 30, use the drawings where provided to solve each problem. Angle measures should be found to the nearest degree; lengths should be found to the nearest tenth of a unit.

21. In the evening, a tree that stands 12 ft tall casts a long shadow. If the angle of depression from the top of the tree to the tip of the shadow is 55°, what is the length of the shadow?

22. A rocket is shot into the air at an angle of 60°. If it is traveling at 200 ft per second, how high in the air is it after 5 seconds? (Ignoring gravity, assume that the path of the rocket is a straight line.)

23. A 4-m beam is used to brace a wall. If the bottom of the beam is 3 m from the base of the wall, what is the angle of elevation to the top of the wall?

24. The basket of a hot-air balloon is 300 ft high. The pilot of the balloon observes a stadium 2200 ft away. What is the measure of the angle of depression?

25. The apothem length of a regular pentagon is approximately 3.44 cm. What is the approximate length of each side of the pentagon?

26. What is the approximate length of the radius of the pentagon in Exercise 25?

27. Each of the legs of an isosceles triangle is 40 cm in length. The base is 30 cm in length. Find the measure of a base angle.

28. The diagonals of a rhombus measure 12 in. and 16 in. Find the measure of the obtuse angle of the rhombus.

29. The term used for measuring the steepness of a hill is the **grade**. A grade of a to b means the hill rises a vertical units for every b horizontal units. If, at some point, the hill is 3 ft above the horizontal and the angle of elevation to that point is 23°, what is the grade of this hill?

30. An observer in a plane 2500 m high sights two ships below. The angle of depression to one ship is 32°, and the angle of depression to the other ship is 44°. How far apart are the ships?

31. If $\sin\theta = \frac{7}{25}$, find $\cos\theta$ and $\sec\theta$.

32. If $\tan\theta = \frac{11}{60}$, find $\sec\theta$ and $\cot\theta$.

33. If $\cot\theta = \frac{21}{20}$, find $\csc\theta$ and $\sin\theta$.

34. In a right circular cone, the radius of the base is 3.2 ft in length and the angle formed by the radius and slant height measures $\theta = 65°$. To the nearest tenth of a foot, find the length of the altitude of the cone. Then use this length of altitude to find the volume of the cone to the nearest tenth of a cubic foot.

Chapter 11 | Test

1. For the right triangle shown, express each of the following in terms of a, b, and c:
  a) $\sin\alpha$
  b) $\tan\beta$ _____

2. For the right triangle shown, express each ratio as a fraction in lowest terms:
  a) $\cos\beta$ _____
  b) $\sin\alpha$ _____

3. Without using a calculator, find the exact value of:
  a) $\tan 45°$ _____ b) $\sin 60°$ _____

4. Use your calculator to find each number correct to four decimal places.
  a) $\sin 23°$ _____ b) $\cos 79°$ _____

5. Using your calculator, find θ to the nearest degree if $\sin\theta = 0.6691.$ _____

6. Without the calculator, determine which number is larger:
  a) $\tan 25°$ or $\tan 26°$ _____
  b) $\cos 47°$ or $\cos 48°$ _____

7. In the drawing provided, find the value of a to the nearest whole number. _____

8. In the drawing provided, find the value of y to the nearest whole number. _____

9. In the drawing provided, find the measure of θ to the nearest degree. _____

10. Using the drawing below, classify each statement as true or false:
  a) $\cos\beta = \sin\alpha$ _____
  b) $\sin^2\alpha + \cos^2\alpha = 1$ _____

11. A kite is flying at an angle of elevation of 67° with the ground. If 100 feet of string have been paid out to the kite, how far is the kite above the ground? Answer to the nearest foot. _____

12. A roofline shows a span of 12 feet across a sloped roof, and this span is accompanied by a 2-foot rise. To the nearest degree, find the measure of θ. _____

13. If $\sin \alpha = \frac{1}{2}$, find:

   a) $\csc \alpha$ _____   b) α _____

14. In a right triangle with acute angles of measures α and β, $\cos \beta = \frac{a}{c}$. Find the following values in terms of the lengths of sides a, b, and c:

   a) $\sin \alpha$ _____   b) $\sec \beta$ _____

15. Use one of the three forms for area (such as the form $A = \frac{1}{2}bc \sin \alpha$) to find the area of the triangle shown. Answer to the nearest whole number. _____

16. On the basis of the drawing provided, complete the Law of Sines.

$$\frac{\sin \alpha}{a} = \underline{\hspace{1.5cm}} = \underline{\hspace{1.5cm}}$$

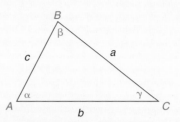

Exercises 16, 17

17. On the basis of the drawing provided, complete this form of the Law of Cosines.

$a^2 = \underline{\hspace{5cm}}$

18. Use the Law of Sines or the Law of Cosines to find α to the nearest degree. _____

19. Use the Law of Sines or the Law of Cosines to find length x to the nearest whole number. _____

20. Each apothem of regular pentagon $ABCDE$ has length a. In terms of a, find an expression for the area A of pentagon $ABCDE$. _____

Algebra Review

A.1 ALGEBRAIC EXPRESSIONS

In algebra, we do not define terms such as *addition, multiplication, number, positive,* and *equality*. However, a *real number* is defined as any number that has a position on the number line, as shown in Figure A.1.

Figure A.1

Any real number positioned to the right of another real number is larger than the number to its left. For example, 4 is larger than -2; equivalently, -2 is less than 4 (smaller numbers are to the left). Numbers such as 3 and -3 are **opposites** or **additive inverses**. Two numerical expressions are **equal** if and only if they have the same value; for example, $2 + 3 = 5$. The axioms of equality are listed in the following list; they are also listed in Section 1.6.

AXIOMS OF EQUALITY

Reflexive $(a = a)$: Any number equals itself.

Symmetric (if $a = b$, then $b = a$): Two equal numbers are equal in either order.

Transitive (if $a = b$ and $b = c$, then $a = c$): If a first number equals a second number and the second number equals a third number, then the first number equals the third number.

Substitution: If one numerical expression equals a second, then it may replace the second.

EXAMPLE 1

Name the axiom of equality illustrated in each case.

a) If AB is the numerical length of the line segment $\overline{AB}$, then $AB = AB$.
b) If $17 = 2x - 3$, then $2x - 3 = 17$.
c) Given that $2x + 3x = 5x$, the statement $2x + 3x = 30$ can be replaced by $5x = 30$.

SOLUTION a) Reflexive   b) Symmetric   c) Substitution

To add two real numbers, think of positive numbers as gains and negative numbers as losses. For instance, $13 + (-5)$ represents the result of combining a gain of $13 with a loss (or debt) of $5; because the outcome is a gain of $8, we have

$$13 + (-5) = 8$$

The answer in addition is the **sum**. Three more examples of addition are

$$13 + 5 = 18 \quad \text{and} \quad (-13) + 5 = -8 \quad \text{and} \quad (-13) + (-5) = -18.$$

If you multiply two *real* numbers, the **product** (answer) will be *positive* if the two numbers have the *same* sign, *negative* if the two numbers have *different signs*, and 0 if either number is 0 or both numbers are 0.

EXAMPLE 2

Simplify each expression:

a) $5 + (-4)$	**b)** $5(-4)$	**c)** $(-7)(-6)$
d) $[5 + (-4)] + 8$	**e)** $7a(0) \cdot 3 \cdot (-5)$	

SOLUTION

a) $5 + (-4) = 1$

b) $5(-4) = -20$

c) $(-7)(-6) = 42$

d) $[5 + (-4)] + 8 = 1 + 8 = 9$

e) $7a(0) \cdot 3(-5) = 0$

Just as $(-3) + 9 = 6$ and $9 + (-3) = 6$, any two sums are equal when the order of the numbers added is reversed. This is often expressed by writing $a + b = b + a$, the property of real numbers known as the Commutative Axiom for Addition. There is also a Commutative Axiom for Multiplication, which is illustrated by the fact that $(6)(-4) = (-4)(6)$; both products are -24.

In a numerical expression, grouping symbols such as brackets and parentheses indicate which operation should be performed first. However, $[5 + (-4)] + 8$ equals $5 + [(-4) + 8]$ because $1 + 8$ equals $5 + 4$. In general, the fact that $(a + b) + c$ equals $a + (b + c)$ is known as the Associative Axiom for Addition. There is also an Associative Axiom for Multiplication, expressed by $a \cdot (b \cdot c) = (a \cdot b) \cdot c$ and illustrated below:

$$(3 \cdot 5)(-2) = 3[5(-2)]$$
$$(15)(-2) = 3(-10)$$
$$-30 = -30$$

SELECTED AXIOMS OF REAL NUMBERS

Commutative Axiom for Addition: $a + b = b + a$

Commutative Axiom for Multiplication: $a \cdot b = b \cdot a$

Associative Axiom for Addition: $a + (b + c) = (a + b) + c$

Associative Axiom for Multiplication: $a \cdot (b \cdot c) = (a \cdot b) \cdot c$

To subtract b from a (to find $a - b$), we change the subtraction problem to the addition problem $a + (-b)$. The answer in subtraction is the **difference** between a and b.

DEFINITION OF SUBTRACTION

$$a - b = a + (-b)$$

where $-b$ is the additive inverse (or opposite) of b.

For $b = 5$, we have $-b = -5$; and for $b = -2$, we have $-b = 2$. For the subtraction $a - (b + c)$, we use the additive inverse of $b + c$, which is $(-b) + (-c)$. That is,

$$a - (b + c) = a + [(-b) + (-c)]$$

EXAMPLE 3

Simplify each expression:

a) $5 - (-2)$ **b)** $(-7) - (-3)$ **c)** $12 - [3 + (-2)]$

SOLUTION
a) $5 - (-2) = 5 + 2 = 7$
b) $(-7) - (-3) = (-7) + 3 = -4$
c) $12 - [3 + (-2)] = 12 + [(-3) + 2] = 12 + (-1) = 11$

NOTE: In part (c), $12 - [3 + (-2)] = 12 - 1$ or 11.

Division can be replaced by multiplication just as subtraction was replaced by addition. We cannot divide by 0! Two numbers whose product is 1 are called **multiplicative inverses** (or **reciprocals**); $-\frac{3}{4}$ and $-\frac{4}{3}$ are multiplicative inverses because $-\frac{3}{4} \cdot -\frac{4}{3} = 1$. The answer in division is the **quotient**.

DEFINITION OF DIVISION

For $b \neq 0$,

$$a \div b = a \cdot \frac{1}{b}$$

where $\frac{1}{b}$ is the multiplicative inverse of b.

NOTE: $a \div b$ is also indicated by a/b or $\frac{a}{b}$.

For $b = 5$ (that is, $b = \frac{5}{1}$), we have $\frac{1}{b} = \frac{1}{5}$; and for $b = -\frac{2}{3}$, we have $\frac{1}{b} = -\frac{3}{2}$.

EXAMPLE 4

Simplify each expression:

a) $12 \div 2$ **b)** $(-5) \div \left(-\frac{2}{3}\right)$

SOLUTION
a) $12 \div 2 = 12 \div \frac{2}{1}$
$= \frac{12}{1} \cdot \frac{1}{2}$ (product of two positive numbers
$= 6$ is a positive number)
b) $(-5) \div \left(-\frac{2}{3}\right) = \left(-\frac{5}{1}\right) \div \left(-\frac{2}{3}\right)$
$= \left(-\frac{5}{1}\right) \cdot \left(-\frac{3}{2}\right)$
$= \frac{15}{2}$ (product of two negative numbers
                is a positive number)

EXAMPLE 5

Morgan works at the grocery store for 3 hours on Friday after school and for 8 hours on Saturday. If he is paid $9 per hour, how much will he be paid in all?

SOLUTION
Method I: Find the total number of hours worked and multiply by 9.

$$9(3 + 8) = 9 \cdot 11 = \$99$$

Method II: Figure the daily wages and add them.

$$(9 \cdot 3) + (9 \cdot 8) = 27 + 72 = \$99$$

Friday's   Saturday's
wages   wages

NOTE: We see that $9(3 + 8) = 9 \cdot 3 + 9 \cdot 8$, where the multiplications on the right are performed before the addition is completed.

The Distributive Axiom was illustrated in Example 5. Because multiplications are performed before additions, we write

$$a(b + c) = a \cdot b + a \cdot c$$
$$2(3 + 4) = 2 \cdot 3 + 2 \cdot 4$$
$$2(7) = 6 + 8$$

The "symmetric" form of the Distributive Axiom is

$$a \cdot b + a \cdot c = a(b + c)$$

This form can be used to combine *like terms* (expressions that contain the same variable factors). A **variable** is a letter that represents a number.

$$4x + 5x = x \cdot 4 + x \cdot 5 \quad \text{(Commutative Axiom for Multiplication)}$$
$$= x(4 + 5) \quad \text{(Symmetric Form of Distributive Axiom)}$$
$$= x(9) \quad \text{(Substitution)}$$
$$= 9x \quad \text{(Commutative Axiom for Multiplication)}$$
$$\therefore 4x + 5x = 9x$$

The Distributive Axiom also distributes multiplication over subtraction.

FORMS OF THE DISTRIBUTIVE AXIOM

$$a(b + c) = a \cdot b + a \cdot c$$
$$a \cdot b + a \cdot c = a(b + c)$$
$$a(b - c) = a \cdot b - a \cdot c$$
$$a \cdot b - a \cdot c = a(b - c)$$

EXAMPLE 6

Combine like terms:

a) $7x + 3x$ b) $7x - 3x$ c) $3x^2y + 4x^2y + 6x^2y$

d) $3x^2y + 4xy^2 + 6xy^2$ e) $7x + 5y$

SOLUTION

a) $7x + 3x = 10x$

b) $7x - 3x = 4x$

c) $3x^2y + 4x^2y + 6x^2y = (3x^2y + 4x^2y) + 6x^2y = 7x^2y + 6x^2y = 13x^2y$

d) $3x^2y + 4xy^2 + 6xy^2 = 3x^2y + (4xy^2 + 6xy^2) = 3x^2y + 10xy^2$

e) $7x + 5y$; cannot combine unlike terms

NOTE: In part (d), $3x^2y$ and $10xy^2$ are not like terms because $x^2y \neq xy^2$.

The statement $4x + 5x = 9x$ says that "the sum of 4 times a number and 5 times the same number equals 9 times the same number." Because x can be any real number, we can also write

$$4\pi + 5\pi = 9\pi$$

in which π is the real number that equals approximately 3.14. Similarly,

$$4\sqrt{3} + 5\sqrt{3} = 9\sqrt{3}$$

in which $\sqrt{3}$ (read "the positive square root of 3") is equal to approximately 1.73.

You may recall the "order of operations" from a previous class; this order is used when simplifying more complicated expressions.

ORDER OF OPERATIONS

1. Simplify expressions within symbols such as parentheses ( ) or brackets [ ], beginning with the innermost symbols of inclusion.

NOTE: The presence of a fraction bar, ——, requires that you simplify a numerator or denominator before dividing.

2. Perform all calculations with exponents.
3. Perform all multiplications and/or divisions in order from left to right.
4. Last, perform all additions and/or subtractions in order from left to right.

EXAMPLE 7

Simplify each numerical expression:

a) $3^2 + 4^2$ **b)** $4 \cdot 7 \div 2$ **c)** $2 \cdot 3 \cdot 5^2$

d) $\dfrac{8 - 6 \div (-3)}{4 + 3(2 + 5)}$ **e)** $2 + [3 + 4(5 - 1)]$

SOLUTION

a) $3^2 + 4^2 = 9 + 16 = 25$
b) $4 \cdot 7 \div 2 = 28 \div 2 = 14$
c) $2 \cdot 3 \cdot 5^2 = 2 \cdot 3 \cdot 25$
$= (2 \cdot 3) \cdot 25 = 6 \cdot 25 = 150$
d) $\dfrac{8 - [6 \div (-3)]}{4 + 3(2 + 5)} = \dfrac{8 - (-2)}{4 + 3(7)} = \dfrac{10}{4 + 21} = \dfrac{10}{25} = \dfrac{2}{5}$
e) $2 + [3 + 4(5 - 1)] = 2 + [3 + 4(4)] = 2 + (3 + 16) = 2 + 19 = 21$

An expression such as $(2 + 5)(6 + 4)$ can be simplified by two different methods. By following the rules of order, we have (7)(10), or 70. An alternative method is described as the FOIL method: First, Outside, Inside, and Last pairs of terms are multiplied and then added. This is how it works:

$$(2 + 5)(6 + 4) = 2 \cdot 6 + 2 \cdot 4 + 5 \cdot 6 + 5 \cdot 4$$
$$= 12 + 8 + 30 + 20$$
$$= 70$$

FOIL is the Distributive Axiom in disguise. We would not generally use FOIL to find the product of $(2 + 5)$ and $(6 + 4)$, but we must use it to find products such as those found in Example 8. Also see Example 2 in Section A.2.

EXAMPLE 8

Use the FOIL method to find the products.

a) $(3x + 4)(2x - 3)$ **b)** $(5x + 2y)(6x - 5y)$

SOLUTION

a) $(3x + 4)(2x - 3)$

$= 3x \cdot 2x + 3x(-3) + 4(2x) + 4(-3)$
$= 6x^2 + (-9x) + 8x + (-12)$
$= 6x^2 - 1x - 12$
$= 6x^2 - x - 12$

b) $(5x + 2y)(6x - 5y) = 5x \cdot 6x + 5x(-5y) + 2y(6x) + 2y(-5y)$
$= 30x^2 + (-25xy) + 12xy + (-10y^2)$
$= 30x^2 - 13xy - 10y^2$

EXAMPLE 9

Use FOIL to express $ab + ac + db + dc$ in factored form as a product.

SOLUTION

$$ab + ac + db + dc = a(b + c) + d(b + c)$$
$$= (b + c)(a + d)$$

Exercises A.1

1 Name the four parts of a mathematical system.

(HINT: See Section 1.3.)

2. Name two examples of mathematical systems.

3. Which axiom of equality is illustrated in each of the following?
a) $5 = 5$
b) If $\frac{1}{2} = 0.5$ and $0.5 = 50\%$, then $\frac{1}{2} = 50\%$.
c) Because $2 + 3 = 5$, we may replace $x + (2 + 3)$ by $x + 5$.
d) If $7 = 2x - 3$, then $2x - 3 = 7$.

4. Give an example to illustrate each axiom of equality:
a) Reflexive   c) Transitive
b) Symmetric   d) Substitution

5. Find each sum:
a) $5 + 7$ c) $(-5) + 7$
b) $5 + (-7)$ d) $(-5) + (-7)$

6. Find each sum:
a) $(-7) + 15$ c) $(-7) + (-15)$
b) $7 + (-15)$ d) $(-7) + [(-7) + 15]$

7. Find each product:
a) $5 \cdot 7$ c) $(-5)7$
b) $5(-7)$ d) $(-5)(-7)$

8. Find each product:
a) $(-7)(12)$ c) $(-7)[(3)(4)]$
b) $(-7)(-12)$ d) $(-7)[(3)(-4)]$

9. The area (the number of squares) of the rectangle in the accompanying drawing can be determined by multiplying the measures of the two dimensions. Will the order of multiplication change the answer? Which axiom is illustrated?

10. Identify the axiom of real numbers illustrated. Give a complete answer, such as Commutative Axiom for Multiplication.
a) $7(5) = 5(7)$
b) $(3 + 4) + 5 = 3 + (4 + 5)$
c) $(-2) + 3 = 3 + (-2)$
d) $(2 \cdot 3) \cdot 5 = 2 \cdot (3 \cdot 5)$

11. Perform each subtraction:
a) $7 - (-2)$ c) $10 - 2$
b) $(-7) - (+2)$ d) $(-10) - (-2)$

12. The temperature changes from $-3°F$ at 2 A.M. to $7°F$ at 7 A.M. Which expression represents the difference in temperatures from 2 A.M. to 7 A.M., $7 - (-3)$ or $(-3) - 7$?

13. Complete each division:
a) $12 \div (-3)$ c) $(-12) \div \left(-\frac{2}{3}\right)$
b) $12 \div \left(-\frac{1}{3}\right)$ d) $\left(-\frac{1}{12}\right) \div \left(\frac{1}{3}\right)$

14. Nine pegs are evenly spaced on a board so that the distance from each end to a peg equals the distance between any two pegs. If the board is 5 feet long, how far apart are the pegs?

15. The four owners of a shop realize a loss of $240 in February. If the loss is shared equally, what number represents the profit for each owner for that month?

16. Bill works at a weekend convention by selling copies of a book. He receives a $2 commission for each copy sold. If he sells 25 copies on Saturday and 30 copies on Sunday, what is Bill's total commission?

17. Use the Distributive Axiom to simplify each expression:
    a) $5(6 + 7)$ c) $\frac{1}{2}(7 + 11)$
    b) $4(7 - 3)$ d) $5x + 3x$

18. Use the Distributive Axiom to simplify each expression:
    a) $6(9 - 4)$ c) $7y - 2y$
    b) $\left(\frac{1}{2}\right) \cdot 6(4 + 8)$ d) $16x + 8x$

19. Simplify each expression:
    a) $6\pi + 4\pi$ c) $16x^2y - 9x^2y$
    b) $8\sqrt{2} + 3\sqrt{2}$ d) $9\sqrt{3} - 2\sqrt{3}$

20. Simplify each expression:
    a) $\pi r^2 + 2\pi r^2$ c) $7x^2y + 3xy^2$
    b) $7xy + 3xy$ d) $x + x + y$

21. Simplify each expression:
    a) $2 + 3 \cdot 4$ c) $2 + 3 \cdot 2^2$
    b) $(2 + 3) \cdot 4$ d) $2 + (3 \cdot 2)^2$

22. Simplify each expression:
    a) $3^2 + 4^2$ c) $3^2 + (8 - 2) \div 3$
    b) $(3 + 4)^2$ d) $[3^2 + (8 - 2)] \div 3$

23. Simplify each expression:
    a) $\dfrac{8 - 2}{2 - 8}$ c) $\dfrac{5 \cdot 2 - 6 \cdot 3}{7 - (-2)}$
    b) $\dfrac{8 - 2 \cdot 3}{(8 - 2) \cdot 3}$ d) $\dfrac{5 - 2 \cdot 6 + (-3)}{(-2)^2 + 4^2}$

24. Use the FOIL method to complete each multiplication:
    a) $(2 + 3)(4 + 5)$ b) $(7 - 2)(6 + 1)$

25. Use the FOIL method to complete each multiplication:
    a) $(3 - 1)(5 - 2)$ b) $(3x + 2)(4x - 5)$

26. Use the FOIL method to complete each multiplication:
    a) $(5x + 3)(2x - 7)$ b) $(2x + y)(3x - 5y)$

27. Using x and y, find an expression for the length of the pegged board shown in the accompanying figure.

28. The cardboard used in the construction of the box shown in the accompanying figure has an area of $xy + yz + xz + xz + yz + xy$. Simplify this expression for the total area of the cardboard.

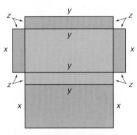

29. A large star is to be constructed, with lengths as shown in the accompanying figure. Give an expression for the total length of the wood strips used in the construction.

30. The area of an enclosed plot of ground that a farmer has subdivided can be found by multiplying $(x + y)$ times $(y + z)$. Use FOIL to complete the multiplication. How does this product compare with the total of the areas of the four smaller plots?

31. The degree measures of the angles of a triangle are $3x$, $5x$, and $2x$. Find an expression for the sum of the measures of these angles in terms of x.

32. The right circular cylinder shown in the accompanying figure has circular bases that have areas of 9π square units. The side has an area of 48π square units. Find an expression for the total surface area.

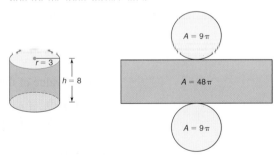

A.2   FORMULAS AND EQUATIONS

A **variable** is a letter used to represent an unknown number. However, the number represented by the Greek letter π is known as a **constant** because it always equals the same number (approximately 3.14); in reality, any real number, such as 5, $\sqrt{2}$, or -1, is a constant. Although we often use x, y, and z as variables, it is convenient to choose r to represent the measure of a radius, h for the measure of height, b for the measure of a base, and so on.

Figure A.2

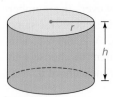

Figure A.3

EXAMPLE 1

For Figure A.2, combine like terms to find the perimeter P (sum of the lengths of all sides) of the figure.

SOLUTION

$$P = (x - 1) + (x + 2) + 2x + (2x + 1)$$
$$= x + (-1) + x + 2 + 2x + 2x + 1$$
$$= 1x + 1x + 2x + 2x + (-1) + 2 + 1$$
$$= 6x + 2$$

When the FOIL method is used with variable expressions, we combine like terms in the simplification.

EXAMPLE 2

Find a simplified expression for the product $(x + 5)(x + 2)$.

SOLUTION

$$(x + 5)(x + 2)$$

$$= x \cdot x + 2 \cdot x + 5 \cdot x + 10$$
$$= x^2 + 7x + 10$$

In Example 2, we multiplied by the FOIL method before adding like terms in accordance with rules of order. When evaluating a variable expression, we must also follow that order. For instance, the value of $a^2 + b^2$ when $a = 3$ and $b = 4$ is given by

$$3^2 + 4^2 \qquad \text{or} \qquad 9 + 16 \qquad \text{or} \qquad 25$$

because exponential expressions represent multiplications and so must be simplified before addition occurs.

EXAMPLE 3

Find the value of the following expressions.

a) $\pi r^2 h$, if $r = 3$ and $h = 4$ (leave π in the answer)
b) $\frac{1}{2}h(b + B)$, if $h = 10$, $b = 7$, and $B = 13$

SOLUTION
a) $\pi r^2 h = \pi \cdot 3^2 \cdot 4$
$$= \pi \cdot 9 \cdot 4 = \pi(36) = 36\pi$$
b) $\frac{1}{2}h(b + B) = \frac{1}{2} \cdot 10(7 + 13)$
$$= \frac{1}{2} \cdot 10(20)$$
$$= \frac{1}{2} \cdot 10 \cdot 20$$
$$= 5 \cdot 20 = 100$$

Many variable expressions are found in formulas. A **formula** is an equation that expresses a rule. For example, $V = \pi r^2 h$ is a formula for calculating the volume V of a right circular cylinder whose altitude has length h and for which the circular base has a radius of length r. See Figure A.3.

EXAMPLE 4

Given the formula $P = 2\ell + 2w$, find the value of P when $\ell = 7$ and $w = 3$.

SOLUTION By substitution, $P = 2\ell + 2w$ becomes

$$P = (2 \cdot 7) + (2 \cdot 3)$$
$$= 14 + 6$$
$$= 20$$

An **equation** is a statement that equates two expressions. Although formulas are special types of equations, most equations are not formulas. Consider the following four examples of equations:

$$x + (x + 1) = 7$$
$$2(x + 1) = 8 - 2x$$
$$x^2 - 6x + 8 = 0$$
$$P = 2\ell + 2w \quad \text{(a formula)}$$

The phrase *solving an equation* means finding the values of the variable that make the equation true when the variable is replaced by those values. These values are known as **solutions** for the equation. For example, 3 is a solution (in fact, the only solution) for the equation $x + (x + 1) = 7$ because $3 + (3 + 1) = 7$ is true.

When each side of an equation is transformed (changed) without having its solution(s) changed, we say that an **equivalent equation** is produced. Some of the properties used to produce equivalent equations when solving an equation are listed in the following box.

PROPERTIES FOR EQUATION SOLVING

Addition Property of Equality (if $a = b$, then $a + c = b + c$): An equivalent equation results when the same number is added to each side of an equation.

Subtraction Property of Equality (if $a = b$, then $a - c = b - c$): An equivalent equation results when the same number is subtracted from each side of an equation.

Multiplication Property of Equality (if $a = b$, then $a \cdot c = b \cdot c$ for $c \neq 0$): An equivalent equation results when each side of an equation is multiplied by the same nonzero number.

Division Property of Equality (if $a = b$, then $\frac{a}{c} = \frac{b}{c}$ for $c \neq 0$). An equivalent equation results when each side of an equation is divided by the same nonzero number.

Operations on numbers are known as inverses if one operation undoes the other. For example, addition and subtraction are **inverse operations**, as are multiplication and division. In problems that involve equation solving, we will utilize inverse operations, as suggested in the following box.

Adding 3	undoes	subtracting 3
Subtracting 5	undoes	adding 5
Multiplying by 2	undoes	dividing by 2
Dividing by 7	undoes	multiplying by 7

EXAMPLE 5

Solve the equation $2x - 3 = 7$.

SOLUTION First add 3 (to eliminate the subtraction of 3 from $2x$):

$$2x - 3 + 3 = 7 + 3 \quad \text{(Add Prop. of Eq.)}$$
$$2x = 10 \quad \text{(simplifying)}$$

Now divide by 2 (to eliminate the multiplication of 2 with x):

$$\frac{2x}{2} = \frac{10}{2} \qquad \text{(Division Prop. of Eq.)}$$

$$x = 5 \qquad \text{(simplifying)}$$

In Example 5, the number 5 is the solution for the original equation. Replacing x in the equation with 5, we confirm this as shown:

$$2x - 3 = 7$$
$$2(5) - 3 = 7$$
$$10 - 3 = 7$$

An equation that can be written in the form $ax + b = c$ for constants a, b, and c is a **linear equation**. Our plan for solving such an equation involves getting variable terms together on one side of the equation and numerical terms together on the other side.

SOLVING A LINEAR EQUATION

1. Simplify each side of the equation; that is, combine like terms.
2. Eliminate additions and/or subtractions by using inverse operations.
3. Eliminate multiplications and/or divisions by using inverse operations.

EXAMPLE 6

Solve the equation $2(x - 3) + 5 = 13$.

SOLUTION

$$2(x - 3) + 5 = 13$$
$$2x - 6 + 5 = 13 \qquad \text{(Distributive Axiom)}$$
$$2x - 1 = 13 \qquad \text{(substitution)}$$
$$2x = 14 \qquad \text{(Addition Prop. of Eq.)}$$
$$x = 7 \qquad \text{(Division Prop. of Eq.)}$$

Some equations involve fractions. To avoid some of the difficulties that fractions bring, we often multiply each side of such equations by the **least common denominator (LCD)** of the fractions involved.

EXAMPLE 7

Solve the equation $\dfrac{x}{3} + \dfrac{x}{4} = 14$.

SOLUTION For the denominators 3 and 4, the LCD is 12. We multiply each side of the equation by 12 and use the Distributive Axiom on the left side.

$$12\left(\frac{x}{3} + \frac{x}{4}\right) = 12 \cdot 14 \qquad \text{(Mult. Prop. of Eq.)}$$

$$\frac{12}{1} \cdot \frac{x}{3} + \frac{12}{1} \cdot \frac{x}{4} = 168 \qquad \text{(Distributive Axiom)}$$

$$4x + 3x = 168 \qquad \text{(substitution)}$$
$$7x = 168 \qquad \text{(substitution)}$$
$$x = 24 \qquad \text{(Division Prop. of Eq.)}$$

To check this result, we replace x with 24 in the given equation $\frac{x}{3} + \frac{x}{4} = 14.$

$$\frac{24}{3} + \frac{24}{4} = 14$$
$$8 + 6 = 14$$

It may happen that the variable appears in the denominator of the only fraction in an equation. In such cases, our method does not change! See Example 8.

EXAMPLE 8

Solve the following equation for n:

$$\frac{360}{n} + 120 = 180$$

SOLUTION Subtracting 120 from each side of the equation, we have

$$\frac{360}{n} = 60$$

Multiplying by n,

$$360 = 60n$$

Dividing by 60,

$$6 = n \text{ (or } n = 6\text{)}.$$

Figure A.4

NOTE: n represents the number of sides shown in the polygon of Figure A.4; $\frac{360}{n}$ and 120 represent the measures of angles in the figure.

Our final example combines many of the ideas introduced in this section and the previous section. Example 9 is based on the formula for the area of a trapezoid.

EXAMPLE 9

See Figure A.5. Use the formula $A = \frac{1}{2} \cdot h \cdot (b + B)$. Given that $A = 77$, $b = 4$, and $B = 7$, find the value of h.

SOLUTION Substitution leads to the equation

$$77 = \frac{1}{2} \cdot h \cdot (4 + 7)$$

$$77 = \frac{1}{2} \cdot h \cdot 11$$

$$2(77) = 2 \cdot \frac{1}{2} \cdot h \cdot 11 \qquad \text{(multiplying by 2)}$$

$$154 = 11h \qquad \text{(simplifying)}$$

$$14 = h \qquad \text{(dividing by 11)}$$

Figure A.5

Exercises A.2

In Exercises 1 to 6, simplify by combining similar terms.

1. $(2x + 3) + (3x + 5)$

2. $(2x + 3) - (3x - 5)$

3. $x + (3x + 2) - (2x + 4)$

4. $(3x + 2) + (2x - 3) - (x + 1)$

5. $2(x + 1) + 3(x + 2)$

   (HINT: Multiply before adding.)

6. $3(2x + 5) - 2(3x - 1)$

In Exercises 7 to 12, simplify by using the FOIL method of multiplication.

7. $(x + 3)(x + 4)$

8. $(x - 5)(x - 7)$

9. $(2x + 5)(3x - 2)$

10. $(3x + 7)(2x + 3)$

11. $(a + b)^2 + (a - b)^2$

12. $(x + 2)^2 - (x - 2)^2$

In Exercises 13 to 16, evaluate each expression.

13. $\ell \cdot w \cdot h$, if $\ell = 4$, $w = 3$, and $h = 5$

14. $a^2 + b^2$, if $a = 5$ and $b = 7$

15. $2 \cdot \ell + 2 \cdot w$, if $\ell = 13$ and $w = 7$

16. $a \cdot b \div c$, if $a = 6$, $b = 16$, and $c = 4$

In Exercises 17 to 20, find the value of the variable named in each formula. Leave π in the answers for Exercises 19 and 20.

17. S, if $S = 2\ell w + 2wh + 2\ell h$, $\ell = 6$, $w = 4$, and $h = 5$

18. A, if $A = \frac{1}{2}a(b + c + d)$, $a = 2$, $b = 6$, $c = 8$, and $d = 10$

19. V, if $V = \frac{1}{3}\pi \cdot r^2 \cdot h$, $r = 3$, and $h = 4$

20. S, if $S = 4\pi r^2$ and $r = 2$

In Exercises 21 to 32, solve each equation.

21. $2x + 3 = 17$

22. $3x - 3 = -6$

23. $-\frac{y}{3} + 2 = 6$

24. $3y = -21 - 4y$

25. $a + (a + 2) = 26$

26. $b = 27 - \frac{b}{2}$

27. $2(x + 1) = 30 - 6(x - 2)$

28. $2(x + 1) + 3(x + 2) = 22 + 4(10 - x)$

29. $\frac{x}{3} - \frac{x}{2} = -5$

30. $\frac{x}{2} + \frac{x}{3} + \frac{x}{4} = 26$

31. $\frac{360}{n} + 135 = 180$

32. $\frac{(n - 2) \cdot 180}{n} = 150$

In Exercises 33 to 36, find the value of the indicated variable for each given formula.

33. w, if $S = 2\ell w + 2wh + 2\ell h$, $S = 148$, $\ell = 5$, and $h = 6$

34. b, if $A = \frac{1}{2} \cdot h \cdot (b + B)$, $A = 156$, $h = 12$, and $B = 11$

35. y, if $m = \frac{1}{2}(x - y)$, $m = 23$, and $x = 78$

36. Y, if $m = \frac{Y - y}{X - x}$, $m = \frac{-3}{2}$, $y = 1$, $X = 2$, and $x = -2$

A.3 INEQUALITIES

In geometry, we sometimes need to work with inequalities. **Inequalities** are statements that involve one of the following relationships:

$<$ means "is less than"
$>$ means "is greater than"
$\leq$ means "is less than or equal to"
$\geq$ means "is greater than or equal to"
$\neq$ means "is not equal to"

The statement $-4 < 7$ is true because negative 4 is less than positive 7. On a horizontal number line, the smaller number is always found to the left of the larger number. An equivalent claim is $7 > -4$, which means positive 7 is greater than negative 4. See the number line at the top of page 541.

Where a and b are real numbers, the statement $a \leq b$ is read "a is less than or equal to b"; that is, $a \leq b$ means "$a < b$ or $a = b$."

Both statements $6 \leq 6$ and $4 \leq 6$ are true. The statement $6 \leq 6$ is true because $6 = 6$ is true. Because $4 < 6$ is true, the statement $4 \leq 6$ is also true. A statement of the form *P or Q* is called a *disjunction*; see Section 1.1 for more information.

EXAMPLE 1

Give two true statements that involve the symbol $\geq$ and the number 5.

SOLUTION

$$5 \geq 5 \qquad \text{because} \qquad 5 = 5 \text{ is true}$$
$$12 \geq 5 \qquad \text{because} \qquad 12 > 5 \text{ is true}$$

The symbol $\neq$ is used to join any two numerical expressions that do not have the same value; for example, $2 + 3 \neq 7$. The following definition is also found in Section 3.5.

DEFINITION

a is **less than** b (that is, $a < b$) if and only if there is a positive number p for which $a + p = b$; a is **greater than** b (that is, $a > b$) if and only if $b < a$.

EXAMPLE 2

Find, if possible, the following:

a) Any number a for which "$a < a$" is true.
b) Any numbers a and b for which "$a < b$ and $b < a$" is true.

SOLUTION

a) There is no such number. If $a < a$, then $a + p = a$ for some positive number p. Subtracting a from each side of the equation gives $p = 0$. This statement ($p = 0$) contradicts the fact that p is positive.
b) There are no such numbers. If $a < b$, then a is to the left of b on the number line. Therefore, $b < a$ is false, because this statement claims that b is to the left of a.

Part (b) of Example 2 suggests the following property.

TRICHOTOMY PROPERTY

If a and b are numerical expressions, then only one of the following statements can be true:

$$a < b, a = b, \text{ or } a > b.$$

EXAMPLE 3

What can you conclude regarding the numbers y and z if $x < y$ and $x > z$?

SOLUTION $x < y$ means that x is to the left of y, as in Figure A.6. Similarly, $x > z$ (equivalently, $z < x$) means that z is to the left of x. With z to the left of x, which is itself to the left of y, we clearly have z to the left of y; thus $z < y$.

Figure A.6

Example 3 suggests a transitive relationship for the inequality "is less than," and this is stated in the following property. The Transitive Property of Inequality can also be stated using $>$, $\leq$, or $\geq$.

TRANSITIVE PROPERTY OF INEQUALITY

For numbers a, b, and c, if $a < b$ and $b < c$, then $a < c$.

This property can be proved as follows:

1. $a < b$ means that $a + p_1 = b$ for some positive number p_1.
2. $b < c$ means that $b + p_2 = c$ for some positive number p_2.
3. Substituting $a + p_1$ for b (from statement 1) into the statement $b + p_2 = c$, we have $(a + p_1) + p_2 = c$.
4. Now $a + (p_1 + p_2) = c$.
5. But the sum of two positive numbers is also positive; that is, $p_1 + p_2 = p$, so statement 4 becomes $a + p = c$.
6. If $a + p = c$, then $a < c$, by the definition of "is less than."

Therefore, $a < b$ and $b < c$ implies that $a < c$.

The Transitive Property of Inequality can be extended to four or more unequal expressions. When a first value is less than a second, the second is less than a third, and so on, then the first is less than the last.

EXAMPLE 4

Two angles are complementary if the sum of their measures is exactly $90°$. If the measure of the first of two complementary angles is more than $27°$, what must you conclude about the measure of the second angle?

SOLUTION
Where $\angle 1$ is the first angle, $m\angle 1 > 27$; it follows that $m\angle 1 = 27 + p$. Also, $m\angle 1 + m\angle 2 = 90$. By substitution,

$$27 + p + m\angle 2 = 90$$
$$m\angle 2 + p = 63$$

By definition, the measure of the second angle must be less than $63°$.

EXAMPLE 5

For the statement $-6 < 9$, determine the inequality that results when each side is changed as follows:

a) Has 4 added to it   **c)** Is multiplied by 3

b) Has 2 subtracted from it   **d)** Is divided by -3

SOLUTION

a) $-6 + 4 \; ? \; 9 + 4$
$$-2 \; ? \; 13 \; \rightarrow \; -2 < 13$$

b) $-6 - 2 \; ? \; 9 - 2$
$$-8 \; ? \; 7 \; \rightarrow \; -8 < 7$$

c) $(-6)(3) \; ? \; 9(3)$
$$-18 \; ? \; 27 \; \rightarrow \; -18 < 27$$

d) $(-6) \div (-3) \; ? \; 9 \div (-3)$
$$2 \; ? \; -3 \; \rightarrow \; 2 > -3$$

As Example 5 suggests, the operations addition and subtraction preserve the order of the inequality. While multiplication and division by a *positive* number preserve the order of the inequality, multiplication and division by a *negative* number reverse the order of the inequality.

OPERATIONS ON INEQUALITIES

Stated for $<$, these properties have counterparts involving $>$, $\leq$, and $\geq$.

Addition:	If $a < b$, then $a + c < b + c$.
Subtraction:	If $a < b$, then $a - c < b - c$.
Multiplication:	i) If $a < b$ and $c > 0$ (c is positive), then $a \cdot c < b \cdot c$.
	ii) If $a < b$ and $c < 0$ (c is negative), then $a \cdot c > b \cdot c$.
Division:	i) If $a < b$ and $c > 0$ (c is positive), then $\frac{a}{c} < \frac{b}{c}$.
	ii) If $a < b$ and $c < 0$ (c is negative), then $\frac{a}{c} > \frac{b}{c}$.

We now turn our attention to solving inequalities such as

$$x + (x + 1) < 7 \qquad \text{and} \qquad 2(x - 3) + 5 \geq 3$$

The method here is almost the same as the one used for equation solving, but there are some very important differences. See the following guidelines.

Warning

Be sure to reverse the inequality symbol upon multiplying or dividing by a *negative* number.

SOLVING AN INEQUALITY

1. Simplify each side of the inequality; that is, combine like terms.
2. Eliminate additions and subtractions.
3. Eliminate multiplications and divisions.

NOTE: For Step 3, see the Warning at the left.

EXAMPLE 6

Solve $2x - 3 \le 7$.

SOLUTION

$$
\begin{aligned}
2x - 3 + 3 &\le 7 + 3 \quad &\text{(adding 3 preserves } \le) \\
2x &\le 10 \quad &\text{(simplify)} \\
\frac{2x}{2} &\le \frac{10}{2} \quad &\text{(division by 2 preserves } \le) \\
x &\le 5 \quad &\text{(simplify)}
\end{aligned}
$$

The possible values of x are shown on a number line in Figure A.7; this picture is the **graph** of the solutions. Note that the point above the 5 is shown solid in order to indicate that 5 is included as a solution.

Figure A.7

EXAMPLE 7

Solve $x(x - 2) - (x + 1)(x + 3) < 9$.

SOLUTION Using the Distributive Axiom and FOIL, we simplify the left side to get

$$(x^2 - 2x) - (x^2 + 4x + 3) < 9$$

Subtraction is performed by adding the additive inverse of each term in $(x^2 + 4x + 3)$.

$$
\begin{aligned}
\therefore (x^2 - 2x) + (-x^2 - 4x - 3) &< 9 \\
-6x - 3 &< 9 \quad &\text{(simplify)} \\
-6x &< 12 \quad &\text{(add 3)} \\
\frac{-6x}{-6} &> \frac{12}{-6} \quad &\text{(divide by } -6 \text{ and reverse the inequality symbol)} \\
x &> -2 \quad &\text{(simplify)}
\end{aligned}
$$

The graph of the solution is shown in Figure A.8. Notice that the circle above the -2 is shown open in order to indicate that -2 is not included as a solution.

Figure A.8

Exercises A.3

1. If line segment AB and line segment CD in the accompanying drawing are drawn to scale, what does intuition tell you about the lengths of these segments?

A ———————————— B

C ———— D

2. Using the number line shown, write two inequalities that relate the values of e and f.

3. If angles *ABC* and *DEF* in the accompanying drawing were measured with a protractor, what does intuition tell you about the degree measures of these angles?

4. Consider the statement $x \geq 6$. Which of the following choices of *x* will make this a true statement?

   $x = -3 \quad x = 0 \quad x = 6 \quad x = 9 \quad x = 12$

5. According to the definition of $a < b$, there is a positive number *p* for which $a + p = b$. Find the value of *p* for the statement given.

   a) $3 < 7$ b) $-3 < 7$

6. Does the Transitive Property of Inequality hold true for four real numbers *a, b, c,* and *d*? That is, is the following statement true?

   If $a < b$, $b < c$, and $c < d$, then $a < d$.

For Exercises 7 and 8, use the Trichotomy Property to decide whether $a < b$, $a = b$, or $a > b$.

7. a) $a = \dfrac{12}{2}$ $b = 4(-2)$

   b) $a = 2^3$ $b = 3^2$

8. a) $a = \dfrac{12 + (-3)}{9}$ $b = \dfrac{5^2 - 4^2}{3^2}$

   b) $a = 2 \cdot 3 + 4$ $b = 2 \cdot (3 + 4)$

9. Of several line segments, $AB > CD$ (the length of segment *AB* is greater than that of segment *CD*), $CD > EF$, $EF > GH$, and $GH > IJ$. What conclusion does the Transitive Property of Inequality allow regarding *IJ* and *AB*?

10. Of several angles, the degree measures are related in this way: $m\angle JKL > m\angle GHI$ (the measure of angle *JKL* is greater than that of angle *GHI*), $m\angle GHI > m\angle DEF$, and $m\angle DEF > m\angle ABC$. What conclusion does the Transitive Property of Inequality allow regarding $m\angle ABC$ and $m\angle JKL$?

11. Classify as true or false.
    a) $5 \leq 4$ c) $5 \leq 5$
    b) $4 \leq 5$ d) $5 < 5$

12. Classify as true or false.
    a) $-5 \leq 4$ c) $-5 \leq -5$
    b) $5 \leq -4$ d) $5 \leq -5$

13. Two angles are supplementary if the sum of their measures is $180°$. If the measure of the first of two supplementary angles is less than $32°$, what must you conclude about the measure of the second angle?

14. Two trim boards need to be used together to cover a 12-ft length along one wall. If Jim recalls that one board is more than 7 ft long, what length must the second board be to span the 12-ft length?

15. Consider the inequality $-3 \leq 5$. Write the statement that results when
    a) each side is multiplied by 4.
    b) -7 is added to each side.
    c) each side is multiplied by -6.
    d) each side is divided by -1.

16. Consider the inequality $-6 > -9$. Write the statement that results when
    a) 8 is added to each side.
    b) each side is multiplied by -2.
    c) each side is multiplied by 2.
    d) each side is divided by -3.

17. Suppose that you are solving an inequality. Complete the chart by indicating whether the inequality symbol should be reversed or kept by writing "change" or "no change."

	Positive	Negative
Add		
Subtract		
Multiply		
Divide		

In Exercises 18 to 28, first solve each inequality. Then draw a number line graph of the solutions.

18. $5x - 1 \leq 29$ 19. $2x + 3 \leq 17$

20. $5 + 4x > 25$ 21. $5 - 4x > 25$

22. $5(2 - x) \leq 30$

23. $2x + 3x < 200 - 5x$

24. $5(x + 2) < 6(9 - x)$

25. $\dfrac{x}{3} - \dfrac{x}{2} \leq 4$

26. $\dfrac{2x - 3}{-5} > 7$

27. $x^2 + 4x \leq x(x - 5) - 18$

28. $x(x + 2) < x(2 - x) + 2x^2$

In Exercises 29 to 32, the claims made are not always true. Cite a counterexample to show why each claim fails.

29. If $a < b$, then $a \cdot c < b \cdot c$.

30. If $a < b$, then $a \cdot c \neq b \cdot c$.

31. If $a < b$, then $a^2 < b^2$.

32. If $a \neq b$ and $b \neq c$, then $a \neq c$.

A.4 FACTORING AND QUADRATIC EQUATIONS

In many geometry applications, it is necessary to use the techniques found in this appendix in order to solve such problems. *Factoring* is the procedure in which a polynomial is replaced by an indicated product of multipliers (factors). Because FOIL can be used to replace the product of $(x - 3)$ and $(2x + 5)$ by $2x^2 - x - 15$, we factor $2x^2 - x - 15$ by replacing it with the factored form $(x - 3)(2x + 5)$.

THE GREATEST COMMON FACTOR

Just as $5x(2x - 3)$ equals $10x^2 - 15x$, we factor $10x^2 - 15x$ as $5x(2x - 3)$ by using the Distributive Property. The individual terms of $10x^2 - 15x$ can be written as follows:

$$10x^2 = 2 \cdot 5 \cdot x \cdot x \text{ and } 15x = 3 \cdot 5 \cdot x.$$

With common factors of 5 and x for the terms of the polynomial $10x^2 - 15x$, the *greatest common factor* (GCF) for $10x^2 - 15x$ is $5x$.

EXAMPLE 1

Using the GCF of its terms, factor each expression.

a) $2x^3 - 6x^2$ b) $ax^2 + ax - 6a$

SOLUTION
a) For $2x^3 - 6x^2$, the GCF is $2x^2$. Then $2x^3 - 6x^2 = 2x^2(x - 3)$.
b) For $ax^2 + ax - 6a$, the GCF is a. Then $ax^2 + ax - 6a = a(x^2 + x - 6)$.

THE DIFFERENCE OF TWO SQUARES

For apparent reasons, expressions such as $a^2 - b^2$ and $4x^2 - 25$ are each known as a *difference of two squares*. Because $(2x + 5)(2x - 5)$ equals $4x^2 - 25$, we can factor this difference of two squares as follows: $4x^2 - 25 = (2x + 5)(2x - 5)$.

In general, $a^2 - b^2 = (a + b)(a - b)$; that is, the two factors of the difference of two squares $a^2 - b^2$ are the sum $(a + b)$ and the difference $(a - b)$.

EXAMPLE 2

Factor each expression:

a) $x^2 - 25$ b) $9x^2 - 16$ c) $8y^2 - 98z^2$

SOLUTION
a) $x^2 - 25 = (x + 5)(x - 5)$
b) $9x^2 - 16 = (3x + 4)(3x - 4)$
c) There is a common factor of 2 for the terms of $8y^2 - 98z^2$, so
$8y^2 - 98z^2 = 2(\underbrace{4y^2 - 49z^2}_{\text{difference of two squares}}) = 2(2y + 7z)(2y - 7z)$

QUADRATIC TRINOMIALS

Consider the following multiplications, each completed by the FOIL method:

$$(x + 3)(x + 7) = x^2 + 7x + 3x + 21 = x^2 + 10x + 21$$
$$(x + 3)(x - 7) = x^2 - 7x + 3x - 21 = x^2 - 4x - 21$$
$$(x - 3)(x - 7) = x^2 - 7x - 3x + 21 = x^2 - 10x + 21$$

Each result, such as $x^2 + 10x + 21$, is known as a *quadratic trinomial* because it has three terms (a trinomial), of which the largest exponent is two (a quadratic term). For quadratic trinomials, sign combinations of the terms are most important; for instance, the trinomial $x^2 + 10x + 21$ has only positive terms due to the multiplication of two sums. Each multiplication shown above can be reversed; that is, each quadratic trinomial shown above at the right can be factored. A polynomial like $x^2 + 2x + 3$ cannot be factored and is said to be *prime*.

Factoring $x^2 - 4x - 21$ would require factors that are a sum and a difference; that is, we know that $x^2 - 4x - 21$ has the form $(x + ?)(x - ?)$. Of course, the final term being numerically 21 suggests that question marks be replaced with 1 and 21 or else 3 and 7. FOIL can be used to eliminate the possible choice of 1 and 21; however, one must be cautious in that the factored form of $x^2 - 4x - 21$ is $(x + 3)(x - 7)$, not $(x - 3)(x + 7)$. Factors can be reversed; for instance, $x^2 - 4x - 21$ equals $(x - 7)(x + 3)$. Multiplication (using FOIL) verifies that $(x + 3)(x - 7)$ or $(x - 7)(x + 3)$ is equal to $x^2 - 4x - 21$.

EXAMPLE 3

Factor each quadratic trinomial:

a) $x^2 + 12x + 32$ **b)** $x^2 + 6x - 40$
c) $x^2 - 13x + 36$ **d)** $ax^2 + ax - 6a$

SOLUTION

a) The factored form of $x^2 + 12x + 32$ is $(x + ?)(x + ?)$. By trial and error, we test combinations that lead to the last term 32, namely $1 \cdot 32$, $2 \cdot 16$, and $4 \cdot 8$.
Now $x^2 + 12x + 32 = (x + 4)(x + 8)$.
b) The factored form of $x^2 + 6x - 40$ is $(x + ?)(x - ?)$;
in turn, $x^2 + 6x - 40 = (x + 10)(x - 4)$
c) The factored form of $x^2 - 13x + 36$ is $(x - ?)(x - ?)$;
in turn, $x^2 - 13x + 36 = (x - 4)(x - 9)$
d) The terms of $ax^2 + ax - 6a$ have the common factor a.
Then $ax^2 + ax - 6a = a(x^2 + x - 6)$ or $a(x + 3)(x - 2)$

NOTE: In part d), we say that the quadratic trinomial is *factored completely*.

The terms of the quadratic trinomial $6x^2 + 7x - 20$ have no GCF other than 1. In the quadratic trinomial $6x^2 + 7x - 20$, the leading term suggests that the factors have a form like $(x + ?)(6x - ?)$ or $(x - ?)(6x + ?)$ or $(2x + ?)(3x - ?)$ or $(2x - ?)(3x + ?)$. The final term of $6x^2 + 7x - 20$ also suggests that we need to consider numerical factors of 20. With some experimentation (trial and error), we find that $6x^2 + 7x - 20 = (2x + 5)(3x - 4)$.

When factoring a polynomial, *always* look first for a common factor (the GCF) for its terms. Factoring $6x^2 + 12x - 18$ is fairly easy when we realize the existence of the GCF of 6. Otherwise, we have to consider lead factors that produce $6x^2$ and numerical factors of 18. The correct factorization of $6x^2 + 12x - 18$ follows:

$$6x^2 + 12x - 18 = 6(x^2 + 2x - 3) = 6(x + 3)(x - 1)$$

EXAMPLE 4

Factor $15x^2 + 22x + 8$.

SOLUTION The terms of this trinomial display a common factor of 1; thus, the factors will have the form $(x + ?)(15x + ?)$ or $(3x + ?)(5x + ?)$. Numerical factors that produce 8 are $1 \cdot 8$ or $2 \cdot 4$. By trial and error, we find that

$$15x^2 + 22x + 8 = (3x + 2)(5x + 4)$$

NOTE: A check of the solution follows:

$$\begin{aligned}
(3x + 2)(5x + 4) &= 3x \cdot 5x + 3x \cdot 4 + 2 \cdot 5x + 2 \cdot 4 \\
&= 15x^2 + 12x + 10x + 8 \\
&= 15x^2 + 22x + 8
\end{aligned}$$

SUMMARY: TO FACTOR A POLYNOMIAL

1. Use the GCF and Distributive Property to replace the polynomial.
2. If the polynomial is a difference of two squares, factor as follows:
$$a^2 - b^2 = (a + b)(a - b)$$
3. For a quadratic trinomial, consider signs as well as factors of the first term and the numerical term to factor the trinomial into two binomial factors.
4. Factor completely.

QUADRATIC EQUATIONS

Where $a \neq 0$, an equation that can be written in the form $ax^2 + bx + c = 0$ is a **quadratic equation**. For example, $x^2 - 7x + 12 = 0$ and $6x^2 = 7x + 3$ are quadratic. Many quadratic equations can be solved by a factoring method that depends on the Zero Product Property.

ZERO PRODUCT PROPERTY

If $a \cdot b = 0$, then $a = 0$ or $b = 0$.

When this property is stated in words, it reads, "If the product of two expressions equals 0, then at least one of the factors must equal 0." This property extends to three or more factors. For instance, if $a \cdot b \cdot c = 0$, then $a = 0$ or $b = 0$ or $c = 0$.

EXAMPLE 5

Solve $x^2 - 7x + 12 = 0$.

SOLUTION First factor the polynomial by reversing the FOIL method of multiplication.

$$(x - 3)(x - 4) = 0 \qquad \text{(factoring)}$$
$$x - 3 = 0 \quad \text{or} \quad x - 4 = 0 \qquad \text{(Zero Product Property)}$$
$$x = 3 \quad \text{or} \quad x = 4 \qquad \text{(Addition Property)}$$

To check $x = 3$, substitute into the given equation:

$$3^2 - (7 \cdot 3) + 12 = 9 - 21 + 12 = 0$$

To check $x = 4$, substitute again:

$$4^2 - (7 \cdot 4) + 12 = 16 - 28 + 12 = 0$$

The solutions are usually expressed as the set $\{3, 4\}$.

If you were asked to use factoring to solve the quadratic equation

$$6x^2 = 7x + 3,$$

it would be necessary to change the equation so that one side would be equal to 0. The form $ax^2 + bx + c = 0$ is the **standard form** of a quadratic equation.

SOLVING A QUADRATIC EQUATION BY THE FACTORING METHOD

1. Be sure the equation is in standard form (one side $= 0$).
2. Factor the polynomial side of the equation.
3. Set each factor *containing the variable* equal to 0.
4. Solve each equation found in step 3.
5. Check solutions by substituting into the original equation.

Step 5, which was shown in Example 5, is omitted in Example 6.

EXAMPLE 6

Solve $6x^2 = 7x + 3$.

SOLUTION First changing to standard form, we have

$$
\begin{array}{lll}
6x^2 - 7x - 3 = 0 & & \text{(standard form)} \\
(2x - 3)(3x + 1) = 0 & & \text{(factoring)} \\
2x - 3 = 0 \quad \text{or} \quad 3x + 1 = 0 & & \text{(Zero Product Property)} \\
2x = 3 \quad \text{or} \quad 3x = -1 & & \text{(Addition-Subtraction Property)} \\
x = \dfrac{3}{2} \quad \text{or} \quad x = \dfrac{-1}{3} & & \text{(division)}
\end{array}
$$

Therefore, $\left\{\frac{3}{2}, -\frac{1}{3}\right\}$ is the solution set.

In some instances, a common factor can be extracted from each term in the factoring step. In the equation $2x^2 + 10x - 48 = 0$, the left side of the equation has the common factor 2. Factoring leads to $2(x^2 + 5x - 24) = 0$ and then to $2(x + 8)(x - 3) = 0$. Of course, only the factors containing variables can equal 0, so the solutions to this equation are -8 and 3.

Equations such as $4x^2 = 9$ and $4x^2 - 12x = 0$ are **incomplete quadratic equations** because one term is missing from the standard form. Either equation can be solved by factoring; in particular, the factoring is given by

$$4x^2 - 9 = (2x + 3)(2x - 3)$$

and

$$4x^2 - 12x = 4x(x - 3)$$

EXAMPLE 7

Solve each incomplete quadratic equation.

   a) $4x^2 - 9 = 0$ **b)** $4x^2 = 12x$

SOLUTION

   a)

$$
\begin{array}{l}
4x^2 - 9 = 0 \\
(2x + 3)(2x - 3) = 0 \quad \text{(factoring)} \\
2x + 3 = 0 \quad \text{or} \quad 2x - 3 = 0 \\
2x = -3 \quad \text{or} \quad 2x = 3 \\
x = -\dfrac{3}{2} \quad \text{or} \quad x = \dfrac{3}{2}
\end{array}
$$

The solution set is $\left\{\dfrac{3}{2}, -\dfrac{3}{2}\right\}$.

b)

$$4x^2 = 12x$$
$$4x^2 - 12x = 0$$
$$4x(x - 3) = 0$$
$$4x = 0 \quad \text{or} \quad x - 3 = 0$$
$$x = 0 \quad \text{or} \quad x = 3$$

The solution set is $\{0, 3\}$.

Exercises A.4

In Exercises 1 to 4, factor by using the GCF.

1. $ax^2 + 5ax + 7a$

2. $5y^3 - 20y^2$

3. $2bx^2 + 4b^2x$

4. $4x + 12y + 8z$

In Exercises 5 to 8, factor each difference of two squares.

5. $y^2 - 9$

6. $16x^2 - 9y^2$

7. $4x^2 - 49y^2$

8. $a^2 - 100$

In Exercises 9 to 16, factor each trinomial product.

9. $x^2 + 7x + 12$

10. $x^2 - 9x + 14$

11. $x^2 + 5x - 24$

12. $y^2 - 4y - 96$

13. $6y^2 + 5y - 6$

14. $12a^2 + 31a + 20$

15. $3x^2 + 11xy - 4y^2$

16. $4a^2 + 12ab + 9b^2$

In Exercises 17 to 22, factor completely.

17. $4x^2 - 16$

18. $6y^2 - 54$

19. $3y^2 + 24y + 45$

20. $30x^2 - 35x + 10$

21. $2ax^2 + 3ax - 35a$

22. $6a^2c^2 + 11abc^2 - 10b^2c^2$

In Exercises 23 and 24, find the three factors for the cubic trinomials.

23. $x^3 + 5x^2 + 4x$

24. $x^3 - 9x$

In Exercises 25 to 32, solve each quadratic equation by factoring.

25. $x^2 - 6x + 8 = 0$

26. $x^2 + 4x = 21$

27. $3x^2 - 51x + 180 = 0$

   (HINT: There is a common factor.)

28. $2x^2 + x - 6 = 0$

29. $3x^2 = 10x + 8$

30. $8x^2 + 10w - 110 = 0$

31. $6x^2 = 5x - 1$

32. $12x^2 + 10x = 12$

In Exercises 33 to 36, solve each incomplete quadratic equation.

33. $2x^2 - 6x = 0$

34. $16x^2 - 9 = 0$

35. $4y^2 = 25$

36. $9y^2 = 18y$

37. Given that the area of the rectangle shown is 66, find the width w and length $w + 5$ by solving the equation $w(w + 5) = 66$.

38. Determine the length x by solving either the equation $\dfrac{x}{4} = \dfrac{6}{x + 5}$ or the equivalent equation $x(x + 5) = 24$.

39. Find the length a by solving the equation $a^2 + (a + 1)^2 = 25$

*40. With the highest degree term being $4x^3$, the equation $4x^3 - 28x^2 + 24x = 0$ is known as a *cubic equation*. Use factoring to find three real solutions for this equation.

*41. With the highest degree term being x^4, the equation $x^4 - 13x^2 + 36 = 0$ is known as a *quartic equation*. Use factoring to find four real solutions for this equation.

A.5 THE QUADRATIC FORMULA AND SQUARE ROOT PROPERTIES

When $ax^2 + bx + c = 0$ cannot be solved by factoring, solutions may be determined by the following formula. In the formula, a is the number multiplied by x^2, b is the number multiplied by x, and c is the constant term. The $\pm$ symbol tells us that there are generally two solutions, one found by adding and one found by subtracting. The symbol $\sqrt{a}$ is read "the square root of a."

QUADRATIC FORMULA

$$x = \frac{-b \pm \sqrt{b^2 - 4ac}}{2a}$$ are solutions for $ax^2 + bx + c = 0$, where $a \neq 0$.

Although the formula may provide two solutions for the equation, an application problem in geometry may have a single positive solution representing the length of a line segment or the degree measure of an angle. Recall that for $a > 0$, $\sqrt{a}$ represents the principal (positive) square root of a.

DEFINITION

Where $a > 0$, the **square root** $\sqrt{a}$ is the positive number for which $\left(\sqrt{a}\right)^2 = a$.

EXAMPLE 1

a) Explain why $\sqrt{25}$ is equal to 5.
b) Without a calculator, find the value of $\sqrt{3} \cdot \sqrt{3}$.
c) Use a calculator to show that $\sqrt{5} \approx 2.236$.

SOLUTION

a) We see that $\sqrt{25}$ must equal 5 because $5^2 = 25$.
b) By definition, $\sqrt{3}$ is the number for which $\left(\sqrt{3}\right)^2 = 3$.
c) By using a calculator, we see that $2.236^2 \approx 5$.

EXAMPLE 2

Simplify each expression, if possible.

a) $\sqrt{16}$ b) $\sqrt{0}$ c) $\sqrt{7}$ d) $\sqrt{400}$ e) $\sqrt{-4}$

SOLUTION

a) $\sqrt{16} = 4$ because $4^2 = 16$.
b) $\sqrt{0} = 0$ because $0^2 = 0$.
c) $\sqrt{7}$ cannot be simplified; however, $\sqrt{7} \approx 2.646$.
d) $\sqrt{400} = 20$ because $20^2 = 400$; a calculator can be used.
e) $\sqrt{-4}$ is not a real number; a calculator gives an "ERROR" message.

Whereas $\sqrt{25}$ represents the principal square root of 25 (namely, 5), the expression $-\sqrt{25}$ can be interpreted as "the negative number whose square is 25"; thus, $-\sqrt{25} = -5$ because $(-5)^2 = 25$. In expressions such as $\sqrt{9 + 16}$ and $\sqrt{4 + 9}$, we first simplify the **radicand** (the expression under the bar of the square root); thus $\sqrt{9 + 16} = \sqrt{25} = 5$ and $\sqrt{4 + 9} = \sqrt{13} \approx 3.606$.

Just as fractions are reduced to lower terms ($\frac{6}{8}$ is replaced by $\frac{3}{4}$), it is also customary to reduce the size of the radicand when possible. To accomplish this, we use the Product Property of Square Roots.

PRODUCT PROPERTY OF SQUARE ROOTS

For $a \geq 0$ and $b \geq 0$, $\sqrt{a \cdot b} = \sqrt{a} \cdot \sqrt{b}$.

When simplifying a square root, we replace the radicand by a product in which the largest possible number (from the list of perfect squares below) is selected as one of the factors:

$$4, 9, 16, 25, 36, 49, 64, 81, 100, 121, \ldots$$

For example,

$$\sqrt{45} = \sqrt{9 \cdot 5}$$
$$= \sqrt{9} \cdot \sqrt{5}$$
$$= 3\sqrt{5}$$

The radicand has now been reduced from 45 to 5. Using a calculator, we see that $\sqrt{45} \approx 6.708$. Also, $3\sqrt{5}$ means 3 times $\sqrt{5}$, and with the calculator we see that $3\sqrt{5} \approx 6.708$.

Leave the smallest possible integer under the square root symbol.

EXAMPLE 3

Simplify the following radicals:

a) $\sqrt{27}$ **b)** $\sqrt{50}$

SOLUTION

a) 9 is the largest perfect square factor of 27. Therefore,
$$\sqrt{27} = \sqrt{9 \cdot 3} = \sqrt{9} \cdot \sqrt{3} = 3\sqrt{3}$$

b) 25 is the largest perfect square factor of 50. Therefore,
$$\sqrt{50} = \sqrt{25 \cdot 2} = \sqrt{25} \cdot \sqrt{2} = 5\sqrt{2}$$

Warning

In Example 5(b), the correct solution is $5\sqrt{2}$, not $2\sqrt{5}$.

The Product Property of Square Roots has a symmetric form that reads $\sqrt{a} \cdot \sqrt{b} = \sqrt{ab}$; for example, $\sqrt{2} \cdot \sqrt{3} = \sqrt{6}$ and $\sqrt{5} \cdot \sqrt{5} = \sqrt{25} = 5$.

The expression $ax^2 + bx + c$ may be **prime** (meaning "not factorable"). Because $x^2 - 5x + 3$ is prime, we solve the equation $x^2 - 5x + 3 = 0$ by using the Quadratic Formula $x = \frac{-b \pm \sqrt{b^2 - 4ac}}{2a}$; see Example 4.

NOTE: When square root radicals are left in an answer, the answer is exact. Once we use the calculator, the solutions are only approximate.

EXAMPLE 4

Find exact solutions for $x^2 - 5x + 3 = 0$. Then use a calculator to approximate these solutions correct to two decimal places.

SOLUTION With the equation in standard form, we see that $a = 1$, $b = -5$, and $c = 3$.

Then
$$x = \frac{-(-5) \pm \sqrt{(-5)^2 - 4(1)(3)}}{2(1)}$$

$$x = \frac{5 \pm \sqrt{25 - 12}}{2} \quad \text{or} \quad x = \frac{5 \pm \sqrt{13}}{2}$$

The exact solutions are $\frac{5 + \sqrt{13}}{2}$ and $\frac{5 - \sqrt{13}}{2}$. Using a calculator, we find that the approximate solutions are 4.30 and 0.70, respectively.

Using the Quadratic Formula to solve the equation $x^2 - 6x + 7 = 0$ yields $x = \frac{6 \pm \sqrt{8}}{2}$. In Example 5, we focus on the simplification of such an expression.

EXAMPLE 5

Simplify $\dfrac{6 \pm \sqrt{8}}{2}$.

SOLUTION Because $\sqrt{8} = \sqrt{4} \cdot \sqrt{2}$ or $2\sqrt{2}$, we simplify the expression as follows:

$$\frac{6 \pm \sqrt{8}}{2} = \frac{6 \pm 2\sqrt{2}}{2} = \frac{\cancel{2}(3 \pm \sqrt{2})}{\cancel{2}} = 3 \pm \sqrt{2}$$

NOTE 1: The number 2 was a common factor for the numerator and the denominator. We then reduced the fraction to lowest terms.

NOTE 2: The approximate values of $3 \pm \sqrt{2}$ are 4.41 and 1.59, respectively. Use your calculator to show that these values are the approximate solutions of the equation $x^2 - 6x + 7 = 0$.

Our final method for solving quadratic equations is used if an incomplete quadratic equation has the form $ax^2 + c = 0$.

SQUARE ROOT PROPERTY

If $x^2 = p$ where $p \geq 0$, then $x = \pm\sqrt{p}$.

According to the Square Roots Property, the equation $x^2 = 6$ has the solutions $\pm\sqrt{6}$.

EXAMPLE 6

Use the Square Roots Property to solve the equation $2x^2 - 56 = 0$.

SOLUTION
$$2x^2 - 56 = 0 \rightarrow 2x^2 = 56 \rightarrow x^2 = 28$$

Then
$$x = \pm\sqrt{28} = \pm\sqrt{4} \cdot \sqrt{7} = \pm 2\sqrt{7}$$

The exact solutions are $2\sqrt{7}$ and $-2\sqrt{7}$; the approximate solutions are 5.29 and -5.29, respectively.

In Example 8, the solutions for the quadratic equation will involve fractions; for this reason, we consider the Quotient Property of Square Roots and Example 7. The Quotient Property enables us to replace the square root of a fraction by the square root of its numerator divided by the square root of its denominator.

QUOTIENT PROPERTY OF SQUARE ROOTS

For $a \geq 0$ and $b > 0$, $\sqrt{\dfrac{a}{b}} = \dfrac{\sqrt{a}}{\sqrt{b}}$.

EXAMPLE 7

Simplify the following square root expressions:

a) $\sqrt{\dfrac{16}{9}}$

b) $\sqrt{\dfrac{3}{4}}$

SOLUTION

a) $\sqrt{\dfrac{16}{9}} = \dfrac{\sqrt{16}}{\sqrt{9}} = \dfrac{4}{3}$

b) $\sqrt{\dfrac{3}{4}} = \dfrac{\sqrt{3}}{\sqrt{4}} = \dfrac{\sqrt{3}}{2}$

EXAMPLE 8

Solve the equation $4x^2 - 9 = 0$.

SOLUTION

$$4x^2 - 9 = 0 \rightarrow 4x^2 = 9 \rightarrow x^2 = \dfrac{9}{4}$$

Then

$$x = \pm\sqrt{\dfrac{9}{4}} = \pm\dfrac{\sqrt{9}}{\sqrt{4}} = \pm\dfrac{3}{2}$$

In summary of Sections A.4 and A.5, quadratic equations have the form $ax^2 + bx + c = 0$ and are solved by one of the following methods:

1. Factoring, when $ax^2 + bx + c$ is easily factored
2. The Quadratic Formula
   $$x = \dfrac{-b \pm \sqrt{b^2 - 4ac}}{2a},$$
   when $ax^2 + bx + c$ is not easily factored or cannot be factored
3. The Square Roots Property, when the equation has the form $ax^2 + c = 0$.

Exercises A.5

1. Use your calculator to find the approximate value of each number, correct to two decimal places:
   a) $\sqrt{13}$ b) $\sqrt{8}$ c) $-\sqrt{29}$ d) $\sqrt{\dfrac{3}{5}}$

2. Use your calculator to find the approximate value of each number, correct to two decimal places:
   a) $\sqrt{17}$ b) $\sqrt{400}$ c) $-\sqrt{7}$ d) $\sqrt{1.6}$

3. Which equations are quadratic?
   a) $2x^2 - 5x + 3 = 0$ d) $\dfrac{1}{2}x^2 - \dfrac{1}{4}x - \dfrac{1}{8} = 0$
   b) $x^2 = x^2 + 4$ e) $\sqrt{2x - 1} = 3$
   c) $x^2 = 4$ f) $(x + 1)(x - 1) = 15$

4. Which equations are incomplete quadratic equations?
   a) $x^2 - 4 = 0$ d) $2x^2 - 4 = 2x^2 + 8x$
   b) $x^2 - 4x = 0$ e) $x^2 = \dfrac{9}{4}$
   c) $3x^2 = 2x$ f) $x^2 - 2x - 3 = 0$

5. Simplify each expression by using the Product Property of Square Roots:
   a) $\sqrt{8}$ c) $\sqrt{900}$
   b) $\sqrt{45}$ d) $(\sqrt{3})^2$

6. Simplify each expression by using the Product Property of Square Roots:
   a) $\sqrt{28}$
   b) $\sqrt{32}$
   c) $\sqrt{54}$
   d) $\sqrt{200}$

7. Simplify each expression by using the Quotient Property of Square Roots:
   a) $\sqrt{\frac{9}{16}}$
   b) $\sqrt{\frac{25}{49}}$
   c) $\sqrt{\frac{7}{16}}$
   d) $\sqrt{\frac{6}{9}}$

8. Simplify each expression by using the Quotient Property of Square Roots:
   a) $\sqrt{\frac{1}{4}}$
   b) $\sqrt{\frac{16}{9}}$
   c) $\sqrt{\frac{5}{36}}$
   d) $\sqrt{\frac{3}{16}}$

9. Use your calculator to verify that the following expressions are equivalent:
   a) $\sqrt{54}$ and $3\sqrt{6}$
   b) $\sqrt{\frac{5}{16}}$ and $\frac{\sqrt{5}}{4}$

10. Use your calculator to verify that the following expressions are equivalent:
    a) $\sqrt{48}$ and $4\sqrt{3}$
    b) $\sqrt{\frac{7}{9}}$ and $\frac{\sqrt{7}}{3}$

In Exercises 11 to 18, determine the values of a, b, and c that are needed in order to use the Quadratic Formula. Choose a > 0.

11. $x^2 - 6x + 8 = 0$

12. $2x^2 - x - 3 = 0$

13. $y^2 - 4y = 12$

14. $y^2 + 6y = 40$

15. $3x^2 = 10x + 25$

16. $5x^2 = 90 - 2x$

17. $(x + 5)(2x - 7) = 117$

18. $(3x + 2)(5x - 4) = 48$

In Exercises 19 to 26, solve each equation by using the Quadratic Formula. Give exact solutions in simplified form. When answers contain square roots, approximate the solutions rounded to two decimal places.

19. $x^2 - 7x + 10 = 0$

20. $x^2 + 7x + 12 = 0$

21. $x^2 + 9 = 7x$

22. $2x^2 + 3x = 6$

23. $x^2 - 4x - 8 = 0$

24. $x^2 - 6x - 2 = 0$

25. $5x^2 = 3x + 7$

26. $2x^2 = 8x - 1$

In Exercises 27 to 32, solve each incomplete quadratic equation. Use the Square Roots Property as needed.

27. $2x^2 = 14$

28. $2x^2 = 14x$

29. $4x^2 - 25 = 0$

30. $4x^2 - 25x = 0$

31. $ax^2 - bx = 0$

32. $ax^2 - b = 0$

33. The length of a rectangle is 3 more than its width. If the area of the rectangle is 40, its dimensions x and $x + 3$ can be found by solving the equation $x(x + 3) = 40$. Find these dimensions.

34. To find the length of $\overline{CP}$ (which is x), one must solve the equation

$$x \cdot (x + 5) = (x + 1) \cdot 4$$

Find the length of $\overline{CP}$.

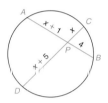

In Exercises 35 and 36, use Theorem 2.5.1 to solve the problem. According to this theorem, the number of diagonals in a polygon of n sides is given by $D = \frac{n(n - 3)}{2}$.

35. Find the number of sides in a polygon that has 9 diagonals.

36. Find the number of sides in a polygon that has the same number of diagonals as it has sides.

In Exercises 37–39, $c^2 = a^2 + b^2$.

37. In the right triangle, find c if $a = 3$ and $b = 4$.

38. In the right triangle, find b if $a = 6$ and $c = 10$.

39. In the right triangle, find a if $b = a + 3$ and $c = a + 4$.

Exercises 37–39

Summary of Constructions, Postulates, Theorems, and Corollaries

▶ Constructions

Section 1.2

1. To construct a segment congruent to a given segment.
2. To construct the midpoint M of a given line segment AB.

Section 1.4

3. To construct an angle congruent to a given angle.
4. To construct the bisector of a given angle.

Section 1.6

5. To construct the line perpendicular to a given line at a specified point on the given line.

Section 2.1

6. To construct the line that is perpendicular to a given line from a point not on the given line.

Section 2.3

7. To construct the line parallel to a given line from a point not on that line.

Section 6.4

8. To construct a tangent to a circle at a point on the circle.
9. To construct a tangent to a circle from an external point.

▶ Postulates

Section 1.3

1. Through two distinct points, there is exactly one line.
2. (Ruler Postulate) The measure of any line segment is a unique positive number.
3. (Segment-Addition Postulate) If X is a point on $\overline{AB}$ and A-X-B, then $AX + XB = AB$.
4. If two lines intersect, they intersect at a point.
5. Through three noncollinear points, there is exactly one plane.
6. If two distinct planes intersect, then their intersection is a line.
7. Given two distinct points in a plane, the line containing these points also lies in the plane.

Section 1.4

8. (Protractor Postulate) The measure of an angle is a unique positive number.
9. (Angle-Addition Postulate) If a point D lies in the interior of angle ABC, then $m\angle ABD + m\angle DBC = m\angle ABC$.

Section 2.1

10. (Parallel Postulate) Through a point not on a line, exactly one line is parallel to the given line.
11. If two parallel lines are cut by a transversal, then the pairs of corresponding angles are congruent.

Section 3.1

12. If the three sides of one triangle are congruent to the three sides of a second triangle, then the triangles are congruent (SSS).
13. If two sides and the included angle of one triangle are congruent to two sides and the included angle of a second triangle, then the triangles are congruent (SAS).
14. If two angles and the included side of one triangle are congruent to two angles and the included side of a second triangle, then the triangles are congruent (ASA).

Section 5.2

15. If the three angles of one triangle are congruent to the three angles of a second triangle, then the triangles are similar (AAA).

Section 6.1

16. (Central Angle Postulate) In a circle, the degree measure of a central angle is equal to the degree measure of its intercepted arc.
17. (Arc-Addition Postulate) If $\overset{\frown}{AB}$ and $\overset{\frown}{BC}$ intersect only at point B, then $m\overset{\frown}{AB} + m\overset{\frown}{BC} = m\overset{\frown}{ABC}$.

Section 8.1

18. (Area Postulate) Corresponding to every bounded region is a unique positive number A, known as the area of that region.
19. If two closed plane figures are congruent, then their areas are equal.
20. (Area-Addition Postulate) Let R and S be two enclosed regions that do not overlap. Then $A_{R \cup S} = A_R + A_S$.
21. The area A of a rectangle whose base has length b and whose altitude has length h is given by $A = bh$.

Section 8.4

22. The ratio of the circumference of a circle to the length of its diameter is a unique positive constant.

Section 8.5

23. The ratio of the degree measure m of the arc (or central angle) of a sector to 360° is the same as the ratio of the area of the sector to the area of the circle; that is, $\frac{\text{area of sector}}{\text{area of circle}} = \frac{m}{360°}$.

Section 9.1

24. (Volume Postulate) Corresponding to every solid is a unique positive number V known as the volume of that solid.
25. The volume of a right rectangular prism is given by

$$V = \ell wh$$

where ℓ measures the length, w the width, and h the altitude of the prism.
26. The volume of a right prism is given by

$$V = Bh$$

where B is the area of a base and h is the length of the altitude of the prism.

▶ Theorems and Corollaries

1.3.1 The midpoint of a line segment is unique.
1.4.1 There is one and only one bisector for a given angle.
1.6.1 If two lines are perpendicular, then they meet to form right angles.
1.6.2 If two lines intersect, then the vertical angles formed are congruent.
1.6.3 In a plane, there is exactly one line perpendicular to a given line at any point on the line.
1.6.4 The perpendicular bisector of a line segment is unique.
1.7.1 If two lines meet to form a right angle, then these lines are perpendicular.
1.7.2 If two angles are complementary to the same angle (or to congruent angles), then these angles are congruent.
1.7.3 If two angles are supplementary to the same angle (or to congruent angles), then these angles are congruent.
1.7.4 Any two right angles are congruent.
1.7.5 If the exterior sides of two adjacent acute angles form perpendicular rays, then these angles are complementary.
1.7.6 If the exterior sides of two adjacent angles form a straight line, then these angles are supplementary
1.7.7 If two line segments are congruent, then their midpoints separate these segments into four congruent segments.
1.7.8 If two angles are congruent, then their bisectors separate these angles into four congruent angles.

2.1.1 From a point not on a given line, there is exactly one line perpendicular to the given line.
2.1.2 If two parallel lines are cut by a transversal, then the pairs of alternate interior angles are congruent.
2.1.3 If two parallel lines are cut by a transversal, then the pairs of alternate exterior angles are congruent.
2.1.4 If two parallel lines are cut by a transversal, then the pairs of interior angles on the same side of the transversal are supplementary.
2.1.5 If two parallel lines are cut by a transversal, then the pairs of exterior angles on the same side of the transversal are supplementary.
2.3.1 If two lines are cut by a transversal so that two corresponding angles are congruent, then these lines are parallel.
2.3.2 If two lines are cut by a transversal so that two alternate interior angles are congruent, then these lines are parallel.
2.3.3 If two lines are cut by a transversal so that two alternate exterior angles are congruent, then these lines are parallel.
2.3.4 If two lines are cut by a transversal so that two interior angles on the same side of the transversal are supplementary, then these lines are parallel.
2.3.5 If two lines are cut by a transversal so that two exterior angles on the same side of the transversal are supplementary, then these lines are parallel.
2.3.6 If two lines are both parallel to a third line, then these lines are parallel to each other.
2.3.7 If two coplanar lines are both perpendicular to a third line, then these lines are parallel to each other.
2.4.1 In a triangle, the sum of the measures of the interior angles is 180°.
2.4.2 Each angle of an equiangular triangle measures 60°.
2.4.3 The acute angles of a right triangle are complementary.
2.4.4 If two angles of one triangle are congruent to two angles of another triangle, then the third angles are also congruent.
2.4.5 The measure of an exterior angle of a triangle equals the sum of the measures of the two nonadjacent interior angles.
2.5.1 The total number of diagonals D in a polygon of n sides is given by the formula $D = \frac{n(n-3)}{2}$.
2.5.2 The sum S of the measures of the interior angles of a polygon with n sides is given by $S = (n-2) \cdot 180°$. Note that $n > 2$ for any polygon.
2.5.3 The measure I of each interior angle of a regular or equiangular polygon of n sides is $I = \frac{(n-2) \cdot 180°}{n}$.
2.5.4 The sum of the measures of the four interior angles of a quadrilateral is 360°.
2.5.5 The sum of the measures of the exterior angles, one at each vertex, of a polygon is 360°.
2.5.6 The measure E of each exterior angle of a regular or equiangular polygon of n sides is $E = \frac{360°}{n}$.

3.1.1 If two angles and a nonincluded side of one triangle are congruent to two angles and a nonincluded side of a second triangle, then the triangles are congruent (AAS).
3.2.1 If the hypotenuse and a leg of one right triangle are congruent to the hypotenuse and a leg of a second right triangle, then the triangles are congruent (HL).
3.3.1 Corresponding altitudes of congruent triangles are congruent.
3.3.2 The bisector of the vertex angle of an isosceles triangle separates the triangle into two congruent triangles.
3.3.3 If two sides of a triangle are congruent, then the angles opposite these sides are also congruent.
3.3.4 If two angles of a triangle are congruent, then the sides opposite these angles are also congruent.
3.3.5 An equilateral triangle is also equiangular.
3.3.6 An equiangular triangle is also equilateral.
3.5.1 The measure of a line segment is greater than the measure of any of its parts.
3.5.2 The measure of an angle is greater than the measure of any of its parts.
3.5.3 The measure of an exterior angle of a triangle is greater than the measure of either nonadjacent interior angle.
3.5.4 If a triangle contains a right or an obtuse angle, then the measure of this angle is greater than the measure of either of the remaining angles.

3.5.5 (Addition Property of Inequality): If $a > b$ and $c > d$, then $a + c > b + d$.

3.5.6 If one side of a triangle is longer than a second side, then the measure of the angle opposite the first side is greater than the measure of the angle opposite the second side.

3.5.7 If the measure of one angle of a triangle is greater than the measure of a second angle, then the side opposite the larger angle is longer than the side opposite the smaller angle.

3.5.8 The perpendicular line segment from a point to a line is the shortest line segment that can be drawn from the point to the line.

3.5.9 The perpendicular line segment from a point to a plane is the shortest line segment that can be drawn from the point to the plane.

3.5.10 (Triangle Inequality) The sum of the lengths of any two sides of a triangle is greater than the length of the third side.

3.5.10 *(Alternative)* The length of one side of a triangle must be between the sum and the difference of the lengths of the other two sides.

4.1.1 A diagonal of a parallelogram separates it into two congruent triangles.

4.1.2 The opposite angles of a parallelogram are congruent.

4.1.3 The opposite sides of a parallelogram are congruent.

4.1.4 The diagonals of a parallelogram bisect each other.

4.1.5 Two consecutive angles of a parallelogram are supplementary.

4.1.6 Two parallel lines are everywhere equidistant.

4.1.7 If two sides of one triangle are congruent to two sides of a second triangle and the included angle of the first triangle is greater than the included angle of the second, then the length of the side opposite the included angle of the first triangle is greater than the length of the side opposite the included angle of the second.

4.1.8 In a parallelogram with unequal pairs of consecutive angles, the longer diagonal lies opposite the obtuse angle.

4.2.1 If two sides of a quadrilateral are both congruent and parallel, then the quadrilateral is a parallelogram.

4.2.2 If both pairs of opposite sides of a quadrilateral are congruent, then the quadrilateral is a parallelogram.

4.2.3 If the diagonals of a quadrilateral bisect each other, then the quadrilateral is a parallelogram.

4.2.4 In a kite, one pair of opposite angles are congruent.

4.2.5 The segment that joins the midpoints of two sides of a triangle is parallel to the third side and has a length equal to one-half the length of the third side.

4.3.1 All angles of a rectangle are right angles.

4.3.2 The diagonals of a rectangle are congruent.

4.3.3 All sides of a square are congruent.

4.3.4 All sides of a rhombus are congruent.

4.3.5 The diagonals of a rhombus are perpendicular.

4.3.6 The diagonals of a rhombus (or square) are perpendicular bisectors of each other.

4.4.1 The base angles of an isosceles trapezoid are congruent.

4.4.2 The diagonals of an isosceles trapezoid are congruent.

4.4.3 The length of the median of a trapezoid equals one-half the sum of the lengths of the two bases.

4.4.4 The median of a trapezoid is parallel to each base.

4.4.5 If two base angles of a trapezoid are congruent, the trapezoid is an isosceles trapezoid.

4.4.6 If the diagonals of a trapezoid are congruent, the trapezoid is an isosceles trapezoid.

4.4.7 If three (or more) parallel lines intercept congruent segments on one transversal, then they intercept congruent segments on any transversal.

5.3.1 If two angles of one triangle are congruent to two angles of another triangle, then the triangles are similar (AA).

5.3.2 The lengths of the corresponding altitudes of similar triangles have the same ratio as the lengths of any pair of corresponding sides.

5.3.3 If an angle of one triangle is congruent to an angle of a second triangle and the pairs of sides including these angles are proportional (in length), then the triangles are similar (SAS~).

5.3.4 If the three sides of one triangle are proportional (in length) to the three corresponding sides of a second triangle, then the triangles are similar (SSS~).

5.3.5 If a line segment divides two sides of a triangle proportionally, then it is parallel to the third side.

5.4.1 The altitude drawn to the hypotenuse of a right triangle separates the right triangle into two right triangles that are similar to each other and to the original right triangle.

5.4.2 The length of the altitude to the hypotenuse of a right triangle is the geometric mean of the lengths of the segments of the hypotenuse.

5.4.3 The length of each leg of a right triangle is the geometric mean of the length of the hypotenuse and the length of the segment of the hypotenuse adjacent to that leg.

5.4.4 (Pythagorean Theorem) The square of the length of the hypotenuse of a right triangle is equal to the sum of the squares of the lengths of the legs.

5.4.5 (Converse of Pythagorean Theorem) If a, b, and c are the lengths of the three sides of a triangle, with c the length of the longest side, and if $c^2 = a^2 + b^2$, then the triangle is a right triangle with the right angle opposite the side of length c.

5.4.6 If the hypotenuse and a leg of one right triangle are congruent to the hypotenuse and a leg of a second right triangle, then the triangles are congruent (HL).

5.4.7 Let a, b, and c represent the lengths of the three sides of a triangle, with c the length of the longest side.
1. If $c^2 > a^2 + b^2$, then the triangle is obtuse and the obtuse angle lies opposite the side of length c.
2. If $c^2 < a^2 + b^2$, then the triangle is acute.

5.5.1 (45-45-90 Theorem) In a right triangle whose angles measure 45°, 45°, and 90°, the legs are congruent and the hypotenuse has a length equal to the product of $\sqrt{2}$ and the length of either leg.

5.5.2 (30-60-90 Theorem) In a triangle whose angles measure 30°, 60°, and 90°, the hypotenuse has a length equal to twice the length of the shorter leg, and the length of the longer leg is the product of $\sqrt{3}$ and the length of the shorter leg.

5.5.3 If the length of the hypotenuse of a right triangle equals the product of $\sqrt{2}$ and the length of either congruent leg, then the angles of the triangle measure 45°, 45°, and 90°.

5.5.4 If the length of the hypotenuse of a right triangle is twice the length of one leg of the triangle, then the angle of the triangle opposite that leg measures 30°.

5.6.1 If a line is parallel to one side of a triangle and intersects the other two sides, then it divides these sides proportionally.

5.6.2 When three (or more) parallel lines are cut by a pair of transversals, the transversals are divided proportionally by the parallel lines.

5.6.3 (The Angle-Bisector Theorem) If a ray bisects one angle of a triangle, then it divides the opposite side into segments whose lengths are proportional to the lengths of the two sides that form the bisected angle.

5.6.4 (Ceva's Theorem) Let D be any point in the interior of $\triangle ABC$. Where E, F, and G lie on $\triangle ABC$, let $\overline{BE}$, $\overline{AF}$, and $\overline{CG}$ be determined by D and the vertices of $\triangle ABC$. Then the product of the ratios of lengths of segments of the sides (taken in order) equals 1; that is, $\frac{AG}{GB} \cdot \frac{BF}{FC} \cdot \frac{CE}{EA} = 1$.

6.1.1 A radius that is perpendicular to a chord bisects the chord.

6.1.2 The measure of an inscribed angle of a circle is one-half the measure of its intercepted arc.

6.1.3 In a circle (or in congruent circles), congruent minor arcs have congruent central angles.

6.1.4 In a circle (or in congruent circles), congruent central angles have congruent arcs.

6.1.5 In a circle (or in congruent circles), congruent chords have congruent minor (major) arcs.

6.1.6 In a circle (or in congruent circles), congruent arcs have congruent chords.

6.1.7 Chords that are at the same distance from the center of a circle are congruent.

6.1.8 Congruent chords are located at the same distance from the center of a circle.

6.1.9 An angle inscribed in a semicircle is a right angle.

6.1.10 If two inscribed angles intercept the same arc, then these angles are congruent.

6.2.1 If a quadrilateral is inscribed in a circle, the opposite angles are supplementary.
(*Alternative*) The opposite angles of a cyclic quadrilateral are supplementary.

6.2.2 The measure of an angle formed by two chords that intersect within a circle is one-half the sum of the measures of the arcs intercepted by the angle and its vertical angle.

6.2.3 The radius (or any other line through the center of a circle) drawn to a tangent at the point of tangency is perpendicular to the tangent at that point.

6.2.4 The measure of an angle formed by a tangent and a chord drawn to the point of tangency is one-half the measure of the intercepted arc.

6.2.5 The measure of an angle formed when two secants intersect at a point outside the circle is one-half the difference of the measures of the two intercepted arcs.

6.2.6 If an angle is formed by a secant and tangent that intersect in the exterior of a circle, then the measure of the angle is one-half the difference of the measures of its intercepted arcs.

6.2.7 If an angle is formed by two intersecting tangents, then the measure of the angle is one-half the difference of the measures of the intercepted arcs.

6.2.8 If two parallel lines intersect a circle, the intercepted arcs between these lines are congruent.

6.3.1 If a line is drawn through the center of a circle perpendicular to a chord, then it bisects the chord and its arc.

6.3.2 If a line through the center of a circle bisects a chord other than a diameter, then it is perpendicular to the chord.

6.3.3 The perpendicular bisector of a chord contains the center of the circle.

6.3.4 The tangent segments to a circle from an external point are congruent.

6.3.5 If two chords intersect within a circle, then the product of the lengths of the segments (parts) of one chord is equal to the product of the lengths of the segments of the other chord.

6.3.6 If two secant segments are drawn to a circle from an external point, then the products of the length of each secant with the length of its external segment are equal.

6.3.7 If a tangent segment and a secant segment are drawn to a circle from an external point, then the square of the length of the tangent equals the product of the length of the secant with the length of its external segment.

6.4.1 The line that is perpendicular to the radius of a circle at its endpoint on the circle is a tangent to the circle.

6.4.2 In a circle (or in congruent circles) containing two unequal central angles, the larger angle corresponds to the larger intercepted arc.

6.4.3 In a circle (or in congruent circles) containing two unequal arcs, the larger arc corresponds to the larger central angle.

6.4.4 In a circle (or in congruent circles) containing two unequal chords, the shorter chord is at the greater distance from the center of the circle.

6.4.5 In a circle (or in congruent circles) containing two unequal chords, the chord nearer the center of the circle has the greater length.

6.4.6 In a circle (or in congruent circles) containing two unequal chords, the longer chord corresponds to the greater minor arc.

6.4.7 In a circle (or in congruent circles) containing two unequal minor arcs, the greater minor arc corresponds to the longer of the chords related to these arcs.

7.1.1 The locus of points in a plane and equidistant from the sides of an angle is the angle bisector.

7.1.2 The locus of points in a plane that are equidistant from the endpoints of a line segment is the perpendicular bisector of that line segment.

7.2.1 The three angle bisectors of the angles of a triangle are concurrent.

7.2.2 The three perpendicular bisectors of the sides of a triangle are concurrent.

7.2.3 The three altitudes of a triangle are concurrent.

7.2.4 The three medians of a triangle are concurrent at a point that is two-thirds the distance from any vertex to the midpoint of the opposite side.

7.3.1 A circle can be circumscribed about (or inscribed in) any regular polygon.

7.3.2 The measure of any central angle of a regular polygon of n sides is given by $c = \frac{360}{n}$.

7.3.3 Any radius of a regular polygon bisects the angle at the vertex to which it is drawn.

7.3.4 Any apothem of a regular polygon bisects the side of the polygon to which it is drawn.

8.1.1 The area A of a square whose sides are each of length s is given by $A = s^2$.

8.1.2 The area A of a parallelogram with a base of length b and with corresponding altitude of length h is given by

$$A = bh$$

8.1.3 The area A of a triangle whose base has length b and whose corresponding altitude has length h is given by

$$A = \frac{1}{2}bh$$

8.1.4 The area A of a right triangle with legs of lengths a and b is given $A = \frac{1}{2}ab$.

8.2.1 (Heron's Formula) If the three sides of a triangle have lengths a, b, and c, then the area A of the triangle is given by

$$A = \sqrt{s(s - a)(s - b)(s - c)}$$

where the semiperimeter of the triangle is

$$s = \frac{1}{2}(a + b + c)$$

8.2.2 (Brahmagupta's Formula) For a cyclic quadrilateral with sides of lengths a, b, c, and d, the area A is given by

$$A = \sqrt{(s - a)(s - b)(s - c)(s - d)}$$

where the semiperimeter of the quadrilateral is

$$s = \frac{1}{2}(a + b + c + d)$$

8.2.3 The area A of a trapezoid whose bases have lengths b_1 and b_2 and whose altitude has length h is given by

$$A = \frac{1}{2}h(b_1 + b_2)$$

8.2.4 The area A of any quadrilateral with perpendicular diagonals of lengths d_1 and d_2 is given by

$$A = \frac{1}{2}d_1d_2$$

8.2.5 The area A of a rhombus whose diagonals have lengths d_1 and d_2 is given by

$$A = \frac{1}{2}d_1d_2$$

8.2.6 The area A of a kite whose diagonals have lengths d_1 and d_2 is given by

$$A = \frac{1}{2}d_1d_2$$

8.2.7 The ratio of the areas of two similar triangles equals the square of the ratio of the lengths of any two corresponding sides; that is,

$$\frac{A_1}{A_2} = \left(\frac{a_1}{a_2}\right)^2$$

8.3.1 The area A of a regular polygon whose apothem has length a and whose perimeter is P is given by

$$A = \frac{1}{2}aP$$

8.4.1 The circumference C of a circle is given by the formula

$$C = \pi d \quad \text{or} \quad C = 2\pi r$$

8.4.2 In a circle whose circumference is C, the length ℓ of an arc whose degree measure is m is given by

$$\ell = \frac{m}{360} \cdot C$$

8.4.3 The area A of a circle whose radius has length r is given by $A = \pi r^2$.

8.5.1 In a circle of radius length r, the area A of a sector whose arc has degree measure m is given by

$$A = \frac{m}{360}\pi r^2$$

8.5.2 The area of a semicircular region of radius length r is $A = \frac{1}{2}\pi r^2$.

8.5.3 Where P represents the perimeter of a triangle and r represents the length of the radius of its inscribed circle, the area A of the triangle is given by

$$A = \frac{1}{2}rP$$

9.1.1 The lateral area L of any prism whose altitude has measure h and whose base has perimeter P is given by $L = hP$.

9.1.2 The total area T of any prism with lateral area L and base area B is given by $T = L + 2B$.

9.2.1 In a regular pyramid, the lengths a of the apothem of the base, the altitude h, and the slant height ℓ satisfy the Pythagorean Theorem; that is, $\ell^2 = a^2 + h^2$ in every regular pyramid.

9.2.2 The lateral area L of a regular pyramid with slant height of length ℓ and perimeter P of the base is given by

$$L = \frac{1}{2}\ell P$$

9.2.3 The total area (surface area) T of a pyramid with lateral area L and base area B is given by $T = L + B$.

9.2.4 The volume V of a pyramid having a base area B and an altitude of length h is given by

$$V = \frac{1}{3}Bh$$

9.2.5 In a regular pyramid, the lengths of altitude h, radius r of the base, and lateral edge e satisfy the Pythagorean Theorem; that is, $e^2 = h^2 + r^2$.

9.3.1 The lateral area L of a right circular cylinder with altitude of length h and circumference C of the base is given by $L = hC$.
(Alternative) Where r is the length of the radius of the base, $L = 2\pi rh$.

9.3.2 The total area T of a right circular cylinder with base area B and lateral area L is given by $T = L + 2B$.
(Alternative) Where r is the length of the radius of the base and h is the length of the altitude, $T = 2\pi rh + 2\pi r^2$.

9.3.3 The volume V of a right circular cylinder with base area B and altitude of length h is given by $V = Bh$.
(Alternative) Where r is the length of the radius of the base, $V = \pi r^2 h$.

9.3.4 The lateral area L of a right circular cone with slant height of length ℓ and circumference C of the base is given by $L = \frac{1}{2}\ell C$.
(Alternative) Where r is the length of the radius of the base, $L = \pi r \ell$.

9.3.5 The total area T of a right circular cone with base area B and lateral area L is given by $T = B + L$.
(Alternative) Where r is the length of the radius of the base and ℓ is the length of the slant height, the total area is $T = \pi r^2 + \pi r \ell$.

9.3.6 In a right circular cone, the lengths of the radius r (of the base), the altitude h, and the slant height ℓ satisfy the Pythagorean Theorem; that is, $\ell^2 = r^2 + h^2$ in every right circular cone.

9.3.7 The volume V of a right circular cone with base area B and altitude of length h is given by $V = \frac{1}{3}Bh$.
(Alternative) Where r is the length of the radius of the base, $V = \frac{1}{3}\pi r^2 h$.

9.4.1 (Euler's Equation) The number of vertices V, the number of edges E, and the number of faces F of a polyhedron are related by the equation $V + F = E + 2$.

9.4.2 The surface area S of a sphere whose radius has length r is given by $S = 4\pi r^2$.

9.4.3 The volume V of a sphere with radius of length r is given by $V = \frac{4}{3}\pi r^3$.

10.1.1 (Distance Formula) The distance d between two points (x_1, y_1) and (x_2, y_2) is given by the formula

$$d = \sqrt{(x_2 - x_1)^2 + (y_2 - y_1)^2}$$

10.1.2 (Midpoint Formula) The midpoint M of the line segment joining (x_1, y_1) and (x_2, y_2) has coordinates x_M and y_M, where

$$(x_M, y_M) = \left(\frac{x_1 + x_2}{2}, \frac{y_1 + y_2}{2}\right)$$

That is, $M = \left(\frac{x_1 + x_2}{2}, \frac{y_1 + y_2}{2}\right)$

10.2.1 If two nonvertical lines are parallel, then their slopes are equal.
(Alternative) If $\ell_1 \parallel \ell_2$, then $m_1 = m_2$

10.2.2 If two lines (neither horizontal nor vertical) are perpendicular, then the product of their slopes is -1.
(Alternative) If $\ell_1 \perp \ell_2$, then $m_1 \cdot m_2 = -1$

10.4.1 The line segment determined by the midpoints of two sides of a triangle is parallel to the third side.

10.4.2 The diagonals of a parallelogram bisect each other.

10.4.3 The diagonals of a rhombus are perpendicular.

10.4.4 If the diagonals of a parallelogram are equal in length, then the parallelogram is a rectangle.

10.5.1 (Slope-Intercept Form of a Line) The line whose slope is m and whose y intercept is b has the equation $y = mx + b$.

10.5.2 (Point-Slope Form of a Line) The line that has slope m and contains the point (x_1, y_1) has the equation

$$y - y_1 = m(x - x_1)$$

10.5.3 The three medians of a triangle are concurrent at a point that is two-thirds the distance from any vertex to the midpoint of the opposite side.

10.6.1 (The Distance Formula) In the xyz coordinate system, the distance d between the points $P_1(x_1, y_1, z_1)$ and $P_2(x_2, y_2, z_2)$ is given by

$$d = P_1 P_2 = \sqrt{(x_2 - x_1)^2 + (y_2 - y_1)^2 + (z_2 - z_1)^2}.$$

10.6.2 (The Midpoint Formula) In the xyz system, the midpoint of the line segment joining the points $P_1(x_1, y_1, z_1)$ and $P_2(x_2, y_2, z_2)$ is given by

$$M = (x_M, y_M, z_M) = \left(\frac{x_1 + x_2}{2}, \frac{y_1 + y_2}{2}, \frac{z_1 + z_2}{2}\right).$$

10.6.3 The equation for the sphere with center (h, k, l) and radius length r is given by the equation $(x - h)^2 + (y - k)^2 + (z - l)^2 = r^2$.

10.6.4 The equation for the sphere with center $(0, 0, 0)$ and radius length r is given by the equation $x^2 + y^2 + z^2 = r^2$.

11.2.1 (The Pythagorean Identity) In any right triangle in which α is the measure of an acute angle,

$$\sin^2\alpha + \cos^2\alpha = 1$$

11.4.1 The area of any acute triangle equals one-half the product of the lengths of two sides and the sine of the included angle. That is,

$$A = \tfrac{1}{2}ab \sin \gamma$$
$$A = \tfrac{1}{2}ac \sin \beta$$
$$A = \tfrac{1}{2}bc \sin \alpha$$

11.4.2 (Law of Sines) In any acute triangle, the three ratios between the sines of the angles and the lengths of the opposite sides are equal. That is,

$$\frac{\sin \alpha}{a} = \frac{\sin \beta}{b} = \frac{\sin \gamma}{c}$$

11.4.3 (Law of Cosines) In acute triangle ABC,

$$c^2 = a^2 + b^2 - 2ab \cos \gamma$$
$$b^2 = a^2 + c^2 - 2ac \cos \beta$$
$$a^2 = b^2 + c^2 - 2bc \cos \alpha$$

or

$$\cos \gamma = \frac{a^2 + b^2 - c^2}{2ab}$$
$$\cos \beta = \frac{a^2 + c^2 - b^2}{2ac}$$
$$\cos \alpha = \frac{b^2 + c^2 - a^2}{2bc}$$

Answers

Selected Exercises and Proofs

CHAPTER 1

1.1 Exercises

1. (a) Not a statement (b) Statement; true (c) Statement; true (d) Statement; false **3.** (a) Christopher Columbus did not cross the Atlantic Ocean. (b) Some jokes are not funny. **5.** Conditional **7.** Simple **9.** Simple **11.** H: You go to the game. C: You will have a great time. **13.** H: The diagonals of a parallelogram are perpendicular. C: The parallelogram is a rhombus. **15.** H: Two parallel lines are cut by a transversal. C: Corresponding angles are congruent. **17.** First write the statement in "If, then" form: If a figure is a square, then it is a rectangle. H: A figure is a square, C: It is a rectangle. **19.** True **21.** True **23.** False **25.** Induction **27.** Deduction **29.** Intuition **31.** None **33.** Angle 1 looks equal in measure to angle 2. **35.** The three angles in one triangle are equal in measure to the corresponding three angles in the other triangle. **37.** *A Prisoner of Society* might be nominated for an Academy Award. **39.** The instructor is a math teacher. **41.** Angles 1 and 2 are complementary. **43.** Alex has a strange sense of humor. **45.** None **47.** June Jesse will be in the public eye. **49.** Marilyn is a happy person. **51.** Valid **53.** Not valid **55.** (a) True (b) True (c) False **57.** (a) True (b) True

1.2 Exercises

1. $AB < CD$ **3.** Two; one **5.** One; none **7.** $\angle ABC$, $\angle ABD$, $\angle DBC$ **9.** Yes; no; yes **11.** $\angle ABC$, $\angle CBA$ **13.** Yes; no **15.** a, d **17.** (a) 3 (b) $2\frac{1}{2}$ **19.** (a) $40°$ (b) $50°$ **21.** Congruent; congruent **23.** Equal **25.** No **27.** Yes **29.** Congruent **31.** $\overline{MN}$ and $\overline{QP}$ **33.** $\overline{AB}$ **35.** 22 **37.** $x = 9$ **39.** $124°$ **41.** $71°$ **43.** $x - 23$ **45.** 10.9 **47.** $x = 102$; $y = 78$ **49.** N 22° E

1.3 Exercises

1. AC **3.** 75 in. **5.** 1.64 ft **7.** 3 mi **9.** (a) A-C-D (b) A, B, C or B, C, D or A, B, D **11.** $\overleftrightarrow{CD}$ means line CD; $\overline{CD}$ means segment CD; CD means the measure or length of $\overline{CD}$; $\overrightarrow{CD}$ means ray CD with endpoint C. **13.** (a) m and t (b) m and p; p and t **15.** $x = 3$; $AM = 7$ **17.** $x = 7$; $AB = 38$ **19.** (a) $\overrightarrow{OA}$ and $\overrightarrow{OD}$ (b) $\overrightarrow{OA}$ and $\overrightarrow{OB}$ (There are other possible answers.) **23.** Planes M and N intersect at $\overleftrightarrow{AB}$. **25.** A **27.** (a) C (b) C (c) H **33.** (a) No (b) Yes (c) No (d) Yes **35.** Six **37.** Nothing **39.** (a) Yes (b) Yes (c) No **41.** $\frac{1}{3}a + \frac{1}{2}b$ or $\frac{2a + 3b}{6}$

1.4 Exercises

1. (a) Acute (b) Right (c) Obtuse **3.** (a) Complementary (b) Supplementary **5.** Adjacent angles **7.** Complementary angles (also adjacent) **9.** Yes; no **11.** (a) Obtuse (b) Straight (c) Acute (d) Obtuse **13.** $m\angle FAC + m\angle CAD = 180$;

$\angle FAC$ and $\angle CAD$ are supplementary. **15.** (a) $x + y = 90$ (b) $x = y$ **17.** $42°$ **19.** $x = 20$; $m\angle RSV = 56°$ **21.** $x = 60$; $m\angle RST = 30°$ **23.** $x = 24$; $y = 8$ **25.** $\angle CAB \cong \angle DAB$ **27.** Angles measure 128° and 52°. **29.** (a) $(180 - x)°$ (b) $(192 - 3x)°$ (c) $(180 - 2x - 5y)°$ **31.** $x = 143$ **37.** It appears that the angle bisectors meet at one point. **39.** It appears that the two sides opposite $\angle$s A and B are congruent. **41.** (a) 90° (b) 90° (c) Equal **43.** $x = 15$; $z = 3$ **45.** 135°

1.5 Exercises

1. Division (or Multiplication) Prop. of Equality **3.** Subtraction Prop. of Equality **5.** Multiplication Prop. of Equality **7.** If 2 $\angle$s are supp., the sum of their measures is 180°. **9.** Angle-Addition Postulate **11.** $AM + MB = AB$ **13.** $\overrightarrow{EG}$ bisects $\angle DEF$. **15.** $m\angle 1 + m\angle 2 = 90°$ **17.** $2x = 10$ **19.** $7x + 2 = 30$ **21.** $6x - 3 = 27$ **23.** 1. Given 2. Distributive Prop. 3. Addition Prop. of Equality 4. Division Prop. of Equality **25.** 1. $2(x + 3) - 7 = 11$ 2. $2x + 6 - 7 = 11$ 3. $2x - 1 = 11$ 4. $2x = 12$ 5. $x = 6$ **27.** 1. Given 2. Segment-Addition Postulate 3. Subtraction Prop. of Equality **29.** 1. Given 2. Definition of angle bisector 3. Angle-Addition Postulate 4. Substitution 5. Substitution (Distribution) 6. Multiplication Prop. of Equality **31.** S1. M-N-P-Q on $\overline{MQ}$ R1. Given 2. Segment-Addition Postulate 3. Segment-Addition Postulate 4. $MN + NP + PQ = MQ$ **33.** $5(x + y)$ **35.** $(-7)(-2) > 5(-2)$ or $14 > -10$ **37.** R1 Given R2 Add. Prop. of Eq. R3 Given R4 Substitution

1.6 Exercises

1. 1. Given 2. If two $\angle$s are $\cong$, then they are equal in measure. 3. Angle-Addition Postulate 4. Addition Property of Equality 5. Substitution 6. If two $\angle$s are equal in measure, then they are $\cong$. **3.** 1. $\angle 1 \cong \angle 2$ and $\angle 2 \cong \angle 3$ 2. $\angle 1 \cong \angle 3$ **11.** 1. Given 3. Substitution 4. $m\angle 1 = m\angle 2$ 5. $\angle 1 \cong \angle 2$ **13.** No; yes; no **15.** No; yes; no **17.** No; yes; yes **19.** (a) Perpendicular (b) Angles (c) Supplementary (d) Right (e) Measure of angle 1 **21.** (a) Adjacent (b) Complementary (c) Ray AB (d) Is congruent to (e) Vertical **23.**

PROOF	
Statements	**Reasons**
1. M-N-P-Q on $\overline{MQ}$	1. Given
2. $MN + NQ = MQ$	2. Segment-Addition Postulate
3. $NP + PQ = NQ$	3. Segment-Addition Postulate
4. $MN + NP + PQ = MQ$	4. Substitution

25.

PROOF	
Statements	**Reasons**
1. ∠TSW with $\overrightarrow{SU}$ and $\overrightarrow{SV}$	1. Given
2. m∠TSW = m∠TSU + m∠USW	2. Angle-Addition Postulate
3. m∠USW = m∠USV + m∠VSW	3. Angle-Addition Postulate
4. m∠TSW = m∠TSU + m∠USV + m∠VSW	4. Substitution

27. In space, there are an infinite number of lines that perpendicularly bisect a given line segment at its midpoint.
29. The sum of measures of ∠s 1, 2, 3, and 4 is the same as the measure of straight angle *AOB*.

1.7 Exercises

1. H: A line segment is bisected. C: Each of the equal segments has half the length of the original segment   **3**   First write the statement in "If, then" form. If a figure is a square, then it is a quadrilateral. H: A figure is a square. C: It is a quadrilateral.
5. H: Each angle is a right angle. C: Two angles are congruent.
7. Statement, Drawing, Given, Prove, Proof   **9.** (a) Given (b) Prove   **11.** After the theorem has been proved.
13. *Given:* $\overleftrightarrow{AB} \perp \overleftrightarrow{CD}$
Prove: ∠AEC is a right angle

15. *Given:* ∠1 is comp. to ∠3; ∠2 is comp. to ∠3
Prove: ∠1 ≅ ∠2

17. *Given:* Lines ℓ and *m* intersect as shown
Prove: ∠1 ≅ ∠2 and ∠3 ≅ ∠4

19. m∠2 = 55°; m∠3 = 125°; m∠4 = 55°   **21.** *x* = 40; m∠1 = 130°
23. *x* = 60; m∠1 = 120°   **25.** *x* = 180; m∠2 = 80°
27. 1. Given   2. If two ∠s are complementary, the sum of their measures is 90. 3. Substitution 4. Subtraction Property of Equality 5. If two ∠s are equal in measure, then they are ≅.
31. 1. Given   2. ∠ABC is a right ∠.   3. The measure of a rt. ∠ = 90.   4. Angle-Addition Postulate 6. ∠1 is comp. to ∠2.

1.7 Selected Proof

33.

PROOF	
Statements	**Reasons**
1. ∠ABC ≅ ∠EFG	1. Given
2. m∠ABC = m∠EFG	2. If two ∠s are ≅, their measures are =
3. m∠ABC = m∠1 + m∠2 m∠EFG = m∠3 + m∠4	3. Angle-Addition Postulate
4. m∠1 + m∠2 = m∠3 + m∠4	4. Substitution
5. $\overrightarrow{BD}$ bisects ∠ABC $\overrightarrow{FH}$ bisects ∠EFG	5. Given
6. m∠1 = m∠2 and m∠3 = m∠4	6. If a ray bisects an ∠, then two ∠s of equal measure are formed
7. m∠1 + m∠1 = m∠3 + m∠3 or 2 · m∠1 = 2 · m∠3	7. Substitution
8. m∠1 = m∠3	8. Division Prop. of Equality
9. m∠1 = m∠2 = m∠3 = m∠4	9. Substitution (or Transitive)
10. ∠1 ≅ ∠2 ≅ ∠3 ≅ ∠4	10. If ∠s are = in measure, then they are ≅

Chapter 1 Review Exercises

1. Undefined terms, defined terms, axioms or postulates, theorems   **2.** Induction, deduction, intuition   **3.** 1. Names the term being defined   2. Places the term into a set or category 3. Distinguishes the term from other terms in the same category 4. Reversible   **4.** Intuition   **5.** Induction   **6.** Deduction
7. H: The diagonals of a trapezoid are equal in length. C: The trapezoid is isosceles.   **8.** H: The parallelogram is a rectangle. C: The diagonals of a parallelogram are congruent.
9. No conclusion   **10.** Jody Smithers has a college degree.
11. Angle *A* is a right angle.   **12.** C   **13.** ∠RST; ∠S; greater than 90°   **14.** Perpendicular   **18.** (a) Obtuse (b) Right **19.** (a) Acute (b) Reflex   **20.** 70°   **21.** 47°
22. 22   **23.** 17   **24.** 34   **25.** 152°   **26.** 39°
27. (a) Point *M* (b) ∠JMH (c) $\overrightarrow{MJ}$ (d) $\overleftrightarrow{KH}$ **28.** $67\frac{1}{2}°$
29. 28° and 152°   **30.** (a) 6*x* + 8 (b) *x* = 4 (c) 11; 10; 11
31. The measure of angle 3 is less than 50°.   **32.** 10 pegs
33. S   **34.** S   **35.** A   **36.** S   **37.** N   **38.** 2. ∠4 ≅ ∠P 3. ∠1 ≅ ∠4   4. If two ∠s are ≅, then their measures are =.   5. Given   6. m∠2 = m∠3 7. m∠1 + m∠2 = m∠4 + m∠3   8. Angle-Addition Postulate   9. Substitution   10. ∠TVP ≅ ∠MVP   **52.** 270°

Chapter 1 Review Exercises Selected Proofs

39.

PROOF

Statements	Reasons
1. $\overline{KF} \perp \overline{FH}$	1. Given
2. $\angle KFH$ is a rt. $\angle$	2. If two segments are $\perp$, then they form a rt. $\angle$
3. $\angle JHF$ is a rt. $\angle$	3. Given
4. $\angle KFH \cong \angle JHF$	4. Any two rt. $\angle$s are $\cong$

40.

PROOF

Statements	Reasons
1. $\overline{KH} \cong \overline{FJ}$ G is the midpoint of both $\overline{KH}$ and $\overline{FJ}$	1. Given
2. $\overline{KG} \cong \overline{GJ}$	2. If two segments are $\cong$, then their midpoints separate these segments into four $\cong$ segments

41.

PROOF

Statements	Reasons
1. $\overline{KF} \perp \overline{FH}$	1. Given
2. $\angle KFJ$ is comp. to $\angle JFH$	2. If the exterior sides of two adjacent $\angle$s form $\perp$ rays, then these $\angle$s are comp.

42.

PROOF

Statements	Reasons
1. $\angle 1$ is comp. to $\angle M$	1. Given
2. $\angle 2$ is comp. to $\angle M$	2. Given
3. $\angle 1 \cong \angle 2$	3. If two $\angle$s are comp. to the same $\angle$, then these angles are $\cong$

43.

PROOF

Statements	Reasons
1. $\angle MOP \cong \angle MPO$	1. Given
2. $\overrightarrow{OR}$ bisects $\angle MOP$; $\overrightarrow{PR}$ bisects $\angle MPO$	2. Given
3. $\angle 1 \cong \angle 2$	3. If two $\angle$s are $\cong$, then their bisectors separate these $\angle$s into four $\cong$ $\angle$s

44.

PROOF

Statements	Reasons
1. $\angle 4 \cong \angle 6$	1. Given
2. $\angle 4 \cong \angle 5$	2. If two angles are vertical $\angle$s, then they are $\cong$
3. $\angle 5 \cong \angle 6$	3. Transitive Property

45.

PROOF

Statements	Reasons
1. Figure as shown	1. Given
2. $\angle 4$ is supp. to $\angle 2$	2. If the exterior sides of two adjacent $\angle$s form a line, then the $\angle$s are supp.

46.

PROOF

Statements	Reasons
1. $\angle 3$ is supp. to $\angle 5$ $\angle 4$ is supp. to $\angle 6$	1. Given
2. $\angle 4 \cong \angle 5$	2. If two lines intersect, the vertical angles formed are $\cong$
3. $\angle 3 \cong \angle 6$	3. If two $\angle$s are supp. to congruent angles, then these angles are $\cong$

Chapter 1 Test

1. Induction [1.1]   **2.** $\angle CBA$ or $\angle B$ [1.4]
3. $AP + PB = AB$ [1.3]   **4.** (a) Point (b) Line [1.3]
5. (a) Right (b) Obtuse [1.4]   **6.** (a) Supplementary
(b) Congruent [1.4]   **7.** m$\angle MNP$ = m$\angle PNQ$ [1.4]
8. (a) Right (b) Supplementary [1.7]   **9.** Kianna will develop
reasoning skills. [1.1]   **10.** 10.4 in. [1.2]   **11.** (a) 11
(b) 16 [1.3]   **12.** 35° [1.4]   **13.** (a) 24° (b) 45° [1.4]

14. (a) 137° (b) 43° [1.4] **15.** (a) 25° (b) 47° [1.7]
16. (a) 23° (b) 137° [1.7] **17.** $x + y = 90$ [1.4]
20. 1. Given 2. Segment-Addition Postulate 3. Segment-Addition Postulate 4. Substitution [1.5] **21.** 1. $2x - 3 = 17$ 2. $2x = 20$ 3. $x = 10$ [1.5] **22.** 1. Given 2. 90°
3. Angle-Addition Postulate 4. 90° 5. Given 6. Definition of Angle-Bisector 7. Substitution 8. $m\angle 1 = 45°$ [1.7]
23. 108° [1.4]

CHAPTER 2

2.1 Exercises

1. (a) 108° (b) 72° **3.** (a) 68.3° (b) 68.3°
5. (a) No (b) Yes (c) Yes **7.** Angle 9 appears to be a right angle. **9.** (a) $m\angle 3 = 87°$ (b) $m\angle 6 = 87°$
(c) $m\angle 1 = 93°$ (d) $m\angle 7 = 87°$ **11.** (a) $\angle 5$ (b) $\angle 5$
(c) $\angle 8$ (d) $\angle 5$ **13.** (a) $m\angle 2 = 68°$ (b) $m\angle 4 = 112°$
(c) $m\angle 5 = 112°$ (d) $m\angle MOQ = 34°$ **15.** $x = 10$;
$m\angle 4 = 110°$ **17.** $x = 12$; $y = 4$; $m\angle 7 = 76°$
19. 1. Given 2. If two parallel lines are cut by a transversal, then the corresponding angles are $\cong$ 3. If two lines intersect, then the vertical angles are $\cong$ 4. $\angle 3 \cong \angle 1$
5. $\angle 1 \cong \angle 4$ **25.** 93° **27.** (a) $\angle 4 \cong \angle 2$ and $\angle 5 \cong \angle 3$
(b) 180° (c) 180° **31.** No

2.1 Selected Proof

21.

PROOF	
Statements	**Reasons**
1. $\overleftrightarrow{CE} \parallel \overleftrightarrow{DF}$; transversal $\overleftrightarrow{AB}$	1. Given
2. $\angle ACE \cong \angle ADF$	2. If two $\parallel$ lines are cut by a transversal, then the corresponding $\angle$s are $\cong$
3. $\overrightarrow{CX}$ bisects $\angle ACE$ $\overrightarrow{DE}$ bisects $\angle CDF$	3. Given
4. $\angle 1 \cong \angle 3$	4. If two $\angle$s are $\cong$, then their bisectors separate these $\angle$s into four $\cong$ $\angle$s

2.2 Exercises

1. *Converse:* If Juan is rich, then he won the state lottery. FALSE.
Inverse: If Juan does not win the state lottery, then he will not be rich. FALSE.
Contrapositive: If Juan is not rich, then he did not win the state lottery. TRUE.
3. *Converse:* If two angles are complementary, then the sum of their measures is 90°. TRUE.
Inverse: If the sum of the measures of two angles is not 90°, then the two angles are not complementary. TRUE.
Contrapositive: If two angles are not complementary, then the sum of their measures is not 90°. TRUE.

5. No conclusion **7.** Alice did not play in the volleyball match.
9. $x = 5$ **11.** (a), (b), and (e) **13.** Suppose that $\angle 1 \cong \angle 2$.
15. If $\angle A$ and $\angle B$ are vertical angles, then $\angle A$ and $\angle B$ are congruent. **17.** If a triangle is equilateral, then all sides of the triangle are congruent. **19.** The areas of $\triangle ABC$ and $\triangle DEF$ are equal. **21.** Parallel

2.2 Selected Proofs

23. Assume that $r \parallel s$. Then $\angle 1 \cong \angle 5$ because they are corresponding angles. But it is given that $\angle 1 \ncong \angle 5$, which leads to a contradiction. Thus, the assumption that $r \parallel s$ is false and it follows that $r \nparallel s$. **25.** Assume that $\overleftrightarrow{FH} \perp \overleftrightarrow{EG}$. Then $\angle 3 \cong \angle 4$ and $m\angle 3 = m\angle 4$. But it is given that $m\angle 3 > m\angle 4$, which leads to a contradiction. Then the assumption that $\overleftrightarrow{FH} \perp \overleftrightarrow{EG}$ must be false and it follows that $\overleftrightarrow{FH}$ is not perpendicular to $\overleftrightarrow{EG}$.
27. Assume that the angles are vertical angles. If they are vertical angles, then they are congruent. But this contradicts the hypothesis that the two angles are not congruent. Hence, our assumption must be false, and the angles are not vertical angles.
31. If M is a midpoint of $\overline{AB}$, then $AM = \frac{1}{2}(AB)$. Assume that N is also a midpoint of $\overline{AB}$ so that $AN = \frac{1}{2}(AB)$. By substitution, $AM = AN$. By the Segment-Addition Postulate, $AM = AN + NM$. Using substitution again, $AN + NM = AN$. Subtracting gives $NM = 0$. But this contradicts the Ruler Postulate, which states that the measure of a line segment is a positive number. Therefore, our assumption is wrong and M is the only midpoint for $\overline{AB}$.

2.3 Exercises

1. $\ell \parallel m$ **3.** $\ell \nparallel m$ **5.** $\ell \nparallel m$ **7.** $p \parallel q$ **9.** None
11. $\ell \parallel n$ **13.** None **15.** $\ell \parallel n$ **17.** 1. Given 2. If two $\angle$s are comp. to the same $\angle$, then they are $\cong$ 3. $\overline{BC} \parallel \overline{DE}$
23. $x = 20$ **25.** $x = 120$ **27.** $x = 9$ **29.** $x = 6$

2.3 Selected Proof

19.

PROOF	
Statements	**Reasons**
1. $\overline{AD} \perp \overline{DC}$ and $\overline{BC} \perp \overline{DC}$	1. Given
2. $\overrightarrow{AD} \parallel \overrightarrow{BC}$	2. If two lines are each $\perp$ to a third line, then these lines are $\parallel$ to each other

2.4 Exercises

1. $m\angle C = 75°$ **3.** $m\angle B = 46°$ **5.** (a) Underdetermined
(b) Determined (c) Overdetermined **7.** (a) Equilateral
(b) Isosceles **9.** (a) Equiangular (b) Right **11.** If two $\angle$s of one triangle are $\cong$ to two $\angle$s of another triangle, then the third $\angle$s of the triangles are $\cong$. **13.** $m\angle 1 = 122°$; $m\angle 2 = 58°$;
$m\angle 5 = 72°$ **15.** $m\angle 2 = 57.7°$; $m\angle 3 = 80.8°$;
$m\angle 4 = 41.5°$ **17.** 35° **19.** 40° **21.** $x = 72$; $m\angle 1 = 72°$;
$m\angle DAE = 36°$ **23.** 360° **25.** $x = 45°$; $y = 45°$
27. $x = 108$ **29.** $y = 20°$; $x = 100°$; $m\angle 5 = 60°$
35. 44° **37.** $m\angle N = 49°$; $m\angle P = 98°$ **39.** 35°
41. 75° **49.** $m\angle M = 84°$

2.5 Exercises

1. Increase   **3.** $x = 113°$; $y = 67°$; $z = 36°$ **5.** (a) 5
(b) 35   **7.** (a) 540° (b) 1440°   **9.** (a) 90° (b) 150°
11. (a) 90° (b) 30°   **13.** (a) 7 (b) 9   **15.** (a) $n = 5$
(b) $n = 10$ **17.** (a) 15 (b) 20   **19.** 135°
21.   **23.**   **25.**

31. Figure (a): 90°, 90°, 120°, 120°, 120° Figure (b): 90°, 90°,
90°, 135°, 135°   **33.** (a) 36° (b) 252°   **35.** The resulting
polygon is also a regular polygon of the same type.   **37.** 150°
39. (a) $n - 3$ (b) $\dfrac{n(n - 3)}{2}$ **41.** 221°   **43.** (a) No (b) Yes
47. 190°

2.5 Selected Proof

29.

PROOF

Statements	Reasons
1. Quad. *RSTV* with diagonals $\overline{RT}$ and $\overline{SV}$ intersecting at W	1. Given
2. $m\angle RWS = m\angle 1 + m\angle 2$	2. The measure of an exterior ∠ of a △ equals the sum of the measures of the nonadjacent interior ∠s of the △
3. $m\angle RWS = m\angle 3 + m\angle 4$	3. Same as 2
4. $m\angle 1 + m\angle 2 = m\angle 3 + m\angle 4$	4. Substitution

2.6 Exercises

1. M, T, X   **3.** N, X   **5.** (a), (c), (b)   **7.** (a), (b)
9. MOM   **11.** (a)

13. (a)   (b)

15. (a) 63° (b) Yes (c) Yes   **17.** WHIM   **19.** SIX   **21.** WOW
23. (a) Clockwise (b) Counterclockwise   **25.** 62,365 kilowatt-
hours   **27.** (a) Line (b) Point (c) Line   **29.** (b), (c)
31. (a) 12 (b) 6 (c) 4 (d) 3   **33.** $x = 50$ **35.** (a) 9 (b) 2

CHAPTER 2 REVIEW EXERCISES

1. (a) $\overline{BC} \parallel \overline{AD}$ (b) $\overline{AB} \parallel \overline{CD}$ **2.** 110°   **3.** $x = 37$
4. $m\angle D = 75°$; $m\angle DEF = 125°$ **5.** $x = 20$; $y = 10$
6. $x = 30$; $y = 35$ **7.** $\overline{AE} \parallel \overline{BF}$ **8.** None   **9.** $\overline{BE} \parallel \overline{CF}$
10. $\overline{BE} \parallel \overline{CF}$ **11.** $\overline{AC} \parallel \overline{DF}$ and $\overline{AE} \parallel \overline{BF}$ **12.** $x = 120°$;
$y = 70°$ **13.** $x = 32°$; $y = 30°$ **14.** $y = -8$; $x = 24$
15. $x = 140°$ **16.** $x = 6$ **17.** $m\angle 3 = 69°$; $m\angle 4 = 67°$;
$m\angle 5 = 44°$ **18.** 110°   **19.** S   **20.** N   **21.** N   **22.** S
23. S   **24.** A
25.

Number of sides	8	12	20	15	10	16	180
Measure of each ext. ∠	45	30	18	24	36	22.5	2
Measure of each int. ∠	135	150	162	156	144	157.5	178
Number of diagonals	20	54	170	90	35	104	15,930

28. Not possible
30. *Statement:* If two angles are right angles, then the angles are
congruent.
Converse: If two angles are congruent, then the angles are right
angles
Inverse: If two angles are not right angles, then the angles are not
congruent.
Contrapositive: If two angles are not congruent, then the angles
are not right angles.
31. *Statement:* If it is not raining, then I am happy.
Converse: If I am happy, then it is not raining.
Inverse: If it is raining, then I am not happy.
Contrapositive: If I am not happy, then it is raining.
32. Contrapositive   **37.** Assume that $x = -3$.
38. Assume that the sides opposite these angles are ≅
39. Assume that $\angle 1 = \angle 2$. Then $m \parallel n$ because congruent corresponding angles are formed. But this contradicts our hypothesis.
Therefore, our assumption must be false, and it follows that
$\angle 1 \not\cong \angle 2$.   **40.** Assume that $m \parallel n$. Then $\angle 1 \cong \angle 3$ because
alternate exterior angles are congruent when parallel lines are cut
by a transversal. But this contradicts the given fact that
$\angle 1 \not\cong \angle 3$. Therefore, our assumption must be false, and it follows that $m \not\parallel n$.   **43.** (a) B, H, W (b) H, S   **44.** (a) Isosceles
Triangle, Circle, Regular Pentagon (b) Circle   **45.** Congruent
46. (a)   (b)   **47.** 90°

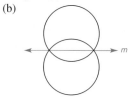

Chapter 2 Review Exercises Selected Proofs

33.

PROOF	
Statements	**Reasons**
1. $\overline{AB} \parallel \overline{CF}$	1. Given
2. $\angle 1 \cong \angle 2$	2. If two $\parallel$ lines are cut by a transversal, then corresponding $\angle$s are $\cong$
3. $\angle 2 \cong \angle 3$	3. Given
4. $\angle 1 \cong \angle 3$	4. Transitive Prop. of Congruence

34.

PROOF	
Statements	**Reasons**
1. $\angle 1$ is comp. to $\angle 2$ $\angle 2$ is comp. to $\angle 3$	1. Given
2. $\angle 1 \cong \angle 3$	2. If two $\angle$s are comp. to the same $\angle$, then these $\angle$s are $\cong$
3. $\overline{BD} \parallel \overline{AE}$	3. If two lines are cut by a transversal so that corresponding $\angle$s are $\cong$, then these lines are $\parallel$

35.

PROOF	
Statements	**Reasons**
1. $\overline{BE} \perp \overline{DA}$ $\overline{CD} \perp \overline{DA}$	1. Given
2. $\overline{BE} \parallel \overline{CD}$	2. If two lines are each $\perp$ to a third line, then these lines are parallel to each other
3. $\angle 1 \cong \angle 2$	3. If two $\parallel$ lines are cut by a transversal, then the alternate interior $\angle$s are $\cong$

36.

PROOF	
Statements	**Reasons**
1. $\angle A \cong \angle C$	1. Given
2. $\overrightarrow{DC} \parallel \overrightarrow{AB}$	2. Given
3. $\angle C \cong \angle 1$	3. If two $\parallel$ lines are cut by a transversal, the alt. int. $\angle$s are $\cong$
4. $\angle A \cong \angle 1$	4. Transitive Prop. of Congruence
5. $\overline{DA} \parallel \overline{CB}$	5. If two lines are cut by a transversal so that corr. $\angle$s are $\cong$, then these lines are $\parallel$

Chapter 2 Test

1. (a) $\angle 5$ (b) $\angle 3$ [2.1]   **2.** (a) r and s (b) ℓ and m [2.3]
3. "not Q" [2.2]   **4.** $\angle R$ and $\angle S$ are not both right angles. [2.2]
5. (a) $r \parallel t$ (b) $a \parallel c$ [2.3]   **7.** (a) $36°$ (b) $33°$ [2.4]
8. (a) Pentagon (b) Five [2.5]   **9.** (a) Equiangular hexagon
(b) $120°$ [2.5]   **10.** A: line; D: line; N: point; O: both;
X: both [2.6]   **11.** (a) Reflection (b) Slide (c) Rotation [2.6]
12. $61°$ [2.1]   **13.** 54 [2.3]   **14.** $50°$ [2.4]   **15.** $78°$ [2.4]
16. 1. Given 2. $\angle 2 \cong \angle 3$ 3. Transitive Prop. of Congruence
4. $\ell \parallel n$ [2.3]   **17.** Assume that $\angle M$ and $\angle Q$ are
complementary. By definition, m$\angle M$ + m$\angle Q$ = $90°$. Also,
m$\angle M$ + m$\angle Q$ + m$\angle N$ = $180°$ because these are the three
angles of $\triangle MNQ$. By substitution, $90°$ + m$\angle N$ = $180°$, so
it follows that m$\angle N$ = $90°$. But this leads to a contradiction
because it is given that m$\angle N$ = $120°$. The assumption must be
false, and it follows that $\angle M$ and $\angle Q$ are *not* complementary. [2.2]
18. 1. Given   2. $180°$ 3. m$\angle 1$ + m$\angle 2$ + $90°$ = $180°$
4. $90°$ S5. $\angle 1$ and $\angle 2$ are complementary. R5. Definition of
complementary angles [2.4]   **19.** $21°$ [2.4]

CHAPTER 3

3.1 Exercises

1. $\triangle BAC \cong \triangle EFD$ (other answers possible)   **3.** m$\angle A$ = $72°$
5. (a) No (b) SSA is not a method for proving triangles
congruent.   **7.** (a) $\cong$ (b) $\cong$ **9.** SSS   **11.** AAS   **13.** ASA
15. ASA   **17.** SSS   **19.** (a) $\angle A \cong \angle A$ (b) ASA
21. $\overline{AD} \cong \overline{EC}$ **23.** $\overline{MO} \cong \overline{MO}$ **25.** 1. Given   2. $\overline{AC} \cong \overline{AC}$
3. SSS   **33.** Yes; SAS or SSS   **35.** No   **37.** (a) $\triangle CBE$,
$\triangle ADE$, $\triangle CDE$ (b) $\triangle ADC$ (c) $\triangle CBD$

3.1 Selected Proofs

27.

PROOF	
Statements	**Reasons**
1. $\overrightarrow{PQ}$ bisects $\angle MPN$	1. Given
2. $\angle 1 \cong \angle 2$	2. If a ray bisects an $\angle$, it forms two $\cong \angle$s
3. $\overline{MP} \cong \overline{NP}$	3. Given
4. $\overline{PQ} \cong \overline{PQ}$	4. Identity
5. $\triangle MQP \cong \triangle NQP$	5. SAS

31.

PROOF	
Statements	**Reasons**
1. $\angle VRS \cong \angle TSR$ and $\overline{RV} \perp \overline{SV}$	1. Given
2. $\overline{RS} \cong \overline{RS}$	2. Identity
3. $\triangle RST \cong \triangle SRV$	3. SAS

3.2 Exercises

1. ASA **3.** SAS **13.** 1. Given 2. If two lines are $\perp$, then they form right $\angle$s 3. Identity 4. $\triangle HJK \cong \triangle HJL$ 5. $\overline{KJ} \cong \overline{JL}$ **21.** $c = 5$ **23.** $b = 8$ **25.** $c = \sqrt{41}$ **35.** (a) 8 (b) 37° (c) 53° **37.** 751 feet

3.2 Selected Proofs

5.

PROOF	
Statements	**Reasons**
1. $\angle 1$ and $\angle 2$ are right $\angle$s $\overline{CA} \cong \overline{DA}$	1. Given
2. $\overline{AB} \cong \overline{AB}$	2. Identity
3. $\triangle ABC \cong \triangle ABD$	3. HL

9.

PROOF	
Statements	**Reasons**
1. $\angle R$ and $\angle V$ are right $\angle$s $\angle 1 \cong \angle 2$	1. Given
2. $\angle R \cong \angle V$	2. All right $\angle$s are $\cong$
3. $\overline{ST} \cong \overline{ST}$	3. Identity
4. $\triangle RST \cong \triangle VST$	4. AAS

17.

PROOF	
Statements	**Reasons**
1. $\angle$s P and R are right $\angle$s	1. Given
2. $\angle P \cong \angle R$	2. All right $\angle$s are $\cong$
3. M is the midpoint of $\overline{PR}$	3. Given
4. $\overline{PM} \cong \overline{MR}$	4. The midpoint of a segment forms two $\cong$ segments
5. $\angle NMP \cong \angle QMR$	5. If two lines intersect, the vertical angles formed are $\cong$
6. $\triangle NPM \cong \triangle QRM$	6. ASA
7. $\angle N \cong \angle Q$	7. CPCTC

27.

PROOF	
Statements	**Reasons**
1. $\overline{DF} \cong \overline{DG}$ and $\overline{FE} \cong \overline{EG}$	1. Given
2. $\overline{DE} \cong \overline{DE}$	2. Identity
3. $\triangle FDE \cong \triangle GDE$	3. SSS
4. $\angle FDE \cong \angle GDE$	4. CPCTC
5. $\overrightarrow{DE}$ bisects $\angle FDG$	5. If a ray divides an $\angle$ into two $\cong \angle$s, then the ray bisects the angle

31.

PROOF	
Statements	**Reasons**
1. $\angle 1 \cong \angle 2$ and $\overline{MN} \cong \overline{QP}$	1. Given
2. $\overline{MP} \cong \overline{MP}$	2. Identity
3. $\triangle NMP \cong \triangle QPM$	3. SAS
4. $\angle 3 \cong \angle 4$	4. CPCTC
5. $\overline{MQ} \parallel \overline{NP}$	5. If two lines are cut by a transversal so that the alt. int. $\angle$s are $\cong$, then the lines are $\parallel$

3.3 Exercises

1. Isosceles **3.** $\overline{VT} \cong \overline{VU}$ **5.** m$\angle U = 69°$
7. m$\angle V = 36°$ **9.** $L = E$ (equivalent) **11.** R and S are disjoint; so $R \cap S = \varnothing$. **13.** Underdetermined
15. Overdetermined **17.** Determined **19.** (a) HL (b) SSS (c) SAS **21.** 55° **23.** m$\angle 2 = 68°$; m$\angle 1 = 44°$
25. m$\angle 5 = 124°$ **27.** m$\angle A = 52°$; m$\angle B = 64°$; m$\angle C = 64°$ **29.** 26 **31.** 12 **33.** Yes **35.** 1. Given
2. $\angle 3 \cong \angle 2$ 3. $\angle 1 \cong \angle 2$ 4. If two $\angle$s of a $\triangle$ are $\cong$, then the opposite sides are $\cong$ **41.** (a) 80° (b) 100° (c) 40°
43. 75° each

3.3 Selected Proof

37.

PROOF	
Statements	**Reasons**
1. ∠1 ≅ ∠3	1. Given
2. $\overline{RU}$ ≅ $\overline{VU}$	2. Given
3. ∠R ≅ ∠V	3. If two sides of a △ are ≅, then the ∠s opposite these sides are also ≅
4. △RUS ≅ △VUT	4. ASA
5. $\overline{SU}$ ≅ $\overline{TU}$	5. CPCTC
6. △STU is isosceles	6. If a △ has two ≅ sides, it is an isosceles △

3.4 Exercises

19. Construct a 90° angle; bisect it to form two 45° ∠s. Bisect one of the 45° angles to get a 22.5° ∠. **31.** 120° **33.** 150°
39. D is on the bisector of ∠A.

3.5 Exercises

1. False **3.** True **5.** True **7.** False **9.** True
11. (a) Not possible (100° + 100° + 60° ≠ 180°)
(b) Possible (45° + 45° + 90° = 180°)
13. (a) Possible (b) Not possible (8 + 9 = 17)
(c) Not possible (8 + 9 < 18) **15.** Scalene right triangle
(m∠ = 90°) **17.** Isosceles obtuse triangle (m∠ = 100°)
19. 4 cm **21.** 72° (two such angles); 36° (one angle only)
23. Nashville **25.** 1. m∠ABC > m∠DBE and
m∠CBD > m∠EBF 3. Angle-Addition Postulate
4. m∠ABD > m∠DBF **29.** BC < EF **31.** 2 < x < 10
33. x + 2 < y < 5x + 12 **35.** Proof: Assume that
PM = PN. Then △MPN is isosceles. But that contradicts the
hypothesis; thus, our assumption must be wrong, and PM ≠ PN.

3.5 Selected Proof

27.

PROOF	
Statements	**Reasons**
1. Quad. RSTU with diagonal $\overline{US}$; ∠R and ∠TUS are right ∠s	1. Given
2. TS > US	2. The shortest distance from a point to a line is the ⊥ distance
3. US > UR	3. Same as (2)
4. TS > UR	4. Transitive Prop. of Inequality

Chapter 3 Review Exercises

15. (a) $\overrightarrow{PR}$ (b) $\overline{PQ}$ **16.** $\overline{BC}, \overline{AC}, \overline{AB}$ **17.** ∠R, ∠Q, ∠P
18. $\overline{DA}$ **19.** (b) **20.** 5, 35 **21.** 20° **22.** 115°
23. m∠C = 64° **24.** Isosceles **25.** The triangle is also
equilateral. **26.** 60°

Chapter 3 Review Exercises Selected Proofs

1.

PROOF	
Statements	**Reasons**
1. ∠AEB ≅ ∠DEC	1. Given
2. $\overline{AE}$ ≅ $\overline{ED}$	2. Given
3. ∠A ≅ ∠D	3. If two sides of a △ are ≅, then the ∠s opposite these sides are also ≅
4. △AEB ≅ △DEC	4. ASA

5.

PROOF	
Statements	**Reasons**
1. $\overline{AB}$ ≅ $\overline{DE}$ and $\overline{AB}$ ∥ $\overline{DE}$	1. Given
2. ∠A ≅ ∠D	2. If two ∥ lines are cut by a transversal, then the alt. int. ∠s are ≅
3. $\overline{AC}$ ≅ $\overline{DF}$	3. Given
4. △BAC ≅ △EDF	4. SAS
5. ∠BCA ≅ ∠EFD	5. CPCTC
6. $\overline{BC}$ ∥ $\overline{FE}$	6. If two lines are cut by a transversal so that alt. int. ∠s are ≅, then the lines are ∥

9.

PROOF	
Statements	**Reasons**
1. $\overline{YZ}$ is the base of an isosceles triangle	1. Given
2. ∠Y ≅ ∠Z	2. Base ∠s of an isosceles △ are ≅
3. $\overrightarrow{XA}$ ∥ $\overline{YZ}$	3. Given
4. ∠1 ≅ ∠Y	4. If two ∥ lines are cut by a transversal, then the corresponding ∠s are ≅
5. ∠2 ≅ ∠Z	5. If two ∥ lines are cut by a transversal, then the alt. int. ∠s are ≅
6. ∠1 ≅ ∠2	6. Transitive Prop. for Congruence

13.

PROOF	
Statements	**Reasons**
1. $\overline{AB} \cong \overline{CD}$	1. Given
2. $\angle BAD \cong \angle CDA$	2. Given
3. $\overline{AD} \cong \overline{AD}$	3. Identity
4. $\triangle BAD \cong \triangle CDA$	4. SAS
5. $\angle CAD \cong \angle BDA$	5. CPCTC
6. $\overline{AE} \cong \overline{ED}$	6. If two $\angle$s of a $\triangle$ are $\cong$, then the sides opposite these $\angle$s are also $\cong$
7. $\triangle AED$ is isosceles	7. If a $\triangle$ has two $\cong$ sides, then it is an isosceles $\triangle$

Chapter 3 Test

1. (a) 75° (b) 17 cm [3.1] **2.** (a) $\overline{XY}$ (b) $\angle Y$ [3.1]
3. (a) SAS (b) ASA [3.1] **4.** Corresponding parts of
congruent triangles are congruent. [3.2] **5.** (a) No (b) Yes [3.2]
6. Yes [3.2] **7.** (a) $c = 10$ (b) $\sqrt{28}$ (or $2\sqrt{7}$) [3.2]
8. (a) $\overline{AM} \cong \overline{MB}$ (b) No [3.3] **9.** (a) 38° (b) 36° [3.3]
10. (a) 7.6 inches (b) 57 [3.3] **13.** (a) $\overline{BC}$ (b) $\overline{CA}$ [3.5]
14. $m\angle V > m\angle U > m\angle T$ [3.5] **15.** $EB > DC$ since
$EB = \sqrt{74}$ and $DC = \sqrt{65}$ [3.2] **16.** $\overline{DA}$ [3.1]
17.

PROOF		
Statements	**Reasons**	
1. $\angle R$ and $\angle V$ are rt $\angle$s	1. Given	
2. $\angle R \cong \angle V$	2. All rt $\angle$s are $\cong$	
3. $\angle 1 \cong \angle 2$	3. Given	
4. $\overline{ST} \cong \overline{ST}$	4. Identity	
5. $\triangle RST \cong \triangle VST$	5. AAS	[3.1]

18. R1. Given R2. If 2 $\angle$s of a $\triangle$ are $\cong$, the opposite sides are
$\cong$ S3. $\angle 1 \cong \angle 3$ R4. ASA S5. $\overline{US} \cong \overline{UT}$ S6. $\triangle STU$ is an
isosceles triangle [3.3] **19.** $a = 10$ cm [3.3]

CHAPTER 4

4.1 Exercises

1. (a) $AB = DC$ (b) $m\angle A = m\angle C$ **3.** (a) 8 (b) 5 (c) 70°
(d) 110° **5.** (a) 6.4 (b) 10.6 **7.** $QR = 11, RN = 11$, and
$QN = 22$ **9.** $AB = DC = 8; BC = AD = 9$
11. $m\angle A = m\angle C = 83°; m\angle B = m\angle D = 97°$
13. $m\angle A = m\angle C = 80°; m\angle B = m\angle D = 100°$
15. $\overline{AC}$ **17.** (a) $\overline{VY}$ (b) 16 **19.** True **21.** True
23. Parallelogram **25.** Parallelogram **27.** 1. Given
2. $\overline{RV} \perp \overline{VT}$ and $\overline{ST} \perp \overline{VT}$ 3. $\overline{RV} \parallel \overline{ST}$ 4. $RSTV$ is a
parallelogram **35.** $\angle P$ is a right angle **37.** $\overline{RT}$ **39.** 255 mph
41. $\overline{AC}$ **43.** $m\angle A = m\angle C = 70°; m\angle B = m\angle D = 110°$;
$ABCD$ is a parallelogram

4.1 Selected Proof

29.

PROOF	
Statements	**Reasons**
1. Parallelogram $RSTV$	1. Given
2. $\overline{RS} \parallel \overline{VT}$	2. Opposite sides of a parallelogram are $\parallel$
3. $\overline{XY} \parallel \overline{VT}$	3. Given
4. $\overline{RS} \parallel \overline{XY}$	4. If two lines are each $\parallel$ to a third line, then the lines are $\parallel$
5. $RSYX$ is a parallelogram	5. If a quadrilateral has opposite sides $\parallel$, then the quadrilateral is a parallelogram
6. $\angle 1 \cong \angle S$	6. Opposite angles of a parallelogram are $\cong$

4.2 Exercises

1. (a) Yes (b) No **3.** Parallelogram **5.** (a) Kite
(b) Parallelogram **7.** $\overline{AC}$ **9.** 6.18 **11.** (a) 8 (b) 7 (c) 6
13. 10 **15.** (a) Yes; diagonal separating kite into 2 $\cong$ $\triangle$s
(b) No **17.** Parallel and Congruent **19.** 1. Given 2. Identity
3. $\triangle NMQ \cong \triangle NPQ$ 4. CPCTC 5. $MNPQ$ is a kite
29. $y = 6; MN = 9; ST = 18$ **31.** $x = 5; RM = 11$
33. $ST = 22$ **35.** $P = 34$ **37.** 270°

4.2 Selected Proofs

21.

PROOF	
Statements	**Reasons**
1. M-Q-T and P-Q-R so that $MNPQ$ and $QRST$ are parallelograms	1. Given
2. $\angle N \cong \angle MQP$	2. Opposite $\angle$s in a parallelogram are $\cong$
3. $\angle MQP \cong \angle RQT$	3. If two lines intersect, the vertical $\angle$s formed are $\cong$
4. $\angle RQT \cong \angle S$	4. Same as (2)
5. $\angle N \cong \angle S$	5. Transitive Prop. for Congruence

23.

PROOF

Statements	Reasons
1. Kite *HJKL* with diagonal $\overline{HK}$	1. Given
2. $\overline{LH} \cong \overline{HJ}$ and $\overline{LK} \cong \overline{JK}$	2. A kite is a quadrilateral with two distinct pairs of ≅ adjacent sides
3. $\overline{HK} \cong \overline{HK}$	3. Identity
4. $\triangle LHK \cong \triangle JHK$	4. SSS
5. $\angle LHK \cong \angle JHK$	5. CPCTC
6. $\overrightarrow{HK}$ bisects $\angle LHJ$	6. If a ray divides an ∠ into two ≅ ∠s, then the ray bisects the ∠

4.3 Exercises

1. (a) rhombus (b) rectangle **3.** (a) rectangle (b) rhombus
5. The parallelogram is a square. **7.** $\overline{MN}$ ∥ to both $\overline{AB}$ and $\overline{DC}$; $MN = AB = DC$ **9.** $x = 5$, $DA = 19$ **11.** $NQ = 10$; $MP = 10$ **13.** $QP = \sqrt{72}$ or $6\sqrt{2}$; $MN = \sqrt{72}$ or $6\sqrt{2}$
15. $\sqrt{41}$ **17.** $\sqrt{34}$ **19.** 5 **21.** True **23.** 1. Given
4. Same as (3) 5. If two lines are each ∥ to a third line, then the two lines are ∥ 6. Same as (2) 7. Same as (3) 8. Same as (4)
9. Same as (5) 10. *ABCD* is a parallelogram **25.** (a)
27. 176 **39.** 20.4 ft **41.** Rhombus **43.** 150°

4.4 Exercises

1. m∠*D* = 122°; m∠*B* = 55° **3.** (a) isosceles (b) isosceles
5. (a) 12.3 cm (b) 12.5 **7.** Trapezoid **9.** (a) Yes (b) No
11. 9.7 **13.** 10.8 **15.** $7x + 2$ **19.** $h = 8$ **21.** 12
23. 22 ft **25.** 14 **35.** (a) 7 (b) 14.2 (c) 10.6 (d) Yes
37. (a) 3 ft (b) 12 ft (c) 13 ft (d) $\sqrt{73}$ ft **39.** 8 ft
41. (a) m∠*P* = 59° (b) *NR* = 4 in. **43.** $x = 144$ or 150

Chapter 4 Review Exercises

1. A **2.** S **3.** N **4.** S **5.** S **6.** A **7.** A **8.** A **9.** A
10. N **11.** S **12.** N **13.** $AB = DC = 17$; $AD = BC = 31$
14. 106° **15.** 52 **16.** m∠*M* = 100°; m∠*P* = 80°
17. $\overline{PN}$ **18.** Kite **19.** m∠*G* = m∠*F* = 72°; m∠*E* = 108°
20. 14.9 cm **21.** *MN* = 23; *PO* = 7 **22.** 26
23. *MN* = 6; m∠*FMN* = 80°; m∠*FNM* = 40°
24. $x = 3$; *MN* = 15; *JH* = 30 **32.** (a) Perpendicular
(b) 13 **33.** (a) Perpendicular (b) 30 **34.** (a) Kites, rectangles, squares, rhombi, isosceles trapezoids
(b) Parallelograms, rectangles, squares, rhombi
35. (a) Rhombus (b) Kite

Chapter 4 Review Exercises Selected Proofs

25.

PROOF

Statements	Reasons
1. *ABCD* is a parallelogram	1. Given
2. $\overline{AD} \cong \overline{CB}$	2. Opposite sides of a parallelogram are ≅
3. $\overline{AD} \parallel \overline{CB}$	3. Opposite sides of a parallelogram are ∥
4. $\angle 1 \cong \angle 2$	4. If two ∥ lines are cut by a transversal, then the alt. int. ∠s are ≅
5. $\overline{AF} \cong \overline{CE}$	5. Given
6. $\triangle DAF \cong \triangle BCE$	6. SAS
7. $\angle DFA \cong \angle BEC$	7. CPCTC
8. $\overline{DF} \parallel \overline{EB}$	8. If two lines are cut by a transversal so that alt. ext. ∠s are ≅, then the lines are ∥

26.

PROOF

Statements	Reasons
1. *ABEF* is a rectangle	1. Given
2. *ABEF* is a parallelogram	2. A rectangle is a parallelogram with a rt. ∠
3. $\overline{AF} \cong \overline{BE}$	3. Opposite sides of a parallelogram are ≅
4. *BCDE* is a rectangle	4. Given
5. ∠*F* and ∠*BED* are rt. ∠s	5. All angles of a rectangle are rt. ∠s
6. $\angle F \cong \angle BED$	6. Any two rt. ∠s are ≅
7. $\overline{FE} \cong \overline{ED}$	7. Given
8. $\triangle AFE \cong \triangle BED$	8. SAS
9. $\overline{AE} \cong \overline{BD}$	9. CPCTC
10. $\angle AEF \cong \angle BDE$	10. CPCTC
11. $\overline{AE} \parallel \overline{BD}$	11. If lines are cut by a transversal so that the corresponding ∠s are ≅, then the lines are ∥

27.

PROOF

Statements	Reasons
1. $\overline{DE}$ is a median of $\triangle ADC$	1. Given
2. E is the midpoint of $\overline{AC}$	2. A median of a $\triangle$ is a line segment drawn from a vertex to the midpoint of the opposite side
3. $\overline{AE} \cong \overline{EC}$	3. Midpoint of a segment forms two $\cong$ segments
4. $\overline{BE} \cong \overline{FD}$ and $\overline{EF} \cong \overline{FD}$	4. Given
5. $\overline{BE} \cong \overline{EF}$	5. Transitive Prop. for Congruence
6. $ABCF$ is a parallelogram	6. If the diagonals of a quadrilateral bisect each other, then the quad. is a parallelogram

28.

PROOF

Statements	Reasons
1. $\triangle FAB \cong \triangle HCD$	1. Given
2. $\overline{AB} \cong \overline{DC}$	2. CPCTC
3. $\triangle EAD \cong \triangle GCB$	3. Given
4. $\overline{AD} \cong \overline{BC}$	4. CPCTC
5. $ABCD$ is a parallelogram	5. If a quadrilateral has both pairs of opposite sides $\cong$, then the quad. is a parallelogram

29.

PROOF

Statements	Reasons
1. $ABCD$ is a parallelogram	1. Given
2. $\overline{DC} \cong \overline{BN}$	2. Given
3. $\angle 3 \cong \angle 4$	3. Given
4. $\overline{BN} \cong \overline{BC}$	4. If two $\angle$s of a $\triangle$ are $\cong$, then the sides opposite these $\angle$s are also $\cong$
5. $\overline{DC} \cong \overline{BC}$	5. Transitive Prop. for Congruence
6. $ABCD$ is a rhombus	6. If a parallelogram has two $\cong$ adjacent sides, then the parallelogram is a rhombus

30.

PROOF

Statements	Reasons
1. $\triangle TWX$ is isosceles with base $\overline{WX}$	1. Given
2. $\angle W \cong \angle X$	2. Base $\angle$s of an isosceles $\triangle$ are $\cong$
3. $\overline{RY} \parallel \overline{WX}$	3. Given
4. $\angle TRY \cong \angle W$ and $\angle TYR \cong \angle X$	4. If two $\parallel$ lines are cut by a transversal, then the corr. $\angle$s are $\cong$
5. $\angle TRY \cong \angle TYR$	5. Transitive Prop. for Congruence
6. $\overline{TR} \cong \overline{TY}$	6. If two $\angle$s of a $\triangle$ are $\cong$, then the sides opposite these $\angle$s are also $\cong$
7. $\overline{TW} = \overline{TX}$	7. An isosceles $\triangle$ has two $\cong$ sides
8. $TR = TY$ and $TW = TX$	8. If two segments are $\cong$, then they are equal in length
9. $TW = TR + RW$ and $TX = TY + YX$	9. Segment Addition Postulate
10. $TR + RW = TY + YX$	10. Substitution
11. $RW = YX$	11. Subtraction Prop. of Equality
12. $\overline{RW} \cong \overline{YX}$	12. If segments are $=$ in length, then they are $\cong$
13. $RWXY$ is an isosceles trapezoid	13. If a quadrilateral has one pair of $\parallel$ sides and the nonparallel sides are $\cong$, then the quad. is an isosceles trapezoid

Chapter 4 Test

1. (a) Congruent (b) Supplementary [4.1] **2.** 18.8 cm [4.1] **3.** $EB = 6$ [4.1] **4.** $\overline{VS}$ [4.1] **5.** $x = 7$ [4.1] **6.** (a) Kite (b) Parallelogram [4.2] **7.** (a) Altitude (b) Rhombus [4.1] **8.** (a) The line segments are parallel. (b) $MN = \frac{1}{2}(BC)$ [4.2] **9.** 15.2 cm [4.2] **10.** $x = 23$ [4.2] **11.** $AC = 13$ [4.3] **12.** (a) $\overline{RV}, \overline{ST}$ (b) $\angle R$ and $\angle V$ (or $\angle S$ and $\angle T$) [4.4] **13.** $MN = 14.3$ in. [4.4] **14.** $x = 5$ [4.4] **15.** S1. Kite $ABCD$; $\overline{AB} \cong \overline{AD}$ and $\overline{BC} \cong \overline{DC}$ R1. Given S3. $\overline{AC} \cong \overline{AC}$ R4. SSS S5. $\angle B \cong \angle D$ R5. CPCTC [4.3] **16.** S1. Trap. $ABCD$ with $\overline{AB} \parallel \overline{DC}$ and $\overline{AD} \cong \overline{BC}$ R1. Given R2. Congruent R3. Identity R4. SAS S5. $\overline{AC} \cong \overline{DB}$ [4.4] **17.** $P = 26$ [4.4]

CHAPTER 5

5.1 Exercises

1. (a) $\frac{4}{5}$ (b) $\frac{4}{5}$ (c) $\frac{2}{3}$ (d) Incommensurable **3.** (a) $\frac{5}{8}$ (b) $\frac{1}{3}$
(c) $\frac{4}{3}$ (d) Incommensurable **5.** (a) 3 (b) 8 **7.** (a) 6 (b) 4
9. (a) $\pm 2\sqrt{7} \approx \pm 5.29$ (b) $\pm 3\sqrt{2} \approx \pm 4.24$
11. (a) 4 (b) $-\frac{5}{6}$ or 3 **13.** (a) $\frac{3 \pm \sqrt{33}}{4} \approx 2.19$ or -0.69
(b) $\frac{7 \pm \sqrt{89}}{4} \approx 4.11$ or -0.61 **15.** 6.3 m/sec **17.** $10\frac{1}{2}$
19. ≈ 29 outlets **21.** (a) $4\sqrt{3} \approx 6.93$ (b) $4\frac{1}{2}$
23. Secretary's salary is \$24,900; salesperson's salary is \$37,350;
vice president's salary is \$62,250. **25.** 60°, 80°, 100°, 120°
27. 40° and 50° **29.** 30.48 cm **31.** $2\frac{4}{7} \approx 2.57$
33. $a = 12$; $b = 16$ **35.** 45° **37.** 4 in. by $4\frac{2}{3}$ in.
39. (a) $\frac{5 + 5\sqrt{5}}{2}$ (b) 8.1

5.2 Exercises

1. (a) Congruent (b) Proportional **3.** (a) Yes (b) No
5. (a) $\triangle ABC \sim \triangle XTN$ (b) $\triangle ACB \sim \triangle NXT$
7. Yes; Yes; Spheres have the same shape; one is an enlargement
of the other unless they are congruent. **9.** (a) 82° (b) 42°
(c) $10\frac{1}{7}$ (d) 8 **11** (a) Yes (b) Yes (c) Yes **13.** $5\frac{1}{3}$
15. 79° **17.** $n = 3$ **19.** 90° **21.** 12 **23.** $10 + 2\sqrt{5}$ or
$10 - 2\sqrt{5}$; ≈ 14.47 or 5.53 **25.** 75 **27.** 2.5 in.
29. 3 ft, 9 in. **31.** 74 ft **33.** No **35.** (a) Yes (b) Yes
37. 50 in. diagonal **39.** 3.75

5.3 Exercises

1. CASTC **3.** (a) True (b) True **5.** SSS~ **7.** SAS~
9. SAS~ **11.** SSS~ **13.** 1. Given 2. If 2 lines are $\perp$, they
form right angles. 3. All right angles are $\cong$. 4. Opposite $\angle$s of
a $\square$ are $\cong$ 5. AA **15.** 1. Given 2. Definition of midpoint
3. If a line segment joins the midpoints of two sides of a $\triangle$, its
length is $\frac{1}{2}$ the length of the third side. 4. Division Prop. of Eq.
5. Substitution 6. SSS~ **17.** 1. $\overline{MN} \perp \overline{NP}$ and $\overline{QR} \perp \overline{RP}$
2. If two lines are $\perp$, then they form a rt. $\angle$. 3. $\angle N \cong \angle QRP$
4. Identity S5. $\triangle MNP \sim \triangle QRP$ R5. AA **19.** 1. $\angle H \cong \angle F$
2. If two $\angle$s are vertical $\angle$s, then they are $\cong$
S3. $\triangle HJK \sim \triangle FGK$ R3. AA **21.** 1. $\frac{RQ}{NM} = \frac{RS}{NP} = \frac{QS}{MP}$
2. $\triangle RQS \sim \triangle NMP$ 3. $\angle N \cong \angle R$ **23.** S1. $\overline{RS} \parallel \overline{UV}$ R1. Given
2. If 2 $\parallel$ lines are cut by a transversal, alternate interior $\angle$s
are $\cong$. 3. $\triangle RST \sim \triangle VUT$ S4. $\frac{RT}{VT} = \frac{RS}{VU}$ R4. CSSTP
25. $4\frac{1}{2}$ **27.** 16 **29.** $EB = 24$ **31.** 27° **37.** $QS = 8$
39. 150 ft

5.3 Selected Proofs

33.

PROOF	
Statements	**Reasons**
1. $\overline{AB} \parallel \overline{DF}$ and $\overline{BD} \parallel \overline{FG}$	1. Given
2. $\angle A \cong \angle FEG$ and $\angle BCA \cong \angle G$	2. If two $\parallel$ lines are cut by a transversal, then the corresponding $\angle$s are $\cong$
3. $\triangle ABC \sim \triangle EFG$	3. AA

35.

PROOF	
Statements	**Reasons**
1. $\triangle DEF \sim \triangle MNP$ $\overline{DG}$ and $\overline{MQ}$ are altitudes	1. Given
2. $\overline{DG} \perp \overline{EF}$ and $\overline{MQ} \perp \overline{NP}$	2. An altitude is a segment drawn from a vertex $\perp$ to the opposite side
3. $\angle DGE$ and $\angle MQN$ are rt. $\angle$s	3. $\perp$ lines form a rt. $\angle$
4. $\angle DGE \cong \angle MQN$	4. Right $\angle$s are $\cong$
5. $\angle E \cong \angle N$	5. If two $\triangle$s are $\sim$ then the corresponding $\angle$s are $\cong$ (CASTC)
6. $\triangle DGE \sim \triangle MQN$	6. AA
7. $\frac{DG}{MQ} = \frac{DE}{MN}$	7. Corresponding sides of $\sim$ $\triangle$s are proportional (CSSTP)

5.4 Exercises

1. $\triangle RST \sim \triangle RVS \sim \triangle SVT$ **3.** $\frac{RT}{RS} = \frac{RS}{RV}$ or $\frac{RV}{RS} = \frac{RS}{RT}$
5. 4.5 **7.** (a) 10 (b) $\sqrt{34} \approx 5.83$ **9.** (a) 8 (b) 4
11. (a) Yes (b) No (c) Yes (d) No **13.** (a) Right
(b) Acute (c) Right (d) No $\triangle$ **15.** 15 ft
17. $6\sqrt{5} \approx 13.4$ m **19.** 20 ft **21.** 12 cm
23. The base is 8; the altitude is 6; the diagonals are 10.
25. $6\sqrt{7} \approx 15.87$ in. **27.** 12 in. **29.** 4 **31.** $9\frac{3}{13}$ in.
33. $5\sqrt{5} \approx 11.18$ **39.** 60° **41.** $TS = 13$;
$RT = 13\sqrt{2} \approx 18.38$ **45.** (a) AA (b) Theorem 5.4.1:
The altitude to the hypotenuse of a rt. $\triangle$ separates it into two
triangles that are similar to each other. (c) Transitive Property
of Similarity

5.5 Exercises

1. (a) a (b) $a\sqrt{2}$ **3.** (a) $a\sqrt{3}$ (b) $2a$
5. $YZ = 8$; $XY = 8\sqrt{2} \approx 11.31$ **7.** $XZ = 10$; $YZ = 10$
9. $DF = 5\sqrt{3} \approx 8.66$; $FE = 10$ **11.** $DE = 12$; $FE = 24$
13. $HL = 6$, $HK = 12$; $MK = 6$ **15.** $AC = 6$;
$AB = 6\sqrt{2} \approx 8.49$ **17.** $RS = 6$; $RT = 6\sqrt{3} \approx 10.39$
19. $DB = 5\sqrt{6} \approx 12.25$ **21.** $6\sqrt{3} + 6 \approx 16.39$
23. 45° **25.** 60°; 146 ft further **27.** $DC = 2\sqrt{3} \approx 3.46$;
$DB = 4\sqrt{3} \approx 6.93$ **29.** $6\sqrt{3} \approx 10.39$ **31.** $4\sqrt{3} \approx 6.93$
33. $6 + 6\sqrt{3} \approx 16.39$ **35.** (a) $6\sqrt{3}$ inches (b) 12 inches
37. $VW = 1.2$

5.6 Exercises

1. 30 oz of ingredient A; 24 oz of ingredient B; 36 oz of
ingredient C **3.** (a) Yes (b) Yes **5.** $EF = 4\frac{1}{6}$, $FG = 3\frac{1}{3}$,
$GH = 2\frac{1}{2}$ **7.** $x = 5\frac{1}{3}$, $DE = 5\frac{1}{3}$, $EF = 6\frac{2}{3}$ **9.** $EC = 16\frac{4}{5}$
11. $a = 5$; $AD = 4$ **13.** (a) No (b) Yes **15.** 9
17. $4\sqrt{6} \approx 9.80$ **19.** 41° **21.** $\frac{AC}{CE} = \frac{AD}{DE}; \frac{DC}{CB} = \frac{DE}{EB}$
23. (a) $\overline{EC}$ (b) $\overline{DB}$ (c) $\overline{FB}$ **25.** $x = \frac{1 + \sqrt{73}}{2}$ or $x = \frac{1 - \sqrt{73}}{2}$;
reject both because each will give a negative number for the
length of a side. **27.** (a) True (b) True **29.** $RK = 1.8$

31. 1. Given 2. Means-Extremes Property 3. Addition Property of Equality 4. Distributive Property 6. Substitution
37. $\frac{-1+\sqrt{5}}{2} \approx 0.62$ **39.** (a) $CD = 2$; $DB = 3$
(b) $CE = \frac{20}{11}$; $EA = \frac{24}{11}$ (c) $BF = \frac{10}{3}$; $FA = \frac{8}{3}$ (d) $\frac{3}{2} \cdot \frac{5}{6} \cdot \frac{4}{5} = 1$

5.6 Selected Proof

33.

PROOF	
Statements	**Reasons**
1. $\triangle RST$ with M the midpoint of $\overline{RS}$; $\overleftrightarrow{MN} \parallel \overline{ST}$	1. Given
2. $RM = MS$	2. The midpoint of a segment divides the segment into two segments of equal measure
3. $\frac{RM}{MS} = \frac{RN}{NT}$	3. If a line is $\parallel$ to one side of a $\triangle$ and intersects the other two sides, then it divides these sides proportionally
4. $\frac{MS}{MS} = 1 = \frac{RN}{NT}$	4. Substitution
5. $RN = NT$	5. Means-Extremes Property
6. N is the midpoint of $\overline{RT}$	6. If a point divides a line segment into two segments of equal measure, then the point is a midpoint

Chapter 5 Review Exercises

1. False **2.** True **3.** False **4.** True **5.** True
6. False **7.** True **8.** (a) $\pm 3\sqrt{2} \approx \pm 4.24$ (b) 26
(c) -1 (d) 2 (e) 7 or -1 (f) $-\frac{9}{7}$ or 4 (g) 6 or -1
(h) -6 or 3 **9.** \$5.28 **10.** Seven packages **11.** \$210
12. The lengths of the sides are 8, 12, 20, and 28. **13.** 18
14. 20 and $22\frac{4}{7}$ **15.** 150° **16.** (a) SSS~ (b) AA
(c) SAS~ (d) SSS~ **19.** $x = 5$; $m\angle F = 97^6$
20. $AB = 6$; $BC = 12$ **21.** 3 **22.** $4\frac{1}{2}$ **23.** $6\frac{1}{4}$
24. $5\frac{3}{5}$ **25.** 10 **26.** 6 **27.** $EO = 1\frac{1}{5}$; $EK = 9$
30. (a) $8\frac{1}{3}$ (b) 21 (c) $2\sqrt{3} \approx 3.46$ (d) 3 **31.** (a) 16
(b) 40 (c) $2\sqrt{5} \approx 4.47$ (d) 4 **32.** (a) 30° (b) 24
(c) 20 (d) 16 **33.** $AE = 20$; $EF = 15$; $AF = 25$;
$m\angle AEF = 90°$ **34.** $4\sqrt{2} \approx 5.66$ in. **35.** $3\sqrt{2} \approx 4.24$ cm
36. 25 cm **37.** $5\sqrt{3} \approx 8.66$ in. **38.** $4\sqrt{3} \approx 6.93$ in.
39. 12 cm **40.** (a) $x = 9\sqrt{2} \approx 12.73$; $y = 9$ (b) $x = 4\frac{1}{2}$;
$y = 6$ (c) $x = 12$; $y = 3$ (d) $x = 2\sqrt{14} \approx 7.48$; $y = 13$
41. 11 km **42.** (a) Acute (b) No $\triangle$ (c) Obtuse (d) Right
(e) No $\triangle$ (f) Acute (g) Obtuse (h) Obtuse

Chapter 5 Review Exercises Selected Proofs

17.

PROOF	
Statements	**Reasons**
1. $ABCD$ is a parallelogram; $\overline{DB}$ intersects $\overline{AE}$ at point F	1. Given
2. $\overline{DC} \parallel \overline{AB}$	2. Opposite sides of a parallelogram are $\parallel$
3. $\angle CDB \cong \angle ABD$	3. If two $\parallel$ lines are cut by a transversal, then the alt. int. $\angle$s are $\cong$
4. $\angle DEF \cong \angle BAF$	4. Same as (3)
5. $\triangle DFE \sim \triangle BFA$	5. AA
6. $\frac{AF}{EF} = \frac{AB}{DE}$	6. CSSTP

18.

PROOF	
Statements	**Reasons**
1. $\angle 1 \cong \angle 2$	1. Given
2. $\angle ADC \cong \angle 2$	2. If two lines intersect, then the vertical $\angle$s formed are $\cong$
3. $\angle ADC \cong \angle 1$	3. Transitive Prop. for Congruence
4. $\angle A \cong \angle A$	4. Identity
5. $\triangle BAE \sim \triangle CAD$	5. AA
6. $\frac{AB}{AC} = \frac{BE}{CD}$	6. CSSTP

Chapter 5 Test

1. (a) $3:5$ (or $\frac{3}{5}$) (b) $\frac{25 \text{ mi}}{\text{gal}}$ [5.1] **2.** (a) $\frac{40}{13}$ (b) 9, -9 [5.1]
3. 15°, 75° [5.1] **4.** (a) 92° (b) 12 [5.2] **5.** (a) SAS~
(b) AA [5.3] **6.** $\triangle ABC \sim \triangle ACD \sim \triangle CBD$ [5.4]
7. (a) $c = \sqrt{41}$ (b) $a = \sqrt{28} = 2\sqrt{7}$ [5.4]
8. (a) Yes (b) No [5.4] **9.** $DA = \sqrt{89}$ [5.4]
10. (a) $10\sqrt{2}$ in. (b) 8 cm [5.5] **11.** (a) 5 m (b) 12 ft [5.5]
12. $EC = 12$ [5.6] **13.** $PQ = 4$; $QM = 6$ [5.6]
14. 1 [5.6] **15.** $DB = 4$ [5.2] **16.** S1. $\overline{MN} \parallel \overline{QR}$
R1. Given 2. Corresponding $\angle$s are $\cong$ 3. $\angle P \cong \angle P$
4. AA [5.3] **17.** 1. Given 2. Identity 3. Given 5. Substitution
6. SAS~ 7. $\angle PRC \cong \angle B$ [5.3]

CHAPTER 6

6.1 Exercises

1. 29° **3.** 47.6° **5.** 56.6° **7.** 313° **9.** (a) 90° (b) 270°
(c) 135° (d) 135° **11.** (a) 80° (b) 120° (c) 160° (d) 80°
(e) 120° (f) 160° (g) 10° (h) 50° (i) 30° **13.** (a) 72°
(b) 144° (c) 36° (d) 72° (e) 18° **15.** (a) 12 (b) $6\sqrt{2}$
17. 3 **19.** $\sqrt{7} + 3\sqrt{3}$ **21.** 90°; square **23.** (a) The
measure of an arc equals the measure of its corresponding central
angle. Therefore, congruent arcs have congruent central angles.
(b) The measure of a central angle equals the measure of its
intercepted arc. Therefore, congruent central angles have
congruent arcs. (c) Draw the radii to the endpoints of the
congruent chords. The two triangles formed are congruent by
SSS. The central angles of each triangle are congruent by
CPCTC. Therefore, the arcs corresponding to the central angles
are also congruent. Hence, congruent chords have congruent arcs.
(d) Draw the four radii to the endpoints of the congruent arcs.
Also draw the chords corresponding to the congruent arcs. The
central angles corresponding to the congruent arcs are also
congruent. Therefore, the triangles are congruent by SAS. The
chords are congruent by CPCTC. Hence, congruent arcs have congruent chords. (e) Congruent central angles have congruent arcs
(from b). Congruent arcs have congruent chords (from d). Hence,
congruent central angles have congruent chords.
(f) Congruent chords have congruent arcs (from c). Congruent
arcs have congruent central angles (from a). Therefore, congruent
chords have congruent central angles. **25.** (a) 15° (b) 70°
27. (a) 72° (b) 60° **29.** 45° **31.** 1. $\overline{MN} \parallel \overline{OP}$ in ⊙O
2. If two ∥ lines are cut by a transversal, then the alt. int. ∠s
arc ≅ 3. If two ∠s are ≅, then their measures are = 4. The
measure of an inscribed ∠ equals $\frac{1}{2}$ the measure of its intercepted
arc 5. The measure of a central ∠ equals the measure of its arc
6. Substitution **39.** If $\overline{ST} \cong \overline{TV}$, then $\overline{ST} \cong \overline{TV}$ (≅ arcs in a
circle have ≅ chords). △STV is an isosceles △ because it has
two ≅ sides. **43.** WZ = 1.75

6.1 Selected Proof

33. *Proof:* Using the chords $\overline{AB}$, $\overline{BC}$, $\overline{CD}$, and $\overline{AD}$ in ⊙O
as sides of inscribed angles, ∠B ≅ ∠D and
∠A ≅ ∠C because they are inscribed angles
intercepting the same arc. △ABE ∼ △CDE
by AA.

6.2 Exercises

1. (a) 8° (b) 46° (c) 38° (d) 54° (e) 126° **3.** (a) 90°
(b) 13° (c) 103° **5.** 18° **7.** (a) 22° (b) 7° (c) 15°
9. (a) 136° (b) 224° (c) 68° (d) 44° **11.** (a) 96° (b) 60°
13. (a) 120° (b) 240° (c) 60° **15.** 28° **17.** $\text{m}\widehat{CE} = 88°$;
$\text{m}\widehat{BD} = 36°$ **19.** (a) Supplementary (b) 107° **21.** 1. $\overline{AB}$
and $\overline{AC}$ are tangents to ⊙O from A 2. The measure of an ∠
formed by a tangent and a chord equals $\frac{1}{2}$ the arc measure
3. Substitution 4. If two ∠s are = in measure, they are ≅
5. $\overline{AB} \cong \overline{AC}$ 6. △ABC is isosceles **27.** ≈ 154.95 mi
29. m∠1 = 36°; m∠2 = 108° **31.** (a) 30° (b) 60°
(c) 150° **33.** ∠X ≅ ∠X; ∠R ≅ ∠W; also, ∠RVX ≅ ∠WSX
35. 10 **37.** $(\sqrt{2} - 1)$ cm

6.2 Selected Proof

23. *Given:* Tangent $\overline{AB}$ to ⊙O at point B; m∠A = m∠B
Prove: $\text{m}\widehat{BD} = 2 \cdot \text{m}\widehat{BC}$
Proof: m∠BCD = m∠A + m∠B; but because
m∠A = m∠B, m∠BCD = m∠B + m∠B
or m∠BCD = 2 · m∠B. m∠BCD also equals
$\frac{1}{2}\text{m}\widehat{BD}$ because it is an inscribed ∠.
Therefore, $\frac{1}{2}\text{m}\widehat{BD} = 2 \cdot \text{m}∠B$ or
$\text{m}\widehat{BD} = 4 \cdot \text{m}∠B$. But if $\overline{AB}$ is a tangent to ⊙O
at B, then $\text{m}∠B = \frac{1}{2}\text{m}\widehat{BC}$. By substitution,
$\text{m}\widehat{BD} = 4(\frac{1}{2}\text{m}\widehat{BC})$ or $\text{m}\widehat{BD} = 2 \cdot \text{m}\widehat{BC}$.

6.3 Exercises

1. 30° **3.** $6\sqrt{5}$ **7.** 3 **9.** DE = 4 and EC = 12 or
DE = 12 and EC = 4 **11.** 4 **13.** x = 6; AE = 3
15. DE = 12; EC = 6 **17.** $9\frac{2}{5}$ **19.** 9 **21.** $5\frac{1}{3}$
23. $3 + 3\sqrt{5}$ **25.** (a) None (b) One (c) 4 **31.** Yes;
$\overline{AE} \cong \overline{CE}$; $\overline{DE} \cong \overline{EB}$ **33.** 20° **35.** AM = 5; PC = 7;
BN = 9 **37.** 12 **39.** 8.7 inches **41.** (a) Obtuse
(b) Equilateral **43.** 45°

6.3 Selected Proofs

27. *Proof:* If $\overline{AF}$ is a tangent to ⊙O and $\overline{AC}$ is a secant to ⊙O,
then $(AF)^2 = AC \cdot AB$. If $\overline{AF}$ is a tangent to ⊙Q
and $\overline{AE}$ is a secant to ⊙Q, then $(AF)^2 = AE \cdot AD$.
By substitution, $AC \cdot AB = AE \cdot AD$.

29. *Proof:* Let M, N, P, and Q be the points of tangency for
$\overline{DC}$, $\overline{DA}$, $\overline{AB}$, and $\overline{BC}$, respectively. Because the
tangent segments from an external point are
congruent, AP = AN, PB = BQ, CM = CQ,
and MD = DN. Thus AP + PB + CM +
MD = AN + BQ + CQ + DN.
Reordering and associating, (AP + PB) +
(CM + MD) = (AN + DN) + (BQ + CQ)
or $\overrightarrow{AB} + CD = DA + BC$.

45. *Proof:* If $\overrightarrow{OM}$ bisects $\overline{RS}$ in ⊙O, then $\overline{RM} \cong \overline{MS}$. Draw $\overline{RO}$
and $\overline{OS}$, which are ≅ because they are radii in the
same circle. Using $\overline{OM} \cong \overline{OM}$, △ROM ≅ △SOM
by SSS. By CPCTC, ∠OMS ≅ ∠OMR, and hence
$\overleftrightarrow{OM} \perp \overline{RS}$.

6.4 Exercises

1. m∠CQD < m∠AQB **3.** QM < QN **5.** CD < AB
7. QM > QN **11.** No; angles are not congruent. **15.** $\overline{AB}$;
$\overline{GH}$; for a circle containing unequal chords, the chord nearest
the center has the greatest length and the chord at the greatest
distance from the center has the least length. **17.** (a) $\overline{OT}$
(b) $\overline{OD}$ **19.** (a) $\text{m}\widehat{MN} > \text{m}\widehat{QP}$ (b) $\text{m}\widehat{MPN} < \text{m}\widehat{PMQ}$
21. Obtuse **23.** (a) m∠AOB > m∠BOC (b) AB > BC
25. (a) $\text{m}\widehat{AB} > \text{m}\widehat{BC}$ (b) AB > BC **27.** (a) ∠C (b) $\overline{AC}$
29. (a) ∠B (b) $\overline{AC}$ **31.** $\overline{AB}$ is $(4\sqrt{3} - 4\sqrt{2})$ units closer
than $\overline{CD}$. **37.** 7

Chapter 6 Review Exercises

1. 9 mm **2.** 30 cm **3.** $\sqrt{41}$ in. **4.** $6\sqrt{2}$ cm **5.** 130°
6. 45° **7.** 80° **8.** 35° **9.** $m\widehat{AC} = m\widehat{DC} = 93\frac{1}{3}°$;
$m\widehat{AD} = 173\frac{1}{3}°$ **10.** $m\widehat{AC} = 110°$ and $m\widehat{AD} = 180°$
11. $m\angle 2 = 44°$; $m\angle 3 = 90°$; $m\angle 4 = 46°$; $m\angle 5 = 44°$
12. $m\angle 1 = 50°$; $m\angle 2 = 40°$; $m\angle 3 = 90°$; $m\angle 4 = 50°$
13. 24 **14.** 10 **15.** A **16.** S **17.** N **18.** S **19.** A
20. N **21.** A **22.** N **23.** (a) 70° (b) 28° (c) 64°
(d) $m\angle P = 21°$ (e) $m\widehat{AB} = 90°$, $m\widehat{CD} = 40°$ (f) 260°
24. (a) 3 (b) 8 (c) 16 (d) 4 (e) 4 (f) 8 or 1 (g) $3\sqrt{5}$
(h) 3 (i) $4\sqrt{3}$ (j) 3 **25.** 29 **26.** If $x = 7$, then $AC = 35$;
$DE = 17\frac{1}{2}$. If $x = -4$, then $AC = 24$; $DE = 12$.
30. $m\angle 1 = 93°$; $m\angle 2 = 25°$; $m\angle 3 = 43°$; $m\angle 4 = 68°$;
$m\angle 5 = 90°$; $m\angle 6 = 22°$; $m\angle 7 = 68°$; $m\angle 8 = 22°$;
$m\angle 9 = 50°$; $m\angle 10 = 112°$ **31.** $24\sqrt{2}$ cm
32. $15 + 5\sqrt{3}$ cm **33.** 14 cm and 15 cm **34.** $AD = 3$;
$BE = 6$; $FC = 7$ **35.** (a) $AB > CD$ (b) $QP < QR$
(c) $m\angle A < m\angle C$

Chapter 6 Review Exercises Selected Proofs

27. *Proof:* If $\overline{DC}$ is tangent to circles B and A at points D
and C, then $\overline{BD} \perp \overline{DC}$ and $\overline{AC} \perp \overline{DC}$. $\angle$s D
and C are congruent because they are right angles.
$\angle DEB \cong \angle CEA$ because they are vertical angles.
$\triangle BDE \sim \triangle ACE$ by AA. It follows that $\frac{AC}{CE} = \frac{BD}{ED}$
because corresponding sides are proportional.
Hence, $AC \cdot ED = CE \cdot BD$.

28. *Proof:* In $\odot O$, if $\overline{EO} \perp \overline{BC}$, $\overline{DO} \perp \overline{BA}$, and $\overline{EO} \cong \overline{OD}$,
$\overline{BC} \cong \overline{BA}$. (Chords equidistant from the center of
the circle are congruent.) It follows that $\widehat{BC} \cong \widehat{BA}$

29. *Proof:* If $\overline{AP}$ and $\overline{BP}$ are tangent to $\odot Q$ at A and B, then
$\overline{AP} \cong \overline{BP}$. $\widehat{AC} \cong \widehat{BC}$ because C is the midpoint
of $\widehat{AB}$. It follows that $\overline{AC} \cong \overline{BC}$ and, using
$\overline{CP} \cong \overline{CP}$, we have $\triangle ACP \cong \triangle BCP$ by SSS.
$\angle APC \cong \angle BPC$ by CPCTC and hence $\overrightarrow{PC}$
bisects $\angle APB$.

Chapter 6 Test

1. (a) 272° (b) 134° [6.1] **2.** (a) 69° (b) 32° [6.1]
3. (a) 48° (b) Isosceles [6.1] **4.** (a) Right (b) Congruent [6.2]
5. (a) 69° (b) 37° [6.2] **6.** (a) 214° (b) 34° [6.2]
7. (a) 226° (b) 134° [6.2] **8.** (a) Concentric (b) 8 [6.1]
9. $2\sqrt{13}$ [6.1] **10.** (a) 1 (b) 2 [6.3] **11.** (a) 10
(b) 5 [6.3] **12.** $2\sqrt{6}$ [6.3] **14.** (a) $m\angle AQB > m\angle CQD$
(b) $AB > CD$ [6.4] **15.** (a) 1 (b) 7 [6.2] **16.** S1. In $\odot O$,
chords $\overline{AD}$ and $\overline{BC}$ intersect at E R1. Given 2. Vertical angles
are congruent 4. AA 5. $\frac{AE}{CE} = \frac{BE}{DE}$ [6.3]

CHAPTER 7

7.1 Exercises

1. A, C, E **11.** The locus of points at a given distance from
a fixed line is two parallel lines on either side of the fixed line
at the same (given) distance from the fixed line. **13.** The locus
of points at a distance of 3 in. from point O is a circle with
center O and radius of length 3 in. **15.** The locus of points

equidistant from points D, E, and F is the point G for which
$DG = EG = FG$. **17.** The locus of the midpoints of the
chords in $\odot Q$ parallel to diameter $\overline{PR}$ is the perpendicular
bisector of $\overline{PR}$. **19.** The locus of points equidistant from
two given intersecting lines is two perpendicular lines that
bisect the angles formed by the two intersecting lines.

21.   **23.**

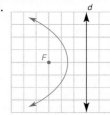

25.

29. The locus of points at a distance of 2 cm from a sphere
whose radius is 5 cm is two concentric spheres (with the same
center). The radius length of one sphere is 3 cm, and the radius
length of the other sphere is 7 cm. **31.** The locus is another
sphere with the same center and a radius of length 2.5 m.
33. The locus of points equidistant from an 8-ft ceiling and the
floor is a plane parallel to the ceiling and the floor and midway
between them.

7.2 Exercises

1. Yes **3.** Incenter **5.** Circumcenter **7.** (a) Angle bisectors
(b) Perpendicular bisectors of sides (c) Altitudes (d) Medians
9. No (need 2) **11.** No: Construct 2 medians to determine the
centroid **13.** Midpoint of the hypotenuse **23.** No **25.** $\frac{10\sqrt{3}}{3}$
27. $RQ = 10$; $SQ = \sqrt{89}$ **29.** (a) 4 (b) 6 (c) 10.5
33. Equilateral **35.** (a) Yes (b) Yes **37.** (a) Yes (b) No
41. 3 in.

7.3 Exercises

1. First, construct the angle bisectors of two consecutive
angles, say A and B. The point of intersection, O, is the center
of the inscribed circle. Second, construct the line segment $\overline{OM}$
perpendicular to $\overline{AB}$. Then, using the radius length $r = OM$,
construct the inscribed circle with center O. **3.** Draw the
diagonals (angle bisectors) $\overline{JL}$ and $\overline{MK}$. These determine center O
of the inscribed circle. Now construct the line segment $\overline{OR} \perp \overline{MJ}$.
Use OR as the length of the radius of the inscribed circle.
9. 27.2 in. **11.** 8.3 cm **13.** $a = 5$ in.; $r = 5\sqrt{2}$ in.
15. $16\sqrt{3}$ ft; 16 ft **17.** (a) 120° (b) 90° (c) 72° (d) 60°
19. (a) 4 (b) 8 (c) 6 (d) 15 **21.** (a) 140° (b) 135°
23. (a) 30° (b) 40° **25.** 6 **27.** (a) Yes (b) No (c) Yes
(d) No **29.** $4 + 4\sqrt{2}$ **31.** 168° **33.** No **35.** 30°

Chapter 7 Review Exercises

7. The locus of points equidistant from the sides of $\angle ABC$ is the bisector of $\angle ABC$.   **8.** The locus of points 1 in. from point B is the circle with center B and radius length 1 in.   **9.** The locus of points equidistant from D and E is the perpendicular bisector of $\overline{DE}$.
10. The locus of points $\frac{1}{2}$ inch from $\overleftrightarrow{DE}$ are two lines parallel to each other and $\overleftrightarrow{DE}$, each line $\frac{1}{2}$ inch from $\overleftrightarrow{DE}$ and on opposite sides of $\overleftrightarrow{DE}$.   **11.** The locus of the midpoints of the radii of a circle is a concentric circle with radius half the length of the given radius.   **12.** The locus of the centers of all circles passing through two given points is the perpendicular bisector of the line segment joining the two given points.   **13.** The locus of the center of a penny that rolls around a half-dollar is a circle.
14. The locus of points in space 2 cm from point A is the sphere with center A and radius 2 cm.   **15.** The locus of points 1 cm from plane P is the two planes parallel to each other and plane P, each plane 1 cm from P and on opposite sides of P.
16. The locus of points in space less than 3 units from a given point is the interior of a sphere.   **17.** The locus of points equidistant from two parallel planes is a parallel plane midway between the two planes.   **24.** (a) 12 (b) 2 (c) $2\sqrt{3}$
25. $DF = 6$, $AE = 9$ **26.** (a) $72°$ (b) $108°$ (c) $72°$
27. (a) $36°$ (b) $144°$ (c) $36°$ **28.** (a) 8 (b) 20 (c) 40 cm
29. (a) 24 in. (b) $3\sqrt{2}$ in.   **30.** (a) No (b) No (c) Yes
(d) Yes   **31.** (a) No (b) Yes (c) No (d) Yes   **32.** 14 in.
33. $40\sqrt{3}$ cm
34.

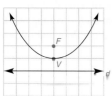

Chapter 7 Test

1. The locus of points equidistant from parallel lines l and m is the line parallel to both l and m and midway between them. [7.1]
2. The locus of points equidistant from the sides of $\angle ABC$ is the bisector of $\angle ABC$. [7.1]   **3.** The locus of points equidistant from D and E is the perpendicular bisector of $\overline{DE}$ [7.1]   **4.** The locus of points 3 cm from point P is the circle with center P and radius length 3 cm. [7.1]   **5.** The locus of points in space 3 cm from point P is the sphere with center P and radius length 3 cm. [7.1]
6. (a) Incenter (b) Centroid [7.2]   **7.** (a) Circumcenter (b) Orthocenter [7.2]   **8.** Equilateral triangle or equiangular triangle [7.2]   **9.** Angle bisectors and medians [7.2]
10. (a) T (b) T (c) F (d) F [7.3]   **11.** (a) 1.5 in.
(b) $3\sqrt{3}$ in. [7.3]   **12.** (a) $72°$ (b) $108°$ [7.3]
13. (a) 10 sides (b) 35 diagonals [7.3]   **14.** 80 cm [7.3]
15. (a) $4\sqrt{3}$ in. (b) 8 in. [7.3]   **16.** (a) rhombus (b) yes [7.1]
17.

CHAPTER 8

8.1 Exercises

1. Two triangles with equal areas are not necessarily congruent. Two squares with equal areas must be congruent because the sides are congruent.   **3.** 37 units^2 **5.** The altitudes to $\overline{PN}$ and to $\overline{MN}$ are congruent. This is because $\triangle$s QMN and QPN are congruent; corresponding altitudes of $\cong \triangle$s are $\cong$.   **7.** Equal
9. 54 cm^2 **11.** 18 m^2 **13.** 72 in^2 **15.** 100 in^2 **17.** 126 in^2
19. 264 units^2 **21.** 144 units^2 **23.** 192 ft^2 **25.** (a) 300 ft^2
(b) 3 gallons   (c) $67.50 **27.** $156 + 24\sqrt{10}$ ft^2
29. (a) 9 sq ft = 1 sq yd (b) 1296 sq in. = 1 sq yd
31. 24 cm^2 **33.** $\overline{MN}$ joins the midpoints of $\overline{CA}$ and $\overline{CB}$, so $MN = \frac{1}{2}(AB)$. Therefore, $\overline{AP} \cong \overline{PB} \cong \overline{MN}$. $\overline{PN}$ joins the midpoints of $\overline{CB}$ and $\overline{AB}$, so $PN = \frac{1}{2}(AC)$. Therefore $\overline{AM} \cong \overline{MC} \cong \overline{PN}$. $\overline{MP}$ joins the midpoints of $\overline{AB}$ and $\overline{AC}$, so $MP = \frac{1}{2}(BC)$. Therefore $\overline{CN} \cong \overline{NB} \cong \overline{MP}$. The four triangles are all $\cong$ by SSS. Therefore, the areas of all these triangles are the same. Hence, the area of the big triangle is equal to four times the area of one of the smaller triangles.   **37.** 8 in.
39. (a) 12 in. (b) 84 in^2 **41.** 56 percent   **43.** By the Area-Addition Postulate, $A_{R \cup S} = A_R + A_S$. Now $A_{R \cup S}$, A_R, and A_S are all positive numbers. Let p represent the area of region S, so that $A_{R \cup S} = A_R + p$. By the definition of inequality, $A_R < A_{R \cup S}$, or $A_{R \cup S} > A_R$.
45. $(a + b)(c + d) = ac + ad + bc + bd$
47. $4\frac{8}{13}$ in.   **49.** 8   **51.** $P = 2x + \frac{96}{x}$ **53.** 48 units^2
55. (a) 10 (b) 26 (c) 18 (d) No

8.1 Selected Proof

53. *Proof:* $A = (LH)(HJ) = s^2$. By the Pythagorean Theorem, $s^2 + s^2 = d^2$.

$$2s^2 = d^2$$
$$s^2 = \frac{d^2}{2}$$

Thus,
$$A = \frac{d^2}{2}$$

8.2 Exercises

1. 30 in.   **3.** $4\sqrt{29}$ m   **5.** 30 ft   **7.** 38   **9.** 84 in^2
11. 1764 mm^2 **13.** 40 ft^2 **15.** 80 units^2
17. $36 + 36\sqrt{3}$ units^2 **19.** 32 in., 16 in., and 28 in.
21. 15 cm   **23.** (a) $\frac{9}{4}$ (b) $\frac{4}{1}$ **27.** $24 + 4\sqrt{21}$ units^2
29. 96 units^2 **31.** 6 yd by 8 yd   **33.** (a) 770 ft (b) $1224.30
35. 624 ft^2 **37.** Square with sides of length 10 in.
39. (a) 52 units (b) 169 units^2 **41.** (a) $\frac{\sqrt{5}}{3}$ (b) $\frac{5}{9}$
43. 60 in^2 **45.** (a) No (b) Yes   **49.** 12 ft^2 **51.** 5 in^2
53. $h = 2.4$ **55.** 2 units

8.2 Selected Proofs

25. Using Heron's Formula, the semiperimeter is $\frac{1}{2}(3s)$, or $\frac{3s}{2}$. Then

$$A = \sqrt{\frac{3s}{2}\left(\frac{3s}{2} - s\right)\left(\frac{3s}{2} - s\right)\left(\frac{3s}{2} - s\right)}$$

$$A = \sqrt{\frac{3s}{2}\left(\frac{s}{2}\right)\left(\frac{s}{2}\right)\left(\frac{s}{2}\right)}$$

$$A = \sqrt{\frac{3s^4}{16}} = \frac{\sqrt{3} \cdot \sqrt{s^4}}{\sqrt{16}}$$

$$A = \frac{s^2\sqrt{3}}{4}$$

47. The area of a trapezoid $= \frac{1}{2}h(b_1 + b_2) = h \cdot \frac{1}{2}(b_1 + b_2)$. The length of the median of a trapezoid is $m = \frac{1}{2}(b_1 + b_2)$. By substitution, the area of a trapezoid is $A = hm$.

Section 8.3

1. (a) 12.25 cm^2 (b) 88.36 in^2 **3.** (a) $1.5625\sqrt{3} \text{ m}^2$
(b) $27\sqrt{3} \text{ in}^2$ **5.** $P = 68.4 \text{ in}$ **7.** $r = \frac{20}{3}\sqrt{3} \text{ cm}$
9. Regular hexagon **11.** Square **13.** 40.96 cm^2
15. $63.48\sqrt{3} \text{ in}^2$ **17.** 97.5 cm^2 **19.** 317.52 in^2
21. $54\sqrt{3} \text{ cm}^2$ **23.** $75\sqrt{3} \text{ in}^2$ **25.** 750 cm^2 **27.** 460.8 ft^2
29. $(24 + 12\sqrt{3}) \text{ in}^2$ **31.** $\frac{2}{1}$ or 2:1 **33.** $(24 + 24\sqrt{2}) \text{ units}^2$
35. $\approx 182 \text{ units}^2$ **37.** $\frac{3}{4}$ or 3.4

8.4 Exercises

1. $C = 16\pi \text{ cm}; A = 64\pi \text{ cm}^2$ **3.** $C = 66 \text{ in.}; A = 346\frac{1}{2} \text{ in}^2$
5. (a) $r = 22 \text{ in.}; d = 44 \text{ in.}$ (b) $r = 30 \text{ ft}; d = 60 \text{ ft}$
7. (a) $r = 5 \text{ in.}; d = 10 \text{ in.}$ (b) $r = 1.5 \text{ cm}; d = 3.0 \text{ cm}$
9. $\frac{8}{3}\pi \text{ in.}$ **11.** $C \approx 77.79 \text{ in.}$ **13.** $r \approx 6.7 \text{ cm}$
15. $\ell \approx 7.33 \text{ in.}$ **17.** 16 in^2 **19.** $0 < b < 20$
21. $5 < AN < 13$ **23.** $(32\pi - 64) \text{ in}^2$
25. $(600 - 144\pi) \text{ ft}^2$ **27.** $\approx 7 \text{ cm}$ **29.** 8 in
31. $A = A_{\text{LARGER CIRCLE}} - A_{\text{SMALLER CIRCLE}}$
$\quad A = \pi R^2 - \pi r^2$
$\quad A = \pi(R^2 - r^2)$
Hint $R^2 - r^2$ is a difference of two squares, so
$\quad A = \pi(R + r)(R - r)$.
33. 3 in. and 4 in. **35.** (a) $\approx 201.06 \text{ ft}^2$ (b) 2.87 quarts. Thus, 3 quarts must be purchased. (c) $47.67 **37.** (a) $\approx 1256 \text{ ft}^2$
(b) 20.93. Thus, 21 lb of seed are needed. (c) $34.65
39. $\approx 43.98 \text{ cm}$ **41.** $\approx 14.43 \text{ in.}$ **43.** $\approx 27,488.94 \text{ mi}$
45. 15.7 ft/s **47.** $12\pi \text{ cm}^2$

8.5 Exercises

1. (a) Sector (b) 25 cm **3.** (a) $(12 + 6\pi) \text{ in.}$ (b) $18\pi \text{ in.}^2$
5. 34 in. **7.** 150 cm^2 **9.** $\frac{3}{2}rs$ **11.** 54 mm **13.** 24 in^2
15. 1 in. **17.** $P = \left(16 + \frac{8}{3}\pi\right) \text{ in. and } A = \frac{32}{3}\pi \text{ in}^2$
19. $\approx 30.57 \text{ in.}$ **21.** $P = (12 + 4\pi) \text{ in.};$
$A = (24\pi - 36\sqrt{3}) \text{ in}^2$ **23.** $\left(25\sqrt{3} - \frac{25}{2}\pi\right) \text{ cm}^2$

25. $\frac{9}{2} \text{ cm}$ **27.** 36π **29.** $90°$ **31.** Cut the pizza into 8 slices.
33. $A = \left(\frac{\pi}{2}\right)s^2 - s^2$ **35.** $r = 3\frac{1}{3} \text{ ft or 3 ft, 4 in.}$
39. (a) 3 (b) 2 **41.** $\frac{308\pi}{3} \approx 322.54 \text{ in}^2$
43. $1875\pi \approx 5890 \text{ ft}^2$ **45.** $\sqrt{5} \text{ cm} \approx 2.2 \text{ cm}$

Chapter 8 Review Exercises

1. 480 **2.** (a) 40 (b) $40\sqrt{3}$ (c) $40\sqrt{2}$ **3.** 50 **4.** 204
5. 336 **6.** 36 **7.** (a) $24\sqrt{2} + 18$ (b) $24 + 9\sqrt{3}$
(c) $33\sqrt{3}$ **8.** $A = 216 \text{ in}^2; P = 60 \text{ in.}$ **9.** (a) $19,000 \text{ ft}^2$
(b) 4 bags (c) $72 **10.** (a) 3 double rolls (b) 3 rolls
11. (a) $\frac{289}{4}\sqrt{3} + 8\sqrt{33}$ (b) $50 + \sqrt{33}$ **12.** 168
13. 5 cm by 7 cm **14.** (a) 15 cm, 25 cm, and 20 cm
(b) 150 cm^2 **15.** 36 **16.** $36\sqrt{3} \text{ cm}^2$ **17.** 20 **18.** (a) $72°$
(b) $108°$ (c) $72°$ **19.** $96\sqrt{3} \text{ ft}^2$ **20.** 6 in. **21.** $162\sqrt{3} \text{ in}^2$
22. (a) 8 (b) $\approx 120 \text{ cm}^2$ **23.** (a) No. $\perp$ bisectors of sides of a parallelogram are not necessarily concurrent. (b) No. $\perp$ bisectors of sides of a rhombus are not necessarily concurrent. (c) Yes. $\perp$ bisectors of sides of a rectangle are concurrent. (d) Yes. $\perp$ bisectors of sides of a square are concurrent.
24. (a) No. $\angle$ bisectors of a parallelogram are not necessarily concurrent. (b) Yes. $\angle$ bisectors of a rhombus are concurrent. (c) No. $\angle$ bisectors of a rectangle are not necessarily concurrent. (d) Yes. $\angle$ bisectors of a square are concurrent.
25. $147\sqrt{3} \approx 254.61 \text{ in}^2$ **26.** (a) 312 ft^2 (b) 35 yd^2
(c) $453.25 **27.** $64 - 16\pi$ **28.** $\frac{49}{2}\pi - \frac{49}{2}\sqrt{3}$
29. $\frac{8}{3}\pi - 4\sqrt{3}$ **30.** $288 - 72\pi$ **31.** $25\sqrt{3} - \frac{25}{3}\pi$
32. $\ell = \frac{2\pi\sqrt{5}}{3} \text{ cm}; A = 5\pi \text{ cm}^2$ **33.** (a) 21 ft
(b) $\approx 346\frac{1}{2} \text{ ft}^2$ **34.** (a) $6\pi \text{ ft}^2$ (b) $(6\sqrt{3} + 4\pi\sqrt{3}) \text{ ft}$
35. $(9\pi - 18) \text{ in}^2$ **39.** (a) $\approx 28 \text{ yd}^2$ (b) $\approx 21.2 \text{ ft}^2$
40. (a) $\approx 905 \text{ ft}^2$ (b) $407.25 (c) Approximately 151 flowers

Chapter 8 Review Exercises Selected Proof

36. *Proof:* By an earlier theorem,
$$A_{\text{RING}} = \pi R^2 - \pi r^2$$
$$= \pi(OC)^2 - \pi(OB)^2$$
$$= \pi[(OC)^2 - (OB)^2]$$
In rt. $\triangle OBC$,
$$(OB)^2 + (BC)^2 = (OC)^2$$
Thus, $\quad (OC)^2 - (OB)^2 = (BC)^2$
In turn, $A_{\text{RING}} = \pi(BC)^2$.

Chapter 8 Test

1. (a) Square inches (b) Equal [8.1] **2.** (a) $A = s^2$
(b) $C = 2\pi r$ [8.4] **3.** (a) True (b) False [8.2]
4. 23 cm^2 [8.1] **5.** 120 ft^2 [8.1] **6.** 24 ft^2 [8.2]
7. 24 cm^2 [8.2] **8.** 6 ft [8.2] **9.** (a) 29 in. (b) 58 in^2 [8.3]
10. (a) $10\pi \text{ in.}$ (b) $25\pi \text{ in}^2$ [8.4] **11.** $\approx 5\frac{1}{2} \text{ in.}$ [8.4]
12. 314 cm^2 [8.4] **13.** $(16\pi - 32) \text{ in}^2$ [8.5]
14. $54\pi \text{ cm}^2$ [8.5] **15.** $(36\pi - 72) \text{ in}^2$ [8.5]
16. $r = 2 \text{ in.}$ [8.3] **17.** (a) 20 sheets (b) $256 [8.1]

CHAPTER 9

9.1 Exercises

1. (a) Yes (b) Oblique (c) Hexagon (d) Oblique hexagonal prism (e) Parallelogram **3.** (a) 12 (b) 18 (c) 8 **5.** (a) cm^2 (b) cm^3 **7.** 132 cm^2 **9.** 120 cm^3 **11.** (a) 16 (b) 8 (c) 16 **13.** (a) $2n$ (b) n (c) $2n$ (d) $3n$ (e) n (f) 2 (g) $n + 2$ **15.** (a) 671.6 cm^2 (b) 961.4 cm^2 (c) 2115.54 cm^3 **17.** (a) 72 ft^2 (b) 84 ft^2 (c) 36 ft^3 **19.** 1728 in^3 **21.** (a) 7500 cm^3 (b) 2350 cm^2 **23.** 6 in. by 6 in. by 3 in. **25.** $x = 3$ **27.** \$4.44 **29.** 640 ft^3 **31.** (a) $T = L + 2B$, $T = hP + 2(e \cdot e)$, $T = e(4e) + 2e^2$, $T = 4e^2 + 2e^2$, $T = 6e^2$ (b) 96 cm^2 (c) $V = Bh$, $V = e^2 \cdot e$, $V = e^3$ (d) 64 cm^3 **33.** 4 cm **35.** $V = 125$ cm^3; $S = 150$ cm^2 **37.** \$210 **39.** 864 in^3 **41.** 10 gal **43.** 720 cm^2 **45.** 2952 cm^3

9.2 Exercises

1. (a) Right pentagonal prism (b) Oblique pentagonal prism **3.** (a) Regular square pyramid (b) Oblique square pyramid **5.** (a) Pyramid (b) F (c) $\overline{EA}, \overline{ED}, \overline{EC}, \overline{ED}$ (d) $\triangle EAB, \triangle EBC, \triangle ECD, \triangle EAD$ (e) No **7.** (a) 5 (b) 8 (c) 5 (d) At the vertex (apex) **9.** 66 in^2 **11.** 32 cm^3 **13.** (a) $n + 1$ (b) n (c) n (d) $2n$ (e) n (f) $n + 1$ **15.** $3a, 4a$ **17.** $3a, 4a$ **19.** (a) Slant height (b) Lateral edge **21.** 4 in. **23.** (a) 144.9 cm^2 (b) 705.18 cm^3 **25.** (a) 60 ft^2 (b) 96 ft^2 (c) 48 ft^3 **27.** $36\sqrt{5} + 36 \approx 116.5$ in^2 **29.** 480 ft^2 **31.** 900 ft^3 **33.** ≈ 24 ft **35.** 336 in^3 **39.** (a) 32 in^3 (b) 8 in **41.** 96 in^2 **43.** $\frac{8}{1}$ or 8:1 **45.** 39.4 in^3

9.3 Exercises

1. (a) Yes (b) Yes (c) Yes **3.** 164.1 cm^3 **5.** (a) $60\pi \approx 188.50$ in^2 (b) $110\pi \approx 345.58$ in^2 (c) $150\pi \approx 471.24$ in^3 **7.** ≈ 54.19 in^2 **9.** 5 cm **11.** The radius has a length of 2 in., and the altitude has a length of 3 in. **13.** $32\pi \approx 100.53$ in^3 **15.** $2\sqrt{13} \approx 7.21$ cm **17.** 2 m **19.** $4\sqrt{3} \approx 6.93$ in. **21.** $3\sqrt{5} \approx 6.71$ cm **23.** (a) $6\pi\sqrt{85} \approx 173.78$ in^2 (b) $6\pi\sqrt{85} + 36\pi \approx 286.88$ in^2 (c) $84\pi \approx 263.89$ in^3 **25.** 54π in^3 **27.** 2000π cm^3 **29.** 1200π cm^3 **31.** $65\pi \approx 204.2$ cm^2 **33.** $192\pi \approx 603.19$ in^3 **37.** $60\pi \approx 188.5$ in^2 **39.** $\frac{4}{1}$ or 4:1 **41.** ≈ 471.24 gal **45.** ≈ 290.60 cm^3 **47.** ≈ 318 gal **49.** ≈ 38 ft^2

9.4 Exercises

1. Polyhedron $EFGHIJK$ is concave.
3. Polyhedron $EFGHIJK$ has 9 faces (F), 7 vertices (V), and 14 edges (E); $V + F = E + 2$ becomes
$$7 + 9 = 14 + 2$$
5. A regular hexahedron has 6 faces (F), 8 vertices (V), and 12 edges (E); $V + F = E + 2$ becomes
$$8 + 6 = 12 + 2$$
7. (a) 8 faces (b) Regular octahedron **9.** 9 faces **11.** (a) $\frac{1}{2}$ (b) $\frac{5}{12}$ (c) $\frac{5}{6}$ **13.** (a) $6\sqrt{2} \approx 8.49$ in. (b) $6\sqrt{3} \approx 10.39$ in. **15.** 44 in^2 **17.** 105.84 cm^2 **19.** (a) 17.64 m^2 (b) 4.2 m **21.** 82.4 cm^3

23. (a) 1468.8 cm^2 (b) \$11.75 **25.** (a) $\frac{3}{2}$ or 3:2 (b) $\frac{3}{2}$ or 3:2 **27.** $r = 3\sqrt{2} \approx 4.24$ in.; $h = 6\sqrt{2} \approx 8.49$ in. **29.** (a) $3\sqrt{3} \approx 5.20$ in. (b) 9 in. **31.** (a) $36\pi \approx 113.1$ m^2 (b) $36\pi \approx 113.1$ m^3 **33.** 1.5 in. **35.** 113.1 ft^2; ≈ 3 pints **37.** $7.4\pi \approx 23.24$ in^3 **39.** (a) Yes (b) Yes **41.** Parallel **43.** Congruent **45.** $S = 36\pi$ units^2; $V = 36\pi$ units^3

Chapter 9 Review Exercises

1. 672 in^2 **2.** 297 cm^2 **3.** Dimensions are 6 in. by 6 in. by 20 in.; $V = 720$ in^3 **4.** $T = 468$ cm^2; $V = 648$ cm^3 **5.** (a) 360 in^2 (b) 468 in^2 (c) 540 in^3 **6.** (a) 624 cm^2 (b) $624 + 192\sqrt{3} \approx 956.55$ cm^2 (c) $1248\sqrt{3} \approx 2161.6$ cm^3 **7.** $\sqrt{89} \approx 9.43$ cm **8.** $3\sqrt{7} \approx 7.94$ in. **9.** $\sqrt{74} \approx 8.60$ in. **10.** $2\sqrt{3} \approx 3.46$ cm **11.** (a) 540 in^2 (b) 864 in^2 (c) 1296 in^3 **12.** (a) $36\sqrt{19} \approx 156.92$ cm^2 (b) $36\sqrt{19} + 36\sqrt{3} \approx 219.27$ cm^2 (c) $96\sqrt{3} \approx 166.28$ cm^3 **13.** (a) 120π in^2 (b) 192π in^2 (c) 360π in^3 **14.** (a) ≈ 351.68 ft^3 (b) ≈ 452.16 ft^2 **15.** (a) $72\pi \approx 226.19$ cm^2 (b) $108\pi \approx 339.29$ cm^2 (c) $72\pi\sqrt{3} \approx 391.78$ cm^3 **16.** $\ell = 10$ in. **17.** ≈ 616 in^2 **18.** ≈ 904.32 cm^3 **19.** 120π units^3 **20.** $\dfrac{\text{surface area of smaller}}{\text{surface area of larger}} = \dfrac{1}{9}$, $\dfrac{\text{volume of smaller}}{\text{volume of larger}} = \dfrac{1}{27}$ **21.** $\approx 183\frac{1}{3}$ in^3 **22.** 288π cm^3 **23.** $\frac{32\pi}{3}$ in^3 **24.** ≈ 1017.36 in^3 **25.** $\left(2744 - \frac{1372}{3}\pi\right)$ in^3 **26.** (a) 8 (b) 4 (c) 12 **27.** 40π mm^3 **28.** (a) $V = 16, E = 24, F = 10$, so $V + F = E + 2$ becomes $16 + 10 = 24 + 2$ (b) $V = 4, E = 6, F = 4$, so $V + F = E + 2$ becomes $4 + 4 = 6 + 2$ (c) $V = 6, E = 12, F = 8$, so $V + F = E + 2$ becomes $6 + 8 = 12 + 2$ **29.** 114 in^3 **30.** (a) $\frac{1}{2}$ (b) $\frac{5}{8}$ **31.** (a) 78 in^2 (b) $16\sqrt{3}$ cm^2 **32.** Right triangle (3, 4, 5)

Chapter 9 Test

1. (a) 15 (b) 7 [9.1] **2.** (a) 16 cm^2 (b) 112 cm^2 (c) 80 cm^3 [9.1] **3.** (a) 5 (b) 4 [9.2] **4.** (a) $32\sqrt{2}$ ft^2 (b) $(16 + 32\sqrt{2})$ ft^2 [9.2] **5.** 15 ft [9.2] **6.** 3 in. [9.2] **7.** 50 ft^3 [9.2] **8.** (a) False (b) True [9.3] **9.** (a) True (b) True [9.3, 9.4] **10.** 12 [9.4] **11.** $3\sqrt{5}$ cm [9.3] **12.** (a) 48π cm^2 (b) 96π cm^3 [9.3] **13.** $h = 6$ in. [9.3] **14.** (a) $\frac{1}{2}$ (b) $\frac{3}{8}$ [9.4] **15.** (a) 1256.6 ft^2 (b) 4188.8 ft^3 [9.4] **16.** 2 hours and 47 minutes [9.4]

CHAPTER 10

10.1 Exercises

3. (a) 4 (b) 8 (c) 5 (d) 9 **5.** $b = 3.5$ or $b = 10.5$ **7.** (a) 5 (b) 10 (c) $2\sqrt{5}$ (d) $\sqrt{a^2 + b^2}$ **9.** (a) $\left(2, \frac{-3}{2}\right)$ (b) $(1, 1)$ (c) $(4, 0)$ (d) $\left(\frac{a}{2}, \frac{b}{2}\right)$ **11.** (a) $(-3, 4)$ (b) $(0, -2)$ (c) $(-a, 0)$ (d) $(-b, -c)$ **13.** (a) $\left(4, -\frac{5}{2}\right)$ (b) $(0, 4)$ (c) $\left(\frac{7}{2}, -1\right)$ (d) (a, b) **15.** (a) $(5, -1)$ (b) $(0, -5)$ (c) $(2, -a)$ (d) $(b, -c)$ **17.** (a) $(-3, -4)$ (b) $(-2, 0)$ (c) $(-a, 0)$ (d) $(-b, c)$ **19.** (a) $x = 4$ (b) $y = b$ (c) $x = 2$ (d) $y = 3$ **21.** $(2.5, -13.7)$ **23.** $(2, 3)$; 16

25. (a) Isosceles (b) Equilateral (c) Isosceles right triangle
27. $x + y = 6$ **29.** $(a, a\sqrt{3})$ or $(a, -a\sqrt{3})$
31. $(0, 1 + 3\sqrt{3})$ and $(0, 1 - 3\sqrt{3})$ **33.** 17   **35.** 9
37. (a) 135π units^3 (b) 75π units^3 **39.** (a) 96π units^3
(b) 144π units^3 **41.** (a) 90π units^2 (b) 90π units^2
43.

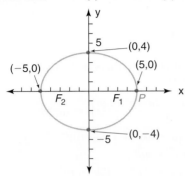

47. (a) $(-3, -1)$ (b) $(1, -3)$ (c) $(3, 1)$ **49.** (a) $(5, 4)$
(b) $(5, 8)$ (c) $(3, 2)$

10.2 Exercises

1. $(4, 0)$ and $(0, 3)$ **3.** $(5, 0)$ and $\left(0, -\frac{5}{2}\right)$ **5.** $(-3, 0)$
7. $(6, 0)$ and $(0, 3)$ **9.** $(4, -3)$ **11.** (a) 4 (b) Undefined
(c) -1 (d) 0 (e) $\frac{d-b}{c-a}$ (f) $-\frac{b}{a}$ **13.** (a) 10 (b) 15
15. (a) Collinear (b) Noncollinear   **17.** (a) $\frac{3}{4}$ (b) $-\frac{5}{3}$
(c) -2 (d) $\frac{a-b}{c}$ **19.** (a) 2 (b) $-\frac{4}{3}$ (c) $-\frac{1}{3}$ (d) $-\frac{h+j}{f+g}$
21. None of these   **23.** Perpendicular   **25.** $\frac{3}{2}$
27. 23 35 Right triangle   **37.** $(4, 7)$; $(0, -1)$; $(10, -3)$
41. $m_{\overline{EH}} = \frac{2c-0}{2b-0} = \frac{2c}{2b} = \frac{c}{b}$
$m_{\overline{FG}} = \frac{c-0}{(a+b)-a} = \frac{c}{b}$
Because of equal slopes, $\overline{EH} \parallel \overline{FG}$. Thus, $EFGH$ is a trapezoid.
45. $-\frac{b^2}{2m}$

10.2 Selected Proof

39. $m_{\overline{VT}} = \dfrac{e - e}{(c - d) - (a + d)} = \dfrac{0}{c - a - 2d} = 0$
$m_{\overline{RS}} = \dfrac{b - b}{c - a} = \dfrac{0}{c - a} = 0$
$\therefore \overline{VT} \parallel \overline{RS}$
$RV = \sqrt{[(a + d) - a]^2 + (e - b)^2}$
$\quad = \sqrt{d^2 + (e - b)^2} = \sqrt{d^2 + e^2 - 2be + b^2}$
$ST = \sqrt{[c - (c - d)]^2 + (b - e)^2}$
$\quad = \sqrt{(d)^2 + (b - e)^2}$
$\quad = \sqrt{d^2 + b^2 - 2be + e^2}$
$\therefore RV = ST$
$RSTV$ is an isosceles trapezoid.

10.3 Exercises

1. (a) $a\sqrt{2}$ if $a > 0$ (b) $\frac{d-b}{c-a}$ **3.** (a) -1 (b) $-\frac{b}{a}$
5. $\overline{AB}$ is horizontal and $\overline{BC}$ is vertical; $\therefore \overline{AB} \perp \overline{BC}$.
Hence, $\angle B$ is a right $\angle$ and $\triangle ABC$ is a right triangle.

7. $m_{\overline{QM}} = \frac{c - 0}{b - 0} = \frac{c}{b}$
$m_{\overline{PN}} = \frac{c - 0}{(a + b) - a} = \frac{c}{b}$
$\therefore \overline{QM} \parallel \overline{PN}$
$m_{\overline{QP}} = \frac{c - c}{(a + b) - b} = \frac{0}{a} = 0$
$m_{\overline{MN}} = \frac{0 - 0}{a - 0} = \frac{0}{a} = 0$
$\therefore \overline{QP} \parallel \overline{MN}$
Because both pairs of opposite sides are parallel, $MNPQ$ is a parallelogram.
9. $m_{\overline{MN}} = 0$ and $m_{\overline{QP}} = 0$; $\therefore \overline{MN} \parallel \overline{QP}$. Also, $\overline{QM}$ and $\overline{PN}$ are both vertical; $\therefore \overline{QM} \parallel \overline{PN}$. Hence, $MQPN$ is a parallelogram. Because $\overline{QM}$ is vertical and $\overline{MN}$ is horizontal, $\angle QMN$ is a right angle. Because parallelogram $MQPN$ has a right $\angle$, it is also a rectangle.   **11.** $A = (0, 0)$; $B = (a, 0)$; $C = (a, b)$
13. $M = (0, 0)$; $N = (r, 0)$; $P = (r + s, t)$
15. $A = (0, 0)$; $B = (a, 0)$; $C = (a - c, d)$
17. (a) Square $A = (0, 0)$; $B = (a, 0)$; $C = (a, a)$; $D = (0, a)$
(b) Square (with midpoints of sides) $A = (0, 0)$; $B = (2a, 0)$;
$C = (2a, 2a)$; $D = (0, 2a)$ **19.** (a) Parallelogram $A = (0, 0)$;
$B = (a, 0)$; $C = (a + b, c)$; $D = (b, c)$
(NOTE: D chosen before C.)
(b) Parallelogram (with midpoints of sides) $A = (0, 0)$;
$B = (2a, 0)$; $C = (2a + 2b, 2c)$; $D = (2b, 2c)$
21. (a) Isosceles triangle $R = (0, 0)$; $S = (2a, 0)$; $T = (a, b)$
(b) Isosceles triangle (with midpoints of sides) $R = (0, 0)$;
$S = (4a, 0)$; $T = (2a, 2b)$ **23.** $r^2 = s^2 + t^2$
25. $c^2 = a^2 - b^2$ **27.** $b^2 = 3a^2$ **29.** (a) Positive
(b) Negative (c) $2a$ **31.** (a) Slope Formula
(b) Distance Formula (c) Midpoint Formula
(d) Slope Formula   **37.** $(6 - 2a, 0)$

10.4 Exercises

21. $m_1 = -\frac{A}{B}$; $m_2 = \frac{B}{A}$; $m_1 \cdot m_2 = -1$, so $\ell_1 \perp \ell_2$.
23. $3x - 2y = 2$ **25.** $x^2 + y^2 = 9$ **27.** $m = -\frac{a}{b}$
29. True. The quadrilateral that results is a parallelogram.

10.4 Selected Proofs

3. The diagonals of a square are perpendicular bisectors of each other.

Proof: Let square $RSTV$ have the vertices shown. Then the midpoints of the diagonals are $M_{\overline{RT}} = (a, a)$ and $M_{\overline{VS}} = (a, a)$. Also, $m_{\overline{RT}} = 1$ and $m_{\overline{VS}} = -1$. Because the two diagonals share the midpoint (a, a) and the product of their slopes is -1, they are perpendicular bisectors of each other.

7. The segments that join the midpoints of the consecutive sides of a quadrilateral form a parallelogram.

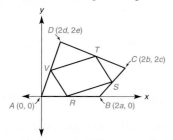

Proof: The midpoints, as shown, of the sides of quadrilateral *ABCD* are

$$R = \left(\frac{0 + 2a}{2}, \frac{0 + 0}{2}\right) = (a, 0)$$

$$S = \left(\frac{2a + 2b}{2}, \frac{0 + 2c}{2}\right) = (a + b, c)$$

$$T = \left(\frac{2d + 2b}{2}, \frac{2e + 2c}{2}\right) = (d + b, e + c)$$

$$V = \left(\frac{0 + 2d}{2}, \frac{0 + 2e}{2}\right) = (d, e)$$

Now we determine slopes as follows:

$$m_{\overline{RS}} = \frac{c - 0}{(a + b) - a} = \frac{c}{b}$$

$$m_{\overline{ST}} = \frac{(e + c) - c}{(d + b) - (a + b)} = \frac{e}{d - a}$$

$$m_{\overline{TV}} = \frac{(e + c) - e}{(d + b) - d} = \frac{c}{b}$$

$$m_{\overline{VR}} = \frac{e - 0}{d - a} = \frac{e}{d - a}$$

Because $m_{\overline{RS}} = m_{\overline{TV}}$, $\overline{RS} \parallel \overline{TV}$. Also $m_{\overline{ST}} = m_{\overline{VR}}$, so $\overline{ST} \parallel \overline{VR}$. Then *RSTV* is a parallelogram.

11. The midpoint of the hypotenuse of a right triangle is equidistant from the three vertices of the triangle.

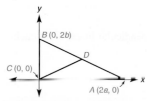

Proof: Let rt. $\triangle ABC$ have vertices as shown. Then *D*, the midpoint of the hypotenuse, is given by

$$D = \left(\frac{0 + 2a}{2}, \frac{2b + 0}{2}\right) = (a, b)$$

Now $BD = DA = \sqrt{(2a - a)^2 + (0 - b)^2}$
$$= \sqrt{a^2 + (-b)^2} = \sqrt{a^2 + b^2}$$

Also, $CD = \sqrt{(a - 0)^2 + (b - 0)^2}$
$$= \sqrt{a^2 + b^2}$$

Then *D* is equidistant from *A*, *B*, and *C*.

15. If the midpoint of one side of a rectangle is joined to the endpoints of the opposite side, an isosceles triangle is formed.

Proof: Let rectangle *ABCD* have vertices as shown above. With *M* the midpoint of $\overline{DC}$,

$$M = \left(\frac{0 + 2a}{2}, \frac{2b + 2b}{2}\right) = (a, 2b)$$

$$MA = \sqrt{(a - 0)^2 + (2b - 0)^2}$$

$$MA = \sqrt{a^2 + 4b^2}$$

$$MB = \sqrt{(a - 2a)^2 + (2b - 0)^2}$$

$$MB = \sqrt{a^2 + 4b^2}$$

Because $MA = MB$, $\triangle AMB$ is isosceles.

10.5 Exercises

1. $x + 2y = 6$; $y = -\frac{1}{2}x + 3$ **3.** $-x + 3y = -40$;
$y = \frac{1}{3}x - \frac{40}{3}$ **9.** $2x + 3y = 15$ **11.** $x + y = 6$
13. $-2x + 3y = -3$ **15.** $bx + ay = ab$
17. $-x + y = -2$ **19.** $5x + 2y = 5$
21. $4x + 3y = -12$ **23.** $-x + 3y = 2$
25. $y = -\frac{b}{a}x + \frac{bg + ha}{a}$ **27.** $(6, 0)$ **29.** $(5, -4)$
31. $(6, -2)$ **33.** $(3, 2)$ **35.** $(6, -1)$ **37.** $(5, -2)$
39. $(6, 0)$ **41.** $\left(\frac{a - c}{b}, a\right)$ **43.** $\left(b, \frac{ab - b^2}{c}\right)$
49. $a = 5, b = 2$

10.5 Selected Proofs

45. The altitudes of a triangle are concurrent.

Proof: For $\triangle ABC$, let $\overline{CH}$, $\overline{AJ}$, and $\overline{BK}$ name the altitudes. Because $\overline{AB}$ is horizontal ($m_{\overline{AB}} = 0$), $\overline{CH}$ is vertical and has the equation $x = b$.
Because $m_{\overline{BC}} = \frac{c - 0}{b - a} = \frac{c}{b - a}$, the slope of altitude $\overline{AJ}$ is $m_{\overline{AJ}} = -\frac{b - a}{c} = \frac{a - b}{c}$. Since $\overline{AJ}$ contains $(0, 0)$, its equation is $y = \frac{a - b}{c}x$.
The intersection of altitudes $\overline{CH}$ ($x = b$) and $\overline{AJ}$ $\left(y = \frac{a - b}{c}x\right)$ is at $x = b$, so $y = \frac{a - b}{c} \cdot b = \frac{b(a - b)}{c} = \frac{ab - b^2}{c}$. That is, $\overline{CH}$ and $\overline{AJ}$ intersect at $\left(b, \frac{ab - b^2}{c}\right)$. The remaining altitude is $\overline{BK}$. Since $m_{\overline{AC}} = \frac{c - 0}{b - 0} = \frac{c}{b}$, $m_{\overline{BK}} = -\frac{b}{c}$. Because $\overline{BK}$ contains $(a, 0)$, its equation is $y - 0 = -\frac{b}{c}(x - a)$ or $y = \frac{-b}{c}(x - a)$.

For the three altitudes to be concurrent, $\left(b, \frac{ab - b^2}{c}\right)$ must lie on the line $y = \frac{-b}{c}(x - a)$. Substituting into $y = -\frac{b}{c}(x - a)$, the equation for altitude $\overline{BK}$, leads to

$$\frac{ab - b^2}{c} = \frac{-b}{c}(b - a)$$
$$= \frac{-b(b - a)}{c}$$
$$= \frac{-b^2 + ab}{c},$$

which is true. Thus, the three altitudes are concurrent.
47. First, find the equation of the line through P that is perpendicular to $Ax + By = C$. Second, find the point of intersection D of the two lines. Finally, use the Distance Formula to find the length of $\overline{PD}$.

10.6 Exercises

1. $(3, 7, 5)$ **3.** $(2, 3, -1)$ **5.** (a) $(2, 3, 4)$
(b) $(3, -2, 5)$ **7.** (a) $(2, 4, 5)$ (b) $(3, -5, 2)$
9. $(x, y, z) = (4, -3, 7) + n(4, 2, -3)$ **11.** $\sqrt{21}$ **13.** 3
15. $(-3, 2, 7)$ **17.** $(3.5, 3, -3)$
19.

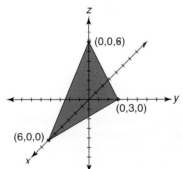

21. a, c **23.** $x^2 + y^2 + z^2 = 25$
25. $(x - 1)^2 + (y - 2)^2 + (z - 3)^2 = 25$ or
$x^2 + y^2 + z^2 - 2x - 4y - 6z - 11 = 0$ **27.** b
29. No common pt; vectors are multiples **31.** Both contain $(0, 0, 0)$, for $n = 0$ and $r = 1$; vectors are multiples **33.** Yes
35. (a) $(7, 1, 1)$ (b) $(-14, 16, -2)$ **37.** $(6, 8, 0), (0, 0, 10)$
39. $S = 400\pi$ units^2; $V = \frac{4000}{3}\pi$ units^3 **41.** Yes; Yes
43. $(5, 6, 5)$ **45.** $(x, y, z) = (12, 6, 0) + n(-2, 3, 1)$ or equivalent **47.** Yes; Yes **49.** $(6, 15, -9)$

Chapter 10 Review Exercises

1. (a) 7 (b) 6 (c) 13 (d) 5 **2.** (a) 8 (b) 10 (c) $4\sqrt{5}$
(d) 10 **3.** (a) $\left(6, \frac{1}{2}\right)$ (b) $(-2, 4)$ (c) $\left(1, -\frac{1}{2}\right)$ (d) $\left(\frac{2x - 3}{2}, y\right)$
4. (a) $(2, 1)$ (b) $(-2, -2)$ (c) $(0, 3)$ (d) $(x + 1, y + 1)$
5. (a) Undefined (b) 0 (c) $-\frac{5}{12}$ (d) $-\frac{4}{3}$ **6.** (a) Undefined
(b) 0 (c) $\frac{1}{2}$ (d) $\frac{4}{3}$ **7.** $(-4, -8)$ **8.** $(3, 7)$ **9.** $x = \frac{4}{3}$
10. $y = -4$ **11.** (a) Perpendicular (b) Parallel (c) Neither
(d) Perpendicular **12.** Noncollinear **13.** $x = 4$
14. $(7, 0)$ and $(0, 3)$ **17.** (a) $3x + 5y = 21$
(b) $3x + y = -7$ (c) $-2x + y = -8$ (d) $y = 5$
18. $m_{\overline{AB}} = \frac{4}{3}$; $m_{\overline{BC}} = \frac{1}{2}$; $m_{\overline{AC}} = -2$. Because $m_{\overline{AC}} \cdot m_{\overline{BC}} = -1$, $\overline{AC} \perp \overline{BC}$ and $\angle C$ is a rt. $\angle$; the triangle is a rt. $\triangle$.
19. $AB = \sqrt{85}$; $BC = \sqrt{85}$. Because $AB = BC$, the triangle

is isosceles. **20.** $m_{\overline{RS}} = -\frac{4}{3}$; $m_{\overline{ST}} = \frac{5}{6}$; $m_{\overline{TV}} = -\frac{4}{3}$; $m_{\overline{RV}} = \frac{5}{6}$. Therefore, $\overline{RS} \parallel \overline{VT}$ and $\overline{RV} \parallel \overline{ST}$ and $RSTV$ is a parallelogram.
21. $(3, 5)$ **22.** $(1, 4)$ **23.** $(3, 5)$ **24.** $(1, 4)$
25. $(16, 11), (4, -9), (-4, 5)$ **26.** (a) $\sqrt{53}$ (b) -4 (c) $\frac{1}{4}$
27. $A = (-a, 0)$; $B = (0, b)$; $C = (a, 0)$
28. $D = (0, 0)$; $E = (a, 0)$; $F = (a, 2a)$; $G = (0, 2a)$
29. $R = (0, 0)$; $U = (0, a)$; $T = (a, a + b)$
30. $M = (0, 0)$; $N = (a, 0)$; $Q = (a + b, c)$; $P = (b, c)$
31. (a) $\sqrt{(a + c)^2 + (b + d - 2e)^2}$ (b) $-\frac{a}{b - e}$ or $\frac{a}{e - b}$
(c) $y - 2d = \frac{a}{e - b}(x - 2c)$ **37.** Yes, because $AB = BC$, an isosceles triangle **38.** (a) $(4, 2, 3)$
(b) $(x, y, z) = (-1, 2, 4) + n(4, 2, 3)$ or
$(x, y, z) = (-1 + 4n, 2 + 2n, 4 + 3n)$
39. (a) $(x, y, z) = (2 + n, -3 + 2n, 5 + 4n)$
(b) $(0, -7, -3)$ **40.** (a) $(1, -2, -4)$ (b) $r = 6$

Chapter 10 Test

1. (a) $(5, -3)$ (b) $(0, -4)$ [10.1] **3.** $CD = 10$ [10.1]
4. $(-3, 5)$ [10.1] **5.**

x	0	3	0	4	[10.2]
y	4	2	4	-2	

6.

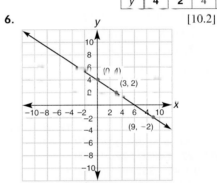

[10.2]

7. (a) -3 (b) $\frac{d - b}{c - a}$ [10.2]
8. (a) $\frac{2}{3}$ (b) $-\frac{3}{2}$ [10.2] **9.** Parallelogram [10.3]
10. $a = \sqrt{b^2 + c^2}$ or $a^2 = b^2 + c^2$ [10.3]
11. (a) Isosceles triangle (b) Trapezoid [10.3]
12. (a) Slope Formula (b) Distance Formula [10.3]
13. $D(0, 0), E(2a, 0), F(a, b)$ [10.3] **14.** (b) [10.4]
15. $m_{\overline{VR}} = m_{\overline{TS}}$ so $-\frac{v}{r} = \frac{v}{t - s}$. Possible answers: $r = s - t$ or $s = r + t$ or equivalent [10.4]
16. (a) $y = x + 4$ (b) $y = \frac{3}{4}x - 3$ [10.5]
17. $y = cx + (b - ac)$ [10.5] **18.** $(4, 1)$ [10.5]
19. $(-1, 4)$ [10.5] **20.** $M = (a + b, c)$ and $N = (a, 0)$. Then $m_{\overline{AC}} = \frac{2c - 0}{2b - 0} = \frac{c}{b}$ and $m_{\overline{MN}} = \frac{c - 0}{a + b - a} = \frac{c}{b}$. With $m_{\overline{AC}} = m_{\overline{MN}}$, it follows that $\overline{AC} \parallel \overline{MN}$. [10.4]
21. (a) $(0, 1, -1)$ (b) $\sqrt{176} = 4\sqrt{1}$
22. (a) $(-1, -9, 20)$ (b) No **23.** (a) Yes (b) Yes

CHAPTER 11

11.1 Exercises

1. $\sin \alpha = \frac{5}{13}$; $\sin \beta = \frac{12}{13}$ **3.** $\sin \alpha = \frac{8}{17}$; $\sin \beta = \frac{15}{17}$
5. $\sin \alpha = \frac{\sqrt{15}}{5}$; $\sin \beta = \frac{\sqrt{10}}{5}$ **7.** 1 **9.** 0.2924
11. 0.9903 **13.** 0.9511 **15.** $a \approx 6.9$ in.; $b \approx 9.8$ in.

17. $a \approx 10.9$ ft; $b \approx 11.7$ ft **19.** $c \approx 8.8$ cm; $d \approx 28.7$ cm
21. $\alpha \approx 29°$; $\beta \approx 61°$ **23.** $\alpha \approx 17°$; $\beta \approx 73°$
25. $\alpha \approx 19°$; $\beta \approx 71°$ **27.** $\alpha \approx 23°$ **29.** $d \approx 103.5$ ft
31. $d \approx 128.0$ ft **33.** $\alpha \approx 24°$
35. (a) ≈ 5.4 ft (b) ≈ 54 ft^2 **37.** $\theta \approx 50°$
39. (a) $h = s \cdot \sin 36°$; (b) $d = 2s \cdot \sin 54°$

11.2 Exercises

1. $\cos \alpha = \frac{12}{13}$; $\cos \beta = \frac{5}{13}$ **3.** $\cos \alpha = \frac{3}{5}$; $\cos \beta = \frac{4}{5}$
5. $\cos \alpha = \frac{\sqrt{10}}{5}$; $\cos \beta = \frac{\sqrt{15}}{5}$ **7.** (a) $\sin \alpha = \frac{a}{c}$; $\cos \beta = \frac{a}{c}$.
Thus, $\sin \alpha = \cos \beta$. (b) $\cos \alpha = \frac{b}{c}$; $\sin \beta = \frac{b}{c}$.
Thus, $\cos \alpha = \sin \beta$. **9.** 0.9205 **11.** 0.9563 **13.** 0
15. 0.1392 **17.** $a \approx 84.8$ ft; $b \approx 53.0$ ft **19.** $a = b = 5$ cm
21. $c \approx 19.1$ in.; $d \approx 14.8$ in. **23.** $\alpha = 60°$; $\beta = 30°$
25. $\alpha \approx 51°$; $\beta \approx 39°$ **27.** $\alpha \approx 65°$; $\beta \approx 25°$
29. $\theta \approx 34°$ **31.** $x \approx 1147.4$ ft **33.** ≈ 8.1 in.
35. ≈ 13.1 cm **37.** $\alpha \approx 55°$ **39.** (a) m$\angle A = 68°$
(b) m$\angle B = 112°$ **43.** 6.8 in^2 **45.** $5r^2 \sin 36° \cos 36°$.

11.3 Exercises

1. $\tan \alpha = \frac{3}{4}$; $\tan \beta = \frac{4}{3}$ **3.** $\tan \alpha = \frac{\sqrt{5}}{1}$; $\tan \beta = \frac{2\sqrt{5}}{5}$
5. $\sin \alpha = \frac{5}{13}$; $\cos \alpha = \frac{12}{13}$; $\tan \alpha = \frac{5}{12}$; $\cot \alpha = \frac{12}{5}$;
$\sec \alpha = \frac{13}{12}$; $\csc \alpha = \frac{13}{5}$ **7.** $\sin \alpha = \frac{a}{c}$; $\cos \alpha = \frac{b}{c}$; $\tan \alpha = \frac{a}{b}$;
$\cot \alpha = \frac{b}{a}$; $\sec \alpha = \frac{c}{b}$; $\csc \alpha = \frac{c}{a}$ **9.** $\sin \alpha = \frac{x\sqrt{x^2+1}}{x^2+1}$;
$\cos \alpha = \frac{\sqrt{x^2+1}}{x^2+1}$; $\tan \alpha = \frac{x}{1}$; $\cot \alpha = \frac{1}{x}$; $\sec \alpha = \sqrt{x^2+1}$;
$\csc \alpha = \frac{\sqrt{x^2+1}}{x}$ **11.** 0.2679 **13.** 1.5399
15. $x \approx 7.5$; $z \approx 14.2$ **17.** $y \approx 5.3$; $z \approx 8.5$ **19.** $d \approx 8.1$
21. $\alpha \approx 37°$; $\beta \approx 53°$ **23.** $\theta \approx 56°$; $\gamma \approx 34°$
25. $\alpha \approx 29°$; $\beta \approx 61°$ **27.** 1.4826 **29.** 2.0000
31. 1.3456 **33.** (b) ≈ 0.4245 **35.** (b) ≈ 7.1853
37. ≈ 1376.8 ft **39.** ≈ 4.1 in. **41.** $\approx 72°$
43. $\alpha \approx 47°$. The heading may be described as N 47° W.
45. $\approx 26,730$ ft **47.** (a) $h \approx 9.2$ ft (b) $V \approx 110.4$ ft^3

11.4 Exercises

1. (a) $\frac{1}{2} \cdot 5 \cdot 6 \cdot \sin 78°$ (b) $\frac{1}{2} \cdot 5 \cdot 7 \cdot \sin 56°$
3. (a) $\frac{\sin 40°}{5} = \frac{\sin \beta}{8}$ (b) $\frac{\sin 41°}{5.3} = \frac{\sin 87°}{c}$
5. (a) $c^2 = 5.2^2 + 7.9^2 - 2(5.2)(7.9) \cos 83°$
(b) $6^2 = 9^2 + 10^2 - 2 \cdot 9 \cdot 10 \cdot \cos \alpha$
7. (a) $\frac{\sin \alpha}{a} = \frac{\sin \beta}{b}$ (b) $a^2 = b^2 + c^2 - 2bc \cos \alpha$
9. (a) (3, 4, 5) is a Pythagorean Triple; γ lies opposite the
longest side and must be a right angle. (b) 90°
11. 8 in^2 **13.** ≈ 11.6 ft^2 **15.** ≈ 15.2 ft^2 **17.** ≈ 11.1 in.
19. ≈ 8.9 m **21.** $\approx 55°$ **23.** $\approx 51°$ **25.** ≈ 10.6
27. ≈ 6.9 **29.** (a) ≈ 213.4 ft (b) $\approx 13,294.9$ ft^2
31. ≈ 8812 m **33.** ≈ 15.9 ft **35.** 6 **37.** ≈ 14.0 ft
41. 51.8 cm^2 **43.** $\frac{1}{2}ab$

Chapter 11 Review Exercises

1. sine; ≈ 10.3 in. **2.** sine; ≈ 7.5 ft **3.** cosine; ≈ 23.0 in.
4. sine; ≈ 5.9 ft **5.** tangent; $\approx 43°$ **6.** cosine; $\approx 58°$
7. sine; $\approx 49°$ **8.** tangent; $\approx 16°$ **9.** ≈ 8.9 units

10. $\approx 60°$ **11.** ≈ 13.1 units **12.** ≈ 18.5 units
13. ≈ 42.7 ft **14.** ≈ 74.8 cm **15.** $\approx 47°$ **16.** $\approx 54°$
17. ≈ 26.3 in^2 **19.** If m$\angle S = 30°$ and m$\angle Q = 90°$, then
the sides of $\triangle RQS$ can be represented by $RQ = x$, $RS = 2x$,
and $SQ = x\sqrt{3}$. $\sin S = \sin 30° = \frac{x}{2x} = \frac{1}{2}$. **21.** ≈ 8.4 ft
22. ≈ 866 ft **23.** $\approx 41°$ **24.** $\approx 8°$ **25.** ≈ 5.0 cm
26. ≈ 4.3 cm **27.** $\approx 68°$ **28.** $\approx 106°$ **29.** 3 to 7 (or 3:7)
30. ≈ 1412.0 m **31.** $\cos \theta = \frac{24}{25}$; $\sec \theta = \frac{25}{24}$
32. $\sec \theta = \frac{61}{60}$; $\cot \theta = \frac{60}{11}$ **33.** $\csc \theta = \frac{29}{20}$; $\sin \theta = \frac{20}{29}$
34. $h \approx 6.9$ ft; $V \approx 74.0$ ft^3

Chapter 11 Test

1. (a) $\frac{a}{c}$ (b) $\frac{b}{a}$ [11.1, 11.3] **2.** (a) $\frac{3}{5}$ (b) $\frac{3}{5}$ [11.1, 11.2]
3. (a) 1 (b) $\frac{\sqrt{3}}{2}$ [11.1, 11.3] **4.** (a) 0.3907
(b) 0.1908 [11.1, 11.2] **5.** $\theta \approx 42°$ [11.1] **6.** (a) $\tan 26°$
(b) $\cos 47°$ [11.2, 11.3] **7.** $a \approx 14$ [11.1] **8.** $y \approx 9$ [11.1]
9. $\theta \approx 56°$ [11.1] **10.** (a) True (b) True [11.2]
11. 92 ft [11.1] **12.** 10° [11.1] **13.** (a) $\csc \alpha$ ▰
(b) $\alpha = 30°$ [11.1, 11.3] **14.** (a) $\frac{a}{c}$ (b) $\frac{c}{a}$ [11.3]
15. ≈ 42 cm^2 [11.4] **16.** $\frac{\sin \alpha}{a} = \frac{\sin \beta}{b} = \frac{\sin \gamma}{c}$ [11.4]
17. $a^2 = b^2 + c^2 - 2bc \cos \alpha$ [11.4] **18.** $\alpha \approx 33°$ [11.4]
19. $x \approx 11$ [11.4] **20.** $5a^2 \tan 54°$ [11.3]

APPENDIX A

A.1 ALGEBRAIC EXPRESSIONS

1. Undefined terms, definitions, axioms or postulates, and
theorems **3.** (a) Reflexive (b) Transitive (c) Substitution
(d) Symmetric **5.** (a) 12 (b) -2 (c) 2 (d) -12
7. (a) 35 (b) -35 (c) -35 (d) 35 **9.** No; Commutative
Axiom for Multiplication **11.** (a) 9 (b) -9 (c) 8 (d) -8
13. (a) -4 (b) -36 (c) 18 (d) $-\frac{1}{4}$ **15.** $-\$60$
17. (a) 65 (b) 16 (c) 9 (d) $8x$ **19.** (a) 10π (b) $11\sqrt{2}$
(c) $7x^2y$ (d) $7\sqrt{3}$ **21.** (a) 14 (b) 20 (c) 14 (d) 38
23. (a) -1 (b) $\frac{1}{9}$ (c) $-\frac{8}{9}$ (d) $-\frac{1}{2}$ **25.** (a) 6
(b) $12x^2 - 7x - 10$ **27.** $5x + 2y$ **29.** $10x + 5y$
31. $10x$

A.2 FORMULAS AND EQUATIONS

1. $5x + 8$ **3.** $2x - 2$ **5.** $5x + 8$ **7.** $x^2 + 7x + 12$
9. $6x^2 + 11x - 10$ **11.** $2a^2 + 2b^2$ **13.** 60
15. 40 **17.** 148 **19.** 12π **21.** 7 **23.** -12
25. 12 **27.** 5 **29.** 30 **31.** 8 **33.** 4 **35.** 32

A.3 INEQUALITIES

1. The length of $\overline{AB}$ is greater than the length of $\overline{CD}$.
3. The measure of angle ABC is greater than the measure of
angle DEF. **5.** (a) 4 (b) 10 **7.** (a) $a > b$ (b) $a < b$
9. $IJ < AB$ **11.** (a) False (b) True (c) True (d) False

13. The measure of the second angle must be greater than 148° and less than 180°. **15.** (a) $-12 \le 20$ (b) $-10 \le -2$ (c) $18 \ge -30$ (d) $3 \ge -5$ **17.**

No change	No change
No change	No change
No change	Change
No change	Change

19. $x \le 7$ **21.** $x < -5$ **23.** $x < 20$ **25.** $x \ge -24$
27. $x \le -2$ **29.** Not true if $c < 0$ **31.** Not true if $a = -3$ and $b = -2$

A.4 FACTORING AND QUADRATIC EQUATIONS

1. $a(x^2 + 5x + 7)$ **3.** $2bx(x + 2b)$ **5.** $(y + 3)(y - 3)$
7. $(2x + 7y)(2x - 7y)$ **9.** $(x + 3)(x + 4)$
11. $(x + 8)(x - 3)$ **13.** $(2y + 3)(3y - 2)$
15. $(x + 4y)(3x - y)$ **17.** $4(x + 2)(x - 2)$
19. $4(y + 5)(y - 3)$ **21.** $x(2x - 7)(x + 5)$
23. $x(x + 1)(x + 4)$ **25.** $\{2, 1\}$ **27.** $\{-1, 7\}$
29. $\left\{-\frac{2}{3}, 4\right\}$ **31.** $\left\{\frac{1}{2}, \frac{1}{3}\right\}$ **33.** $\{0, 3\}$ **35.** $\left\{\frac{5}{2}, -\frac{3}{2}\right\}$
37. $w = 6, w + 5 = 11$ **39.** $a = 3$ **41.** $\{2, -2, 3, -3\}$

A.5 THE QUADRATIC FORMULA AND SQUARE ROOT PROPERTIES

1. (a) 3.61 (b) 2.83 (c) -5.39 (d) 0.77 **3.** a, c, d, f
5. (a) $2\sqrt{2}$ (b) $3\sqrt{5}$ (c) 30 (d) 3 **7.** (a) $\frac{3}{4}$ (b) $\frac{5}{7}$ (c) $\frac{\sqrt{7}}{4}$
(d) $\frac{\sqrt{6}}{3}$ **9.** (a) $\sqrt{54} \approx 7.35$ and $3\sqrt{6} \approx 7.35$
(b) $\sqrt{\frac{5}{16}} \approx 0.56$ and $\frac{\sqrt{5}}{4} \approx 0.56$ **11.** $a = 1, b = -6, c = 8$
13. $a = 1, b = -4, c = -12$
15. $a = 3, b = -10, c = -25$
17. $a = 2, b = 3, c = -152$ **19.** $x = 5$ or $x = 2$
21. $x = \frac{7 \pm \sqrt{13}}{2} \approx 5.30$ or 1.70
23. $x = 2 \pm 2\sqrt{3} \approx 5.46$ or -1.46
25. $x = \frac{3 \pm \sqrt{149}}{10} \approx 1.52$ or -0.92
27. $x = \pm\sqrt{7} \approx \pm 2.65$ **29.** $x = \pm\frac{5}{2}$
31. $x = 0$ or $x = \frac{b}{a}$ **33.** 5 by 8 **35.** $n = 6$ **37.** $c = 5$
39. $1 + 2\sqrt{2}$

Glossary

acute angle an angle whose measure is between $0°$ and $90°$

acute triangle a triangle whose three interior angles are all acute

adjacent angles two angles that have a common vertex and a common side between them

altitude of cone (pyramid) the line segment from the vertex (apex) of the cone perpendicular to the plane of the base

altitude of cylinder (prism) a line segment between and perpendicular to each of the two bases

altitude of parallelogram a line segment drawn perpendicularly from one side to a nonadjacent side (or extension of that side)

altitude of trapezoid a line segment drawn perpendicularly from a vertex to the remaining parallel side

altitude of triangle a line segment drawn perpendicularly from a vertex of the triangle to the opposite side of the triangle; the length of the altitude is the height of the triangle

angle the plane figure formed by two rays that share a common endpoint

angle bisector see bisector of angle

angle of depression (elevation) acute angle formed by a horizontal ray and a ray determined by a downward (an upward) rotation

apex of pyramid (cone) same as vertex of pyramid (cone)

apothem of regular polygon any line segment drawn from the center of the regular polygon perpendicular to one of its sides

arc the segment (part) of a circle determined by two points on the circle and all points between them

area the measurement, in square units, of the amount of region within an enclosed plane figure

auxiliary line a line (or part of a line) added to a drawing to help complete a proof or solve a problem

axiom *see* postulate

base a side (of a plane figure) or face (of a solid figure) to which an altitude is drawn

base angles of isosceles triangle the two congruent angles of the isosceles triangle

base of isosceles triangle the side of the triangle whose length is unique; the side opposite the vertex

bases of trapezoid the two parallel sides of the trapezoid

bisector of angle a ray that separates the given angle into two smaller, congruent angles

Cartesian plane the two-dimensional coordinate system determined by x and y axes

Cartesion space the three-dimensional coordinate system determined by x, y, and z axes

center of circle the interior point of the circle whose distance from all points on the circle is the same

center of regular polygon the common center of the inscribed and circumscribed circles of the regular polygon

center of sphere the interior point of the sphere whose distance from all points on the sphere is the same

central angle of circle an angle whose vertex is at the center of the circle and whose sides are radii of the circle

central angle of regular polygon an angle whose vertex is at the center of the regular polygon and whose sides are two consecutive radii of the polygon

centroid of triangle the point of concurrence for the three medians of the triangle

chord of circle any line segment that joins two points on the circle

circle the set of points in a plane that are at a fixed distance from a given point (the center of the circle) in the plane

circumcenter of triangle the center of the circumscribed circle of a triangle; the point of concurrence for the perpendicular bisectors of the three sides of the triangle

circumference the linear measure of the distance around a circle

circumscribed circle a circle that contains all vertices of a polygon whose sides are chords of the circle

circumscribed polygon a polygon whose sides are all tangent to a circle in the interior of the polygon

collinear points points that lie on the same line

common tangent a line (or segment) that is tangent to more than one circle; can be a common external tangent or a common internal tangent

complementary angles two angles whose sum of measures is $90°$

concave polygon a polygon in which at least one diagonal lies in the exterior of the polygon

concentric circles (spheres) two or more circles (spheres) that have the same center

conclusion the "then" clause of an "If, then" statement; the part of a theorem indicating the claim to be proved

concurrent lines (planes) three or more lines (planes) that contain the same point

congruent refers to figures (such as angles) that can be made to coincide

converse relative to the statement "If P, then Q," this statement has the form "If Q, then P"

convex polygon a polygon in which all diagonals lie in the interior of the polygon

coplanar points points that lie in the same plane

corollary a theorem that follows from another theorem as a "by-product"; a theorem that is easily proved as the consequence of another theorem

cosecant in a right triangle, the ratio $\frac{\text{hypotenuse}}{\text{opposite}}$

cosine in a right triangle, the ratio $\frac{\text{adjacent}}{\text{hypotenuse}}$

cotangent in a right triangle, the ratio $\frac{\text{adjacent}}{\text{opposite}}$

cube a right square prism whose edges are congruent

cyclic polygon a polygon that can be inscribed in a circle

cylinder (circular) the solid generated by all line segments parallel to the axis of the cylinder and that contain corresponding endpoints on the two congruent circular bases

decagon a polygon with exactly 10 sides

deduction a form of reasoning in which specific conclusions are reached through the use of established principles

degree the unit of measure that corresponds to $\frac{1}{360}$ of a complete revolution; used with angles and arcs

diagonal of polygon a line segment that joins two nonconsecutive vertices of a polygon

diameter any line segment that joins two points on a circle (or sphere) and that also contains the center of the circle (or sphere)

dodecagon a polygon that has exactly 12 sides

dodecahedron a polyhedron that has exactly 12 faces

dodecahedron (regular) a polyhedron that has exactly 12 faces that are congruent regular pentagons

edge of polyhedron any line segment determined by the intersection of two faces of the polyhedron (includes prisms and pyramids)

equiangular polygon a type of polygon whose angles are congruent (equal)

equilateral polygon a type of polygon whose sides are congruent (equal)

equivalent equations equations for which the solutions are the same

extended proportion a proportion that has three or more members, such as $\frac{a}{b} = \frac{c}{d} = \frac{e}{f}$

extended ratio a ratio that compares three or more numbers, such as $a:b:c$

exterior refers to all points that lie outside an enclosed (bounded) plane or solid figure

exterior angle of polygon an angle formed by one side of the polygon and an extension of a second side that has a common endpoint with the first side

extremes of a proportion the first and last terms of a proportion; in $\frac{a}{b} = \frac{c}{d}$, a and d are the extremes

face of polyhedron any one of the polygons that lies in a plane determined by the vertices of the polyhedron; includes base(s) and lateral faces of prisms and pyramids

geometric mean the repeated second and third terms of certain proportions; in $\frac{a}{b} = \frac{b}{c}$, b is the geometric mean of a and c

height the length of the altitude of a geometric figure

heptagon a polygon that has exactly seven sides

heptahedron a polyhedron that has exactly seven faces

hexagon a polygon that has exactly six sides

hexahedron a polyhedron that has exactly six faces

hexahedron (regular) a polyhedron that has six congruent square faces; also called a cube

hypotenuse of right triangle the side of a right triangle that lies opposite the right angle

hypothesis the "if" clause of an "If, then" statement; the part of a theorem providing the given information

icosahedron (regular) a polyhedron with 20 congruent faces that are equilateral triangles

incenter of triangle the center of the inscribed circle of a triangle; the point of concurrence for the three bisectors of the angles of the triangle

induction a form of reasoning in which a number of specific observations are used to draw a general conclusion

inscribed angle of circle an angle whose vertex is on a circle and whose sides are chords of the circle

inscribed circle a circle that lies inside a polygon in such a way that the sides of the polygon are tangents of the circle

inscribed polygon a polygon whose vertices all lie on a circle in such a way that the sides of the polygon are chords of the circle

intercepted arc the arc (an arc) of a circle that is cut off in the interior of an angle (or related angle)

intercepts the points at which the graph of an equation intersects the axes

interior refers to all points that lie inside an enclosed (bounded) plane or solid figure

interior angle of polygon any angle formed by two consecutive sides of the polygon in such a way that the angle lies in the interior of the polygon

intersection the elements that two sets have in common; the points that two geometric figures share

intuition drawing a conclusion through insight

inverse relative to the statement "If P, then Q," this statement has the form "If not P, then not Q"

isosceles trapezoid a trapezoid that has two congruent legs (its nonparallel sides)

isosceles triangle a triangle that has two congruent sides

kite a quadrilateral that has two distinct pairs of congruent adjacent sides

lateral area the sum of areas of the lateral faces of a solid or the area of the curved lateral surface, excluding the base area(s) (as in prisms, pyramids, cylinders, and cones)

legs of isosceles triangle the two congruent sides of the triangle

legs of right triangle the two sides that form the right angle of the triangle

legs of trapezoid the two nonparallel sides of the trapezoid

lemma a theorem that is introduced and proved so that a later theorem can be proved

line of centers the line (or line segment) that joins the centers of two coplanar circles

line segment the part of a line determined by two points and all points on the line that lie between those two points

locus the set of all points that satisfy a given condition or conditions

major arc an arc whose measure is between 180° and 360°

mean proportional *see* geometric mean

means of a proportion the second and third terms of a proportion; in $\frac{a}{b} = \frac{c}{d}$, b and c are the means

median of trapezoid the line segment that joins the midpoints of the two legs (nonparallel sides) of the trapezoid

median of triangle the line segment joining a vertex of the triangle to the midpoint of the opposite side

midpoint the point on a line segment (or arc) that separates the line segment (arc) into two congruent parts

minor arc an arc whose measure is between 0° and 180°

nonagon a polygon that has exactly nine sides

noncollinear points three or more points that do not lie on the same line

noncoplanar points four or more points that do not lie in the same plane

obtuse angle an angle whose measure is between 90° and 180°

obtuse triangle a triangle that has exactly one interior angle that is obtuse

octagon a polygon that has exactly eight sides

octahedron a polyhedron that has exactly eight faces

octahedron (regular) a polyhedron with eight congruent faces that are equilateral triangles

opposite rays two rays having a common endpoint and that together form a line

orthocenter of triangle the point of concurrence for the three altitudes of the triangle

parallel lines (planes) two lines in a plane (or two planes) that do not intersect

parallelogram a quadrilateral that has two pairs of parallel sides

parallelepiped a right rectangular prism; a box

pentagon a polygon that has exactly five sides

pentahedron a polyhedron that has exactly five faces

perimeter of polygon the sum of the lengths of the sides of the polygon

perpendicular bisector of line segment a line (or part of a line) that is both perpendicular to a given line segment and bisects that line segment

perpendicular lines two lines that intersect to form congruent adjacent angles

pi (π) the constant ratio of the circumference of a circle to the length of its diameter; this ratio is commonly approximated by the fraction $\frac{22}{7}$ or the decimal 3.1416

point of tangency (contact) the point at which a tangent to a circle touches the circle

polygon a plane figure whose sides are line segments that intersect only at their endpoints

polyhedron a solid figure whose faces are polygons that intersect other faces along common sides of the polygons

postulate a statement that is assumed to be true

Quadratic Formula the formula $x = \frac{-b \pm \sqrt{b^2 - 4ac}}{2a}$, which provides solutions for the equation $ax^2 + bx + c = 0$, where a, b, and c are real numbers and $a \neq 0$

quadrilateral a polygon that has exactly four sides

radian the measure of a central angle of a circle whose intercepted arc has a length equal to the radius of the circle

radius any line segment that joins the center of a circle (or sphere) to a point on the circle (or sphere)

radius of regular polygon any line segment that joins the center of the polygon to one of its vertices.

ratio a comparison between two quantities a and b, generally written $\frac{a}{b}$ or $a{:}b$

ray the part of a line that begins at a point and extends infinitely far in one direction

rectangle a parallelogram that contains a right angle

reflex angle an angle whose measure is between 180° and 360°

regular polygon a polygon whose sides are congruent and whose interior angles are congruent

regular polyhedron a polyhedron whose edges are congruent and whose faces are congruent

regular prism a right prism whose bases are regular polygons

regular pyramid a pyramid whose base is a regular polygon and whose lateral faces are congruent isosceles triangles

rhombus a parallelogram with two congruent adjacent sides

right angle an angle whose measure is exactly 90°

right circular cone a cone in which the line segment joining the vertex of the cone to the center of the circular base is perpendicular to the base

right circular cylinder a cylinder in which the line segment joining the centers of the circular bases is perpendicular to the plane of each base

right prism a prism in which lateral edges are perpendicular to the base edges that they intersect

right trapezoid a trapezoid that contains a right angle

right triangle a triangle in which exactly one interior angle is a right angle

scalene triangle a triangle in which no two sides are congruent

secant in a right triangle, the ratio $\frac{\text{hypotenuse}}{\text{adjacent}}$

secant of circle a line (or part of a line) that intersects a circle at two points

sector of circle the plane region bounded by two radii of the circle and the arc that is intercepted by the central angle formed by those radii

segment of circle the plane region bounded by a chord and a minor arc (major arc) that has the same endpoints as that chord

semicircle the arc of a circle determined by a diameter; an arc of a circle whose measure is exactly 180°

set any collection of objects, numbers, or points

similar polygons polygons that have the same shape

sine in a right triangle, the ratio $\frac{\text{opposite}}{\text{hypotenuse}}$

skew quadrilateral a quadrilateral whose sides do not all lie in one plane

slant height of cone any line segment joining the vertex (apex) of the cone to a point on the circular base

slant height of regular pyramid a line segment joining the vertex (apex) of the pyramid to the midpoint of a base edge of the pyramid

slope a measure of the steepness of a line; in the rectangular coordinate system, the slope m of the line through (x_1, y_1) and (x_2, y_2) is $m = \frac{y_2 - y_1}{x_2 - x_1}$ where $x_1 \neq x_2$

sphere the set of points in space that are at a fixed distance from a given point (the center of the sphere)

straight angle an angle whose measure is exactly 180°; an angle whose sides are opposite rays

straightedge an idealized instrument used to construct parts of lines

supplementary angles two angles whose sum of measures is 180°

surface area the measure of the total area of a solid; the sum of the lateral area and base area in many solid figures.

symmetry with respect to a line (ℓ) figure for which every point A on the figure has a second point B on the figure for which ℓ is the perpendicular bisector of $\overline{AB}$

symmetry with respect to a plane (R) solid figure for which every point A on the figure has a second point B on the figure for which plane R is the perpendicular bisector of $\overline{AB}$.

symmetry with respect to a point (P) figure for which every point A on the figure has a second point C on the figure for which P is the midpoint of $\overline{AC}$.

tangent in a right triangle, the ratio $\frac{\text{opposite}}{\text{adjacent}}$

tangent circles two circles that have one point in common; the circles may be externally tangent or internally tangent

tangent line of circle a line that touches a circle at only one point

tetrahedron a polyhedron that has exactly four faces

tetrahedron (regular) a four-faced solid in which the faces are congruent equilateral triangles

theorem a statement that follows logically from previous definitions and principles; a statement that can be proved

torus a three-dimensional solid that has a "doughnut" shape

transversal a line that intersects two or more lines, intersecting each at one point

trapezoid a quadrilateral having exactly two parallel sides

triangle a polygon that has exactly three sides

triangle inequality a statement that the sum of the lengths of two sides of a triangle cannot be greater than the length of the third side

union the joining together of any two sets, such as geometric figures

valid argument an argument in which the conclusion follows logically from previously stated (and accepted) premises or assumptions

vertex angle of isosceles triangle the angle formed by the two congruent sides of the triangle

vertex of angle the point at which the two sides of the angle meet

vertex of isosceles triangle the point at which the two congruent sides of the triangle meet

vertex of polygon any point at which two sides of the polygon meet

vertex of polyhedron any point at which three (or more) edges of the polyhedron meet

vertical angles a pair of angles that lie in opposite positions when formed by two intersecting lines

volume the measurement, in cubic units, of the amount of space within a bounded region of space

Index

Abbreviations

AA	angle-angle (proves $\triangle$s similar)
ASA	angle-side-angle (proves $\triangle$s congruent)
AAS	angle-angle-side (proves $\triangle$s congruent)
add.	addition
adj.	adjacent
alt.	altitude, alternate
CASTC	Corresponding angles of similar triangles are congruent.
cm	centimeters
cm^2	square centimeters
cm^3	cubic centimeters
comp.	complementary
corr.	corresponding
cos	cosine
cot	cotangent
CPCTC	Corresponding parts of congruent triangles are congruent.
csc	cosecant
CSSTP	Corresponding sides of similar triangles are proportional.
eq.	equality
exs.	exercises
ext.	exterior
ft	foot (or feet)
gal	gallon
HL	hypotenuse-leg (proves $\triangle$s congruent)
hr	hour
in.	inch (or inches)
ineq.	inequality

int.	interior
isos.	isosceles
km	kilometers
m	meters
mi	miles
mm	millimeters
n-gon	polygon of n sides
opp.	opposite
pent.	pentagon
post.	postulate
prop.	property
pt.	point
quad.	quadrilateral
rect.	rectangle
rt.	right
SAS	side-angle-side (proves $\triangle$s congruent)
SAS$\sim$	side-angle-side (proves $\triangle$s similar)
sec	secant
sin	sine
SSS	side-side-side (proves $\triangle$s congruent)
SSS$\sim$	side-side-side (proves $\triangle$s similar)
st.	straight
supp.	supplementary
tan	tangent
trans.	transversal, transitive
trap.	trapezoid
vert.	vertical (angle or line)
yd	yards